Contemporary Mathematicians

Gian-Carlo Rota
Editor

Nathan Jacobson, circa 1940

Nathan Jacobson
Collected Mathematical Papers

Volume 1
(1934–1946)

Birkhäuser
Boston · Basel · Berlin
1989

Nathan Jacobson
Department of Mathematics
Yale University
New Haven, CT 06520

Library of Congress Cataloging-in-Publication Data
Jacobson, Nathan, 1910–
[Selections. 1989]
Collected mathematical papers / Nathan Jacobson.
p. cm.—(Contemporary mathematicians)
Includes bibliographies.
ISBN 0-8176-3410-X (v. 1)
1. Mathematics. I. Title. II. Series.
QA3.J3325 1989
510—dc20 89-14889
CIP

Printed and bound by Edwards Brothers Inc., Ann Arbor, Michigan.
Printed in the United States of America.

ISBN 0-8176-3410-X
ISBN 3-7643-3410-X

Bibliography of Nathan Jacobson's Books and Papers

Note: **Boldface** numbers at the end of each entry denote the volume in which the entry appears.

Books

The Theory of rings: *Mathematical Surveys*, *No. II*, Amer. Math. Soc., 1943 (Russian Translation, 1947).

Lectures in Abstract Algebra: *Vol. 1*, *Basic Concepts*, D. Van Nostrand Co. Inc., 1951 (Springer-Verlag reprint, 1975; Chinese translation, 1966).

Lectures in Abstract Algebra: *Vol. 2*, *Linear Algebra*, D. Van Nostrand Co. Inc., 1953 (Springer-Verlag reprint, 1975; Chinese translation, 1960).

Lectures in Abstract Algebra: *Vol. 3*, *Theory of Fields and Galois Theory*, D. Van Nostrand Co. Inc., 1964 (Springer-Verlag reprint, 1975).

Structure of Rings, Amer. Math. Soc. Colloquium Publications, Vol. 37, 1956, 1964 (Russian translation, 1961).

Lie Algebras, Interscience Publishers (John Wiley and Sons), 1962 Interscience Tracts in Pure and Applied Mathematics, No. 10 (Dover reprint, 1979; Russian translation, 1964; Chinese translation, 1964).

Structure and Representations of Jordan Algebras, Amer. Math. Soc. Colloquium Publications, Vol. 39, 1968.

Lectures on Quadratic Jordan Algebras, Tata Institute of Fundamental Research, Bombay, 1969.

Exceptional Lie Algebras, Lecture Notes in Pure and Applied Mathematics, Marcel Dekker Inc., New York, 1971.

Basic Algebra I, W. H. Freeman and Co., New York, 1974; second edition, 1985.

Pi-Algebras: *An Introduction*, Springer Verlag, 1975.

Basic Algebra II, W. H. Freeman and Co., New York, 1980; second edition, 1989.

Structure Theory of Jordan Algebras, University of Arkansas Lecture Notes in Mathematics, 1981.

Finite Dimensional Division Algebras (with David Saltman), Springer-Verlag Grundlehre Series, in press.

Papers

[1] "Non-commutative polynomials and cyclic algebras", (Princeton University dissertation) *Annals of Math.* **35** (1934) 197–208. **(1)**

[2] "A note on non-commutative polynomials", *Annals of Math.* **35** (1934) 209–210. **(1)**

[3] "Locally compact rings" (with O. Taussky), *Proc. Nat. Acad. Sci.* **21** (1935) 106–108. **(1)**

[4] "Rational methods in the theory of Lie algebras", *Annals of Math.* **36** (1935) 875–881. **(1)**

[5] "On pseudo-linear transformations", *Proc. Nat. Acad. Sci.* **21** (1935) 667–670. **(1)**

[6] "Totally disconnected locally compact rings", *Amer. J. Math.* **58** (1936) 433–449. **(1)**

[7] "Simple Lie algebras of type A", *Proc. Nat. Acad. Sci.* **23** (1937) 240–242. **(1)**

[8] "Pseudo-linear transformations", *Annals of Math.* **38** (1937) 484–507. **(1)**

[9] "A class of normal simple Lie algebras of characteristic zero", *Annals of Math.* **38** (1937) 508–517. **(1)**

[10] "A note on non-associative algebras", *Duke Math. J.* **3** (1937) 544–548. **(1)**
[11] "Abstract derivation and Lie algebras", *Trans. Amer. Math. Soc.* **42** (1937) 206–224. **(1)**
[12] "p-Algebras of exponent p", *Bull. Amer. Math. Soc.* **43** (1937) 667–670. **(1)**
[13] "A note on topological fields", *Amer. J. Math.* **59** (1937) 889–894. **(1)**
[14] "Simple Lie algebras of type A", *Annals of Math.* **39** (1938) 181–188. **(1)**
[15] "Simple Lie algebras over a field of characteristic zero", *Duke Math. J.* **4** (1938) 534–551. **(1)**
[16] "Normal semi-linear transformations", *Amer. J. Math.* **61** (1939) 45–58. **(1)**
[17] "An application of E. H. Moore's determinant of a Hermitian matrix", *Bull. Amer. Math. Soc.* **45** (1939) 745–848. **(1)**
[18] "Structure and automorphisms of semi-simple Lie groups in the large", *Annals of Math.* **40** (1939) 755–763. **(1)**
[19] "Cayley numbers and normal simple Lie algebras of type G", *Duke Math. J.* **5** (1939) 775–783. **(1)**
[20] "The fundamental theorem of the Galois theory for quasi-fields", *Annals of Math.* **41** (1940) 1–7. **(1)**.
[21] "A note on hermitian forms", *Bull. Amer. Math. Soc.* **46** (1940) 264–268. **(1)**
[22] "Restricted Lie algebras of characteristic p", *Trans. Amer. Math. Soc.* **50** (1941) 15–25. **(1)**
[23] "Classes of restricted Lie algebras of characteristic p I", *Amer. J. Math.* **63** (1941) 481–515. **(1)**
[24] "Classes of restricted Lie algebras of characteristic p II", *Duke Math. J.* **10** (1943) 107–121. **(1)**
[25] "An extension of Galois theory to non-normal and non-separable fields", *Amer. J. Math.* **66** (1944) 1–29. **(1)**
[26] "Schur's theorems on commutative matrices", *Bull. Amer. Math. Soc.* **50** (1944) 431–436. **(1)**
[27] "Relations between the composites of a field and those of a subfield", *Amer. J. Math.* **66** (1944) 636–644. **(1)**
[28] "Galois theory of purely inseparable fields of exponent one", *Amer. J. Math.* **66** (1944) 645–648. **(1)**
[29] "Construction of central simple associative algebras", *Annals of Math.* **45** (1944) 658–666. **(1)**
[30] "The equation $x' = xd - dx = b$", *Bull. Amer. Math. Soc.* **50** (1944) 902–905. **(1)**
[31] "Structure theory of simple rings without finiteness assumptions", *Trans. Amer. Math. Soc.* **57** (1945) 228–245. **(1)**
[32] "The radical and semi-simplicity for arbitrary rings", *Amer. J. Math.* **67** (1945) 300–320. **(1)**
[33] "Structure theory for algebraic algebras of bounded degree", *Annals of Math.* **46** (1945) 695–707. **(1)**
[34] "A topology for the set of primitive ideals in an arbitrary ring", *Proc. Nat. Acad. Sci.* **31** (1945) 333–338. **(1)**
[35] "On the theory of primitive rings", *Annals of Math.* **48** (1947) 8–21. **(1)**
[36] "A note on division rings", *Amer. J. Math.* **69** (1947) 27–36. **(1)**
[37] "Isomorphisms of Jordan rings", *Amer. J. Math.* **70** (1948) 317–326. **(1)**
[38] "The center of a Jordan ring", *Bull. Amer. Math. Soc.* **54** (1948) 316–322. **(1)**
[39] "Lie and Jordan triple systems", *Amer. J. Math.* **71** (1949) 149–170. **(2)**
[40] "Classification and representation of semi-simple Jordan algebras" (with F. D. Jacobson), *Trans. Amer. Math. Soc.* **65** (1949) 141–169. **(2)**
[41] "Derivation algebras and multiplication algebras of semi-simple Jordan algebras", *Annals of Math.* **50** (1949) 866–874. **(2)**
[42] "Enveloping algebras of semi-simple Lie algebras", *Can. J. Math.* **2** (1950) 257–266. **(2)**
[43] "Some remarks on one-sided inverses", *Proc. Amer. Math. Soc.* **1** (1950) 352–355. **(2)**
[44] "Jordan homomorphisms of rings" (with C. E. Rickart), *Trans. Amer. Math. Soc.* **69** (1950) 479–502. **(2)**

[45] "Completely reducible Lie algebras of linear transformations", *Proc. Amer. Math. Soc.* **2** (1951) 105–113. **(2)**
[46] "General representation theory of Jordan algebras", *Trans. Amer. Math. Soc.* **70** (1951) 509–530. **(2)**
[47] "Une généralization du théorème d'Engel", *C. R. Acad. Sci.* **234** (1952) 679–681. **(2)**
[48] "Homomorphisms of Jordan rings of self-adjoint elements" (with C. E. Rickart), *Trans. Amer. Math. Soc.* **72** (1952) 310–322. **(2)**
[49] "A note on Lie algebras of characteristic p", *Amer. J. Math.* **74** (1952) 357–359. **(2)**
[50] "Operator commutativity in Jordan algebras", *Proc. Amer. Math. Soc.* **3** (1952) 973–976. **(2)**
[51] "A Kronecker factorization theorem for Cayley algebras and the exceptional simple Jordan algebra", *Amer. J. Math.* **76** (1954) 447–452. **(2)**
[52] "Structure of alternative and Jordan bimodules", *Osaka Math. J.* **6** (1954) 1–71. **(2)**
[53] "A note on automorphisms and derivations of Lie algebras", *Proc. Amer. Math. Soc.* **6** (1955) 281–283. **(2)**
[54] "Commutative restricted Lie algebras", *Proc. Amer. Math. Soc.* **6** (1955) 476–481. **(2)**
[55] "A note on two dimensional division ring extensions", *Amer. J. Math.* **77** (1955) 593–599. **(2)**
[56] "A theorem on the structure of Jordan algebras", *Proc. Nat. Acad. Sci.* **42** (1956) 140–147. **(2)**
[57] "Generation of separable and central simple algebras", *J. Math. Pures Appl.* **36** (1957) 217–227. **(2)**
[58] "On reduced exceptional simple Jordan algebras" (with A. A. Albert), *Annals of Math.* **66** (1957) 400–417. **(2)**
[59] "On Jordan algebras with two generators" (with L. J. Paige), *J. Rat. Mech. Anal.* **6** (1957) 895–906. **(2)**
[60] "Composition algebras and their automorphisms", *Rend. Circ. Math. Palermo* **7 (Series II)** (1958) 1–26. **(2)**
[61] "Nilpotent elements in semi-simple Jordan algebras", *Math. Ann.* **136** (1958) 375–386. **(2)**
[62] "A note on three dimensional simple Lie algebras", *J. Math. Mech.* **7** (1958) 823–832. **(2)**
[63] "Some groups of transformations defined by Jordan algebras I", *J. Reine Angew. Math.* **201** (1959) 178–195. **(2)**
[64] "Some groups of transformations defined by Jordan algebras II", *J. Reine Angew. Math.* **204** (1960) 74–98. **(2)**
[65] "Some groups of transformations defined by Jordan algebras III", *J. Reine Angew. Math.* **207** (1961) 61–85. **(2)**
[66] "Macdonald's theorem on Jordan algebras", *Archiv Math.* **13** (1962) 241–250. **(2)**
[67] "A coordinatization theorem for Jordan algebras", *Proc. Nat. Acad. Sci.* **48** (1962) 1154–1160. **(2)**
[68] "A note on automorphisms of Lie algebras", *Pac. J. Math.* **12** (1962) 303–315. **(2)**
[69] "Generic norm of an algebra", *Osaka Math. J.* **15** (1963) 25–50. **(2)**
[70] "Clifford algebras for algebras with involution of type D", *J. Algebra* **1** (1964) 288–300. **(2)**
[71] "Triality and Lie algebras of type D_4", *Rend. Circ. Math. Palermo* **13 (Series II)** (1964) 1–25. **(2)**
[72] "Cartan subalgebras of Jordan algebras", *Nagoya Math. J.* **27** (1966) 591–609. **(3)**
[73] "Structure theory for a class of Jordan algebras", *Proc. Nat. Acad. Sci.* **55** (1966) 243–251. **(3)**
[74] "Quadratic Jordan algebras of quadratic forms with base points" (with K. McCrimmon), *Indian Math. Soc.* **35** (1971) 1–45. **(3)**
[75] "Generally algebraic quadratic Jordan algebras" (with J. Katz), *Scri. Math.* **29** (1971) 215–227. **(3)**

[76] "Structure groups and Lie algebras of Jordan algebras of symmetric elements of associative algebras with involution", *Adv. Math.* **20** (1976) 106–150. **(3)**
[77] "Localization of Jordan algebras" (with K. McCrimmon and M. Parvathi), *Commun. Algebra* **6** **(9)** (1978) 911–958. **(3)**
[78] "Bimodule structure of certain Jordan algebras relative to subalgebras with one generator", *Hokkaido Math. J.* **10** (1981) 333–342. **(3)**
[79] "Some application of Jordan norms to associative algebras", *Adv. Math.* **48** (1983) 149–165. **(3)**
[80] "Forms of the generic norm of a separable algebra", *J. Algebra* **86** (1984) 76–84. **(3)**
[81] "Some projective varieties defined by Jordan algebras", *J. Algebra* **97** (1985) 565–598. **(3)**
[82] "Jordan algebras of real symmetric matrices", *Algebras, Groups and Geometries*, **4** (1987) 291–304. **(3)**

Expository Papers

[83] "*The Classical Groups* by Hermann Weyl" (A book review), *Bull. Amer. Math. Soc.* **46** (1940) 592–595. **(3)**
[84] "Representation theory for Jordan rings", *Proc. Intl. Congr. Math.* (1950) **2** 37–43. **(3)**
[85] "Le problème de Kurosch", *Séminaire Bourbaki* **64** (1951–1952) 295–303. **(3)**
[86] "Some aspects of the theory of representations of Jordan algebras", *Proc. Intl. Congr. Math.* (1954) **3** 28–33. **(3)**
[87] "Jordan algebras", Report of a Conference on Linear Algebras, *Nat. Acad. Sci.–Nat. Research Council* **502** (1957) 12–19. **(3)**
[88] "Representation theory of Jordan algebras, Some Aspects of Ring Theory", *Cen. Int. Mat. Est.* 1966. **(3)**
[89] "Forms of algebras; Some Recent Advances in Basic Sciences", *Academic Press* (1966) 41–71. **(3)**
[90] "Associative algebras with involution and Jordan algebras", *Proc. K. Akad. van Wetenschappen* **69** (1966) 202–212. **(3)**
[91] "Connections between associative and Jordan rings", *Ist. Naz. Alta Math. Symp. Math.* **9** (1972) 261–268. **(3)**
[92] "Abraham Adrian Albert" (an obituary), *Bull. Amer. Math. Soc.* **80** (1974) 1076–1093. **(3)**
[93] "PI-Algebras, Ring Theory", Proceedings of Oklahoma Conference, edited by B. R. MacDonald, Marcel Dekker (1974) 1–30. **(3)**
[94] "Some recent developments in the theory of PI-algebras", Proceedings Winter School, Reinhardsbrunn, German Democratic Republic (1976) 17–21. **(3)**
[95] "Some recent developments in the theory of algebras with polynomial identities, I. Razmyslov's central polynomial, II. The Artin-Procesi theorem, III. On Shirshov's local finiteness theorems", *Proceedings* 18*th SRI* Springer-Verlag Canberra, 1978, 8–46. **(3)**
[96] "Survey of Jordan structure theory", *Southeast Asian Bull. Math.* **5** (1981) 27–38. **(3)**
[97] *Emmy Noether Collected Papers* (Introduction), Springer-Verlag, 1983, 16–26. **(3)**
[98] "Brauer factor sets, Noether factor sets and crossed products, *Emmy Noether in Bryn Mawr*, edited by Srinivasan and Sally, Springer-Verlag, 1983, 1–20. **(3)**

Papers Listed by Topic

(Ore) Skew Polynomial Domains: 1, 2, 5, 8, 16.

Topological Algebra: 3, 6, 13.

Lie Algebras and General Non-associative Algebras: 4, 7, 9, 10, 11, 14, 15, 19, 22, 23, 24, 39, 42, 45, 47, 49, 53, 54, 62, 68, 69, 71.

Galois Theory: 20, 25, 27, 28, 29, 36, 55.

General Structure Theory of Rings: 31, 32, 33, 34, 35, 43, 55, 93, 94, 95.

Jordan Rings: 37, 38, 39, 40, 41, 44, 46, 48, 50, 51, 61, 63, 64, 65, 66, 67, 69, 72, 73, 74, 75, 76, 77, 78, 79, 80, 81, 82, 84, 86, 87, 88, 90, 91, 96.

Associative Algebras: 1, 12, 29, 57, 70, 79, 85, 98.

Miscellaneous: 17, 18, 21, 26, 30, 60, 92, 97.

Doctoral Students of Nathan Jacobson

University of North Carolina

Charles L. Carroll, Jr. Normal simple Lie algebras of type D and order 28 over a field of characteristic zero (1945).

Yale University

Eugene Schenkman A theory of subinvariant Lie algebras (1950).
Charles W. Curtis Additive ideal theory in general rings (1951).
William G. Lister A structure theory of Lie triple systems (1951).
Henry G. Jacob A theorem on Kronecker products (1953).
George B. Seligman Lie algebras of prime characteristic (1954).
Morris Weisfeld Derivations in division rings (1954).
Bruno Harris Galois theory of Jordan algebras (1956).
Earl J. Taft Invariant Wedderburn factors (1956).
Dallas W. Sasser On Jordan matrix algebras (1957).
Maria J. Woneburger On the group of similitudes and its projective group (1957).
Tae-Il Suh On isomorphisms of little projective groups of Cayley planes (1961).
Herbert F. Kreimer, Jr. Differential, difference, and related operational rings (1962).
Charles M. Glennie Identities in Jordan algebras (1963).
David A. Smith On Chevalley's method in the theory of Lie algebras and linear groups of prime characteristic (1963).
Dominic C. Soda Groups of type D_4 defined by Jordan algebras (1964).
Harry P. Allen Jordan algebras and Lie algebras of type D_4 (1965).
Eugene A. Klotz Isomorphisms of simple Lie rings (1965).
Kevin M. McCrimmon Norms and noncommutative Jordan algebras (1965).
Joseph C. Ferrar On Lie algebras of type E_6 (1966).
Daya-Nand Verma Structure of certain induced representations of complex semi-simple Lie algebras (1966).
Lynn Barnes Small Mapping theorems in simple rings with involution (1967).
John R. Faulkner Octonion planes defined by quadratic Jordan algebras (1969).
Samuel R. Gordon On the automorphism group of a semi-simple Jordan algebra of characteristic zero (1969).
Michel Racine The Arithmetics of quadratic Jordan algebras (1971).
Jerome M. Katz Automorphisms of the lattice of inner ideals of certain quadratic Jordan algebras (1972).
Ronald Infante Strongly normal difference extensions (1973).
Louis H. Rowen On algebras with polynomial identity (1973).
Georgia M. Benkart Inner ideals and the structure of Lie algebras (1974).
David J. Saltman Azumaya algebras over rings of characteristic p (1976).
Robert A. Bix Separable Jordan algebras over commutative rings (1977).
Leslie Hogben Radical classes of Jordan algebras (1978).
Craig L. Huneke Determinantal ideals and questions related to factoriality (1978).

Chronology

September 8, 1910	Born, Warsaw, Poland (U.S. Citizen)
1930	A.B. University of Alabama
1934	Ph.D. Princeton University
1934–1935	Assistant, Institute for Advanced Study
1935–1936	Lecturer, Bryn Mawr College
1936–1937	National Research Council Fellow, University of Chicago
1937–1938	Instructor, University of North Carolina
1938–1940	Assistant Professor, University of North Carolina
1940–1941	Visiting Associate Professor, Johns Hopkins University
1941–1942	Associate Professor, University of North Carolina
1942–1943	Associate Ground School Instructor, University of North Carolina
1943–1947	Associate Professor, Johns Hopkins University
1947–1949	Associate Professor, Yale University
1949–1961	Professor, Yale University
1961–1963	James E. English Professor, Yale University
1963–1981	Henry Ford II Professor, Yale University
1981–	Henry Ford II Professor Emeritus, Yale University
Summer, 1947	Visiting Professor, University of Chicago
1951–1952	Fulbright Scholar, University of Paris
1957–1958	Visiting Professor, University of Paris
Oct. 1964–Jan. 1965	Visiting Professor, University of Chicago
Spring, 1965	Lecturer, Mathematical Society of Japan
Spring, 1969	Visiting Professor, Tata Institute of Fundamental Research
Oct. 1981–Jan. 1982	Visiting Professor, ETH, Zurich
Sept.–Nov. 1983	Visiting Professor, Nanjing University, People's Republic of China
Nov.–Dec. 1983	Visiting Professor, Taiwan National University, Republic of China
Feb.–March 1984	Visitor, Center for Advanced Studies, University of Virginia
Sept.–Oct. 1985	Visiting Professor, Pennsylvania State University
April 1988	John Hasbrouck Van Vleck Distinguished Visiting Professor, Wesleyan University
Visiting Lecturer	Australia, Israel, Italy, People's Republic of China, University of Texas
1951–1952	Guggenheim Fellow
1954	Elected Member of National Academy of Sciences

1960	Elected Member of American Academy of Arts and Sciences
1971–1973	President of American Mathematical Society
1972	Honorary D.Sc., University of Chicago
1972	Honorary Member of London Mathematical Society
1972–1974	Vice President of International Mathematical Union
1981	Sesquicentennial Honorary Professor, University of Alabama

Preface

This collection contains all my published papers, both research and expository, that were published from 1934 to 1988. The research papers arranged in chronological order appear in Volume I and II and in the first part of Volume III. The expository papers, which are mainly reports presented at conferences, appear in chronological order in the last part of Volume III. Volume I covers the period 1910 to 1947, the year I moved to Yale, Volume II covers the period 1947 to 1965 when I became Chairman of the Department at Yale and Volume III covers the period from 1965 to 1989, which goes beyond my assumption of an emeritus status in 1981. I have divided the time interval covered in each volume into subintervals preceded by an account of my personal history during this period, and a commentary on the research papers published in the period. I have omitted commentaries on the expository papers and have sorted out the commentaries on the research papers according to the principal fields of my research.

The personal history has been based on my recollections, checked against written documentation in my file of letters as well as diaries. One of these was a diary I kept of my trip to the USSR in 1961; the others were diaries Florie (Florence) kept during other major visits abroad. I have also consulted Professor A. W. Tucker on historical details on Princeton during the 1930's.

The material on personal history and commentary has had critical readings by Florie, by George Seligman and by Kevin McCrimmon. I take this opportunity to express my sincere thanks to them for numerous suggestions which have resulted in an improved manuscript.

Nathan Jacobson
Hamden, Connecticut
September 25, 1989

Table of Contents

A Personal History and Commentary

1910-1943

I was born in the Jewish ghetto of Warsaw on October 5, 1910.[1] When I was five years old, my father, Charles Jacobson (his "Ellis Island name") emigrated to the United States where he settled in Nashville, Tennessee. He owned a small grocery store in the rear of which there were living quarters that he occupied. After two years he was able to save enough money to pay for the passage of his family (my mother, née Pauline Rosenberg, my brother, Sol (Solomon) who was a year and a half older than I, and myself).

It was a long journey from Warsaw to New York: in a sealed freight car through Germany to Rotterdam where we had to wait four months for a Dutch ship to take us across the U-boat-infested Atlantic. The trip had been organized by HIAS (Hebrew Immigrant Aid Society) which looked after our physical and spiritual needs during the stay in Rotterdam. After a brief visit with relatives in New York, our family was re-united in Nashville where we squeezed into the modest quarters that my father occupied.

Though I had begun elementary school in Warsaw (where I recall the Jewish students were required to sit on special benches to segregate them from their Polish peers), most of my elementary and secondary schooling took place in Birmingham, Alabama where we had moved after a few months in Nashville. For my last two years in high school I attended the S. D. Lee High School in Columbus, Mississippi, a pleasant southern town[2] where we had moved from Birmingham. After graduation I entered the University of Alabama in 1926 and graduated with an A.B. degree in 1930. I had intended to study law to follow in the footsteps of a maternal uncle who had come to the United States a few years before my father and had obtained a degree in law at the University of Alabama. My initial objective of law as well as my interests at the time led me to a curriculum that was heavily weighted in the humanities, especially in history. However, I also took all the mathematics courses that were offered (not a large number). Apparently I had distinguished myself in these, for I was offered a teaching assistantship in mathematics in my junior year. This marked a turning point in my college career since I decided then to major in mathematics and to pursue this study beyond college. I had two inspiring teachers at Alabama, Professor William P. Ott, who had received a doctorate at the University of Chicago in the calculus of variations and Professor Fred Lewis, who later received a doctorate at Ohio State University with a dissertation on linear groups. Many years later, in 1982, at a ceremony in which I was designated a "Sesquicentennial Honorary Professor" of the University of Alabama, I was able to acknowledge in a brief speech my indebtedness to Professors Lewis and Ott.

During my senior year at Alabama I applied for admission and financial aid to the three top graduate schools in the country: Princeton, Harvard and Chicago. I was

---

[1] On the Jewish calendar the date was the second day of the high holy day of "Rosh Hashanah". Through an error this was converted to September 8th which is the date that now appears on all official documents.

[2] This did not appear to be so pleasant when we visited it briefly in 1960.

fortunate enough to be awarded a research assistantship at Princeton with a stipend of $500. Tuition at that time was $100 and my room and board at the Graduate College amounted to $540. After the first year I was a part-time instructor for two years at a salary of $1100, and during my fourth year a Procter Fellow with a stipend of $1400. These salaries and stipends provided a substantial surplus over living expenses. This enabled me to make a grand tour of Europe by car in the summer of 1935 with two companions from Princeton: H. F. Bohnenblust, who had been a student and was then an assistant professor, and Robert J. Walker, who had also entered Princeton in 1930. During this trip, I made a side trip alone to visit my relatives in Warsaw (an uncle, aunts and cousins). This was the last time I would see them. All were killed by the Nazis in the death camps after the German invasion of Poland in 1939.

Princeton was a very exciting place in the thirties, especially for a young man from Alabama. The graduate school was small; about two hundred students in all of whom about twenty were in mathematics Enrollment was restricted to young, unmarried males. The mathematics students formed a closely knit group. We all lived in the Graduate College and we ate together, particularly at dinner in Procter Hall where academic gowns were required attire. Much to the annoyance of the other students, our dinner talk was almost exclusively confined to mathematics. This was an important part of our learning process.

The list of students awarded Ph.D. degrees in 1934 was exceptionally large: J. L. Barnes, N. Jacobson, S. C. Kleene, J. Levine, M. I. S. Robertson, J. B. Rosser, J. L. Vanderslice, R. J. Walker. The Ph.D.'s in the years 1930–1933 included J. H. C. Whitehead, H. F. Bohnenblust, B. Hoffman, A. W. Tucker (chronological order). The roster for 1935–1938 included A. H. Taub, J. Bardeen, N. E. Steenrod, A. M. Turing. During the early thirties the Princeton mathematics department had almost a monopoly on the post-graduate Procter fellowships. These were fellowships awarded to students at Oxford, Cambridge and the University of Paris (one to each annually) and were open to all departments of the university. Among the recipiants in mathematics were E. Linfoot, A. M. Turing, C. Ehresmann and C. Chabauty.

The Princeton mathematics faculty during this time was outstanding. In topology its senior members were J. W. Alexander and S. Lefschetz. Lefschetz also still had an active interest in algebraic geometry. Topology and differential geometry were the strongest fields. The faculty in the latter consisted of L. P. Eisenhart, O. Veblen, T. Y. Thomas and M. S. Knebelman. For a while, E. Hille, who left Princeton for Yale in 1933, was the only analyst. A junior member, H. F. Bohnenblust was added and S. Bochner was appointed after Hille resigned. The only algebraist was J. H. M. Wedderburn and the only logician was A. Church. The members in mathematical physics were H. P. Robertson and E. U. Condon. J. von Neumann and E. Wigner shared a visiting professorship of mathematical physics. Besides the regular faculty, Princeton had a number of foreign visiting professors and lecturers: P. Alexandroff (from whom I took a semester's course in topology), W. Blaschke, H. Bohr, R. Courant, W. V. D. Hodge, G. Polya, J. A. Schouten.

During the thirties, Princeton's mathematics journal, *Annals of Mathematics*, published some of the most important papers of that period. Among these was Haar's paper on measures on locally compact groups. This led to von Neumann's paper that gave an affirmative answer to Hilbert's fifth problem for the case of compact groups. Another early application of Haar's measure was contained in Pontrjagin's paper on structure and duality of commutative groups (given in full generality by van Kampen).

All of these papers appeared in the *Annals* as did Pontrjagin's paper on duality for topological spaces.

My dissertation advisor was Wedderburn. I was fascinated by the material of his course on matrices that became his book, *Lectures on Matrices*. The last part of the course was devoted to an exposition of Wedderburn's classical structure theory of finite dimensional algebras over arbitrary fields (dating from 1908). In some sense his results reduced the study of a finite dimensional algebra to that of a division algebra. These became a focal point of research during the late 1920's and early 1930's. The most general construction of finite dimensional division algebras was that of crossed products, which included cyclic algebras. An open question that was settled in the affirmative by S. A. Amitsur in 1972 was: Do there exist non-crossed product central division algebras? Wedderburn suggested that I study division algebras that could be defined as idealizers of one-sided ideals of rings. I undertook such a study for skew polynomial rings. This could give only cyclic algebras, but it led to some new results on these that constituted my dissertation. This was published in the Annals [1]. During this period I became acquainted also with Wedderburn's paper, "Algebras which do not possess a finite basis" that had been published in the 1924 *Transactions of the American Mathematical Society*. This paper was one of those that inspired my later work on structure theory of rings ([31], [32], [33], [34]).

The advent of the Institute for Advanced Study in 1932 had great potential, but in the beginning it amounted only to the transfer of several members of the Princeton department to the new Institute and the addition of Einstein—who did not lecture. Gradually new members were added: Hermann Weyl in 1933 and Marston Morse in 1935. From the time of his arrival, Weyl gave a number of important courses, perhaps the most influential of which was his first course, Continuous Groups I. and II. given in 1934. Richard Brauer had been designated as Weyl's research assistant, but he was unable to arrive in Princeton until the fall of 1934. I was asked to fill the gap and I wrote up the lecture notes for I. given in the spring of 1934. This part of the course was largely preparatory to the structure and representation theory of semi-simple Lie groups presented in II. The last part of II. was devoted to a "Seminar on various topics in group theory." This included lectures by Brauer on the determination of the Betti numbers of the classical simple groups and by H. S. M. Coxeter on groups generated by reflections. The achievement of Lie had been to reduce the study of local properties of an analytic group to the study of corresponding properties of a certain non-associative algebra called its infinitesimal group. In his lectures, Weyl proposed an independent study of such algebras, which became known as Lie algebras. He felt that it would be of interest to study these algebras over arbitrary fields of characteristic 0 and that it would be desirable to derive their properties "rationally", that is, without recourse to extension of the base field to its algebraic closure—as had been done by Wedderburn in 1908 for associative algebras. These goals attracted A. A. Albert and myself. My first paper on Lie algebras, "Rational methods in the theory of Lie algebras" [4] was directed toward the second goal.

During the spring semester of 1935, Emmy Noether, who had left Germany and had taken a position as professor at Bryn Mawr College, came to the Institute once a week to lecture on class field theory. I had met her the summer before at a meeting of the American Mathematical Society at which I had presented an abstract of my dissertation, and she had made a comment on my paper. I became better acquainted with her during her visits to Princeton and especially at occasional dinners at the

Brauers'. These contacts with her were a memorable experience, as was attending her course. It was a terrible shock when we learned of her passing on April 14 as a result of an operation. On April 26 a Memorial Service was held for her at Goodhart Hall of Bryn Mawr College that I and many of her friends and students from Princeton attended. It was a moving occasion at which we were enthralled to hear the Memorial Address by Professor Weyl that is reproduced in his *Gesammelte Abhandlungen* (Springer-Verlag, 1968, vol. 3, pp. 425–444).

Since a number of courses by Emmy had been announced for the following year, Professor Anna Pell Wheeler, the chairperson of the department at Bryn Mawr, had to find someone who could give these courses. She offered me a lectureship for the year 1935–1936. As I recall, my teaching duties amounted to three courses: one undergraduate course, one beginning graduate course and one advanced graduate course. This was my first full-time teaching experience.

The following year I held a National Research Council fellowship. Most of that academic year I spent at the University of Chicago working with Albert. Then I returned to Princeton since this appeared to be a better vantage point for seeking a regular teaching position.

Perhaps it is appropriate to describe the employment situation at that time. First, this was in the depth of the Great Depression. Salaries declined in some instances and there were very few new positions. Moreover, for the new Jewish Ph.D's the situation was further aggravated by anti-Semitism that was prevalent, especially in the top universities—the only ones that had any interest in fostering research. There were no Jews on the mathematics faculties of Yale or Harvard, one at Chicago and three at Princeton, and it had been decided either by the department or by the administration that these two departments could hire no more Jews. The chairman of the department at Cornell, who happened to be Jewish, came to the annual meeting of the American Mathematical Society held that year at St. Louis to interview candidates for four openings at Cornell. He interviewed no Jews, and he filled the positions with four young Ph.D's, three of whom had top credentials and all of whom were non-Jewish.

For me there was one mitigating circumstance. Professor Archibald Henderson, a Southern gentleman and scholar[3], was head of the department at the University of North Carolina. At the same time the university had an extraordinary president, Dr. Frank Graham, who was noted for his espousal of liberal causes. Professor Henderson decided to upgrade his department by hiring two research mathematicians and giving them light teaching loads so that they would have ample time for research. He hired Reinhold Baer, a recent emigré from Nazi Germany, as an assistant professor and me as an instructor. After a year, in 1938, I was promoted to assistant professor. In 1940 I accepted a position as visiting associate professor at Johns Hopkins University for one year to replace Oscar Zariski who had been invited to Harvard as visiting professor. My understanding at the time was that it was likely that he would stay at Harvard, in which case, I would be offered a tenure position at Hopkins. However, at the end of the year he returned to Hopkins and I returned to North Carolina—with a promotion to associate professor.

---

[3] He had written a Cambridge tract *The Twenty-Seven Lines Upon a Cubic Surface* in 1911, had written an authorized biography of Bernard Shaw and numerous articles on drama and books on American history, especially the history of North Carolina.

In the summer of 1938 Albert organized a conference on algebra at the University of Chicago. I was invited to participate and to be a visiting lecturer for the summer quarter. As such I gave a course entitled *Continuous Groups* that was perhaps more notable for its audience, that included Irving Kaplansky and George Whitehead, than for its content: Topological Spaces, Topological Groups, Local Isomorphism, Elements of Lie Theory. I prepared a set of lecture notes for the course, including exercises. An account of the algebra conference is given in the paper "Some recent advances in algebra" by MacLane in the *American Mathematical Monthly*, XLVI, 1939, pp. 3–19.

Soon after the United States was drawn into World War II by the Japanese attack on Pearl Harbor, the Navy established a Pre-Flight School for training pilots on the campus of the University of North Carolina. The mathematics department of this school was staffed by younger members of the faculty of the university including myself and Ralph Boas from Duke. During the spring of 1942, prior to assuming our teaching duties at the Pre-Flight School, we were sent to Chicago for a month's course in teacher training. We had daily homework assignments. For me these became a pleasant task, thanks to the help of Florence Dorfman, a student at the University of Chicago, whom I had met the previous summer. She had obtained a masters degree with Otto Schilling and was working for a doctorate with Adrian Albert. Florie and I were married August 25, 1942. The Navy granted me an "emergency leave" for the wedding. (It was not an emergency; our first-born, Michael Sidney, was born in 1944.)

My title at the Pre-Flight school was Associate Ground School Instructor. The teaching duties consisted of several sections of arithmetic and of a course called Principles of Flying. At the end of the year the Navy decided to convert its civilian instructors to reserve officers. All except Boas and myself became officers.

The elimination of civilian instructors at the Pre-Flight School and the fact that I was not needed at the University made it advisable for me to seek a temporary position elsewhere. The most attractive possibility appeared to be an offer from Johns Hopkins of an associative professorship for a two year term. A substantial part of the teaching there would be in an army training program and while there was no guarantee that the position would become a permanent one, Professor Murnaghan, the chairman, wrote that I shouldn't worry "about how permanent the appointment is. The facts are that we have no algebraist here and both Zariski and I want one very much and at present you are the man we want". I accepted the Hopkins offer and we moved to Baltimore in September of 1943.

In February of 1943 I gave an invited address, Some Aspects of the Theory of Semi-Linear Transformations, at the February meeting in New York of the American Mathematical Society.

The main topics of my research prior to the move to Hopkins were: (Ore) Skew Polynomial Domains, Topological Algebra, Lie Algebras and General Non-Associative Algebras, Galois Theory. The papers dealing with the first topic are [1], [2], [5], [8], [16]. I have already indicated the background of [1], which was my dissertation. The main result of this paper is a determination of the index of a cyclic algebra by a norm condition in an Ore polynomial ring. An improved derivation of this result is given in section 1.7 of *Finite Dimensional Division Algebras*. The paper [2] gives a characterization of skew polynomial rings by properties of the division algorithm that can defined in such a ring. Paper [5] is an abstract of [8]. In these papers I defined pseudo-linear transformation of a vector space, a concept that encompasses semi-linear transformations as well as differential transformations. The theory of a single pseudo-linear transformation is equivalent to that of a module over a skew polynomial ring. The ring of linear transformations that commute with a given pseudo-linear transformation T is isomorphic to the ring of endomorphisms of the module over the skew polynomial ring associated with T. If T is a differential transformation in a vector space $\mathfrak{R}$ over a field $\mathfrak{F}$, that is, T is additive and $(x\alpha)T = (xT)\alpha + x\alpha'$ where $\alpha \to \alpha'$ is a derivation d in $\mathfrak{F}$, then the ring of linear transformations commuting with T is a finite dimensional algebra over the field of d-constants. Division algebras obtained in this way were studied in Amitsur's paper "Differential polynomials and division algebras" (*Annals of Mathematics* 59, 1954, pp. 245–278).

Classically one defines a normal linear transformation T in a finite dimensional vector space over $\mathbb{R}$ or $\mathbb{C}$ equipped with a positive definite symmetric or hermitian scalar product by the condition $TT^* = T^*T$, where T^* is the adjoint of T relative to the given scalar product. In [16] I considered a generalization of this concept to semi-linear transformations in a vector space over a division ring with involution equipped with a hermitian or skew hermitian scalar product that is totally regular in the sense that $(x, x) = 0 \Rightarrow x = 0$. The results obtained have applications to projective geometry.

The papers on topological algebra are [3] (jointly with O. Taussky), [6], [13]. In 1932 Pontrjagin, in a paper, "Über stetige algebraische Körper," published in the *Annals of Mathematics*, proved that any connected locally-compact (bicompact in his terminology) division ring is topologically isomorphic to either $\mathbb{R}$, $\mathbb{C}$ or Hamilton's quaternion algebra $\mathbb{H}$. After the appearance in 1934 of the Pontrjagin-van Kampen structure theory of locally-compact abelian groups it appeared natural to derive Pontrjagin's theorem on connected locally-compact division rings by applying the structure theory of locally-compact abelian groups to the additive group of the division ring. This was done in [3] and this method was applied to obtain results on connected locally compact rings in general. The thrust of these was that with mild restrictions such rings are finite dimensional algebras over $\mathbb{R}$. We showed also that a separable locally-compact (not necessarily associative) field is either connected or totally-disconnected.

Paper [6] has two parts: the first part deals with the structure of the additive group of a totally-disconnected locally-compact separable (non-discrete) ring. The second part gives a sharp determination of the totally disconnected locally compact separable associative division rings $\mathfrak{F}$. It is shown that any such $\mathfrak{F}$ is a special type of cyclic algebra over its center $\mathfrak{C}$ which, by a result of D. van Dantzig, is either an extension of finite degree of a field of p-adic numbers or a field of Laurent series over a finite field. An improved treatment of the structure theory of totally-disconnected locally-compact division rings is given in *Basic Algebra* II, pp. 599–608.

Paper [13], "A note on topological fields" arose in connection with the problem of classifying simple Lie algebras over a $\mathfrak{P}$-adic field. The result proved is the following: Let $\mathfrak{F}$ be a locally compact separable (non-discrete) totally disconnected division ring with an involution F of second kind. Then $\mathfrak{F}$ is commutative.

I have already mentioned my first paper on Lie algebras: [4] "Rational methods in the theory of Lie algebras". Between 1937 and 1943 I wrote a number of papers on Lie algebras dealing with the problem of the classification of simple Lie algebras over arbitrary fields of characteristic 0 and the beginnings of a similar study for Lie algebras over fields of characteristic p ($\neq 0$). The problem of classifying the simple Lie algebras over $\mathbb{R}$ was solved by Elie Cartan in 1914. He showed that the problem could be reduced to the sub-problem of classifying the normal (=central) simple ones, defined to be those that remain simple on extension of the base field $\mathbb{R}$ to $\mathbb{C}$ and hence became one of the algebras A_n, B_n, C_n, D_n, G_2, F_4, E_6, E_7, or E_8 in the Killing-Cartan list of simple algebras over $\mathbb{C}$ (Cartan's Thèse, 1894).

In 1935, W. Landherr, who was a doctoral student of Artin's, wrote a dissertation on the problem of classifying the simple Lie algebras over arbitrary fields of characteristic 0. He showed that, as in Cartan's case, reduction could be made to algebras that are normal simple in the sense that they remain simple on extensions of the base field Φ to its algebraic closure Ω. Moreover, he studied the Lie algebras $\mathfrak{L}$ of type A, that is, the $\mathfrak{L}$ such that $\mathfrak{L}_\Omega \equiv \Omega \otimes_\Phi \mathfrak{L} \cong M_n(\Omega)'$, the Lie algebra of $n \times n$ matrices of trace 0 over Ω. He distinguished two types, A_{I}, in which the elements of $\mathfrak{L}$ regarded as $n \times n$ matrices with entries in Ω all have characteristic polynomials in $\Phi[\lambda]$, and A_{II}, in which there are elements of $\mathfrak{L}$ whose characteristic polynomials are not in $\Phi[\lambda]$. He obtained some general theorems on the Lie algebras of type A_{I}, namely, if $\mathfrak{A}$ is a central simple associative algebra over Φ and $\mathfrak{A}_l$ is the corresponding Lie algebra obtained by replacing the associative product ab by the Lie product $[ab] = ab - ba$ then the derived Lie algebra $\mathfrak{A}_l'$ is of type A_{I} and every Lie algebra of type A_{I} is obtained in this way. He showed also that if $\mathfrak{A}$ and $\mathfrak{B}$ are central simple associative then $\mathfrak{A}_l' \cong \mathfrak{B}_l'$ if and only if $\mathfrak{A}$ and $\mathfrak{B}$ are either isomorphic or anti-isomorphic. He gave a complete classification of the Lie algebras of type A over any $\mathfrak{P}$-adic field, and in his Habilitatsionschrift (1938) he obtained a similar result for Lie algebras of type A over any number field.

My first paper on the classification problem was [7], which is an announcement of the results of [14]. These papers give a determination of the simple Lie algebras of type A_{II} over a field Φ of characteristic 0: Let $\mathfrak{A}$ be an involutorial simple associative algebra of second kind, J the involution in $\mathfrak{A}$. Let $\mathfrak{S}_J$ be the set of J-skew elements of $\mathfrak{A}$. Then $\mathfrak{S}_J$ is a subalgebra of the Lie algebra $\mathfrak{A}_l$ and the derived algebra $\mathfrak{S}_J'$ is a simple Lie algebra of type A_{II}. Moreover, any simple Lie algebra of type A_{II} over Φ can be obtained in this way and, if $\mathfrak{A}_1$ and $\mathfrak{A}_2$ are involutorial simple of second kind with involutions J_1 and J_2, respectively, then $\mathfrak{S}_{J_1}' \cong \mathfrak{S}_{J_2}'$ implies $\mathfrak{A}_1 \cong \mathfrak{A}_2$, and if $\mathfrak{A}_1$ and $\mathfrak{A}_2$ are identified, then J_1 and J_2 are cogredient ($J_2 = S^{-1}J_1 S$ for an automorphism S).

Similar results were obtained for Lie algebras of type B, C and D in [9]. A Lie algebra $\mathfrak{L}$ is of type B (resp. D) if $\mathfrak{L}_\Omega \cong \mathfrak{S}_t(M_n(\Omega))$ the subalgebra of $M_n(\Omega)_l$ of t-skew matrices where t is the transpose involution $X \to X'$ and n is odd (resp. even). The Lie algebra $\mathfrak{L}$ is of type C if $\mathfrak{L}_\Omega \cong \mathfrak{S}_{t_S}(M_n(\Omega))$ where $n = 2\nu$ and t_S is the involution $X \to S^{-1}X'S$ in $M_n(\Omega)$), $S = \begin{pmatrix} 0 & 1_\nu \\ -1_\nu & 0 \end{pmatrix}$. To insure simplicity and to avoid duplication one assumes $n > 5$ for type B, $n > 2$ for type C, and $n > 6$ for type D. Also it is necessary to assume $n \neq 8$ (so $n > 8$) for type D. This is due to the fact that the group of automorphisms for $\mathfrak{S}_t(M_8(\Omega))$ is more complicated than for other Lie algebras of type D, and this implies that the results obtained for Lie algebras of type D in general fail in the case in which $n = 8$ (and the dimensionality of $\mathscr{L}$ over Φ is 28. Some results in this case are given in [71]).

In the short paper [10], "A note on non-associative algebras", I considered arbitrary non-associative algebras that are finite dimensional over a field Φ. For such an algebra

$\mathfrak{R}$, relations between $\mathfrak{R}$ and the associative algebra $\mathfrak{A}$ of linear transformations of the vector space generated by the left and right multiplications determined by the elements of $\mathfrak{R}$ were studied. I showed that $\mathfrak{R}$ is a direct sum of simple ideals if and only if $\mathfrak{A}$ is a semi-simple associative algebra, and $\mathfrak{R}$ is simple if and only if $\mathfrak{A}$ is simple. I defined the centroid $\mathfrak{C}$ ("centrum" in the terminology of [10]) as the center of the algebra $\mathfrak{A}$. If $\mathfrak{R}$ is simple then $\mathfrak{C}$ is a field and $\mathfrak{R}$ can be regarded in a natural way as an algebra over $\mathfrak{C}$. Defining $\mathfrak{R}$ to be normal simple if $\mathfrak{R}_\Omega$ is simple for Ω the algebraic closure of Φ, I showed that $\mathfrak{R}$ is normal simple if and only if $\mathfrak{R}$ is simple and its centroid $\mathfrak{C}$ is $\Phi 1$, 1, the identity map. I showed also that if $\mathfrak{R}$ is simple and n is its dimensionality over $\mathfrak{C}$ then $\mathfrak{A} \cong M_n(\mathfrak{C})$. If $\mathfrak{R}_1$ and $\mathfrak{R}_2$ are simple and $\mathfrak{C}_i$ is the centroid of $\mathfrak{R}_i$ then any isomorphism S of $\mathfrak{R}_1/\Phi$ onto $\mathfrak{R}_2/\Phi$ defines in a natural way an isomorphism s of $\mathfrak{C}_1$ onto $\mathfrak{C}_2$ and S is an s-semi-linear isomorphism of $\mathfrak{R}_1/\mathfrak{C}_1$ onto $\mathfrak{R}_2/\mathfrak{C}_2$.

Paper [15] uses the results of [10] to extend Landherr's result on Lie Algebras of type $A_{\rm I}$ and my results on types $A_{\rm II}$, B, C and D to non-normal algebras. I also showed that the isomorphism problems for the Lie algebras could be reduced to problems on equivalence of hermtian or skew hermitian bilinear forms over involutorial division algebras. The latter problems were solved for the case in which the base field was a $\mathfrak{P}$-adic field. This gave a complete classification of the simple Lie algebras of types $A - D$ ($n \neq 8$ for type D) over a $\mathfrak{P}$-adic field.

An improved derivation of the general results of this paper is given in Chapter X of *Lie Algebras*.

If $\mathfrak{R}$ is a finite dimensional algebra (not necessarily associative) over $\mathbb{C}$, the group of automorphisms of $\mathfrak{R}$ is a Lie group whose Lie algebra (=infinitesimal group) is the derivation algebra $\mathfrak{D}(\mathfrak{R})$ of $\mathfrak{R}$. This is the set of derivations of $\mathfrak{R}$, that is, the linear transformations D of $\mathfrak{R}$ such that $(xy)D = x(yD) + (xD)y$ for $x, y \in \mathfrak{R}$. Now replace $\mathfrak{C}$ by any field Φ. Then we obtain the Lie algebra $\mathfrak{D}(\mathfrak{R})$ of derivations of $\mathfrak{R}$. Something new occurs if the characteristic of Φ is a prime p: If D is a derivation then so is D^p. Hence $\mathfrak{D}(\mathfrak{R})$ is a restricted Lie algebra of linear transformations, that is, $\mathfrak{D}(\mathfrak{R})$ is a linear space of linear transformations closed under the binary composition $(D, E) \rightarrow [D, E] = DE - ED$ and the unary composition $D \rightarrow D^p$. These concepts were introduced in [11], as well as that of an abstract restricted Lie algebra over a field of characteristic p (whose definition was changed slightly in [22]; see below). The main results of [11] are: 1) the determination of the structure of $\mathfrak{D}(\mathfrak{R})$ for $\mathfrak{R}$ semi-simple associative (Theorems 6, 13); 2) a Galois correspondence between the subfields of $\mathfrak{R} = \mathfrak{F}(c_1, \ldots, c_m)$, $c_i^p = \gamma_i \in \mathfrak{F}$ and certain objects in $\mathfrak{D}(\mathfrak{R})$ (Theorem 12); 3) an analogue for derivations of Hilbert's Satz 90 (Theorem 15).

A Lie algebra $\mathscr{L}$ over Φ is said to be of *type G* if $\mathscr{L}_\Omega$ is isomorphic to the Lie algebra G_2 of 14 dimensions in the Killing-Cartan list. For $\Phi = \Omega$, Cartan had indicated in his paper on classification over $\mathbb{R}$ that $\mathscr{L} = \mathfrak{D}(\mathfrak{A})$ where $\mathfrak{A}$ is the (unique) Cayley (or octonion) algebra over Ω. In [19], I showed that if $\mathfrak{A}$ is any Cayley algebra over a field Φ (of characteristic 0) then $\mathfrak{D}(\mathfrak{A})$ is a Lie algebra of type G and every Lie algebra of type G is obtained in this way. If $\mathfrak{A}_1$ and $\mathfrak{A}_2$ are Cayley algebras then $\mathfrak{D}(\mathfrak{A}_1) \cong \mathfrak{D}(\mathfrak{A}_2)$ if and only if $\mathfrak{A}_1 \cong \mathfrak{A}_2$. These results reduce the problem of classifying the Lie algebras of type G to the more tractable problem of classifying the Cayley algebras over Φ. Substantially the same results were published in a paper by E. Bannow, "Die Automorphismengruppen der Cayley Zahlen", *Abhandlungen Mathematischen Seminar Hansischen Universität* 13, 1940, pp. 240–256.

If $\mathfrak{A}$ is an associative algebra over a field Φ of characteristic $p \neq 0$, one has the following identity

$$(a + b)^p = a^p + b^p + s(a, b)$$

where $s(a, b) = \sum_{i=0}^{p-1} s_i(a, b)$ and $is_i(a, b)$ is the coefficient of λ^{i-1} in $a(ad\ (\lambda a + b))^{p-1}$ and $a(adc) = [a, c]$. Thus $s(a, b)$ is a linear combination of Lie products in a and b of degree p. A formula like this is given as (15) in [11], and the explicit formula we have displayed appears in sec. 6 of H. Zassenhaus' paper, "Über Liesche Ringe mit Primzahlcharacteristik" (*Abhandlungen Mathematischen Seminar Hansischen Universität* 13, 1940, pp. 1–100; see also *Lie Algebras*, pp. 185 f.) Also, as shown in [11], $a(ad\ b)^p = a(ad\ b^p)$ and a^p is homogeneous of degree p. These observations led me to define in [22] the concept of a *restricted Lie algebra* $\mathscr{L}$ *over a field* Φ *of characteristic* p as a Lie algebra endowed with a unary composition $a \to a^{[p]}$ such that

$$(a + b)^{[p]} = a^{[p]} + b^{[p]} + s(a, b)$$

where $s(a, b)$ is as above and

$$(\alpha a)^{[p]} = \alpha^p a^{[p]}, \qquad \alpha \in \Phi$$

$$a(ad\ b)^p = a(ad\ b^{[p]}).$$

This paper is devoted to the basic concepts and elementary results on restricted Lie algebras. If $\mathfrak{A}$ is an associative algebra over a field of characteristic p, $\mathfrak{A}_l$ will now denote the restricted Lie algebra obtained from $\mathfrak{A}$ by using the Lie product $[ab] = ab - ba$ and taking $a^{[p]} = a^p$. Given any restricted Lie algebra, there exists an associative algebra $\mathfrak{U}$, called the u-algebra of $\mathscr{L}$ that is the analogue for restricted Lie algebras of the universal enveloping algebra: We have an isomorphism $l \to \{l\}$ of $\mathscr{L}$ onto a subalgebra $\{\mathscr{L}\}$ of $\mathfrak{U}$ and $\mathfrak{U}$ is generated by $\{\mathscr{L}\}$. Moreover, if θ is a homomorphism of $\mathscr{L}$ into an $\mathfrak{A}_l$, $\mathfrak{A}$ associative, then the homomorphism $\{\theta\}$: $\{l\} \to \theta l$ has a unique extension to a homomorphism of $\mathfrak{U}$ into $\mathfrak{A}$. If the dimensionality $[\mathscr{L} : \Phi] = m < \infty$ then $[\mathfrak{U} : \Phi] = p^m - 1$. This implies that in this case there are only a finite number of inequivalent restricted irreducible representations (homomorphisms of $\mathscr{L}$ as restricted Lie algebra into matrix Lie algebras $M_n(\Phi)_l$). As is known from Cartan's theory of representations of Lie algebras, this is not the case for ordinary Lie algebras.

Papers [23] and [24] are concerned with some classes of simple restricted Lie algebras over a field Φ of characteristic p. The first of these deals with the "classical" Lie algebras of types A–D that are the counterparts of the algebras treated in [15]. The second paper introduces a class of restricted Lie algebras that has no counterpart for finite dimensional Lie algebras of characteristic 0. These restricted Lie algebras are generalizations of the derivation algebras of purely inseparable field extensions of Φ defined in [11]. Let $\xi_1, \ldots, \xi_m \in \Phi$ and let $\mathfrak{A} = \Phi[x_1, \ldots, x_m]$ the commutative associative algebra with base $x_1^{\alpha_1} \ldots x_m^{\alpha_m}$, $0 \leq \alpha_i < p$, where $x_i^0 = 1$, $x_i^p = \xi_i$. Then $\mathscr{L} = \mathfrak{D}(\mathfrak{A})$ is a normal simple Lie algebra of dimension mp^m if either $p > 2$ or $m > 1$. If $m = 1$ and the $\xi_i = 1$ then $\mathscr{L}$ has a base $(e_0, e_1, \ldots, e_{p-1})$ with the multiplication table

$$[e_\alpha e_\beta] = (\alpha - \beta) e_{\alpha+\beta}$$

where the subscripts are reduced modulo p. This Lie algebra was discovered by Witt and was first published by Zassenhaus in the paper referred to above. If Φ is algebraically closed then by changing the set of generators x_i we may assume the $\xi_i = 1$ or 0. The algebra $\Phi[x_1, \ldots, x_m]$ with $x_i^p = 1$ is the group algebra over Φ of the direct product of m cyclic groups of order p. The derivation algebra $\mathscr{L}$ of this algebra is often called the *Jacobson-Witt algebra.* The following problem on descent was considered in [24]: Suppose $\mathscr{L}$ is a restricted Lie algebra over a field Φ of characteristic p such that for some extension field P/Φ, $\mathscr{L}_{\mathrm{P}}$ is a Jacobson-Witt algebra over P. Does there exist an algebra $\mathfrak{A}/\Phi$ of the form $\mathfrak{A} = \Phi[y_1, \ldots, y_m]$ with base $(y_1^{\alpha_1}, \ldots, y_m^{\alpha_m}, 0 \leq \alpha_i < p, y_i^p = \xi_i \in \Phi)$ such that $\mathscr{L} = \mathfrak{D}(£)$? An affirmative answer was given in [24] for P/Φ separable algebraic or purely inseparable of exponent one. An affirmative answer for arbitrary P was given by Harry Allen and Moss Sweedler in "A theory of linear descent based upon Hopf algebra techniques" (*Journal of Algebra* 11, 1969, pp. 242–294; especially pp. 288–294) and by William C. Waterhouse in "Automorphism schemes and forms of Witt Lie algebras" (*Journal of Algebra* 17, 1971, pp. 34–40).

A result attributed to me and to Y. Manin by Seligman (*Modular Lie Algebras*, Springer-Verlag, p. 106) is that any finite dimensional restricted Lie algebra $\mathscr{L}$ of characteristic p can be imbedded in a Jacobson-Witt algebra. Manin's proof appears in *Sibirskii Matematicheskii Zhurnal*, 3, 1962, pp. 479–480. Perhaps it is appropriate to sketch here my proof that has not been published previously.[4] We begin with the fact that $\mathscr{L}$ has a faithful representation $l \to L$ by linear transformations in a finite dimensional vector space V. Let $(\xi_1, \ldots, \xi_m)$ be a base for V and let $\mathfrak{O} = \Phi[\xi_1, \ldots, \xi_m]$ be the group algebra of the direct product of the groups $\langle \xi_i \rangle$ of order p, so $\mathfrak{D}(\mathfrak{O})$ is a Jacobson-Witt algebra. If $l \in \mathscr{L}$ the corresponding L has a unique extension to a derivation $d(l)$ of $\mathfrak{A}$. Then $l \to d(l)$ is an isomorphism of $\mathscr{L}$ onto a subalgebra of $\mathfrak{D}(\mathfrak{O})$.

In her paper "Nichtkommutative Algebra" (*Mathematische Zeitschrift*, 37, 1933, pp. 514–541; especially pp. 529–533), Emmy Noether developed a Galois theory of finite dimensional normal division algebras and normal simple algebras based on inner automorphisms. In [20] I considered the opposite case in which we have a finite group G of automorphisms of a quasi-field (=division ring or skew field) P such that every $S \neq 1$ in G is an outer automorphism. If Φ is the division subring of G-fixed elements then the left and right dimensionality of P over Φ is n and one has the usual Galois correspondence between the set of subgroups of G and the set of division subrings of P containing Φ. This was my first paper on Galois theory. It was followed by [25], [27] and [36] that were published in the period 1943–1946.

The three papers [17], [18] and [21] are somewhat isolated, though the theorem proved in [17] on symplectic symmetric matrices (matrices B such that $R^{-1}B'R = B$ where R is a non-singular skew symmetric matrix) can be viewed as the determination of the generic minimum polynomial of symplectic symmetric matrices (cf. *Structure and Representations of Jordan algebras*, p. 230 f.). The proof is based on a definition of the determinant of a hermitian quaternionic matrix due to E. H. Moore.

If $\mathfrak{R}$ is a vector space over a quasi-field Φ with an involutorial anti-automorphism $\alpha \to \bar{\alpha}$ then one can define hermitian forms on $\mathfrak{R}$ over Φ with respect to $\alpha \to \bar{\alpha}$. In [21] I consider such forms in which Φ is either a quadratic extension of Φ_0 of characteristic $\neq 2$ or is a quaternion algebra over Φ_0. In the first case $\alpha \to \bar{\alpha}$ is the automorphism $\neq 1_{\Phi}$ in Φ and in the second it is the standard map $\alpha \to tr(\alpha)1 - \alpha$ in Φ. Given a hermitian form (x, y) in $\mathfrak{R}$ over Φ I associated to (x, y) the symmetric bilinear form $\{x, y\} = \frac{1}{2}tr(x, y)$. The main results of [21] is a characterization of the $\{x, y\}$ that can be obtained in this way and the theorem that equivalence holds for two her-

[4] I have not been able to date this proof.

mitian forms if and only if it holds for the associated symmetric bilinear forms. This result reduced the study of hermitian forms to that of symmetric bilinear forms. A number of important special cases are considered in [21]. (cf. *Symmetric Bilinear Forms* by J. Milnor and D. Husemoller, Springer Ergeb. Series, v. 73, 1973, p. 114 f.).

Paper [18] contains a number of remarks on the connection between Lie groups and the corresponding Lie algebras.

ANNALS OF MATHEMATICS
Vol. 35, No. 2, April, 1934

NON-COMMUTATIVE POLYNOMIALS AND CYCLIC ALGEBRAS[1]

BY NATHAN JACOBSON

(Received October 18, 1933)

Introduction

Consider the cyclic algebra $\mathfrak{F}'(\Pi)$ whose basis consists of the products $a^j x^k$ ($j, k = 0, 1, \cdots, r-1$) over a field $\mathfrak{F}$ where a generates a cyclic field $\mathfrak{F}' = \mathfrak{F}(a)$ over $\mathfrak{F}$ and

$$xa = a^{(1)}x \tag{1}$$

$$\Pi = x^r - \gamma = 0 \qquad (\gamma \neq 0) \tag{2}$$

where a is an element of $\mathfrak{F}'$ such that the correspondence a into $a^{(1)}$ generates the Galois group of $\mathfrak{F}'$ over $\mathfrak{F}$. We consider $\mathfrak{F}'(\Pi)$ as a residue algebra of the polynomial Π in the domain $\mathfrak{F}'[x]$ of all polynomials in x whose coefficients $\subset \mathfrak{F}'$ and for which multiplication is defined by (1). The main purpose of the present paper is to show that the problem of determining the division algebra part of $\mathfrak{F}'(\Pi)$ in the Wedderburn decomposition may be reduced to the problem of finding a single irreducible factor of Π in $\mathfrak{F}'[x]$. In §1 we discuss the factorization theory in $\mathfrak{F}'[x]$. In §2 and §3 we determine the relation between $\mathfrak{F}'(\Pi)$ and an algebra associated with a factor of Π. Finally, in §4 we derive a "norm" condition in order that Π have an irreducible factor of degree t and from this condition the usual norm conditions for cyclic algebras follow as simple corollaries. The methods used are applicable to the algebras defined by (1) and (2) where $\mathfrak{F}'$ is any division algebra of finite order over $\mathfrak{F}$. We hope to consider this case in a later paper.

In the course of the preparation of this paper, I had the privilege of discussing its details with Professor Wedderburn. I am very grateful to him for the stimulus of these discussions.

1. Cyclic Polynomial Domains

Let $\mathfrak{F}' = \mathfrak{F}(a)$ be a cyclic field of order r generated by a over the field $\mathfrak{F}$ and $a^{(1)} = \theta(a)$ be another root of the minimal equation satisfied by a in $\mathfrak{F}$ such that the correspondence a into $a^{(1)}$ is a generating automorphism of the cyclic Galois group of $\mathfrak{F}'$ over $\mathfrak{F}$. We define the *cyclic polynomial domain*, $\mathfrak{F}'[x]$, as the set of expressions of the form

$$P = a_0 + a_1 x + \cdots + a_n x^n$$

---

[1] Presented to the American Mathematical Society, June 19, 1933.

where the $a_i \subset \mathfrak{F}'$ and x is an indeterminate. Addition of such expressions is defined in the usual manner. Multiplication is defined by (1) together with the distributive and associative laws. From (1) one obtains

$$x^k b = b^{(k)} x^k$$

where $b = b(a)$ is a polynomial of degree $< r$ in a and $b^{(k)} = b(\theta^{(k)}(a))$, $\theta^{(k)}$ being the k-th iterative of θ. It is easily verified that $\mathfrak{F}'[x]$ is a domain of integrity. It may also be considered as an algebra with an infinite basis over $\mathfrak{F}$. In the present section we take the former viewpoint and give an outline of the theory of factorization in $\mathfrak{F}'[x]$. We mention first certain well known properties of $\mathfrak{F}'[x]$ referring to the papers of Wedderburn and of Ore for the proofs.[2]

For any two polynomials P and Q there exists a division process leading to polynomials S, R and S_1, R_1 such that

$$P = SQ + R \qquad P = QS_1 + R_1$$

where the degree of R, $\delta(R) < \delta(Q)$ and $\delta(R_1) < \delta(Q)$. Further, S, R and S_1, R_1 are unique.

The existence of a highest common right (left) factor of P and Q, $(P, Q)_R$ $((P, Q)_L)$ and a least common left (right) multiple, $[P, Q]_L$ $([P, Q]_R)$ follows from the division process.[3] These are unique except for left (right) unit multipliers. We have

$$\delta(PQ) = \delta[P, Q]_L + \delta(P, Q)_R = \delta[P, Q]_R + \delta(P, Q)_L. \tag{3}$$

There exists A_1, B_1 and A_2, $B_2 \subset \mathfrak{F}'[x]$ such that

$$(P, Q)_R = A_1P + B_1Q \qquad (P, Q)_L = PA_2 + QB_2.$$

We recall the definition of a *generalized left-transform* of P by Q as $P^* = [P, Q]_L Q^{-1}$.[4] If $(P, Q)_R = 1$, then (3) shows that $\delta(P) = \delta(P^*)$ and P^* is said to be *left-equivalent*[5] to P. Right-transforms and right-equivalence may be defined in an analogous way. We refer to Ore's paper (O. pp. 488–491) for a discussion of transforms, recalling merely that right- and left-equivalence imply each other and hence we drop the modifiers "right" and "left" in what follows.

The fundamental factorization theorem for $\mathfrak{F}'[x]$ is the following

THEOREM 1. *A polynomial A may be expressed as $P_1P_2 \cdots P_s$ where the P's are irreducible. If $Q_1Q_2 \cdots Q_t$ is another factorization of A into irreducible factors, then $s = t$ and the P's and Q's are equivalent in pairs.*

We refer to Ore's paper (O. p. 494) for the proof.

[2] J. H. M. Wedderburn, On continued fractions in non-commutative quantities, Ann. of Math., **15**, (1913–1914), pp. 101–105 and Non-commutative domains of integrity, Jour. für Math., **167**, (1932), pp. 129–141. The latter paper will be referred to as W. Ö. Ore, Theory of non-commutative polynomials, Ann. of Math., **34**, (1933), pp. 480–508, referred to as O.

[3] W. pp. 132–135 or O. pp. 483–486.

[4] W. p. 135 or O. p. 488.

[5] Ore (O. p. 488) uses the term *similar*. We prefer *equivalent* since *similar* has a different meaning in connection with the theory of algebras. See p. 204.

The preceding properties hold for a very wide class of domains of integrity, namely, the proper Euclidean domains as defined by Wedderburn.[6] We shall now discuss certain properties peculiar to $\mathfrak{F}'[x]$.

Let $\mathfrak{C}[x]$ denote the *centrum* of $\mathfrak{F}'[x]$ i.e. the set of elements of $\mathfrak{F}'[x]$ commutative with every element in $\mathfrak{F}'[x]$. If a polynomial Π is commutative with x, its coefficients $\subset \mathfrak{F}$ and if it commutes with a, it contains only terms in the powers of x^r. It follows that $\mathfrak{C}[x]$ consists of all polynomials in x^r with coefficients $\subset \mathfrak{F}$.

A domain of integrity has been called *Hamiltonian* by Wedderburn[7] if for every element P there exists a $\bar{P}$, an *adjoint* of P such that $P\bar{P} = \Pi \subset$ centrum of the domain. It follows that $\bar{P}P = \Pi$.

THEOREM 2. *$\mathfrak{F}'[x]$ is a Hamiltonian domain of integrity.*

For any P,

$$1P = P1, \qquad x^rP = Px^r, \cdots, x^{kr}P = Px^{kr} \cdots.$$

By division obtain

$$x^{kr} = PU_k + V_k. \tag{4}$$

Since the polynomials $V_0, V_1, V_2, \cdots$ all have $\delta < \delta(P)$, they can not be linearly independent with respect to $\mathfrak{F}$. There exists $\alpha_0, \alpha_1, \cdots \alpha_f$ not all zero such that

$$\sum_{k=0}^{f} \alpha_k V_k = 0. \tag{5}$$

From (4) and (5) we have

$$\Pi = \Sigma\alpha_k x^{kr} = P(\Sigma\alpha_k U_k) = P\bar{P}. \tag{6}$$

Since there exists a division process in $\mathfrak{C}[x]$, it follows easily that the polynomial Π of lowest degree (and hence the adjoint of lowest degree) is unique except for a multiplier $\subset \mathfrak{F}$. If the inverses of the polynomials $\mathfrak{C}[x]$ are adjoined to $\mathfrak{F}'[x]$, we obtain a division algebra containing $\mathfrak{F}'[x]$.[8]

THEOREM 3. *Let $\Pi \subset \mathfrak{C}[x]$ and be irreducible in $\mathfrak{C}[x]$. If $\Pi = P_1P_2 \cdots P_s$ where the P's are irreducible in $\mathfrak{F}'[x]$, then all the P's are equivalent.*

Suppose Π has the form (6) and $\alpha_f = 1$, $\alpha_0 \neq 0$.[9] Let Q be an irreducible right factor of Π with leading coefficient 1, i.e.

$$\Pi = RQ. \tag{7}$$

We may transform (7) by an element of $\mathfrak{F}'$ or by x obtaining

$$\Pi = b\Pi b^{-1} = (bRb^{-1})(bQb^{-1}) = x\Pi x^{-1} = (xRx^{-1})(xQx^{-1}), \quad b \subset \mathfrak{F}'. \tag{8}$$

[6] W. p. 132.
[7] W. p. 138.
[8] W. p. 138.
[9] If $\alpha_0 = 0$, $\Pi = x^r$ and the theorem is trivial.

The transform xRx^{-1} is obtained by taking conjugates of the coefficients of R. We indicate generally the result of any transformation as (8) by $\Pi = R'Q'$. It follows that $Q^* = [Q', Q]_L Q^{-1}$ is a right factor of R. Q^* is a unit if and only if Q' and Q differ by unit factors, which means in this case that $Q' = Q$. But, if $Q' = Q$ for every transformation, then $Q \subset \mathfrak{C}[x]$ and from the assumption of the irreducibility of Π in $\mathfrak{C}[x]$ it follows that $\Pi = Q$. In the contrary case we obtain a new factor Q^*, equivalent to Q' and hence to Q such that

$$\Pi = R_1Q^*Q = R_1Q_1.$$

We suppose Q_1 normalized so that its leading coefficient is 1 and then we use the same process on it as used before on Q. We obtain $Q_1^* = [Q_1, Q_1']_L Q^{-1}$ which has factors equivalent to Q unless $Q_1 = Q_1'$.[10] Continuing in this way we finally reach a stage at which $Q_k = Q_k'$ for every transform. As before, it follows that $\Pi = Q_k$. Thus, we obtain a factorization of the type desired. It follows then from Theorem 1 that any factorization has the desired properties.

2. Residue Classes in $\mathfrak{F}'[x]$

In this section we consider the properties of $\mathfrak{F}'[x]$ as an algebra over $\mathfrak{F}$.

The sub-algebra $\mathfrak{F}'[x]P$ (the set of elements of the form KP, $K \subset \mathfrak{F}'[x]$) is left-invariant in $\mathfrak{F}'[x]$. One may divide the elements of $\mathfrak{F}'[x]$ into classes modulo $\mathfrak{F}'[x]P$ or (mod P). The set of these classes forms a vector space Σ_P whose order is tr over $\mathfrak{F}$ where $t = \delta(P)$. In order that Σ_P be an algebra it must be closed under multiplication, i.e. from

$$A \equiv A_1 \ (\text{mod } P), \qquad B \equiv B_1 \ (\text{mod } P) \tag{9}$$

must follow

$$BA \equiv B_1A_1 \ (\text{mod } P). \tag{10}$$

But (9) and (10) imply

$$PA \equiv 0 \ (\text{mod } P) \quad \text{or} \quad PA = A'P. \tag{11}$$

Since A is arbitrary, it follows that a necessary and sufficient condition that Σ_P be an algebra is that (11) hold for every $A \subset \mathfrak{F}'[x]$. However, if we restrict ourselves to the sub-set, $\mathfrak{F}'[P]$, of elements A which satisfy (11), we may associate with every P a unique difference or residue algebra $\mathfrak{F}'(P) = \mathfrak{F}'[P] - \mathfrak{F}'[x]P$. $\mathfrak{F}'[P]$ is a domain of integrity and an algebra over $\mathfrak{F}$. It is in fact the maximal sub-algebra of $\mathfrak{F}'[x]$ which contains $\mathfrak{F}'[x]P$ as an invariant sub-algebra. We shall call $\mathfrak{F}'[P]$ the *normalizer*[11] of P and $\mathfrak{F}'(P)$ the *residue algebra* of P.

---

[10] See O. p. 488–491.

[11] Ore, in Formale Theorie der linearen Differentialgleichungen II, Jour. fur Math., **168**, (1932) p. 242, uses the term "*Eigenring*." We use *normalizer* because of the analogy with the theory of groups. We refer to this paper of Ore's as O′.

Suppose that $A \subset \mathfrak{F}'[P]$ and $(A, P)_R = 1$. There exist an A_1 and a P_1 such that

$$A_1A + P_1P = 1. \tag{12}$$

By division of PA_1 by P on the right we get $PA_1 = A_1'P + A_2$ where $\delta(A_2) < \delta(P)$. From (12) we get

$$PA_1A = P(1 - P_1P) = (1 - PP_1)P \tag{13}$$

and from (11)

$$PA_1A = A_1'PA + A_2A = A_1'A'P + A_2A \tag{14}$$

and from (13) and (14)

$$A_2A = (1 - P_1P - A_1'A')P.$$

Then $P^* = [P, A]_L A^{-1}$ is a right factor of A, which is impossible because of the degrees of P^* and A_2 unless $A_2 = 0$, or $PA_1 = A_1'P$. Since $\mathfrak{F}'(P)$ is an algebra it follows that there is a P_2 such that $AA_1 + P_2P = 1$. We have immediately

THEOREM 4. *If P is irreducible, $\mathfrak{F}'(P)$ is a division algebra.*[12]

In the same way we may treat the algebra $P\mathfrak{F}'[x]$ and we may define the left-normalizer $\mathfrak{F}_L'[P]$ as the set of elements A' which satisfy (11) and we may define the left-residue algebra as $\mathfrak{F}_L'(P) = \mathfrak{F}'[P] - P\mathfrak{F}'[x]$. If we let A and A' be corresponding elements, (11) gives an isomorphism between $\mathfrak{F}'[P]$ and $\mathfrak{F}_L'[P]$ and also between $\mathfrak{F}'(P)$ and $\mathfrak{F}_L'(P)$.

We shall require later the trivial remark that $\mathfrak{F}'(P) \cong \mathfrak{F}'(bPc)$ where $b, c \subset \mathfrak{F}'$. This is verified immediately.

3. The Structure of Cyclic Algebras

A *cyclic algebra* is defined to be an algebra $\mathfrak{F}'(\Pi)$ where $\Pi = x^r - \gamma$ $(\gamma \neq 0)$. $\mathfrak{F}'(\Pi)$ is a normal simple algebra. We shall now discuss the structure of $\mathfrak{F}'(\Pi)$ and its relation to the factorization of Π in $\mathfrak{F}'[x]$.

Suppose that $\Pi = P\bar{P}$ where $(P, \bar{P})_R = 1$. Then from (3) we have $[P, \bar{P}]_L = \Pi$. There exists a Q and an R' such that

$$Q\bar{P} + R'P = 1 \tag{15}$$

where we may take $\delta(Q) < \delta(P)$. From (11) we obtain

$$A\bar{P} = \bar{P}A'. \tag{16}$$

Since $Q \subset \mathfrak{F}'[P]$, we may set

$$Q\bar{P} = \bar{P}Q' = e \tag{17}$$

---

[12] This theorem has been given by Ore (O′ p. 242). Our proof, however, is different from his.

and since the rôles of P and $\bar{P}$ may be interchanged,

$$R'P = PR = \bar{e}. \tag{18}$$

Now, (15), (17) and (18) give $(P, \bar{P})_L = 1$ and as before

$$[P, \bar{P}]_R = \Pi. \tag{19}$$

(15), (17) and (18) imply also

$$e^2 \equiv e \pmod{\Pi}, \qquad \bar{e}^2 \equiv \bar{e} \pmod{\Pi}, \qquad e\bar{e} \equiv \bar{e}e \equiv 0 \pmod{\Pi}. \tag{20}$$

If $A \subset \mathfrak{F}'[P]$ from (15) and (17) follows

$$A = \bar{P}(Q'A) + (R'A')P = \bar{P}Y + ZP \tag{21}$$

and conversely any polynomial of the form $\bar{P}Y + ZP \subset \mathfrak{F}'[P]$. Thus, $\mathfrak{F}'[P]$ is identical with the set of these polynomials. If $\delta(A) < \delta(P)$, we may take $\delta(Y) < \delta(P)$ in (21).

As in §2 let Σ_P denote the set of elements of $\mathfrak{F}'[x]$ of $\delta < \delta(P) = t$. We have seen that Σ_P is a vector space whose order is rt over $\mathfrak{F}$. Let $\Sigma_P^{(1)} = \Sigma_P \wedge \mathfrak{F}'[P]$. $\Sigma_P^{(1)}$ is a vector space whose elements may be used as representatives of the elements of the residue algebra $\mathfrak{F}'(P)$. Finally, let $\Sigma_P^{(2)}$ be the sub-space of elements W of Σ_P such that $\bar{P}W = W^*P$ or, what amounts to the same thing, such that

$$W\bar{P} = PW^*. \tag{22}$$

We shall require later

LEMMA 1. $\Sigma_P = \Sigma_P^{(1)} + \Sigma_P^{(2)}$, $\Sigma_P^{(1)} \wedge \Sigma_P^{(2)} = 0$.

$\Sigma_P^{(1)} \wedge \Sigma_P^{(2)} = 0$. For, if W satisfies (22) and (16), then $\bar{P}W' = PW^*$ which is impossible because of (19) unless $W' = W^* = W = 0$. Now, (21) with $\delta(Y) < t$ gives a linear transformation of Σ_P into $\Sigma_P^{(1)}$ determined by dividing $\bar{P}Y$ on the right by P and letting Y correspond to A. $\Sigma_P^{(2)}$ is precisely the sub-space which goes into 0 under this transformation. Hence, order of Σ_P = order of $\Sigma_P^{(1)}$ + order of $\Sigma_P^{(2)}$ and from the previous result, $\Sigma_P = \Sigma_P^{(1)} + \Sigma_P^{(2)}$.

We prove next the fundamental

THEOREM 5. *If* $\Pi = P\bar{P} = \bar{P}P$, $(P, \bar{P})_R = 1$, *then* $\mathfrak{F}'(P) \cong e\mathfrak{F}'(\Pi)e$ *where* e *is given by* (17).

We recall the Peirce decomposition of $\mathfrak{F}'(\Pi)$ relative to the idempotent elements e and $\bar{e}$:

$$\mathfrak{F}'(\Pi) = e\mathfrak{F}'(\Pi)e + e\mathfrak{F}'(\Pi)\bar{e} + \bar{e}\mathfrak{F}'(\Pi)e + \bar{e}\mathfrak{F}'(\Pi)\bar{e}$$

or

$$S \equiv eSe + eS\bar{e} + \bar{e}Se + \bar{e}S\bar{e} \pmod{\Pi}.$$

eSe is divisible on both sides by $\bar{P}$ and if

$$eSe \equiv S_{11} \pmod{\Pi} \tag{23}$$

where $\delta(S_{11}) < r$, S_{11} is divisible on both sides by $\bar{P}$ i.e.

$$S_{11} = T_{11}\bar{P} = \bar{P}T'_{11}. \tag{24}$$

and so $S_{11} \subset \mathfrak{F}'[P]$. Similarly, we obtain

$$S_{12} = T^*_{12}P = \bar{P}T_{12}, \qquad S_{21} = T_{21}\bar{P} = PT^*_{21}, \qquad S_{22} = T'_{22}P = PT_{22} \tag{25}$$

where $\delta(S_{ik}) < r$ and

$$S \equiv S_{11} + S_{12} + S_{21} + S_{22} \pmod{\Pi}.$$

Now, if $\delta(S) < r$, from considerations of degree follows

$$S = S_{11} + S_{12} + S_{21} + S_{22}. \tag{26}$$

Conversely, suppose S is any element of $\delta < r$ and let (26) be an expression for S where we assume merely that the S_{ik} satisfy (24) and (25) and $\delta(S_{ik}) < r$. Then, this expression is the Peirce decomposition of S. To show this we have merely to show that an expansion of this type is unique. Suppose then $S = 0$ where S has the form (26) the S_{ik} satisfying (24) and (25). Then

$$\bar{P}(T'_{11} + T_{12}) = -P(T^*_{21} + T_{22})$$

which according to (19) implies $S_{11} + S_{12} = 0 = S_{21} + S_{22}$ and by the same kind of argument applied to $S_{11} + S_{12}$, $S_{11} = S_{12} = S_{21} = S_{22} = 0$ or (26) is unique.

We shall show that the correspondence between the class of $eSe \pmod{\Pi}$ and the class of $S_{11} \pmod{P}$, where S_{11} and eSe satisfy (23), is an isomorphism between $e\mathfrak{F}'(\Pi)e$ and $\mathfrak{F}'(P)$.

In the first place, the elements S_{11} obtained by (23) are a complete set of representatives of the elements of $\mathfrak{F}'(P)$. For, let T_{11} be any element of $\Sigma^{(1)}_P$. Then $S_{11} = T_{11}\bar{P} = \bar{P}T'_{11}$ has $\delta < r$ and according to the Peirce decomposition, (26), $S_{11} \subset e\mathfrak{F}'(\Pi)e$. As T_{11} takes on all values in $\Sigma^{(1)}_P$, S_{11} generates a complete set of residues $\pmod{P}$ since $\bar{P}$ has an inverse, Q, $\pmod{P}$. Hence, the correspondence between classes under consideration is $(1-1)$. It is evidently preserved under addition and scalar multiplication. Let

$$S^{(1)}_{11}S^{(2)}_{11} \equiv S^{(3)}_{11} \pmod{P}.$$

From (17) we have

$$eS^{(1)}_{11}S^{(2)}_{11}e \equiv eS^{(3)}_{11}e \pmod{\Pi} \tag{27}$$

and by (15)

$$eS^{(1)}_{11}S^{(2)}_{11}e = eS^{(1)}_{11}(e + \bar{e})S^{(2)}_{11}e = eS^{(1)}_{11}eS^{(2)}_{11}e + eS^{(1)}_{11}\bar{e}S^{(2)}_{11}e. \tag{28}$$

Further,

$$eS^{(1)}_{11}\bar{e}S^{(2)}_{11}e = \bar{P}Q'S^{(1)}_{11}R'PS^{(2)}_{11}\bar{P}Q' = \bar{P}Q'S^{(1)}_{11}R'S^{(2)}_{11}\Pi Q' \equiv 0 \pmod{\Pi} \tag{29}$$

and so from (27), (28), (29) and (20) follows

$$eS_{11}^{(3)}e \equiv (eS_{11}^{(1)}e)\ (eS_{11}^{(2)}e)\ (\text{mod }\Pi)$$

i.e. the correspondence is an isomorphism.

If we allow S_{12} in (25) to correspond to T_{12}, we obtain a non-singular linear transformation of the elements of $e\mathfrak{F}'(\Pi)\bar{e}$ into the elements of $\Sigma_P^{(2)}$. For, if $T_{12} \subset \Sigma_P^{(2)}$, as before, $S_{12} = \bar{P}T_{12} \subset e\mathfrak{F}'(\Pi)\bar{e}$ so that the correspondence is defined over the whole of $\Sigma_P^{(2)}$. It is evidently $(1-1)$ and linear with respect to $\mathfrak{F}$. Hence, we have

Lemma 2. *The orders of $e\mathfrak{F}'(\Pi)\bar{e}$ and $\Sigma_P^{(2)}$ over $\mathfrak{F}$ are equal.*

It follows from Theorem 5 that $\mathfrak{F}'(P)$ is a normal simple algebra *similar to* $(\sim)$ $\mathfrak{F}'(\Pi)$, i.e. their division algebras in the Wedderburn decomposition are isomorphic. If P is irreducible, $\mathfrak{F}'(P)$ is the division algebra part of $\mathfrak{F}'(\Pi)$ (in the sense of isomorphism) and e is a primitive idempotent element.

Theorem 5 holds also in the more general case in which $\Pi \subset \mathfrak{C}[x]$ and is irreducible there. However, it can be shown without difficulty that $\mathfrak{F}'(\Pi)$ is cyclic over its centrum and so from the point of view of the theory of algebras there is no gain in generality in this case. We return to our previous consideration of $\mathfrak{F}'(\Pi)$, $\Pi = x^r - \gamma$.

Lemma 3. *If $\Pi = P\bar{P}$, P irreducible, then there exists an associate, P_1, of P such that $\Pi = P_1\bar{P}_1$ and $(P_1, \bar{P}_1)_R = 1$.*

If $(P, \bar{P})_R \neq 1$, then $\bar{P} = TP = PT$ and

$$\bar{P}^2 \equiv 0\ (\text{mod }\Pi). \tag{30}$$

Let b and c be arbitrary elements of $\mathfrak{F}'$. Then

$$\Pi = (bPc)\ (c^{-1}Pb^{-1}) = \bar{P}_1P_1 = P_1\bar{P}_1.$$

If the lemma is false, then (30) holds for every $\bar{P}_1$. A direct calculation shows that $\bar{P}_1$ is divisible on both sides by P and so

$$\bar{P}_1\bar{P} \equiv \bar{P}\bar{P}_1 \equiv 0\ (\text{mod }\Pi) \tag{31}$$

and in fact (31) holds for any pair of associates of P. If x^k is a power of x actually present in $\bar{P}$, $a^{(k)}\bar{P} - \bar{P}a$ has one term less than P and by (30) and (31) is nilpotent. Continuing in this way we may obtain a linear combination of terms as $\bar{P}_1 = x^l (l < r)$ which is nilpotent. But this is impossible since $\gamma \neq 0$.

It follows from Theorem 5, Lemma 3 and the remark at the end of §2 that if $\Pi = P\bar{P}$, P irreducible, $\mathfrak{F}'(P) \cong$ division algebra part of $\mathfrak{F}'(\Pi)$.

Theorem 6. *If P is an irreducible factor of Π of degree t, the order of the division algebra part of $\mathfrak{F}'(\Pi)$ is t^2.*

It is only necessary to compute the order of $\mathfrak{F}'(P)$ where P is irreducible and $(P, \bar{P})_R = 1$. Let n^2 denote this order. By Lemma 1 $n^2 = rt - h$ where $h =$ order of $\Sigma_P^{(2)}$. The order of the matric part (e_{ik}) of $\mathfrak{F}'(\Pi)$ is then

$$m^2 = \frac{r^2}{n^2}. \tag{32}$$

h is also the order of $e\mathfrak{F}'(\Pi)\bar{e}$ according to Lemma 2. If e is taken as e_{11} in the matric part of $\mathfrak{F}'(\Pi)$, then $e\mathfrak{F}'(\Pi)\bar{e}$ is the direct product of the division algebra part $\mathfrak{F}'(\Pi)$ and the algebra whose basis is $(e_{12}, e_{13}, \cdots, e_{1m})$ over $\mathfrak{F}$. It follows that

$$m^2 = \left(\frac{h}{rt - h} + 1\right)^2 = \left(\frac{rt}{rt - h}\right)^2 = \frac{r^2t^2}{n^4}. \tag{33}$$

From (32) and (33) we find $n = t$.

Theorem 6 gives an indirect proof that the degrees of the irreducible factors of Π are equal. This is also a consequence of Theorem 3.

We summarize the results of this section in the following

THEOREM 7. *If $\Pi = P_1P_2 \cdots P_s$ where the P's are irreducible, all the P's have the same degree. If this common degree is t, $\mathfrak{F}'(\Pi)$ has division algebra of order t^2 and matric algebra of order s^2. Further, $\mathfrak{F}'(\Pi) \sim \mathfrak{F}'(P_i)$.*

4. Norm Conditions

We shall now derive a criterion based on Theorem 7 in order that the division algebra part of $\mathfrak{F}'(\Pi)$ have order t^2. From this criterion the norm conditions for cyclic algebras follow as immediate corollaries.

If $P \subset \mathfrak{F}'[x]$, evidently

$$P = p_{11} + p_{12}x + \cdots + p_{1r}\, x^{r-1}$$

where the p_{1i} are polynomials in x^r with coefficients $\subset \mathfrak{F}'$. This expression for P is unique. In the same way let

$$x^{k-1}P = p_{k1} + p_{k2}x + \cdots + p_{kr}\, x^{r-1} \qquad (k = 1, \cdots r). \tag{34}$$

If $\mathfrak{X}$ denotes the matrix

$$\begin{pmatrix} 1 \\ x \\ \vdots \\ x^{r-1} \end{pmatrix}$$

and $\mathfrak{P}$ the matrix (p_{ij}), then the equations (34) may be written as

$$\mathfrak{X}P = \mathfrak{P}\mathfrak{X}. \tag{35}$$

The correspondence of P into $\mathfrak{P}$ is readily verified to be an isomorphism.

THEOREM 8. *The characteristic equation of $\mathfrak{P}$ has coefficients which are polynomials in $\mathfrak{C}[x]$.*

If we allow the p_{1i} in the expression for P to have denominators $\subset \mathfrak{C}[x]$, then P represents a general element of the quotient algebra of $\mathfrak{F}'[x]$. This algebra has as its centrum the field $\mathfrak{K} = \mathfrak{F}(x^r = z)$ and is in fact the cyclic algebra $\mathfrak{K}'(\Omega)$ where $\mathfrak{K}' = \mathfrak{K}(a)$ and $\Omega = x^r - z$. The matrices $\mathfrak{P}$ give a representation of $\mathfrak{K}'(\Omega)$. The characteristic equation of $\mathfrak{P}$ is the minimal equation of the algebra $\mathfrak{K}'(\Omega)$. Hence, the coefficients of this equation $\subset \mathfrak{K}$.

Obviously, if $P \subset \mathfrak{F}'[x]$, the characteristic equation of $\mathfrak{P}$ has integral coefficients in $\mathfrak{R}$, i.e. $\subset \mathfrak{C}[x]$.

In particular, the determinant of $\mathfrak{P} \subset \mathfrak{C}[x]$. We shall call this determinant the *norm* of P, $N(P)$. If $b \subset \mathfrak{F}'$, $N(b)$ is the ordinary norm, the product of the conjugates of b. For $\Pi \subset \mathfrak{C}[x]$, $N(\Pi) = \Pi^r$. Finally, if

$$P = a_0 + a_1 x + \cdots + a_t x^t, \qquad a_i \subset \mathfrak{F}', \qquad t < r,$$

then

$$\mathfrak{P} = \begin{bmatrix} a_0 & a_1 & \cdot & \cdot & \cdot & \cdot & \cdot & \cdot & \cdot & a_t & \cdot & \cdot & 0 \\ 0 & a_0^{(1)} & & & & & & & & & a_t^{(1)} & & \cdot \\ \cdot & \cdot & \cdot & & & & & & & & & & \cdot \\ \cdot & \cdot & & \cdot & & & & & & & & & \cdot \\ \cdot & \cdot & & & \cdot & & & & & & & & \cdot \\ 0 & \cdot & \cdot & \cdot & 0 & a_0^{(r-t-1)} & & & & & & & a_t^{(r-t-1)} \\ a_t^{(r-t)}x^r & 0 & & & & & a_0^{(r-t)} & & & & \cdot & \cdot & a_{t-1}^{(r-t)} \\ & \cdot & & & & & & & & & \cdot & \cdot & \cdot \\ & & \cdot & & & & & & & & & \cdot & \cdot \\ & & & a_t^{(r-1)}x^r & & \cdot & \cdot & \cdot & \cdot & \cdot & \cdot & \cdot & a_0^{(r-1)} \end{bmatrix}$$

and so

$$N(P) = (-1)^t N(-1)^t N(a_t) x^{tr} + \cdots + N(a_0). \tag{36}$$

For any P we have

$$(\text{adjoint } \mathfrak{P})\mathfrak{P} = N(P)\mathfrak{J}$$

where $\mathfrak{J}$ is the unit matrix and hence from (35)

$$(\text{adjoint } \mathfrak{P})\mathfrak{X}P = N(P)\mathfrak{X}. \tag{37}$$

By comparing elements in the first row of the matrices on the two sides of the equality sign in (37), we find that P is a factor of $N(P)$. Further

$$N(PQ) = N(P)N(Q). \tag{38}$$

We may now prove the following

THEOREM 9. *If t is the smallest positive integer such that $\alpha\Pi^t = N(P)$ for some $\alpha \subset \mathfrak{F}$ and $P \subset \mathfrak{F}'[x]$, then the division algebra part of $\mathfrak{F}'(\Pi)$ has order t^2 and is isomorphic to $\mathfrak{F}'(P)$.*

Let P, α and t satisfy the conditions of the theorem. Then P is irreducible. For, suppose $P = P_1P_2$ where $\delta(P_1) = t_1 < t$. From (38) follows $\alpha\Pi^t = N(P_1)N(P_2)$. $N(P_1)$ and $N(P_2) \subset \mathfrak{C}[x]$ and have P_1 and P_2 respectively as factors. Since Π is irreducible in $\mathfrak{C}[x]$, we must have $N(P_1) = \beta\Pi^u$ where $u < t$ and $\beta \subset \mathfrak{F}$. This contradicts our assumption about t.

P is a right factor of $\alpha\Pi^t$ and if not a factor of Π, $(P, \Pi)_R = 1$. Hence, $P =$

$[P, \Pi]_L \Pi^{-1}$ is a factor of $\alpha \Pi^{t-1}$. By repeating this argument sufficiently often, we are lead to a contradiction.

Thus, P is an irreducible factor of Π and hence $\delta(P) \leqq r$. From (36) we have $\delta(P) = t$ and the theorem follows from Theorem 7.

If the leading coefficient of P is a_t, then $(-1)^t a_t^{-1} P$ also satisfies the condition of the theorem and has $(-\Pi)^t$ for its norm.

THEOREM 10. *A necessary and sufficient condition that $\mathfrak{F}'(\Pi)$ be matric is that γ be the norm of an element of $\mathfrak{F}'$.*[13]

By Theorem 9 the condition is that

$$\gamma - x^r = N(m - x) = N(m) - x^r.$$

Hence, $\gamma = N(m)$. By retracing the steps we get the converse.

THEOREM 11. *If q is the smallest positive integer for which γ^q is the norm of an element of $\mathfrak{F}'$, then the division algebra part of $\mathfrak{F}'(\Pi)$ has order $\geqq q^2$.*

If the order of the division algebra is t^2, then according to Theorem 9, there exists a P such that $N(P) = (\gamma - x^r)^t$ where we assume that the leading coefficient of P is $(-1)^t$ and last coefficient is a_0. From (36) follows

$$(-1)^t x^{tr} + \cdots + N(a_0) = (\gamma - x^r)^t$$

and hence $\gamma^t = N(a_0)$ so that $t \geqq q$.[14]

If $r = st$, there is a unique sub-field $\mathfrak{F}_t$ of $\mathfrak{F}'$ of order t over $\mathfrak{F}$. $\mathfrak{F}_t$ is in fact the sub-field of $\mathfrak{F}'$ which is invariant under the automorphism which sends a into $a^{(t)}$. If $b \subset \mathfrak{F}_t$, we denote $bb^{(1)} \cdots b^{(t-1)}$ by $N_t(b)$. By means of $\mathfrak{F}_t$ and the automorphism which sends any b of $\mathfrak{F}_t$ into $b^{(1)}$ we can define a cyclic polynomial domain $\mathfrak{F}_t[x]$. If P is a polynomial in $\mathfrak{F}_t[x]$, we denote its residue algebra in $\mathfrak{F}_t[x]$ by $\mathfrak{F}_t(P)$. We require the following

LEMMA. *If $P = x^t - \delta$, $\delta \subset \mathfrak{F}$, then $\mathfrak{F}'(P) \cong \mathfrak{F}_t(P)$ and is a cyclic algebra.*

For, if $A = a_0 + a_1 x + \cdots + a_{t-1} x^{t-1}$ and $PA = A'P$, then the $a_i \subset \mathfrak{F}_t$.

We may now prove a result of Albert's:[15]

THEOREM 12. *If $r = st$ and $\gamma^s = N(c)$, $c \subset \mathfrak{F}'$, then there exists an element $d \subset \mathfrak{F}_t$ such that $\gamma = N_t(d)$.*

By the hypothesis and by Theorem 10 we conclude that the cyclic algebra $\mathfrak{F}'(x^r - \gamma^s)$ is a matric algebra. But

$$x^r - \gamma^s = (x^t - \gamma)(x^{t(s-1)} + \gamma x^{t(s-2)} + \cdots + \gamma^{s-1}) = P\bar{P}$$

where $P = x^t - \gamma$. Evidently $(P, \bar{P}) = 1$ and hence by Theorem 5, $\mathfrak{F}'(x^r - \gamma^s) \sim \mathfrak{F}'(x^t - \gamma) \cong \mathfrak{F}_t(x^t - \gamma)$. Thus $\mathfrak{F}_t(x^t - \gamma)$ is matric and so by Theorem 10

$$\gamma = N_t(d) \qquad d \subset \mathfrak{F}_t.$$

[13] cf. Helmut Hasse, Theory of cyclic algebras over an algebraic number field, Trans. Am. Math. Soc., **34**, (1932), p. 199, or A. A. Albert, On the construction of cyclic algebras with a given exponent, Am. Jour. of Math., **54**, (1932), p. 10. These papers will be denoted by H. and by A. respectively.

[14] cf. H. p. 194 and A. p. 10.

[15] A. p. 8.

The following theorem also has been given by Albert:[16]

THEOREM 13. *A cyclic algebra is a direct product of cyclic algebras of prime power orders.*

We recall the well known theorem that the direct product $\mathfrak{F}'(x^r - \gamma_1) \times \mathfrak{F}'(x^r - \gamma_2) \sim \mathfrak{F}'(x^r - \gamma_1\gamma_2)$.[17] Now, suppose that $r = st$, $(s, t) = 1$ and determine l, m and k so that

$$ls + mt = 1 + kr \qquad l, m, k > 0.$$

Then

$$\mathfrak{F}'(x^r - \gamma^{ls}) \times \mathfrak{F}'(x^r - \gamma^{mt}) \sim \mathfrak{F}'(x^r - \gamma) \times \mathfrak{F}'(x^r - \gamma^{kr}).$$

But by Theorem 10, $\mathfrak{F}'(x^r - \gamma^{kr})$ is matric and hence

$$\mathfrak{F}'(x^r - \gamma) \sim \mathfrak{F}'(x^r - \gamma^{ls}) \times \mathfrak{F}'(x^r - \gamma^{mt}).$$

On the other hand, since

$$x^r - \gamma^{ls} = (x^t - \gamma^l)(x^{t(s-1)} + \gamma^l x^{t(s-2)} + \cdots + \gamma^{l(s-1)}) = P\bar{P}$$

where $P = x^t - \gamma^l$ and $(P, \bar{P}) = 1$, it follows as in the proof of Theorem 12 that $\mathfrak{F}'(x^r - \gamma^{ls}) \sim \mathfrak{F}_t(x^t - \gamma^l)$ and similarly $\mathfrak{F}'(x^r - \gamma^{mt}) \sim \mathfrak{F}_s(x^s - \gamma^m)$. Hence

$$\mathfrak{F}'(x^r - \gamma) \sim \mathfrak{F}_t(x^t - \gamma^l) \times \mathfrak{F}_s(x^s - \gamma^m).$$

By comparing the degrees of the algebras involved we have

$$\mathfrak{F}'(x^r - \gamma) \cong \mathfrak{F}_t(x^t - \gamma^l) \times \mathfrak{F}_s(x^s - \gamma^m).$$

If s or t is not a power of a prime we repeat the argument using $\mathfrak{F}_s(x^s - \gamma^m)$ or $\mathfrak{F}_t(x^t - \gamma^l)$ in place of $\mathfrak{F}'(x^r - \gamma)$. Continuing in this way we obtain the theorem.

---

[16] A. p. 9.

[17] A. p. 6.

PRINCETON UNIVERSITY.

ANNALS OF MATHEMATICS
Vol. 35, No. 2, April, 1934

A NOTE ON NON-COMMUTATIVE POLYNOMIALS

BY NATHAN JACOBSON

(Received October 18, 1933)

Ö. Ore[1] has recently considered polynomials in an indeterminate x whose coefficients (taken on the left) lie in a non-commutative field or division algebra $\mathfrak{A}$ where

$$xa = \bar{a}x + a' \qquad a, \bar{a}, a' \subset \mathfrak{A} \tag{1}$$

He has shown that if the associative and distributive postulates are assumed, the set of these polynomials constitutes a domain of integrity, $\mathfrak{A}[x]$, and the operations $(^-)$ and $(')$ in $\mathfrak{A}$ satisfy the conditions:

$$\overline{(a + b)} = \bar{a} + \bar{b} \qquad \overline{(ab)} = \bar{a}\bar{b}$$

$$(a + b)' = a' + b' \qquad (ab)' = \bar{a}b' + a'b.$$

For every element $A \neq 0, \subset \mathfrak{A}[x]$ there is defined a real non-negative number $\delta(A)$, (the degree of A) such that

(I) $\delta(AB) = \delta(A) + \delta(B) \qquad \delta(A + B) \leqq \text{Max.}\ (\delta(A), \delta(B))$

(II) No infinite sequence $A_1, A_2, A_3 \cdots$ can exist such that

$$\delta(A_1) > \delta(A_2) > \delta(A_3) > \cdots$$

(III) For every pair A, B there exists a Q and an R such that $A = QB + R$ where either $\delta(R) < \delta(B)$ or $R = 0$.

A domain of integrity satisfying (II) and (III) will be called *right-Euclidean.*[2] The property (I) will be referred to as the *non-Archimedian* property of δ.[3]

It is the purpose of the present note to prove the converse of these results and thus obtain an abstract characterization of polynomial domains.

THEOREM. *A domain of integrity in which there is defined a non-negative function δ having the properties (I), (II) and (III) is a polynomial domain.*

The condition $\delta(a) = 0$ is necessary and sufficient in order that a be a unit.[4]

---

[1] Ö. Ore, Theory of non-commutative polynomials, Annals of Math., **34** (1933) pp. 480–508.

[2] J. H. M. Wedderburn, in Non-commutative domains of integrity, Jour. für Math., **167** (1932) p. 132, calls a domain Euclidean if two sided division exists and (II) is satisfied in the domain.

[3] A. Ostrowski, Über einige Lösungen der Funktionalgleichung $\varphi(x)\varphi(y) = \varphi(xy)$, Acta Math., **41** (1916–1917), p. 272.

[4] cf. Wedderburn, loc. cit.

For it follows from (I) that $\delta(1) = 0$ and since $\delta(aa^{-1}) = \delta(a) + \delta(a^{-1}) = 0$ then $\delta(a) = 0 = \delta(a^{-1})$. Conversely, if $\delta(a) = 0$, (III) gives $1 = qa + r$ and since $\delta(a) = 0$, we must have $r = 0$ so that $q = a^{-1}$. (I) shows that the set of units is closed under rational operations and thus forms a non-commutative field $\mathfrak{A}$.

Let x be an element of the domain for which $\delta(x)$ has the minimum positive value. The existence of such an element follows from (II). Now, there is no element P for which

$$n\delta(x) < \delta(P) < (n + 1)\delta(x) \tag{2}$$

where n is a positive integer. For, if we obtain by division

$$P = a_0x^n + P_1 \tag{3}$$

where $\delta(P_1) < \delta(x^n)$, then it follows from (I) that $\delta(P) \leqq \delta(a_0x^n)$. But if P_1 is transposed to the side of P in (3) and (I) is again applied, we obtain $\delta(a_0x^n) \leqq \delta(P)$ and hence

$$\delta(P) = \delta(a_0x^n) = \delta(a_0) + n\delta(x) \tag{4}$$

and from (2) follows $0 < \delta(a_0) < \delta(x)$ contrary to the assumption on x.

Now, if $\delta(P) = n\delta(x)$, (4) shows that $\delta(a_0) = 0$ and so $a_0 \subset \mathfrak{A}$. By repeating the argument for P_1 and using induction as allowed by (II), we obtain finally

$$P = a_0x^n + a_1x^{n-1} + \cdots + a_n \tag{5}$$

where the $a_i \subset \mathfrak{A}$. It is evident from considerations of degree that the expression for P given by (5) is unique, i.e. x is transcendental over $\mathfrak{A}$.

The conditions (I) imply (1) and the proof is complete.

PRINCETON UNIVERSITY.

Reprinted from the Proceedings of the NATIONAL ACADEMY OF SCIENCES,
Vol. 21, No. 2, pp. 106–108. February, 1935.

LOCALLY COMPACT RINGS

BY N. JACOBSON AND O. TAUSSKY*

PRINCETON UNIVERSITY AND BRYN MAWR COLLEGE

Communicated January 10, 1935

1. L. Pontrjagin[1] has given a topological characterization of the fields of real numbers, complex numbers and quaternions. Recently[2] he described the structure of locally compact and connected abelian groups satisfying the second axiom of countability (separable). As every field and every ring (commutative or not) is an abelian group with respect to addition, it is possible to derive the previous results of Pontrjagin from his theorems on groups and it is possible to obtain by this method far more general results concerning topological rings.

E. R. van Kampen[3] extended Pontrjagin's main theorems on groups to the non-connected case. The following is the Pontrjagin-van Kampen theorem which is fundamental for the present paper:

Any locally compact and separable abelian group is the direct sum of a vector group N and a group C in which the component C_1 of the 0*-element is compact.*

From this it follows that C_1 is the maximal compact and connected subgroup of the whole group.

We are concerned with topological rings, i.e., having an infinite number of elements and for which $a + b$ and $a \cdot b$ are continuous functions of both variables.

By means of the above theorem we shall extend Pontrjagin's theorem on fields to the non-connected and non-associative case:[4]

THEOREM I. *A locally compact and separable (not necessarily associative or commutative) field F is either a hypercomplex system over the real field or is totally disconnected.*

2. LEMMA 1. *A field satisfying the first axiom of countability cannot be compact.*

If the field F were compact it would contain a sequence $a_n \longrightarrow 0$. Then the sequence a_n^{-1} diverges. For if a_{nj}^{-1} is a sub-sequence converging to a limit a, then $a_{nj}a_{nj}^{-1} = 1 \longrightarrow 0$, which is impossible.

LEMMA 2. *Let N be a topological ring which is—considered as a group with respect to addition—continuously isomorphic to an n-dimensional real vector group. Then N is a hypercomplex system over the field of real numbers.*

From the isomorphism and continuity it follows that N contains n elements such that every element of N is a linear form with real coefficients in these elements. It is merely necessary to show that the automorphism $x \longrightarrow \alpha x$ where α is a real number, commutes with multiplication: $x(\alpha y) = \alpha(xy) = (\alpha x)y$. This holds for integral values of α by the distributive law. Since there are no elements of finite order in N, it holds for rational values. By continuity it holds for all real α.

3. We now prove Theorem I. As F is a locally compact and separable abelian group when regarded additively, it is a direct sum $F = N + C$ where N is a vector group and C is a group in which the component C_1 of 0 is compact. Now C_1 is an ideal: If x is any element of F the set xC_1 is a compact and connected sub-group and hence $xC_1 \subset C_1$; in the same way it follows that $C_1 x \subset C_1$. Since F is a field it contains only the 0-ideal and the 1-ideal. Thus either $C_1 = 0$ or $C_1 = F$. But by Lemma 1, $C_1 = F$ is impossible and so $C_1 = 0$.

Now $F = N + C$ where N is a vector group and C a group in which the component of 0 is 0 itself, i.e., C is totally disconnected. We prove now that either $F = N$ or $F = C$. N is an ideal. For in the group F, N is the component of 0 and xN is a connected set containing 0. Hence $xN \subset N$ and similarly $Nx \subset N$. Since N is an ideal either $F = N$ and the field is, according to Lemma 2, a real hypercomplex system; or, $F = C$ and the field is totally disconnected.

If F is associative we obtain from Theorem I and the Frobenius theorem the

THEOREM II. *A locally compact and separable associative field is isomorphic to the field of real numbers, of complex numbers or of quaternions, or is totally disconnected.*

4. THEOREM III. *A locally compact, separable and connected (not necessarily associative or commutative) ring R which contains no absolute zero-divisor* (i.e., *an element c such that $cR = Rc = 0$*) *is a hypercomplex system over the real field.*

The assumption that R contains no absolute zero-divisor holds, for instance, if R has an identity or if R has no nilpotent ideal.

The original Pontrjagin result on connected groups shows that $R = N + C$ where N is a real vector group and C is compact and connected. Therefore it follows as before that C is an ideal. We proceed to show that $C = 0$.

LEMMA 3. *C cannot have all its elements of finite order.*

Let χ be a character of C. If c is of order m, $m\chi(c) = \chi(mc) = 0$ and $\chi(c)$ is a rational point on the circle. χ has only rational values. Since χ is continuous and C connected $\chi(c) = 0$ for all c. This contradicts the fact that non-zero characters exist for C.[5]

LEMMA 4. *$C = mC$ where m is a positive integer and mC is the set of elements mc, c in C.*

mC is closed; hence C is continuously homomorphic with the quotient group $C - mC$. The latter is compact and connected and has only elements of finite order. By Lemma 3, $C - mC = 0$, or $C = mC$.

LEMMA 5. *If C' is the closure of the group generated by the elements of infinite order of C, then $C' = C$.*

This follows immediately from the fact that $C - C'$ has only elements of finite order.

Let us now prove Theorem III. We shall show that $RC = CR = 0$ and from the assumption that R contains no absolute zero-divisor will follow that $C = 0$.

(1) Let $x \epsilon N$ and $xC = C'$. Then $\left(\frac{1}{m}x\right)C = \left(\frac{1}{m}x\right)(mC) = xC = C'$ (m a positive integer). If $u \epsilon C'$ for every m we have a c_m such that $\left(\frac{1}{m}x\right)c_m = u$. A sub-sequence of $\{c_m\}$ converges to a limit c and hence $u = 0 \cdot c = 0$ and $C' = 0$.

(2) Let $x \epsilon C$ and have infinite order. For every integral m the relation $(mx)C = x(mC) = xC = C'$ holds. There exists a sequence of integers m_j such that $m_j x \longrightarrow 0$. It follows then as in (1) that $xC = 0$. By Lemma 5 we have for *every* $x \epsilon C$, $xC = 0$.

5. In the case of non-connected rings we may apply the above arguments to C_1 the compact component of 0 in place of C. We may show that $C_1C_1 = 0$ and $C_1N = NC_1 = 0$ (not necessarily CC_1 or C_1C, however). If we assume, for example, that the ring has no nilpotent ideal, then $C_1 = 0$ and $R = N + C$ where N is an ideal and a real hypercomplex system. Again, if we suppose that R is *simple*, i.e., has only the 0- and 1-ideals then R is either a real hypercomplex system or totally disconnected.

* Alfred Jarrow Scientific Research Fellow.

[1] L. Pontrjagin, "Über stetige algebraische Körper," *Ann. Math.*, **33**, 163–174 (1932).

[2] L. Pontrjagin, "The Theory of Topological Commutative Groups," *Ibid.*, **35**, 361–388 (1934).

[3] E. R. van Kampen, "Locally Compact Abelian Groups," *Proc. Nat. Acad. Sci.*, **20**, 434–436 (1934).

[4] D. van Dantzig considers non-connected fields in "Topologische Algebra I," *Math. Ann.*, **107**, 589 (1932). An example of a connected non-associative field is the field of the Cayley numbers.

[5] This has been proved by Pontrjagin by means of the Peter-Weyl theory of representations and the Haar group measure, l. c. in footnote 2.

ANNALS OF MATHEMATICS
Vol. 36, No. 4, October, 1935

RATIONAL METHODS IN THE THEORY OF LIE ALGEBRAS

BY NATHAN JACOBSON

(Received May 20, 1934)

Introduction. The present note gives a development by rational methods of that part of the theory of Lie algebras (infinitesimal groups) which centers around the Lie-Engel theorems.[1] The main results are for the most part well known. The chief interest lies in the method, which consists of studying by elementary and direct means the relation between a Lie algebra and an enveloping associative algebra generated by it. The incidental machinery developed should be useful in rationalizing other parts of the theory.

My interest in this subject was aroused by the lectures of Prof. Weyl. I am indebted also to Prof. Albert for collaborating with me in the early stages of this work.

1. **Preliminaries.** We suppose that the underlying field F has characteristic 0. A *Lie algebra* over F is a linear space having a finite basis with respect to F, and having defined in it a composition of pairs of elements a, b resulting in the *commutator* $[a, b]$ such that

$$(1)\quad \alpha [a, b] = [\alpha a, b] = [a, \alpha b], \qquad [a + b, c] = [a, c] + [b, c], \qquad \alpha \in F.$$

$$(2)\quad [a, b] = -[b, a], \qquad [a, [b, c]] + [b, [c, a]] + [c, [a, b]] = 0.$$

If B and C are linear sub-spaces of L, the linear space generated by the commutators $[b, c]$, $b \in B$, $c \in C$ will be called the *commutator* of B and C, and denoted by $[B, C]$. In addition to the usual rules of calculation with linear spaces, we note, as a consequence of (2):

$$(3)\quad [B, C] = [C, B], \qquad [B, [C, D]] \subset [C, [D, B]] + [D, [B, C]].$$

B is a Lie *sub-algebra* of L if $B' = [B, B] \subset B$. B is *invariant* if $[L, B] \subset B$. The projection space L (mod B) of an invariant B is a Lie algebra. $L' = [L, L]$ is the first *derived* algebra of L, $L'' = [L', L']$ the second derived, etc. L is *solvable* if its derived sequence leads to 0, i.e. there exists an integer t such that $L^{(t)} = 0$. The nth *power* of L is defined by induction as $L^n = [L, L^{n-1}]$. If $L^s = 0$, L is said to be *nilpotent*. Evidently $L^{k+1} \supset L^{(k)}$ and hence if L is nilpotent, it is solvable. A Lie algebra is *semi-simple* if it contains no solvable invariant sub-algebra other than 0. If B and C are invariant, it is easily seen that their commutator $[B, C]$ is also invariant. If S is solvable and invariant, and $S^{(t-1)} \neq 0$ while $S^{(t)} = 0$, then $S^{(t-1)}$ is an abelian invariant sub-algebra of L.

[1] For the usual proofs of the Lie-Engel theorems see H. Weyl, *Darstellung kontinuerlicher halb-einfacher Gruppen* II, Math. Zeits., **24** (1925), p. 375.

Hence we may say L is semi-simple if it contains no abelian invariant sub-algebra $\neq 0$. The sum of two solvable invariant sub-algebras is easily seen to be solvable and invariant. Thus there exist a unique maximal solvable invariant sub-algebra of L. We call it the *Lie radical* of L. If $L = L_1 + L_2$ where $L_1 \frown L_2 = 0$ and $[L_1, L_2] = 0$, then L is said to be the *direct sum* of L_1 and L_2, $L = L_1 (+) L_2$. It follows that L_1 and L_2 are invariant in L.

We recall that every Lie algebra admits a representation, the *adjoint representation*, defined by associating with every element a of L the following linear transformation of L:

$$x \rightarrow x' = [a, x] = (a)x \qquad \text{where} \qquad x \in L .$$

This is a representation in the sense that the correspondence $a \rightarrow (a)$ is linear and $[a, b] \rightarrow (a)(b) - (b)(a)$. If we have any representation of a Lie algebra, we may consider the associative algebra generated by the transformations of this representation. We are led in this way to study the following situation, of which the above is a special case: a Lie algebra L is embedded in an algebra[2] A in such a way that the commutator $[a, b]$ is realized by means of $ab - ba$ which is defined in A. (1) and (2) then follow from the postulates for an algebra. We suppose also, as we may without loss of generality, that the sub-algebra generated by multiplication and linear combination of the elements of L is the whole algebra A. A will then be called an *enveloping algebra* of L.

2. **Semi-simple enveloping algebras and Lie's theorems.** Let L be a Lie algebra whose enveloping algebra is A. In this section we propose to investigate the structure of L when A is semi-simple and we shall apply the results to obtain Lie's theorems. Only the most elementary facts in the theory of associative algebras will be required. Besides the definitions, we need refer only to the theorem that an algebra, all of whose elements are nilpotent, is itself nilpotent.[3] This result will be used for commutative algebras in this section, and for these a trivial direct proof may be given. In the next section, however, we require the result for arbitrary algebras.

LEMMA 1. *If M is an invariant Lie sub-algebra of L, and B and A respectively are their enveloping algebras, then AB and BA are invariant. If B is nilpotent, so are AB and BA.*

We wish to show first that

$$AB \subset BA + B, \qquad BA \subset AB + B. \tag{4}$$

It is sufficient to show the first of these and because of the distributive law, we need to prove only that $l_1 l_2 \cdots l_s m_1 m_2 \cdots m_t \in BA + B$. Let $[l_s, m_1] = m' \in M$, then $l_s m_1 = m' + m_1 l_s$ and

[2] The term *algebra* will mean linear associative algebra with a finite basis over F in contrast with the term Lie algebra.

[3] For a proof of this theorem see J. H. Maclagan Wedderburn, *On hypercomplex numbers*, Proc. London Math. Soc., series 2, **6** (1908), p. 91 or L. E. Dickson, *Algebras and their arithmetics*, p. 46.

$$l_1 \cdots l_s m_1 \cdots m_t = l_1 \cdots l_{s-1} m' m_2 \cdots m_t + l_1 \cdots l_{s-1} m_1 l_s m_2 \cdots m_t .$$

If we use induction on s and t, we get the desired result. (4) implies that $ABA \subset AB \frown BA$ and hence AB and BA are invariant. From (4) follows also

$$(5) \qquad AB^k \subset B^k A + B^k, \qquad B^k A \subset AB^k + B^k$$

and from (5)

$$(6) \qquad (AB)^k \subset AB^k, \qquad (BA)^k \subset B^k A .$$

This equation shows that if B is nilpotent, so are AB and BA.

LEMMA 2. *If $[l, m] = m'$ is commutative with m, then m' is nilpotent.*

The operation (′) defined by l has the formal properties of a derivative. It is linear and

$$(m_1 m_2)' = [l, m_1 m_2] = [l, m_1] m_2 + m_1 [l, m_2] = m_1' m_2 + m_1 m_2' .$$

From the commutativity of m and m', we have $\varphi(m)' = \varphi'(m) m'$ where $\varphi(\lambda)$ is a polynomial with coefficients in F and $\varphi'(\lambda)$ is the derivative of $\varphi(\lambda)$. Now suppose $\varphi(m) = 0$. There exists such a polynomial since A has a finite basis. It follows that

$$\varphi(m) = 0, \qquad \varphi'(m) m' = 0, \cdots, \qquad \varphi^{(k)}(m)(m')^{2k-1} = 0, \cdots .$$

If the degree of $\varphi(\lambda)$ is h, then $\varphi^{(h)}(m) = h!$ and $(m')^{2h-1} = 0$.

Any associative algebra A defines a Lie algebra in which $[a, b] = ab - ba$. A is its own enveloping algebra. We use the Lie algebra notation A' for the derived algebra of A and denote the centrum of A by C.

LEMMA 3. *$C_1 = A' \frown C$ is a nilpotent algebra.*

For, if c and $c' \in C_1$, and $c' = [x, y] + [z, w] + \cdots$, then

$$cc' = [cx, y] + [cz, w] + \cdots \in A' .$$

But $cc' \in C$ also and hence C_1 is closed under multiplication. Let tr (c) be the trace of c in the regular representation of A. Since $c, c^2, c^3, \cdots \in A'$, tr $(c) =$ tr $(c^2) \cdots = 0$. Hence every element of C_1 is nilpotent and the commutative algebra C_1 is nilpotent.

LEMMA 4. *If A is semi-simple, $C_1 = A' \frown C = 0$.*

By Lemma 3, C_1 is nilpotent algebra. $AC_1 = C_1 A$ is nilpotent and invariant in A. Because of the semi-simplicity, $AC_1 = C_1 A = 0$. But then C_1 is nilpotent and invariant. Hence $C_1 = 0$.

Let S denote the Lie radical of L and A the enveloping algebra of L. In this notation we have

THEOREM 1. *If A is semi-simple, $L = S\ (+)\ L_1$ where S is abelian and L_1 is semi-simple.*

We wish to show first that $[S, L] = S^* = 0$. $S^* \subset L' \subset A'$. If S^* is $\neq 0$, it is a solvable invariant Lie sub-algebra of L and it contains $S_1 \neq 0$ an abelian invariant sub-algebra of L. (S_1 may be taken as one of the algebras of the derived sequence of S^*.) $S_2 = [S_1, L]$ has a basis of elements of the form $s' = [s, l]$ where $s \in S$ and $l \in L$. By Lemma 2, these elements are nilpotent. Hence the commutative enveloping algebra B of S_2 is nilpotent. By Lemma 1, AB is nilpotent and invariant in A. Hence $B = 0$ and $S_2 = 0$. Thus S_1 consists exclusively of elements of the centrum. This is impossible because of Lemma 5. Hence $S^* = 0$.

On account of Lemma 4, we may obtain an L^* such that $L = S + (L' + L^*) = S + L_1$, where the spaces S, L' and L^* are independent. L_1 is a Lie algebra and $[S, L_1] = [L_1, S] = 0$. Hence $L = S\ (+)\ L_1$ q.e.d.

$L_1 \cong L - S$ is a semi-simple Lie algebra. It is a consequence of the structure theorem of Cartan that such an algebra is equal to its derived algebra. If we apply this result to L_1, we have $L_1 = L_1' \subset L'$ and hence $L_1 = L'$, $L = S\ (+)\ L'$. We shall not require this stronger form of Theorem 1 in the sequel.

The following is a generalization of a theorem of Cartan's on absolutely irreducible Lie algebras of linear transformations:[4]

THEOREM 2. *If L is a completely reducible Lie algebra of linear transformations, then L $S\ (+)\ L'$, S abelian and L' semi-simple.*

This follows directly from Theorem 1 and the proposition from general representation theory that the enveloping algebra of a completely reducible set of linear transformations is semi-simple.

The following result is fundamental for Lie's theorems on solvable Lie algebras:

THEOREM 3. *If L is solvable and N is the radical of the enveloping algebra A of L, then A (mod N) is abelian. L' is a nilpotent Lie algebra contained in N.*

If $l_1 \equiv m_1$, $l_2 \equiv m_2$ (mod N), then $[l_1, l_2] \equiv [m_1, m_2]$ (mod N). Thus the elements of L taken mod N define a Lie algebra which is isomorphic with L (not necessarily $(1 - 1)$). We denote this algebra as L (mod N). It is solvable and its enveloping algebra is A (mod N), which is semi-simple. By Theorem 1, L (mod N) is abelian and hence so is A (mod N). For any pair of elements $l_1, l_2 \in L$, $[l_1, l_2] \equiv 0$ (mod N), or $L' \subset N$. Hence L' is a nilpotent Lie algebra.

Now suppose that the elements of the solvable Lie algebra are linear transformations of a vector space $\mathfrak{R}$. In the notation of Theorem 3 we have A (mod N) is a commutative semi-simple algebra. Suppose $N^s = 0$ but $N^{s-1} \neq 0$. If B is any sub-set of A, we denote the sub-space of $\mathfrak{R}$ generated by the transforms of the vectors of $\mathfrak{R}$ by the elements of B by $B\mathfrak{R}$. Denote $N^k\mathfrak{R}$ by $\mathfrak{R}_k$. We have $\mathfrak{R}_{s+1} = 0$, $\mathfrak{R}_s \neq 0$. From the associative law, $N^{l-k}\mathfrak{R}_k = \mathfrak{R}_l$ $(l > k)$. Hence

$$\mathfrak{R} > \mathfrak{R}_1 > \mathfrak{R}_2 > \cdots > \mathfrak{R}_s > \mathfrak{R}_{s+1} = 0.$$

The spaces $\mathfrak{R}_k$ are invariant with respect to A. For,

$$A\mathfrak{R}_k = AN^k\mathfrak{R} = (AN)N^{k-1}\mathfrak{R} \subset N^k\mathfrak{R} = \mathfrak{R}_k$$

since $AN \subset N$.

[4] E. Cartan, *Les groupes de transformations etc.*, Annales de l'Ecole Normale, 3rd ser., **26** (1909), p. 148.

Consider the transformations of A in the projection space $\mathfrak{R}_k$ (mod $\mathfrak{R}_{k+1}$). If $a \equiv a'$ (mod N), $a, a' \in A$, then $a\mathfrak{x} \equiv a'\mathfrak{x}$ (mod $\mathfrak{R}_{k+1}$) where $\mathfrak{x} \in \mathfrak{R}_k$. Thus the representation of A determined by $\mathfrak{R}_k$ (mod $\mathfrak{R}_{k+1}$) is also a representation of A (mod N). In particular the transformations of N are 0 in $\mathfrak{R}_k$ (mod $\mathfrak{R}_{k+1}$). Let us choose a basis for $\mathfrak{R}_s$ and supplement it to obtain a basis for $\mathfrak{R}_{s-1}$, supplement this to obtain a basis for $\mathfrak{R}_{s-2}$, etc. With respect to this basis the matrices of the transformations of A all have the form

$$a = \begin{bmatrix} a_1 & a_{12} & \cdots & a_{1r} \\ & a_2 & \cdots & a_{2r} \\ & & \ddots & \vdots \\ 0 & & & a_r \end{bmatrix} \tag{7}$$

where the a_i are the representation matrices of A (mod N) in $\mathfrak{R}_{i-1}$ (mod $\mathfrak{R}_i$). In particular if $a \in N$, $a_1 = a_2 = \cdots a_r = 0$.

It follows from the general theory of representations of algebras that there exists a fixed algebraic field Z of finite degree over F such that any representation of the semi-simple commutative algebra A (mod N) can be completely reduced into l-dimensional representations in Z. It follows that by choosing a suitable basis in the space obtained from $\mathfrak{R}_{i-1}$ (mod $\mathfrak{R}_i$) by extending the field to Z, the matrices a may be taken to have the form

$$a_i = \begin{bmatrix} \alpha_1^{(i)} & & & 0 \\ & \alpha_2^{(i)} & & \\ & & \ddots & \\ 0 & & & \alpha_{r_i}^{(i)} \end{bmatrix}. \tag{8}$$

We have proved in this way the results which are essentially due to Lie:

THEOREM 4. *If L is a solvable Lie algebra of transformations, its matrices may be taken in the form* (7). *The matrices of L' then have $a_1 = a_2 = \cdots = a_r = 0$. By transforming in an algebraic field Z over F we may take the a_i to have the form* (8).

3. **Engel's theorem.** We suppose again that the elements of the Lie algebra are linear transformations of a vector space $\mathfrak{R}$. The characteristic polynomial $f(\lambda | l) = \lambda^n - \psi_1(l)\lambda^{n-1} + \psi_2(l)\lambda^{n-2} - \cdots$ of a general element l of L will be called the *characteristic polynomial of L in $\mathfrak{R}$*. In a similar fashion we define the characteristic polynomial $f(\lambda \mid a)$ of the enveloping algebra A of L. The first coefficient $\psi_1(a)$ is called the *trace* of a in $\mathfrak{R}$.

The trace of the elements of A can be computed from the characteristic polynomial of L. Since the trace is linear, one needs to compute only the trace of a single term such as $l_1 l_2 \cdots l_r$, where the $l_i \in L$. We may suppose that the trace has already been determined for terms of $(r - 1)$ factors or less. The form $\frac{1}{r!}$ tr $(\Sigma^* l_{k_1} l_{k_2} \cdots l_{k_r})$, the summation extending over all permutations of

$l_1, l_2, \cdots, l_r$, is a symmetric multilinear form associated with $\text{tr}(l^r)$: If $m_1, m_2, \cdots, m_t$ is a basis for L, and $l = \lambda_1 m_1 + \lambda_2 m_2 + \cdots + \lambda_t m_t$, $l_i = \lambda_1^{(i)} m_1 + \lambda_2^{(i)} m_2 + \cdots + \lambda_t^{(i)} m_t$, and

$$\text{tr}(l^r) = \frac{1}{r!} \sum \alpha_{i_1 i_2 \cdots i_r} \lambda_{i_1} \lambda_{i_2} \cdots \lambda_{i_r}$$

then

$$\frac{1}{r!} \text{tr} \left(\sum{}^* l_{k_1} l_{k_2} \cdots l_{k_r}\right) = \frac{1}{r!} \sum \alpha_{i_1 i_2 \cdots i_r} \lambda_{i_1}^{(1)} \lambda_{i_2}^{(2)} \cdots \lambda_{i_r}^{(r)}$$

$\text{tr}(l^r)$ can be computed from $\psi_1(l), \psi_2(l), \cdots$ and hence also the associated multilinear form can be computed. On the other hand, if $[l_{k+1}, l_k] = l'$,

$$\text{tr}(l_1 \cdots l_{k-1} l_{k+1} l_k l_{k+2} \cdots l_r - l_1 l_2 \cdots l_r) = \text{tr}(l_1 \cdots l_{k-1} l' l_{k+2} \cdots l_r) = \mu$$

and this involves only products of $(r - 1)$ terms and so can be computed. Since any permutation is a product of transpositions, all the terms $\text{tr}(l_{k_1} \cdots l_{k_r})$ can be expressed in terms of $\text{tr}(l_1 l_2 \cdots l_r)$ and traces of terms involving $(r - 1)$ factors. Hence

$$\text{tr}(l_1 l_2 \cdots l_r) = \frac{1}{r!} \text{tr}\left(\sum{}^* l_{k_1} l_{k_2} \cdots l_{k_r}\right) + \omega \tag{9}$$

where ω can be determined.[5]

We apply this method to obtain a generalization of Engel's theorem.

THEOREM 5. *If the characteristic polynomial of L in $\mathfrak{R}$ is λ^n, then the enveloping algebra of L is nilpotent, and L is a nilpotent Lie algebra.*

We shall show that the trace of every element of the enveloping algebra A is 0. Assume this true for elements which are linear combinations of products of $r - 1$ terms. We have $\omega = 0$ in (9). But $\text{tr}(l^r) = 0$ and hence the associated form $\frac{1}{r!} \text{tr}(\Sigma^* l_{k_1} \cdots l_{k_r}) = 0$. By (9) we have $\text{tr}(l_1 \cdots l_r) = 0$. If a is any element of A, $\text{tr}(a) = \text{tr}(a^2) = \cdots = 0$, and hence a is nilpotent. By the associative algebra theorem quoted earlier, A is nilpotent. It follows that L is a nilpotent Lie algebra.

4. **Applications to abstract Lie algebras.** By means of the adjoint representation we may apply the above results to abstract Lie algebras (not necessarily contained in associative algebras).

Theorem 3 yields

THEOREM 6. *The first derived algebra of a solvable Lie algebra is nilpotent.*

The adjoint representation of the solvable Lie algebra L is a solvable Lie

[5] This trace argument was given by Prof. Weyl in his course on Continuous Groups, Institute for Advanced Study, Spring term 1934. It was also found by Dr. van Kampen at Hamburg, 1928, but has not been published previously.

algebra (L) of transformations. The elements of L' correspond in this representation to the elements of $(L)'$ the derived of (L). By Theorem 2 $(L)'$ is contained in an associative nilpotent algebra. Hence we have for some integer t, $(l'_1)(l'_2) \cdots (l'_t) = 0$ for arbitrary $l'_i \in (L)'$. According to the definition of the adjoint representation, if $l'_i \to (l'_i)$, then $[l'_1[l'_2[\cdots[l'_t x]\cdots] = 0$, where x is any element of L. Using this equation for $x \in L'$, we have $L'^{t+1} = 0$.

An element l of L is *nilpotent* if there exists an integer s such that the term of s brackets $[l[l[\cdots[lx]\cdots] = 0$ for any $x \in L$.

Theorem 5 becomes

THEOREM 7. *If every element of a Lie algebra L is nilpotent, then L is nilpotent.*

For if $l \to (l)$ in the adjoint representation, then

$$[l[l[\cdots[lx]\cdots] = (l)^s x = 0.$$

Hence all the transformations of the adjoint representation are nilpotent and so by Theorem 4, the enveloping algebra of the Lie algebra of this representation is nilpotent, i.e. there exists an integer t such that $(l_1)(l_2) \cdots (l_t) = 0$ for arbitrary l_i. Thus $[l_1[l_2[\cdots[l_t x]\cdots] = 0$, or $L^{t+1} = 0$.

THE INSTITUTE FOR ADVANCED STUDY,
PRINCETON, N. J.

Reprinted from the Proceedings of the NATIONAL ACADEMY OF SCIENCES,
Vol. 21, No 12, pp. 667–670. December, 1935.

ON PSEUDO-LINEAR TRANSFORMATIONS

BY N. JACOBSON

DEPARTMENT OF MATHEMATICS, BRYN MAWR COLLEGE

Communicated November 13, 1935

The present note contains the principal results of a paper which I hope to publish in detail at some later date.

1. Let $\mathfrak{R}$ be a vector space of finite dimension n over an arbitrary field $\mathfrak{F}$ (not necessarily commutative). A *pseudo-linear transformation* (p. l. t.) $\mathbf{T}$ of $\mathfrak{R}$ is defined by the conditions

$$(x + y)\mathbf{T} = x\mathbf{T} + y\mathbf{T} \qquad x, y \in \mathfrak{R} \tag{1}$$

$$(x\alpha)\mathbf{T} = (x\mathbf{T})\bar{\alpha} + x\alpha' \qquad \alpha, \bar{\alpha}, \alpha' \in \mathfrak{F} \tag{2}$$

where the correspondence $\alpha \longrightarrow \bar{\alpha}$ is a 1-*automorphism* and $\alpha \longrightarrow \alpha'$ a *differentiation* in $\mathfrak{F}$, i.e.,

$$(\overline{\alpha + \beta}) = \bar{\alpha} + \bar{\beta} \qquad (\overline{\alpha\beta}) = \bar{\alpha}\,\bar{\beta} \tag{3}$$

$$(\alpha + \beta)' = \alpha' + \beta' \qquad (\alpha\beta)' = \alpha'\bar{\beta} + \alpha\beta' \tag{4}$$

and $\alpha \longrightarrow \bar{\alpha}$ is $(1 - 1)$.

The importance of these transformations is indicated by the following examples:

(*a*) $\mathfrak{F}$, any field; $\bar{\alpha} \equiv \alpha$; $\alpha' \equiv 0$. In this case **T** is an ordinary linear transformation of a vector space over a non-commutative field.

(*b*) $\mathfrak{F}$, the complex field; $\bar{\alpha}$, the conjugate complex of α; $\alpha' \equiv 0$. **T** is then an *anti-linear transformation* as defined by v. d. Waerden.[1] Such transformations are on almost an equal footing with linear transformations in complex projective geometry. We may generalize this type and obtain *semi-linear transformations*, i.e., p. l. t. for which $\mathfrak{F}$ is any field, $\alpha \longrightarrow \bar{\alpha}$ any 1-automorphism, $\alpha' \equiv 0$. Another instance of these can be found in linear difference equations (cf. (*c*)).

(*c*) The differential equations

$$u_i' \equiv \frac{du_i}{d\tau} = \sum_{j=1}^{n} u_j\alpha_{ji}(\tau) \quad (i = 1, \ldots n) \tag{5}$$

where the α's are analytic in τ, define a p. l. t. in the following way: Let $\mathfrak{R}$ be the n-space of forms $u = \sum u_i\alpha_i(\tau)$ in the independent vectors u_i with coefficients $\alpha_i(\tau)$ in the field of analytic functions of τ. We may associate with (5) the transformation **T** of $\mathfrak{R}$ defined by

$$u \longrightarrow u\mathbf{T} = \sum u_i\alpha_i'(\tau) + \sum u_i'\alpha_i(\tau)$$

where u_i' is given by (5). **T** is p. l.

2. If $(e_1, \ldots, e_n)$ is a basis for $\mathfrak{R}$, **T** is determined by its effect on the e_i. Suppose $e_i\mathbf{T} = \sum e_j\tau_{ji}$ or

$$(e_1, \ldots, e_n)\mathbf{T} = (e_1, \ldots, e_n)T \qquad T = (\tau_{ij}).$$

T is called *the matrix of* **T** in $(e_1, \ldots, e_n)$.

Let $\mathfrak{F}[\mathbf{T}]$ denote the ring of transformations in $\mathfrak{R}$ generated by **T** and the elements of $\mathfrak{F}$. The p. l. t $\mathbf{T}_1$ and $\mathbf{T}_2$ with the same automorphism and differentiation are *similar* if the abelian group $\mathfrak{R}$ with the operators $\mathfrak{F}[\mathbf{T}_1]$, $(\mathfrak{R}, \mathfrak{F}[\mathbf{T}_1])$, is operator isomorphic with $(\mathfrak{R}, \mathfrak{F}[\mathbf{T}_2])$. This means that the matrices T_i of $\mathbf{T}_i$ are related thus:

$$T_2 = A^{-1}T_1\bar{A} + A^{-1}A'$$

where $A = (\alpha_{ij})$, $\bar{A} = (\bar{\alpha}_{ij})$. $A' = (\alpha'_{ij})$. Similarly we may employ the concept of group with operators to define *reducibility, decomposability, complete reducibility* of p. l. t.

It is well known that if $\mathfrak{F}$ is commutative and $\mathbf{T}_i$ $(i = 1, 2)$ are linear, then $\mathbf{T}_1$ and $\mathbf{T}_2$ are similar if and only if certain polynomials—the invariant factors or the elementary divisors—determined by them are the same. We prove an analogous result here. The method is briefly this: We introduce the ring $\mathfrak{F}[t]$ of *non-commutative polynomials*[2] in the indeterminate t where

$$\alpha t = t\bar{\alpha} + \alpha' \tag{6}$$

and the space $\mathfrak{R}[t]$ of all forms $\sum e_i\alpha_i(t)$, $\alpha_i(t)\epsilon\mathfrak{F}[t]$. $\mathfrak{R}[t]$ is a group $(\mathfrak{R}[t], \mathfrak{F}[t])$ with operators $\mathfrak{F}[t]$. The correspondences $\sum e_i\alpha_i(t) \longrightarrow \sum e_i\alpha_i(\mathbf{T})$, $\beta(t) \longrightarrow \beta(\mathbf{T})$ define an operator homomorphism of $(\mathfrak{R}[t], \mathfrak{F}[t])$ and $(\mathfrak{R}, \mathfrak{F}[\mathbf{T}])$ By the usual method[3] of reducing a matrix with elements in $\mathfrak{F}[t]$ to diagonal form by elementary transformations we may obtain a basis for $\mathfrak{R}$ such that the matrix of $\mathbf{T}$ has the form

$$\begin{Bmatrix} T_1 & 0 & \dots & 0 \\ 0 & T_2 & & \vdots \\ \vdots & \vdots & \ddots & \cdot \\ 0 & 0 & \cdot & T_r \end{Bmatrix} \tag{7}$$

where

$$T_i = \begin{Bmatrix} 0 & 0 & \dots & & \tau_{n_i}^{(i)} \\ 1. & 0. & & & \cdot \\ \vdots & \cdot & \ddots & & \cdot \\ \cdot & & & \cdot 0 & \cdot \\ 0 & \dots & & \cdot 1 & \tau_1^{(i)} \end{Bmatrix} \tag{8}$$

The invariant subspace $(\mathfrak{R}_i, \mathfrak{F}[\mathbf{T}])$ corresponding to $\mathbf{T}_i$ is said to be *cyclic* and $\mathbf{T}$ acting in it, *cyclic* also. $(\mathfrak{R}_i, \mathfrak{F}[\mathbf{T}])$ is generated by a single vector g_i and its transforms $g_i\alpha_i(\mathbf{T})$ by the elements of $\mathfrak{F}[\mathbf{T}]$. The polynomial $\mu_i(t) = t^{n_i} - t^{n_i-1}\tau_1^{(i)} - \dots - \tau_{n_i}^{(i)}$ is called the *order* of g_i. We have

THEOREM 1. *$(\mathfrak{R}, \mathfrak{F}[\mathbf{T}])$ is a direct sum of the cyclic subspaces $(\mathfrak{R}_i, \mathfrak{F}[\mathbf{T}]) \equiv (g_i)$ where the order $\mu_i(t)$ of g_i is both a right factor and a left factor of $\mu_j(t)$ for $j > i$.*

THEOREM 2. *Two cyclic subspaces are operator isomorphic if and only if their generators are similar in the sense of Ore.*[4]

The undecomposable parts of $(\mathfrak{R}, \mathfrak{F}[\mathbf{T}])$ are cyclic. The orders of the generators of these parts will be called the *elementary divisors* of $\mathbf{T}$. With the aid of a theorem of Krull[5] we may prove

THEOREM 3. *The p. l. t. $\mathbf{T}_1$ and $\mathbf{T}_2$ are similar if and only if their elementary divisors may be paired off into similar pairs.*

3. The orders $\mu_i(t)$ in Theorem 1 are not unique in general (even in the sense of similarity). Nevertheless they give criteria for the reducibility, decomposability and full reducibility of $\mathbf{T}$.

An extension of Theorem 1 may be given. To describe this and for other purposes we require the following definitions: A polynomial $\mu^*(t)$ is *finite* if the left ideal generated by it is two sided. It follows for our particular ring that the right ideal generated by $\mu^*(t)$ is also two sided. $\mu(t)$ is *bounded* if it is a left factor of a finite polynomial $\mu^*(t)$. It follows that $\mu(t)$ is also a right factor of $\mu^*(t)$. The finite $\mu^*(t)$ of minimum degree and leading coefficient 1 of which the bounded $\mu(t)$ is a factor is called *the bound* of $\mu(t)$. It is unique. A p. l. t. $\mathbf{T}$ is *finite* if $\mathfrak{F}[\mathbf{T}]$ has a finite basis over $\mathfrak{F}$. We prove

THEOREM 4. *The finite polynomials with leading coefficients one constitute a commutative system* $\mathfrak{B}^*$ *closed under multiplication and in which unique factorization holds.*

THEOREM 5. **T** *is finite if and only if the* $\mu_i(t)$ *are bounded.*

If **T** is finite, the bound $\mu_r^*(t)$ of the last $\mu_i(t)$ is called the *minimum function* of **T**. It is analogous to the minimum function of an ordinary linear transformation. We have for instance

THEOREM 6. *A finite p. l. t. is completely reducible if and only if its minimum function has no multiple factors in* $\mathfrak{B}^*$.

The following extension of Theorem 1 may be proved.

THEOREM 7. *The invariant subspaces in Theorem 1 may be chosen so that each* $\mu_i(t)$, $i < r$, *is bounded and* $\mu_i^*(t)$ *is a factor of* $\mu_j(t)$ *for* $j > i$.

In most cases the $\mu_i(t)$ thus normalized are unique to within similarity. However, I have been unable to prove this as yet for the general case.

4. We consider next the automorphisms of $(\mathfrak{R}, \mathfrak{F}[\mathbf{T}])$. These give rise to matrices A such that $A\mathbf{T} = \mathbf{T}\bar{A} + A'$ and conversely. For cyclic transformations the ring of automorphisms $\mathfrak{A}$ is isomorphic to the invariant ring[6] of its order. The general theorems of v. d. Waerden,[7] Fitting[8] and others are applicable here and give the structure of certain hypercomplex systems. My former results[9] on cyclic algebras are obtainable in this way.

5. Non-trivial information about **T** can be obtained just from the nature of $\mathfrak{F}[t]$.

If by a suitable choice of t we may obtain $\alpha' \equiv 0$, $\mathfrak{F}[t]$ is called *semi-linear*, and if by a suitable t we obtain $\bar{\alpha} \equiv \alpha$, $\mathfrak{F}[t]$ is called *differential*.

THEOREM 8. *If* $\mathfrak{F}$ *is commutative,* $\mathfrak{F}[t]$ *is either semi-linear or differential.*

It is not difficult to determine the finite elements of all differential and semi-linear rings. In most cases these consist of the elements of $\mathfrak{F}$ alone. It follows that *the corresponding p. l. t. are cyclic.* There are also interesting special rings $\mathfrak{F}[t]$ in which every element is bounded. *The corresponding p. l. t. are finite.*

There are numerous applications of these results to complex projective geometry, differential and difference equation and hypercomplex numbers. I shall give these in the subsequent fuller exposition of this subject.

[1] v. d. Waerden, "Gruppen von lineare Transformationen," p. 4.

[2] Ore, "Theory of Non-Commutative Polynomials," *Ann. Math.*, **34**, 480–508 (1933).

[3] v. d. Waerden, *Moderne Algebra II*, pp. 135–142.

[4] Ore, loc. cit. in footnote 2, p. 488.

[5] Krull, "Über verallgemeinerte Abelsche Gruppen," *Math. Zeits.*, **23**, 182–187 (1925).

[6] Ore, "Formale Theorie der linearen Differentialgleichungen II," *Crelle Jour.*, **168**, 241–243 (1932)

[7] v. d. Waerden, *Moderne Algebra II*, pp. 165–169.

[8] Fitting, "Die Theorie der Automorphismenringe," etc., *Math. Annalen*, **107**, 514–542 (1932).

[9] Jacobson, "Non-Commutative Polynomials and Cyclic Algebras," *Annals of Math.*, **35**, 197–208 (1934).

TOTALLY DISCONNECTED LOCALLY COMPACT RINGS.[1]

By N. Jacobson.

Introduction.

1. Miss Taussky and I[2] have recently considered the theory of locally compact separable (l. c. s.) rings and shown that it could be reduced for the most part to the consideration of two essentially different topological types, the connected and the totally disconnected. Furthermore the connected l. c. s. rings satisfying mild algebraic conditions were shown to be hypercomplex systems with finite bases over the field of real numbers and hence their structure could be given by means of the classical results on hypercomplex numbers.

In this paper I consider the totally disconnected (t. d.) l. c. s. rings restricting myself in the main to simple rings and particularly to fields. The work is divided into two parts which are quite independent of each other. Part I is concerned with the nature of the additive abelian group of a locally-compact separable totally-disconnected (l. c. s. t. d.) simple ring $\mathfrak{S}$. $\mathfrak{S}$ may be non-commutative and non-associative. It is necessary to distinguish two cases: (1) the characteristic of $\mathfrak{S}$, $\chi(\mathfrak{S}) = p$ a prime and (2) $\chi(\mathfrak{S}) = 0$. In (1) the additive group of $\mathfrak{S}$ is a direct sum of cyclic groups of order p. (The precise meaning of this type of direct sum usually involving an infinite number of summands is given below (§ 5).) In (2) $\mathfrak{S}$ is additively a finite dimensional vector space over the field of p-adic numbers P_p. It follows that $\mathfrak{S}$ is a hypercomplex system with a finite basis relative to P_p.

Now if $\mathfrak{S}$ is associative and $\chi(\mathfrak{S}) = 0$, the above result together with the known theory of hypercomplex systems over a p-adic field gives a complete solution of the problem of determining the algebraic structure of $\mathfrak{S}$. For, by Wedderburn's theorem $\mathfrak{S}$ is a system consisting of all matrices of a fixed finite degree with coefficients in a p-adic division algebra $\mathfrak{F}$. By the results of Hasse[3] $\mathfrak{F}$ is cyclic over its centrum $\mathfrak{C}$ and the precise nature of the latter can be described.

On the other hand there is no such well-developed theory which will apply directly to solve the structure problem for the case $\chi(\mathfrak{S}) = p$. It is necessary

[1] Presented to the Society, April 19, 1935.

[2] "Locally compact rings," *Proceedings of the National Academy of Sciences*, vol. 21 (1935), pp. 106-108.

[3] H. Hasse, "Über $\mathfrak{P}$-adische Schiefkörper etc.," *Mathematische Annalen*, vol. 104 (1931), referred to below as H.

to develop new methods for the solution of this problem. This has been done in part II under the further assumption that $\mathfrak{S} = \mathfrak{F}$ is a field (not necessarily commutative). Moreover the methods apply at the same time to fields of characteristic 0 and thus afford a new treatment of p-adic division algebras. The main idea underlying this treatment is that the orders (Ordnungen) of a p-adic division algebra are compact and open (c. o.) subrings and conversely. Since the existence of c. o. subrings of $\mathfrak{F}$ can be shown directly the procedure is as follows: We first consider the ideal theory of any c. o. subring $\mathfrak{R}$ of $\mathfrak{F}$ and then prove the existence of a unique maximal c. o. subring $\mathfrak{R}_\omega$. By means of $\mathfrak{R}_\omega$ a valuation (Bewertung) of $\mathfrak{F}$ is defined. We then use the methods of Hasse to show that $\mathfrak{F}$ is a cyclic algebra over its centrum $\mathfrak{C}$. The structure of the valued commutative field $\mathfrak{C}$ can be determined as, indeed, it has been by v. Dantzig [4] and by Hasse and Schmidt.[5]

2. For a comprehensive account of the foundations of topological algebra the reader is referred to the paper by v. Dantzig (D. l. c. in footnote 4). We recall a few of the important definitions at this point.

By a *space* $\mathfrak{S}$ we shall always mean a Hausdorff space.[6] $\mathfrak{S}$ is said to be *compact* if every infinite sequence of points in it has a limit point. $\mathfrak{S}$ is *locally compact* (l. c.) if every point has a neighborhood whose closure is compact. $\mathfrak{S}$ is *separable* (s.) if it contains a denumerable set of neighborhoods which generate all the open sets by logical addition. $\mathfrak{S}$ is *connected* if it is impossible to decompose it into a logical sum $\mathfrak{S}_1 \cup \mathfrak{S}_2$ of the open and non-intersecting sets $\mathfrak{S}_1$ and $\mathfrak{S}_2$. $\mathfrak{S}$ is *totally disconnected* (t. d.) if for every pair of distinct points a_1, a_2 there is a decomposition of the space into a sum of non-intersecting open sets $\mathfrak{S}_1$ and $\mathfrak{S}_2$ such that $\mathfrak{S}_i \supset a_i$. If $\mathfrak{S}$ is l. c. and t. d. then it is *zero-dimensional*, i. e., every point has arbitrarily small open and closed neighborhoods.[7]

A *locally compact separable ring* $\mathfrak{S}$ is a l. c. s. space in which there is defined two binary operations $+$ and $\cdot$ such that

1. $\mathfrak{S}$ is an abelian group under $+$.

---

[4] D. van Dantzig in *Studien over topologische Algebra*, Dissertation, Amsterdam, H. J. Paris, 1931, determines the structure of all l. c. commutative and associative fields. The foundations of topological algebra may be found in this paper (in Dutch) or in van Dantzig, "Zur topologische Algebra," *Mathematische Annalen*, vol. 107 (1933), referred to as D.

[5] H. Hasse and F. K. Schmidt, "Die Struktur diskret bewerteter Körper," *Journal f. d. reine u. angew. Math.*, vol. 170 (1933), pp. 4-63.

[6] F. Hausdorff, *Grundzüge der Mengenlehre*, 1914, p. 213.

[7] K. Menger, *Dimensionstheorie*, Teubner, 1928, p. 207.

2. $a(b+c)=ab+ac$, $(b+c)a=ba+ca$.

3. $+$, $-$, $\cdot$, and $\div$ (when it exists) are continuous operations: If $a_\mu \to a$ (converges to a) and $b_\mu \to b$, then $a_\mu \pm b_\mu \to a \pm b$ and $a_\mu b_\mu \to ab$. If a_μ^{-1} and a^{-1} exist, then $a_\mu^{-1} \to a^{-1}$.

I. The additive group of a simple ring.

3. A l. c. s. t. d. ring $\mathfrak{S}$ is a l. c. s. t. d. abelian group under addition. It is therefore natural to begin our investigation with an account of the theory of l. c. s. t. d. abelian groups (§ 3 and § 4). It may be noted in the sequel that for our present purpose (the classification of simple rings) only a portion of the group theoretic results are needed. However, we may justify the more complete discussion on the ground that it shows up the precise extent to which the ring restrictions effect the structure of its additive group, and it should be useful for future generalizations. The main results on l. c. s. t. d. abelian groups are due to v. Dantzig (dissertation) and independently to Alexander and Cohen.[8]

In the remainder of this section $\mathfrak{S}$ will denote a l. c. s. t. d. abelian group with addition as the group operation.

THEOREM 3. 1. *$\mathfrak{S}$ is complete.*

The completeness is meant in the following sense: if $a_1, a_2, \cdots$ is a Cauchy sequence, i. e. has the property that for any given neighborhood U of 0 there exists a positive integer $N(U)$ such that $a_\mu - a_\nu \in U$ if μ and $\nu > N$, then $a_1, a_2, \cdots$ converges. To prove this property suppose U is compact, i. e., has a compact closure. Since $a_{N+1} - a_N, a_{N+2} - a_N, \cdots$ all belong to U, they have a limit b. It follows easily that $a_\mu \to a_N + b$.

THEOREM 3. 2. *If U is an open and closed compact neighborhood of* 0, *then there exists an open and closed compact subgroup $\mathfrak{G} \subset U$.*[9]

Denote the set of elements $\{-u\}$ where $u \in U$ by $-U$ and $V = U \frown (-U)$ (logical intersection). Since $-U$ and U are open and closed, so is their intersection V. We denote the complement of V in $\mathfrak{S}$ by $\mathfrak{S} \mid V$.

---

[8] J. W. Alexander and L. W. Cohen in "A classification of the homology groups of compact spaces," *Annals of Mathematics*, vol. 33 (1932), pp. 538-566, deal with groups with generators, which may be shown by virtue of Theorem 3. 2 to include the c. s. t. d. groups.

[9] This theorem is due to v. Dantzig cf. his dissertation, p. 18, and also E. R. v. Kampen, "Locally compact abelian groups," *Proceedings of the National Academy of Sciences*, vol. 20 (1934).

If $U_1 \supset U_2 \supset U_3 \supset \cdots$ is a decreasing sequence of neighborhoods of 0 whose intersection is 0 ($U_\mu \to 0$) then for μ sufficiently large $(U_\mu + V) \cap (\mathfrak{S} | V)$ is vacuous. Otherwise we have for every μ a $u_\mu \in U_\mu$ and $v_\mu \in V$ such that $u_\mu + v_\mu = h_\mu \in \mathfrak{S}|V$. Suppose $v_{\mu_k} \to v \in V$. Since $u_{\mu_k} \to 0$, $h_{\mu_k} \to v$ which is impossible since $\mathfrak{S}|V$ is closed.

Let $\mathfrak{G}$ denote the set of elements g of $\mathfrak{S}$ such that $g + V \subset V$. $\mathfrak{G}$ is a group $\subset V$ and by the above remark $\mathfrak{G} \supset U_\mu$ for μ sufficiently large. Hence $\mathfrak{G}$ is open. Being a group $\mathfrak{G}$ is also closed.[10]

By Theorem 3. 2 there exists a sequence of compact open and closed subgroups $\mathfrak{G}_\mu \to 0$. The cosets of $\mathfrak{G}_\mu$ constitute a fundamental set of neighborhoods for $\mathfrak{S}$. In the rest of I we will therefore mean by a neighborhood of 0 a compact open and closed group neighborhood.

Corollary. $\sum_{\mu=1}^{\infty} a_\mu$ *exists if and only if* $a_\mu \to 0$.

Set $\sum_{\mu=1}^{n} a_\mu = s_n$. By the completeness of $\mathfrak{S}$, $\sum_{\mu=1}^{\infty} a_\mu$ exists if and only if $\{s_n\}$ is a Cauchy sequence. Let U be a (compact group) neighborhood of 0. If $\mu > N(U)$, $a_\mu \in U$ and hence also for $n, m > N$

$$s_n - s_m = \begin{cases} a_{m+1} + a_{m+2} + \cdots + a_n & \text{if} \quad n > m \\ 0 & \text{if} \quad n = m \\ a_{n+1} + a_{n+2} + \cdots + a_m & \text{if} \quad n < m \end{cases}$$

is contained in U. Thus $\{s_n\}$ is a Cauchy sequence and Σa_μ exists. The converse is obvious.

4. A group $\mathfrak{S}$ is called a *p-group* (p a prime) if for every $a \in \mathfrak{S}$, $p^\mu a \to 0$. Let $\mathfrak{G}$ be a compact t. d. group and

$$\mathfrak{G}^{(p)} = \underset{\mu}{I}(p^\mu \mathfrak{G}) \equiv \mathfrak{G} \cap p\mathfrak{G} \cap p^2\mathfrak{G} \cap \cdots.$$

Lemma 4. 1. *A necessary and sufficient condition that an element* $h \in \mathfrak{G}^{(p)}$ *is that there exist a sequence of integers* $\mu_i \to \infty$ *such that* $p^{\mu_i}h \to h$.

(1) Suppose $p^{\mu_i}h \to h$. Since $p^{\mu_i}h \in p^\mu\mathfrak{G}$ for $\mu_i \geqq \mu$ and $p^\mu\mathfrak{G}$ is closed, we have $p^\mu\mathfrak{G} \supset h$. Hence $h \in \underset{\mu}{I} p^\mu\mathfrak{G} = \mathfrak{G}^{(p)}$.

(2) Let $h \in \mathfrak{G}^{(p)}$ so that $h = p^\nu g_\nu$ $(\nu = 1, 2, \cdots)$. If $g_{\nu_k} \to g$ then $h = \lim_{k\to\infty} p^{\nu_k} g$. For if k is sufficiently large $(g_{\nu_k} - g) \in U$ where U is an arbitrary (group) neighborhood of 0. Hence

[10] D. p. 609. That $\mathfrak{G}$ is closed may also be seen directly.

$$p^{\nu_k}(g_{\nu_k} - g) = (h - p^{\nu_k}g) \in U$$

or $\lim_{k\to\infty} p^{\nu_k}g = h$. Set $\mu_k = \nu_{k+1} - \nu_k$. By dropping enough terms of the sequence $\{\nu_k\}$ we may suppose that $\mu_k \to \infty$. We assert that $p^{\mu_k}h \to h$. If k is large enough, $(h - p^{\nu_k}g) \in U$ and hence

$$p^{\mu_k}(h - p^{\nu_k}g) = (p^{\mu_k}h - p^{\nu_{k+1}}g) \in U$$
$$(h - p^{\mu_k}h) = (h - p^{\nu_{k+1}}g) - (p^{\mu_k}h - p^{\nu_{k+1}}g) \in U$$

i. e. $p^{\mu_k}h \to h$.

LEMMA 4. 2. *If* $p^{\mu_k}g \to 0$ *then* $p^{\mu}g \to 0$ $(\mu = 1, 2, \cdots)$.

If k is large enough so that $p^{\mu_k}g \in U$ then $p^{\mu}g \in U$ also for $\mu \geqq \mu_k$.

According to this lemma we have that the elements g for which the sequence $\{p^{\mu}g\}$ has 0 for limit point form a closed subgroup $\mathfrak{G}_p$.

LEMMA 4. 3. $\mathfrak{G} = \mathfrak{G}^{(p)} \oplus \mathfrak{G}_p$.

Let g be an arbitrary element of $\mathfrak{G}$ and $p^{\nu_k}g \to h$. Since $p^{\mu_k}h \to h$ where $\mu_k = \nu_{k+1} - \nu_k$, $h \in \mathfrak{G}^{(p)}$. Suppose $p^{\mu'_k}g \to h' \in \mathfrak{G}^{(p)}$. Then $p^{\nu'_k}h' \to h$. For if U is a neighborhood of 0 and $k > K_1(U)$, $(h' - p^{\mu'_k}g) \in U$ and hence also $(p^{\nu'_k}h' - p^{\nu'_{k+1}}g) \in U$. If $k > K_2(U)$ we have besides $(p^{\nu'_{k+1}}g - h) \in U$. Hence

$$(h - p^{\nu'_k}h') \in U, \quad \text{or} \quad p^{\nu'_k}h' \to h.$$

Thus $p^{\nu'_k}(g - h') \to 0$ and by Lemma 4. 2 $p^{\nu}(g - h') \to 0$.

Set $g = h' + (g - h')$. $h' \in \mathfrak{G}^{(p)}$ and $(g - h') \in \mathfrak{G}_p$. Evidently this decomposition is unique, i. e., $\mathfrak{G}^{(p)} \cap \mathfrak{G}_p = 0$ and so $\mathfrak{G} = \mathfrak{G}^{(p)} \oplus \mathfrak{G}_p$. (This shows that h' determined above is unique and in particular $p^{\mu_k}g \to h'$ not merely $p^{\mu'_k}g$).

LEMMA 4. 4. *There exist primes p such that* $\mathfrak{G}^{(p)} \neq \mathfrak{G}$ *and hence such that* $\mathfrak{G}_p \neq 0$.

If $\mathfrak{H}_0$ is an open subgroup of $\mathfrak{G}$ then $\mathfrak{G} - \mathfrak{H}_0$ is compact and discrete and hence it is finite. Let p be a divisor of its order. $\mathfrak{G} - \mathfrak{H}_0$ has a subgroup $\mathfrak{H} - \mathfrak{H}_0$ of index p. Hence $\mathfrak{G} - \mathfrak{H} \cong (\mathfrak{G} - \mathfrak{H}_0) - (\mathfrak{H} - \mathfrak{H}_0)$ has order p and $p\mathfrak{G} \subset \mathfrak{H} < \mathfrak{G}$ ($<$ means properly contained in). It follows from Lemma 4. 3 that $\mathfrak{G}_p \neq 0$.

THEOREM 4. 1. $\mathfrak{G}$ *is a direct sum of p-groups.*[11]

[11] Alexander and Cohen, *loc. cit.*,[8] p. 557.

Let p_1 be a prime for which $\mathfrak{G}_{p_1} \neq 0$. We have $\mathfrak{G} = \mathfrak{G}_{p_1} \oplus \mathfrak{G}^{(p_1)}$. Since $\mathfrak{G}^{(p_1)}$ is closed, it is compact and may be treated as $\mathfrak{G}$, i. e. $\mathfrak{G}^{(p_1)} = \mathfrak{G}^{(p_1)}_{p_2} \oplus \mathfrak{G}^{(p_1 p_2)}$. Continuing this process we obtain

$$\mathfrak{G} = \mathfrak{G}_{p_1} \oplus \mathfrak{G}^{(p_1)}_{p_2} \oplus \cdots \oplus \mathfrak{G}^{(p_1 \cdots p_{n-1})}_{p_n} \oplus \mathfrak{G}^{(p_1 \cdots p_n)}$$

where $\mathfrak{G}^{(p_1 \cdots p_i)}_{p_{i+1}} \neq 0$ for $i < n$ if $\mathfrak{G}^{(p_1 \cdots p_n)} \neq 0$. We must show that this process is uniformly convergent in the sense that $\mathfrak{G}^{(p_1 \cdots p_n)} \subset U$ where U is any neighborhood of 0 and $n > N(U)$. Choose N so that $\{p_1, p_2, \cdots, p_n\}$ for $n > N$ includes all the distinct prime factors of the order m of $\mathfrak{G} - U$. Let $h \in \mathfrak{G}^{(p_1 \cdots p_n)}$. Then there exist sequences of integers $\{\mu_k\}, \{\nu_k\}, \cdots, \{\rho_k\}$ such that $p_1^{\mu_k} h \to h$, $p_2^{\nu_k} h \to h, \cdots, p_n^{\rho_k} h \to h$. It follows that $p_1^{\mu_k} p_2^{\nu_k} \cdots p_n^{\rho_k} h \to h$. If $\mu_k, \nu_k, \cdots, \rho_k$ are sufficiently large $p_1^{\mu_k} p_2^{\nu_k} \cdots p_n^{\rho_k} h = p_1^{\mu'_k} p_2^{\nu'_k} \cdots p_n^{\rho'_k} m h \in m\mathfrak{G}$ and hence $h \in m\mathfrak{G} \subset U$ or $\mathfrak{G}^{(p_1 \cdots p_n)} \subset U$. Hence we may write

$$\mathfrak{G} = \mathfrak{G}_{p_1} \oplus \mathfrak{G}^{(p_1)}_{p_2} \oplus \mathfrak{G}^{(p_1 p_2)}_{p_3} \oplus \cdots.$$

The components $\mathfrak{G}^{(p_1 \cdots p_i)}_{p_{i+1}}$ are *characteristic subgroups* of $\mathfrak{G}$, i. e. they are carried into themselves by every automorphisms of $\mathfrak{G}$.

5. Let $\mathfrak{S}$ be a l. c. s. t. d. simple[12] ring (not necessarily associative or commutative). Since $\mathfrak{S}$ is homogeneous it is either discrete or dense in itself. The first case is, of course, uninteresting for the present considerations and hence we restrict ourselves to the second.

THEOREM 5. 1. *$\mathfrak{S}$ is a p-group.*

Let $\mathfrak{S}_p$ denote the subset of elements a of $\mathfrak{S}$ such that $p^\nu a \to 0$. $\mathfrak{S}_p$ is an ideal[13] and hence either $\mathfrak{S}_p = \mathfrak{S}$ or $\mathfrak{S}_p = 0$. $\mathfrak{S}$ contains a compact and open subgroup $\mathfrak{G}$. Since $\mathfrak{S}$ is not discrete $\mathfrak{G} \neq 0$. By Lemma 4. 4 $\mathfrak{G}$ contains elements $g \neq 0$ such that $p^\nu g \to 0$ for some p. Thus $\mathfrak{S}_p \neq 0$ and so $\mathfrak{S}_p = \mathfrak{S}$.

p is unique. For, suppose that $p_1^\nu a \to 0$ and $p_2^\nu a \to 0$ where $(p_1, p_2) = 1$. Integers α_ν and β_ν can be determined so that $\alpha_\nu p_1^\nu + \beta_\nu p_2^\nu = 1$ $(\nu = 1, 2, 3, \cdots)$. If U is an arbitrary neighborhood of 0, there exists an $N(U)$ such that $p_1^\nu a$ and $p_2^\nu a \in U$ for all $\nu \geqq N$. Then $a = \alpha_\nu p_1^\nu a + \beta_\nu p_2^\nu a \in U$. Since U is arbitrary a must be 0.

We now distinguish two cases:

(1) *Characteristic p.* $p\mathfrak{S} = 0$.

Let $\mathfrak{G}_1 > \mathfrak{G}_2 > \mathfrak{G}_3 > \cdots$ be a sequence of compact and open subgroups of $\mathfrak{S}$ whose intersection is 0. $\mathfrak{G}_\mu - \mathfrak{G}_{\mu+1}$ has order a power of p and hence

[12] A ring $\mathfrak{S}$ is said to be *simple* if (0) and (1) $\equiv \mathfrak{S}$ are its only (two-sided) ideals.

[13] The term *ideal* will refer exclusively to two-sided ideals.

by inserting a finite number of compact and open subgroups between $\mathfrak{G}_\mu$ and $\mathfrak{G}_{\mu+1}$ and changing the notation, we may suppose that $\mathfrak{G}_\mu - \mathfrak{G}_{\mu+1}$ has order p. If x_μ is an element of $\mathfrak{G}_\mu | \mathfrak{G}_{\mu+1}$ then $\mathfrak{G}_\mu = (x^\mu) \oplus \mathfrak{G}_{\mu+1}$ where (x_μ) denotes the group of order p generated by x_μ. Hence $\mathfrak{G}_1 = (x_1) \oplus (x_2) \oplus \cdots \oplus x_{n-1} \oplus \mathfrak{G}_n$ and since $\mathfrak{G}_n \to 0$

$$\mathfrak{G}_1 = (x_1) \oplus (x_2) \oplus (x_3) \oplus \cdots.$$

Suppose $y_1, y_2, y_3, \cdots$ is a denumerable dense set in $\mathfrak{S}$. Then $\mathfrak{S} = (\mathfrak{G}_1 + y_1) \cup (\mathfrak{G}_1 + y_2) \cup \cdots$.[14] If y_{μ_1} is the first $y \notin \mathfrak{G}_1$ set $y_{\mu_1} = x_{-1}$ and form $\mathfrak{G}_{-1} = (x_{-1}) \oplus \mathfrak{G}_1$. Again if y_{μ_2} is the first $y \notin \mathfrak{G}_{-1}$ set $y_{\mu_2} = x_{-2}$ and form $\mathfrak{G}_{-2} = (x_{-2}) \oplus \mathfrak{G}_{-1}$. Continuing in this way we obtain the theorem:

THEOREM 5.2. *$\mathfrak{S}$ has a denumerable set of generators $x_1, x_2, \cdots$ and $x_{-1}, x_{-2}, \cdots$ such that every element is expressible uniquely as an infinite linear combination of the x's with coefficients mod p and in which only a finite number of the $x_{-\mu}$'s occur. Every such expression is an element of $\mathfrak{S}$.*

(2) *Characteristic* 0. $p\mathfrak{S} \neq 0$.

LEMMA 5.1. *$\mathfrak{S}$ is a hypercomplex system over the field of rational numbers* P.[15]

If m is a positive integer, then $m\mathfrak{S}$ is an ideal and hence is either $= \mathfrak{S}$ or $= 0$. But $m\mathfrak{S} = 0$ is impossible: If $m = kl$, $m\mathfrak{S} = k(l\mathfrak{S}) = 0$ and hence either $k\mathfrak{S} = 0$ or $l\mathfrak{S} = 0$. We obtain in this way that a prime q exists for which $q\mathfrak{S} = 0$. Since p is the only prime such that $p^\nu a \to 0$ for every a in $\mathfrak{S}$, $p = q$ and we have a contradiction of the assumption $p\mathfrak{S} \neq 0$.

The elements of order m form an ideal $\neq \mathfrak{S}$. Hence this ideal is $= (0)$, i. e., every element of $\mathfrak{S}$ has infinite order. From $m\mathfrak{S} = \mathfrak{S}$ follows that for every a there exists a unique a' such that $ma' = a$. We denote a' by $(1/m)a$ and define $(n/m)a = na'$ where n is any integer. If $n/m = n_1/m_1$, then $(n/m)a = (n_1/m_1)a$. Using the fact that $\mathfrak{S}$ has no elements of finite order we can easily verify the following rules:

$$(*) \qquad \begin{gathered} \alpha(a+b) = \alpha a + \alpha b \\ (\alpha+\beta)a = \alpha a + \beta a \qquad (\alpha\beta)a = \alpha(\beta a) \\ \alpha(ab) = (\alpha a)b = a(\alpha b) \end{gathered}$$

where α and β are rational numbers.

---

[14] This is the only point in the entire discussion at which the assumption of separability seems to be necessary. In the rest of the paper it may be replaced by the weaker Hausdorff first denumerability axiom.

[15] $\mathfrak{S}$ does not have a finite basis over P, however.

Let $\mathfrak{G}$ be a compact and open subgroup of $\mathfrak{S}$. Then $\mathfrak{G} > p\mathfrak{G} > p^2\mathfrak{G} > \cdots$ and $I_\nu(p^\nu\mathfrak{G}) = 0$. For otherwise by Lemma 4.1 $\mathfrak{G}$ would contain an element $h \neq 0$ for which the sequence $\{p^\nu h\}$ has h for a limit point. Let $p^{-\mu}\mathfrak{G}$ denote the compact and open subgroup of p^μ-th parts of the elements of $\mathfrak{G}$ and $\Gamma_\mu = p^\mu\mathfrak{G} | p^{\mu+1}\mathfrak{G}$ $(\mu = 0, \pm 1, \pm 2, \cdots)$. Then $\mathfrak{G} < p^{-1}\mathfrak{G} < p^{-2}\mathfrak{G} < \cdots$. Since $p^\nu a \to 0$ for every a, there exists an integer N such that $p^N a \in \mathfrak{G}$ and hence $a \in p^{-N}\mathfrak{G}$. Thus $\mathfrak{G} \cup p^{-1}\mathfrak{G} \cup p^{-2}\mathfrak{G} \cup \cdots = \mathfrak{S}$.

If $a \in \Gamma_\mu$, evidently $p^\nu a \in \Gamma_{\mu+\nu}$ for any integral ν. If α is a rational number prime to p (numerator and denominator in the reduced form of α are prime to p), then $\alpha a \in \Gamma_\mu$. For, if α is an integer prime to p, $\alpha a \in \Gamma_{\mu+\epsilon}$ where $\epsilon \geqq 0$. But $\epsilon > 0$ is impossible since in this case $\alpha a \in p^{\mu+1}\mathfrak{G}$ contradicting the fact that the order of the coset of a in $p^\mu\mathfrak{G} - p^{\mu+1}\mathfrak{G}$ is p. The result for reciprocals of integers follows from symmetry and hence it follows for all rational α's by combining the two special cases.

If $a \in \Gamma_\mu$ we denote the real-valued function $p^{-\mu}$ of a by $\| a \|$. We have then:

$$\| a + b \| \leqq \max (\| a \|, \| b \|)$$
$$\| p^\nu a \| = p^{-\nu} \| a \|$$
$$\| \alpha a \| = \| a \| \quad \text{if} \quad (\alpha, p) = 1.$$

If we set dist $(a, b) = \| a - b \|$, we obtain a metric which gives the topology of $\mathfrak{S}$.

LEMMA 5.2. *If $\{\alpha_\nu\}$ is a sequence of rational numbers, $\{\alpha_\nu a\}$ $(a \neq 0)$ converges if and only if $\{\alpha_\nu\}$ converges in the p-adic topology.*[16]

Because of the completeness of $\mathfrak{S}$ and of the p-adic field P_p it is necessary to show only that $\alpha_\nu a \to 0$ if and only if $\alpha_\nu \underset{p}{\to} 0$ (in the p-adic topology). This is evident from the above considerations.

LEMMA 5.3. *$\mathfrak{S}$ is a hypercomplex system over the field of p-adic numbers.*

If α is a p-adic number, say $\alpha = \lim_p \alpha_\nu$ where the α_ν are rational numbers, then we define $\alpha a = \lim \alpha_\nu a$. This is independent of the particular sequence $\{\alpha_\nu\}$ approaching α. From the continuity of addition and multiplication we obtain the validity of equations (*) for all p-adic α, β. Finally, if $\{\alpha_\nu\}$ is a sequence of p-adic numbers converging to α and $a_\nu \to a$ then $\alpha_\nu a_\nu \to \alpha a$, i. e. αa is a continuous function of α and a.

[16] For a definition of the p-adic field P_p and the p-adic topology of P see v. d. Waerden, *Moderne Algebra* I, Springer, 1930, pp. 218-220.

We denote the p-adic vector space determined by the elements $a_1, a_2, \cdots, a_r$ of $\mathfrak{S}$ by $(a_1, a_2, \cdots, a_r)$.

LEMMA 5.4. *$(a_1, a_2, \cdots, a_r)$ is closed.*

We may suppose that $a_1, \cdots, a_r$ are linearly independent (with p-adic coefficients). Let $b_\nu = \beta_{\nu 1}a_1 + \cdots + \beta_{\nu r}a_r \rightarrow b$. We must show that $b \in (a_1, a_2, \cdots, a_r)$. For each ν choose $\lambda(\nu) = 1, 2, \cdots, r$ such that $\| \beta_{\nu,\lambda(\nu)}a_{\lambda(\nu)} \| \geqq \| \beta_{\nu,\mu}a_\mu \|$ for $\mu \neq \lambda$. Since $\lambda(\nu)$ has only a finite range one of its values, say $\lambda(\nu) = 1$ occurs infinitely often. By restricting ourselves to a subsequence we may suppose that $\| \beta_{\nu 1}a_1 \| \geqq \| \beta_{\nu\mu}a_\mu \|$ for all ν and μ. Then $\| (\| \beta_{\nu 1}a_1 \|)\beta_{\nu\mu}a_\mu \| \leqq 1$ and hence we may suppose that $\lim_{\nu\to\infty} \| \beta_{\nu 1}a_1 \| \beta_{\nu\mu}a_\mu$ exists and $= \gamma_\mu a_\mu$. Since $\| (\| \beta_{\nu 1}a_1 \|)\beta_{\nu 1}a_1 \| = 1$, $\gamma_1 \neq 0$. If $\lim_p \| \beta_{\nu 1}a_1 \| = \infty$, then $\| \beta_{\nu 1}a_1 \| b \rightarrow 0$ and hence $\gamma_1 a_1 + \gamma_2 a_2 + \cdots + \gamma_r a_r = 0$ contrary to the linear independence of the a's. Thus $\lim_p \| \beta_{\nu 1}a_1 \|$ exists and since $\| \beta_{\nu 1}a_1 \| \geqq \| \beta_{\nu\mu}a_\mu \|$ we may suppose that $\beta_{\nu\mu}a_\mu \rightarrow \beta_\mu a_\mu$. Then

$$b = \beta_1 a_1 + \beta_2 a_2 + \cdots + \beta_r a_r \in (a_1, a_2, \cdots, a_r).$$

THEOREM 5.3. *$\mathfrak{S}$ is a hypercomplex system with a finite basis over the field of p-adic numbers.*

Choose a_1 in Γ_0. If $\mathfrak{S} \neq (a_1)$, then there exists an $a_2 \in \Gamma_0$ such that $a_2 \not\in (a_1) + p\mathfrak{G}$. For if every b in Γ_0, $\in (a_1) + p\mathfrak{G}$ then every b_ν in $\Gamma_\nu \in (a_1) + p^{\nu+1}\mathfrak{G}$. Thus

$$\begin{array}{lll} b = \alpha_1 a_1 + b_1 & b_1 \in \Gamma_{m_1} & m_1 > 0 \\ b_1 = \alpha_2 a_1 + b_2 & b_2 \in \Gamma_{m_2} & m_2 > m_1 \\ \cdot\ \cdot\ \cdot & \cdot\ \cdot\ \cdot & \cdot\ \cdot\ \cdot \\ b_\nu = \alpha_\nu a_1 + b_{\nu+1} & b_{\nu+1} \in \Gamma_{m_{\nu+1}} & m_{\nu+1} > m_\nu \end{array}$$

and hence $b = (\alpha_1 + \cdots + \alpha_\nu)a_1 + b_{\nu+1}$. Since $\lim b_\nu = 0$ and (a_1) is closed, $b \in (a_1)$ contradicting $(a_1) \neq \mathfrak{S}_1$. Now choose $a_2 \begin{cases} \in \Gamma_0 \\ \not\in (a_1) + p\mathfrak{G} \end{cases}$. The subgroup generated by a_1 and a_2 mod $p\mathfrak{G}$ then has order p^2. If $\mathfrak{S} \neq (a_1, a_2)$, we repeat the process and obtain an $a_3 \in \Gamma_0$ such that the order of the group generated by a_1, a_2, a_3 mod $p\mathfrak{G}$ is p^3 and so on. Since $\mathfrak{G} - p\mathfrak{G}$ is finite this process breaks off and we obtain a finite basis for $\mathfrak{S}$.

Before leaving this part we note that the above methods are applicable to cases other than the one we have considered. We shall not attempt to give the maximum generality of this group-theoretic method but mention certain extensions which are readily made.

A compact t. d. ring $\mathfrak{G}$ is a direct sum of p-rings.

This follows at once from the remark that the $\mathfrak{G}^{(p_1\cdots p_i)}_{p_{i+1}}$ in Theorem 4. 1 are characteristic subgroups. For, then they are ideals and the sum is direct in the sense of rings as well as of groups.

Theorem 5. 3 may be stated under the more general assumptions that $\mathfrak{S}$ is a hypercomplex system over P and is a p-group.

II. The structure of associative fields.

6. Let $\mathfrak{F}$ be a l. c. s. t. d. associative field (not necessarily commutative). For the present we may drop the distinction between the two types characteristic 0 and characteristic p. We assume, of course, that $\mathfrak{F}$ is not discrete.

There exists a $y \neq 0$ sufficiently near 0 such that $y\bar{U} < U$ where U is a compact neighborhood of 0 and $\bar{U}$ is its closure. Hence $U > y\bar{U} > y^2\bar{U} > \cdots$. Suppose $y^{\nu_k} \rightarrow z \neq 0$. Then $y^{\mu_k} \rightarrow 1$ where $\mu_k = \nu_{k+1} - \nu_k$. If u is any element of U, $y^{\mu}u \in y^{\mu_k}\bar{U}$ for $\mu \geqq \mu_k$ and hence $u = \lim_{i\rightarrow\infty} y^{\mu_i}u$ is contained in the closed set $y^{\mu_k}\bar{U}$. Thus $U \subset y^{\mu_k}\bar{U}$ which is impossible. Hence $y^{\nu} \rightarrow 0$.

We require the existence of such a y in proving

LEMMA 6. 1. *If $a_\nu \rightarrow \infty$ then $a_\nu^{-1} \rightarrow 0$.*[17]

Evidently $\{a_\nu^{-1}\}$ can have no limit point other than 0 and hence it is sufficient to show that 0 is really a limit point of this sequence. Let $y \neq 0$ be an element such that $y^\nu \rightarrow 0$ and U a compact neighborhood of 0. If infinitely many $a_\nu y^j$ for $\nu = 1, 2, \cdots$ and j fixed were contained in U, then the corresponding a_ν would be contained in the compact set $\bar{U}y^{-j}$ and hence would have a limit point. It follows that a sequence $\{b_\nu\}$ may be extracted from $\{a_\nu\}$ so that $b_\nu y^\nu \in \mathfrak{F}|U$. Since $y^\mu \rightarrow 0$ we may determine for each ν an integer k_ν such that $b_\nu y^{k_\nu} \in \mathfrak{F}|U$ but $b_\nu y^{k_\nu+1} \in U$. Evidently k_ν $(\geqq \nu) \rightarrow \infty$. Because of the compactness of U we may suppose that $b_\nu y^{k_\nu+1} \rightarrow z$. Then $b_\nu y^{k_\nu} \rightarrow zy^{-1} = w \in \mathfrak{F}|U$ and hence is $\neq 0$. From $y^{-k_\nu}b_\nu^{-1} \rightarrow w^{-1}$ and $y^{k_\nu} \rightarrow 0$ follows $b_\nu^{-1} \rightarrow 0$.

Let $\mathfrak{G}$ be a compact and open subgroup of $\mathfrak{F}$ and $\mathfrak{R}(\mathfrak{G})$ the set of elements a of $\mathfrak{F}$ such that $a\mathfrak{G} \subset \mathfrak{G}$.

---

[17] $a_\nu \rightarrow \infty$ means that $\{a_\nu\}$ is absolutely divergent, i. e., has no convergent subsequence. Lemma 6. 1 is the "Perfektisierungsaxiom" for fields of van Dantzig (D, p. 612). The present proof holds for all locally compact fields satisfying the first denumerability axiom.

LEMMA 6.2. *$\mathfrak{R}(\mathfrak{G})$ is a compact and open domain of integrity (d. o. i.).*

That $\mathfrak{R}(\mathfrak{G})$ is a ring may be verified directly. Since $\mathfrak{F}$ is a field $\mathfrak{R}$ has no zero-divisors. By the continuity of multiplication we have that $\mathfrak{R}$ contains a neighborhood of 0 and hence is open. If $\{a_\nu\}$ is an infinite sequence of elements of $\mathfrak{R}$ and $g \neq 0$ belongs to $\mathfrak{G}$ then $\{a_\nu g\}$ is a sequence in $\mathfrak{G}$ and hence has a limit point h. Thus $\{a_\nu\}$ has hg^{-1} as limit point and $\mathfrak{R}$ is compact.

7. In this section we consider the ideal theory of a compact and open (c. o.) d. o. i. $\mathfrak{R}$ contained in $\mathfrak{F}$. The existence of such subrings has just been shown.

THEOREM 7.1. *$\mathfrak{R}$ satisfies the chain conditions for left-(right-) ideals.*[18]

If $\mathfrak{J}$ is a left-(right-) ideal then $\mathfrak{J} \supset \mathfrak{R}b$ where $b \in \mathfrak{J}$. If $b \neq 0$, $\mathfrak{R}b$ is open and hence $\mathfrak{J} = \sum_{b \neq 0} \mathfrak{R}b$ is open. Let $\mathfrak{J}_1 < \mathfrak{J}_2 < \mathfrak{J}_3 < \cdots$ be an increasing sequence of ideals and $\nu_1, \nu_2, \nu_3, \cdots$ the orders of the finite quotient groups $\mathfrak{R} - \mathfrak{J}_1, \mathfrak{R} - \mathfrak{J}_2, \mathfrak{R} - \mathfrak{J}_3, \cdots$. Then $\nu_1 > \nu_2 > \nu_3 > \cdots$. Hence the sequence of $\mathfrak{J}$'s is finite. Similarly, every decreasing sequence of left-(right-) ideals $\mathfrak{J}_1 > \mathfrak{J}_2 > \mathfrak{J}_3 > \cdots > \mathfrak{J}_\omega$ containing a fixed ideal $\mathfrak{J}_\omega$ is finite.

Let $\mathfrak{P}$ be the totality of elements b of $\mathfrak{R}$ for which $b\mathfrak{R} < \mathfrak{R}$. We propose to show that $\mathfrak{P}$ is a prime ideal in $\mathfrak{R}$ and that $\mathfrak{P} > \mathfrak{P}^2 > \cdots \to 0$.

LEMMA 7.1. *$\mathfrak{R}|\mathfrak{P}$ contains arbitrarily small ideals.*

If U is any neighborhood of 0, there exists a $z \neq 0$ sufficiently near 0 such that $\mathfrak{R}z\mathfrak{R} \subset U$. $\mathfrak{R}z\mathfrak{R}$ is evidently an ideal.

LEMMA 7.2. *$\mathfrak{R}|\mathfrak{P}$ is a compact group relative to multiplication.*

$\mathfrak{R}|\mathfrak{P}$ consists of those elements $a, a', \cdots$ of $\mathfrak{F}$ which satisfy $a\mathfrak{R} = \mathfrak{R}$, $a'\mathfrak{R} = \mathfrak{R}, \cdots$. It follows then that $a'a\mathfrak{R} = \mathfrak{R}$, $a^{-1}\mathfrak{R} = \mathfrak{R}$, i. e., $aa', a^{-1} \in \mathfrak{R}|\mathfrak{P}$. Hence $\mathfrak{R}|\mathfrak{P}$ is a group. Since $\mathfrak{R}|\mathfrak{P}$ is closed and contained in $\mathfrak{R}$ it is compact.

LEMMA 7.3. *An element b of $\mathfrak{R}$ belongs to $\mathfrak{P}$ if and only if $b^\nu \to 0$.*

Since b has an inverse and $\mathfrak{R} > b\mathfrak{R}$ we have $\mathfrak{R} > b\mathfrak{R} > b^2\mathfrak{R} > \cdots$. If $b^{\nu_k} \to c \neq 0$, then $b^{\mu_k} \to 1$ where $\mu_k = \nu_{k+1} - \nu_k$. But this is impossible for the reasons given in § 6. The converse is obvious.

[18] The conditions referred to are the "Teilerkettensatz" and the "eingeschrankte Vielfachenkettensatz" of E. Noether, cf. "Abstrakter Aufbaue der Idealtheorie in algebraischen Zahl-und Funktionsenkorper," *Mathematische Annalen*, vol. 96 (1926), p. 26.

If $\mathfrak{P}_1$ is the set of b's such that $\mathfrak{R}b < \mathfrak{R}$, we may show as in Lemma 7.3 that $\mathfrak{P}_1$ consists of those elements b of $\mathfrak{R}$ whose power sequence $b^\nu \to 0$. Hence $\mathfrak{P}_1 = \mathfrak{P}$.

LEMMA 7.4. *If $\mathfrak{A}$ is a finite ring, its radical $\mathfrak{N}$ consists of the properly nilpotent elements,* i.e., the elements z such that za and az are nilpotent for all a in $\mathfrak{A}$.

This lemma is well known for rings with a finite basis over a field and it may be proved in exactly the same manner for our case of finite rings.[19]

THEOREM 7.2. *$\mathfrak{P}$ is a prime ideal in $\mathfrak{R}$ and* $\underset{\nu}{I}\mathfrak{P}^\nu = 0$.[20]

If $b \in \mathfrak{P}$ and $a \in \mathfrak{R}$ then ab and $ba \in \mathfrak{P}$. Hence $\mathfrak{P}$ is invariant. Since $\mathfrak{P}$ contains a sufficiently small neighborhood of 0, it contains an ideal $\mathfrak{J}$ of $\mathfrak{R}$. Consider the finite ring $\mathfrak{A} = \mathfrak{R} - \mathfrak{J}$. In the isomorphism $\mathfrak{R} \sim \mathfrak{A}$ the elements of $\mathfrak{P}$ are precisely those elements of $\mathfrak{R}$ which correspond to properly nilpotent elements of $\mathfrak{A}$. Hence $\mathfrak{P}$ corresponds to the radical $\mathfrak{N}$ of $\mathfrak{A}$. It follows that $\mathfrak{P}$ is an ideal and $\mathfrak{P}^n \equiv 0(\mathfrak{J})$ for sufficiently high n. Since $\mathfrak{J}$ is arbitrarily small, $\underset{\nu}{I}\mathfrak{P}^\nu = 0$. $\mathfrak{P}$ is evidently prime.

From this theorem and Wedderburn's theorem on finite fields [21] we obtain

THEOREM 7.3. *$\mathfrak{R} - \mathfrak{P}$ is a commutative field.*

8. Let $\mathfrak{R}_1$ be a c. o. d. o. i. contained in $\mathfrak{F}$ and $\mathfrak{P}_1$ its prime ideal defined as in the last section. (It is easily seen that $\mathfrak{P}_1$ is the only prime ideal in $\mathfrak{R}_1$). Consider $\mathfrak{R}_2 = \mathfrak{R}(\mathfrak{P}_1)$ the largest d. o. i. in which $\mathfrak{P}_1$ is contained as a left-ideal, i.e., $\mathfrak{R}_2$ consists of the elements a of $\mathfrak{F}$ such that $a\mathfrak{P}_1 \subset \mathfrak{P}_1$. Evidently $\mathfrak{R}_2 \supset \mathfrak{R}_1$ and is c. o. Determine $\mathfrak{R}_3 = \mathfrak{R}(\mathfrak{P}_2)$ where $\mathfrak{P}_2$ is the prime ideal of $\mathfrak{R}_2$ and so on. We thus obtain a sequence of c. o. d. o. i. $\mathfrak{R}_1 \subset \mathfrak{R}_2 \subset \mathfrak{R}_3 \subset \cdots$. Then $\mathfrak{R} = \lim \mathfrak{R}_\nu = \mathfrak{R}_1 \cup \mathfrak{R}_2 \cup \mathfrak{R}_3 \cup \cdots$ is an open d. o. i.

We wish to show that $\mathfrak{R}$ *is compact.* This will follow from the following definition and lemma.

Definition. An element $a \in \mathfrak{F}$ is *integral* if its power sequence $\{a^\nu\}$ has no divergent subsequence.

[19] For a proof for hypercomplex systems with a finite basis see L. E. Dickson, *Algebras and their arithmetics*, Chicago, p. 47, or German edition, p. 97.

[20] $\mathfrak{P}^\nu$ is the smallest ring containing all the products $b_1 b_2 \cdots b_\nu$ where $b_\mu \in \mathfrak{P}$.

[21] J. H. M. Wedderburn, "A theorem on finite algebras," *Transactions of the American Mathematical Society*, vol. 6 (1905), pp. 349-352.

LEMMA 8.1. *A subring $\mathfrak{R}$ of $\mathfrak{F}$ containing only integral elements is compact.*

If $\{a_\nu\} \subset \mathfrak{R}$ and $a_\nu \to \infty$ then $a_\nu^{-1} \to 0$. Hence for a sufficiently large N, $a_N^{-1} \in \mathfrak{P}_1$ and so $\lim_{\mu\to\infty} a_N^{-\mu} = 0$. But this implies that $\lim_{\mu\to\infty} a_N^{\mu} = \infty$ which is impossible.

Since $\mathfrak{R} = \lim \mathfrak{R}_\nu$ as well as the $\mathfrak{R}_\nu$ contains integral elements only, it is compact. Since $\mathfrak{R}$ is c. o., it follows by the argument of Theorem 7.1 that $\mathfrak{R} = \mathfrak{R}_N$ for N sufficiently large and hence $\mathfrak{R}(\mathfrak{P}) = \mathfrak{R}$ where $\mathfrak{P}$ is the prime ideal of $\mathfrak{R}$.

Let $\mathfrak{P}^*$ denote the set of inverses of the elements of $\mathfrak{P}$.

THEOREM 8.1. $\mathfrak{F} = \mathfrak{R} \cup \mathfrak{P}^*$.

If $b \in \mathfrak{F}|\mathfrak{R} = \mathfrak{F}|\mathfrak{R}(\mathfrak{P})$, then there exists an element c_1 of $\mathfrak{P}$ such that $b_1 = bc_1 \in \mathfrak{F}|\mathfrak{P}$. If $b_1 \in \mathfrak{F}|\mathfrak{R}$, we may repeat this process and obtain a c_2 in $\mathfrak{P}$ such that $b_1c_2 = bc_1c_2 \in \mathfrak{F}|\mathfrak{P}$ and so on. Since $\mathfrak{P}^\nu \to 0$ the sequence $c_1, c_1c_2, c_1c_2c_3, \cdots$, if infinite, converges to 0 contradicting $bc_1c_2 \cdots c_\nu \in \mathfrak{F}|\mathfrak{P}$. Thus the sequence of c's breaks off after, say, r steps and we obtain $bc_1c_2 \cdots c_r = u \in \mathfrak{R}|\mathfrak{P}$. Hence $b^{-1} = c_1c_2 \cdots c_r u^{-1} \in \mathfrak{P}$, since $u^{-1} \in \mathfrak{R}|\mathfrak{P}$. Thus $\mathfrak{F}|\mathfrak{R} = \mathfrak{P}^*$.

This theorem shows that $\mathfrak{R}$ may be characterized topologically as the totality of integral elements of $\mathfrak{F}$. We have then

THEOREM 8.2. *$\mathfrak{R}$ is the only maximal c. o. d. o. i. in $\mathfrak{F}$.*

If $f \in \mathfrak{P}^\nu|\mathfrak{P}^{\nu+1}$, we say that f has *exponent* ν and if $f^{-1} \in \mathfrak{P}^\nu|\mathfrak{P}^{\nu+1}$, f has *exponent* $-\nu$. Define the value $|f| = \gamma^{\exp f}$ where $f \neq 0$ and $0 < \gamma < 1$ and $|0| = 0$.

THEOREM 8.3. *$|f|$ gives a non-archimedean valuation (Bewertung) of $\mathfrak{F}$: $|f+g| \leqq \max(|f|, |g|)$, $|fg| = |f| \cdot |g|$.*

We choose an element x in $\mathfrak{P}|\mathfrak{P}^2$ and we shall show that if f has exponent ν then $f = f_0x^\nu$ where $f_0 \in \mathfrak{R}|\mathfrak{P}$. First if f and g have exponent ν, fg^{-1} and $g^{-1}f$ have exponent 0. It is sufficient to prove this for $\nu > 0$. Now $fg^{-1} \in \mathfrak{P}$ implies $f \in \mathfrak{P}g \subset \mathfrak{P}^{\nu+1}$ which is contrary to $\exp f = \nu$. Similarly $fg^{-1} \in \mathfrak{P}^*$ is impossible and hence $fg^{-1} \in \mathfrak{R}|\mathfrak{P}$ and $\exp fg^{-1} = 0$. For the same reasons $\exp g^{-1}f = 0$. We observe next that $\exp x^\nu = \nu$. For if $\nu > 0$ $\exp x^\nu \geqq \nu$ and if $> \nu$ then $x^\nu = \Sigma y_1y_2 \cdots y_\mu$ where $y \in \mathfrak{P}$ and $\mu > \nu$. Then $1 = \Sigma x^{-\nu}y_1y_2 \cdots y_\mu$ and since $x^{-1}y \in \mathfrak{R}$ this implies that $1 \in \mathfrak{P}$ which is impossible. Thus $\exp(fx^{-\nu} = f_0) = 0$ and $f = f_0x^\nu$, $f_0 \in \mathfrak{R}|\mathfrak{P}$.

Suppose $f = f_0 x^\nu$, $g = g_0 x^\mu$ where $f_0, g_0 \in \mathfrak{R}|\mathfrak{P}$ and $\nu \leqq \mu$. Then

$$f + g = (f_0 + g_0 x^{\mu-\nu}) x^\nu \qquad (f_0 + g_0 x^{\mu-\nu}) \in \mathfrak{R}.$$

Hence

$$\exp(f+g) \geqq \min(\exp f, \exp g) \qquad |f+g| \leqq \max(|f|, |g|).$$

$$fg = (f_0 x^\nu g_0 x^{-\nu}) x^{\nu+\mu} \qquad (f_0 x^\nu g_0 x^{-\nu}) \in \mathfrak{R}|\mathfrak{P}$$

and so

$$\exp(fg) = \exp f + \exp g \qquad |fg| = |f| \cdot |g|.$$

$\mathfrak{J}$ is a *fractional left-(right-) ideal in* $\mathfrak{F}$ if it is a subgroup and $\mathfrak{R}\mathfrak{J} \subset \mathfrak{J}$ $(\mathfrak{J}\mathfrak{R} \subset \mathfrak{J})$. It follows easily that $\mathfrak{J}$ is open. The inverse $\mathfrak{J}^{-1}$ of $\mathfrak{J}$ is the set of elements a such that $a\mathfrak{J} \subset \mathfrak{R}$, or $\mathfrak{J}^{-1}\mathfrak{J} \subset \mathfrak{R}$. $\mathfrak{J}^{-1}$ is evidently a fractional left-ideal. If $\mathfrak{J} \neq (0), \neq 1$ then $\mathfrak{J}$ is called *proper.*

THEOREM 8.4. *Every proper fractional left-(right-) ideal is two-sided, principal and a power of* $\mathfrak{P}$.

If $b \in \mathfrak{J}$ and c has exponent $\geqq \exp b$, then $\exp cb^{-1} \geqq 0$, and $cb^{-1} \in \mathfrak{R}$. Hence $c \in \mathfrak{R}b \subset \mathfrak{J}$. If $\mathfrak{J}$ contains elements of arbitrarily small exponent, $\mathfrak{J} = \mathfrak{F}$. Otherwise let b have the smallest exponent of the elements of $\mathfrak{J}$. It follows that $\mathfrak{J}$ is the set of elements of $\mathfrak{F}$ of exponent $\geqq \exp b = k$ and hence $\mathfrak{J} = \mathfrak{P}^k = (b)$.

It follows directly from this theorem that $\mathfrak{J}\mathfrak{J}^{-1} = \mathfrak{R}$ as well as $\mathfrak{J}^{-1}\mathfrak{J}$. If $\mathfrak{F}$ has characteristic 0, $p\mathfrak{R}$ is an ideal $\neq 0$ and hence is $= \mathfrak{P}^e$ where $e \geqq 1$.

9. In this final section we shall apply the arithmetic results obtained in § 8 to obtain the structure of $\mathfrak{F}$.

Let p^n be the order of the finite field $\mathfrak{R} - \mathfrak{P}$ of characteristic p. $\mathfrak{R} - \mathfrak{P}$ contains a primitive q-th root of unity where $q = p^n - 1$, i. e., there exists a $u_0 \in \mathfrak{R}$ such that $u_0{}^q \equiv 1(\mathfrak{P})$ and q is the smallest power for which such a congruence holds. Let $x \equiv 0(\mathfrak{P})$ but $\not\equiv 0(\mathfrak{P}^2)$. Then the correspondence $a \leftrightarrow xax^{-1}$ where $a \in \mathfrak{R}$ determines an automorphism of the Galois group of $\mathfrak{R} - \mathfrak{P}$. It follows that

$$xu_0x^{-1} \equiv u_0{}^s(\mathfrak{P}), \quad \text{where} \quad s = p^t.$$

THEOREM 9.1. $\mathfrak{F}$ *contains a primitive* q-*th root of unity.*

We shall determine $u \in \mathfrak{R}|\mathfrak{P}$ such that $u^q = 1$ and $u \equiv u_0(\mathfrak{P})$. Since u_0 is a primitive q-th root of unity mod $\mathfrak{P}$ it will follow that u is primitive. Let r be the exponent of s mod q, i. e., r is the least positive integer for which $s^r \equiv 1(q)$. Then r is also the least positive integer satisfying

$$(**) \qquad x^r u_0 x^{-r} \equiv u_0(\mathfrak{P}).$$

If $u_0{}^q \not\equiv 1$, suppose $u_0{}^q \equiv 1(\mathfrak{P}^k)$, $\not\equiv 1(\mathfrak{P}^{k+1})$. Then $u_0{}^q - 1 = v_0 x^k$ where $v_0 \in \mathfrak{R}|\mathfrak{P}$. Since $v_0 u_0 \equiv u_0 v_0(\mathfrak{P})$ and $(u_0{}^q - 1)u_0 = u_0(u_0{}^q - 1)$, we have $x^{-k} u_0 x^k \equiv u_0(\mathfrak{P})$ and hence by (**) $k \equiv 0(r)$, say $k = rl$. Thus

$$u_0{}^q \equiv 1 + w x^{rl}(\mathfrak{P}^{rl+1}) \qquad w \not\equiv 0(\mathfrak{P}).$$

Set $u_1 = u_0 + y x^{rl}$ where y is to be determined $\in \mathfrak{R}$ so that $u_1{}^q \equiv 1(\mathfrak{P}^{rl_1})$ for $l_1 > l$. This requires in particular that

$$({}_*{}^*{}_*) \qquad u_1{}^q \equiv 1 + (w + q u_0{}^{q-1} y) x^{rl} \equiv 1(\mathfrak{P}^{rl+1})$$

by (**) and the commutativity of u_0 and y mod $\mathfrak{P}$. Since q and $u_0{}^{q-1} \not\equiv 0(\mathfrak{P})$, y may be chosen so that $(w + q u_0{}^{q-1} y) \equiv 0(\mathfrak{P})$ and will then satisfy $({}_*{}^*{}_*)$. But then we will necessarily have $u_1{}^q \equiv 1(\mathfrak{P}^{rl_1})$ for $l_1 > l$. If u_1 is not a q-th root of one we may repeat this process with it in place of u_0. In this way we either obtain a q-th root of one after a finite number of steps or else an infinite sequence $\{u_\nu\}$ such that

$$u_\nu \equiv u_{\nu-1}(\mathfrak{P}^{rl_{\nu-1}}) \qquad u_\nu{}^q \equiv 1\,(\mathfrak{P}^{rl_\nu})$$

where $l < l_1 < l_2 < \cdots$. The $\{u_\nu\}$ converge to a limit u having the desired properties:

$$u^q = 1 \qquad u \equiv u_\nu(\mathfrak{P}^{rl_{\nu-1}}) \qquad x u x^{-1} \equiv u^s(\mathfrak{P}).$$

THEOREM 9. 2. *x may be normalized so that $xux^{-1} = u^s$, $x \in \mathfrak{P}|\mathfrak{P}^2$.*

This theorem has been given by Hasse for p-adic fields. Moreover his proof (H. p. 511) goes over word for word for the present more general case.

THEOREM 9. 3. *Every element of $\mathfrak{F}$ may be represented uniquely in the form $v = (v_0 + v_1 x + v_2 x^2 + \cdots) x^{-k}$ where $k \geqq 0$ and $v_\nu = u^{k_\nu}$ or 0.*

If $v \equiv 0(\mathfrak{P}^{-k})$ for $k \geqq 0$, then $vx^k \equiv 0(\mathfrak{R})$ and hence $vx^k \equiv v_0(\mathfrak{P})$ where $v_0 = u^{k_0}$ or 0. Now $(vx^k - v_0)x^{-1} \equiv v_1(\mathfrak{P})$ where $v_1 = u^{k_1}$ or 0 and so on. We obtain thus a sequence $v_\nu = u^{k_\nu}$ or 0 such that

$$vx^k \equiv v_0 + v_1 x + v_2 x^2 + \cdots + v_\nu x^\nu(\mathfrak{P}^{\nu+1})$$

and hence

$$v = (v_0 + v_1 x + v_2 x^2 + \cdots) x^{-k}.$$

If $(\sum_{\nu=0}^{\infty} v_\nu x^\nu) x^{-k} = (\sum_{\nu=0}^{\infty} v'_\nu x^\nu) x^{-l}$, evidently $k = l$ so that $\Sigma v_\nu x^\nu = \Sigma v'_\nu x^\nu$. Hence

$v_0 \equiv v'_0(\mathfrak{P})$ and so $v_0 = v'_0$ since v_0 and v'_0 are members of the complete set of incongruent residues $0, u, u^2, \cdots, u^q \bmod \mathfrak{P}$. Likewise $v_1 = v'_1, v_2 = v'_2, \cdots$.

THEOREM 9.4. *The centrum $\mathfrak{C}$ of $\mathfrak{F}$ consists of all elements of the form $(c_0 + c_1x^r + c_2x^{2r} + \cdots)x^{-kr}$ where $c_\nu = 0$ or $u^{k\nu}$ and commutes with x.*

If $c = \sum_{\nu=-k}^{\infty} c_\nu x^\nu$, $u^{-1}cu = c$ implies $c_\nu = c_\nu u^{s^\nu - 1}$. Hence if $c_\nu \neq 0$, $u^{s^\nu-1} = 1$ and $s^\nu \equiv 1(q)$ or $\nu \equiv 0(r)$. Further $xcx^{-1} = c$ implies that $c_\nu{}^s = c_\nu$ and hence c_ν commutes with x. Thus the two conditions that c commutes with x and with u imply that c has the form $\sum_{\nu=-k}^{\infty} c_\nu x^{\nu r}$, $c_\nu x = xc_\nu$. Since these two conditions insure that c belongs to the centrum $\mathfrak{C}$, we have the theorem.

The above proof shows also that the elements of $\mathfrak{F}$ commutative with u are all of the form $\sum_{\nu=-k}^{\infty} v_\nu x^{r\nu}$ where $v_\nu = u^{k\nu}$ or 0. These elements form a commutative subfield $\mathfrak{C}' \supset \mathfrak{C}$. Since $\mathfrak{C}' = \mathfrak{C}(u)$ it has a finite basis over $\mathfrak{C}$.

THEOREM 9.5. *$\mathfrak{C}'$ is a cyclic field of degree r over $\mathfrak{C}$.*

The automorphism $a \leftrightarrow xax^{-1}$ of $\mathfrak{F}$ leaves $\mathfrak{C}'$ invariant (though not element-wise) and hence induces an automorphism S in $\mathfrak{C}'$. The elements of $\mathfrak{C}'$ invariant under S are precisely the elements commutative with x and hence belonging to $\mathfrak{C}$. Since x^r is the smallest power of x commutative with u, the order of S is r. Hence $\mathfrak{C}'$ is cyclic of degree r over $\mathfrak{C}$ with S as generating automorphism of its Galois group relative to $\mathfrak{C}$.

THEOREM 9.6. *$\mathfrak{F}$ is a cyclic algebra over its centrum.*

For, by Theorem 9.5, u satisfies an irreducible equation of degree r, $\phi_r(u) = 0$ having coefficients in $\mathfrak{C}$. Thus

$$\phi_r(u) = 0 \qquad xu = u^s x \qquad (s = p^t) \qquad x^r = z \in \mathfrak{C}$$

gives a description of the algebra $\mathfrak{F}$ relative to its centrum $\mathfrak{C}$.

To complete the description of $\mathfrak{F}$ it is necessary to give the structure of $\mathfrak{C}$. Since $\mathfrak{C}$ is closed it is a l. c. s. t. d. commutative field and hence as mentioned in the introduction its form has been described by v. Dantzig and by Hasse and Schmidt. We shall merely state the results here and sketch briefly their derivation.

For the case $\chi(\mathfrak{F}) = 0$ we have seen that $\mathfrak{F}$ and hence also $\mathfrak{C}$ has finite order over the p-adic field P_p. From considerations analogous to those of

Theorems 9.1 and 9.2 it follows that $\mathfrak{C}$ is generated by a $p^m - 1$ $(m \geqq 1)$ root of one and a second element z $(= x^r)$. Certain additional normalizations may be made. The reader is referred to Hasse's paper (H. p. 514) for these.

If $\chi(\mathfrak{F}) = p$, we have

THEOREM 9.7. *The powers of u and 0 form a finite field K_{p^n} of p^n elements.*

It suffices to show that $v \equiv u^j + u^k = u^l$ or 0 and $-u^k = u^m$. Suppose $v \neq 0$. Since v is algebraic mod p (satisfies an algebraic equation whose coefficients are in the field K_p of residues mod p), it is a root of unity and hence $\epsilon\, \mathfrak{R}|\mathfrak{P}$. If $v \equiv u^l$ (mod $\mathfrak{P}$), $v - u^l$ is algebraic mod p and $\equiv 0$ (mod $\mathfrak{P}$). This is impossible unless $v - u^l = 0$. Similarly we may show that $-u^k$ is $=$ some u^m.

By the same argument we see that x is transcendental mod p. Thus $\mathfrak{F}$ is uniquely determined by the field K_{p^n} and its automorphism $a \leftrightarrow a^s$ $(s = p^t)$. For when these are given we have $\mathfrak{F}$ determined as the set of integral power series in x with a finite number of negative powers and coefficients in K_{p^n} where the multiplication is determined by $xa = a^s x (a \,\epsilon\, K_{p^n})$.

PRINCETON UNIVERSITY.

Corrections

Professor Irving Kaplansky has pointed out two errors. On page 457 of his paper "Locally compact rings" (*Amer. J. Math.* **70** (1948) 447–459). First, he notes that on page 440 of this paper there is a tacit assumption that the mapping $x \to p^{-1}x$ is continuous. This can not be justified without further assumptions. This invalidates the proof of Theorem 5.3. The theorem is correct for rings with unit, since in this case p^{-1} is an element of the ring and hence $x \to p^{-1}x$ is continuous. Also Kaplansky notes that the assertion "made on page 442 that if S is a locally compact p-group which is an algebra over the rational numbers, then S is an algebra of finite order over the field of p-adic numbers" is false. He gives a counter-example in which the multiplication in S is trivial ($S^2 = 0$).

Reprinted from
American Journal of Mathematics
April 1936.

Reprinted from the Proceedings of the NATIONAL ACADEMY OF SCIENCES,
Vol. 23, No. 4, pp. 240–242 April, 1937.

SIMPLE LIE ALGEBRAS OF TYPE A

BY N. JACOBSON[1]

DEPARTMENT OF MATHEMATICS, UNIVERSITY OF CHICAGO

Communicated March 8, 1937

If $\mathfrak{A}$ is an associative algebra over Φ it becomes a Lie algebra[2] when $[a, b]$ is defined as $ab - ba$ where ab is the associative product originally given in $\mathfrak{A}$. If in addition $\mathfrak{A}$ is self-reciprocal, i.e., there is defined a correspondence $a \longrightarrow a^J \epsilon \mathfrak{A}$ such that

$$(a + b)^J = a^J + b^J \quad (\alpha a)^J + \alpha a^J \quad (ab)^J = b^J a^J, \qquad \alpha \epsilon \Phi$$

then the elements a such that $a^J = -a$ are called *J-skew* and their totality

$\mathfrak{S}_J$ is easily seen to be a Lie subalgebra of $\mathfrak{A}$. If $a^{J^2} = a$ for all a, J is called an *involutorial anti-automorphism* (i.a.a.). Then $K = S^{-1}JS$ is also an i.a.a. for S an arbitrary automorphism of the associative algebra $\mathfrak{A}$ over Φ and we call K *cogredient* to J. It is easily seen that $\mathfrak{S}_K$ consists of the elements a^S where a varies in $\mathfrak{S}_J$ and so $\mathfrak{S}_J$ and $\mathfrak{S}_K$ are isomorphic. Hence the derived algebras $\mathfrak{S}_J'$ and $\mathfrak{S}_K'$ are isomorphic also. If $J = K$, i.e., $SJ = JS$ then S is an automorphism of $\mathfrak{A}$ inducing an automorphism in the Lie algebras $\mathfrak{S}_J$ and $\mathfrak{S}_J'$.

In a paper[3] which will appear shortly in the *Annals of Mathematics* I discussed the Lie algebras $\mathfrak{S}_J$ obtainable in this way from a normal simple algebra $\mathfrak{A}$ over Φ of characteristic 0 and showed their connection with the theory of abstract Lie algebras of types B, C and D. The present note will give an outline of an extension of these results to involutorial simple algebras of second kind as defined by Albert.[4]

Following Landherr[5] we say that a Lie algebra $\mathfrak{B}$ over Φ of characteristic 0 has *type A* if $\mathfrak{B}_\Omega \cong \Omega_n'$ where Ω is the algebraic closure of Φ and Ω_n the algebra of all n-rowed matrices with coördinates in Ω. $\mathfrak{B}$ has *type* $A_{\rm I}$ or $A_{\rm II}$ according as the coefficients of the characteristic equations of the matrices representing the elements of $\mathfrak{B}$ in Ω_n all belong to Φ or not. In either case $\mathfrak{B}$ is simple since Ω_n' is.

Now suppose $\mathfrak{A}$ is simple of order $2n^2$ over Φ of characteristic 0 with $P = \Phi(q)$, $q^2 = \mu$ in Φ as centrum and J an i.a.a. of second kind in $\mathfrak{A}$, i.e., $\xi^J = \bar{\xi} = \alpha - \beta q$ for $\xi = \alpha + \beta q$ in P. We prove

THEOREM 1. *If* $n > 2$ $\mathfrak{S}_J'$ *is a Lie algebra of type* A_{II}.

An important lemma used in the proof is that the enveloping algebra over Φ of $\mathfrak{S}_J'$ is $\mathfrak{A}$. The classification of the Lie algebras $\mathfrak{S}_J'$ and their automorphisms are given by

THEOREM 2. *If* $\mathfrak{A}_1$ *and* $\mathfrak{A}_2$ *are involutorial simple algebras of order* $2n^2$ *over* Φ, $n > 2$, *and* J_1 *and* J_2 *are i.a.a. of second kind in* $\mathfrak{A}_1$ *and* $\mathfrak{A}_2$, *respectively, such that* $\mathfrak{S}_{J_1}' \cong \mathfrak{S}_{J_2}'$ *then* $\mathfrak{A}_1 \cong \mathfrak{A}_2$ *and when* $\mathfrak{A}_1$ *and* $\mathfrak{A}_2$ *are identified* J_1 *and* J_2 *are cogredient.*

THEOREM 3. *If* $\mathfrak{A}$ *is involutorial simple of order* $2n^2$, $n > 2$, *over* Φ *and* J *an i.a.a. of second kind then any automorphism in* $\mathfrak{S}_J'$ *may be extended to a unique automorphism of the associative algebra* $\mathfrak{A}$.

The extended automorphism commutes with J. The inner automorphism $a \longrightarrow g^{-1}ag$ commutative with J are given by *J-orthogonal* elements, i.e., $gg^J = \gamma \neq 0$ in Φ. These form an invariant subgroup of index 1 or 2 in the group of all automorphisms of $\mathfrak{A}$ commutative with J. It follows that if $\mathfrak{g}_J$ is the group of J-orthogonal elements and $\mathfrak{d}$ the set of multiples $\delta 1$, $\delta \neq 0$ in P then $\mathfrak{g}_J/\mathfrak{d}$ is isomorphic either to the complete group of automorphisms of $\mathfrak{S}_J'$ or to an invariant subgroup of index 2 of this group.

As a converse of Theorem 1 we have

THEOREM 4. *If* $\mathfrak{B}$ *is a Lie algebra of type* A_{II} *then* $n > 2$ *and* $\mathfrak{B} \cong \mathfrak{S}_J'$ *where* J *is an i.a.a. of second kind in a simple algebra* $\mathfrak{A}$.

Thus a complete classification of Lie algebras of type A_{II} depends on the classification of simple associative algebras of second kind and of i.a.a.'s relative to cogredience in these algebras. It can be shown that the latter problem is essentially one of ordinary cogredience of hermitian matrices with coördinates in an involutorial division algebra of second kind. These questions are largely arithmetic in character and depend on the structure of Φ.

A detailed account of this theory will appear in another journal.

[1] NATIONAL RESEARCH FELLOW.

[2] For definitions of the principal concepts in the theory of Lie algebras, see N. Jacobson, "Rational Methods in the Theory of Lie Algebras," *Ann. Math.*, **36**, 875–881 (1935).

[3] "A Class of Normal Simple Lie Algebras of Characteristic 0." An abstract of this paper appeared in the *Bull. Am. Math. Soc.*, **43**, Jan. (1937).

[4] A. A. Albert, "Involutorial Simple Algebras and Riemann Matrices," *Ann. Math.*, **36**, 894 (1935).

[5] W. Landherr, "Ueber einfache Liesche Ringe," *Hamb. Abhandlungen*, **11**, 50 (1935). $\mathfrak{B}_\Omega$ denotes the algebra obtained from $\mathfrak{B}$ by extending the field Φ to Ω. The Lie algebras of type A_I were determined completely by Landherr.

ANNALS OF MATHEMATICS
Vol. 38, No. 2, April, 1937

PSEUDO-LINEAR TRANSFORMATIONS[1]

BY N. JACOBSON

(Received May 29, 1936)

It is well known that two matrices T_1 and T_2 with coördinates in a commutative field are similar, i.e. there exists a non-singular matrix A such that $T_2 = A^{-1}T_1A$ if and only if they have the same elementary divisors. In a number of problems occurring in complex projective geometry, differential and difference equations, and hypercomplex numbers a more general problem of similarity is encountered: namely, when are two matrices with coördinates in a non-commutative field $\mathfrak{F}$ related in the fashion

$$(*)\qquad T_2 = A^{-1}T_1A^S + A^{-1}A_S$$

where $A = (\alpha_{ij})$ is a matrix in $\mathfrak{F}$, $A^S = (\alpha_{ij}^S)$, $A_S = (\alpha_{ijS})$ and the α^S and α_S are obtained respectively from the α by a fixed 1-automorphism and differentiation in $\mathfrak{F}$? (§1 for definitions). In analogy with the special case of matrices in a commutative field under ordinary similarity, we define the elementary divisors of a matrix subject to the transformation (*) as certain non-commutative polynomials and prove that a necessary and sufficient (n.a.s.c.) that T_1 and T_2 be similar is that their elementary divisors be similar (§8).

As in the usual theory the present notion of similarity carries with it notions of reducibility, decomposability, and complete reducibility (§3). Criteria for these may be given in terms of the elementary divisors. In the important special case of finite matrices these criteria may be sharpened (§9).

We consider also the automorphism ring $\mathfrak{A}$ of T, i.e. the ring of all matrices A such that $AT = TA^S + A_S$. This is a generalization of the ring of matrices commutative with T and has many interesting special cases. For example, cyclic algebras and the invariant ring of a differential polynomial as defined by Ore are included here. The general theory of the ring $\mathfrak{A}$ may be deduced from results due to v. d. Waerden, Fitting and others. A more detailed discussion is given of some of the special cases which have application to hypercomplex numbers and differential equations.

In the body of the paper we take the abstract point of view, formulating the problems in terms of linear transformations or groups with operators. The concepts of similarity, reducibility, etc. appear naturally as special cases of the concepts of isomorphism, existence of allowable subgroups, etc. of groups with operators. This procedure has the added advantage of enabling us to apply directly the general theorems on groups to our case. The reader is

[1] An abstract of this paper appeared in the Proc. Nat. Acad. Sci. [7]. A portion of this work was done by the author as National Research Fellow at the University of Chicago.

referred to v. d. Waerden ([15] vol. I, pp. 132–144 and II, Chap. 15) for a discussion of these notions.

1. **Definition of pseudo-linear transformations.** Let $\mathfrak{R}$ be a vector space of finite dimensionality n over an arbitrary field $\mathfrak{F}$ (not necessarily commutative). A *pseudo-linear transformation* (p.l.t.) $\mathbf{T}$ of $\mathfrak{R}$ is defined by the following conditions:

$$(x + y)\mathbf{T} = x\mathbf{T} + y\mathbf{T} \qquad x, y \in \mathfrak{R} \tag{1}$$

$$(x\alpha)\mathbf{T} = (x\mathbf{T})\alpha^S + x\alpha_S \tag{2}$$

where the correspondence $\alpha \to \alpha^S$ is a 1-*automorphism* and $\alpha \to \alpha_S$ an *S-differentiation* in $\mathfrak{F}$, i.e.

$$(\alpha + \beta)^S = \alpha^S + \beta^S \qquad (\alpha\beta)^S = \alpha^S\beta^S \tag{3}$$

$$(\alpha + \beta)_S = \alpha_S + \beta_S \qquad (\alpha\beta)_S = \alpha_S\beta^S + \alpha\beta_S \tag{4}$$

and $\alpha \to \alpha^S$ is $(1 - 1)$. Equation (1) states that $\mathbf{T}$ is an automorphism of the abelian group $\mathfrak{R}$ and (2) gives the commutation relation

$$\alpha\mathbf{T} = \mathbf{T}\alpha^S + \alpha_S \tag{2'}$$

between $\mathbf{T}$ and the automorphism $x \to x\alpha$ of $\mathfrak{R}$. We note the following examples of p.l.t.

(a) $\alpha^S \equiv \alpha$, $\alpha_S \equiv 0$. In this case $\mathbf{T}$ is a linear transformation (l.t.) in a vector space over any (non-commutative) field. The case $\mathfrak{F}$ commutative is classical.

(b) $\mathfrak{F}$ the field of complex numbers, α^S the conjugate complex of α, $\alpha_S \equiv 0$. P.l.t. of this type were first defined by Segre[2] and are of interest in complex projective geometry. If we suppose that $\mathfrak{F}$ is any field, $\alpha \to \alpha^S$ any 1-automorphism, $\alpha_S \equiv 0$, we obtain as a generalization of the transformations of Segre a type of p.l.t. which we shall call *semi-linear* (s.l.t.). Thus the transformation $x \to x\mu \equiv x\mathbf{M}$ is an s.l.t. with automorphism $\alpha \to \alpha^S = \mu^{-1}\alpha\mu$.

(c) The differential equations

$$u_i' = \frac{du_i}{d\tau} = \sum_{j=1}^{n} u_j\alpha_{ji}(\tau) \qquad (i = 1, \cdots, n) \tag{5}$$

where the $\alpha(\tau)$ are analytic functions of τ, define a p.l.t. in the following way: For $\mathfrak{R}$ we take the space of forms $u = \sum u_i\alpha_i(\tau)$ in the independent vectors u_i with coördinates $\alpha_i(\tau)$ in the field $\mathfrak{F}$ of analytic functions of τ. We define $\mathbf{T}$ by

$$u \to u\mathbf{T} = \sum u_i'\alpha_i(\tau) + \sum u_i\alpha_i'(\tau)$$

where u_i is given by (5).[3] $\mathbf{T}$ is easily seen to be pseudo-linear with the identity as its automorphism. More generally if $\mathfrak{F}$ is any field $\alpha^S \equiv \alpha$ we shall call the corresponding p.l.t. *differential* (d.t.).

---

[2] C. Segre [14] and v. d. Waerden [16] p. 4.

[3] Cf. Krull [8] p. 188.

2. **The matrices of a p.l.t.** Let $(e_1, \cdots, e_n)$ be a basis for $\mathfrak{R}$ and suppose that the p.l.t. $\mathbf{T}$ sends e_k into $e_k\mathbf{T}$ where

$$e_k\mathbf{T} = e_1\tau_{1k} + e_2\tau_{2k} + \cdots + e_n\tau_{nk} \qquad (k = 1, \cdots, n)$$

or, using the usual rule of matrix multiplication

$$(6) \qquad (e_1\mathbf{T}, \cdots, e_n\mathbf{T}) = (e_1, \cdots, e_n)T \qquad T = (\tau_{ij}).$$

If x is the vector $\sum e_i\xi_i \equiv \begin{pmatrix} \xi_1 \\ \xi_2 \\ \vdots \\ \xi_n \end{pmatrix}$, then according to (1) and (2)

$$(7) \qquad x\mathbf{T} = T\begin{pmatrix} \xi_1^S \\ \xi_2^S \\ \vdots \\ \xi_n^S \end{pmatrix} + \begin{pmatrix} \xi_{1S} \\ \xi_{2S} \\ \vdots \\ \xi_{nS} \end{pmatrix}$$

Thus the p.l.t. $\mathbf{T}$ is completely determined by its matrix in a fixed coördinate system, its automorphism $\alpha \to \alpha^S$ and its differentiation $\alpha \to \alpha_S$. Conversely, if T is any matrix and $\alpha \to \alpha^S$ and $\alpha \to \alpha_S$ any 1-automorphism and differentiation, the correspondence $x \to x\mathbf{T}$ defined by (7) is a p.l.t. We restrict ourselves in the present paper to the study of p.l.t.'s with the same automorphism and differentiation and so, for the sake of simplicity, we write $\alpha^S \equiv \bar{\alpha}$, $\alpha_S \equiv \alpha'$.

Let $(e_1^*, \cdots, e_n^*)$ be a second basis for $\mathfrak{R}$ related to $(e_1, \cdots, e_n)$ by

$$(e_1^*, \cdots, e_n^*) = (e_1, \cdots, e_n)A \qquad (e_1, \cdots, e_n) = (e_1^*, \cdots, e_n^*)A^{-1}$$

where $A = (\alpha_{ij})$. If T and T^* are respectively the matrices of $\mathbf{T}$ relative to the bases $(e_1, \cdots, e_n)$ and $(e_1^*, \cdots, e_n^*)$ a simple computation shows that

$$(8) \qquad T^* = A^{-1}T\bar{A} + A^{-1}A'.$$

3. **Similarity, reducibility, decomposability and complete reducibility.** If $\mathbf{T}$ is a p.l.t. we denote the ring of transformations[4] in $\mathfrak{R}$ generated by $\mathbf{T}$ and the elements of $\mathfrak{F}$ by $\mathfrak{F}[\mathbf{T}]$. From the definition (1) and (2) it follows that the elements of $\mathfrak{F}[\mathbf{T}]$ may be represented as polynomials in $\mathbf{T}$ with coefficients (on the right of the powers of $\mathbf{T}$) in $\mathfrak{F}$. We shall consider $\mathfrak{R}$ as a group $(\mathfrak{R}, \mathfrak{F}[\mathbf{T}])$ with operators $\mathfrak{F}[\mathbf{T}]$.

We say that the p.l.t.'s $\mathbf{T}_1$ and $\mathbf{T}_2$ are *similar* if $(\mathfrak{R}, \mathfrak{F}[\mathbf{T}_1])$ is isomorphic as a group with operators to $(\mathfrak{R}, \mathfrak{F}[\mathbf{T}_2])$ i.e., if there exists a $(1 - 1)$ mapping $x \to x\mathbf{A}$ of $\mathfrak{R}$ on itself such that

$$(9) \qquad (x + y)\mathbf{A} = x\mathbf{A} + y\mathbf{A} \qquad (x\mathbf{O}_1)\mathbf{A} = (x\mathbf{A})\mathbf{O}_2$$

[4] The sum $\mathbf{T}_1 + \mathbf{T}_2$ of two transformations is defined as $x(\mathbf{T}_1 + \mathbf{T}_2) \equiv x\mathbf{T}_1 + x\mathbf{T}_2$, the product as $x(\mathbf{T}_1\mathbf{T}_2) \equiv (x\mathbf{T}_1)\mathbf{T}_2$.

where $\mathbf{O}_k = \mathbf{T}_k^m\alpha_0 + \mathbf{T}_k^{m-1}\alpha_1 + \cdots + \alpha_m$ $(k = 1, 2)$. In order that (9b) hold it is sufficient with (9a) to require merely that $(x\alpha)\mathbf{A} = (x\mathbf{A})\alpha$ and $(x\mathbf{T}_1)\mathbf{A} = (x\mathbf{A})\mathbf{T}_2$. Thus $\mathbf{A}$ is a non-singular l.t. such that

$$\mathbf{T}_1\mathbf{A} = \mathbf{A}\mathbf{T}_2, \qquad \mathbf{T}_2 = \mathbf{A}^{-1}\mathbf{T}_1\mathbf{A}. \tag{10}$$

It follows readily that the matrices T_1, T_2, A of $\mathbf{T}_1$, $\mathbf{T}_2$, $\mathbf{A}$ satisfy the equation

$$T_1 = A^{-1}T_2\bar{A} + A^{-1}A'. \tag{11}$$

Conversely, if the matrices of two p.l.t. (with the same automorphism and differentiation, are related according to (11), the p.l.t. are similar.

If $\mathfrak{R}_1$ is an allowable subgroup, or *invariant subspace* of $\mathfrak{R}$, i.e. a subspace which is transformed into itself by $\mathbf{T}$, then $\mathbf{T}$ induces a p.l.t. $\mathbf{T}_1$ in $\mathfrak{R}_1$ which may then also be considered as a group with operators $(\mathfrak{R}_1, \mathfrak{F}[\mathbf{T}_1])$. We call $\mathbf{T}_1$ a *contraction* and denote its relation to $\mathbf{T}$ by $\mathbf{T} \geq \mathbf{T}_1$, or $\mathbf{T}_1 \leq \mathbf{T}$. In this case $\mathbf{T}$ also induces a p.l.t. in the *projection space* (factor group) $\mathfrak{R}/\mathfrak{R}_1$ which we denote as $\mathbf{T}/\mathbf{T}_1$. If $\mathfrak{R}_1$ and $\mathfrak{R}_2$ are two invariant subspaces, then so is their vector sum $\mathfrak{R}_1 \cup \mathfrak{R}_2$ and their intersection $\mathfrak{R}_1 \cap \mathfrak{R}_2$. If $\mathbf{T}_1$ and $\mathbf{T}_2$ are the contractions of $\mathbf{T}$ in $\mathfrak{R}_1$ and $\mathfrak{R}_2$, we denote the p.l.t.'s corresponding to $\mathfrak{R}_1 \cup \mathfrak{R}_2$ and $\mathfrak{R}_1 \cap \mathfrak{R}_2$ by $\mathbf{T}_1 \cup \mathbf{T}_2$ and $\mathbf{T}_1 \cap \mathbf{T}_2$ respectively. The set of contractions of $\mathbf{T}$ considered relative to the operations $\cup$ and $\cap$ is an instance of an abstract system of the type called a modular lattice by G. Birkhoff and a Dedekind structure by O. Ore.[5] Its characteristic property is the Dedekind law:

If $\mathbf{T}_1 \geqq \mathbf{T}_3$, then $\mathbf{T}_1 \cap (\mathbf{T}_2 \cup \mathbf{T}_3) = (\mathbf{T}_1 \cap \mathbf{T}_2) \cup \mathbf{T}_3$.

If $\mathbf{T} > \mathbf{T}_1$ ($\mathbf{T} \geqq \mathbf{T}_1$ but $\mathbf{T} \neq \mathbf{T}_1$) where $\mathbf{T}_1 \neq 0$, then $\mathbf{T}$ is *reducible*. If a basis of $\mathfrak{R}_1$ is supplemented to give a basis for $\mathfrak{R}$, it follows that the matrix of $\mathbf{T}$ relative to this basis has the form

$$\begin{pmatrix} T_1 & Q \\ O & T_2 \end{pmatrix} \tag{12}$$

where T_1 is a matrix of $\mathbf{T}_1$ and T_2 of $\mathbf{T}/\mathbf{T}_1$. The sequence of contractions

$$\mathbf{T} = \mathbf{T}_1 > \mathbf{T}_2 > \cdots > \mathbf{T}_l > \mathbf{T}_{l+1} = 0$$

is a *composition series* for $\mathbf{T}$ if each $\mathbf{T}_i/\mathbf{T}_{i+1}$ is irreducible. l is the *length* of the series and $\mathbf{T}_i/\mathbf{T}_{i+1}$ $(i = 1, \cdots, l)$ the *composition factors*. The Jordan-Hölder theorem[6] asserts that any two composition series for $\mathbf{T}$ have the same length and that their composition factors are similar in pairs. $\mathbf{T}$ is *decomposable*, $\mathbf{T} = \mathbf{T}_1 \oplus \mathbf{T}_2$, if $\mathbf{T} = \mathbf{T}_1 \cup \mathbf{T}_2$ where $\mathbf{T}_1 \cap \mathbf{T}_2 = 0$ and $\mathbf{T}_i \neq 0$. If a basis for $\mathfrak{R}$ is chosen to consist of a basis for $\mathfrak{R}_1$ plus a basis for $\mathfrak{R}_2$, then $\mathbf{T}$ has a matrix of the form (12) with $Q = 0$. Conversely, if $\mathbf{T}$ has a matrix of the form (12) ((12) with $Q = 0$), then $\mathbf{T}$ is reducible (decomposable). The decomposition of $\mathbf{T}$ may be continued to obtain $\mathbf{T} = \mathbf{T}_1 \oplus \mathbf{T}_2 \oplus \cdots \oplus \mathbf{T}_u$ where

[5] G. Birkhoff [2] p. 445 and Ore [13] p. 411.

[6] v. d. Waerden [15] I, p. 140.

each $\mathbf{T}_i$ is indecomposable. In this connection we have the theorem of Krull[7] that any two decompositions of $\mathbf{T}$ into indecomposable parts have the same number of parts which may be paired off into similar pairs. If the indecomposable parts of $\mathbf{T}$ are irreducible, then $\mathbf{T}$ is said to be *completely reducible*. It follows that $\mathbf{T}$ has a matrix of the form

$$\begin{pmatrix} T_1 & & & & \\ & T_2 & & & \\ & & \cdot & & \\ & & & \cdot & \\ & & & & T_u \end{pmatrix}$$

where the T_i are irreducible and conversely.

4. **Non-commutative polynomials.** Let $\mathfrak{F}[t, \bar{\ }, '] \equiv \mathfrak{F}[t]$ be the ring of non-commutative polynomials in the indeterminate t with coefficients taken on the right in $\mathfrak{F}$ and in which multiplication is defined by

$$\alpha t = t\bar{\alpha} + \alpha' \tag{13}$$

$\mathfrak{F}[t]$ is a domain of integrity with Euclidean algorithm. The ring $\mathfrak{F}[\mathbf{T}]$ associated with the p.l.t. is in a sense a representation of $\mathfrak{F}[t]$ and so it is natural to expect that the properties of the latter will be useful in studying $\mathbf{T}$. In this section we shall give a resumé of those properties of $\mathfrak{F}[t]$ which will be required below referring the reader to a paper of Ore's for a complete discussion.[8]

From the Euclidean algorithm follows the existence of a h.c.l.f. (h.c.r.f. $(\alpha(t), \beta(t))_L$ $((\alpha(t), \beta(t))_R)$ and a l.c.r.m. (l.c.l.m.) $[\alpha(t), \beta(t)]_R$ $([\alpha(t), \beta(t)]_L)$ of any two polynomials $\alpha(t)$, $\beta(t)$ in $\mathfrak{F}[t]$. We suppose these to be *normalized*, i.e. to have leading coefficient $= 1$. It follows that they are unique. The *left-transform* of $\alpha(t)$ by $\beta(t)$ is defined as

$$\alpha_1(t) \equiv \beta(t)^{-1}\cdot\alpha(t)\cdot\beta(t) = \beta(t)^{-1}[\alpha(t), \beta(t)]_R\bar{\beta}_0^{(r-d)}\alpha_0$$

where r is the degree $\sigma\alpha(t)$, $d = \sigma(\alpha(t), \beta(t))_R$, α_0 and β_0 the leading coefficients of $\alpha(t)$ and $\beta(t)$ respectively. Since

$$\sigma\alpha(t) + \sigma\beta(t) = \sigma[\alpha(t), \beta(t)]_R + \sigma(\alpha(t), \beta(t))_L$$

we have $\sigma\alpha_1(t) = r - d$ and so the factor $\bar{\beta}_0^{(r-d)}\alpha_0$ gives $\alpha_1(t)$ the leading coefficient α_0. If $d = 0$, $\sigma\alpha(t) = \sigma\alpha_1(t)$ and then $\alpha(t)$ and $\alpha_1(t)$ are said to be *left-similar*. We note the following rules for transforms:

$$\beta_1(t)^{-1}\cdot(\beta_2(t)^{-1}\cdot\alpha(t)\cdot\beta_2(t))\cdot\beta_1(t) = (\beta_2(t)\beta_1(t))^{-1}\cdot\alpha(t)\cdot(\beta_2(t)\beta_1(t)) \tag{14}$$

$$\beta(t)^{-1}\cdot[\alpha_1(t), \alpha_2(t)]_R\cdot\beta(t) = [\beta(t)^{-1}\cdot\alpha_1(t)\cdot\beta(t), \beta(t)^{-1}\cdot\alpha_2(t)\cdot\beta(t)]_R \tag{15}$$

$$\beta(t)^{-1}\cdot\alpha_1(t)\alpha_2(t)\cdot\beta(t) \equiv 0\ (\beta(t)^{-1}\cdot\alpha_1(t)\cdot\beta(t))_R. \tag{16}$$

[7] Krull [8] p. 186.

[8] Ore [12].

Similarly we define right-transforms and right-similarity. Right-similarity and left-similarity are coexistential and so we shall speak simply of *similarity* without the modifiers and denote it by the symbol $\simeq$. If $\mathfrak{F}[t]$ is commutative, i.e. $\mathfrak{F}$ is commutative and $\bar{\alpha} \equiv \alpha$, $\alpha' \equiv 0$, then similarity implies identity.

$\alpha(t)$ is *reducible* if it has a proper factor. If $\alpha(t) = [\alpha_1(t), \alpha_2(t)]_R$, $(\alpha_1(t), \alpha_2(t))_L = 1$ and $\sigma\alpha_i(t) > 0$, then $\alpha(t)$ is said to be *decomposable* into the *left-components* $\alpha_1(t)$, $\alpha_2(t)$. In this case $\sigma\alpha(t) = \sigma\alpha_1(t) + \sigma\alpha_2(t)$. If $\alpha(t) = \alpha_1(t)\beta_2(t) = \alpha_2(t)\beta_1(t)$, then $\alpha(t) = [\beta_1(t), \beta_2(t)]_L$, $(\beta_1(t), \beta_2(t))_R = 1$ and so $\alpha(t)$ is also decomposable into the *right-components* $\beta_1(t)$, $\beta_2(t)$ and conversely. $\alpha(t)$ is *completely reducible* if $\alpha(t) = [\pi_1(t), \cdots, \pi_u(t)]_R$ where the $\pi_i(t)$ are irreducible. It may be shown that this is equivalent to $\alpha(t) = [\rho_1(t), \cdots, \rho_u(t)]_L$ where the $\rho_i(t)$ are irreducible and that every factor of a completely reducible polynomials is also completely reducible.[9]

If the right-ideal $(\mu^*(t))_R$ (the right multiples of $\mu^*(t)$) is two sided, i.e. for every $\alpha(t)$ there exists a $\beta(t)$ such that

$$\alpha(t)\mu^*(t) = \mu^*(t)\beta(t) \tag{17}$$

then we say that $\mu^*(t)$ is *finite*. Another formulation of this condition is

$$\mu^*(t) \equiv 0 \ (\alpha(t)^{-1}\cdot\mu^*(t)\cdot\alpha(t))_R \tag{17'}$$

for every $\alpha(t)$. The correspondence $\mathbf{A}_{\mu^*}: \alpha(t) \to \beta(t)$ is a homomorphic mapping of $\mathfrak{F}[t]$ onto a subring $\mathfrak{G}[t_1]$ where $\mathfrak{F} \to \mathfrak{G}$, $t \to t_1$. However, it is easily seen that $\mathfrak{F} = \mathfrak{G}$. For the correspondent of α is $\beta = \rho^{-1}\bar{\alpha}^{(s)}\rho$ where $s = \sigma\mu^*(t)$, and ρ is the leading coefficient of $\mu^*(t)$ and so as α varies over the whole of $\mathfrak{F}$, so does β. Since t_1 is of the first degree in t, $\mathfrak{F}[t_1] = \mathfrak{F}[t]$ and so $\mathbf{A}_{r^*}$ is a 1-automorphism of $\mathfrak{F}[t]$. Thus for every $\beta(t)$ there corresponds an $\alpha(t)$ satisfying (17), or

$$\mu^*(t) \equiv 0 \ (\beta(t)\cdot\mu^*(t)\cdot\beta(t)^{-1})_L \tag{18}$$

for every $\beta(t)$.

If $\mathfrak{J}$ is a two-sided ideal in $\mathfrak{F}[t]$ and $\mu^*(t)$ an element of $\mathfrak{J}$ having minimum degree and leading coefficient 1, it is easily seen from the division process that every element of $\mathfrak{J}$ is both a r.m. and a l.m. of $\mu^*(t)$. Hence $\mu^*(t)$ is uniquely determined by $\mathfrak{J}$ and if $\alpha(t)$ is arbitrary $\alpha(t)\mu^*(t) = \mu^*(t)\beta(t)$, i.e. $\mu^*(t)$ is a normalized finite polynomial. Thus we have a $(1 - 1)$ correspondence between the two-sided ideals and the elements $\mu^*(t)$ and the theory of the latter which we shall develop here is equivalent to that of two-sided ideals.

If $\mu_1^*(t)$ and $\mu_2^*(t)$ are finite, then so is $\mu_3^*(t) = \mu_1^*(t)\mu_2^*(t)$ and $\mathbf{A}_{\mu_3^*} = \mathbf{A}_{\mu_1^*}\mathbf{A}_{\mu_2^*}$. If $\mu_3^*(t) = \mu_1^*(t)\mu_2^*(t)$ where $\mu_3^*(t)$ and $\mu_1^*(t)$ are finite, then so is $\mu_2^*(t)$. Let $\mathfrak{B}^*$ denote the totality of normalized finite elements. $\mathfrak{B}^*$ is closed under multiplication.

[9] Up to this point we have given an abstract of Ore's results. The rest of this section I believe is new.

If $\mu(t)$ is a l.f. of a $\mu^*(t) \in \mathfrak{B}^*$, say $\mu^*(t) = \mu(t)\nu(t)$, then $\mu(t)$ is *bounded*. Suppose

$$\delta(t) = (\mu(t), \mu^*(t))_R = \alpha_1(t)\mu(t) + \alpha_2(t)\mu^*(t),$$

then

$$\delta(t)\nu(t) = \mu^*(t)[\beta_1(t) + \beta_2(t)\nu(t)]$$

where $\beta_1(t)$, $\beta_2(t)$ are given by (17). Thus $\sigma\delta(t) = \sigma\mu(t)$ and $\mu^*(t) = \nu_1(t)\mu(t)$: *If $\mu(t)$ is a l.f. (r.f.) of a finite polynomial $\mu^*(t)$, then it is also a r.f. (l.f.) of $\mu^*(t)$.*

Let $\mu^*(t)$ be a polynomial of least degree in $\mathfrak{B}^*$ divisible by the bounded polynomial $\mu(t)$. Equation (16) shows that $\mu^*(t)$ is divisible by

$$\mu_1(t) = [\mu(t), \alpha(t)^{-1}\cdot\mu(t)\cdot\alpha(t), \beta(t)^{-1}\cdot\mu(t)\cdot\beta(t), \cdots]_R$$

where $1, \alpha(t), \beta(t), \cdots$ exhaust all the elements of $\mathfrak{F}[t]$. It follows from (14) and (15) that

$$\mu_1(t) \equiv 0 \ (\alpha(t)^{-1}\cdot\mu_1(t)\cdot\alpha(t))_R$$

for every $\alpha(t)$ and hence $\mu_1(t) \in \mathfrak{B}^*$. Since $\mu^*(t)$ was assumed to be minimal, $\mu_1(t) = \mu^*(t)$ and this polynomial is unique. We shall call it the *bound* of $\mu(t)$. It may be characterized also as

$$[\mu(t), \alpha(t)\cdot\mu(t)\cdot\alpha(t)^{-1}, \beta(t)\cdot\mu(t)\cdot\beta(t)^{-1}, \cdots]_L .$$

It is easily seen that the product of two bounded polynomials $\mu_1(t)\mu_2(t)$ is bounded with bound of degree $\sigma \leq \sigma\overset{*}{\mu}_1(t)\overset{*}{\mu}_2(t)$ where $\overset{*}{\mu}_i(t)$ is the bound of $\mu_i(t)$. We denote the multiplicative system of normalized bounded polynomials by $\mathfrak{B}$.

Lemma 1. *If $\pi(t)$ is an irreducible bounded polynomial, then its bound $\pi^*(t)$ is irreducible in $\mathfrak{B}^*$.*

If $\pi^*(t) = \overset{*}{\pi}_1(t)\overset{*}{\pi}_2(t)$, $\sigma\overset{*}{\pi}_i(t) > 0$, then $\pi(t)$ can not be a factor of $\overset{*}{\pi}_1(t)$ because of the minimality of $\pi^*(t)$. Hence $(\pi(t), \overset{*}{\pi}_1(t)) = 1$ and $\pi(t) = \overset{*}{\pi}_1(t)^{-1}\cdot\pi(t)\overset{*}{\pi}_1(t)$ is a factor of $\overset{*}{\pi}_2(t)$ again contrary to the minimality of $\pi^*(t)$.

Lemma 2. *$\mathfrak{B}^*$ is a commutative multiplicative system.*

Any polynomial in $\mathfrak{B}^*$ may be factored into a product of polynomials which are irreducible in $\mathfrak{B}^*$. It is therefore sufficient to prove that every pair of irreducible polynomials in $\mathfrak{B}^*$ are commutative. If $\overset{*}{\pi}_1(t) = \overset{*}{\pi}_2(t)$, this is evident. If $\overset{*}{\pi}_1(t) \neq \overset{*}{\pi}_2(t)$, then $(\overset{*}{\pi}_1(t), \overset{*}{\pi}_2(t)) = 1$ for otherwise their common factor would have two distinct bounds. Hence

$$\overset{*}{\pi}_1(t)\overset{*}{\pi}_2(t) = \overset{*}{\pi}_2(t)\overset{*}{\rho}_1(t) = [\overset{*}{\pi}_1(t), \overset{*}{\pi}_2(t)]_R$$

and similarly $\overset{*}{\pi}_2(t)\overset{*}{\pi}_1(t) = [\overset{*}{\pi}_1(t), \overset{*}{\pi}_2(t)]_R$. Thus $\overset{*}{\pi}_1(t)\overset{*}{\pi}_2(t) = \overset{*}{\pi}_2(t)\overset{*}{\pi}_1(t)$.

Corollary. *If $\overset{*}{\mu}_1(t)$ and $\overset{*}{\mu}_2(t) \in \mathfrak{B}^*$, then* $\mathbf{A}_{\overset{*}{\mu}_1}\mathbf{A}_{\overset{*}{\mu}_2} = \mathbf{A}_{\overset{*}{\mu}_2}\mathbf{A}_{\overset{*}{\mu}_1}$.

Theorem 1. *The polynomials in $\mathfrak{B}^*$ may be factored uniquely into irreducible polynomials in $\mathfrak{B}^*$.*

Let

$$\mu^*(t) = \pi_1^*(t)\pi_2^*(t) \cdots \pi_k^*(t) = \rho_1^*(t)\rho_2^*(t) \cdots \rho_l^*(t)$$

be two factorizations of $\mu^*(t)$ into factors irreducible in $\mathfrak{B}^*$. If $\pi(t)$ is a divisor of $\mu^*(t)$ irreducible in $\mathfrak{F}[t]$, then by the argument in Lemma 1, $\pi(t)$ is a divisor of one of the $\pi_i^*(t)$ and of one of the $\rho_1^*(t)$. Since we can permute the $\pi^*(t)$ and the $\rho^*(t)$, we may assume that $i = j = 1$. By the uniqueness of the bound of $\pi(t)$, we can conclude that $\pi_1^*(t) = \rho_1^*(t)$. We cancel off this factor and repeat the process to obtain the theorem.

We conclude this section with an explicit determination of the elements of $\mathfrak{B}^*$ in certain polynomial rings.

1. $\mathfrak{F}[t]$ *semi-linear,* i.e. has a generator t such that $\alpha t = t\bar{\alpha}$. Let

$$\mu^*(t) = t^m + t^{m-1}\mu_1 + \cdots + \mu_m \tag{19}$$

$\epsilon\, \mathfrak{B}^*$. The condition $t\mu^*(t) = \mu^*(t)t_1$ gives $t_1 = t$ and hence $\bar{\mu}_i = \mu_i$; $\alpha\mu^*(t) = \mu^*(t)\alpha_1$ implies $\alpha_1 = \bar{\alpha}^{(m)}$ and hence $\bar{\alpha}^{(m-i)}\mu_i = \mu_i\bar{\alpha}^{(m)}$. Thus if $\mu_i \neq 0$

$$\bar{\alpha}^{(m)} = \mu_i^{-1}\bar{\alpha}^{(m-i)}\mu_i \qquad \text{or} \qquad \bar{\alpha}^{(i)} = \mu_i^{-1}\alpha\mu_i$$

i.e. $\alpha \to \bar{\alpha}^{(i)}$ is an inner automorphism. If q is the smallest integer such that $\alpha \to \bar{\alpha}^{(q)}$ is an inner automorphism, then $i \equiv 0 \pmod{q}$. q will be called the *relative order* of the automorphism. It is easily seen that $\mu^*(t)$ has the form

$$\mu^*(t) = t^m + t^{m-q}\nu_1 + t^{m-2q}\nu_2 + \cdots \tag{20}$$

where

$$\bar{\nu}_j = \nu_j \qquad \nu_j\bar{\alpha}^{(jq)} = \alpha\nu_j \tag{21}$$

for all α. Conversely these conditions are sufficient to insure that $\mu^*(t)\ \epsilon\ \mathfrak{B}^*$.

THEOREM 2. *Let $\mathfrak{F}[t]$ be semi-linear. If the automorphism of $\mathfrak{F}[t]$ has relative infinite order, $\mathfrak{B}^*$ consists of the powers of t only. If the automorphism has relative finite order q, then* (20) *with* (21) *gives the form of the elements of $\mathfrak{B}^*$.*

2. $\mathfrak{F}[t]$ *differential,* i.e. $\alpha t = t\alpha + \alpha'$. If there exists a ρ such that $\alpha' = \rho\alpha - \alpha\rho$ for all α, then we shall call $\alpha \to \alpha'$ an *inner differentiation.* In this case we may replace t by $t_1 = t + \rho$ and obtain $\alpha t_1 = t_1\alpha$. $\mathfrak{F}[t]$ is then of the type considered in 1. We therefore assume that is an *outer differentiation.*

Let $\mu^*(t)$ of the form (19) be an element of $\mathfrak{B}^*$. $t\mu^*(t) = \mu^*(t)t_1$ implies $t_1 = t$ and $\mu_i' = 0$; $\alpha\mu^*(t) = \mu^*(t)\alpha_1$ implies $\alpha_1 = \alpha$ and since

$$\alpha t^j = t^j\alpha + \binom{j}{1}t^{j-1}\alpha' + \binom{j}{2}t^{j-2}\alpha'' + \cdots + \alpha^{(j)} \tag{22}$$

we have among other conditions

$$m\alpha' = \mu_1\alpha - \alpha\mu_1.$$

If the characteristic of $\mathfrak{F}$, $\chi(\mathfrak{F}) = 0$, it follows from the assumption that $\alpha \to \alpha'$ is outer that $m = 0$.

THEOREM 3. *If $\mathfrak{F}[t]$ is differential of characteristic* 0 *and with an outer differentiation, then* $\mathfrak{B}^*$ *contains the identity element only.*

If $\chi(\mathfrak{F}) = p \neq 0$, the conditions that $\mu^*(t) \in \mathfrak{B}^*$ are somewhat complicated but may be satisfied non-trivially. We hope to discuss this case in a later paper.

5. **Matrices with elements in $\mathfrak{F}[t]$.** Let $\mathfrak{F}[t]_n$[10] denote the ring of matrices of n rows and columns with coordinates in $\mathfrak{F}[t]$. If $U(t) \in \mathfrak{F}[t_n]$ and has an inverse $U^{-1}(t)$ in $\mathfrak{F}[t]_n$, then $U(t)$ is a *unit* in $\mathfrak{F}[t]_n$. If $B(t) = U(t)A(t)V(t)$ where $U(t)$, $V(t) \in \mathfrak{U}_n$ the multiplicative group of units, then $B(t)$ and $A(t)$ are *associates.* This relation is evidently reflexive, symmetric and transitive.

We consider the following problem: Given $A(t)$, to select an associate of $A(t)$ which has a simple normal form and to determine to what extent this normal form is unique. The result[11] given by Wedderburn and by van der Waerden in this connection is the following

LEMMA 3. *There exists an associate $B(t)$ of $A(t)$ having the form*

$$B(t) = \begin{pmatrix} \beta_1(t) & & & & & & \\ & \beta_2(t) & & & & & \\ & & \ddots & & & & \\ & & & \beta_r(t) & & & \\ & & & & 0 & & \\ & & & & & \ddots & \\ & & & & & & 0 \end{pmatrix} \tag{23}$$

where $\beta_i(t)$ is both a l.f. and a r.f. of $\beta_j(t)$ for $j > i$.

We shall prove an extension of this, namely

THEOREM 4. *The associate $B(t)$ of the diagonal form* (23) *may be chosen so that each $\beta_i(t)$ for $i < r$ is bounded and the bound $\beta_i^*(t)$ is a factor of $\beta_j(t)$ for $j > i$.*

(1) We begin with the associate $B(t)$ in (23). If every left-transform of $\beta_1(t)$ is a l.f. of $\beta_2(t)$, $\beta_1(t)$ is bounded and its bound $\beta_1^*(t)$ divides $\beta_2(t)$ and hence also $\beta_j(t)$, $j > 2$. Consider the matrix

$$B_{n-1}(t) = \begin{pmatrix} \beta_2(t) & & & & & \\ & \ddots & & & & \\ & & \beta_r(t) & & & \\ & & & 0 & & \\ & & & & \ddots & \\ & & & & & 0 \end{pmatrix}$$

[10] In general if $\mathfrak{A}$ is a ring $\mathfrak{A}_n$ will denote the ring of all matrices of n rows and columns with coördinates in $\mathfrak{A}$.

[11] Wedderburn [17] pp. 139–141 and v. d. Waerden [15] II pp. 120–125. Wedderburn's result is in fact an extension of Lemma 3. However, it too is included in Theorem 4.

of $\mathfrak{F}[t]_{n-1}$. Using induction on the number of rows we may conclude that $B_{n-1}(t)$ has an associate

$$C_{n-1}(t) = U_{n-1}(t)B_{n-1}(t)V_{n-1}(t) = \begin{pmatrix} \gamma_2(t) & & & & & \\ & \ddots & & & & \\ & & \gamma_r(t) & & & \\ & & & 0 & & \\ & & & & \ddots & \\ & & & & & 0 \end{pmatrix}$$ [12]

where $\gamma_i^*(t)$ is a factor of $\gamma_j(t)$ for $i < j, < r$. Then

$$(24) \quad C(t) = \begin{pmatrix} 1 & \\ & U_{n-1}(t) \end{pmatrix} B(t) \begin{pmatrix} 1 & \\ & V_{n-1}(t) \end{pmatrix} = \begin{pmatrix} \gamma_1(t) & & & & & & \\ & \gamma_2(t) & & & & & \\ & & \ddots & & & & \\ & & & \gamma_r(t) & & & \\ & & & & 0 & & \\ & & & & & \ddots & \\ & & & & & & 0 \end{pmatrix}$$

where $\gamma_1(t) = \beta_1(t)$. Since $\gamma_i(t)$ for $i > 2$ are linear combinations of

$$\beta_2(t), \cdots, \beta_r(t),$$

it follows that $\gamma_1^*(t)$ is a factor of $\gamma_i(t)$ and hence the associate (24) of $A(t)$ has the required normal form.

(2) Suppose next that the transform $\lambda(t)^{-1} \cdot \beta_1(t) \cdot \lambda(t)$ is not a l.f. of $\beta_2(t)$. Then $\beta_1(t)$ is not a l.f. of $\lambda(t)\beta_2(t) = \xi(t)$. Let $\delta(t) = (\beta_1(t), \xi(t))_L$ and $\beta_1(t) = \delta(t)\mu(t)$, $\xi(t) = \delta(t)\nu(t)$ where $(\mu(t), \nu(t))_L = 1$. It follows that $\eta_1(t)$, $\zeta_1(t)$, $\eta_2(t)$, $\zeta_2(t)$ can be determined so that

$$\mu(t)\eta_1(t) + \nu(t)\zeta(t) = 1 \qquad \mu(t)\eta_2(t) + \nu(t)\zeta_2(t) = 0$$

and so that

$$W_2(t) = \begin{pmatrix} \eta_1(t) & \eta_2(t) \\ \zeta_1(t) & \zeta_2(t) \end{pmatrix}$$

---

[12] The rank $r - 1$ is invariant: Wedderburn [17] p. 140.

is a unit in $\mathfrak{F}[t]_2$.[13] Thus

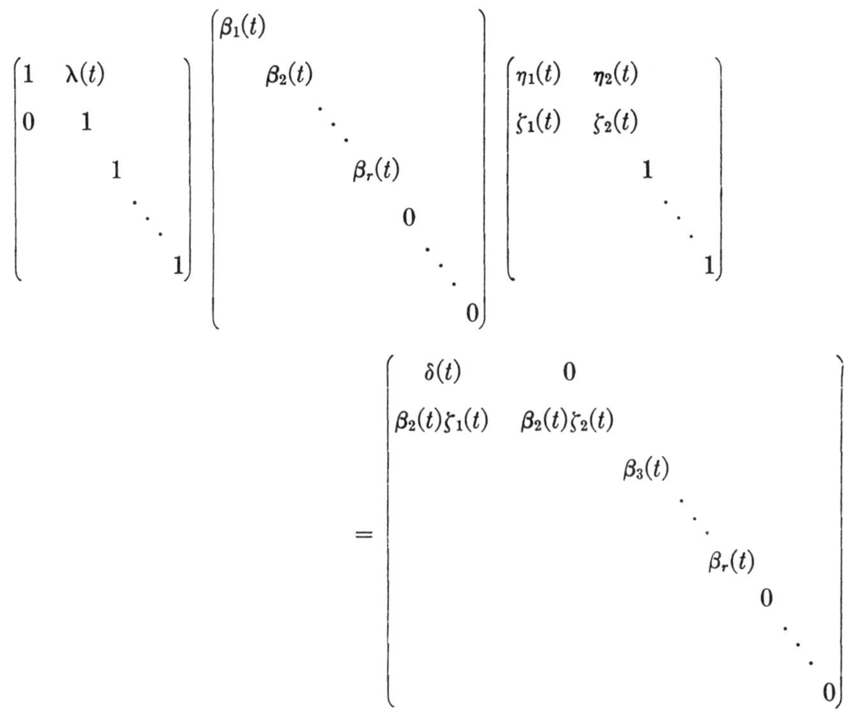

is an associate of $A(t)$. Since the degree $\sigma\delta(t) < \sigma\beta_1(t)$, it follows from the usual proof of Lemma 3 that there exists an associate $C(t)$ of the form (23) with $\sigma\gamma_1(t) < \sigma\beta_1(t)$. We repeat the argument with $C(t)$ in place of $B(t)$. After a finite number of repetitions of this process we arrive at an associate of $A(t)$ of the form (23) in which the first element has a bound which divides the succeeding elements. The theorem then follows from (1).

An associate (23) satisfying the conditions of Theorem 4 is a *normal form* for $A(t)$. The polynomials $\beta_i(t)$ may be taken to have leading coefficients $= 1$. When this has been done the $\beta_i(t) \neq 1$ are called the *invariant factors* of the normal form.

6. **The invariant factors of a p.l.t.** We return to the discussion of the p.l.t. **T** whose matrix T relative to the basis $(e_1, \cdots, e_n)$ of $\mathfrak{R}$ is given by (6).

Let $\mathfrak{R}[t]$ be the extension of $\mathfrak{R}$ obtained by allowing the coefficients to range in $\mathfrak{F}[t]$: $\mathfrak{R}[t]$ is the totality of forms $x(t) = \sum e_i \xi_i(t)$, $\xi_i(t) \in \mathfrak{F}[t]$. This definition is evidently independent of the choice of basis of $\mathfrak{R}$. $\mathfrak{R}[t]$ is an abelian group under addition. The correspondence $x(t) \rightarrow x(t)\alpha(t)$, $\alpha(t) \in \mathfrak{F}[t]$ is an automorphism of this group. Thus $\mathfrak{R}[t]$ may be looked upon as an abelian group with operators $\mathfrak{F}[t]$, $(\mathfrak{R}[t], \mathfrak{F}[t])$.

[13] Wedderburn [17] p. 134.

With $x(t) = \sum e_i\xi_i(t)$ we associate the vector $x = \sum e_i\xi_i(\mathbf{T})$ and with the automorphism $x(t) \to x(t)\alpha(t)$ we associate the automorphism $x \to x\alpha(\mathbf{T})$. These correspondences define an operator homomorphism of $(\mathfrak{R}[t], \mathfrak{F}[t])$ into $(\mathfrak{R}, \mathfrak{F}[\mathbf{T}])$.[14] Let $\mathfrak{N}$ denote the allowable subgroup sent into 0 by this homomorphism and define $f_1(t), \cdots, f_n(t)$ by

$$(f_1(t), \cdots, f_n(t)) = (e_1, \cdots, e_n)(T - t1) \tag{25}$$

where 1 denotes the unit matrix.

LEMMA 4. *$(f_1(t), \cdots, f_n(t))$ is an independent basis for $\mathfrak{N}$.*

The correspondent of $f_j(t)$ in the homomorphism of $(\mathfrak{R}[t], \mathfrak{F}[t])$ and $(\mathfrak{R}, \mathfrak{F}[\mathbf{T}])$ is

$$e_1\tau_{1j} + \cdots + e_{j-1}\tau_{j-1,j} + e_j(\tau_{jj} - \mathbf{T}) + e_{j+1}\tau_{j+11,j} + \cdots + e_n\tau_{nj} = 0$$

because of (6). Thus $f_j(t) \in \mathfrak{N}$. If $\sum f_j(t)\delta_j(t) = 0$, then

$$t\delta_j(t) = \sum \tau_{ji}\delta_i(t) \qquad (j = 1, \cdots, n) \tag{26}$$

since the e's are independent. If any $\delta(t) \neq 0$, let $\delta_s(t)$ be one of maximum degree. The equation (26) for $j = s$ is impossible. Hence all $\delta(t)$ must be 0 and the $f_j(t)$ are independent. Now let $x(t) = \sum e_i\xi_i(t)$ be any element of $\mathfrak{R}[t]$. By means of the relations

$$e_jt = -f_j(t) + \sum e_i\tau_{ij} \qquad (j = 1, \cdots, n) \tag{25'}$$

we may express $x(t)$ in the form

$$x(t) = \sum f_i(t)\psi_i(t) + \sum e_i\xi_i$$

where $\xi_i \in \mathfrak{F}$. If $x(t) \in \mathfrak{N}$, we have $x = \sum e_i\xi_i(\mathbf{T}) = 0 = \sum e_i\xi_i$ since $f_i(\mathbf{T}) = 0$. Thus $\xi_i = 0$ and $x(t) = \sum f_i(t)\psi_i(t)$ as was to be shown.

We may replace the bases $(e_1, \cdots, e_n)$ and $(f_1(t), \cdots, f_n(t))$ of $\mathfrak{R}[t]$ and $\mathfrak{N}$ respectively by

$$(e_1^*(t), \cdots, e_n^*(t)) = (e_1, \cdots, e_n)U^{-1}(t),$$

$$(f_1^*(t), \cdots, f_n^*(t)) = (f_1(t), \cdots, f_n(t))V(t),$$

where $V(t)$, $U^{-1}(t)$ are units in $\mathfrak{F}[t]_n$. Then (25) becomes

$$(f_1^*(t), \cdots, f_n^*(t)) = (e_1^*(t), \cdots, e_n^*(t))T^*(t)$$

where

$$T^*(t) = U(t)(T - t1)V(t).$$

In view of §5 we may choose $U(t)$ and $V(t)$ so that $T^*(t)$ has the form

$$T^*(t) = \begin{pmatrix} 1 & & & & & \\ & \ddots & & & & \\ & & 1 & & & \\ & & & \mu_1(t) & & \\ & & & & \ddots & \\ & & & & & \mu_r(t) \end{pmatrix}$$

[14] This is essentially the method used by Krull [9] pp. 22–32 to obtain the canonical form of ordinary linear transformations.

where the invariant factors $\mu_i(t)$ are bounded with bounds $\mu_i^*(t)$ dividing $\mu_j(t)$ if $i < r, < j$. We shall call the set $\{\mu_k(t)\}$ $(k = 1, \cdots, r)$ a set of *invariant factors of the p.l.t.* $\mathbf{T}$.

Referring back to $(\mathfrak{R}, \mathfrak{F}[\mathbf{T}])$ which is operator isomorphic to the factor group $(\mathfrak{R}[t], \mathfrak{F}[t])/\mathfrak{N}$, we may conclude that the vectors $e_1^*, \cdots, e_n^*$ corresponding to $e_1^*(t), \cdots, e_n^*(t)$ generate $(\mathfrak{R}, \mathfrak{F}[\mathbf{T}])$ in the sense each of its vectors may be written as $\sum e_i^* \xi_i(\mathbf{T})$. Since $e_1^*(t), \cdots, e_{n-r}^*(t) \in \mathfrak{N}$, $e_1^* = 0, \cdots, e_{n-r}^* = 0$ and so may be discarded. We denote e_{n-r+k}^* by g_k $(k = 1, \cdots, r)$. If $\sum g_k \delta_k(\mathbf{T}) = 0$, we have, since

$$f_1^*(t) = e_1^*(t), \cdots, f_{n-r}^*(t) = e_{n-r}^*(t), f_{n-r+1}^*(t) = e_{n-r+1}^*(t)\mu_1(t), \cdots, f_n^*(t) = e_n^*(t)\mu_r(t)$$

form a basis for $\mathfrak{N}$, that $\delta_k(t)$ is a r.m. of $\mu_k(t)$. It follows that if

$$\mu_k(t) = t^{n_k} - t^{n_k-1}\tau_1^{(k)} - \cdots - \tau_{n_k}^{(k)}$$

then

$$(g_1, g_1\mathbf{T}, \cdots, g_1\mathbf{T}^{n_1-1}; \cdots; g_r, g_r\mathbf{T}, \cdots, g_r\mathbf{T}^{n_r-1})$$

is an ordinary independent basis for $\mathfrak{R}$. Thus $\sum n_i = n$. Relative to this basis T the matrix of $\mathbf{T}$ has the form

$$\begin{pmatrix} T_1 & & & \\ & T_2 & & \\ & & \ddots & \\ & & & T_r \end{pmatrix} \tag{27}$$

where

$$T_k = \begin{pmatrix} 0 & \cdot & \cdot & \cdot & \tau_{n_k}^{(k)} \\ 1 & 0 & & & \cdot \\ & \ddots & \ddots & & \cdot \\ & & \ddots & \ddots & \cdot \\ & & & 1 & \tau_1^{(k)} \end{pmatrix} \tag{28}$$

The invariant subspace $\mathfrak{R}_k = (g_k, g_k\mathbf{T}, \cdots)$ is generated by the single vector g_k (acted upon by $\mathfrak{F}[\mathbf{T}]$). Let $\mathbf{T}_k$ denote the contraction of $\mathbf{T}$ in $\mathfrak{R}$. Thus if $g \in \mathfrak{R}_k$, $g\mathbf{T}_k \equiv g\mathbf{T}$. We shall call $\mathfrak{R}_k$ a *cyclic subspace* and $\mathbf{T}_k$ a *cyclic p.l.t.* The polynomial $\mu_k(t)$ is the *order* of the *generator* g_k of $\mathfrak{R}_k$. The above result may be stated as

THEOREM 5. $\mathbf{T} = \mathbf{T}_1 \oplus \mathbf{T}_2 \oplus \cdots \oplus \mathbf{T}_r$ *where the* $\mathbf{T}_k$ *are cyclic with generators* g_k *whose orders* $\mu_k(t)$ *are the invariant factors of a normal form of the matrix* $T - t1$.

7. Cyclic p.l.t. In this section we suppose that $\mathbf{T}$ acting in $\mathfrak{R}$ is cyclic and that g is a generator of $\mathfrak{R}$ whose order is $\mu(t)$. $\sigma\mu(t) = n$. If $h = g\alpha(\mathbf{T})$, its order is readily seen to be $\nu(t) = \alpha(t)^{-1} \cdot \mu(t) \cdot \alpha(t)$. Thus h is a generator of $\mathfrak{R}$ if and only if $\sigma\nu(t) = \sigma\mu(t)$, i.e. $\mu(t) \simeq \nu(t)$. Hence

LEMMA 5. *A n.a.s.c. that two cyclic p.l.t. be similar is that the orders of their corresponding generators be similar polynomials.*[15]

The vectors of $\mathfrak{R}$ may be represented uniquely in the form $g\lambda(\mathbf{T})$ where $\lambda(t)$ is of degree $<n$. Now let $\mathfrak{R}_1$ be an invariant subspace of $\mathfrak{R}$ and $\mathbf{T}_1$ the contraction of $\mathbf{T}$ in $\mathfrak{R}_1$. Let $g_1 = g\lambda_1(\mathbf{T}_1)$ be a vector of $\mathfrak{R}_1$ for which the polynomial $\lambda_1(t)$ has minimum degree for the vectors of $\mathfrak{R}_1$. It is an immediate consequence of the division process that every vector of $\mathfrak{R}_1$ is a "multiple" of g_1, i.e. $\mathbf{T}_1$ is cyclic with generator g_1. Further, the order $\mu(t)$ is a r.m. of $\lambda_1(t)$, $\mu(t) = \lambda_1(t)\mu_1(t)$. The right associate of $\mu_1(t)$ whose leading coefficient is 1 is the order of g_1.

LEMMA 6. *A n.a.s.c. that the cyclic* $\mathbf{T}$ *be irreducible is that the order* $\mu(t)$ *be an irreducible polynomial.*

The above results establish a correspondence between a factorization of $\mu(t)$ into irreducible factors and a composition series of $\mathbf{T}$. From this correspondence it is easily seen that Ore's theorem on the unique factorization (in the sense of similarity) of polynomials in $\mathfrak{F}[t]$ is precisely the Jordan-Hölder theorem applied to cyclic p.l.t. A similar connection exists between the concepts of decomposability applied to polynomials and to p.l.t.

Suppose that the order $\mu(t)$ is decomposable, say

$$\mu(t) = \lambda_1(t)\mu_1(t) = \lambda_2(t)\mu_2(t)$$

where

$$(\lambda_1(t), \lambda_2(t))_L = 1 = (\mu_1(t), \mu_2(t))_R$$

and all of these polynomials are normalized. Set $g_1 = g\lambda_1(\mathbf{T})$, $g_2 = g\lambda_2(\mathbf{T})$ and let $\mathfrak{R}_1$ and $\mathfrak{R}_2$ be the cyclic subspaces generated by these vectors. We may determine $\nu_1(t)$, $\nu_2(t)$ so that

$$\lambda_1(t)\nu_1(t) + \lambda_2(t)\nu_2(t) = 1$$

and hence

$$g = g_1\nu_1(\mathbf{T}) + g_2\nu_2(\mathbf{T}).$$

By comparing the dimensions of $\mathfrak{R}$, $\mathfrak{R}_1$, and $\mathfrak{R}_2$ we obtain $\mathfrak{R}_1 \cap \mathfrak{R}_2 = 0$ and hence $\mathbf{T} = \mathbf{T}_1 \oplus \mathbf{T}_2$. Conversely, if $\mathbf{T} = \mathbf{T}_1 \oplus \mathbf{T}_2$ and correspondingly $\mathfrak{R} = \mathfrak{R}_1 \oplus \mathfrak{R}_2$, let $g_1 = g\lambda_1(\mathbf{T})$, $g_2 = g\lambda_2(\mathbf{T})$ be generators of $\mathfrak{R}_1$ and $\mathfrak{R}_2$ such that $\lambda_1(t)$ and $\lambda_2(t)$ are normalized l.f.'s of $\mu(t)$. The orders $\mu_1(t)$, $\mu_2(t)$ of g_1 and g_2 are then normalized r.f.'s of $\mu(t)$. Since $g = g_1\nu_1(\mathbf{T}) + g_2\nu_2(\mathbf{T})$, we have

$$(\lambda_1(t), \lambda_2(t))_L = 1, \qquad [\mu_1(t), \mu_2(t)]_L = \mu(t).$$

Since $\mathfrak{R}_1 \cap \mathfrak{R}_2 = 0$,

$$\sigma\mu(t) = \sigma\mu_1(t) + \sigma\mu_2(t)$$

and hence

$$(\mu_1(t), \mu_2(t))_R = 1.$$

Thus we have proved

[15] Noether-Schmeidler [10], p. 21.

Lemma 7. *A n.a.s.c. that a cyclic* $\mathbf{T}$ *be indecomposable is that* $\mu(t)$ *be an indecomposable polynomial*

Lemma 6 and 7 give

Lemma 8. *A n.a.s.c. that a cyclic* $\mathbf{T}$ *be completely reducible is that* $\mu(t)$ *be a completely reducible polynomial.*

8. **Criteria for similarity, reducibility, etc.** Theorem 5 enables us to apply the lemmas of the preceding section to arbitrary p.l.t. For according to this theorem a p.l.t. $\mathbf{T}$ is decomposable unless it is cyclic and the condition that it be cyclic is that the normal form of $T - t1$ have but one invariant factor, i.e. $r = 1$. Thus Lemma 6 and 7 give respectively

Theorem 6. *A n.a.s.c. that* $\mathbf{T}$ *be irreducible is that it be cyclic and its order* $\mu(t)$ *be an irreducible polynomial.*

Theorem 7. *A n.a.s.c. that* $\mathbf{T}$ *be indecomposable is that it be cyclic and* $\mu(t)$ *be an indecomposable polynomial.*

The condition that $\mathbf{T}$ be completely reducible is, according to Lemma 8 that all of the invariant factors $\mu_k(t)$ be completely reducible. However, if we recall that every factor of a completely reducible polynomial is completely reducible, we obtain

Theorem 8. *A n.a.s.c. that* $\mathbf{T}$ *be completely reducible is that the last invariant factor* $\mu_r(t)$ *be a completely reducible polynomial.*

Let

$$\mathbf{T} = \mathbf{T}_1 \oplus \mathbf{T}_2 \oplus \cdots \oplus \mathbf{T}_u \tag{29}$$

be a decomposition of $\mathbf{T}$ into indecomposable parts. By Theorem 3 each $\mathbf{T}_i$ is cyclic and the order $\rho_i(t)$ of its generator is indecomposable. $\rho_i(t)$ is uniquely determined in the sense of similarity by $\mathbf{T}_i$. We shall call these polynomials the *elementary divisors of* $\mathbf{T}$. In virtue of the theorem of Krull that the indecomposable parts of an abelian group with operators are unique in the sense of isomorphism, we have that the $\mathbf{T}_i$ in (29) are unique in the sense of similarity and hence the elementary divisors are determined in the same sense by $\mathbf{T}$. Hence

Theorem 9. *A n.a.s.c. that two p.l.t. be similar is that their elementary divisors be similar in pairs.*

The following theorem gives some additional information regarding the invariance of the invariant factors $\mu_i(t)$.

Theorem 10. $\mathbf{T}_r$ *is a cyclic contraction of maximum dimension* n_r *of* $\mathbf{T}$. *If* $\lambda(t)$ *is the order of any cyclic contraction of dimension* n_r, *then* $\lambda(t) = \lambda^{(1)}(t)\mu_{r-1}^*(t)$, $\mu_r(t) = \mu^{(1)}(t)\mu_{r-1}^*(t)$ *where* $\mu_{r-1}^*(t)$ *is the bound of* $\mu_{r-1}(t)$ *and* $\lambda^{(1)}(t) \simeq \mu^{(1)}(t)$.

In the notation of Theorem 5 suppose $\mathbf{T} = \mathbf{T}_1 \oplus \mathbf{T}_2 \oplus \cdots \oplus \mathbf{T}_r$ where $\mathbf{T}_i$ has generator g_i of order $\mu_i(t)$. Let $\lambda(t)$ denote the order of

$$g = g_1\alpha_1(\mathbf{T}) + \cdots + g_r\alpha_r(\mathbf{T}).$$

The order of $g_i\alpha_i(\mathbf{T})$ is $\lambda_i(t) = \alpha_i(t)^{-1}\cdot\mu_i(t)\cdot\alpha_i(t)$ and $\lambda(t) = [\lambda_1(t), \cdots, \lambda_r(t)]_R$. Since $[\lambda_1(t), \cdots, \lambda_{r-1}(t)]_R$ is a factor of $\mu_{r-1}^*(t)$, $\lambda(t)$ is a l.f. of $[\mu_{r-1}^*(t), \lambda_r(t)]_R$.

Since $\mu_r(t) = \mu^{(1)}(t)\mu^*_{r-1}(t)$, $\lambda_r(t)$ is a l.f. of $\lambda^{(1)}(t)\mu^*_{r-1}(t)$ where $\lambda^{(1)}(t) = \alpha_r(t)^{-1}\cdot\mu^{(1)}(t)\cdot\alpha_r(t)$. Thus $\lambda(t)$ is a l.f. of $\lambda^{(1)}(t)\mu^*_{r-1}(t)$. Since $\sigma\lambda^{(1)}(t) \leqq \sigma\mu^{(1)}(t)$ we have $\sigma\lambda(t) \leqq \sigma\mu_r(t)$. If equality is to hold we must have $\sigma\lambda^{(1)}(t) = \sigma\mu^{(1)}(t)$, i.e. $\lambda^{(1)}(t) \simeq \mu^{(1)}(t)$.

Special cases.

(1). If $\mathbf{T}$ is an l.t. and $\mathfrak{F}$ is commutative, $\mathfrak{F}[t]$ is commutative and similarity implies identity. Theorem 9 shows that the elementary divisors are a complete set of invariants of $\mathbf{T}$ and Theorem 10 may be extended to give the same result for the invariant factors.

(2). If $\mathbf{T}$ is an s.l.t. with automorphism of relative infinite order, the finite elements of $\mathfrak{F}[t]$ consist of the powers t^m. Hence the bounded elements are also of this form. Thus in the decomposition $\mathbf{T} = \mathbf{T}_1 \oplus \cdots \oplus \mathbf{T}_{r-1} \oplus \mathbf{T}_r$ of Theorem 5 $\mathbf{T}_1 \oplus \cdots \oplus \mathbf{T}_{r-1}$ is nilpotent and $\mathbf{T}_r$ is cyclic.

(3). If $\mathbf{T}$ is a d.t. and $\mathfrak{F}$ has characteristic 0, 1 is the only normalized bounded element in $\mathfrak{F}[t]$. Thus $r = 1$ and $\mathbf{T}$ is cyclic. This result gives the well-known equivalence between the theory of systems of n linear homogeneous differential equations (analytic case) and that of a single n-th order linear homogeneous differential equation.

9. **Finite p.l.t.** The homomorphism $\mathfrak{F}[t] \sim \mathfrak{F}[\mathbf{T}]$ between the polynomial ring and the ring of a p.l.t. defines an isomorphism $\mathfrak{F}[\mathbf{T}] \simeq \mathfrak{F}[t]/\mathfrak{J}$ where $\mathfrak{J}$ is a two-sided ideal. If $\mathbf{T} \neq 0$, then $\mathfrak{J} \neq (1)$. Thus either $\mathfrak{J} = (0)$ or $\mathfrak{J} = (\mu^*(t))$ where $\mu^*(t)$ is a normalized finite polynomial $\neq 0$, $\neq 1$. In the latter case we shall call $\mathbf{T}$ *finite* and $\mu^*(t)$ its *minimum function.* $\mu^*(t)$ may be defined also as the normalized polynomial of least degree such that $x\mu^*(\mathbf{T}) = 0$ for all x in $\mathfrak{R}$.

Let $\mathbf{T}$ be finite and decomposed as $\mathbf{T}_1 \oplus \cdots \oplus \mathbf{T}_r$ into the cyclic parts $\mathbf{T}_i$ with generators g_i whose orders $\mu_i(t)$ are chosen as in Theorem 5. We have $g_r\alpha(\mathbf{T})\mu^*(\mathbf{T}) = 0$ and hence $\mu^*(t)$ is a r.m. of $\alpha(t)^{-1}\cdot\mu_r(t)\cdot\alpha(t)$. Since $\alpha(t)$ is arbitrary, it follows that $\mu_r(t)$ is bounded with bound $\mu_r^*(t)$ a factor of $\mu^*(t)$. On the other hand it is easily seen that $x\mu_r^*(\mathbf{T}) = 0$ for all x. Thus $\mu_r^*(t) = \mu^*(t)$.

THEOREM 11. *A n.a.s.c. that* $\mathbf{T}$ *be finite is that* $\mu_r(t)$ *be bounded. If the condition is satisfied, the minimum function of* $\mathbf{T}$ *is* $\mu_r^*(t)$.

Suppose $\mu^*(t) = \pi_1^*(t)^{e_1}\,\pi_2^*(t)^{e_2}\cdots\pi_m^*(t)^{e_m}$ where $\pi_i^*(t)$ are distinct and irreducible in $\mathfrak{B}^*$. Set $\pi_i^*(t)^{e_i} = \rho_i^*(t)$ and $\mu^*(t)\rho_i^*(t)^{-1} = \tau_i^*(t)$. Let $\mathfrak{R}_i^*$ denote the subspace of $\mathfrak{R}$ consisting of all vectors of the form $x\tau_i^*(\mathbf{T})$. Since $\tau_i^*(t)$ is finite $\mathfrak{R}_i^*$ is an invariant subspace. If $x_i \in \mathfrak{R}_i^*$ then $x_i\rho_i^*(\mathbf{T}) = 0$ and since $(\rho_i^*(t), \tau_i^*(t)) = 1$,

$$\mathfrak{R}_i^* \cap (\mathfrak{R}_1^* \cup \cdots \cup \mathfrak{R}_{i-1}^* \cup \mathfrak{R}_{i+1}^* \cup \cdots \mathfrak{R}_m^*) = 0.$$

Since

$$(\tau_1^*(t), \cdots, \tau_m^*(t)) = 1,$$

$$\mathfrak{R} = \mathfrak{R}_1^* \cup \cdots \cup \mathfrak{R}_m^*$$

and correspondingly

$$\mathbf{T} = \mathbf{T}_1 \oplus \mathbf{T}_2 \oplus \cdots \oplus \mathbf{T}_m .$$

Thus we have proved

THEOREM 12. *If* $\mathbf{T}$ *is finite and indecomposable, then* $\mu^*(t)$ *is a power of a polynomial irreducible in* $\mathfrak{B}^*$.

The following is an extension of a well-known result on ordinary linear transformations.

THEOREM 13. *A n.a.s.c. that a finite* $\mathbf{T}$ *be completely reducible is that its minimum function have no multiple factors in* $\mathfrak{B}^*$.

If $\mathbf{T}$ is completely reducible, its elementary divisors are irreducible polynomials and hence their bounds are irreducible in $\mathfrak{B}^*$. It is readily seen that $\mu^*(t)$ is the l.c.m. of these bounds and hence has no multiple factors in $\mathfrak{B}^*$. To prove the converse we note that if $\pi(t)$ is an elementary divisor of $\mathbf{T}$, its bound $\pi^*(t)$ is a factor of $\mu^*(t)$ and hence is irreducible in $\mathfrak{B}^*$. Since $\pi^*(t)$ can be defined as a l.c.m. of irreducible polynomials, it is completely reducible. Hence its factor $\pi(t)$ is also completely reducible, but being indecomposable, it must be irreducible. Thus every elementary divisor of $\mathbf{T}$ is irreducible and so $\mathbf{T}$ is completely reducible.[16]

LEMMA 9. *If* $\rho(t)$ *is a normalized indecomposable bounded polynomial and* $\rho^*(t)$ *its bound, then the indecomposable parts of* $\rho^*(t)$ *are similar to* $\rho(t)$.

Suppose $\lambda_k(t)$ is a normalized polynomial decomposable into k parts similar to $\rho(t)$. $\lambda_k(t)$ is a factor of $\rho^*(t)$. Suppose $\lambda_k(t) \neq \rho^*(t)$ and consider the p.l.t. $\mathbf{T}$ defined as the direct sum of (decomposable into) the two cyclic p.l.t.'s whose orders are $\rho(t)$ and $\lambda_k(t)$. Thus $\mathbf{T}$ has a matrix of the form

$$T = \begin{pmatrix} R & 0 \\ 0 & L_k \end{pmatrix}$$

where R and L_k are associated with $\rho(t)$ and $\lambda_k(t)$ as in (28). We assert that $\mathbf{T}$ is cyclic. For otherwise the number r of invariant factors $\mu_j(t)$ in Theorem 5 is >1. By Theorem 10 $\mu_r(t)$ is the order of a cyclic subspace of maximum dimension of $\mathbf{T}$ and if $\sigma\mu_r(t) = \sigma\lambda_k(t)$, $\lambda_k(t)$ is divisible by $\overset{*}{\mu}_{r-1}(t)$. We obtain a decomposition of $\mathbf{T}$ into indecomposable parts by decomposing the $\mu_j(t)$. Thus $\mu_{r-1}(t)$ is divisible by a polynomial similar to $\rho(t)$ and so $\lambda_k(t)$ is divisible by $\rho^*(t)$ contrary to the assumption that $\rho^*(t) \neq \lambda_k(t)$. On the other hand if $\sigma\mu_r(t) > \sigma\lambda_k(t)$ we must have $\sigma\mu_{r-1}(t) < \sigma\rho(t)$ and hence $\mathbf{T}$ has indecomposable parts not similar to $\rho(t)$ contrary to Krull's theorem. Thus $r > 1$ is impossible and $\mathbf{T}$ is cyclic. Let $\lambda_{k+1}(t)$ denote its order. $\lambda_{k+1}(t)$ is decomposable into $k+1$ parts similar to $\rho(t)$ and hence divides $\rho^*(t)$. If $\lambda_{k+1}(t) \neq \rho^*(t)$ we repeat the process. If we start with $\lambda_1(t) = \rho(t)$ we arrive in this way after a finite

[16] The remainder of this section was added Aug. 14, 1936. These results are related in part to some recent work by Prof. M. H. Ingraham and Dr. M. C. Wolf on matrices with elements in a division algebra. I am indebted to Prof. Ingraham for communicating a copy of their manuscript to me.

number of steps at $\rho^*(t)$ and obtain a decomposition for it into parts similar to $\rho(t)$.

An immediate consequence of Lemma 9 and Krull's theorem is

LEMMA 10. *If $\rho_1(t)$ and $\rho_2(t)$ are indecomposable bounded polynomials with the same bound, then $\rho_1(t) \simeq \rho_2(t)$.*

From this and Theorem 9 we have

THEOREM 14. *A n.a.s.c. that two finite p.l.t. be similar is that the bounds of their elementary divisors be identical.*

Let $\pi(t)$ be any irreducible bounded polynomial with bound $\pi^*(t)$ and $\rho_e(t) = \pi_1(t) \cdots \pi_e(t)$ where the $\pi_i(t) \simeq \pi(t)$. Each $\pi_i(t)$ divides $\pi^*(t)$, say $\pi^*(t) = \pi_i(t)\gamma_i(t) = \delta_i(t)\pi_i(t)$. Suppose

$$\pi^*(t)^{e-1} = \rho_{e-1}(t)\alpha(t) \qquad \rho_{e-1}(t) = \pi_1(t) \cdots \pi_{e-1}(t).$$

Then

$$\pi^*(t)^e = \rho_{e-1}(t)\alpha(t)\pi^*(t) = \rho_{e-1}(t)\pi^*(t)\beta(t) = \rho_e(t)\gamma_e(t)\beta(t).$$

Hence $\rho_e^*(t)$ the bound of $\rho_e(t)$ is $\pi^*(t)^f$ where $f \leqq e$. We shall show next that $\pi_1(t), \cdots, \pi_e(t)$ may be so chosen that $\rho_e^*(t) = \pi^*(t)^e$. Suppose $\rho_{e-1}(t)$ has already been found so that $\rho_{e-1}^*(t) = \pi^*(t)^{e-1}$. There exists an irreducible $\pi_e(t) \simeq \pi(t)$ such that the bound of $\rho_e(t) = \rho_{e-1}(t)\pi_e(t)$ is $\pi^*(t)^e$. For otherwise all the $\rho_e(t)$ divide $\pi^*(t)^{e-1}$ and hence their l.c.r.m. $\rho_{e-1}(t)\pi^*(t)$ divides $\pi^*(t)^{e-1}$. But then $\rho_{e-1}(t)$ divides $\pi^*(t)^{e-2}$ contrary to assumption. We require these remarks to prove the following extensions of Theorem 12.

THEOREM 15. *A n.a.s.c. that the bounded $\rho_e(t) = \pi_1(t) \cdots \pi_e(t)$ ($\pi_i(t)$ irreducible and $\simeq$ to $\pi(t)$) be indecomposable is that its bound $\rho_e^*(t) = \pi^*(t)^e$.*

Sufficiency. If $\rho_e(t) = [\rho_{e_1}^{(1)}(t), \rho_{e_2}^{(2)}(t)]_R$ where $\rho_{e_i}^{(i)}(t) = \pi_1^{(i)}(t) \cdots \pi_{e_i}^{(i)}(t)$ and $\pi_j^{(i)}(t) \simeq \pi(t)$ then $\rho_{e_i}^{(i)}(t)$ divides $\pi^*(t)^{e_i}$ and hence $\rho_e(t)$ divides $\pi^*(t)^f$ where $f = \max(e_1, e_2)$. Hence we can not have both e_1 and $e_2 < e$.

Necessity. Suppose $\rho_e^*(t) = \pi^*(t)^f$ where $f \leqq e$. We have seen that a polynomial $\rho_f'(t) = \pi_1'(t) \cdots \pi_f'(t)$ exists having the bound $\pi^*(t)^f$. It follows that $\rho_f'(t)$ is indecomposable and by Lemma 10 is $\simeq \rho_e(t)$. Thus $e = f$.

THEOREM 16. *A n.a.s.c. that a finite polynomial be indecomposable is that it be divisible on the left (right) by only one normalized irreducible polynomial.*

The condition is evidently sufficient. Now if $\rho_e(t)$ is indecomposable and divisible by the distinct normalized irreducible polynomials $\pi_1(t)$ and $\pi_2(t)$, then $\rho_e(t)$ is divisible by their l.c.r.m. $\mu(t)$, i.e. $\rho_e(t) = \mu(t)\pi_3(t) \cdots \pi_e(t)$. Since $\mu(t)$ divides $\pi^*(t)$ it follows easily that $\rho_e(t)$ divides $\pi^*(t)^{e-1}$ contrary to Theorem 15.

As a consequence of this theorem we note that if $\rho_e(t) = \pi_1(t) \cdots \pi_e(t)$ is indecomposable, so is $\rho_k(t) = \pi_1(t) \cdots \pi_k(t)$, $1 \leqq k \leqq e$.

If $x(\mathbf{T})$ is a variable element of $\mathfrak{F}[\mathbf{T}]$, the correspondence $x(\mathbf{T}) \rightarrow x(\mathbf{T})\mathbf{T}$ is a p.l.t. $\tilde{\mathbf{T}}$ determined by $\mathbf{T}$ in the vector space $\mathfrak{F}[\mathbf{T}]$. The correspondence $\mathbf{T} \rightarrow \tilde{\mathbf{T}}$ defines an isomorphism between $\mathfrak{F}[\mathbf{T}]$ and $\mathfrak{F}[\tilde{\mathbf{T}}]$ analogous to the regular representation of a hypercomplex system. By means of this isomorphism Theorems 12 and 13 yield results on the structure of $\mathfrak{F}[\mathbf{T}]$. Thus it follows that a n.a.s.c.

that $\mathfrak{F}[\mathbf{T}]$ be a semi-simple ring is that $\mu^*(t)$ have no multiple factors in $\mathfrak{B}^*$. This will also be a consequence of a general theorem of the next section.

10. **The automorphism ring of a p.l.t.** An (*operator*) *automorphism* $\mathbf{A}$ of $(\mathfrak{R}, \mathfrak{F}[\mathbf{T}])$ (or of $\mathbf{T}$) is defined by the conditions

$$(x + y)\mathbf{A} = x\mathbf{A} + y\mathbf{A} \tag{30a}$$

$$(x\mathbf{A})\mathbf{O} = (x\mathbf{O})\mathbf{A} \tag{30b}$$

for every $\mathbf{O}$ in $\mathfrak{F}[\mathbf{T}]$. It is sufficient to require

$$(x\mathbf{A})\alpha = (x\alpha)\mathbf{A} \qquad (x\mathbf{A})\mathbf{T} = (x\mathbf{T})\mathbf{A}$$

in place of (30b). Thus $\mathbf{A}$ is an l.t. commutative with $\mathbf{T}$ and conversely.

Let $\mathbf{A}$ be an arbitrary l.t., $\mathbf{T}$ a p.l.t. and x a vector such that $(x\mathbf{A})\mathbf{T} = (x\mathbf{T})\mathbf{A}$. Then

$$(x\alpha)\mathbf{TA} = (x\mathbf{TA})\bar{\alpha} + (x\mathbf{A})\alpha' \qquad (x\alpha)\mathbf{AT} = (x\mathbf{AT})\bar{\alpha} + (x\mathbf{A})\alpha'$$

so that $(x\alpha)\mathbf{TA} = (x\alpha)\mathbf{AT}$. Thus the totality of vectors for which $\mathbf{A}$ and $\mathbf{T}$ commute constitutes a vector subspace of $\mathfrak{R}$. Hence $\mathbf{A}$ is an automorphism of $\mathbf{T}$ is and only if $e_k\mathbf{TA} = e_k\mathbf{AT}$ $(k = 1, \cdots, n)$ where $(e_1, \cdots, e_n)$ is a basis for $\mathfrak{R}$. This condition in terms of the matrices A, T of $\mathbf{A}$ and $\mathbf{T}$ is

$$AT = T\bar{A} + A'. \tag{31}$$

In a similar fashion we define an *automorphism* $\mathbf{A}(t)$ of $(\mathfrak{R}[t], \mathfrak{F}[t])$ by

$$[x(t) + y(t)]\mathbf{A}(t) = x(t)\mathbf{A}(t) + y(t)\mathbf{A}(t)$$
$$[x(t)\alpha(t)]\mathbf{A}(t) = [x(t)\mathbf{A}(t)]\alpha(t).$$

If $(e_1(t), \cdots, e_n(t))$ is a basis of $\mathfrak{R}[t]$ then $\mathbf{A}(t)$ is determined by the matrix $A(t)$ given by

$$(e_1(t)\mathbf{A}(t), \cdots, e_n(t)\mathbf{A}(t)) = (e_1(t), \cdots, e_n(t))A(t)$$

and conversely any matrix defines an automorphism. If $\mathbf{A}(t)$ leaves $\mathfrak{N}$ invariant where $(\mathfrak{R}, \mathfrak{F}[\mathbf{T}]) \simeq (\mathfrak{R}[t], \mathfrak{F}[t])/\mathfrak{N}$, then $\mathbf{A}(t)$ induces an automorphism in the quotient group $(\mathfrak{R}, \mathfrak{F}[\mathbf{T}])$, the *projection* of $\mathbf{A}(t)$. Suppose

$$(f_1(t), \cdots, f_n(t)) = (e_1(t), \cdots, e_n(t))T(t)$$

is a basis of $\mathfrak{N}$. We recall that $T(t)$ is an associate of $T - t1$. The condition that $A(t)$ leave $\mathfrak{N}$ invariant is

$$A(t)T(t) = T(t)A_1(t). \tag{32}$$

The projection $\mathbf{A}$ of $\mathbf{A}(t)$ is 0 if and only if

$$A(t) \equiv 0 \qquad (T(t))_R$$

where $(T(t))_R$ denotes the ideal of right multiples of $T(t)$.

Conversely if $\mathbf{A}$ is any automorphism of $(\mathfrak{R}, \mathfrak{F}[\mathbf{T}])$, it may be extended to an automorphism $\mathbf{A}(t)$ of $(\mathfrak{R}[t], \mathfrak{F}[t])$. For if $x(t) = e_1\xi_1(t) + \cdots + e_n\xi_n(t)$, then we may define $A(t)$ by

$$x(t)\mathbf{A}(t) = e_1\mathbf{A}\xi_1(t) + \cdots + e_n\mathbf{A}\xi_n(t).$$

This is independent of the choice of the basis $(e_1, \cdots, e_n)$. Now if $x(t) \in \mathfrak{N}$, i.e. $\sum e_i\xi_i(\mathbf{T}) = 0$, then since $\mathbf{A}$ commutes with all $\xi_i(\mathbf{T})$, $x(t)\mathbf{A}(t)$ also $\in \mathfrak{N}$. Thus $\mathbf{A}$ may be obtained as a projection of an $\mathbf{A}(t)$ leaving $\mathfrak{N}$ invariant.

Let $\mathfrak{A}(\mathbf{T}) = \mathfrak{A}$ denote the ring of automorphisms of $\mathbf{T}$. The above discussion shows that $\mathfrak{A}$ is inverse isomorphic to the ring of matrices with elements in $\mathfrak{F}$ satisfying (31) and also inverse isomorphic to the ring of matrices with elements in $\mathfrak{F}[t]$ satisfying (32) taken mod $(T(t))_R$. The latter will be called the *invariant ring* of $T(t)$.[17]

THEOREM 17. *If* $\mathbf{T}_1 \simeq \mathbf{T}_2$ *then* $\mathfrak{A}(\mathbf{T}_1) \simeq \mathfrak{A}(\mathbf{T}_2)$.

For $\mathbf{T}_2 = \mathbf{A}^{-1}\mathbf{T}_1\mathbf{A}$ and hence if $\mathbf{A}_1 \in \mathfrak{A}(\mathbf{T}_1)$, $\mathbf{A}_2 = \mathbf{A}^{-1}\mathbf{A}_1\mathbf{A} \in \mathfrak{A}(\mathbf{T}_2)$ and conversely.

THEOREM 18. *If* $\mathbf{T}$ *is irreducible,* $\mathfrak{A}(\mathbf{T})$ *is a field.*

If $\mathbf{A} \in \mathfrak{A}(\mathbf{T})$, the space $\mathfrak{R}\mathbf{A}$ consisting of all vectors $x\mathbf{A}$ is an invariant subspace of $\mathfrak{R}$. Hence either $\mathfrak{R}\mathbf{A} = 0$ or $\mathfrak{R}\mathbf{A} = \mathfrak{R}$, i.e. $\mathbf{A} = 0$ or $\mathbf{A}$ has an inverse $\mathbf{A}^{-1}$. Evidently $\mathbf{A}^{-1}$ also $\in \mathfrak{A}(\mathbf{T})$.

The following theorem, a special case of a theorem on the automorphisms of abelian groups due to v. d. Waerden, gives the structure of $\mathfrak{A}(\mathbf{T}) = \mathfrak{A}$ for a completely reducible p.l.t.

THEOREM 19. *Let*

$$\mathbf{T} = (\mathbf{T}_1^{(1)} \oplus \cdots \oplus \mathbf{T}_{n_1}^{(1)}) \oplus \cdots \oplus (\mathbf{T}_1^{(m)} \oplus \cdots \oplus \mathbf{T}_{n_m}^{(m)})$$

where each $\mathbf{T}_j^{(i)}$ *is irreducible and* $\mathbf{T}_k^{(i)} \simeq \mathbf{T}_l^{(i)}$ *but* $\mathbf{T}_k^{(i)} \not\simeq \mathbf{T}_l^{(j)}$ *if* $j \neq i$. *Then*

$$\mathfrak{A} = \mathfrak{A}^{(1)} \oplus \cdots \oplus \mathfrak{A}^{(m)}, \qquad \mathfrak{A}^{(i)} \simeq \mathfrak{D}_{n_i}^{(i)} \tag{33}$$

where $\mathfrak{D}^{(i)}$ *is a field, the automorphism ring of* $\mathbf{T}_j^{(i)}$.

For a proof of this theorem the reader is referred to v.d. Waerden [15] vol. II, pp. 165–168.

Consider the case of a finite completely reducible $\mathbf{T}$. Its minimum function $\mu^*(t) = \overset{*}{\pi}_1(t) \cdots \overset{*}{\pi}_m(t)$ where the $\overset{*}{\pi}_i(t)$ are irreducible in $\mathfrak{B}^*$ and $\overset{*}{\pi}_i(t) \neq \overset{*}{\pi}_j(t)$ $(i \neq j)$. Any two irreducible factors of $\overset{*}{\pi}_i(t)$ are similar. Let $\pi_i(t)$ be such a factor. Then $\pi_i(t) \not\simeq^1 \pi_j(t)$ and (33) gives the structure of $\mathfrak{A}$ where $\mathfrak{D}^{(i)}$ is inverse isomorphic to the invariant ring of $\pi_i(t)$. In particular if $m = 1$, i.e. $\mu^*(t)$ is irreducible in $\mathfrak{B}^*$, $\mathfrak{A}$ is a simple ring $\simeq \mathfrak{D}_{n_1}$.

11. **Periodic s.l.t. and generalized cyclic algebras.** An s.l.t. will be called *periodic* if some power of it, say $\mathbf{T}^m = \mu \neq 0$, i.e. $\mathbf{T}^m$ is the s.l.t. $x \to x\mu = x\mathbf{M}$

[17] Cf. Ore [11] p. 241. The connection between the ring of matrices commutative with a matrix in a commutative $\mathfrak{F}$ and the invariant ring was first noted by Frobenius. Cf. Wedderburn [18] p. 106.

whose automorphism is $\alpha \to \mu^{-1}\alpha\mu$ and matrix (in any coordinate system) is $\mu 1$. If $\alpha \to \bar{\alpha}$ and T are the automorphism and matrix of $\mathbf{T}$ relative to $(e_1, \cdots, e_n)$, $\mathbf{T}^m$ is the s.l.t. whose automorphism and matrix are

$$\alpha \to \bar{\alpha}^{(m)} \qquad \text{and} \qquad T\bar{T} \cdots \bar{T}^{(m-1)}.$$

Hence n.a.s.c.'s that an s.l.t. be periodic are

$$\bar{\alpha}^{(m)} = \mu^{-1}\alpha\mu \qquad T\bar{T} \cdots \bar{T}^{(m-1)} = \mu 1. \tag{34}$$

Thus $\alpha \to \bar{\alpha}$ has relative finite order, say q and $m = qr$. (34) implies.

$$\bar{\mu}1 = \bar{T} \cdots \bar{T}^{(m)} = T^{-1}(T \cdots \bar{T}^{(m-1)})\bar{T}^{(m)} = T^{-1}\mu 1\bar{T}^{(m)} = \mu 1$$

and so $\mu^*(t) = t^m - \mu \in \mathfrak{B}^*$. $\mathbf{T}$ is a finite s.l.t. whose minimum function is a factor of $\mu^*(t)$.

THEOREM 20. *If $\chi(\mathfrak{F}) = p$ is not a divisor of $m/q = r$, $\mathbf{T}$ is completely reducible.*

If $\bar{\alpha}^{(q)} = \nu\alpha\nu^{-1}$, $u = t^q\nu$ commutes with every α. $t^m - \mu = u^r\rho - \mu$ where $\rho^{-1} = \bar{\nu}^{(q(r-1))} \cdots \bar{\nu}^{(q)}\nu$. Since $\mu \neq 0$, the factors of $u^r\rho - \mu$ in $\mathfrak{B}^*$ are polynomials in u. Suppose

$$u^r\rho - \mu = \overset{*}{\pi}_1(u)^2\overset{*}{\pi}_2(u) \qquad \overset{*}{\pi}_i(u) \in \mathfrak{B}^*. \tag{35}$$

Since u commutes with the coefficients of these polynomials, we may differentiate (35) formally with respect to u and obtain

$$ru^{r-1}\rho = [\overset{*}{\pi}_1(u)\overset{*}{\pi}_1(u)' + \overset{*}{\pi}_1(u)'\overset{*}{\pi}_1(u)]\overset{*}{\pi}_2(u) + \overset{*}{\pi}_1(u)^2\overset{*}{\pi}_2(u)' \equiv 0 \; (\overset{*}{\pi}_1(u))$$

since $\overset{*}{\pi}_1(u) \in \mathfrak{B}^*$. If $r \not\equiv 0 \pmod p$, this implies $(u^{r-1}, u^r\rho - \mu) \neq 1$ which is impossible since $\mu \neq 0$. Hence $t^m - \mu$ has no multiple factors in $\mathfrak{B}^*$ and $\mathbf{T}$ is completely reducible.

If $m = q$, $\mathbf{T}$ will be called *normal*. In the remainder of this section we suppose that $\mathbf{T}$ is a normal periodic s.l.t. In this case $\mu^*(t) = t^q - \mu$ is irreducible in $\mathfrak{B}^*$. Hence $\mu^*(t)$ is the minimum function of $\mathbf{T}$ which is then completely reducible. If

$$\mu(t) = t^l - t^{l-1}\mu_1 - \cdots - \mu_l$$

is an irreducible factor of $\mu^*(t)$ in $\mathfrak{F}[t]$, then $l \mid m$, $m = m'l$ and a matrix of $\mathbf{T}$ has the form

$$\begin{pmatrix} U & & & \\ & U & & \\ & & \ddots & \\ & & & U \end{pmatrix} \tag{36}$$

where

$$(37)\qquad \begin{pmatrix} 0 & \cdot & \cdot & \cdot & \cdot & \mu_l \\ 1 & & & & & \cdot \\ & \cdot & & & & \cdot \\ & & \cdot & & & \cdot \\ & & & \cdot & & \mu_2 \\ & & & & 1 & \mu_1 \end{pmatrix}$$

Thus $l \mid n$, $n = n'l$. We shall call l the *index* of **T**.

T is a multiple of the identity matrix if and only if $l = 1$ i.e. $\mu(t) = t - \mu_l \equiv t - \nu$. Then

$$t^q - \mu = (t - \nu)(t^{q-1} + t^{q-2}\bar{\nu}^{(q-1)} + \cdots + \bar{\nu}\bar{\nu}^{(2)} \cdots \bar{\nu}^{(q-1)})$$

and hence

$$\mu = N(\nu) = \nu\bar{\nu} \cdots \bar{\nu}^{(q-1)}$$

and conversely if $\mu = N(\nu)$, **T** has index $= 1$.

THEOREM 21. *A n.a.s.c. that a normal periodic s.l.t. have index* 1 *is that* $\mu = N(\nu)$ *for some* ν *in* $\mathfrak{F}$.

In particular if $\mu = 1$, we may take $\nu = 1$ and then (36) is the identity matrix. In the language of matrices this becomes the following extension of Hilbert's theorem on cyclic fields.[18]

COROLLARY. *Let* $\mathfrak{F}$ *be any field and* $\alpha \rightarrow \bar{\alpha}$ *any automorphism whose order and relative order are finite and* $= q$. *If* T *is a matrix with elements in* $\mathfrak{F}$ *such that* $T\bar{T} \cdots \bar{T}^{(q-1)} = 1$, *then there exists a matrix* A *in* $\mathfrak{F}$ *such that* $T = A^{-1}\bar{A}$.

The automorphism ring $\mathfrak{A}$ of **T** with matrix (36) is according to the general theory a simple algebra $\mathfrak{D}_{n'}$ where $\mathfrak{D}$ is inverse isomorphic to the invariant ring of $\mu(t)$. If **T** has the index 1 ($\mu(t) = t - \nu$). It follows from the definition of the invariant ring that $\mathfrak{D}$ is inverse isomorphic to the subfield of $\mathfrak{F}$ of elements β such that $\bar{\beta} = \nu^{-1}\beta\nu$.

If $n = q$, **T** is cyclic and $\mathfrak{A}$ is inverse isomorphic to the invariant ring of $\mu^*(t) = t^q - \mu$. The latter is a generalization of cyclic algebras[19] first considered by Dickson. The above results are applicable to the study of its structure. Thus we have the following extension of a theorem conjectured by Dickson and proved by Albert.[20]

THEOREM 22. *If* q *is a prime, the invariant ring* $\mathfrak{A}$ *of* $(t^q - \mu) \in \mathfrak{B}^*$ *is a field if* $\mu \neq N(\nu)$ *for any* $\nu \in \mathfrak{F}$. *If* $\mu = N(\nu)$, $\mathfrak{A} \simeq \mathfrak{D}_q$ *where* $\mathfrak{D}$ *is the subfield of* β*'s such that* $\bar{\beta} = \nu^{-1}\beta\nu$.

12. **Differential transformations.** In this section we suppose that $\mathfrak{F}$ is commutative and **T** is a d.t. in $\mathfrak{R}$ over $\mathfrak{F}$. We denote the subfield of *constants* (elements β such that $\beta' = 0$) of $\mathfrak{F}$ by $\mathfrak{K}$.

---

[18] Hilbert [5] p. 272.

[19] A discussion of cyclic algebras from the point of view of the invariant ring of non-commutative polynomials is given in Jacobson [6].

[20] Dickson [3] p. 227 and Albert [1] p. 624. Our extension consists in removing the assumption that $\mathfrak{F}$ has a finite basis over its centrum.

We have seen that the automorphism ring $\mathfrak{A}$ of $\mathbf{T}$ is given by the matrices A such that

$$(38) \qquad [A, T] \equiv AT - TA = A'$$

where T is a matrix of $\mathbf{T}$. This matrix equation is equivalent to a system of n^2 linear homogeneous differential equations in the elements of A. Hence the number of solutions of (38) which are linearly independent relative to $\mathfrak{K}$ is finite. We therefore have the following theorem due to Ore[21] for cyclic d.t.

THEOREM 23. *If* $\mathbf{T}$ *is a d.t. in* $\mathfrak{R}$ *over a commutative field* $\mathfrak{F}$, *its automorphism ring has a finite basis over the field of constants* $\mathfrak{K}$.

Suppose $\mathfrak{K}$ is perfect and let $\mathfrak{K}(\theta)$ be an algebraic extension of $\mathfrak{K}$ of degree r. Consider $\mathfrak{F}(\theta)$. Since θ is a separable element, we may extend the derivative defined in $\mathfrak{F}$ to $\mathfrak{F}(\theta)$ by defining $\theta' = 0$. It follows that $\mathfrak{F}(\theta)$ has degree r over $\mathfrak{F}$ also; for if $\theta^m + \alpha_1\theta^{m-1} + \cdots + \alpha_m = 0$ is the minimum equation of θ over $\mathfrak{F}$, we obtain by differentiation $\alpha_1'\theta^{m-1} + \cdots + \alpha_m' = 0$. Hence each $\alpha_i' = 0$ or $\alpha_i \in \mathfrak{K}$ and $m = r$.

Let $\mathfrak{R}_{\mathfrak{F}(\theta)}$ be the extension of $\mathfrak{R}$ obtained by letting the coefficients vary in $\mathfrak{F}(\theta)$. If $\mathbf{T}$ is a d.t. in $\mathfrak{R}$, it has a unique extension $\mathbf{T}(\theta)$ in $\mathfrak{R}_{\mathfrak{F}(\theta)}$: The matrix T of $\mathbf{T}$ relative to $(e_1, \cdots, e_n)$ is also the matrix of $\mathbf{T}(\theta)$ relative to this basis. It is easily seen that *the automorphism ring of* $\mathbf{T}(\theta)$ *is* $\mathfrak{A}_{\mathfrak{F}(\theta)}$ *where* $\mathfrak{A}$ *is that of* $\mathbf{T}$; for an automorphism of $\mathbf{T}(\theta)$ is determined by $A(\theta)$ such that

$$(39) \qquad [A(\theta), T] = A(\theta)'.$$

If $\theta_1, \cdots, \theta_r$ is a basis for $\mathfrak{F}(\theta)$ over $\mathfrak{F}$, we may write $A(\theta)$ uniquely in the form

$$A_1\theta_1 + A_2\theta_2 + \cdots + A_r\theta_r \qquad A_i \in \mathfrak{F}_n .$$

Since $\theta_i' = 0$, (39) gives $[A_i, T] = A_i'$, i.e. A_i determine automorphism of $\mathbf{T}$. The converse of this is obvious.

Now suppose $\mathbf{T}$ is an irreducible d.t. which decomposes in every algebraic extension $\mathfrak{F}(\theta)$ of $\mathfrak{F}$ into parts all of which are similar. Then the automorphism ring $\mathfrak{A}$ of $\mathbf{T}$ is a normal division algebra over $\mathfrak{K}$. For if the centrum of $\mathfrak{A}$ has degree m over $\mathfrak{K}$, it follows from the theory of hypercomplex numbers that there exists a field $\mathfrak{K}(\theta)$ such that $\mathfrak{A}_{\mathfrak{F}(\theta)}$ is a direct sum of m matrix algebras. But then $\mathbf{T}(\theta)$ would have dissimilar components unless $m = 1$. This criterion may be used to construct examples of d.t.'s whose automorphism rings are normal division algebras.

BIBLIOGRAPHY

1. A. A. ALBERT, *On normal simple algebras*, Trans. Am. Math. Soc., **34** (1932), pp. 620–625.
2. G. BIRKHOFF, *On the combination of subalgebras*, Proc. Camb. Phil. Soc., **29** (1933), pp. 441–464.
3. L. E. DICKSON, *New division algebras*, Trans. Am. Math. Soc., **28** (1926), pp. 207–234.
4. H. FITTING, *Die Theorie der Automorphismenringe Abelscher Gruppen* etc., Math. Annalen, **107** (1932), pp. 514–542.

[21] Ore [11] p. 236.

5. D. Hilbert, *Die Theorie der algebraische Zahlkorper*, Jahr. Deutsch. Math. Verein., 4 (1897).
6. N. Jacobson, *Non-commutative polynomials and cyclic algebras*, Annals of Math. **35** (1934), pp. 197–209.
7. N. Jacobson, *On pseudo-linear transformations*, Proc. Nat. Acad. Sci., **21** (1935), pp. 667–670.
8. W. Krull, *Über verallgemeinerte endliche Abelsche Gruppen*, Math. Zeitsch., **23** (1925), pp. 160–196.
9. W. Krull, *Theorie und Anwendung der verallgemeinerten Abelschen Gruppen*, Sitz. Heidelberg Akad., **17** (1926), pp. 3–32.
10. E. Noether and W. Schmeidler, *Moduln in nichtkommutative Bereichen* etc., Math. Zeitsch., **8** (1920), pp. 1–35.
11. O. Ore, *Formale theorie lineare differentialgleichungen* II, Jour. für Math., **168** (1932), pp. 233–252.
12. O. Ore, *Theory of non-commutative polynomials*, Annals of Math., **34** (1933), pp. 480–508.
13. O. Ore, *On the foundations of abstract algebra* I, Annals of Math., **36** (1935), pp. 406–437.
14. C. Segre, *Un nuovo campo di ricerche geometriche*, Atti Torino **25** (1889), pp. 276–301.
15. B. L. van der Waerden, *Moderne Algebra* I and II, J. Springer, Berlin, 1930.
16. B. L. van der Waerden, *Gruppen von lineare Transformationen*, Ergebnisse der Math., J. Springer, Berlin, 1935.
17. J. H. M. Wedderburn, *Non-commutative domains of integrity*, Jour. für Math., **167** (1932), pp. 129–141.
18. J. H. M. Wedderburn, *Lectures on matrices*, Am. Math. Soc. Colloquium Publications, vol. XVII, New York, 1934.

Bryn Mawr College.

ANNALS OF MATHEMATICS
Vol. 38, No. 2, April, 1937

A CLASS OF NORMAL SIMPLE LIE ALGEBRAS OF CHARACTERISTIC ZERO[1]

BY N. JACOBSON[2]

(Received March 24, 1937)

Introduction. The classification of simple Lie algebras (infinitesimal groups) over the field of complex numbers is due to Killing and to Cartan.[3] In a second paper[4] Cartan determined the simple Lie algebras over the real field. The methods used in these papers are purely algebraic and so it is clear that the results hold equally well for arbitrary algebraically closed fields of characteristic zero and arbitrary real closed fields. More recently W. Landherr[5] considered this problem for arbitrary fields of characteristic zero and showed that it could be reduced to that of determining all *normal simple algebras* i.e. algebras which remain simple when the underlying field is extended to its algebraic closure. It therefore suffices to classify those Lie algebras which become a fixed Lie algebra in the Killing-Cartan list when the field is extended. Landherr has obtained a partial solution of this problem for the algebras which lead to Cartan's type A.[6] In this paper we determine the algebras which lead to Cartan's types B, C, D.[7] We give an invariantive realization of these systems and also obtain their groups of automorphisms. The relation between the Lie algebra and this group seems to be an algebraic substitute for the relation between an infinitesimal group and the integrated group given by Lie's theory.

1. **Anti-automorphisms in an algebra.** Let $\mathfrak{A}$ be an associative algebra with a finite basis over a commutative field Φ. It is readily verified that if we define in $\mathfrak{A}$ the new composition $[a, b] = ab - ba$, $\mathfrak{A}$ becomes a Lie algebra i.e.

$$\alpha[a, b] = [\alpha a, b] = [a, \alpha b] \qquad [a + b, c] = [a, c] + [b, c]$$

$$[a, b] = -[b, a] \qquad [[a, b], c] + [[b, c], a] + [[c, a], b] = 0.$$

Suppose $\mathfrak{A}$ is *self-reciprocal* i.e. there is defined a correspondence (an *anti-automorphism*) J, $a \rightarrow a^J$ in $\mathfrak{A}$ such that

$$(a + b)^J = a^J + b^J \qquad (\alpha a)^J = \alpha a^J \qquad (ab)^J = b^J a^J.$$

---

[1] Presented to the Society, Dec. 29, 1936.
[2] National Research Fellow.
[3] E. Cartan, *Thèse*, Paris (1894) Chapter 5.
[4] *Les groupes réels simples et continus*, Annales de l'École Normale, **31** (1914) pp. 263–355.
[5] *Über einfache Liesche Ringe*, Hamb. Abhandlungen **11** (1935) pp. 41–64.
[6] His classification is complete for p-adic fields.
[7] With the exception of type D and order 28.

We shall call a *J-skew* if $a^J = -a$. Let $\mathfrak{S}_J$ denote the set of J-skew elements. If $a, b \in \mathfrak{S}_J$ so do $a + b$, αa and $[a, b]$ so that $\mathfrak{S}_J$ is a Lie subalgebra of $\mathfrak{A}$.

A second anti-automorphism K is *cogredient* to J if $K = S^{-1}JS$ where S is an automorphism of $\mathfrak{A}$. In this case the elements of $\mathfrak{S}_K$ are a^S, a in $\mathfrak{S}_J$ ($\mathfrak{S}_K = \mathfrak{S}_J^S$) and the correspondence $a \rightarrow a^S$ establishes an isomorphism between the Lie algebras $\mathfrak{S}_J$ and $\mathfrak{S}_K$.

The elements g of $\mathfrak{A}$ such that $gg^J = \gamma 1 = g^J g$ where $\gamma \neq 0 \in \Phi$ will be called *J-orthogonal*. Their totality is a group $\mathfrak{G}_J$ under multiplication. $\mathfrak{G}_J$ contains the invariant subgroup $\mathfrak{D}$ consisting of the Φ-multiples of 1. We note that $g^{J^2} = (\gamma g^{-1})^J = \gamma(g^J)^{-1} = \gamma\gamma^{-1}g = g$ and $(g^{-1}ag)^J = g^J a^J (g^J)^{-1} = g^{-1}a^J g$. Hence if a is J-skew so is $g^{-1}ag$ and so if g is a fixed element of $\mathfrak{G}_J$ the correspondence $a \rightarrow g^{-1}ag$ is an automorphism of the Lie algebra $\mathfrak{S}_J$. If S is an automorphism of $\mathfrak{A}$ then $\gamma 1 = g^S g^{JS} = g^S g^{S(S^{-1}JS)} = g^S g^{SK}$, $K = S^{-1}JS$. Thus the correspondence $g \rightarrow g^S$ is an isomorphism between $\mathfrak{G}_J$ and $\mathfrak{G}_K$.

Let $\tilde{\Phi}$ be an over-field of Φ and $\xi_1, \xi_2, \cdots$ be a basis of $\tilde{\Phi}$ over Φ. The elements of the algebra $\tilde{\mathfrak{A}} = \mathfrak{A}_{\tilde{\Phi}}$ obtained from $\mathfrak{A}$ by extending Φ to $\tilde{\Phi}$ are uniquely expressible in the form $a = \xi_1 a_1 + \xi_2 a_2 + \cdots$ (finite sums) where $a_i \in \mathfrak{A}$. If M is an automorphism or anti-automorphism of $\mathfrak{A}$, M has a unique extension defined by $a^M = \xi_1 a_1^M + \xi_2 a_2^M + \cdots$ in $\tilde{\mathfrak{A}}$. If $M = J$ is an anti-automorphism and a is J-skew,

$$\xi_1 a_1^J + \xi_2 a_2^J + \cdots = -(\xi_1 a_1 + \xi_2 a_2 + \cdots)$$

then $a_i^J = -a_i$. Thus the J-skew elements of $\tilde{\mathfrak{A}}$ are the linear combinations with coefficients in $\tilde{\Phi}$ of the J-skew elements of $\mathfrak{A}$ i.e. $\mathfrak{S}_{\tilde{\Phi}J} = \mathfrak{S}_{J\tilde{\Phi}}$.

J is *involutorial* if $J^2 = I$ the identity mapping. If K is cogredient to J then K is involutorial also.

2. Skew elements of a matric algebra. In the remainder of the paper J, K will denote involutorial anti-automorphism (i. a. a.), $\mathfrak{A}$ a normal simple algebra over Φ of characteristic 0.

Suppose first that $\mathfrak{A} \cong \Phi_n$ the matrix algebra of order n^2 over Φ. The correspondence J_0: $A = (\alpha_{ij}) \rightarrow A' = (\alpha_{ji})$ is an i. a. a.[8] If J is any anti-automorphism $J = SJ_0$ where S is an automorphism and since any automorphism of a normal simple algebra is inner[9] we have $A^J = S^{-1}A'S$ where $A \rightarrow S^{-1}AS$ is the automorphism S. The condition $J^2 = I$ is $S^{-1}S'A(S^{-1}S') = A$ for all A. Hence $S^{-1}S' = \sigma 1$ and $S' = \sigma S$ and since $S'' = S$, $\sigma^2 = 1$ and $\sigma = \pm 1$. Thus S is either symmetric or skew-symmetric (relative to J_0).[10] The i. a. a. $K = V^{-1}JV$ is given as $A^K = T^{-1}A'T$ where $T = V'SV$ and $A^V =$

---

[8] J_0 is defined by means of a fixed basis E_{ij}, of Φ_n such that $E_{ij}E_{kl} = \delta_{jk}E_{il}$: we have $E_{ij}^{J_0} = E_{ji}$.

[9] For a proof of this theorem see M. Deuring, *Algebren*, Springer 1935, p. 42.

[10] Cf. A. A. Albert, *Involutorial simple algebras and real Riemann matrices*, Annals of Math. **36** (1935) p. 897.

$V^{-1}AV$ as is readily verified. It should be noted that the matrix S associated with J is determined only to within a multiple in Φ. Hence we have

THEOREM 1. *The i. a. a. J and K are cogredient if and only if the matrix S of J is cogredient (in the usual sense) to a multiple of the matrix T of K.*

(1) If $S' = -S$, since S is non-singular $n = 2\nu$, ν an integer and S is cogredient to

$$\begin{pmatrix} 0 & 1_\nu \\ -1_\nu & 0 \end{pmatrix}$$

(1_ν the identity matrix of Φ_ν).[11] The corresponding i. a. a. K defines the Lie algebra $\mathfrak{S}_K$ consisting of the matrices of the form

$$\begin{pmatrix} A_{11} & A_{12} \\ A_{21} & A_{22} \end{pmatrix}$$

where $A_{ij} \epsilon \Phi_\nu$ and $A_{11}' = -A_{22}$, $A_{12}' = A_{12}$, $A_{21}' = A_{21}$.[12] Thus $\mathfrak{S}_K$ has order $n(n+1)/2$ and is generated by the abelian subalgebra $\mathfrak{H}$ of matrices

$$H = \begin{pmatrix} \lambda_1 & & & & & \\ & \ddots & & & & \\ & & \lambda_\nu & & & \\ & & & -\lambda_1 & & \\ & & & & \ddots & \\ & & & & & -\lambda_\nu \end{pmatrix} \tag{1}$$

and

$$E_{\lambda_i-\lambda_k} = \begin{pmatrix} E_{ik} & 0 \\ 0 & -E_{ki} \end{pmatrix} \qquad E_{-\lambda_i+\lambda_k} = \begin{pmatrix} E_{ki} & 0 \\ 0 & -E_{ik} \end{pmatrix}$$

$$E_{\lambda_i+\lambda_k} = \begin{pmatrix} 0 & E_{ik} + E_{ki} \\ 0 & 0 \end{pmatrix} \qquad E_{-\lambda_i-\lambda_k} = \begin{pmatrix} 0 & 0 \\ E_{ik} + E_{ki} & 0 \end{pmatrix}$$

$$E_{2\lambda_i} = \begin{pmatrix} 0 & E_{ii} \\ 0 & 0 \end{pmatrix} \qquad E_{-2\lambda_i} = \begin{pmatrix} 0 & 0 \\ E_{ii} & 0 \end{pmatrix}$$

where $i < k$ and E_{ij} is the matrix in Φ_ν having a 1 in the ith row and j^{th} column. The linear forms $\alpha = \pm\lambda_i \pm \lambda_k$, $\pm 2\lambda_i$ are the *roots* of $\mathfrak{S}_K$. The multiplication table of this algebra is

---

[11] See, for example J. H. M. Wedderburn, *Lectures on matrices*, New York, 1934, p. 96.

[12] $\mathfrak{S}_K$ is the "infinitesimal group" of the complex group when Φ is the field of complex numbers. Cf. H. Weyl, *Theorie der Darstellung kontinuierlicher halbeinfacher Gruppen* II, Math. Zeitschrift **24** (1926) p. 333.

$$(2)\qquad \begin{aligned} [H_1, H_2] &= 0 \text{ for any } H_1, H_2 \text{ in } \mathfrak{H} \\ [H, E_\alpha] &= \alpha E_\alpha \qquad [E_\alpha, E_{-\alpha}] = H_\alpha \in \mathfrak{H} \\ [E_\alpha, E_\beta] &= \begin{cases} 0 \text{ if } \alpha + \beta \text{ is not a root} \\ N_{\alpha\beta} E_{\alpha+\beta} \neq 0 \;\; N_{\alpha\beta} \text{ an integer if } \alpha + \beta \text{ is a root.} \end{cases} \end{aligned}$$

It follows that $\mathfrak{S}_K$ *is a simple Lie algebra.* For if $\mathfrak{J} \neq 0$ is an ideal in $\mathfrak{S}_K$ and $B = H_0 + \sum \kappa_\alpha E_\alpha \neq 0$ is in $\mathfrak{J}$ then so is

$$(3)\qquad \underbrace{[H, [H, \cdots [H}_{r}, B] \cdots]] = \sum \kappa_\alpha \alpha^r E_\alpha \qquad r = 1, 2, \cdots$$

for all values of $\lambda_1, \cdots, \lambda_\nu$. The determinant of $m = 2\nu^2$ rows (m, the number of roots)

$$V(\lambda_1, \cdots, \lambda_\nu) = \begin{vmatrix} \alpha_1 & \alpha_2 & \cdots & \alpha_m \\ \alpha_1^2 & \alpha_2^2 & \cdots & \alpha_m^2 \\ \cdot & \cdot & & \cdot \\ \cdot & \cdot & & \cdot \\ \cdot & \cdot & & \cdot \\ \alpha_1^m & \alpha_2^m & \cdots & \alpha_m^m \end{vmatrix} = \prod_j \alpha_j \prod_{i<k} (\alpha_i - \alpha_k)$$

where the α_i are the roots is not identically 0 and so $\lambda_1^0, \cdots, \lambda_\nu^0$ can be chosen in Φ so that $V(\lambda_1^0, \cdots, \lambda_\nu^0) \neq 0$. By multiplying the relations (3) by the cofactors of the column containing α in V and adding we obtain that

$$V(\lambda_1^0, \cdots, \lambda_\nu^0)\kappa_\alpha E_\alpha \in \mathfrak{J}$$

and hence also $\kappa_\alpha E_\alpha$. Thus if $\kappa_\alpha \neq 0$, $E_\alpha \in \mathfrak{J}$. If $\nu > 1$ every root may be obtained from a fixed one by adding roots and hence by the third relation (2) all E_α belong to $\mathfrak{J}$. It follows that all $H_\alpha \in \mathfrak{J}$ and since the H_α generate $\mathfrak{H}$, $\mathfrak{J} = \mathfrak{S}_K$. If $\nu = 1$, $\mathfrak{J} \supset [E_\alpha, E_{-\alpha}]$ which forms a basis for $\mathfrak{H}$ and hence $\mathfrak{J} \supset E_{-\alpha}$ also. If all $\kappa_\alpha = 0$, $H_0 \neq 0$ and some $[H_0, E_\alpha]$ is a non-zero multiple of E_α. As before it follows that $\mathfrak{J} = \mathfrak{S}_K$.

(2) $S' = S$. In this case S is cogredient to

$$(4)\qquad \begin{pmatrix} \rho_1 & & & \\ & \rho_2 & & \\ & & \ddots & \\ & & & \rho_n \end{pmatrix}$$

and if Φ contains $\sqrt{\rho_i}$, S is cogredient to 1_n and also to either

$$(5)\qquad \begin{pmatrix} 1 & & \\ & 0 & 1_\nu \\ & 1_\nu & 0 \end{pmatrix} \quad \text{or} \quad \begin{pmatrix} 0 & 1_\nu \\ 1_\nu & 0 \end{pmatrix}$$

according as $n = 2\nu + 1$ or $n = 2\nu$.[13] If we use the i. a. a. corresponding to 1_n we see that $\mathfrak{S}_J$ is isomorphic to the Lie algebra of all skew-symmetric matrices in Φ_n and so $\mathfrak{S}_J$ has order $n(n - 1)/2$. For our purposes it is more convenient to take the i. a. a. defined by the matrices (5) and to distinguish the two cases odd and even order. A discussion analogous to that given in (1) may be made. We shall merely state the results referring the reader to Prof. Weyl's paper[14] for the details.

(a) $n = 2\nu + 1$. $\mathfrak{H}$ is the subalgebra of matrices

$$H = \begin{pmatrix} 0 & & & & & & \\ & \lambda_1 & & & & & \\ & & \ddots & & & & \\ & & & \lambda_\nu & & & \\ & & & & -\lambda_1 & & \\ & & & & & \ddots & \\ & & & & & & -\lambda_\nu \end{pmatrix}$$

and there are E_α's such that (2) holds. The roots α are $\pm\lambda_i \pm\lambda_k$ $(i < k)$ and $\pm\lambda_i$. As before $\mathfrak{S}_K$ is simple.

(b) $n = 2\nu$. $\mathfrak{H}$ is the subalgebra

$$H = \begin{pmatrix} \lambda_1 & & & & & \\ & \ddots & & & & \\ & & \lambda_\nu & & & \\ & & & -\lambda_1 & & \\ & & & & \ddots & \\ & & & & & -\lambda_\nu \end{pmatrix}$$

and the roots α corresponding to the E_α are $\pm\lambda_i \pm\lambda_k$ $(i < k)$. The proof given above for the simplicity of $\mathfrak{S}_K$ in (1) is valid here for $\nu > 2$. If $\nu = 1$ $\mathfrak{S}_K$ is abelian and if $\nu = 2$ $\mathfrak{S}_K$ is a direct sum of two algebras of type $2a$ in which $\nu = 1$.

It can be shown that $\mathfrak{S}_K$ of type 2a with $\nu = 1$ is isomorphic to $\mathfrak{S}_K$ of type 1 for $\nu = 1$ and type 2a with $\nu = 2$ is isomorphic to type 1 with $\nu = 2$. No other algebra of type 2 is isomorphic with one of type 1.[15]

In either (1) or (2) the set of matrices $\mathfrak{S}_K$ are an absolutely irreducible system.[16] By Burnside's theorem the enveloping algebra of $\mathfrak{S}_K$ (the smallest subalgebra of Φ_n containing $\mathfrak{S}_K$) is Φ_n.

[13] Wedderburn, *Lectures on matrices*, p. 95.

[14] Loc. cit. in 12, pp. 342–345.

[15] Cartan, *Thèse*, p. 78.

[16] Weyl II, p. 334 and p. 344.

3. **Skew elements of a normal simple algebra.** Now suppose that $\mathfrak{A}$ is a normal simple algebra of degree n (order n^2) and Ω is the algebraic closure of Φ. Then $\mathfrak{A}_\Omega \cong \Omega_n$ and $\mathfrak{A}$ has a representation $a \to A$ by matrices in Ω such that the linear combinations $\sum \tilde{\alpha}_i A_i$ with coefficients $\tilde{\alpha}_i$ in Ω of the matrices (representing the elements) of $\mathfrak{A}$ comprise all the elements of Ω_n. The i. a. a. J in $\mathfrak{A}$ induces an i. a. a. J in Ω_n $A \to A^J$ where A^J corresponds to a^J, $\sum \tilde{\alpha}_i A_i \to \sum \tilde{\alpha}_i A_i^J$. The J-skew elements of Ω_n are the linear combinations with coefficients in Ω of the matrices of $\mathfrak{S}_J$. Now let G be a matrix of $\mathfrak{A}$ such that G is J-orthogonal in Ω_n i.e. $GG^J = \tilde{\gamma}1$, $\tilde{\gamma} \neq 0$ in Ω. Then GG^J is a matrix of $\mathfrak{A}$ and commutes with all the matrices of $\mathfrak{A}$. Hence $\tilde{\gamma} = \gamma \in \Phi$ and G is the matrix of a J-orthogonal element of $\mathfrak{A}$.

We have seen that there is an automorphism S in Ω_n such that $K = S^{-1}JS$ is one of the normalized i. a. a.'s defined in the last section and $\tilde{\mathfrak{S}}_K$ the set of K-skew elements of Ω_n is one of the systems (1), (2a) or (2b). The correspondence $a \to A^S$ is a second representation of $\mathfrak{A}$ by matrices in Ω_n and the i. a. a. J in $\mathfrak{A}$ induces by means of this representation the i. a. a. K in Ω_n since $A^S \to A^{JS} = A^{SK}$. We may restrict our attention to the second representation and write $a \to A$ in place of $a \to A^S$ and J in place of K. We shall say that $\mathfrak{S}_J$ has type (1), (2a) or (2b) according as $\tilde{\mathfrak{S}}_J$ is the system (1), (2a) or (2b). If $\mathfrak{S}_J$ has type (1) its order is $n(n + 1)/2$ and if it has types (2) its order is $n(n - 1)/2$. In either case since the enveloping algebras (1), (2a) or (2b) is Ω_n, the enveloping algebra of $\mathfrak{S}_J$ is $\mathfrak{A}$.

A Lie algebra $\mathfrak{L}$ over Φ will be called *normal simple* if $\mathfrak{L}_\Omega$ where Ω is the algebraic closure of Φ is simple. This of course implies that $\mathfrak{L}$ is simple. For if $\mathfrak{J}$ is an ideal in $\mathfrak{L}$, $\mathfrak{J}_\Omega$ is one in $\mathfrak{L}_\Omega$. We have seen that all the algebras $\mathfrak{S}_J$ with the exception of type 2 for $n = 4$ are normal simple. We summarize our results in the theorem:

THEOREM 2. *If $\mathfrak{A}$ is a normal simple algebra of degree n and J an i. a. a. in $\mathfrak{A}$, $\mathfrak{S}_J$ the set of J-skew elements is a normal simple algebra except when it has type* 2 *and $n = 4$. The order of $\mathfrak{S}_J$ is $n(n + 1)/2$ (type* 1*) or $n(n - 1)/2$ (type* 2*). In either case the enveloping algebra of $\mathfrak{S}_J$ is $\mathfrak{A}$.*

COROLLARY. *If J and K are i. a. a. in a normal simple algebra and $\mathfrak{S}_J = \mathfrak{S}_K$ then $J = K$.*

For the elements a of $\mathfrak{S} = \mathfrak{S}_J = \mathfrak{S}_K$ we have $a^J = -a = a^K$. Since the enveloping algebra of $\mathfrak{S}$ is $\mathfrak{A}$ any element b in $\mathfrak{A}$ has the form $\sum a_1 a_2 \cdots a_r$, $a_i \in \mathfrak{S}$. Hence $b^J = \sum (-1)^r a_r a_{r-1} \cdots a_1 = b^K$, i.e. $J = K$.

4. **Automorphisms of the Lie algebra $\mathfrak{S}_J$.** From now on we suppose when $\mathfrak{S}_J$ has type 2 that $n \geqq 6$ and $\neq 8$. In view of the isomorphisms noted in §2 the only loss in generality here is in the restriction $n \neq 8$.

Let $G: a \to a^G$ be an automorphism $\neq 0$ of $\mathfrak{S}_J$. G is $(1 - 1)$ since $\mathfrak{S}_J$ is simple and the elements mapped into 0 by an automorphism form an ideal. We shall show that G may be realized as a similarity transformation by a J-orthogonal element. For this purpose we suppose first that $\Phi = \Omega$ is algebraically closed. Then $\mathfrak{A} \cong \Omega_n$ and $\mathfrak{S}_J$ may be taken to be one of the systems (1), (2a) or (2b).

If A is a matrix of $\mathfrak{S}_J$ the correspondence $A \to A^G$ is a second representation of $\mathfrak{S}_J$ and since the totality of matrices $\{A^G\}$ coincides with the totality $\{A\}$ this representation is irreducible also. If, as above, $H = H(\lambda_1, \cdots, \lambda_\nu)$ denotes the general element of the maximal abelian subalgebra $\mathfrak{H}$, by Cartan's theory of representations there exists a matrix Q in Ω_n such that

$$K = Q^{-1}H^G Q = \begin{pmatrix} \Lambda_1 & & & \\ & \Lambda_2 & & \\ & & \cdot & \\ & & & \cdot \end{pmatrix}$$

where the Λ's the weights of the representation are linear forms in the λ's with coefficients all integers in (1) and either all integers or all halves of odd integers in (2).[17] Since every A of $\mathfrak{S}_J$ is similar to $-A'$, if Λ_0 is a root of the characteristic equation of A so is $-\Lambda_0$. It follows that the negative of every weight is a weight. For if no $\Lambda_k = -\Lambda_p$, λ_i may be chosen $= \lambda_i^0$ in Ω so that

$$\Lambda_p(\lambda_1^0, \cdots, \lambda_\nu^0) \neq 0, \neq -\Lambda_k(\lambda_1^0, \cdots, \lambda_\nu^0).$$

Suppose $Q^{-1}E_\alpha^G Q = F_\alpha = (\varphi_{ij\alpha})$ where E_α is defined as in §2. Since $[H, E_\alpha] = \alpha E_\alpha$, $[K, F_\alpha] = \alpha F_\alpha$ and so $\alpha\varphi_{ij\alpha} = (\Lambda_i - \Lambda_j)\varphi_{ij\alpha}$. Since $F_\alpha \neq 0$ it follows that every root α is expressible as a difference of weights. Thus the number of linearly independent weights is ν. We order them lexicographically[18] and may suppose that $\Lambda_1 > \Lambda_2 > \cdots > \Lambda_\nu > 0 > -\Lambda_\nu > \cdots > -\Lambda_1$. We recall also that the set of weights is invariant under arbitrary permutations of the λ's.[19]

(1). In this case the weights are $\pm\Lambda_1, \cdots, \pm\Lambda_\nu$ and the possible differences $\neq 0$ are $\pm\Lambda_p \pm \Lambda_q$, $p \neq q$ and $\pm 2\Lambda_r$. Since their number is equal to the number of roots all of these differences are roots. Evidently $2\Lambda_1 > \pm\Lambda_p \pm \Lambda_q > \pm 2\Lambda_r$ if $r \neq 1$ and so $2\Lambda_1 = 2\lambda_1$ the highest root and $\Lambda_1 = \lambda_1$.

(2a). The weights here are $0, \pm\Lambda_1, \cdots, \pm\Lambda_\nu$ and the possible differences are $\pm\Lambda_p \pm \Lambda_q$, $\pm\Lambda_r$, $\pm 2\Lambda_r$. If $2\Lambda_r$ is a root $\Lambda_r = \lambda_i/2$ or $= (\lambda_i \pm \lambda_k)/2$ $(i < k)$ and since the coefficients must all be halves of odd integers, $\nu = 1$ or 2 which are cases outside of the present consideration. Thus no $\pm 2\Lambda_r$ is a root and so all $\pm\Lambda_p \pm \Lambda_q$, $\pm\Lambda_r$ are. Comparing highest linear forms we have $\Lambda_1 + \Lambda_2 = \lambda_1 + \lambda_2$. If $\Lambda_1 - \Lambda_2 = \lambda_i$, $\Lambda_1 = (\lambda_1 + \lambda_2 + \lambda_i)/2$ and in order that the coefficients all be halves of odd integers $i \neq 1, 2$ and so $\nu = 3$, $i = 3$. Thus $\Lambda_2 = (\lambda_1 + \lambda_2 - \lambda_3)/2$ and by permuting the λ's and changing all signs we see that $(\lambda_1 - \lambda_2 + \lambda_3)/2$ and $(\lambda_1 - \lambda_2 - \lambda_3)/2$ are also weights. But this is impossible since there are only $\nu = 3$ positive weights. It follows that $\Lambda_1 - \Lambda_2 = \lambda_i \pm \lambda_k$ and $i < k$ since $\Lambda_1 - \Lambda_2$ is positive. If $k \neq 2$, $i \neq 1$ for otherwise the coefficients of $\Lambda_1 = (\lambda_1 + \lambda_2 + \lambda_i \pm \lambda_k)/2$ would not satisfy our arithmetic condition. Hence $\nu = 4$, $\Lambda_1 = (\lambda_1 + \lambda_2 + \lambda_3 \pm \lambda_4)/2$ and

[17] Cf. Weyl, *Darstellung kontinuierlicher halb-einfacher Gruppen* I, Math. Zeitschr. **23** (1925) p. 278 and Weyl II p. 334 and p. 344.

[18] We say that $\Lambda_i > \Lambda_j$ or $\Lambda_i - \Lambda_j = m_1\lambda_1 + \cdots + m_\nu\lambda_\nu > 0$ if the first $m_k \neq 0$ is positive.

[19] Weyl II, p. 334 and p. 344.

$\Lambda_2 = (\lambda_1 + \lambda_2 - \lambda_3 \mp \lambda_4)/2$. Then $\Lambda_3 = (\lambda_1 - \lambda_2 + \lambda_3 \mp \lambda_4)/2$ and $\Lambda_4 = (\lambda_1 - \lambda_2 - \lambda_3 \pm \lambda_4)/2$ are the remaining positive weights. But then $\pm\Lambda_p \pm \Lambda_q$, $\pm\Lambda_r$ do not exhaust all the roots and so this case is ruled out also. Hence $k = 2$, $i = 1$ and $\Lambda_1 - \Lambda_2 = \lambda_1 \pm \lambda_2$. Since $\Lambda_1 \neq 0$ $\Lambda_1 - \Lambda_2 = \lambda_1 - \lambda_2$ and $\Lambda_1 = \lambda_1$.

(2b). The weights are $\pm\Lambda_1, \cdots, \pm\Lambda_\nu$ and the possible differences are $\pm\Lambda_p \pm \Lambda_q$, $\pm 2\Lambda_r$. As in (2a) we may rule out the possibility that $\pm 2\Lambda_r$ are roots and so all $\pm\Lambda_p \pm \Lambda_q$ are. Again $\Lambda_1 + \Lambda_2 = \lambda_1 + \lambda_2$. As above $\Lambda_1 - \Lambda_2 = \lambda_i$ is impossible and $\Lambda_1 - \Lambda_2 = \lambda_i \pm \lambda_k$ $i \neq 1$, $k \neq 2$ gives $\nu = 4$ a case excluded by our assumption.[20] Hence $\Lambda_1 - \Lambda_2 = \lambda_1 - \lambda_2$ and again $\Lambda_1 = \lambda_1$.

Since the highest weight $\Lambda_1 = \lambda_1$ the representation by A^G is similar to that by A.[21] Thus there is a fixed matrix G in Ω_n such that $A^G = G^{-1}AG$ for all A. Since A^G and A are J-skew we have also $A^G = G^J A (G^J)^{-1}$ and hence GG^J commutes with all A. Since the system $\{A\}$ is irreducible $GG^J = \gamma 1$, $\gamma \neq 0$.

We may now prove the following theorem.

THEOREM 3. *If $\mathfrak{A}$ is a normal simple algebra, J an i. a. a. in $\mathfrak{A}$ and G an automorphism of the Lie algebra $\mathfrak{S}_J$ then there is a J-orthogonal element g such that $a^G = g^{-1}ag$ for all a.*[22]

We have seen in §3 that $\mathfrak{A}$ may be represented by matrices in Ω_n where Ω is the algebraic closure of Φ the field of $\mathfrak{A}$, in such a fashion that J induces in Ω_n a normalized i. a. a. of §2 and the linear combinations with coefficients in Ω of the matrices of $\mathfrak{S}_J$ is the system (*) = (1), (2a) or (2b). The automorphism G extends to an automorphism $\tilde{A} \to \tilde{A}^G$ of (*). We have seen that $\tilde{A}^G = \tilde{G}^{-1}\tilde{A}\tilde{G}$ where $\tilde{G}\tilde{G}^J = \tilde{\gamma}1$, $\tilde{\gamma} \neq 0$ in Ω. In particular for A of $\mathfrak{S}_J$ we have $A^G = \tilde{G}^{-1}A\tilde{G}$. Since the enveloping algebra of the matrices of $\mathfrak{S}_J$ is the algebra of matrices of $\mathfrak{A}$ we have $\tilde{G}^{-1}B\tilde{G}$ is a matrix of $\mathfrak{A}$ if B is and so $B \to \tilde{G}^{-1}B\tilde{G}$ is an automorphism of $\mathfrak{A}$. But the automorphisms of $\mathfrak{A}$ are inner and hence there is a matrix G of $\mathfrak{A}$ such that $\tilde{G}^{-1}B\tilde{G} = G^{-1}BG$. It follows that $\tilde{G} = \tilde{\rho}G$ and since $\tilde{G}$ is J-orthogonal, so is G.

THEOREM 4. *The group of automorphisms $\mathfrak{H}_J$ of $\mathfrak{S}_J \cong \mathfrak{G}_J/\mathfrak{D}$ where $\mathfrak{D}$ is the set of multiples $\delta 1$, $\delta \neq 0$ in Φ.*

Theorem 3 shows that $\mathfrak{G}_J$ is homomorphic to $\mathfrak{H}_J$. It is easily seen that the elements of $\mathfrak{G}_J$ which define the identity automorphism of $\mathfrak{S}_J$ are those of $\mathfrak{D}$. Hence $\mathfrak{H}_J \cong \mathfrak{G}_J/\mathfrak{D}$.

---

[20] Our proof breaks down for $\nu = 4$ since $\Lambda_1 = (\lambda_1 + \lambda_2 + \lambda_3 + \lambda_4)/2$, $\Lambda_2 = (\lambda_1 + \lambda_2 - \lambda_3 - \lambda_4)/2$, $\Lambda_3 = (\lambda_1 - \lambda_2 + \lambda_3 - \lambda_4)/2$ and $\Lambda_4 = (\lambda_1 - \lambda_2 - \lambda_3 + \lambda_4)/2$ are weights satisfying our conditions and $\pm\Lambda_p \pm \Lambda_q$ give all the roots $\pm\lambda_i \pm \lambda_j$. The result does not hold in this case as has been shown by E. Cartan: *Le principe de dualité*, Bull. Sci. Math. **49** (1925) p. 367.

[21] Weyl I, p. 281.

[22] This theorem gives an algebraic connection between $\mathfrak{S}_J$ and $\mathfrak{G}_J$. If Φ is the field of complex numbers an analytic connection between these systems is known by Lie's theory.

5. **Isomorphism of distinct $\mathfrak{S}_J$'s.** Let $\mathfrak{A}_1$ and $\mathfrak{A}_2$ be two normal simple algebras and J_1, J_2 i. a. a. in $\mathfrak{A}_1$ and $\mathfrak{A}_2$ respectively such that $\mathfrak{S}_{J_1} \cong \mathfrak{S}_{J_2}$. Then $\mathfrak{S}_{J_1\Omega} \cong \mathfrak{S}_{J_2\Omega}$ and since the algebras of type 2 that we are considering are not isomorphic to any of type 1, $\mathfrak{S}_{J_1}$ and $\mathfrak{S}_{J_2}$ have the same type and hence $\mathfrak{A}_1$ and $\mathfrak{A}_2$ have the same degree n. We have seen that $\mathfrak{A}_1$ and $\mathfrak{A}_2$ have representations in Ω_n such that the linear combinations in Ω of the matrices of $\mathfrak{S}_{J_1}$ or $\mathfrak{S}_{J_2}$ comprise the elements of one of the systems (1), (2a) or (2b) denoted as (*). The isomorphism $A_1 \to A_2$ defines an automorphism in (*) and hence by Theorem 3 $A_2 = \tilde{G}^{-1}A_1\tilde{G}$. Since the enveloping algebras of the matrices of $\mathfrak{S}_{J_1}$ and $\mathfrak{S}_{J_2}$ using linear combinations in Φ only are the matrices of $\mathfrak{A}_1$ and $\mathfrak{A}_2$, the correspondence $B_1 \to \tilde{G}^{-1}B_1\tilde{G} = B_2$ for any matrix B_1 of $\mathfrak{A}_1$ defines an isomorphism between $\mathfrak{A}_1$ and $\mathfrak{A}_2$. If we identify $\mathfrak{A}_1$ and $\mathfrak{A}_2$ and repeat the argument we see that there is an automorphism of the matrices of $\mathfrak{A}$, $B \to \tilde{G}^{-1}B\tilde{G}$ sending the J_1-skew matrices into the J_2-skew matrices. It follows from the corollary to Theorem 2 that J_1 and J_2 are cogredient.

THEOREM 5. *A necessary and sufficient condition that $\mathfrak{S}_{J_1} \cong \mathfrak{S}_{J_2}$ where J_1 and J_2 are i. a. a. in the normal simple algebras $\mathfrak{A}_1$ and $\mathfrak{A}_2$ respectively is that $\mathfrak{A}_1 \cong \mathfrak{A}_2$ and J_1 and J_2 be cogredient.*

6. **Connection with the theory of abstract Lie algebras.** We have seen that any normal simple algebra with an i. a. a. J leads to a simple Lie algebra which becomes one of the systems (*) when the field is sufficiently extended. In this section we shall prove the converse, that any Lie algebra $\mathfrak{L}$ that has the type (*) when the field is extended is isomorphic to an algebra $\mathfrak{S}_J$.

Let $\mathfrak{L}$ be a Lie algebra over Φ and Ω the algebraic closure of Φ. We suppose that $\mathfrak{L}_\Omega$ has one of the types (*). Thus if $a_1, a_2, \cdots, a_m$ is a basis for $\mathfrak{L}$ and hence also for $\mathfrak{L}_\Omega$ there exist elements h, e_α linear combinations of the a_i in Ω satisfying the relations (2). But these linear combinations involve only a finite number of elements of Ω and so there is an algebraic extension $\tilde{\Phi}$ of finite degree over Φ such that $\tilde{\mathfrak{L}} = \mathfrak{L}_{\tilde{\Phi}}$ has the type (*). We may suppose also that $\tilde{\Phi}$ is a Galois field with the Galois group $\mathfrak{G} = (i, s, t \cdots)$. Thus $\mathfrak{L}$ has a representation by matrices in $\tilde{\Phi}_n$ such that the linear combination of these matrices with coefficients in $\tilde{\Phi}$ constitute the system (*). The matrices A_i corresponding to a_i form a basis for (*) over Φ and if $[a_i, a_j] = \sum \gamma_{ijk} a_k$ $(\gamma_{ijk} \in \Phi)$ then $[A_i, A_j] = \sum \gamma_{ijk} A_k$. It is clear from the definition of the system (*) that if $\tilde{A} = (\tilde{\alpha}_{ij})$ belongs to the system then so does $\tilde{A}^s = (\alpha_{ij}^s)$ for any $s \in \mathfrak{G}$. In particular A_i^s belongs to the system, $A_1^s, \cdots, A_m^s$ are linearly independent relative to $\tilde{\Phi}$ and $[A_i^s, A_j^s] = \sum \gamma_{ijk} A_k^s$ since $\gamma_{ijk}^s = \gamma_{ijk} \in \Phi$. It follows that for each $s \in \mathfrak{G}$ the correspondence $\tilde{A} = \sum \tilde{\alpha}_i A_i \to \tilde{A}^S = \sum \tilde{\alpha}_i A_i^s$ is an automorphism of the Lie algebra (*). By Theorem 3 there exists a matrix $\tilde{G}_s$ in $\tilde{\Phi}_n$ such that $\tilde{A}^S = G_s^{-1}\tilde{A}G_s$. For the matrices $A = \sum \alpha_i A_i$ $(\alpha_i \in \Phi)$ representing the elements of $\mathfrak{L}$ we have $A^s = A^S$ and hence

$$A^s = G_s^{-1} A G_s \qquad s \in \mathfrak{G}. \tag{6}$$

The totality of matrices satisfying the equations (6) is an algebra $\mathfrak{A}$ over Φ containing the matrices of $\mathfrak{L}$. Since the enveloping algebra of the matrices of $\tilde{\mathfrak{L}}$ is $\tilde{\Phi}_n$ there are n^2 linearly independent matrices $A_1, \cdots, A_{n^2}$ of the form $A_{i_1}A_{i_2} \cdots A_{i_k}$ $(i_\beta = 1, \cdots, m)$ and these belong to $\mathfrak{A}$. Now suppose $\tilde{B} = \sum_1^n \beta_i A_i$ is in $\mathfrak{A}$. Then

$$\sum \beta_i^s A_i^s = \tilde{B}^s = \tilde{G}_s^{-1}\tilde{B}\tilde{G}_s = \sum \beta_i \tilde{G}_s^{-1} A_i \tilde{G}_s = \sum \beta_i A_i^s$$

and hence $\beta_i^s = \beta_i = \beta_i \in \Phi$. Thus $\mathfrak{A}$ consists of the totality of linear sums of $A_1, \cdots, A_{n^2}$ with coefficients in Φ, or $\mathfrak{A}$ is the enveloping algebra of the matrices of $\mathfrak{L}$ using coefficients in Φ only. Since $\tilde{\mathfrak{A}} = \mathfrak{A}_{\tilde{\Phi}} = \tilde{\Phi}_n$, $\mathfrak{A}$ is normal simple. Since $(A_{i_1}A_{i_2} \cdots A_{i_k})^J = (-1)^k A_{i_k} \cdots A_{i_2}A_{i_1} \in \mathfrak{A}$ $(i_\beta = 1, 2, \cdots, m)$, $\mathfrak{A}$ is carried into itself by the i. a. a. J defined in $\tilde{\mathfrak{A}}$ and hence J is an i. a. a. in $\mathfrak{A}$. The matrices of $\mathfrak{L}$ are precisely the J-skew elements of $\mathfrak{A}$. We have therefore proved:

THEOREM 6. *If $\mathfrak{L}$ is a Lie algebra such that $\mathfrak{L}_\Omega$ is isomorphic to one of the algebras* (1), (2a) *or* (2b) *then $\mathfrak{L}$ may be realized as the set of J-skew elements of a normal simple algebra $\mathfrak{A}$.*

7. Cogredience in a normal simple algebra. It has been shown by Albert[23] that if $\mathfrak{A}$ is an involutorial normal simple algebra, i.e. has an i. a. a. defined in it then $\mathfrak{A} = \mathfrak{F}_m$ where $\mathfrak{F}$ is an involutorial normal division algebra. Thus $\mathfrak{F}$ has exponent 2 and its order is a power of 2 and if Φ is suitably restricted (e.g. an algebraic field of the rationals, a p-adic field, or the real field) then $\mathfrak{F}$ is a quaternion algebra.[24] We consider, however, the general case of Φ arbitrary and shall investigate the condition that two i. a. a.'s in $\mathfrak{F}_n$ be cogredient. The discussion is similar to that given in the proof of Theorem 1.

We begin with a fixed i. a. a. $A = (a_{ij}) \rightarrow (\bar{a}_{ji}) = \bar{A}'$ where $a_{ij} \in \mathfrak{F}$ and $a \rightarrow \bar{a}$ is an i. a. a. in $\mathfrak{F}$. As before an arbitrary i. a. a. J in $\mathfrak{F}_n$ has the form $A^J = S^{-1}\bar{A}'S$ where $\bar{S}' = \pm S$, i.e. S is a hermitian or quasi-hermitian matrix. If $A^K = T^{-1}\bar{A}'T$ we have as before that K is cogredient to J if and only if $\tau T = \bar{V}'SV$, $\tau \in \Phi$. Thus the question of cogredience of i. a. a. is one of ordinary cogredience of hermitian or quasi-hermitian matrices with elements in a normal division algebra. We hope to discuss this problem in a later paper.

UNIVERSITY OF CHICAGO.

[23] Loc. cit. in 10, pp. 894–901.

[24] These results are due to Brauer, Albert, Hasse and E. Noether. For proofs and references see Deuring's *Algebren*.

Reprinted from DUKE MATHEMATICAL JOURNAL
Vol. 3, No. 3, September, 1937

A NOTE ON NON-ASSOCIATIVE ALGEBRAS

BY N. JACOBSON

It is the purpose of this note to obtain relations between an arbitrary algebra $\mathfrak{R}$ (not necessarily associative) and an algebra $\mathfrak{A}$ (necessarily associative) of linear transformations determined by $\mathfrak{R}$. If $\mathfrak{A}$ is simple, the centrum $\mathfrak{C}$ of $\mathfrak{A}$ is an algebraic field and $\mathfrak{R}$ may be regarded as an algebra over $\mathfrak{C}$. When this is done $\mathfrak{R}$ becomes a *normal simple algebra*, i.e., remains simple when this field is extended to its algebraic closure. A field having this property for algebras of characteristic 0 has been defined previously but less directly by Landherr.[1] Some of our results have been announced for Lie algebras of characteristic 0 by Albert.[2]

1. Let $\mathfrak{R}$ be an arbitrary algebra (not necessarily associative or commutative) with a finite basis over a commutative field Φ; $\mathfrak{R}$ is a finite dimensional vector space over Φ in which there is defined a composition xy of pairs of elements x, y such that

$$(x + y)z = xz + yz, \qquad z(x + y) = zx + zy, \tag{1}$$

$$(xy)\alpha = x(y\alpha) = (x\alpha)y, \qquad \alpha \in \Phi. \tag{2}$$

The mapping $x \to xa \equiv xA_r$ of $\mathfrak{R}$ on itself will be called the *right multiplication* determined by a. Equations (1) and (2) show that A_r is a linear transformation in the vector space $\mathfrak{R}$. Similarly we define the *left multiplication* determined by a as $x \to ax \equiv xA_l$. Let $\mathfrak{A}$ be the enveloping algebra of the left and right multiplications of $\mathfrak{R}$, i.e., the smallest algebra of linear transformations in $\mathfrak{R}$ containing all the left and right multiplications. The elements of $\mathfrak{A}$ are sums of terms of the type $A_{1i_1} \cdots A_{si_s}$ ($i_\alpha = r$ or l) where A_{ji_j} is a multiplication determined by a_j. We shall therefore denote an arbitrary element of $\mathfrak{A}$ by $\Sigma A_{1i_1} \cdots A_{si_s}$ (not summed on i_α!). Thus $\mathfrak{A}$ may also be defined as the smallest ring of linear transformations containing all the multiplications.

If $a_1, \cdots, a_n$ is a basis for $\mathfrak{R}$ over Φ and A is a linear transformation in this vector space, then A is completely determined by the matrix (α_{ij}) such that $a_jA = \Sigma a_i\alpha_{ij}$. The correspondence between A and the matrix (α_{ij}) determines, as is well known, a reciprocal isomorphism between the ring of all linear transformations in $\mathfrak{R}$ over Φ and the matrix ring Φ_n of all n-rowed square matrices

Received February 11, 1937; presented to the American Mathematical Society, April 9, 1937. The author is a National Research Fellow.

[1] W. Landherr, *Über einfache Liesche Ringe*, Hamb. Abhandlungen, vol. 11 (1935), pp. 41–64.

[2] Bull. Am. Math. Soc., vol. 41 (1935), p. 344.

with coördinates in Φ. In particular $\mathfrak{A}$ may be represented (reciprocally) as a subring of Φ_n.

If Σ is an extension of the field Φ, we define $\mathfrak{R}_\Sigma$ to be the algebra over Σ having the same basis as $\mathfrak{R}$ has over Φ, i.e., if $\mathfrak{R} = a_1\Phi + \cdots + a_n\Phi$, then $\mathfrak{R}_\Sigma = a_1\Sigma + \cdots + a_n\Sigma$. Any linear transformation A in $\mathfrak{R}$ over Φ has a unique extension to a linear transformation in $\mathfrak{R}_\Sigma$ over Σ. The matrix of A (relative to the basis $a_1, \cdots, a_n$) and of its extension are, of course, identical. It follows readily from the matrix representation that the enveloping algebra of the multiplications of $\mathfrak{R}_\Sigma$ is the extension algebra $\mathfrak{A}_\Sigma$.

As usual we define a (two-sided) *ideal* $\mathfrak{S}$ of $\mathfrak{R}$ as a subspace of $\mathfrak{S}$ such that $\mathfrak{S} \supset zx, xz$ for all $z \in \mathfrak{S}$ and $x \in \mathfrak{R}$. Thus $\mathfrak{S}$ is a subspace invariant under all the left and right multiplications and hence under all the transformations of $\mathfrak{A}$. $\mathfrak{R}$ is a *direct sum* of the ideals $\mathfrak{R}_1, \cdots, \mathfrak{R}_k$ ($\mathfrak{R} = \mathfrak{R}_1 \oplus \cdots \oplus \mathfrak{R}_k$) if every x in $\mathfrak{R}$ is expressible uniquely as $x_1 + \cdots + x_k$, x_i in $\mathfrak{R}_i$. This notion coincides with that of decomposability of $\mathfrak{R}$ relative to the system $\mathfrak{A}$. Since $x_ix_j \in$ the intersection $\mathfrak{R}_i \cap \mathfrak{R}_j = 0$, $x_ix_j = 0$ for any $x_i \in \mathfrak{R}_i$, $x_j \in \mathfrak{R}_j$. $\mathfrak{R}$ is *simple* if it has no proper ideal, or in other words, if $\mathfrak{A}$ is an irreducible system of linear transformations.[3]

THEOREM 1. *A necessary and sufficient condition that $\mathfrak{R}$ be a direct sum of simple algebras is that $\mathfrak{A}$ be a completely reducible system.*

If $\mathfrak{R} = \mathfrak{R}_1 \oplus \cdots \oplus \mathfrak{R}_k$, where the $\mathfrak{R}_i$ are simple, then the $\mathfrak{R}_i$ are irreducible subspaces and $\mathfrak{A}$ is completely reducible. Conversely if the $\mathfrak{R}_i$ are irreducible, they are simple. For let $\mathfrak{S}_i$ be an ideal relative to $\mathfrak{R}_i$, i.e., $z_ix_i, x_iz_i \in \mathfrak{S}_i$ for all $x_i \in \mathfrak{R}_i$, $z_i \in \mathfrak{S}_i$. Since $x_jz_i = z_ix_j = 0$ for $x_j \in \mathfrak{R}_j$ ($j \neq i$), we have xz_i, $z_ix \in \mathfrak{S}_i$, and so $\mathfrak{S}_i$ is an ideal of $\mathfrak{R}$ and hence an invariant subspace relative to $\mathfrak{A}$. This contradicts the irreducibility of $\mathfrak{R}_i$.

We recall that an algebra $\mathfrak{A}$ of linear transformations in a vector space $\mathfrak{R}$ is completely reducible if it is semi-simple. Suppose conversely that $\mathfrak{A}$ is completely reducible, say, $\mathfrak{R} = \mathfrak{R}_1 \oplus \cdots \oplus \mathfrak{R}_k$, where the $\mathfrak{R}_i$ are irreducible invariant subspaces, and let $\mathfrak{N}$ be a nilpotent ideal of $\mathfrak{A}$. If $\mathfrak{B}$ is any subalgebra of $\mathfrak{A}$, and $\mathfrak{S}$ a subspace of $\mathfrak{R}$, we denote the subspace of elements ΣyB, $y \in \mathfrak{S}$, $B \in \mathfrak{B}$ by $\mathfrak{S}\mathfrak{B}$. Since $(\mathfrak{S}\mathfrak{B})\mathfrak{A} = \mathfrak{S}(\mathfrak{B}\mathfrak{A})$, $\mathfrak{S}\mathfrak{B}$ is invariant if $\mathfrak{B}$ is a right ideal. In particular $\mathfrak{R}_i\mathfrak{N}$ is invariant, and since the $\mathfrak{R}_i$ are irreducible, either $\mathfrak{R}_i\mathfrak{N} = 0$ or $\mathfrak{R}_i\mathfrak{N} = \mathfrak{R}_i$. But $\mathfrak{R}_i\mathfrak{N} = \mathfrak{R}_i$ implies $\mathfrak{R}_i = \mathfrak{R}_i\mathfrak{N}^\rho = 0$ if ρ is sufficiently high. Thus $\mathfrak{R}_i\mathfrak{N} = 0$ and $\mathfrak{R}\mathfrak{N} = 0$, i.e., $\mathfrak{N} = 0$ and so $\mathfrak{A}$ is semi-simple. By Theorem 1 we have therefore

THEOREM 2. *A necessary and sufficient condition that $\mathfrak{R}$ be a direct sum of simple algebras is that $\mathfrak{A}$ be semi-simple.*

[3] For definitions of *irreducibility, direct sum, complete reducibility, equivalence* (operator-isomorphism) for systems of linear transformations, see van der Waerden's *Moderne Algebra*, vol. II, 1931, §108. We shall also require a number of results on the structure and representation of semi-simple algebras. These may be found in §§115, 116, 118, 119, 121 of van der Waerden's book.

An element z is an *absolute zero-divisor* if $z \neq 0$ and $zx = 0 = xz$ for all x in $\mathfrak{R}$. Consider $\mathfrak{R}_\Sigma$ where Σ is an extension of Φ and suppose that $z' = a_1\zeta_1' + \cdots + a_n\zeta_n'$ is an absolute zero-divisor. If $a_ia_j = \Sigma a_k\gamma_{kij}$ $(\gamma \in \Phi)$, then $z'a_i = a_iz' = 0$ implies that $\sum_i \gamma_{kij}\zeta_i' = 0$ and $\sum_j \gamma_{kij}\zeta_j' = 0$. Since these linear homogeneous equations have a non-trivial solution $\zeta_1', \cdots, \zeta_n'$ in Σ, they also have one, say, $\zeta_1, \cdots, \zeta_n$ in Φ, and so $\Sigma a_i\zeta_i$ is an absolute zero-divisor in $\mathfrak{R}$. Thus $\mathfrak{R}_\Sigma$ has absolute zero-divisors if and only if $\mathfrak{R}$ has. We note also that a simple algebra $\mathfrak{R}$ has no absolute zero-divisors unless $\mathfrak{R} = z\Phi$ where $z^2 = 0$. We suppose from now on that $\mathfrak{R}$ has no absolute zero-divisors. With this restriction we have

THEOREM 3. *If $\mathfrak{R} = \mathfrak{R}_1 \oplus \cdots \oplus \mathfrak{R}_k$ and the $\mathfrak{R}_i$ are simple, then $\mathfrak{A} = \mathfrak{A}_1 \oplus \cdots \oplus \mathfrak{A}_k$ and the $\mathfrak{A}_i$ are simple, and conversely. $\mathfrak{A}_i$ is the enveloping algebra of the left and right multiplications of $\mathfrak{R}_i$ (acting in $\mathfrak{R}$).*

Let $\mathfrak{R} = \mathfrak{R}_1 \oplus \cdots \oplus \mathfrak{R}_k$ and $\mathfrak{A}_i$ be the enveloping algebra of the multiplications of $\mathfrak{R}_i$. The elements of $\mathfrak{A}_i$ map $\mathfrak{R}_j$ $(j \neq i)$ on 0 and $\mathfrak{R}$ on a subspace $\neq 0$ of $\mathfrak{R}_i$. It follows directly that $\mathfrak{A} = \mathfrak{A}_1 \oplus \cdots \oplus \mathfrak{A}_k$. $\mathfrak{A}_i \neq 0$ since the elements of $\mathfrak{R}_i$ are not absolute zero-divisors. Since the transformations of $\mathfrak{A}$ map $\mathfrak{R}_j$ on 0, the algebra $\mathfrak{A}_i$ is isomorphic to the enveloping algebra of the multiplications of $\mathfrak{R}_i$ acting in $\mathfrak{R}_i$. The latter is simple since it is an irreducible system of linear transformations and hence $\mathfrak{A}_i$ is simple also. Conversely if $\mathfrak{A} = \mathfrak{A}_1 \oplus \cdots \oplus \mathfrak{A}_k$, where the $\mathfrak{A}_i$ are simple, $\mathfrak{A}$ is completely reducible and hence $\mathfrak{R} = \mathfrak{R}_1 \oplus \cdots \oplus \mathfrak{R}_{k'}$, where the $\mathfrak{R}_i$ are simple algebras. By the first part and the uniqueness of the decomposition of an algebra as a direct sum of simple algebras we conclude that $k = k'$ and $\mathfrak{A}_i$ is the enveloping algebra of the multiplications of $\mathfrak{R}_i$.

COROLLARY. *$\mathfrak{R}$ is simple if and only if $\mathfrak{A}$ is.*

2. Let $\mathfrak{C}$ denote the centrum of $\mathfrak{A}$. If $\mathfrak{R}$ is itself associative and has an identity, $\mathfrak{C}$ coincides with the multiplications determined by the elements of the centrum $\mathfrak{C}'$ of $\mathfrak{R}$. For if $c \in \mathfrak{C}'$, $C_r = C_l \in \mathfrak{C}$, and if $C \in \mathfrak{C}$ and $1C = c$, then $xC = (1x)C = (1C)x = cx = (x1)C = x(1C) = xc$, so that $c \in \mathfrak{C}'$ and $C = C_r = C_l$. If $\mathfrak{R}$ is associative but has no identity, we may adjoin an identity to it and repeat the argument. We then obtain the fact that $\mathfrak{C}$ is the algebra of linear transformations determined by the multiplications of $\mathfrak{C}'$ plus the identity mapping. When $\mathfrak{R}$ is arbitrary we shall call $\mathfrak{C}$ the *extended centrum* of $\mathfrak{R}$.

If $\mathfrak{R}$ is simple, so is $\mathfrak{A}$, and hence $\mathfrak{C} = \mathrm{P}$ is an algebraic field of finite order over Φ. If $\xi \in \mathrm{P}$,

$$(xy)\xi = (x\xi)y = x(y\xi),$$

and so $\mathfrak{R}$ may be regarded as an algebra over P.

An algebra $\mathfrak{R}$ over Φ will be called *normal simple* if $\mathfrak{R}_\Omega$, the algebra obtained by extending Φ to its algebraic closure Ω, is simple.

THEOREM 4. *$\mathfrak{R}$ is normal simple if and only if it is simple and its extended centrum consists of the multiples of the identity transformation.*

By hypothesis the centrum of the simple algebra $\mathfrak{A}$ consists of the Φ-multiples of 1. It is a well-known result that $\mathfrak{A}_\Omega$ is simple in this case. But $\mathfrak{A}_\Omega$ is the enveloping algebra of the multiplications of $\mathfrak{R}_\Omega$, and hence by the corollary to Theorem 3 the latter is simple. On the other hand, if $\mathfrak{C}$ is larger than Φ, $\mathfrak{A}_\Omega$ is not simple[4] and hence $\mathfrak{R}$ is not normal simple.

Thus if $\mathfrak{R}$ is an arbitrary simple algebra, it becomes normal simple when regarded as an algebra over its extended centrum.

THEOREM 5. *If $\mathfrak{R}$ is simple and has order n over its extended centrum* P, *then $\mathfrak{A} \cong \mathrm{P}_n$, the algebra of n-rowed square matrices with coefficients in* P.

We regard P as the underlying field. Since $\mathfrak{R}_\Omega$ is simple, $\mathfrak{A}$ is an absolutely irreducible system of linear transformations. It follows from Burnside's theorem that $\mathfrak{A}$ contains n^2 linearly independent linear transformations and hence is isomorphic to P_n.

More generally the structure of $\mathfrak{A}$ when $\mathfrak{R}$ is a direct sum of simple algebras may be deduced from Theorems 3 and 5.

3. Now suppose that $\mathfrak{R}$ is an associative algebra with an identity. It is well known that the right multiplications form an algebra $\mathfrak{R}_r$ isomorphic to $\mathfrak{R}$ and the left multiplications form an algebra $\mathfrak{R}_l$ reciprocally isomorphic to $\mathfrak{R}$. $\mathfrak{R}_l$ ($\mathfrak{R}_r$) is the totality of linear transformations in the vector space $\mathfrak{R}$ commutative with those of $\mathfrak{R}_r$ ($\mathfrak{R}_l$). Thus $\mathfrak{R}_r \cap \mathfrak{R}_l = \mathfrak{C}$.

If $\mathfrak{R}$ is normal simple, $\mathfrak{R}_r \cap \mathfrak{R}_l = 1\Phi$. If the order of $\mathfrak{R}$ over Φ, $(\mathfrak{R}:\Phi) = n$ by Theorem 5, $(\mathfrak{A}:\Phi) = n^2 = (\mathfrak{R}_r:\Phi)(\mathfrak{R}_l:\Phi)$. Thus $\mathfrak{A}$ is a direct product of $\mathfrak{R}_r$ and $\mathfrak{R}_l$ and so we have obtained an elementary proof of the following theorem due to Brauer:[5]

THEOREM 6. *The direct product of a normal simple algebra and its reciprocal algebra is a complete matric algebra.*

4. We return to the general case in which $\mathfrak{R}$ is not necessarily associative and suppose that $x \rightarrow x^S$ is an automorphism of $\mathfrak{R}$ over Φ, i.e.,

$$(x + y)^S = x^S + y^S, \qquad (x\alpha)^S = x^S\alpha, \qquad (xy)^S = x^S y^S,$$

and $x \rightarrow x^S$ is (1-1). If $P = \sum A_{1i_1} \cdots A_{si_s}$ ($i_\alpha = r$ or l) is an element of $\mathfrak{A}$, we define $P^S = \sum A^S_{1i_1} \cdots A^S_{si_s}$, where A^S_i is the right or left multiplication determined by a^S. P^S is independent of the representation of P. For if $\sum A_{1i_1} \cdots A_{si_s} = \sum B_{1j_1} \cdots B_{tj_t}$ ($j_\beta = r, l$), i.e.,

$$x(\sum A_{1i_1} \cdots A_{si_s}) = x(\sum B_{1j_1} \cdots B_{tj_t})$$

[4] If $\mathfrak{C}$ is separable, $\mathfrak{A}_\Omega$ is semi-simple though not simple, and if $\mathfrak{C}$ is inseparable, $\mathfrak{A}_\Omega$ has a radical. Cf. van der Waerden, loc. cit., §119.

[5] R. Brauer, *Über Systeme hyperkomplexer Zahlen*, Math. Zeits., vol. 30 (1929), p. 103.

for all x, then

$$x^S(\sum A^S_{1i_1} \cdots A^S_{si_s}) = x^S(\sum B^S_{1j_1} \cdots B^S_{tj_t})$$

for all x^S. Since x^S ranges over all of $\mathfrak{R}$ when x does, we have $\sum A^S_{1i_1} \cdots A^S_{si_s} = \sum B^S_{1j_1} \cdots B^S_{tj_t}$. It follows that the correspondence $P \to P^S$ is an automorphism of $\mathfrak{A}$ and $(xP)^S = x^S P^S$.

Any automorphism of an associative algebra induces an automorphism in its centrum. Hence if $\mathfrak{R}$ is simple, an automorphism of $\mathfrak{R}$ defines an automorphism in the field $\mathfrak{C} = \mathrm{P}$, the extended centrum.

THEOREM 7. *If $\mathfrak{R}$ is normal simple over* P *and* $\mathrm{P} \supset \Phi$ *such that* $(\mathrm{P}:\Phi)$ *is finite, then the automorphisms of $\mathfrak{R}$ over Φ have the property* $(x\xi)^S = x^S \xi^S$, *where* $\xi \in \mathrm{P}$ *and* $\xi \to \xi^S$ *is an automorphism of* P.

Let $\mathfrak{G}$ be the group of automorphisms of $\mathfrak{R}$ over Φ and $\mathfrak{X}$ the subgroup consisting of the automorphisms of $\mathfrak{R}$ over P. Theorem 7 shows that $\mathfrak{X}$ is an invariant subgroup of $\mathfrak{G}$ and that $\mathfrak{G}/\mathfrak{X}$ is isomorphic to a subgroup of the Galois group $\mathfrak{g}$ of P over Φ.

Now suppose that $\mathfrak{R} = \mathfrak{R}_0 \times \mathrm{P}$ ($= \mathfrak{R}_{0\mathrm{P}}$ regarded as an algebra over Φ) where $\mathfrak{R}_0$ is a normal simple algebra over Φ. If $a_1, \cdots, a_n$ is a basis for $\mathfrak{R}_0$ over Φ or for $\mathfrak{R}$ over P and S is an element of $\mathfrak{g}$, then the correspondence $x = \sum a_i \xi_i \to \sum a_i \xi_i^S = x^{S_1}$ is an automorphism of $\mathfrak{R}$ such that $(x\xi)^{S_1} = x^{S_1}\xi^{S_1} = x^{S_1}\xi^S$. Let $\mathfrak{G}_1$ denote the subgroup of $\mathfrak{G}$ consisting of the elements S_1. Evidently $\mathfrak{G}_1 \cong \mathfrak{g}$ and $\mathfrak{G}_1 \cap \mathfrak{X} = I$ the identity mapping. By Theorem 7 any element of $\mathfrak{G}$ has the form $S_1 H$ where $S_1 \in \mathfrak{G}_1$ and $H \in \mathfrak{X}$. Hence $\mathfrak{G}/\mathfrak{X} \cong \mathfrak{G}_1 \cong \mathfrak{g}$. This result may be used, as we shall show in another paper, to determine the automorphisms of simple Lie algebras and simple continuous groups.

UNIVERSITY OF CHICAGO

ABSTRACT DERIVATION AND LIE ALGEBRAS*

BY

NATHAN JACOBSON†

The purpose of this paper is the investigation of the algebraic properties of the set of operations mapping an algebra on itself and having the formal character of derivation in the field of analytic functions. Some of the results obtained are analogous to well-known theorems on automorphisms of algebras.‡ The considerations in I are general and quite elementary. In II and III we restrict ourselves to the derivations of an associative algebra having a finite basis and in the main to semi-simple algebras. A number of results of the theory of algebras are presupposed. These may be found in Deuring's *Algebren*, Springer, 1935.

I. Derivations in an arbitrary algebra

1. Let $\mathfrak{R}$ be an arbitrary algebra (hypercomplex system not necessarily commutative or associative, or of finite order) over a commutative field $\mathfrak{F}$. Then $\mathfrak{R}$ is a vector space (with elements $x, y, \cdots$) over $\mathfrak{F}$ (with elements $\alpha, \beta, \cdots$) in which a composition $xy \epsilon \mathfrak{R}$ is defined such that

$$(1)\quad (x+y)z = xz + yz,\quad z(x+y) = zx + zy,\quad (xy)\alpha = (x\alpha)y = x(y\alpha).$$

A *derivation* D of $\mathfrak{R}$ is a single valued mapping of $\mathfrak{R}$ on itself such that

$$(2)\ \text{(a)}\ (x+y)D = xD + yD,\ \text{(b)}\ (x\alpha)D = (xD)\alpha,\ \text{(c)}\ (xy)D = (xD)y + x(yD).$$

Thus D is a linear transformation in the vector space $\mathfrak{R}$ satisfying the special condition (2c). It is well known that the sum D_1+D_2, difference D_1-D_2, scalar product $D\alpha$ and product D_1D_2 (defined respectively by $x(D_1 \pm D_2) \equiv xD_1 \pm xD_2$, $x(D\alpha) \equiv (xD)\alpha$, $x(D_1D_2) \equiv ((xD_1)D_2)$ of linear transformations are linear transformations. If D, D_1, D_2 are derivations we have besides

$$(3)\quad \begin{aligned}(xy)(D_1 \pm D_2) &= (xy)D_1 \pm (xy)D_2 = (xD_1)y + x(yD_1) \pm (xD_2)y \pm x(yD_2)\\ &= (x(D_1 \pm D_2))y + x(y(D_1 \pm D_2)),\end{aligned}$$

$$(4)\quad (xy)D\alpha = ((xy)D)\alpha = ((xD)y + x(yD))\alpha = (xD\alpha)y + x(yD\alpha),$$

* Presented to the Society, December 31, 1936; received by the editors November 6, 1936.

† National Research Fellow.

‡ A direct connection between derivations and automorphisms may sometimes be established. For example if $\mathfrak{R}$ is the ring of polynomials $\mathfrak{F}[x]$ where $\mathfrak{F}$ is a field of characiterstic 0, and D is defined by $f(x)D=f'(x)$ the usual derivative then $\exp D = 1+D+D^2/2!+\cdots$ is an automorphism since $f(x)\exp D = f(x+1)$.

$$(5) \quad \begin{aligned}(xy)D_1D_2 &= ((xy)D_1)D_2 = ((xD_1)y + x(yD_1))D_2 \\ &= (xD_1D_2)y + x(yD_1D_2) + (xD_1)(yD_2) + (xD_2)(yD_1).\end{aligned}$$

Thus $D_1 \pm D_2$, $D\alpha$ are derivations, but not in general D_1D_2. However (5) shows that the commutator $[D_1, D_2] = D_1D_2 - D_2D_1$ does satisfy (2c) and so is a derivation. We recall the relations

$$(6) \quad [D_1, D_2] = -[D_2, D_1], \; [[D_1, D_2], D_3] + [[D_2, D_3], D_1] + [[D_3, D_1], D_2] = 0.$$

As a consequence of (2) we have Leibniz's formula:

$$(7) \quad (xy)D^k = (xD^k)y + C_{k,1}(xD^{k-1})(yD) + C_{k,2}(xD^{k-2})(yD^2) + \cdots + x(yD^k).$$

Hence if $\mathfrak{F}$ has characteristic $p \neq 0$ we have

$$(8) \quad (xy)D^p = (xD^p)y + x(yD^p);$$

i.e., D^p is a derivation.

By *a restricted Lie algebra of linear transformations* we shall mean a system of linear transformations closed relative to the operations of addition, subtraction, scalar multiplication, commutation, and taking pth powers, if p ($=0$ or a prime) is the characteristic of the field over which the vector space is defined.* With this definition we have

THEOREM 1. *The derivations of an algebra $\mathfrak{R}$ over $\mathfrak{F}$ constitute a restricted Lie algebra $\mathfrak{D}$ of linear transformations in $\mathfrak{R}$.*

We call $\mathfrak{D}$ the *derivation algebra* or, more briefly, the *d-algebra* of $\mathfrak{R}$ over $\mathfrak{F}$. It should be noted that we are regarding $\mathfrak{D}$ as an algebra over $\mathfrak{F}$.

2. Suppose $D, E, D_1, D_2, \cdots$ are elements of any associative algebra $\mathfrak{A}$. As a generalization of the multinomial theorem in a commutative algebra we have

$$(9) \quad (D_1 + D_2 + \cdots + D_r)^k = \sum \begin{Bmatrix} D_1 & D_2 & \cdots & D_r \\ j_1 & j_2 & \cdots & j_r \end{Bmatrix},$$

where the summation is extended over $j_1, \cdots, j_r$ such that $j_\alpha \geqq 0$ and $j_1 + \cdots + j_r = k$ and where $\{D_1 \cdots D_r / j_1 \cdots j_r\}$ denotes the sum of the $(j_1 + \cdots + j_r)!/(j_1! \cdots j_r!)$ terms obtained by multiplying j_1 of the D_1's, j_2 of the D_2's, $\cdots, j_r$ of the D_r's together in every possible order. Let $D_{i_1} + D_{i_2} + \cdots + D_{i_s} = D_{i_1 i_2 \cdots i_s}$ where $i_1, i_2, \cdots, i_s$ are distinct and have values in the range $1, 2, \cdots, k$. Consider

$$D^k_{i_1 \cdots i_k} - \sum_C D^k_{i_1 \cdots i_{k-1}} + \cdots + (-1)^{k-1} \sum_C D^k_{i_1} = Q,$$

* We use the convention $D^0 = 0$.

where $\sum_C D^k_{i_1\cdots i_s}$ denotes the sum of the $C_{k,s}$ terms obtained by letting $i_1, \cdots, i_s$ run through all the combinations of $1, 2, \cdots, k$ taken s at a time. By (9), Q is a sum of terms of the form $\{D_{m_1}\cdots D_{m_t}/j_1\cdots j_t\}$ where $j_\alpha > 0$ and $j_1+j_2+\cdots+j_t=k$. Since

$$\begin{Bmatrix} D_{m_1} & \cdots & D_{m_t} \\ j_1 & \cdots & j_t \end{Bmatrix} = \begin{Bmatrix} D_{m_1} & \cdots & D_{m_t} & D_{n_1} & \cdots & D_{n_r} \\ j_1 & \cdots & j_t & 0 & \cdots & 0 \end{Bmatrix},$$

where $n_1, n_2, \cdots, n_r$ are distinct indices different from $m_1, m_2, \cdots, m_r$, the term $\{D_{m_1}\cdots D_{m_t}/j_1\cdots j_t\}$ has the coefficient $C_{k-t,r}$ in $\sum_C D^k_{i_1\cdots i_{t+r}}$ and hence the coefficient of this term in Q is

$$C_{k-t,k-t} - C_{k-t,k-t-1} + \cdots + (-1)^{k-t}C_{k-t,0} = \delta_{kt},$$

i.e., $=0$ or 1 according as $k\neq t$ or $k=t$. Hence

$$(10)\qquad D^k_{i_1\cdots i_k} - \sum_C D^k_{i_1\cdots i_{k-1}} + \cdots + (-1)^k \sum_C D^k_{i_1} = \begin{Bmatrix} D_1 & \cdots & D_k \\ 1 & \cdots & 1 \end{Bmatrix}.^*$$

Since

$$\begin{Bmatrix} D_1+\cdots+D_r & D \\ k & 1 \end{Bmatrix} = \sum \begin{Bmatrix} D_1 & \cdots & D_r & D \\ j_1 & \cdots & j_r & 1 \end{Bmatrix},$$

where $j_\alpha \geqq 0$ and $j_1+\cdots+j_r=k$, we may derive the following formula similar to (10):

$$(11)\qquad \begin{Bmatrix} D_{i_1\cdots i_k} & D \\ k & 1 \end{Bmatrix} - \cdots + (-1)^{k-1}\sum_C \begin{Bmatrix} D_{i_1} & D \\ k & 1 \end{Bmatrix} = \begin{Bmatrix} D_1 & \cdots & D_k & D \\ 1 & \cdots & 1 & 1 \end{Bmatrix}.$$

If in (10) and (11) we set j_1 of the D's equal to D_1, j_2 equal to $D_2, \cdots, j_l$ equal to D_l then $\{D_1\cdots D_k/1\cdots 1\}$ and $\{D_1\cdots D_kD/1\cdots 11\}$ become respectively $(j_1!\cdots j_l!)\{D_1\cdots D_l/j_1\cdots j_l\}$ and $(j_1!\cdots j_l!)\{D_1\cdots D_lD/j_1\cdots j_l1\}$ and we obtain expressions for these as sums of kth powers and as sums of terms of the type $\{ED/k1\}$.

An analogue of (7) is

$$(12)\qquad DE^k = E^kD + C_{k,1}E^{k-1}D' + \cdots + D^{(k)}$$

where $D'=[D, E], \cdots, D^{(j)}=[D^{(j-1)}, E]$. Hence

$$E^lDE^{k-l} = E^kD + C_{k-l,1}E^{k-1}D' + \cdots + C_{k-l,j}E^{k-j}D^{(j)} + \cdots + E^lD^{(k-l)},$$

and summing on $l=0, 1, \cdots, k$ we have

* If we set $D_1=D_2=\cdots=D_k=1$ in (10) we obtain the identity
$$k^k - C_{k,1}(k-1)^k + C_{k,2}(k-2)^k - \cdots + (-1)^{k-1}C_{k,k-1}1^k = k!.$$

$$\left\{\begin{matrix} E & D \\ k & 1 \end{matrix}\right\} = C_{k+1,1}E^kD + \cdots + C_{k+1,j+1}E^{k-j}D^{(j)} + \cdots + D^{(k)}, \tag{13}$$

since

$$C_{k,j} + C_{k-1,j} + \cdots + C_{j,j} = C_{k+1,j+1}.$$

If the characteristic of $\mathfrak{A}$ is $p \neq 0$ special cases of (12) and (13) are

$$\text{(a)} \quad [D, E^p] = D^{(p)}, \qquad \text{(b)} \quad \left\{\begin{matrix} E & D \\ p-1 & 1 \end{matrix}\right\} = D^{(p-1)}. \tag{14}$$

Equations (11) and (14b) show that $\{D_1 \cdots D_p/1 \cdots 1\}$ is expressible as a linear combination of $(p-1)$-fold commutators, i.e., of the type $D^{(p-1)}$ where $D = D_p$ and E is a sum of the other D_i's. Hence we see also that $(j_1! \cdots j_l!)\ \{D_1 \cdots D_l/j_1 \cdots j_l\}$ where $j_1 + \cdots + j_l = p$ is a linear sum of $(p-1)$-fold commutators. If no $j_i = p$, $(j_1! \cdots j_l!) \not\equiv 0 \pmod p$ and so $\{D_1D_2 \cdots D_l/j_1j_2 \cdots j_l\}$ is a linear sum of $(p-1)$-fold commutators and (9) becomes

$$(D_1 + D_2 + \cdots + D_r)^p = D_1^p + D_2^p + \cdots + D_r^p + S, \tag{15}$$

where S is a linear sum of $(p-1)$-fold commutators.

3. If $\mathfrak{D}$ is any system of linear transformations we define the *enveloping algebra* $\mathfrak{A}$ of $\mathfrak{D}$ to be the totality of linear combinations of products of a finite number of elements of $\mathfrak{D}$. We call k the degree of the monomial $D_1D_2 \cdots D_k$, $D_i \epsilon \mathfrak{D}$. Suppose $\mathfrak{D}$ is a Lie algebra of linear transformations and consider $D_1D_2 \cdots D_k$ where $k < p$ if $p \neq 0$ and arbitrary if $p = 0$. We have

$$D_1 \cdots D_{i-1}D_{i+1}D_iD_{i+2} \cdots D_k = D_1D_2 \cdots D_k + D_1 \cdots D_{i-1}D'D_{i+2} \cdots D_k,$$

where $D' = [D_{i+1}, D_i] \epsilon \mathfrak{D}$. Since any arrangement $i_1i_2 \cdots i_k$ of $1, 2, \cdots, k$ may be obtained from $1, 2, \cdots, k$ by a sequence of transpositions of adjacent indices

$$D_{i_1}D_{i_2} \cdots D_{i_k} = D_1D_2 \cdots D_k + R,$$

where R is a sum of terms of degree $< k$. Hence

$$\left\{\begin{matrix} D_1 & D_2 & \cdots & D_k \\ 1 & 1 & \cdots & 1 \end{matrix}\right\} = (k!)D_1D_2 \cdots D_k + S,$$

where degree of $S < k$. Since the left-hand side of this equation is expressible by (10) as a sum of kth powers of elements in $\mathfrak{D}$ and $k! \not\equiv 0 \pmod p$, we have by induction that $D_1D_2 \cdots D_k$ is a linear combination of lth powers of elements of $\mathfrak{D}$ where $l \leq k$.

THEOREM 2. *If $\mathfrak{D}$ is a Lie algebra of linear transformations the elements in the enveloping algebra $\mathfrak{A}$ of degree $k<p$ if $p\neq 0$ and of arbitrary degree if $p=0$ are expressible as linear combinations of lth powers $l\leqq k$, of elements of $\mathfrak{D}$.**

If $\mathfrak{D}$ is restricted (10) shows that $\{D_1D_2\cdots D_r/j_1j_2\cdots j_r\}\epsilon\mathfrak{D}$ if $j_1+j_2+\cdots+j_r=p$ and $D_i\epsilon\mathfrak{D}$. This transformation is also expressible as a sum of $(p-1)$-fold commutators of elements of $\mathfrak{D}$. Since $(D_1+D_2+\cdots+D_r)^{p^k}=((D_1+D_2+\cdots+D_r)^{p^{k-1}})^p$ for $j_1, j_2, \cdots, j_r$ such that $j_1+j_2+\cdots+j_r=p^k$, we have

$$\begin{Bmatrix} D_1 & D_2 & \cdots & D_r \\ j_1 & j_2 & \cdots & j_r \end{Bmatrix} = \sum \left\{ \begin{matrix} \begin{Bmatrix} D_1 & D_2 & \cdots & D_r \\ k_{11} & k_{12} & \cdots & k_{1r} \end{Bmatrix} & \begin{Bmatrix} D_1 & D_2 & \cdots & D_r \\ k_{21} & k_{22} & \cdots & k_{2r} \end{Bmatrix} & \cdots \\ m_1 & m_2 & \cdots \end{matrix} \right\},$$

where the summation is extended over the non-negative integers such that the ordered set $(k_{l1}, k_{l2}, \cdots, k_{lr})\neq(k_{m1}, k_{m2}, \cdots, k_{mr})$ for $l\neq m$ and

$$k_{l1}+k_{l2}+\cdots+k_{lr}=p^{k-1} \qquad (l=1, 2, \cdots),$$

$$m_1+m_2+\cdots=p,$$

$$k_{1i}m_1+k_{2i}m_2+\cdots=j_i \qquad (i=1, 2, \cdots, r).$$

Hence we see by induction on k that $\{D_1D_2\cdots D_r/j_1j_2\cdots j_r\}\epsilon\mathfrak{D}$ *for all* $j_1, j_2, \cdots$ *such that* $j_1+j_2+\cdots+j_r=p^k$.

4. Because of (14a) we are led to the definition: A *restricted Lie Algebra* $\mathfrak{R}$ *of characteristic* p ($=0$ or not) is an algebra (i.e., satisfies (1)) in which the composition $[x, y]$ (in place of xy) satisfies

$$[x, y] \doteq -[y, x], \tag{16}$$

$$[[x, y], z]+[[y, z], x]+[[z, x], y]=0, \tag{17}$$

for every y there exists an element denoted as y^p such that

$$[\cdots[[x, \overbrace{y]y]\cdots y}^{p}]=[x, y^p] \tag{18}$$

for all x. A *restricted subalgebra* $\mathfrak{S}$ of $\mathfrak{R}$ is a subalgebra containing y^p for every y in $\mathfrak{S}$. Similarly we define restricted ideal, etc.†

Suppose $\mathfrak{R}$ is an associative algebra. We may define a new composition $[x, y]=xy-yx$ in terms of xy defined in $\mathfrak{R}$. It is readily verified that $\mathfrak{R}$ is a

* This is a slight extension of a result announced recently by M. Zorn (Bulletin of the American Mathematical Society, vol. 42 (1936), p. 485). Cf. H. Poincaré, *Sur les groupes continus*, Cambridge Philosophical Transactions, vol. 18 (1899), pp. 220–255.

† For definitions of the important concepts in the theory of Lie algebras the reader is referred to Jacobson, *Rational methods in the theory of Lie algebras*, Annals of Mathematics, vol. 36 (1935), pp. 875–881.

restricted Lie algebra if y^p is defined as the pth power of y in $\mathfrak{R}$. We shall call this Lie algebra *the restricted Lie algebra determined by the associative* $\mathfrak{R}$.

5. If $\mathfrak{R}$ is any algebra the mapping a_r: $x \rightarrow xa$ is a linear transformation and will be called the *right multiplication* determined by a. Suppose D is a derivation in $\mathfrak{R}$. Equation (2c) gives the commutation relation

$$[a_r, D] = (aD)_r. \tag{19}$$

Similarly we define a_l as $x \rightarrow ax$ and call this mapping the *left multiplication* determined by a. In place of (19) we have $[a_l, D] = (aD)_l$. If $\mathfrak{R}$ is a Lie algebra $a_r = -a_l$ and, by (16) and (17),

$$[x, y]a_r = [xa_r, y] + [x, ya_r].$$

Thus a_r is a derivation which we call *inner*.

THEOREM 3. *The totality of inner derivations of a (restricted) Lie algebra* $\mathfrak{R}$ *is a (restricted) ideal* $\mathfrak{J}$ *in the d-algebra* $\mathfrak{D}$ *of* $\mathfrak{R}$. $\mathfrak{J} \cong \mathfrak{R}/\mathfrak{C}$ *where* $\mathfrak{C}$ *is the centrum of* $\mathfrak{R}$.*

If a_r and b_r are multiplications associated with a and b it follows directly from the definition of $\mathfrak{R}$ that $a_r \pm b_r = (a \pm b)_r$, $a_r\alpha = (a\alpha)_r$, $[a_r, b_r] = [a, b]_r$ and if $\mathfrak{R}$ is restricted $(a_r)^p = (a^p)_r$. Hence $\mathfrak{J}$ is a subalgebra of $\mathfrak{D}$ and is restricted if $\mathfrak{R}$ is. Furthermore the correspondence $a \rightarrow a_r$ is a homomorphism between $\mathfrak{R}$ and $\mathfrak{J}$. Since the elements of $\mathfrak{C}$ are the ones corresponding to 0 in this homomorphism $\mathfrak{R}/\mathfrak{C} \cong \mathfrak{J}$. Equation (19) shows that $\mathfrak{J}$ is an ideal.

Suppose $\mathfrak{R}$ is associative and D a derivation. D is also a derivation in the restricted Lie algebra determined by $\mathfrak{R}$. Hence the d-algebra of $\mathfrak{R}$ as an associative algebra is a restricted subalgebra of the d-algebra of $\mathfrak{R}$ as a Lie algebra. Moreover the inner derivations $x \rightarrow [x, a]$ are derivations of the associative $\mathfrak{R}$ since

$$[xy, a] = [x, a]y + x[y, a].$$

Thus $\mathfrak{J}$ is a restricted ideal in the d-algebra of the associative $\mathfrak{R}$.

If $\mathfrak{R}$ is associative, $\mathfrak{D}$ its d-algebra, $D \epsilon \mathfrak{D}$ and $c \epsilon \mathfrak{C}$ the centrum of $\mathfrak{R}$ then $c_r = c_l \equiv c$ and it is easily verified that $Dc \epsilon \mathfrak{D}$ also. Hence $\mathfrak{D}$ has $\mathfrak{C}$ as well as $\mathfrak{F}$ as a set of multipliers under which it is invariant. A subalgebra $\mathfrak{E}$ of $\mathfrak{D}$ which contains with every element E also Ec for every c in $\mathfrak{C}$ will be called a $\mathfrak{C}$-*subalgebra* of $\mathfrak{D}$.

If $\mathfrak{R}$ is arbitrary, $D \epsilon \mathfrak{D}$ the elements $k \epsilon \mathfrak{R}$ such that $kD = 0$ are called *D-constants*. Their totality is a subalgebra. If $kD = 0$ for all D then k is a *constant*. If $\mathfrak{R}$ has an identity 1 we have $1^2 = 1$ and hence $1(1D) + (1D)1 = 1D$ or $1D = 0$

* The *centrum* is the set of elements c such that $[c, x] = 0$ for all x in $\mathfrak{R}$.

so that 1 is a constant. More generally if $\mathfrak{D}_1$ is a subalgebra of $\mathfrak{D}$ we denote the set of elements k in $\mathfrak{R}$ such that $kD_1=0$ for all $D_1\epsilon\mathfrak{D}_1$ by $\mathfrak{R}(\mathfrak{D}_1)$. $\mathfrak{R}(\mathfrak{D}_1)$ is a subalgebra. On the other hand if $\mathfrak{R}_1$ is a subalgebra of $\mathfrak{R}$ we define $\mathfrak{D}(\mathfrak{R}_1)$ to be the set of derivations E such that $x_1E=0$ for all $x_1\epsilon\mathfrak{R}_1$. $\mathfrak{D}(\mathfrak{R}_1)$ is a restricted subalgebra of $\mathfrak{D}$. Evidently $\mathfrak{D}(\mathfrak{R}(\mathfrak{D}_1))\supset\mathfrak{D}_1$ and $\mathfrak{R}(\mathfrak{D}(\mathfrak{R}_1))\supset\mathfrak{R}_1$. If $\mathfrak{R}$ is associative with centrum $\mathfrak{C}$, $\mathfrak{D}(\mathfrak{R}_1)$ is a restricted $\mathfrak{C}$-subalgebra of $\mathfrak{D}$.

$\mathfrak{S}$ is a *characteristic subalgebra* of $\mathfrak{R}$ if it is mapped on itself by every element of $\mathfrak{D}$. The subalgebra of constants $\mathfrak{R}_0$, the centrum $\mathfrak{C}$ and the powers of $\mathfrak{R}$ are characteristic. If $\mathfrak{S}$ is characteristic, $\mathfrak{D}(\mathfrak{S})$ is an ideal. In particular $\mathfrak{D}(\mathfrak{C})$ is an ideal containing $\mathfrak{J}$ if $\mathfrak{R}$ is associative or a Lie algebra. The derivations mapping $\mathfrak{R}$ on the characteristic subalgebra $\mathfrak{S}$ also form a restricted ideal $\mathfrak{G}$. In the case of a Lie algebra or an associative algebra the ideal associated in this way with $\mathfrak{C}$ is the *annihilator* of $\mathfrak{J}$, i.e., the set of elements G such that $[a_r, G]=0$ for all a_r. This is an immediate consequence of (19).

II. Derivations in an associative algebra with a finite basis

6. In the remainder of the paper $\mathfrak{R}$ will denote an associative algebra with a finite basis over $\mathfrak{F}$. We propose to study the d-algebra $\mathfrak{D}$ of $\mathfrak{R}$.

Theorem 4. *If* $\mathfrak{R}=\mathfrak{R}_1\oplus\mathfrak{R}_2$ *and* $\mathfrak{R}_1^2=\mathfrak{R}_1$, $\mathfrak{R}_2^2=\mathfrak{R}_2$ *then* $\mathfrak{D}=\mathfrak{D}_1\oplus\mathfrak{D}_2$ *where* $\mathfrak{D}_i$ *is isomorphic to the d-algebra of* $\mathfrak{R}_i$.

$\mathfrak{R}_1$ is characteristic; for $\mathfrak{R}_1^2=\mathfrak{R}_1$ and so the arbitrary element x_1 of $\mathfrak{R}_1$ has the form $\sum y_1z_1$, $y_1, z_1\epsilon\mathfrak{R}_1$. Hence $x_1D=\sum(y_1z_1)D=\sum(y_1D)z_1+\sum y_1(z_1D)\epsilon\mathfrak{R}_1$ since this is an ideal. Similarly $\mathfrak{R}_2$ is characteristic. Let $\mathfrak{D}_i$ be the ideals mapping $\mathfrak{R}$ onto $\mathfrak{R}_i$. Since $\mathfrak{R}_1\cap\mathfrak{R}_2=0$, $\mathfrak{D}_1\cap\mathfrak{D}_2=0$ and hence $[\mathfrak{D}_1, \mathfrak{D}_2]\subset\mathfrak{D}_1\cap\mathfrak{D}_2=0$.† If $x=x_1+x_2$, $x_i\epsilon\mathfrak{R}_i$ and D any derivation, the mappings $x\rightarrow x_1D\equiv xD_1$ and $x\rightarrow x_2D\equiv xD_2$ are derivations in $\mathfrak{D}_1$ and $\mathfrak{D}_2$ respectively. Since $D=D_1+D_2$, $\mathfrak{D}=\mathfrak{D}_1\oplus\mathfrak{D}_2$. The isomorphism between $\mathfrak{D}_1$ and the d-algebra of $\mathfrak{R}_1$ follows directly from the fact that the transformations of $\mathfrak{D}_1$ induce all the derivations in $\mathfrak{R}_1$ and map $\mathfrak{R}_2$ into 0. Similarly $\mathfrak{D}_2$ is isomorphic to the d-algebra of $\mathfrak{R}_2$.

Let $x_1, x_2, \cdots, x_r$ be a basis for $\mathfrak{R}$ over $\mathfrak{F}$ ($\mathfrak{R}=x_1\mathfrak{F}+x_2\mathfrak{F}+\cdots+x_r\mathfrak{F}$) and suppose $x_ix_j=\sum x_p\gamma_{pij}$, $\gamma_{pij}\epsilon\mathfrak{F}$. If D is a derivation in $\mathfrak{R}$ and

$$(x_1D, x_2D, \cdots, x_rD) = (x_1, x_2, \cdots, x_r)\Delta,\ \Delta = (\alpha_{ij}),\ \alpha_{ij}\epsilon\mathfrak{F},$$

then the condition $(x_ix_j)D=(x_iD)x_j=x_i(x_jD)$ gives

$$\sum_p \alpha_{kp}\gamma_{pij} = \sum_p \gamma_{kpj}\alpha_{pi} + \sum_p \gamma_{kip}\alpha_{pj} \qquad (i, j, k = 1, 2, \cdots, r), \tag{20}$$

a set of n^3 linear homogeneous equations for the coordinates α_{ij} of Δ. Con-

† $[\mathfrak{A},\mathfrak{B}]$ denotes the smallest subspace of $\mathfrak{D}$ containing all the elements $[A, B]$, where $A\epsilon\mathfrak{A}, B\epsilon\mathfrak{B}$.

versely if Δ is any matrix whose coordinates satisfy (20) the linear transformation D determined by Δ satisfies $(x_i x_j)D=(x_iD)x_j+x_i(x_jD)$ for all i, j and hence $(xy)D=(xD)y+x(yD)$ for all x, y, i.e., D is a derivation. Now suppose $\mathfrak{K}$ is a field containing $\mathfrak{F}$ and let $\mathfrak{N}_{\mathfrak{K}}=x_1\mathfrak{K}+x_2\mathfrak{K}+\cdots+x_r\mathfrak{K}$ and $\mathfrak{D}^*$ be the d-algebra of $\mathfrak{N}_{\mathfrak{K}}$ (over $\mathfrak{K}$). Evidently the matrix Δ also determines a derivation D^* in $\mathfrak{N}_{\mathfrak{K}}$. Furthermore since the maximum number of linearly independent solutions of (20) in $\mathfrak{K}$ is the same as in $\mathfrak{F}$ it follows that if $D_1, D_2, \cdots, D_s$ is a basis for $\mathfrak{D}$ then $D_1^*, D_2^*, \cdots, D_s^*$ is a basis for $\mathfrak{D}^*$, and if $[D_i, D_j]=\sum D_\rho \mu_{\rho ij}$, $D_i^p=\sum D_\rho \nu_{\rho i}$ ($\mu_{\rho ij}$, $\nu_{\rho i}\epsilon\mathfrak{F}$), then $[\Delta_j, \Delta_i]=\sum \Delta_\rho \mu_{\rho ij}$, $\Delta_i^p=\sum\Delta_\rho \nu_{\rho i}$ and hence $[D_i^*, D_j^*]=\sum D_\rho^* \mu_{\rho ij}$, $(D_i^*)^p=\sum D_\rho^* \nu_{\rho i}$. Thus we have proved

THEOREM 5. *If $\mathfrak{D}$ is the d-algebra of $\mathfrak{N}$ then $\mathfrak{D}_{\mathfrak{K}}$ is the d-algebra of $\mathfrak{N}_{\mathfrak{K}}$.*

7. We now consider the d-algebra of a semi-simple algebra $\mathfrak{N}$. Since $\mathfrak{N}=\mathfrak{N}_1\oplus\mathfrak{N}_2\oplus\cdots\oplus\mathfrak{N}_t$ where $\mathfrak{N}_i$ are simple and $\mathfrak{N}_i^2=\mathfrak{N}_i$ we have as a consequence of Theorem 4

THEOREM 6. *The d-algebra of a semi-simple algebra is a direct sum of algebras isomorphic to the d-algebras of its simple components.*

We suppose therefore that $\mathfrak{N}$ is simple and let $\mathfrak{C}$ denote its centrum. $\mathfrak{C}$ is an algebraic field over $\mathfrak{F}$ and is characteristic. Let $\mathfrak{C}_0$ be the subfield of constants of $\mathfrak{C}$. Because of (19),

$$[D_1, D_2]c_0 = [D_1c_0, D_2] = [D_1, D_2c_0],$$

where c_0 here denotes the multiplication determined by the element c_0 of $\mathfrak{C}_0$. Thus $\mathfrak{D}$ as well as $\mathfrak{N}$ may be regarded as an algebra over $\mathfrak{C}_0$. We may therefore suppose that $\mathfrak{C}_0=\mathfrak{F}$, i.e., the only constants in $\mathfrak{C}$ are the multiples of 1 by elements of $\mathfrak{F}$. In this case we shall show that $\mathfrak{C}$ is an inseparable field of a simple type over $\mathfrak{F}$.

Let c be any element of $\mathfrak{C}$ not in $\mathfrak{F}$. Since $cD\epsilon\mathfrak{C}$, we have

$$\begin{aligned} \phi(c)D &\equiv (c^r + c^{r-1}\gamma_1 + \cdots + \gamma_r)D \\ &= (rc^{r-1} + (r-1)c^{r-2}\gamma_1 + \cdots + \gamma_{r-1})(cD) \qquad (21) \\ &= \phi'(c)(cD), \end{aligned}$$

where $\phi'(\lambda)$ is the formal derivative of the polynomial $\phi(\lambda)$ in the polynomial ring $\mathfrak{F}[\lambda]$. If $\phi(c)=0$ is the minimum equation of c and D is chosen so that $cD\neq 0$, (21) gives $\phi'(c)=0$ and hence $\phi'(\lambda)=0$. Thus c is inseparable. In particular if the characteristic $p=0$, $\mathfrak{C}=\mathfrak{F}$ and $\mathfrak{N}$ is a normal simple algebra. If $p\neq 0$, $c^p=\gamma\epsilon\mathfrak{F}$ since $c^pD=pc^{p-1}(cD)=0$ for all D.

LEMMA 1. *If $\mathfrak{F}$ is a field of characteristic $p\neq 0$, the polynomial $\lambda^p-\alpha$ is either irreducible or a pth power of a linear factor in $\mathfrak{F}[\lambda]$.*

Suppose $\lambda^p-\alpha$ is reducible and $\phi(\lambda)$ of degree $<p$ is an irreducible factor, say

$$\lambda^p - \alpha = \phi(\lambda)^r\psi(\lambda), \qquad (\phi(\lambda), \psi(\lambda)) = 1.$$

Differentiating we obtain

$$0 = r\phi(\lambda)^{r-1}\phi'(\lambda)\psi(\lambda) + \phi(\lambda)^r\psi'(\lambda).$$

$\psi'(\lambda)\neq 0$ implies that $\phi(\lambda)^r$ divides $r\phi(\lambda)^{r-1}\phi'(\lambda)\psi(\lambda)$ and $\phi(\lambda)$ divides $r\phi'(\lambda)\psi(\lambda)$. Since $\phi'(\lambda)\neq 0$ and $(\phi(\lambda), \psi(\lambda))=1$, it follows that $r=p$ and hence $\psi(\lambda)$ has degree 0 contrary to the assumption $\psi'(\lambda)\neq 0$. Hence $\psi'(\lambda)=0$ or $\psi(\lambda)$ has degree 0 and may be taken to be 1. Then $r\phi(\lambda)^{r-1}\phi'(\lambda)=0$ and so $r=p$, $\lambda^p-\alpha=\phi(\lambda)^p$.

We return to the consideration of the structure of $\mathfrak{C}$ in the case $p\neq 0$. If $\mathfrak{C}\neq\mathfrak{F}$ choose $c_1\epsilon\mathfrak{C}$, $\notin\mathfrak{F}$. $c_1^p=\gamma_1\epsilon\mathfrak{F}$. The polynomial $\lambda^p-\gamma_1$ is irreducible in $\mathfrak{F}[\lambda]$. For otherwise $\lambda^p-\gamma_1=(\lambda-\delta)^p$, $\delta\epsilon\mathfrak{F}$ and $\lambda^p-\gamma_1=(\lambda-c_1)^p=(\lambda-\delta)^p$, $c_1=\delta\epsilon\mathfrak{F}$ contrary to the choice of c_1. The order of $\mathfrak{F}^1=\mathfrak{F}(c_1)$ over $\mathfrak{F}$ is therefore p. If $\mathfrak{C}\neq\mathfrak{F}^1$ choose $c_2\epsilon\mathfrak{C}$, $\notin\mathfrak{F}^1$. $c_2^p=\gamma_2\epsilon\mathfrak{F}$ and the polynomial $\lambda^p-\gamma_2$ is irreducible in $\mathfrak{F}^1$. Hence $\mathfrak{F}^2=\mathfrak{F}^1(c_2)=\mathfrak{F}(c_1, c_2)$ has order p over $\mathfrak{F}^1$ and consequently p^2 over $\mathfrak{F}$. Continuing in this way we prove that $\mathfrak{C}=\mathfrak{F}(c_1, c_2, \cdots, c_m)$, $c_i^p=\gamma_i$ and $\mathfrak{C}$ has order p^m over $\mathfrak{F}$.

8. We determine first the structure of the d-algebra $\mathfrak{D}$ of a normal simple algebra $\mathfrak{R}$, i.e., $\mathfrak{C}=\mathfrak{F}$. The following theorem is fundamental.

THEOREM 7. *If $\mathfrak{S}$ is a semi-simple subalgebra of $\mathfrak{R}$, any derivation in $\mathfrak{S}$ may be extended to an inner derivation in $\mathfrak{R}$.*†

By Wedderburn's theorem $\mathfrak{R}$ is the totality of $t\times t$ matrices with coordinates in a normal division algebra $\mathfrak{G}$. In particular the elements z of $\mathfrak{S}$ are such matrices and we have a representation $z\rightarrow z$ of $\mathfrak{S}$ by matrices in $\mathfrak{G}$. We suppose first that this representation is irreducible. If D is any derivation in $\mathfrak{S}$ it is readily verified that

$$z\rightarrow\begin{pmatrix} z & 0\\ 0 & z\end{pmatrix}, \qquad z\rightarrow\begin{pmatrix} z & 0\\ zD & z\end{pmatrix} \tag{22}$$

are also representations of $\mathfrak{S}$ by matrices $(2t\times 2t)$ in $\mathfrak{G}$. Since, as E. Noether‡ has shown, every representation of a semi-simple algebra by matrices in a normal division algebra is completely reducible, any two representations with

† This proof is an extension of an argument communicated to me by R. Brauer.

‡ E. Noether, *Nichtkommutative Algebra*, Mathematische Zeitschrift, vol. 37 (1933), pp. 514–541. The theorem is stated here only for simple algebras but the proof given is also valid for semi-simple algebras.

the same irreducible parts are similar. Thus the two representations in (22) are similar, i.e., there exists a fixed non-singular matrix

$$A = \begin{pmatrix} a_{11} & a_{21} \\ a_{12} & a_{22} \end{pmatrix}, \qquad a_{ij} \epsilon \mathfrak{R}$$

such that

$$\begin{pmatrix} z & 0 \\ zD & z \end{pmatrix} \begin{pmatrix} a_{11} & a_{12} \\ a_{21} & a_{22} \end{pmatrix} = \begin{pmatrix} a_{11} & a_{12} \\ a_{21} & a_{22} \end{pmatrix} \begin{pmatrix} z & 0 \\ 0 & z \end{pmatrix}$$

for all $z \epsilon \mathfrak{S}$. Hence

$$za_{11} = a_{11}z, \qquad za_{12} = a_{12}z, \qquad (zD)a_{11} + za_{21} = a_{21}z,$$
$$(zD)a_{12} + za_{22} = a_{22}z.$$

By Schur's lemma, a_{11} and a_{12} are either 0 or non-singular and both cannot be 0 since A is non-singular. If $a_{11} \neq 0$, we set $a = -a_{21}a_{11}^{-1}$ and if $a_{11} = 0$, we set $a = -a_{22}a_{12}^{-1}$. Then $a \epsilon \mathfrak{R}$ and $zD = [z, a]$ as was to be shown.

If $z \to z$ is not irreducible it is completely reducible and so there exists a fixed matrix b in $\mathfrak{R}$ such that

$$b^{-1}zb = \begin{pmatrix} z_1 & & & \\ & z_2 & & \\ & & \ddots & \\ & & & z_l \end{pmatrix}$$

and $z \to z_i$ are irreducible representations of $\mathfrak{S}$. As before

$$z \to \begin{pmatrix} z_i & 0 \\ (zD)_i & z_i \end{pmatrix}, \qquad z \to \begin{pmatrix} z_i & 0 \\ 0 & z_i \end{pmatrix}$$

are similar representations of $\mathfrak{S}$ and there exists a matrix a_i such that $(zD)_i = [z_i, a_i]$. Then if

$$a = \begin{pmatrix} a_1 & & & \\ & a_2 & & \\ & & \ddots & \\ & & & a_l \end{pmatrix},$$

$b^{-1}(zD)b = [b^{-1}zb, a]$ and $zD = [z, bab^{-1}]$, $bab^{-1} \epsilon \mathfrak{R}$.

As a special case we have

THEOREM 8. *The d-algebra of a normal simple algebra contains only inner derivations.*

COROLLARY. *If $\mathfrak{R}$ is simple, $\mathfrak{D}(\mathfrak{C}) = \mathfrak{J}$.*

If $D \epsilon \mathfrak{D}(\mathfrak{C})$, $(xc)D = (xD)c$ for all x and all $c \epsilon \mathfrak{C}$. Thus D is a derivation of $\mathfrak{R}$ considered as an algebra over $\mathfrak{C}$. By Theorem 8, D is inner and so $\mathfrak{D}(\mathfrak{C}) \subset \mathfrak{J}$ Since $\mathfrak{J} \supset \mathfrak{D}(\mathfrak{C})$ we have equality.

Suppose again that $\mathfrak{R}$ is normal simple. Theorem 8 implies that $\mathfrak{D} = \mathfrak{J} \cong \mathfrak{R}/\mathfrak{F}$ where $\mathfrak{R}$ is the restricted Lie algebra determined by the associative $\mathfrak{R}$ and $\mathfrak{F}$ is the centrum consisting of the multiples of 1. We may extend $\mathfrak{F}$ to the field $\mathfrak{K}$ such that $\mathfrak{R}_{\mathfrak{K}} = \mathfrak{K}_n$ is the complete matrix algebra of order n^2 over $\mathfrak{K}$, i.e., $\mathfrak{K}_n$ has a basis e_{ij} $(i, j = 1, 2, \cdots, n)$ such that $e_{ij}e_{kl} = \delta_{jk}e_{il}$. We consider the structure of the Lie algebra $\mathfrak{K}_n$ having basis e_{ij} also and multiplication table

$$[e_{ij}, e_{kl}] = \delta_{jk}e_{il} - \delta_{il}e_{kj}. \tag{23}$$

The centrum of $\mathfrak{K}_n$ is $\mathfrak{K}$ the totality of multiples of $1 = e_{11} + e_{22} + \cdots + e_{nn}$. This is an ideal as is $\mathfrak{K}_n' = [\mathfrak{K}_n, \mathfrak{K}_n]$. From (23) follows that e_{ts}, $e_{tt} - e_{ss} \epsilon \mathfrak{K}_n'$ if $t \neq s$. Evidently every element of $\mathfrak{K}_n'$ has trace 0. Conversely if $a = \sum e_{ij}\alpha_{ij}$ and $\mathrm{tr}(a) = \alpha_{11} + \alpha_{22} + \cdots + \alpha_{nn} = 0$,

$$a = (e_{11} - e_{nn})\alpha_{11} + (e_{22} - e_{nn})\alpha_{22} + \cdots + (e_{n-1,n-1} - e_{nn})\alpha_{n-1,n-1} + \sum_{t \neq s} e_{ts}\alpha_{ts} \epsilon \mathfrak{K}_n'$$

and so $\mathfrak{K}_n'$ is the set of matrices of trace 0 and is generated by $e_{11} - e_{nn}$, $e_{22} - e_{nn}, \cdots, e_{n-1,n-1} - e_{nn}$, e_{ts} $(t \neq s)$. These $n^2 - 1$ elements are evidently linearly independent and hence form a basis for $\mathfrak{K}_n'$. Since $(e_{ss} - e_{tt})^p = e_{ss} - e_{tt}$, $e_{st}^p = 0$ if $p \neq 0$ is the characteristic of $\mathfrak{K}$, $\mathfrak{K}_n'$ by (15) contains the pth power of every element belonging to it, i.e., $\mathfrak{K}_n'$ is a restricted ideal. $\mathfrak{K}_n'$ contains 1 if and only if $\mathrm{tr}(1) = n \equiv 0 \pmod{p}$.

Suppose $\mathfrak{B}$ is an ideal $\neq \mathfrak{K}$ in $\mathfrak{K}_n$ and $b = \sum e_{ij}\beta_{ij} \epsilon \mathfrak{B}$, $\notin \mathfrak{K}$. Suppose first $\beta_{uv} \neq 0$ for some pair $u, v, u \neq v$. If $n > 2$, choose $t \neq v, \neq u$ and then $[[[b, e_{vu}], e_{tu}], e_{tt}]\beta_{uv}^{-1} = e_{tu} \epsilon \mathfrak{B}$. If $p \neq 2$, $[[b, e_{vu}], e_{vu}]\ (-2\beta_{uv})^{-1} = e_{vu} \epsilon \mathfrak{B}$. If all $\beta_{uv} = 0$ then $b = e_{11}\beta_{11} + e_{22}\beta_{22} + \cdots + e_{nn}\beta_{nn}$ and since $b \notin \mathfrak{K}$, $\beta_{uu} \neq \beta_{vv}$ for some pair $u \neq v$ and hence $[b, e_{uv}]\ (\beta_{uu} - \beta_{vv})^{-1} = e_{uv} \epsilon \mathfrak{B}$. Thus in any case unless $n = p = 2$ $\mathfrak{B}$ contains an e_{st}, $s \neq t$ and since by (23), $[e_{st}, \mathfrak{K}_n] = \mathfrak{K}_n'$, $\mathfrak{B} \supset \mathfrak{K}_n'$. If $\mathfrak{B} \neq \mathfrak{K}_n$, $\mathfrak{B} = \mathfrak{K}_n'$.

Any ideal of $\mathfrak{K}_n/\mathfrak{K}$ the derivation ring of the associative algebra $\mathfrak{K}_n$ has the form $\mathfrak{B}/\mathfrak{K}$ where $\mathfrak{B}$ is an ideal in the Lie algebra $\mathfrak{K}_n$ containing $\mathfrak{K}$. If $p \nmid n$ the only such ideals are $\mathfrak{K}$ and $\mathfrak{K}_n$. Hence $\mathfrak{K}_n/\mathfrak{K}$ is a simple Lie algebra, i.e., has no proper ideals.

If $p | n$ and either $p \neq 2$ or $n > 2$, $\mathfrak{K}_n/\mathfrak{K}$ has one proper ideal $\mathfrak{K}_n'/\mathfrak{K}$ and this is restricted. It may be shown by a direct argument similar to the above that $\mathfrak{K}_n'/\mathfrak{K}$ is simple except when $p = n = 2$ and hence the Lie algebra $\mathfrak{K}_n/\mathfrak{K}$ is semi-simple. Since $\mathfrak{K}_n'/\mathfrak{K}$ is the only proper ideal in $\mathfrak{K}_n/\mathfrak{K}$ the latter is not a direct sum of simple ideals.†

THEOREM 9. *If $\mathfrak{R}$ is a normal simple algebra of order n^2 and $p \nmid n^2$ then the d-algebra $\mathfrak{D}$ of $\mathfrak{R}$ is simple.*

THEOREM 10. *If $\mathfrak{R}$ is normal simple and $p \mid n^2$ but either $p \neq 2$ or $n > 2$ $\mathfrak{D}$ is semi-simple though not simple.*

To prove these theorems we note that a proper ideal $\mathfrak{B}$ of $\mathfrak{D}$ becomes a proper ideal $\mathfrak{B}_{\mathfrak{K}}$ of $\mathfrak{D}_{\mathfrak{K}}$ the d-algebra of $\mathfrak{R}_{\mathfrak{K}}$ when $\mathfrak{F}$ is extended to $\mathfrak{K}$. By choosing $\mathfrak{K}$ so that $\mathfrak{R}_{\mathfrak{K}} = \mathfrak{K}_n$, $\mathfrak{D}_{\mathfrak{K}} \cong \mathfrak{K}_n/\mathfrak{K}$ it follows that $\mathfrak{D}$ has no such ideals if $p \nmid n^2$. If $p \mid n^2$ and either $p \neq 2$ or $n > 2$, $[\mathfrak{D}, \mathfrak{D}]_{\mathfrak{K}} \cong \mathfrak{K}_n'/\mathfrak{K}$ is a proper restricted ideal of $\mathfrak{D}_{\mathfrak{K}}$ and hence $\mathfrak{D}' = [\mathfrak{D}, \mathfrak{D}]$ is a proper restricted ideal in $\mathfrak{D}$. $\mathfrak{D}'$ is simple since $\mathfrak{D}_{\mathfrak{K}}'$ is.

If $n = p = 2$ it is easily seen that $\mathfrak{K}_2/\mathfrak{K}$ and hence $\mathfrak{D}$ is solvable.

9. We consider next the d-algebra $\mathfrak{D}$ of the other extreme case, namely, $\mathfrak{R} = \mathfrak{C} = \mathfrak{F}(c_1, c_2, \cdots, c_m)$ where $c_i^p = \gamma_i$ and the order of $\mathfrak{R}$ over $\mathfrak{F}$ is p^m, $p \neq 0$. Let D be any element of $\mathfrak{D}$ and consider the correspondence $D \rightarrow (c_1D, c_2D, \cdots, c_mD)$ mapping $\mathfrak{D}$ on the space $\mathfrak{R}^{(m)}$ of ordered m-tuples of elements of $\mathfrak{R}$. This correspondence is linear relative to $\mathfrak{F}$ and since $c_1D = c_2D = \cdots = c_mD = 0$ implies that $D = 0$ it is $(1-1)$. Moreover if $(d_1, d_2, \cdots, d_m)$ is an arbitrary element of $\mathfrak{R}^{(m)}$ there is a $D \epsilon \mathfrak{D}$ such that $c_iD = d_i$. For $\mathfrak{R} \cong \mathfrak{F}[\lambda_1, \lambda_2, \cdots, \lambda_m]/\mathfrak{P}$ where $\mathfrak{P}$ is the ideal having the basis $\lambda_1^p - \gamma_1, \lambda_2^p - \gamma_2, \cdots, \lambda_m^p - \gamma_m$. If $d_1(\lambda_1, \cdots, \lambda_m)$, $d_2(\lambda_1, \cdots, \lambda_m), \cdots, d_m(\lambda_1, \cdots, \lambda_m)$ are arbitrary polynomials, then the transformation D defined by

$$c(\lambda_1, \lambda_2, \cdots, \lambda_m)D = \sum_i \frac{\partial c(\lambda_1, \lambda_2, \cdots, \lambda_m)}{\partial \lambda_i} d_i(\lambda_1, \lambda_2, \cdots, \lambda_m)$$

is easily verified to be a derivation in $\mathfrak{F}[\lambda_1, \lambda_2, \cdots, \lambda_m]$. If $z(\lambda_1, \lambda_2, \cdots, \lambda_m) \epsilon \mathfrak{P}$ then $zD \epsilon \mathfrak{P}$ also. It follows that D induces a derivation in $\mathfrak{F}[\lambda_1, \lambda_2, \cdots, \lambda_m]/\mathfrak{P}$, i.e., in $\mathfrak{R}$ and since $d_i(\lambda_1, \cdots, \lambda_m)$ were arbitrary, D may be chosen so that $c_iD = d_i$. We have therefore established an isomorphism between $\mathfrak{D}$ and $\mathfrak{R}^{(m)}$ considered as vector spaces over $\mathfrak{F}$. The order of $\mathfrak{R}^{(m)}$ is mp^m and hence the order of $\mathfrak{D}$ is mp^m also.

LEMMA 2. *If $\mathfrak{R}$ is any commutative field, D a derivation in it, and $\mathfrak{F}$ the subfield of D-constants, a necessary and sufficient condition that the elements $y_1, y_2, \cdots, y_r$ be linearly dependent over $\mathfrak{F}$ is that the Wronskian*

† If $p = 0$ a fundamental theorem due to E. Cartan, *Thèse*, Paris, 1894, states that a semi-simple Lie algebra is a direct sum of simple algebras. The algebras $\mathfrak{K}_n/\mathfrak{K}$ for $p \mid n$ show that this does not hold for $p \neq 0$. A second example of this type will be given below.

$$\begin{vmatrix} y_1 & y_2 & \cdots & y_r \\ y_1 D & y_2 D & \cdots & y_r D \\ \cdot & \cdot & \cdots & \cdot \\ y_1 D^{r-1} & y_2 D^{r-1} & \cdots & y_r D^{r-1} \end{vmatrix} = 0.$$

The usual proof of this result for analytic functions is valid here.† As a consequence we have

LEMMA 3. *The differential equation* $y(D^r + D^{r-1}a_1 + \cdots + a_r) = 0$, $a_i \epsilon \mathfrak{R}$, *has at most r solutions* $y_1, y_2, \cdots, y_r$ *in* $\mathfrak{R}$ *linearly independent over* $\mathfrak{F}$.

It has been shown by R. Baer‡ that if $\mathfrak{R}$ is a field of the type $\mathfrak{F}(c_1, c_2, \cdots, c_m)$, $c_i^p = \gamma_i \epsilon \mathfrak{F}$, there exists a derivation D such that the D-constants are precisely the elements of $\mathfrak{F}$. Let D denote a fixed derivation of this type and set $cD = c'$ for any $c \epsilon \mathfrak{R}$. $D^p, D^{p^2}, \cdots$ are derivations and since $\mathfrak{R}$ is commutative the transformation $Da_0 + D^p a_1 + \cdots + D^{p^{m-1}} a_{m-1}$ is a derivation for arbitrary right multiplication a_i $(= a_{ir})$ in $\mathfrak{R}$. If $Da_0 + D^p a_1 + \cdots + D^{p^{m-1}} a_{m-1} = 0$, i.e., $y(Da_0 + D^p a_1 + \cdots + D^{p^{m-1}} a_{m-1}) = 0$ for all y in $\mathfrak{R}$, it follows by Lemma 3 and the fact that $\mathfrak{F}$ is the set of D-constants that all $a_i = 0$. Thus as the a_i vary in $\mathfrak{R}$ we obtain in this way mp^m linearly independent (over $\mathfrak{F}$) derivations and hence the complete algebra $\mathfrak{D}$. We shall therefore call D a *generator* of $\mathfrak{D}$. Since D^{p^m} is a derivation we have

$$D^{p^m} = D^{p^{m-1}} b_{m-1} + D^{p^{m-2}} b_{m-2} + \cdots + Db_0.$$

Taking commutators with D we have by (19),

$$0 = D^{p^{m-1}} b'_{m-1} + D^{p^{m-2}} b'_{m-2} + \cdots + Db'_0,$$

and hence $b'_i = 0$, i.e., $b_i = \beta_i \epsilon \mathfrak{F}$, and

$$D^{p^m} = D^{p^{m-1}} \beta_{m-1} + D^{p^{m-2}} \beta_{m-2} + \cdots + D\beta_0. \tag{24}$$

As a consequence of (19), we note also

$$[D^{p^k} a, D^{p^j} b] = D^{p^k} a^{(p^j)} b - D^{p^j} b^{(p^k)} a, \tag{25}$$

where $a^{(p^j)} = aD^{p^j}$. If $E = Da \neq 0$ then the E-constants are the same as the D-constants since the multiplication a is non-singular and hence E is a generator of $\mathfrak{D}$ also.

THEOREM 11. *The d-algebra of the field* $\mathfrak{R} = \mathfrak{F}(c_1, c_2, \cdots, c_m)$, $c_i^p = \gamma_i$ *is simple except when* $p = 2$, $m = 1$.

---

† See, for example, T. Chaundy, *Differential Calculus*, Oxford, 1935, p. 106.

‡ R. Baer, *Algebraische Theorie der differentiierbaren Funktionenkörper*. I, Sitzungsberichte, Heidelberger Akademie, 1927, pp. 15–32.

Let $\mathfrak{B}\neq 0$ be an ideal in $\mathfrak{D}$ and $E=Db_0+D^pb_1+\cdots+D^{p^j}b_j$, $b_j\neq 0, j<m$, belong to $\mathfrak{B}$. We call j the *length* of E and suppose E chosen in $\mathfrak{B}$ so that j is minimal. We assert that $j=0$. For if $j>0$ we may suppose $b_j=1$. This is evident if $b_j'=0$ or $b_j=\beta_j\epsilon\mathfrak{F}$ and if $b_j'\neq 0$, $[E, D(b_j')^{-1}]=Db_0^*+D^pb_1^*+\cdots+D^{p^j}\epsilon\mathfrak{B}$. But if $E=Db_0+D^pb_1+\cdots+D^{p^{j-1}}b_{j-1}+D^{p^j}$ then

$$[E, Da] = D(ab_0' - a'b_0 - \cdots - a^{(p^{j-1})}b_{j-1} - a^{(p^j)}) + D^pab_1' + \cdots + D^{p^{j-1}}ab_{j-1}'.$$

$[E, Da]$ has length $<j$ and may be chosen $\neq 0$ since by Lemma 3 a may be chosen so that $ab_0'-a'b_0-\cdots-a^{(p^{j-1})}b_{j-1}-a^{(p^j)}\neq 0$. This contradicts the minimality of j in $\mathfrak{B}$ and shows that $E=Db_0$, $b_0\neq 0$. Since E as well as D is a generator of $\mathfrak{D}$, by changing the notation we may suppose that $\mathfrak{B}\supset D$. Then $\mathfrak{B}\supset [Da, D]=Da'$ also. Since the null space of the linear transformation D in $\mathfrak{R}$ has order 1, the order of $\mathfrak{R}'$ the set of all a' is p^m-1 over $\mathfrak{F}$. If $1\notin\mathfrak{R}'$ the smallest space containing all a' and 1 is $\mathfrak{R}$. Since $\mathfrak{B}\supset D$ and Da', $\mathfrak{B}$ will then contain Da for all a in $\mathfrak{R}$. Also if $p\neq 2$, $\mathfrak{B}\supset\frac{1}{2}[Da', Db]+\frac{1}{2}[Da'b, D]=Da''b$ and since a'' is not identically 0 and b is arbitrary, $\mathfrak{B}\supset Da$ for all a. Suppose finally that $p=2$ and $\mathfrak{R}'\supset 1$, say $u'=1$. Here $\mathfrak{B}\supset [[D^2a, D], Db]=D^2a''b+Da'b''$. If $m>1$, a'' is not identically 0 and hence b may be chosen so that $a''b=u$. Set $a'b''=v$. $\mathfrak{B}\supset [D^2u+Dv, Da]+D(va+ua'+a)'=D^2a$ and $[D^2a, Db]+D^2a'b=Dab''$. Thus in any case unless $p=2$, $m=1$, $\mathfrak{B}\supset Da$ for all a and since $[D^{p^k}b, Da]+Da^{(p^k)}b=D^{p^k}b'a$, $\mathfrak{B}\supset$ all $D^{p^k}a$ so that $\mathfrak{B}=\mathfrak{D}$.

If $p=2$, $m=1$, $\mathfrak{D}$ has order 2 and hence is solvable. In all other cases the algebras $\mathfrak{D}$ are simple algebras which, like inseparable fields, have no counterparts for $p=0$.

If E is any derivation, the totality of expressions $E^{p^k}a_0+E^{p^{k-1}}a_1+\cdots+Ea_{\mathfrak{R}}$, $a_i\epsilon\mathfrak{R}$ is, by virtue of (15) and (19), a restricted $\mathfrak{R}$-($=\mathfrak{E}$-)subalgebra $\mathfrak{E}$ of $\mathfrak{D}$. Conversely if $\mathfrak{E}$ is any restricted $\mathfrak{R}$-subalgebra of $\mathfrak{D}$, $\mathfrak{E}$ is generated in this fashion. To prove this let $E=D^{p^e}g_0+D^{p^{e-1}}g_1+\cdots+Dg_e$, $g_0\neq 0$, be an element of smallest length in $\mathfrak{E}$. Since $\mathfrak{E}$ is an $\mathfrak{R}$-algebra we may suppose that $g_0=1$ and then $E=D^{p^e}+D^{p^{e-1}}g_1+\cdots+Dg_e$ is unique. If $F=D^{p^f}h_0+D^{p^{f-1}}h_1+\cdots+Dh_f\epsilon\mathfrak{E}$, $f\geqq e$ and

$$F_1 = F - E^{p^{f-e}}h_0 = D^{p^{f-1}}k_0 + \cdots + Dk_{f-1}$$

by (15) and (25). F_1 has length $\leqq f-1$ and belongs to $\mathfrak{E}$. Repeating this process we obtain an expression for F of the form $E^{p^{f-e}}a_0+\cdots+Ea_{f-e}$.

If as in I we denote the elements of $\mathfrak{R}$ which are constants for all the derivations in $\mathfrak{E}$ by $\mathfrak{R}(\mathfrak{E})$ it is clear that $\mathfrak{R}(\mathfrak{E})$ coincides with the subfield $\mathfrak{S}$ of E-constants. On the other hand if E is any derivation, $\mathfrak{S}$ the set of E-constants, then the argument at the beginning of this section shows that E generates the d-algebra $\mathfrak{D}(\mathfrak{S})$ of $\mathfrak{R}$ considered as a field over $\mathfrak{S}$. Hence $\mathfrak{D}(\mathfrak{S})=\mathfrak{E}$, i.e.,

$$\mathfrak{D}(\mathfrak{R}(\mathfrak{E})) = \mathfrak{E}.$$

If $\mathfrak{S}$ is any subfield of $\mathfrak{R}$, $\mathfrak{D}(\mathfrak{S})$ contains an element E such that the E-constants are precisely $\mathfrak{S}$. Hence $\mathfrak{R}(\mathfrak{D}(\mathfrak{S}))$ cannot be larger than $\mathfrak{S}$ and so

$$\mathfrak{R}(\mathfrak{D}(\mathfrak{S})) = \mathfrak{S}.$$

We have therefore proved

THEOREM 12. *There is a (1–1) correspondence between the subfields $\mathfrak{S}$ of $\mathfrak{R}$ containing $\mathfrak{F}$ and the restricted $\mathfrak{R}$-subalgebras $\mathfrak{E}$ of the d-algebra $\mathfrak{D}$ of $\mathfrak{R}$ over $\mathfrak{F}$. The correspondence is given by either $\mathfrak{E}=\mathfrak{D}(\mathfrak{S})$ or $\mathfrak{S}=\mathfrak{R}(\mathfrak{E})$.*

10. We now suppose that $\mathfrak{R}$ is simple and that $\mathfrak{R}\supset\mathfrak{C}\supset\mathfrak{F}$ where $\mathfrak{C}=\mathfrak{F}(c_1, c_2, \cdots, c_m)$, $c_i^p=\gamma_i$ and $p\neq 0$. Let $\mathfrak{D}$ denote the d-algebra of $\mathfrak{R}$ over $\mathfrak{F}$ and $\mathfrak{E}$ that of $\mathfrak{C}$ over $\mathfrak{F}$. If $D\epsilon\mathfrak{D}$, D induces a derivation in $\mathfrak{C}$ and hence $\mathfrak{D}$ is homomorphic with a subalgebra $\mathfrak{E}_1$ of $\mathfrak{E}$. Since $\mathfrak{D}(\mathfrak{C})$ is the set of elements corresponding to 0 in this homomorphism, we have $\mathfrak{E}_1\cong\mathfrak{D}/\mathfrak{D}(\mathfrak{C})$. But by the corollary to Theorem 8, $\mathfrak{D}(\mathfrak{C})=\mathfrak{J}$ and hence $\mathfrak{E}_1\cong\mathfrak{D}/\mathfrak{J}$. We wish to show that $\mathfrak{E}_1=\mathfrak{E}$.

$\mathfrak{R}$ may be regarded as a normal simple algebra over $\mathfrak{C}$ and there exists a separable field $\mathfrak{C}(s)$ over $\mathfrak{C}$ such that $\mathfrak{R}\times\mathfrak{C}(s)=\mathfrak{C}(s)_n$ the matrix algebra of order n^2 with elements in $\mathfrak{C}(s)$. As has been shown by Albert† the separable extension $\mathfrak{C}(s)$ of the inseparable field $\mathfrak{C}$ has the form $\mathfrak{K}(c_1, \cdots, c_m)=\mathfrak{C}_{\mathfrak{K}}$ where $\mathfrak{K}$ is a separable field over $\mathfrak{F}$. Now consider $\mathfrak{R}_{\mathfrak{K}}$. The centrum of this algebra is $\mathfrak{C}_{\mathfrak{K}}=\mathfrak{C}(s)$ and if $x_1, x_2, \cdots, x_{n^2}$ form a basis of $\mathfrak{R}$ over $\mathfrak{C}$ they are also a basis for $\mathfrak{R}_{\mathfrak{K}}$ over $\mathfrak{C}_{\mathfrak{K}}=\mathfrak{C}(s)$. It follows that $\mathfrak{R}_{\mathfrak{K}}=\mathfrak{R}\times\mathfrak{C}(s)=\mathfrak{C}(s)_n=(\mathfrak{C}_{\mathfrak{K}})_n$.

The d-algebra of $\mathfrak{R}_{\mathfrak{K}}$ is $\mathfrak{D}_{\mathfrak{K}}$ and the ideal of inner derivations of $\mathfrak{D}_{\mathfrak{K}}$ is $\mathfrak{J}_{\mathfrak{K}}$. If E^* is any derivation in $\mathfrak{C}_{\mathfrak{K}}$ over $\mathfrak{K}$, the correspondence $\sum e_{ij}c_{ij}^*\rightarrow\sum e_{ij}(c_{ij}^*E^*)$, $c_{ij}^*\epsilon\mathfrak{C}_{\mathfrak{K}}$ is readily verified to be a derivation in $\mathfrak{R}_{\mathfrak{K}}$ inducing E^* in $\mathfrak{C}_{\mathfrak{K}}$. Hence $\mathfrak{D}_{\mathfrak{K}}/\mathfrak{J}_{\mathfrak{K}}$ is isomorphic to the complete d-algebra of $\mathfrak{C}_{\mathfrak{K}}$ and so has order mp^m over $\mathfrak{K}$. Since $\mathfrak{D}_{\mathfrak{K}}/\mathfrak{J}_{\mathfrak{K}}\cong(\mathfrak{D}/\mathfrak{J})_{\mathfrak{K}}$, $\mathfrak{D}/\mathfrak{J}$ has order mp^m over $\mathfrak{F}$. Comparing orders we have $\mathfrak{E}\cong\mathfrak{D}/\mathfrak{J}$.

THEOREM 13. *Suppose $\mathfrak{R}$ is a simple algebra of order n^2 over its centrum $\mathfrak{C}=\mathfrak{F}(c_1, c_2, \cdots, c_m)$, $c_i^p=\gamma_i$, $p\neq 0$. Then the d-algebra of $\mathfrak{D}$ over $\mathfrak{R}$ is semi-simple unless $p=2$ and either $n\leqq 2$ or $m=1$.*

Let $\mathfrak{B}$ be a solvable ideal in $\mathfrak{D}$. $\mathfrak{B}+\mathfrak{J}$‡ is an ideal and $(\mathfrak{B}+\mathfrak{J})/\mathfrak{J}$ is a solva-

† A. A. Albert, *Simple algebras of degree p^e over a centrum of characteristic p*, these Transactions, vol. 40 (1936), p. 113.

‡ $\mathfrak{B}+\mathfrak{J}$ denotes the smallest space containing $\mathfrak{B}$ and $\mathfrak{J}$.

ble ideal in $\mathfrak{D}/\mathfrak{J}\cong\mathfrak{E}$. But by Theorem 11, $\mathfrak{E}$ is simple. Hence $(\mathfrak{B}+\mathfrak{J})/\mathfrak{J}=0$ or $\mathfrak{B}+\mathfrak{J}=\mathfrak{J}$ and $\mathfrak{B}\subset\mathfrak{J}$. However, by Theorem 10, $\mathfrak{J}$ is semi-simple and so $\mathfrak{B}=0$.

$\mathfrak{D}$ is not a direct sum of $\mathfrak{J}$ and a second ideal. For we have seen (§5) that the elements commutative with all elements in $\mathfrak{J}$ are those mapping $\mathfrak{R}$ into $\mathfrak{C}$. If F is such an element, then F^* the extension of F maps $\mathfrak{R}_{\mathfrak{R}}$ into $\mathfrak{C}_{\mathfrak{R}}$ (cf. Theorem 5). If $e_{ij}F^*=c_{ij}^*\epsilon\mathfrak{C}_{\mathfrak{R}}$, it follows from $e_{ij}e_{kl}=\delta_{jk}e_{il}$ that $c_{ij}^*=0$. Hence $(e_{ij}c^*)F^*=e_{ij}(c^*F^*)$ for $c^*\epsilon\mathfrak{C}_{\mathfrak{R}}$. If this belongs to $\mathfrak{C}_{\mathfrak{R}}$ we must have $c^*F^*=0$. Thus $F^*=0$, $F=0$, and $\mathfrak{D}$ is not a direct sum.

III. Theory of D-fields

11. In this part we propose to study $\mathfrak{R}=\mathfrak{F}\,(c_1, c_2, \cdots, c_m)$, $c_i^p=\gamma_i$, $p\neq 0$ relative to the fixed derivation D and shall obtain several analogues of theorems on automorphisms of cyclic fields. Without loss of generality we may assume that $\mathfrak{F}$ is the field of D-constants and hence D is a generator of the d-algebra of $\mathfrak{R}$. We have seen that D satisfies (24),

$$D^{p^m} = D^{p^{m-1}}\beta_1 + D^{p^{m-2}}\beta_2 + \cdots + D\beta_m,$$

and no equation of lower degree of the form $D^r+D^{r-1}a_1+\cdots+a_r$, $a_i\epsilon\mathfrak{R}$.

Suppose $y_1, y_2, \cdots, y_{pm}$ is a basis for $\mathfrak{R}$ and

$$(y_1D, y_2D, \cdots, y_{p^m}D) = (y_1, y_2, \cdots, y_{p^m})\Delta \qquad \Delta = (\alpha_{ij}).$$

If $f(\lambda)$ is the characteristic function $|\lambda 1-\Delta|$, then by the Hamilton-Cayley theorem, $f(D)=0$. Since the degree of $f(\lambda)$ is p^m we have

$$f(\lambda) = |\lambda 1 - \Delta| = \lambda^{p^m} - \lambda^{p^{m-1}}\beta_1 - \cdots - \lambda\beta_m. \tag{26}$$

Since the characteristic and minimum equations of A are identical, A is similar to

$$B = \begin{pmatrix} 0 & & & & & & & 0 \\ 1 & & & & & & & \beta_m \\ & 1 & & & & & & 0 \\ & & \ddots & & & & & \vdots \\ & & & & & & & \beta_1 \\ & & & & \ddots & & & 0 \\ & & & & & & & \vdots \\ & & & & & & 1 & 0 \end{pmatrix}.$$

It follows that $\mathfrak{R}$ has a basis of the form $z, zD, zD^2, \cdots, zD^{p^m-1}$, i.e., $\mathfrak{R}$ is a cyclic space relative to the linear transformation D.†

A polynomial of the form $\lambda^{p^e}+\lambda^{p^{e-1}}\rho_1+\cdots+\lambda\rho_e$ will be called a *p-polynomial*.‡ A subfield $\mathfrak{S}$ of $\mathfrak{R}$ containing $\mathfrak{F}$ and vD for every v in $\mathfrak{S}$ will be called a *D-subfield* of $\mathfrak{R}$. Thus $\mathfrak{S}$ is a space invariant under the transformation D.

THEOREM 14. *There is a* (1–1) *correspondence between the D-subfields of* $\mathfrak{R}$ *and the p-polynomial factors of* $f(\lambda)$.

Any subspace $\mathfrak{S}$ of $\mathfrak{R}$ is cyclic with generator w. If $g(\lambda)$ is a polynomial of least degree such that $wg(D)=0$ then $g(\lambda)$ is the minimum function of D acting in $\mathfrak{S}$ and the order of $\mathfrak{S}$ = degree of $g(\lambda)$. $g(\lambda)$ is therefore uniquely determined by $\mathfrak{S}$ and is a factor of $f(\lambda)$. For if $h(\lambda)=f(\lambda)q(\lambda)+g(\lambda)r(\lambda)=(g(\lambda), f(\lambda))$ then $wh(D)=0$ and since $g(\lambda)$ is minimal, $g(\lambda)=h(\lambda)$. Conversely if $g(\lambda)$ is a factor of $f(\lambda)$, $f(\lambda)=g(\lambda)k(\lambda)$, the vectors v such that $vg(D)=0$ form an invariant subspace $\mathfrak{S}$. $\mathfrak{S}\supset zk(D), zd(D)D, \cdots$ if z is a generator of $\mathfrak{R}$, and if the degree of $g(\lambda)$ is r, $zk(D), zk(D)D, \cdots, zk(D)D^{r-1}$ are linearly independent. Hence the order of $\mathfrak{S}$ is $\geqq r$. On the other hand the minimum function of D in $\mathfrak{S}$ is $g(\lambda)$ so that order of $\mathfrak{S}$ is r, $\mathfrak{S}=(zk(D), zk(D)D, \cdots)$. Thus we have a (1–1) correspondence between the invariant subspaces $\mathfrak{S}$ of $\mathfrak{R}$ and the factors $g(\lambda)$ of $f(\lambda)$. If $\mathfrak{S}$ is a field, D is a generator of the d-algebra of $\mathfrak{S}$ over $\mathfrak{F}$ and hence $g(\lambda)$ is a p-polynomial. Conversely if $g(\lambda)$ is a p-polynomial and $v_1, v_2\epsilon\mathfrak{S}$, i.e., $v_1g(D)=v_2g(D)=0$, then since $g(D)$ is a derivation, $v_1v_2g(D)=(v_1g(D))v_2+v_1(v_2g(D))=0$ so that $\mathfrak{S}$ is closed under multiplication and hence is a D-subfield of $\mathfrak{R}$.

Suppose $g(\lambda)=\lambda^{p^e}+\lambda^{p^{e-1}}\rho_1+\cdots+\lambda\rho_e$, $h(\lambda)=\lambda^{p^f}+\lambda^{p^{f-1}}\sigma_1+\cdots+\lambda\sigma_f$ and $e\leqq f$. Then $g(\lambda)-h(\lambda)^{p^{e-f}}=\lambda^{p^{e-1}}\tau_1+\cdots+\lambda\tau_{e-1}$. By repeating this process we may express $g(\lambda)$ in the form

$$g(\lambda) = h(\lambda)^{p^{e-f}} + h(\lambda)^{p^{e-f-1}}\omega_1 + \cdots + h(\lambda)\omega_{e-f} + r(\lambda) \qquad (\omega_1 = \tau_1),$$

where $r(\lambda)$ is a p-polynomial of degree $<p^f$. Since $r(\lambda)$ is the remainder obtained by dividing $g(\lambda)$ by $h(\lambda)$, by continuing the euclid algorithm we find that $(g(\lambda), h(\lambda))$ *is a p-polynomial.*

If $k(\lambda)$ is any polynomial (coefficients in $\mathfrak{F}$), then

$$\lambda^{p^j} = k(\lambda)q_j(\lambda) + s_j(\lambda) \qquad (j = 0, 1, 2, \cdots),$$

where degree $s_j(\lambda)<$degree $k(\lambda)=r$. Since there are at most r independent polynomials of degree $<r$ there exist elements, $\alpha_0, \alpha_1, \cdots, \alpha_r$ not all 0 such that $s_r(\lambda)\alpha_0+s_{r-1}(\lambda)\alpha_1+\cdots+s_0(\lambda)\alpha_r=0$ and hence

† For a discussion of cyclic spaces see Jacobson, *Pseudo-linear transformations*, Annals of Mathematics, vol. 38 (1937), p. 496.

‡ This term is due to O. Ore, *On a special class of polynomials*, these Transactions, vol. 35 (1933), p. 560.

$$h(\lambda) = \lambda^{p^r}\alpha_0 + \lambda^{p^{r-1}}\alpha_1 + \cdots + \lambda\alpha_r = k(\lambda)\sum q_j(\lambda)\alpha_j,$$

i.e., *any polynomial is a factor of some p-polynomial.*† Since the h.c.f. of p-polynomials is a p-polynomial, the p-polynomial of least degree divisible by $k(\lambda)$ is unique. We denote it by $\{k(\lambda)\}$.

Now suppose $\mathfrak{S}$ is a subspace of $\mathfrak{R}$ invariant under D and $k(\lambda)$ is the minimum function of D in $\mathfrak{S}$. Let $\{\mathfrak{S}\}$ denote the enveloping field of $\mathfrak{S}$. $\{\mathfrak{S}\}$ is a D-field and D has minimum function $\{k(\lambda)\}$ in $\{\mathfrak{S}\}$. If $\mathfrak{S}_1$ and $\mathfrak{S}_2$ are invariant subspaces, $k_1(\lambda)$, $k_2(\lambda)$ the corresponding minimum functions, then $\mathfrak{S}_1+\mathfrak{S}_2$ and $\mathfrak{S}_1 \cap \mathfrak{S}_2$ are invariant and the associated functions are respectively $[k_1(\lambda), k_2(\lambda)]$ and $(k_1(\lambda), k_2(\lambda))$.

12. Let $\mathfrak{M}$ denote the algebra of linear transformations generated by D and the multiplications of $\mathfrak{R}$. Since $D^{p^m-1}a_1+D^{p^m-2}a_2+\cdots+a_{p^m}=0$ implies all $a_i=0$, $\mathfrak{M}$ has order p^{2m} over $\mathfrak{F}$ and hence is isomorphic to $\mathfrak{F}_{p^m}$ the algebra of all $p^m\times p^m$ matrices in $\mathfrak{F}$. The multiplication of the elements of $\mathfrak{M}$ may be ascertained from the multiplications of the elements of $\mathfrak{R}$ and the rules

$$\text{(27)} \qquad \text{(a) } aD = Da + a', \qquad \text{(b) } f(D) = D^{p^m} - D^{p^{m-1}}\beta_1 - \cdots - D\beta_m = 0.$$

Let c be an arbitrary element of $\mathfrak{R}$ and consider the powers of $D_1=D+c$. From (27a) we obtain by induction

$$\text{(28)} \qquad D_1{}^k = (D+c)^k = D^k + C_{k,1}D^{k-1}V_1(c) + C_{k,2}D^{k-2}V_2(c) + \cdots + V_k(c),$$

where

$$\text{(29)} \qquad V_1(c) = c, \qquad V_j(c) = V_{j-1}(c)' + V_{j-1}(c)c.$$

For $k=p^i$, (28) specializes to

$$\text{(30)} \qquad D_1{}^{p^i} = (D+c)^{p^i} = D^{p^i} + V_{p^i}(c).$$

D_1 evidently satisfies (27a), and from (30) and (27b) we have as the condition that D_1 also satisfies (27b),

$$\text{(31)} \qquad V(c) = V_{p^m}(c) - V_{p^{m-1}}(c)\beta_1 - \cdots - V_1(c)\beta_m = 0.$$

On the other hand if D_1 satisfies (27) the correspondence $D\rightarrow D_1$, $a\rightarrow a$ defines an automorphism of $\mathfrak{M}$ and conversely. Since every automorphism of $\mathfrak{M}\cong\mathfrak{F}_{p^m}$ is inner there exists an element $B\epsilon\mathfrak{M}$ such that

$$B^{-1}aB = a, \qquad B^{-1}D_1B = D$$

for all a in $\mathfrak{R}$. Since $\mathfrak{R}$ has maximum order for a commutative subfield of $\mathfrak{M}$, $B=b\epsilon\mathfrak{R}$ and hence the second condition gives

† This result is due to Ore, loc. cit., p. 581.

$$c = b^{-1}b', \tag{32}$$

i.e., c is a *logarithmic derivative*. We have therefore proved

THEOREM 15. *A necessary and sufficient condition that $c\epsilon\mathfrak{R}$ be a logarithmic derivative is that* (31) *hold.*

This is an analogue of Hilbert's theorem on the elements of norm 1 in a cyclic field. $V(c)$ takes the part of the norm and derivation that of the generating automorphism of the cyclic field.

We denote the set of logarithmic derivatives by $\mathfrak{L}$. Since $-b'/b = (b^{-1})'/(b^{-1})$and $b'/b+c'/c=(bc)'/bc$, $\mathfrak{L}$ is a group under addition and the correspondence $b \rightarrow b'/b$ establishes a homomorphism between the multiplicative group of $\mathfrak{R}$ and $\mathfrak{L}$. The elements corresponding to 0 here are those of $\mathfrak{F}$. Hence $\mathfrak{L} \cong \mathfrak{R}/\mathfrak{F}$.

By means of the recursion formula (29) we may prove by induction

$$V_j(c) = \sum_{i=1}^{j} P_{ij}, \qquad P_{ij} = \sum \frac{j!}{\alpha!\beta!\cdots}\left(\frac{c}{1!}\right)^{\alpha}\left(\frac{c'}{2!}\right)^{\beta}\cdots, \tag{33}$$

where the summation in P_{ij} is extended over all non-negative integers such that

$$\alpha + \beta + \gamma + \cdots = i, \qquad \alpha + 2\beta + 3\gamma + \cdots = j.$$

(The coefficients in P_{ij} are understood to be the integers obtained by cancelling the common factors in $j!/(\alpha!\beta!\cdots)(1!)^{\alpha}(2!)^{\beta}\cdots$.) By (33) it is easily seen that $V_p(c) = c^p + c^{(p-1)}$. Since

$$D_1^{p^j} = (D_1^{p^{j-1}})^p = (D^{p^{j-1}} + V_{p^{j-1}}(c))^p = D^{p^j} + V_{p^j}(c),$$

we have

$$V_{p^j}(c) = (V_{p^{j-1}}(c))^p + (V_{p^{j-1}}(c))^{(p^j - p^{j-1})}, \tag{34}$$

and hence

$$V_{p^j}(c) = c^{p^j} + (c^{(p-1)})^{p^{j-1}} + (c^{(p^2-1)})^{p^{j-2}} + \cdots + c^{(p^j-1)}. \tag{35}$$

Then $V_{p^j}(c)' = c^{(p^j)}$ and so by (27b), $(V(c))'=0$, i.e., $V(c)\epsilon\mathfrak{F}$ *for any c in* $\mathfrak{R}$. Also by (35), or more directly by (30),

$$V(b + c) = V(b) + V(c). \tag{36}$$

UNIVERSITY OF CHICAGO,
CHICAGO, ILL.

Reprinted from
Transactions of the American Mathematical Society
September 1937.

p-ALGEBRAS OF EXPONENT p*

BY NATHAN JACOBSON†

A. A. Albert and O. Teichmüller have recently investigated the structure of p-algebras, that is, normal simple algebras of degree p^e and characteristic p.‡ In particular they showed that a necessary and sufficient condition that such an algebra have exponent p is that it be similar to an algebra A having a maximal subfield $C=F(c_1, c_2, \cdots, c_m)$, where $c_i^p=\gamma_i \epsilon F$, the underlying field. The latter algebra is cyclic. It is the purpose of this note to apply some results of my paper *Abstract derivation and Lie algebras*§ to obtain a new generation of A. For $m=1$ this generation is more symmetric than the cyclic generation. We obtain a condition that A be a matrix algebra in terms of the new generation, and when $m=1$ we have as a consequence a reciprocity law for fields of characteristic p.

Let A be a normal simple algebra of degree p^m (order p^{2m}) over a field F of characteristic p and suppose A contains the maximal subfield $C=F(c_1, c_2, \cdots, c_m)$, $c_i^p=\gamma_i \epsilon F$. Let D be an arbitrary derivation of C over F, that is, a mapping $x \rightarrow xD$ of C into itself such that

$$(x+y)D = xD + yD, \qquad (x\alpha)D = (xD)\alpha,$$
$$(xy)D = (xD)y + x(yD), \qquad \alpha \epsilon F.$$

It is known that D may be chosen so that the only elements z such that $zD=0$ are those of F,‖ and for a D of this type I have shown that

$$(1) \qquad x(D^{p^m} + D^{p^{m-1}}\tau_1 + \cdots + D\tau_m) = 0, \qquad \tau_i \epsilon F,$$

* Presented to the Society, April 10, 1937.

† National Research Fellow.

‡ A. A. Albert, *On normal division algebras of degree p^e over F of characteristic p*, Transactions of this Society, vol. 39 (1936), pp. 183–188, and *Simple algebras of degree p^e over a centrum of characteristic p*, Transactions of this Society, vol. 40 (1936), pp. 112–126. O. Teichmüller, *p Algebren*, Deutsche Mathematik, vol. 1 (1936), pp. 362–388.

§ Transactions of this Society, vol. 42 (1937), pp. 206–224, referred to as J.

‖ R. Baer, *Algebraische Theorie der differentierbaren Funktionenkörper*. I, Sitzungsberichte Heidelberger Akademie, 1927, pp. 15–32.

or

$$x^{(p^m)} + x^{(p^{m-1})}\tau_1 + \cdots + x'\tau_m = 0$$

for all $x\epsilon C$, but no equation of the form

$$x^{(r)} + x^{(r-1)}b_1 + \cdots + x'b_{r-1} + b_r = 0, \qquad b_i\epsilon C,$$

can hold if $r<p^m$.* I have shown also that any derivation in a simple subalgebra of a normal simple algebra may be extended to an inner derivation in the latter.† Thus there exists an element d in A such that $[x, d]\equiv xd-dx=xD$ for all $x\epsilon C$.

We note that

$$(2) \qquad xd^k = d^kx + C_{k,1}d^{k-1}x' + \cdots + x^{(k)},$$

where the coefficients are those of the binomial theorem, and hence $xd^{p^j}=d^{p^j}x+x^{(p^j)}$. It follows from (1) that $(d^{p^m}+d^{p^{m-1}}\tau_1 + \cdots +d\tau_m)$ commutes with every x, and since C is a maximal subfield of A, $(d^{p^m}+d^{p^{m-1}}\tau_1+ \cdots +d\tau_m)=c\epsilon C$. Deriving with respect to d (taking commutators), we have $[c, d]=0$, and so $c=\delta\epsilon F$ and

$$(3) \qquad d^{p^m} + d^{p^{m-1}}\tau_1 + \cdots + d\tau_m = \delta.$$

We assert that C and d generate the whole of A. Suppose

$$(4) \qquad d^r + d^{r-1}b_1 + \cdots + b_r = 0, \qquad b_i\epsilon C,$$

is an equation of least degree having coefficients in C and satisfied by d. If $x\epsilon C$ by (2)

$$d^{r-1}x_1 + d^{r-2}x_2 + \cdots + x_r = 0,$$

where, if we use the $C_{r,k}$ notation for binomial coefficients,

$$x_1 = C_{r,1}x', \qquad x_2 = C_{r,2}x'' + C_{r-1,1}x'b_1, \cdots,$$
$$x_r = x^{(r)} + x^{(r-1)}b_1 + \cdots + x'b_{r-1}.$$

Since (4) has minimum degree, $x_1=x_2= \cdots =x_r=0$. But by (1) $x_r=0$ is impossible for all x unless $r\geqq p^m$. It follows that $r=p^m$ and $1, d, \cdots, d^{p^m-1}$ are (right) independent over C. Thus C and d generate an algebra of order p^m over C and hence p^{2m} over F, and so C and d generate all of A. The field C, the derivation

* J, p. 218.

† J, p. 214.

D, and the equation (3) give a complete description of A.

Let $V(x)$ for $x \epsilon C$ be the function

$$V_{p^m}(x) + V_{p^{m-1}}(x)\tau_1 + \cdots + V_1(x)\tau_m,$$

where

$$V_{p^i}(x) = x^{p^i} + (x^{(p-1)})^{p^{i-1}} + (x^{(p^2-1)})^{p^{i-2}} + \cdots + x^{(p^i-1)}.$$

I have shown that $V(x) \epsilon F$, and that it has properties analogous to the norm in cyclic fields.* Now suppose $\delta = V(x_0)$. If $d_1 = d - x_0$, then $[x, d_1] = xD$ for all x, and since

$$d_1^{p^i} = (d - x_0)^{p^i} = d^{p^i} - V_{p^i}(x_0),$$

we have

$$d_1^{p^m} + d_1^{p^{m-1}}\tau_1 + \cdots + d_1\tau_m = 0,$$

and so $A \cong F_{p^m}$, the algebra of all p^m-rowed square matrices with elements in F.†

Conversely suppose that $A \cong F_{p^m}$. Then there exists in A a field $\tilde{C} \cong C$ and an element $\tilde{d}_1$ such that $[\tilde{x}, \tilde{d}_1] = \widetilde{xD}$ where $x \longleftrightarrow \tilde{x}$ in the isomorphism between C and $\tilde{C}$ and

$$\tilde{d}_1^{p^m} + \tilde{d}_1^{p^{m-1}}\tau_1 + \cdots + \tilde{d}_1\tau_m = 0.$$

This isomorphism between C and $\tilde{C}$ may be extended to an automorphism in A.‡ Hence there exists an element d_1 corresponding to $\tilde{d}_1$ such that $[x, d_1] = xD$ and

$$d_1^{p^m} + d_1^{p^{m-1}}\tau_1 + \cdots + d_1\tau_m = 0.$$

We observe that $d - d_1$ commutes with all the elements of C, and hence $d_1 = d - x_0$, $x_0 \epsilon C$. It follows as before that $\delta = V(x_0)$.

THEOREM. *A necessary and sufficient condition that A be $\cong F_{p^m}$ is that $\delta = V(x_0)$, $x_0 \epsilon C$.*

We now consider the special case where $m = 1$, $C = F(c)$, $c^p = \gamma$. Let D be the derivation such that $cD = 1$. It is easily seen that $D^p = 0$ and hence A is generated by c and d such that

---

* See J, p. 224.

† The symbol $\cong$ denotes isomorphism. For the above equations and result see J, p. 223.

‡ M. Deuring, *Algebren*, 1935, p. 42.

$[c, d]=1$ and $d^p=\delta$. Thus A has the basis $d^i c^j$ $(i, j=0, 1, \cdots, p-1)$ such that

$$c^p = \gamma, \qquad d^p = \delta, \qquad cd - dc = 1. \tag{5}$$

The condition that $A \cong F_p$ is $\delta = V(x_0)$, $x_0 \epsilon F(c)$. Here $V(x)=x^p + x^{(p-1)}$, and so if $x=\xi_0+c\xi_1+\cdots+c^{p-1}\xi_{p-1}$, then

$$V(x) = (\xi_0^p - \xi_{p-1}) + \gamma\xi_1^p + \cdots + \gamma^{p-1}\xi_{p-1}^p. \tag{6}$$

If δ is not a p-th power, (5) is essentially symmetric in c and d. We define the derivation $d \rightarrow dE = -1$ in $F(d)$. Since $E^p=0$, the condition that $A \cong F_p$ is that

$$\gamma = V(y_0) = y_0 E^{p-1} + y_0^p, \quad y_0 \epsilon F(d).$$

But if

$$y = \eta_0 + d\eta_1 + \cdots + d^{p-1}\eta_{p-1},$$

then

$$V(y) = (\eta_0^p - \eta_{p-1}) + \delta\eta_1^p + \cdots + \delta^{p-1}\eta_{p-1}^p.$$

Thus we have the following reciprocity theorem for arbitrary fields of characteristic p.

THEOREM. *If γ and δ are not p-th powers in F, then $(\xi_0^p - \xi_{p-1}) + \gamma\xi_1^p + \cdots + \gamma^{p-1}\xi_{p-1}^p = \delta$ is solvable for $\xi_i \epsilon F$ if and only if $(\eta_0^p - \eta_{p-1}) + \delta\eta_1^p + \cdots + \delta^{p-1}\eta_{p-1}^p = \gamma$ is solvable for $\eta_i \epsilon F$.*

UNIVERSITY OF CHICAGO

Reprinted from
Bulletin of the American Mathematical Society
October 1937.

A NOTE ON TOPOLOGICAL FIELDS.*

By N. JACOBSON.[1]

In a recent paper [2] the author showed that a locally compact, separable, totally disconnected (l. c. s. t. d.) field $\mathfrak{F}$ has a valuation determined as follows: Let $\mathfrak{J}$ be the totality of elements a, called integers, such that $\{a^\nu\}$ is bounded (contained in a compact and closed set)[3] and $\mathfrak{P}$ the subset of elements b such that $b^\nu \to 0$. Then $\mathfrak{J}$ is a compact and open domain of integrity and $\mathfrak{P}$ a two-sided principal prime ideal $= (x)$ in $\mathfrak{J}$. Every element y in $\mathfrak{F}$ has the form ux^k where k is a rational integer and $u \in \mathfrak{J} \mid \mathfrak{P}$ the part of $\mathfrak{J}$ not in $\mathfrak{P}$. If we define $\exp y$ to be k and $|y| = \gamma^{\exp y}$ where $0 < \gamma < 1$ then y is a non-archimedean valuation of $\mathfrak{F}$, i. e.,

$$|y_1 y_2| = |y_1|\,|y_2| \qquad |y_1 + y_2| \leqq \max(|y_1|, |y_2|).$$

We showed also that $\mathfrak{F}$ is a cyclic algebra over its centrum $\mathfrak{C}$ and the latter is either a $\mathfrak{p}$-adic field [4] or a field of power series in an indeterminate z with coefficients in a finite field. The former case obtains if the characteristic $\chi(\mathfrak{F}) = 0$ and the latter, which we shall call a z-adic field, if $\chi(\mathfrak{F}) = p \neq 0$.

Now if $\mathfrak{C}$ is any commutative $\mathfrak{p}$-adic or z-adic field it may be topologized by means of the neighborhoods $\{a + U_k\}$ of a where U_k is the set of points $(\alpha_0 + \alpha_1\pi + \cdots)\pi^k$, $k = 0, \pm 1, \pm 2, \cdots$, $(\mathfrak{p} = (\pi))$ in the $\mathfrak{p}$-adic field or $(\alpha_0 + \alpha_1 z + \cdots)z^k$ in the z-adic field. In this topology a sequence of elements $a^{(\nu)} = (\alpha_0^{(\nu)} + \alpha_1\pi^{(\nu)} + \cdots)\pi^{k_\nu}$ $((\alpha_0^{(\nu)} + \alpha_1^{(\nu)}z + \cdots)z^{k_\nu})$ converges to 0 if and only if $k_\nu \to \infty$. It follows that the neighborhoods $a + U_k$ are closed and, since the α_i have only a finite range, it is easily seen that they are also compact. The set of elements of finite length, e. g. $(\alpha_0 + \alpha_1\pi + \cdots \alpha_m\pi^m)\pi^k$ is denumerable and everywhere dense in $\mathfrak{C}$. Hence $\mathfrak{C}$ is a l. c. s. t. d. field. If $\mathfrak{F}$ is an algebra with a finite basis over $\mathfrak{C}$ it is a vector space over $\mathfrak{C}$ and hence it is l. c. s. t. d. in the usual vector space topology. Thus we see that the theory of l. c. s. t. d. fields is equivalent to that of division algebras over a $\mathfrak{p}$-adic or a z-adic field. In this note we shall apply the topological methods

---

* Presented to the Society September 7, 1937. Received by the Editors June 15, 1937.

[1] National Research Fellow.

[2] "Totally disconnected locally compact rings," *American Journal of Mathematics*, vol. 58 (1936), pp. 433-449. This paper will be referred to as T.

[3] An equivalent condition is that $\{a^\nu\}$ have no divergent subsequence: T, p. 444.

[4] i. e. an algebraic extension of finite order of a field of Hensel p-adic numbers.

of our earlier paper to investigate the automorphisms and anti-automorphisms of $\mathfrak{F}$.

Let J be an automorphism or an anti-automorphism of the l. c. s. t. d. field $\mathfrak{F}$, i. e. a $(1-1)$ correspondence of $\mathfrak{F}$ on itself such that $(a+b)^J = a^J + b^J$ and $(ab)^J = a^J b^J$ (automorphism) or $(ab)^J = b^J a^J$ (anti-automorphism). If $c \in$ the centrum $\mathfrak{C}$, $c^J \in \mathfrak{C}$ also and hence the correspondence $c \rightarrow c^J$ is an automorphism of $\mathfrak{C}$. By a theorem of F. K. Schmidt [5] J is a continuous transformation in $\mathfrak{C}$.

Suppose $b_1, b_2, \cdots, b_m$ $(m = r^2)$ [6] is a basis for $\mathfrak{F}$ over $\mathfrak{C}$. Then every y may be written uniquely in the form $\Sigma c_i b_i$, $c_i \in \mathfrak{C}$. Set $c_i = E_i(y)$.

LEMMA 1. *$E_i(y)$ is a continuous function of y.*

Since $E_i(y)$ is linear it suffices to show that if $y_\nu = \Sigma c_i^{(\nu)} b_i \rightarrow 0$ then $E_i(y_\nu) = c_i^{(\nu)} \rightarrow 0$. If $c_i^{(\nu)} \nrightarrow 0$ there exists a subsequence which does not have 0 as a limit point and we may suppose that this is the whole sequence $\{c_i^{(\nu)}\}$. For each ν let $\lambda(\nu) = 1, \cdots, m$ be an index such that $| c_{\lambda(\nu)}^{(\nu)} | \geqq | c_j^{(\nu)} |$ for all $j = 1, \cdots, m$. Since $\lambda(\nu)$ has a finite range, one of its values, say $\lambda(\nu) = 1$, occurs infinitely often and hence by restricting ourselves to a subsequence we may suppose that $| c_1^{(\nu)} | \geqq | c_j^{(\nu)} |$ for all ν. Since $\lim c_i^{(\nu)} \neq 0$, $\lim c_1^{(\nu)} \neq 0$ and there is a subsequence such that $\lim (c_1^{(\nu_k)})^{-1} = d \in \mathfrak{C}$.[7] Again we write ν for ν_k Then

$$\lim (c_1^{(\nu)})^{-1} y_\nu = 0 = \lim (b_1 + (c_1^{(\nu)})^{-1} c_2^{(\nu)} b_2 + \cdots + (c_1^{(\nu)})^{-1} c_m^{(\nu)} b_m).$$

Since $|(c_1^{(\nu)})^{-1} c_j^{(\nu)} | \leqq 1$ we may suppose that $c_1^{(\nu)} c_j^{(\nu)} \rightarrow c_j \in \mathfrak{C}$. It follows that $b_1 + c_2 b_2 + \cdots + c_m b_m = 0$ contrary to the linear independence of the b's.

LEMMA 2. *J is a continuous mapping of $\mathfrak{F}$ on itself.*

Suppose

$$y_\nu = c_1^{(\nu)} b_1 + \cdots + c_m^{(\nu)} b_m \rightarrow y = c_1 b_1 + \cdots + c_m b_m.$$

By Lemma 1, $c_i^{(\nu)} \rightarrow c_i$ and since J is continuous in $\mathfrak{C}$, $c_i^{(\nu)J} \rightarrow c_i^J$. Hence

$$y_\nu^J = c_1^{(\nu)J} b_1^J + \cdots + c_m^{(\nu)J} b_m^J \rightarrow y^J = c_1^J b_1^J + \cdots + c_m^J b_m^J.$$

---

[5] F. K. Schmidt, "Mehrfach perfekte Körper," *Mathematische Annalen*, vol. 108 (1933), pp. 1-25.

[6] The order of a division algebra over its centrum is the square of an integer called its degree. See, for example, Deuring's "Algebren," *Ergebnisse der Mathematik*, p. 47.

[7] We require here the property of a locally compact field that $a_\nu \rightarrow 0$ if $\{a_\nu\}$ diverges: T, p. 442.

Since J^{-1} is also an automorphism or anti-automorphism it is continuous and hence J is a homeomorphism of $\mathfrak{F}$.

THEOREM 1. *Any automorphism or anti-automorphism J of $\mathfrak{F}$ is isometric, i. e., $|y^J| = |y|$ for every y in $\mathfrak{F}$.*

The condition that $\exp u = 0$ is that $\{u^\nu\}$ and $\{u^{-\nu}\}$ be bounded (or have no divergent subsequence). Since J is a homeomorphism $\{(u^J)^\nu\} = \{(u^\nu)^J\}$ and $\{(u^J)^{-\nu}\}$ are bounded also. Hence $\exp u^J = 0$. Now let x be an element such that $\mathfrak{P} = (x)$, or $\exp x = 1$. Consider x^J. Since $(x^J)^\nu = (x^\nu)^J \to 0, x^J \epsilon \mathfrak{P}$ and hence $\exp x^J = l > 0$. Similarly $\exp x^{J^{-1}} = k > 0$. If $x^J = ux^l$ where $\exp u = 0$, $x = u^{J^{-1}}(x^{J^{-1}})^l$ or $(x^{J^{-1}})^l u^{J^{-1}}$. Hence $1 = \exp x = lk$ and so $l = k = 1$. Thus $\exp x^J = 1$. If $\exp y = n$ an integer, $y = ux^n$ and $y^J = u^J(x^J)^n$ or $= (x^J)^n u^J$ and hence $\exp y^J = n$. Hence for any y, $|y^J| = |y|$.

The condition that $\mathfrak{F}$ have an automorphism is evidently no restriction since the identity transformation is such a correspondence. On the other hand we shall see that the existence of an anti-automorphism strongly restricts the structure of $\mathfrak{F}$.

We recall that $\mathfrak{F}$ is cyclic over $\mathfrak{C}$. Its degree $r(= \sqrt{m})$ over $\mathfrak{C}$ may be characterized as the order of the automorphism $u \to xux^{-1}$ (mod $\mathfrak{P}$) of the finite field $\mathfrak{J} - \mathfrak{P}$ where x is determined as above.[8] We have for any u in $\mathfrak{J}$

$$xux^{-1} \equiv u^s \pmod{\mathfrak{P}} \qquad (s = p^t)$$

where p is the characteristic of $\mathfrak{J} - \mathfrak{P}$. It follows that

$$(x^J)^{-1}u^J x^J \equiv (u^s)^J \pmod{\mathfrak{P}}$$
$$\equiv (u^J)^s \pmod{\mathfrak{P}}.$$

Since $|x^J| = |x|$, $x^J = vx$ where $|v| = 1$. Then

$$(u^J)^s \equiv x^{-1}(v^{-1}u^J v)x \equiv x^{-1}u^J x \pmod{\mathfrak{P}}$$

since $v^{-1}u^J v \equiv u^J$ (mod $\mathfrak{P}$). Thus

$$x^{-1}u^J x \equiv xu^J x^{-1} \pmod{\mathfrak{P}}$$

for all u^J. Since u^J varies over a complete set of residues mod $\mathfrak{P}$ when u does we have

$$x^{-1}ux \equiv xux^{-1} \pmod{\mathfrak{P}}$$
$$x^2ux^{-2} \equiv u \pmod{\mathfrak{P}}.$$

Thus the order of the automorphism $u \to xux^{-1}$ (mod $\mathfrak{P}$) is either 1 or 2 and hence $r = 1$ or 2.

[8] T, pp. 446-448.

THEOREM 2. *The degree of an anti-automorphic l. c. s. t. d. field is either* 1 *or* 2.

The condition $r = 1$ or 2 is also sufficient that $\mathfrak{F}$ have an anti-automorphism. For if $r = 1$, $\mathfrak{F}$ is commutative and any automorphism is an anti-automorphism. If $r = 2$, $\mathfrak{F}$ is a quaternion algebra over its centrum and it is well known that $\mathfrak{F}$ has anti-automorphisms J. In fact J may be chosen so that $J^2 = I$ the identity and $c^J = c$ for all c in $\mathfrak{C}$.

We suppose now that J is an involutorial anti-automorphism ($J^2 = I$) and the characteristic $\chi(\mathfrak{F}) \neq 2$. Let $\mathfrak{K}_J$ denote the symmetric or invariant elements ($h^J = h$), $\mathfrak{S}_J$ the set of skew elements ($s^J = - s$) and $\mathfrak{C}_0 = \mathfrak{K}_J \frown \mathfrak{C}$. Evidently $\mathfrak{C}_0$ is a closed subfield of $\mathfrak{C}$ and hence is l. c. s. t. d. For any y we have $y = \frac{1}{2}(y + y^J) + \frac{1}{2}(y - y^J) = h + s$ where $h = \frac{1}{2}(y + y^J) \in \mathfrak{K}_J$ and $s = \frac{1}{2}(y - y^J) \in \mathfrak{S}_J$. In particular this holds for y in $\mathfrak{C}$ and in this case $h \in \mathfrak{C}_0$, $s \in \mathfrak{S}_J \frown \mathfrak{C}$. Hence if $\mathfrak{C} \neq \mathfrak{C}_0$ there exists an element $q \neq 0$ in $\mathfrak{S}_J \frown \mathfrak{C}$. Then $y = h + (sq^{-1})q$ and $sq^{-1} \in \mathfrak{K}_J$. For c in $\mathfrak{C}$ we obtain a quadratic extension $\mathfrak{C}_0(q)$ of $\mathfrak{C}_0$. It follows that $\mathfrak{C}_0$ is infinite and therefore is either a $\mathfrak{p}$-adic or a z-adic field.[9] We note also that $q^2 = q_0$ is symmetric

We wish to show that if $\mathfrak{C} \neq \mathfrak{C}_0$ then $\mathfrak{F} = \mathfrak{C}$, i. e., $\mathfrak{F}$ is commutative. For the present we drop the restriction that $\mathfrak{C}_0$ is l. c. s .t. d. and allow it to denote any commutative field of characteristic $\neq 2$. We suppose that $\mathfrak{F}$ is a division algebra of order 8 over $\mathfrak{C}_0$ with centrum $\mathfrak{C} = \mathfrak{C}_0(q)$ and J is an involutorial anti-automorphism of $\mathfrak{F}$ leaving the elements of $\mathfrak{C}_0$ invariant and mapping q into $- q$. The order of $\mathfrak{F}$ over $\mathfrak{C}$ is of course 4.

It is easily seen that $\mathfrak{K}_J$ and $\mathfrak{S}_J$ defined as above are vector spaces over $\mathfrak{C}_0$. If $h \in \mathfrak{K}_J$, $hq \in \mathfrak{S}_J$ and if $s \in \mathfrak{S}_J$, $sq \in \mathfrak{K}_J$. Hence these spaces have the same order over $\mathfrak{C}_0$. Evidently $\mathfrak{S}_J \frown \mathfrak{K}_J = 0$ and as we showed before $\mathfrak{F} = \mathfrak{S}_J + \mathfrak{K}_J$. It follows that $\mathfrak{K}_J$ and $\mathfrak{S}_J$ have order 4 over $\mathfrak{C}_0$ and since $\mathfrak{K}_J = \mathfrak{S}_J q$, $\mathfrak{F} = \mathfrak{S}_J \mathfrak{C}$.[10]

Let $T(y)$ and $N(y)$ denote respectively the trace and norm of y relative to $\mathfrak{C}$. Then $y^2 - T(y)y + N(y) = 0$ and if $y \not\in \mathfrak{C}$ this is the equation of least degree having coefficients in $\mathfrak{C}$ and satisfied by y. Then

$$(y^J)^2 - T(y)^J y^J + N(y)^J = 0 = (y^J)^2 - T(y^J) y^J + N(y^J)$$

and hence for $y \not\in \mathfrak{C}$

[9] We note that any infinite closed subfield $\mathfrak{C}_0$ of a l. c. s. t. d. field is not discrete. For if $\{a_\nu\}$ is a divergent sequence in $\mathfrak{C}_0$ then $a_\nu^{-1} \to 0$.
and hence $\in \mathfrak{C}_0$.

[10] If $\mathfrak{A}$ and $\mathfrak{B}$ are vector spaces over $\mathfrak{C}_0$, $\mathfrak{A}\mathfrak{B}$ denotes the smallest vector space over $\mathfrak{C}_0$ containing all the products ab where $a \in \mathfrak{A}$ and $b \in \mathfrak{B}$.

$$(*) \qquad T(y^J) = T(y)^J \qquad N(y^J) = N(y)^J.$$

But for $y \in \mathfrak{C}$, $T(y) = 2y$, $N(y) = y^2$ and so (*) holds for every y in $\mathfrak{F}$. If $y \in \mathfrak{S}_J$, $T(y)^J = T(y^J) = T(-y) = -T(y)$ and $N(y)^J = N(-y) = N(y)$, i. e.. $T(y) \in \mathfrak{S}_J \frown \mathfrak{C}$ and $N(y) \in \mathfrak{C}_0$. $\mathfrak{S}_J \frown \mathfrak{C}_0$ consists of the $\mathfrak{C}_0$-multiples of q.

Let $\mathfrak{S}'_J$ be the totality of elements of trace 0 in $\mathfrak{S}_J$. $\mathfrak{S}'_J$ is a vector space over $\mathfrak{C}_0$ and if $y, t \in \mathfrak{S}'_J$ so does $[y, t] = yt - ty$. If y is any element of $\mathfrak{S}_J$, $y = \frac{1}{2}T(y) + y_0$ where $y_0 = y - \frac{1}{2}T(y) \in \mathfrak{S}'_J$. Thus $\mathfrak{S}_J = \mathfrak{S}'_J + \mathfrak{C}_0 q$ and $\mathfrak{S}'_J \frown \mathfrak{C}_0 q = 0$. It follows that the order of $\mathfrak{S}'_J$ over $\mathfrak{C}_0$ is 3. Since $1 \not\in \mathfrak{C}'_J$, the order of $\mathfrak{F}_0 = \mathfrak{S}'_J + \mathfrak{C}_0 1$ is 4 over $\mathfrak{C}_0$.

For y, t in $\mathfrak{S}'_J$ we have

$$\begin{aligned} yt &= \tfrac{1}{2}[(y+t)^2 - y^2 - t^2 + yt - ty] \\ &= \tfrac{1}{2}[-N(y+t) + N(y) + N(t) + yt - ty] \in \mathfrak{F}_0. \end{aligned}$$

Hence $\mathfrak{F}_0$ is an algebra over $\mathfrak{C}_0$. Since $\mathfrak{F}_0 = \mathfrak{S}'_J + \mathfrak{C}_0 1$, $\mathfrak{S}_J = \mathfrak{S}'_J + \mathfrak{C}_0 q$ and $\mathfrak{F} = \mathfrak{S}_J \mathfrak{C}$, we have $\mathfrak{F} = \mathfrak{F}_0 \mathfrak{C}$. We have therefore proved the following theorem.

THEOREM 3. *If $\mathfrak{F}$ is a division algebra of order 8 over $\mathfrak{C}_0$ of characteristic $\neq 2$ with $\mathfrak{C} = \mathfrak{C}_0(q)$, $q^2 \in \mathfrak{C}_0$, as centrum and J an involutorial anti-automorphism in $\mathfrak{F}$ such that $q^J = -q$ then $\mathfrak{F}$ is a direct product $\mathfrak{C} \times \mathfrak{F}_0$ of $\mathfrak{C}$ and $\mathfrak{F}_0$ the join of $\mathfrak{C}_0 1$ and the space of skew elements of trace 0.*[11]

Since the centrum of $\mathfrak{F}$ is $\mathfrak{C}$ that of $\mathfrak{F}_0$ must be $\mathfrak{C}_0$.

If $\mathfrak{C}_0$ is a commutative l. c. s. t. d. field it is well known that an algebra $\mathfrak{F}_0$ of degree 2 over $\mathfrak{C}_0$ has every quadratic field as a splitting field, i. e., if $\mathfrak{C} = \mathfrak{C}_0(q)$ then $\mathfrak{C} \times \mathfrak{F}_0$ is a complete matrix algebra over $\mathfrak{C}$.[12] Hence

THEOREM 4. *If $\mathfrak{F}$ is an l. c. s. t. d. field of characteristic $\neq 2$ and J an involutorial anti-automorphism such that $c^J \neq c$ for some c in the centrum $\mathfrak{C}$, then $\mathfrak{F} = \mathfrak{C}$ is commutative.*

In the remainder of the paper we shall extend the above results to $\mathfrak{F}_n$ a complete matrix algebra of n rows and columns over an l. c. s. t. d. field $\mathfrak{F}$. Let e_{ij} $(i, j = 1, \cdots, n)$ be a set of matrix units in $\mathfrak{F}_n$; $e_{ij} e_{kl} = \delta_{jk} e_{il}$ and every element b in $\mathfrak{F}_n$ has the form $\Sigma y_{ij} e_{ij}$ where $y_{ij} \in \mathfrak{F}$. As before let $\mathfrak{C}$

[11] Cf. A. A. Albert, "Involutorial simple algebras and real Riemann matrices," *Annals of Mathematics*, vol. 36 (1935), p. 909.

[12] Deuring's *Algebren*, p. 113. The proof given there for $\mathfrak{C}_0$ a $\mathfrak{p}$-adic field holds also for a z-adic field. See *Algebren*, p. 137.

denote the centrum of $\mathfrak{F}$ or of $\mathfrak{F}_n$. We recall that $\mathfrak{F}$ may be characterized as the totality of elements commutative with the e_{ij}.

Let J be an anti-automorphism of $\mathfrak{F}_n$ and $e_{ji}{}^J = f_{ij}$. Then $f_{ij}f_{kl} = \delta_{jk}f_{il}$ and it follows that there exists a non-singular element s in $\mathfrak{F}_n$ such that $e_{ij} = s^{-1}f_{ij}s$. The correspondence $b \to b^K = s^{-1}b^Js$ is an anti-automorphism also and $e_{ij}{}^K = e_{ji}$. Hence if $y \in \mathfrak{F}$, y^K which commutes with the e_{ij} belongs to $\mathfrak{F}$ also and $\mathfrak{F}$ itself is anti-automorphic. By Theorem 1 we have

THEOREM 5. *If $\mathfrak{F}$ is an l. c. s. t. d. field such that $\mathfrak{F}_n$ is anti-automorphic then $\mathfrak{F}$ has degree* 1 *or* 2 *over its centrum.*

If J is involutorial, then

$$e_{ij} = e_{ij}{}^{J^2} = f_{ji}{}^J = (se_{ji}s^{-1})^J = (s^{-1})^J se_{ij}s^{-1}s^J.$$

Thus $s^{-1}s^J$ commutes with all e_{ij} and $s^J = su$ where $u \in \mathfrak{F}$. If $u = -1$ it is easily seen that the correspondence $b \to b^K = s^{-1}b^Js$ is an involutorial anti-automorphism in $\mathfrak{F}_n$ mapping $\mathfrak{F}$ into itself. If $u \neq 1$, $s^J + s = s(u+1) = t$ is non-singular and $t^J = t$. Hence the correspondence $b \to b^K = t^{-1}b^Jt$ is an involutorial anti-automorphism mapping $\mathfrak{F}$ into itself.[13] In either case $\mathfrak{F}$ is involutorial anti-automorphic. Evidently if $c^J \neq c$ then $c^K \neq c$ for c in $\mathfrak{C}$. Hence by Theorem 4 we have

THEOREM 6. *If $\mathfrak{F}$ is an l. c. s. t. d. field of characteristic $\neq 2$ and $\mathfrak{F}_n$ has an involutorial anti-automophism J mapping an element c of the centrum into $c^J \neq c$ then $\mathfrak{F} = \mathfrak{C}$ is commutative.*

PRINCETON UNIVERSITY AND
THE INSTITUTE FOR ADVANCED STUDY.

Reprinted from
American Journal of Mathematics
October 1937.

[13] Cf. Albert, *loc. cit.* in 11, p. 897.

ANNALS OF MATHEMATICS
Vol. 39, No. 1, January, 1938

SIMPLE LIE ALGEBRAS OF TYPE A[1]

BY N. JACOBSON[2]

(Received March 24, 1937)

In a recent paper[3] the author discussed the Lie algebras of characteristic 0 obtainable as the set of skew elements of an involutorial normal simple associative algebra. The present paper gives an extension of these results to simple associative algebras of second kind as defined by Albert.[4] The resulting Lie algebras in this case constitute Landherr's class A_{II}.[5] We reduce the problem of classifying these Lie algebras, as in our earlier work on Lie algebras of types B, C, D,[6] to two standard problems in associative algebra, namely, classification of involutorial simple algebras and of generalized hermitian matrices relative to cogredience. In this sense a complete determination of normal simple Lie algebras of characteristic 0 except those of a finite number of orders[7] results from Landherr's work on the algebras of type A_I and our own on types A_{II}, B, C, D.

1. Let $\mathfrak{A}$ be a simple associative algebra of order $2n^2$ over the field Φ of characteristic 0 and $P = \Phi(q)$, $q^2 = \mu \in \Phi$, be its centrum. Suppose $\mathfrak{A}$ is involutorial of the second kind i.e. there is defined a $(1 - 1)$ correspondence J in $\mathfrak{A}$ such that

$$(a + b)^J = a^J + b^J \qquad (ab)^J = b^J a^J \qquad (\xi a)^J = \xi^J a^J \equiv \bar{\xi} a^J \qquad a^{J^2} = a$$

where $\xi^J \equiv \bar{\xi} = \alpha - \beta q$ if $\xi = \alpha + \beta q$, $\alpha, \beta \in \Phi$. We shall call J an involutorial anti-automorphism (i.a.a.) of *second kind.* It is easily seen that $\mathfrak{S}_J$ the set of J-skew elements ($a^J = -a$) forms a Lie algebra over Φ relative to commutator multiplication $[a, b] \equiv ab - ba$.[8] $\mathfrak{H}_J$ the set of J-symmetric elements ($a^J = a$) is a vector space over Φ and if $a \in \mathfrak{H}_J$, $qa \in \mathfrak{S}_J$. On the other

---

[1] Presented to the Society March 26, 1937. An abstract of this paper appeared in the Proc. Nat. Acad. Sci. **23** (1937) pp. 240–242.

[2] National Research Fellow.

[3] *A class of normal simple Lie algebras of characteristic* 0, Annals of Math. **38** (1937) pp. 508–517, referred to hereafter as J.

[4] A. A. Albert, *Involutorial simple algebras and real Riemann matrices*, Annals of Math. **36** (1935) pp. 894–910.

[5] W. Landherr, *Über einfache Liesche Ringe*, Hamb. Abhandlungen **11** (1935) pp. 41–64, referred to as L.

[6] J.

[7] The exceptional orders are those of the five exceptional Lie algebras of Killing and Cartan and order 28 corresponding to an algebra of type D. Cf. E. Cartan, *Thèse* Paris 1894, Chap. 5 and J. p. 513. The Lie algebras of type A_{II} over a p-adic field were determined prior to the present work by Landherr loc. cit.

[8] Cf. J. p. 508.

hand if $a \in \mathfrak{S}_J$, $qa \in \mathfrak{H}_J$. Thus the two vector spaces $\mathfrak{S}_J$ and $\mathfrak{H}_J$ have the same order over Φ. Evidently the intersection $\mathfrak{S}_J \frown \mathfrak{H}_J = 0$ and if a is any element of $\mathfrak{A}$, $a = \frac{1}{2}(a + a^J) + \frac{1}{2}(a - a^J) = b + c$ where $b \in \mathfrak{H}_J$ and $c \in \mathfrak{S}_J$. Thus $\mathfrak{A} = \mathfrak{S}_J + \mathfrak{H}_J$ and hence the order of $\mathfrak{S}_J$ over Φ, $(\mathfrak{S}_J : \Phi) = (\mathfrak{H}_J : \Phi) = n^2$. If $a_0, a_1, \cdots, a_m$ $(m = n^2 - 1)$ is a basis for $\mathfrak{S}_J$ over Φ, $qa_0, qa_1, \cdots, qa_m$ is one for $\mathfrak{H}_J$ over Φ and so $a_0, a_1, \cdots, a_m$ is a basis for $\mathfrak{A}$ over P. Hence $\mathfrak{S}_{JP}$[9] is isomorphic to $\mathfrak{A}$ over P regarded as a Lie algebra relative to commutator multiplication. If $\mathfrak{B}$ is any Lie algebra we denote its derived algebra $[\mathfrak{B}, \mathfrak{B}]$ by $\mathfrak{B}'$. Evidently $(\mathfrak{B}_\Lambda)' = (\mathfrak{B}')_\Lambda$ for Λ any extension field of the field of $\mathfrak{B}$. In particular $\mathfrak{S}'_{JP} \cong \mathfrak{A}'$ (over P).

If Ω is the algebraic closure of P (or of Φ), for $\mathfrak{A}$ over P we have $\mathfrak{A}_\Omega \cong \Omega_n$ the matrix algebra of n rows and columns over Ω and hence $\mathfrak{S}'_{J\Omega} \cong \Omega'_n$.[10] Since Ω'_n is simple,[11] $\mathfrak{S}'_J$ is a normal simple Lie algebra. We note also that $\mathfrak{S}'_J$ has order m over Φ since Ω'_n has this order over Ω and we may suppose that in the basis $a_0, a_1, \cdots, a_m$ for $\mathfrak{A}$ over P, $a_1, \cdots, a_m$ form a basis for $\mathfrak{S}'_J$ over Φ. Since $[q, a] = 0$, $q \notin \mathfrak{S}'_J$ for otherwise $\mathfrak{S}'_J$ would not be simple. Hence we may suppose also that $a_0 = q$ in the basis $a_0, a_1, \cdots, a_m$ for $\mathfrak{A}$ over P or $\mathfrak{S}_J$ over Φ.

Considering $\mathfrak{A}$ over P, the condition $\mathfrak{A}_\Omega \cong \Omega_n$ means that $\mathfrak{A}$ over P may be represented by matrices in Ω_n such that the linear combinations of these matrices with coefficients in Ω is Ω_n. Hence if $a_i \to A_i$ in the representation $A_0, A_1, \cdots, A_m$ form a basis for Ω_n and $A_1, \cdots, A_m$ for Ω'_n over Ω. Employing a matric basis E_{ij} such that $E_{ij}E_{kl} = \delta_{jk}E_{il}$ we see that Ω'_n is generated by E_{ij} $(i \neq j)$ and $E_{ii} - E_{jj}$ and hence consists of all the matrices of trace 0 and so $A_1, \cdots, A_m$ is a basis over Ω for these matrices.

The enveloping algebra over Φ of $\mathfrak{S}_J$ (smallest algebra over Φ contained in $\mathfrak{A}$ and containing $\mathfrak{S}_J$) is $\mathfrak{A}$ itself since it contains $q, a_1, \cdots, a_m$ and $q^2, qa_1, \cdots, qa_m$. We shall require the stronger result:

LEMMA 1. *If $n > 2$ the enveloping algebra over Φ of $\mathfrak{S}'_J$ is $\mathfrak{A}$.*

Let $\mathfrak{B}$ denote this algebra. Since $\mathfrak{B}$ contains $a_1, \cdots, a_m$ it suffices to show that $\mathfrak{B}$ contains q also and by the above discussion $\mathfrak{B}$ will contain q if it contains any element of $\mathfrak{S}_J$ not in $\mathfrak{S}'_J$. Now there exists a matrix W in Ω_n such that tr $W = 0$ but tr $W^3 \neq 0$. For example we may take

$$W = \rho_1 E_{11} + \rho_2 E_{22} + \cdots + \rho_n E_{nn}$$

where $\rho_1 + \rho_2 + \cdots + \rho_n = 0$ but $\rho_1^3 + \rho_2^3 + \cdots + \rho_n^3 \neq 0$. Then $W = \omega_1 A_1 + \cdots + \omega_m A_m$, $\omega_i \in \Omega$. It follows that there are elements $\alpha_1, \cdots, \alpha_m$ in Φ also such that tr $A^3 \neq 0$ where $A \doteq \alpha_1 A_1 + \cdots + \alpha_m A_m$. Then the

[9] If $\mathfrak{A}$ is a Lie algebra or an associative algebra with basis $a_1, a_2, \cdots$ over Φ and Σ an extension field of Φ then $\mathfrak{A}_\Sigma$ denotes the algebra having the basis $a_1, a_2, \cdots$, but with coefficients in Σ.

[10] In general we shall denote the algebra of n-rowed matrices with coördinates in an algebra $\mathfrak{A}$ by $\mathfrak{A}_n$.

[11] Ω'_n is one of the simple algebras (class A) in Killing-Cartan's classification. If Ω is the field of complex numbers Ω'_n is the infinitesimal group of the unimodular linear group.

element $a = \alpha_1 a_1 + \cdots + \alpha_m a_m \in \mathfrak{S}'_J$ and a^3 is an element of $\mathfrak{B}$ contained in $\mathfrak{S}_J$ but not in $\mathfrak{S}'_J$.

It is well known that the field of the coefficients of the characteristic equations of the matrices (representing the elements) of $\mathfrak{A}$ over P in Ω_n is P.[12] By Newton's identities this field is the same as the field of the traces of the matrices of $\mathfrak{A}$. It follows that if $n > 2$ the field Λ of the characteristic equations of the matrices of $\mathfrak{S}'_J$ is also P. For by a result of an earlier paper[13] the traces of the matrices of an enveloping algebra of a Lie algebra $\mathfrak{L}$ are expressible rationally in terms of the coefficients of the characteristic equations of the elements of $\mathfrak{L}$. Hence by Lemma 1 $\Lambda = \text{P}$. Following Landherr[14] we shall say that a Lie algebra $\mathfrak{L}$ over Φ has *type* A_{II} if $\mathfrak{L}_\Omega \cong \Omega'_n$ and in the representation of $\mathfrak{L}$ by matrices in Ω'_n the field of the characteristic equations of the matrices of $\mathfrak{L}$ is a proper overfield of Φ. With this definition we have

THEOREM 1. *If $\mathfrak{A}$ is an involutorial simple algebra of second kind and order $2n^2$, $n > 2$, over Φ and J is an* i.a.a. *of second kind in $\mathfrak{A}$ then $\mathfrak{S}'_J$ is a normal simple Lie algebra of type A_{II}.*

If $n = 2$ we have for any a in $\mathfrak{A}$ that $a^2 - \text{tr}\,(a)a + N(a) = 0$ where $N(a)$ is the determinant in the representation in Ω_n. Hence if $a \in \mathfrak{S}'_J$, $a^2 = \alpha$ and since a^2 is J-symmetric $\alpha \in \Phi$. Thus the field $\Lambda = \Phi$ in this case and the enveloping algebra over Φ of $\mathfrak{S}'_J$ is $\mathfrak{B} \neq \mathfrak{A}$.

2. If G is an automorphism of $\mathfrak{A}$ over Φ it induces an automorphism in P. Hence we have either $\xi^G = \xi$ for all ξ in P or $\xi^G = \bar{\xi}$. In the former case G is an automorphism of $\mathfrak{A}$ over P and hence is inner i.e. there exists an element g in $\mathfrak{A}$ such that $a^G = g^{-1}ag$ for all a.[15] If S is a second automorphism then $S^{-1}GS$ is the inner automorphism $a \to (g^S)^{-1}ag^S$. Thus the inner automorphisms constitute an invariant subgroup of the complete group of automorphisms. If S_1 and S_2 are outer automorphisms we must have $\xi^{S_1} = \bar{\xi} = \xi^{S_2}$ and hence $\xi^{S_1 S_2^{-1}} = \xi$. It follows that $S_1 S_2^{-1}$ is inner and so if there exist outer automorphisms the index of the group of inner automorphisms in the complete group is 2.

We note that if S is an outer automorphism and J an i.a.a. of second kind then $\xi^{SJ} = \xi$ and hence $V = SJ$ is an anti-automorphism of $\mathfrak{A}$ over P. Since $\mathfrak{A}$ over P is normal simple it follows from a theorem of Brauer's that $\mathfrak{A}$ over P has exponent 1 or 2 and if $\mathfrak{A} = \mathfrak{F}_r$ where $\mathfrak{F}$ is a normal division algebra then the degree of $\mathfrak{F}$ is 2^e, $e \geqq 0$.[16] Conversely if $\mathfrak{A}$ over P has exponent 1 or 2 it has an anti-automorphism V and hence $\mathfrak{A}$ over Φ has an outer automorphism $S = VJ$.

---

[12] See for example M. Deuring, *Algebren*, Ergebnisse der Math. Springer 1935, pp. 50–52.

[13] N. Jacobson, *Rational methods in the theory of Lie algebras*, Annals of Math. **36** (1935) pp. 879–880.

[14] L. p. 50.

[15] For a proof of this theorem see Deuring's *Algebren*, p. 43.

[16] R. Brauer, *Über Systeme hyperkomplexer Zahlen*, Math. Zeitsch. **30** (1929) p. 103 or Deuring's Algebren p. 45 and pp. 58–59.

If K is a second i.a.a. (not necessarily distinct from J) cogredient to J i.e. $K = S^{-1}JS$ for some automorphism S then K is also of the second kind and it is easily seen that $\mathfrak{S}_J \cong \mathfrak{S}_K = \mathfrak{S}_J^S$ and hence also $\mathfrak{S}_J' \cong \mathfrak{S}_K'$.[17] The following criterion for cogredience will be used below:

LEMMA 2. *Suppose $n > 2$ and J and K are i.a.a. of the second kind such that there is an automorphism S of $\mathfrak{A}$ over Φ carrying $\mathfrak{S}_J'$ into $\mathfrak{S}_K'$. Then $K = S^{-1}JS$.*

Since $\mathfrak{S}_{S^{-1}JS}' = (\mathfrak{S}_J')^S = \mathfrak{S}_K' = \mathfrak{S}'$ we have for $a \in \mathfrak{S}'$ that $a^{S^{-1}JS} = -a = a^K$. But by Lemma 1 any element b of $\mathfrak{A}$ has the form $\sum \beta_{i_1 \cdots i_r} a_{i_1} a_{i_2} \cdots a_{i_r}$ $(i_\alpha = 1, \cdots, m;\ \beta \in \Phi)$ and hence

$$b^{S^{-1}JS} = \sum (-1)^r \beta_{i_1 \cdots i_r} a_{i_r} a_{i_{r-1}} \cdots a_{i_1} = b^K$$

and $K = S^{-1}JS$.

As a special case of Lemma 2 we note that if S is an automorphism mapping $\mathfrak{S}_J'$ into itself then $S^{-1}JS = J$ i.e. S commutes with J. The converse of this is immediate. If $S = G$ is inner, say $a^G = g^{-1}ag$, then $a^{G^{-1}JG} = (g^Jg)^{-1}a^J(g^Jg)$ and the condition $GJ = JG$ is that $g^Jg = \gamma \in \mathrm{P}$. Then $gg^J = \gamma$ also and $(gg^J)^J = \bar{\gamma} = g^Jg = \gamma \in \Phi$ i.e. g is a J-orthogonal element.[18]

3. The following lemma, proved by Landherr,[19] is fundamental for the remainder of the paper:

LEMMA 3. *Let $A \rightarrow A^G$ be an automorphism of the Lie algebra Φ_n' over Φ, then there exists a matrix G in Φ_n such that either $A^G = G^{-1}AG$ or $A^G = -G^{-1}A'G$ where A' is the transposed of A.*

The automorphisms which may be expressed in the form $A \rightarrow G^{-1}AG$ will be called *inner*. As in the preceding section we may show that they form an invariant subgroup $\mathfrak{G}_0$ of the complete group of automorphisms $\mathfrak{G}$. If $n = 2$ we have $-A' = Q^{-1}AQ$ where $Q = \begin{pmatrix} 0 & 1 \\ -1 & 0 \end{pmatrix}$ and A is any matrix of trace 0 i.e. in Φ_n'. Hence $\mathfrak{G}_0 = \mathfrak{G}$. On the other hand if $n > 2$ there exists no such Q since this condition implies tr $A^3 = 0$. Hence in this case $\mathfrak{G}_0 \neq \mathfrak{G}$ and it is easily seen that $\mathfrak{G}_0$ has index 2 in $\mathfrak{G}$.

Now suppose $\mathfrak{A}_1$ and $\mathfrak{A}_2$ are two involutorial simple algebras of order $2n^2$, $n > 2$ over Φ, $\mathrm{P}_1 = \Phi(q_1)$, $\mathrm{P}_2 = \Phi(q_2)$ their centrums and J_1 and J_2 i.a.a. of second kind such that the Lie algebras over Φ, $\mathfrak{S}_{J_1}'$ and $\mathfrak{S}_{J_2}'$ are isomorphic. By passing to isomorphic fields if necessary we may suppose that P_1 and P_2 have the same algebraic closure Ω. Then $\mathfrak{A}_1$ and $\mathfrak{A}_2$ have representations by matrices in Ω_n such that the linear combinations in Ω of the matrices of $\mathfrak{S}_{J_1}'$ or of $\mathfrak{S}_{J_2}'$ constitute Ω_n'. Let $a_0^{(1)}, a_1^{(1)}, \cdots, a_m^{(1)}$ be a basis of $\mathfrak{S}_{J_1}$ over Φ and hence of $\mathfrak{A}_1$ over P_1 such that $a_0^{(1)} = q_1$ and $a_1^{(1)}, \cdots, a_m^{(1)}$ is a basis of $\mathfrak{S}_{J_1}'$ over Φ.

[17] J. p. 509.

[18] J. p. 509.

[19] L. p. 59. Cf. A. Weinstein, *Fundementalsatz der Tensorrechnung*, Math. Zeitsch. **16** (1923) pp. 78-91.

If $a_1^{(2)}, \cdots, a_m^{(2)}$ are the elements corresponding to $a_1^{(1)}, \cdots, a_m^{(1)}$ in the isomorphism between $\mathfrak{S}'_{J_1}$ and $\mathfrak{S}'_{J_2}$ and $a_0^{(2)} = q_2$, then $a_i^{(2)}$, $i = 0, \cdots, m$ is a basis of $\mathfrak{S}_{J_2}$ over Φ and of $\mathfrak{A}_2$ over P_2. If $a_i^{(1)} \to A_i^{(1)}$ and $a_i^{(2)} \to A_i^{(2)}$ in the representations of $\mathfrak{A}_1$ and $\mathfrak{A}_2$ then the correspondence $\sum_{j=1}^m \omega_j A_j^{(1)} \to \sum \omega_j A_j^{(2)}$ where the ω's are arbitrary in Ω is an automorphism of Ω'_n over Ω. Hence by Lemma 3 there exists a matrix G in Ω_n such that either $A_j^{(2)} = G^{-1}A_j^{(1)}G$ or $A_j^{(2)} = -G^{-1}(A_j^{(1)})'G$. By Lemma 1 the matrices of $\mathfrak{A}_1$ and $\mathfrak{A}_2$ have the form $B_1 = \sum \beta_{i_1 \cdots i_r} A_{i_1}^{(1)} \cdots A_{i_r}^{(1)}$ and $B_2 = \sum \beta_{i_1 \cdots i_r} A_{i_1}^{(2)} \cdots A_{i_r}^{(2)}$ $(i_\alpha = 1, \cdots, m;$ $\beta \in \Phi)$ respectively. Thus if $A_j^{(2)} = G^{-1}A_j^{(1)}G$ then the correspondence G defined by $B_1^G = G^{-1}B_1G$ is an isomorphism between the matrices of $\mathfrak{A}_1$ and those of $\mathfrak{A}_2$ which extends the isomorphism $\sum \alpha_j A_j^{(1)} \to \sum \alpha_j A_j^{(2)}$ $(\alpha \in \Phi)$ between the matrices of $\mathfrak{S}'_{J_1}$ and those of $\mathfrak{S}'_{J_2}$. On the other hand if $A_j^{(2)} = -G^{-1}(A_j^{(1)})'G$ then G defined by $B_1^G = (G^{-1}B_1'G)^{J_2}$ has this property. In any case $\mathfrak{A}_1$ and $\mathfrak{A}_2$ are isomorphic.

If $\mathfrak{A}_1$ and $\mathfrak{A}_2$ are identical, say $= \mathfrak{A}$, the argument shows that the isomorphism between $\mathfrak{S}'_{J_1}$ and $\mathfrak{S}'_{J_2}$ may be extended to an automorphism G of $\mathfrak{A}$ over Φ. Evidently the correspondence $B_1 \to G^{-1}B_1G$ and $B_1 \to G^{-1}B_1'G$ leave the matrices of P (scalar matrices) unaltered. Hence in the first case G is an inner automorphism of $\mathfrak{A}$ and in the second case it is outer. As noted earlier the latter possibility is excluded if the exponent of $\mathfrak{A}$ over $\mathrm{P} > 2$. By Lemma 2 we have

THEOREM 2. *If $\mathfrak{A}_1$ and $\mathfrak{A}_2$ are involutorial simple algebras of order $2n^2$, $n > 2$, over Φ and J_1 and J_2 i.a.a. of second kind in $\mathfrak{A}_1$ and $\mathfrak{A}_2$ respectively such that $\mathfrak{S}'_{J_1} \cong \mathfrak{S}'_{J_2}$ then $\mathfrak{A}_1 \cong \mathfrak{A}_2$ and when $\mathfrak{A}_1$ and $\mathfrak{A}_2$ are identified J_1 and J_2 are cogredient.*

THEOREM 3. *If $\mathfrak{A}$ is involutorial simple of order $2n^2$, $n > 2$, over Φ and J an i.a.a. of second kind in $\mathfrak{A}$ then any automorphism in $\mathfrak{S}'_J$ may be extended to an automorphism of the associative algebra $\mathfrak{A}$.*

This results by noting that the above argument is valid even when $\mathfrak{S}'_{J_1}$ and $\mathfrak{S}'_{J_2}$ are the same, say, $= \mathfrak{S}'_J$.

The automorphism G of $\mathfrak{A}$ extending the isomorphism between $\mathfrak{S}'_{J_1}$ and $\mathfrak{S}'_{J_2}$ is unique. For by Lemma 1 we have $b = \sum \beta_{i_1 \cdots i_r} a_{i_1}^{(1)} \cdots a_{i_r}^{(1)}$ for any b in $\mathfrak{A}$ and since G is an automorphism $B^G = \sum \beta_{i_1 \cdots i_r} a_{i_1}^{(1)G} \cdots a_{i_r}^{(1)G} = \sum \beta_{i_1 \cdots i_r} a_{i_1}^{(2)} \cdots a_{i_r}^{(2)}$. If $J_1 = J_2 = J$ we have seen that G commutes with J. Thus the group of automorphism $\mathfrak{A}_J$ of $\mathfrak{S}'_J$ is isomorphic to the subgroup of the automorphism group of $\mathfrak{A}$, consisting of those automorphisms commutative with J. The elements of $\mathfrak{A}_J$ which correspond to inner automorphisms of $\mathfrak{A}$ form an invariant subgroup $\mathfrak{A}_J^0$ of index 1 or 2 in $\mathfrak{A}_J$. If $G \in \mathfrak{A}_J^0$ then $a^G = g^{-1}ag$ where $g \in \mathfrak{G}_J$ the group of J-orthogonal elements. The correspondence $g \to G$ is a homomorphism between $\mathfrak{G}_J$ and $\mathfrak{A}_J^0$ and it is easily seen that the elements of $\mathfrak{G}_J$ mapped into the identity of $\mathfrak{A}_J^0$ i.e. such that $g^{-1}ag = a$ for all a have the form $\delta 1$, $\delta \in \mathrm{P}$. If we denote the subgroup ofthese elements by $\mathfrak{D}$ we obtain the isomorphism $\mathfrak{A}_J^0 = \mathfrak{G}_J/\mathfrak{D}$.

4. Now suppose that $\mathfrak{L}$ is a normal simple Lie algebra over Φ of type A i.e. $\mathfrak{L}_\Omega \cong \Omega_n'$. Then $\mathfrak{L}$ has a representation by matrices in Ω_n' such that the linear combinations in Ω of these matrices constitute Ω_n'. Let $a_1, \cdots, a_m$ $(m = n^2 - 1)$ be a basis of $\mathfrak{L}$ over Φ, $a_j \rightarrow A_j$ in our representation, and $\mathfrak{A}$ denote the enveloping algebra over Φ of the matrices of $\mathfrak{L}$. The A_j have coördinates in an algebraic field Σ of finite degree over Φ. Hence the coördinates of all the matrices of $\mathfrak{A}$ are in Σ. Since $A_1, \cdots, A_m$ are linearly independent over Σ every element of Σ_n' is a Σ-linear combination of $A_1, \cdots, A_m$ and so $\mathfrak{L}_\Sigma \cong \Sigma_n'$. By passing to an algebraic extension we may suppose that Σ is a Galois field of finite degree over Φ with $\mathfrak{g} = (i, s, t, \cdots)$ as Galois group. If $M = (\mu_{ij})$ is any element of Σ_n we denote (μ_{ij}^s) by M^s and note that $M \rightarrow M^s$ is an automorphism of Σ_n regarded as an algebra over Φ and $(M^s)' = (M')^s$. Thus if $[a_i, a_j] = \sum \gamma_{ijk} a_k$ $(\gamma \in \Phi)$ then $[A_i^s, A_j^s] = \sum \gamma_{ijk} A_k^s$ for all s in $\mathfrak{g}$ and the correspondence $M = \sum \mu_j A_j \rightarrow M^S = \sum \mu_j A_j^s$ is an automorphism S of Σ_n' over Σ. By Lemma 3 S is either inner i.e. there exists a G_s such that $A_j^S = A_j^s = G_s^{-1} A_j G_s$ or else there exists a G_s such that $A_j^s = -G_s^{-1} A_j' G_s$. As shown above, if $n = 2$ the automorphisms of Σ_n' are all inner but if $n > 2$ the automorphisms of the form $A \rightarrow -GA'G$ are outer. Let G_s be a fixed matrix satisfying our conditions. Then any other matrix H_s having this property is a multiple of G_s. For $G_s^{-1} H_s$ commutes with all A_j and hence with all the elements of Σ_n. It follows that $G_s^{-1} H_s = \rho_s 1$, $H_s = \rho_s G_s$.

Let $\mathfrak{g}_0$ be the totality of elements s of $\mathfrak{g}$ such that the corresponding S is inner. Then the following table holds:

1. If $s, t \in \mathfrak{g}_0$ then $st \in \mathfrak{g}_0$ and $G_{st} = \rho_{s,t} G_t G_s^t$, $\rho_{s,t} \neq 0$ in Σ.
2. If $s \in \mathfrak{g}_0$, $t \notin \mathfrak{g}_0$ then $st \notin \mathfrak{g}_0$ and $G_{st} = \rho_{s,t} G_t G_s^t$.
3. If $s \notin \mathfrak{g}_0$, $t \in \mathfrak{g}_0$ then $st \notin \mathfrak{g}_0$ and $G_{st} = \rho_{s,t} (G_t')^{-1} G_s^t$.
4. If $s \notin \mathfrak{g}_0$, $t \notin \mathfrak{g}_0$ then $st \in \mathfrak{g}_0$ and $G_{st} = \rho_{s,t} (G_t')^{-1} G_s^t$.

We shall prove 1 and leave the others, which are derived in a similar fashion, to the reader. Assume $A_j^s = G_s^{-1} A_j G_s$ and $A_j^t = G_t^{-1} A_j G_t$. Then $A_j^{st} = (A_j^s)^t = (G_s^t)^{-1} A_j^t G_s^t = (G_s^t)^{-1} G_t^{-1} A_j G_t G_s^t$. Hence $st \in \mathfrak{g}_0$ and G_{st} is a multiple of $G_t G_s^t$. Our table shows that $\mathfrak{g}_0$ is an invariant subgroup of index 1 or 2 in $\mathfrak{g}$.

Suppose first that $\mathfrak{g}_0 = \mathfrak{g}$. Set $A_0 = 1$ and $A = \sum_{i=0}^m \alpha_i A_i$ where $\alpha_i \in \Phi$. Then

$$A^s = G_s^{-1} A G_s \qquad s \in \mathfrak{g}$$

The matrices satisfying these equations form an algebra $\mathfrak{B}$ over Φ containing the algebra $\mathfrak{A}$. On the other hand if $N = \sum \nu_i A_i$ is in $\mathfrak{B}$ then

$$\sum \nu_i^s A_i^s = N^s = G_s^{-1} N G_s = \sum \nu_i A_i^s$$

and so $\nu_i^s = \nu_i \in \Phi$. Thus $\mathfrak{B}$ has order $m + 1 = n^2$ and $\mathfrak{B} \supset \mathfrak{A} \supset$ the matrices $\sum_{j=1}^m \alpha_j A_j$ of $\mathfrak{L}$. Hence either $\mathfrak{A} = \mathfrak{B}$ or $\mathfrak{A}$ is the set of matrices of $\mathfrak{L}$. But the latter is impossible since the matrices of trace 0 do not form an algebra under ordinary multiplication and so $\mathfrak{A} = \mathfrak{B}$. $\mathfrak{A}' \supset$ the matrices $\sum \alpha_j A_j$ of $\mathfrak{L}' = \mathfrak{L}$ and since tr $A_0 = n$, $A_0 \notin \mathfrak{A}'$ and $\mathfrak{A}'$ is the totality of matrices

$\sum \alpha_j A_j$ i.e. $\mathfrak{A}' \cong \mathfrak{L}$. Since $\mathfrak{A}_\Sigma \cong \Sigma_n$, $\mathfrak{A}$ is normal simple with Σ as a splitting field. It follows that the traces of the elements of $\mathfrak{A}$ and the coefficients of the characteristic equations of the matrices of $\mathfrak{L}$ are all in Φ and hence $\mathfrak{L}$ has type A_{I} in the sense of Landherr.[20]

If $\mathfrak{L}$ is of type A_{II} $\mathfrak{g}/\mathfrak{g}_0$ has order 2 and $n > 2$. Corresponding to $\mathfrak{g}_0$ there is a quadratic subfield $\mathrm{P} = \Phi(q)$ such that the group of Σ over P is $\mathfrak{g}_0$. $\mathfrak{L}_{\mathrm{P}}$ is isomorphic to the set of matrices $\sum_{j=1}^m \xi_j A_j$, $\xi \in \mathrm{P}$ and $(\mathfrak{L}_\rho)_\Sigma \cong \Sigma_n'$. Since the automorphisms s of the Galois group of Σ over P induce inner automorphisms S in Σ_n' we conclude by the above argument $\mathfrak{L}_{\mathrm{P}} \cong \mathfrak{B}'$ where $\mathfrak{B}$ is a normal simple algebra over P, namely, the set of matrices $\sum_{i=0}^m \xi_i A_i$, $\xi \in \mathrm{P}$. Now let t be an element of $\mathfrak{g}$ not in $\mathfrak{g}_0$. Then t induces an automorphism distinct from the identity in P and so if $\xi \in \mathrm{P}$, $\xi^t = \bar{\xi}$. Since $t \notin \mathfrak{g}_0$, $A_j^t = -G_t^{-1} A_j' G_t$ and hence for $H = G_t'$ we have $-A_j = H^{-1}(A_j')^t H$. Let $A_0 = q1$. Then the correspondence $X = \sum \xi_i A_i \to X^J = H^{-1}(X')^t H = -\sum \bar{\xi}_i A_i$ is readily seen to be an i.a.a. of second kind in $\mathfrak{B}$ regarded as an algebra over Φ. $\mathfrak{S}_J$ is the set of matrices $\sum \alpha_i A_i$, $\alpha \in \Phi$ and $\mathfrak{S}_J'$ whose order is m is the set $\sum_1^m \alpha_j A_j$ i.e. $\mathfrak{S}_j' \cong \mathfrak{L}$. Hence by Lemma 1 $\mathfrak{B} = \mathfrak{A}$, the enveloping algebra of $\mathfrak{S}_J'$ the set of matrices of $\mathfrak{L}$.

THEOREM 4. *If $\mathfrak{L}$ is a Lie algebra of type A_{II} then $n > 2$ and $\mathfrak{L} \cong \mathfrak{S}_J'$ where J is an i.a.a. of second kind in a simple algebra $\mathfrak{A}$.*

5. In this section we consider the question of cogredience of i.a.a.'s in $\mathfrak{A}$.

By Wedderburn's theorem $\mathfrak{A} \cong \mathfrak{F}_r$ the r-rowed matrix algebra with coördinates in a division algebra, and as has been shown by Albert[21] there is an i.a.a. $a \to \bar{a}$ in $\mathfrak{F}$ such that $J_0 : A = (a_{ij}) \to (\bar{a}_{ji}) = \bar{A}' = A^{J_0}$ $(a_{ij} \in \mathfrak{F})$ is an i.a.a. of second kind in $\mathfrak{A}$. Since $J_0^{-1}J = P$ is an automorphism leaving the elements of P invariant it is inner, say $A^P = P^{-1}AP$ where the matrix P is determined only to within a multiple $\neq 0$ in P. Thus $J = J_0 P$ and $A^J = P^{-1}\bar{A}'P$. The condition that J is involutorial is $A^{J^2} = P^{-1}\bar{P}'A(\bar{P}')^{-1}P = A$ i.e. $\bar{P}' = \rho P$, $\rho \neq 0$ in P. But then $P = \bar{\rho}\bar{P}' = \bar{\rho}\rho P$ so that $\bar{\rho}\rho = 1$. Hence there exists a σ in P such that $\rho = \sigma(\bar{\sigma})^{-1}$ and σP is hermitian.[22] We may therefore suppose that P is hermitian in the expression $P^{-1}\bar{A}'P$ for A^J. P will then be determined to within a multiple $\neq 0$ in Φ. Thus associated with the i.a.a. we have a unique ray $\{P\}$ of non-singular hermitian matrices consisting of all Φ-multiples $\neq 0$ of a fixed one of the set.

We shall say that the rays $\{P\}$ and $\{Q\}$ are *cogredient* if for any $P \in \{P\}$ and $Q \in \{Q\}$ here is an S in $\mathfrak{F}_r$ and a μ in Φ such that $Q = \mu\bar{S}'PS$ i.e. Q is cogredient in the usual sense to a multiple of P.

If $K = S^{-1}JS$ where $A^S = S^{-1}AS$ and $\{Q\}$ is the ray of K then $A^K = Q^{-1}\bar{A}'Q = (\bar{S}'PS)^{-1}\bar{A}'(\bar{S}'PS)$ and hence $\mu\bar{S}'PS = Q$ or $\{P\}$ and $\{Q\}$ are cogredient. The converse of this is immediate. If there exist outer automor-

[20] L. p. 50.

[21] Albert loc cit. in (4) p. 897.

[22] Cf. Albert p. 897.

phisms of $\mathfrak{A}$ and S_0 is a fixed one then any other outer automorphism is the product of S_0 by an inner automorphism. The mapping $S_0^{-1}J_0S_0$ is an i.a.a. of second kind which we may suppose has the form $A \to U_0^{-1}\bar{A}'U_0$ where U_0 is a fixed hermitian matrix in the ray $\{U_0\}$. Consider $S_0^{-1}JS_0 = S_0^{-1}J_0PS_0 = (S_0^{-1}J_0S_0)(S_0^{-1}PS_0)$. As we have seen $S_0^{-1}PS_0$ is the inner automorphism $A \to (P^{S_0})^{-1}AP^{S_0}$. Hence $A^{S_0^{-1}JS_0} = (P^{S_0})^{-1}U_0^{-1}\bar{A}'U_0P^{S_0}$. Thus the $\{U_0P^{S_0}\}$ determines an i.a.a. cogredient to J also and the ray of any i.a.a. cogredient to J is cogredient either to $\{P\}$ or to $\{U_0P^{S_0}\}$.

In particular if $\mathfrak{F} = \mathrm{P}$ $(r = n)$ we define J_0 as before and let S_0 be the automorphism $A \to \bar{A}$. Then $S_0^{-1}J_0S_0 = J_0$ and we may suppose that $U_0 = 1$. Hence the i.a.a. cogredient to J have rays cogredient to either $\{P\}$ or $\{\bar{P}\}$. However we note that P and $\bar{P}$ are cogredient matrices. For P is cogredient to a diagonal matrix D, say $P = \bar{V}'DV$ and since D is hermitian $\bar{D} = D$. Then $\bar{P} = V'D\bar{V}$ and hence $\bar{P} = V'(\bar{V}')^{-1}PV^{-1}\bar{V} = \bar{W}'PW$ where $W = V^{-1}\bar{V}$. It follows in this case that a necessary and sufficient condition that J and K be cogredient is that $\{P\}$ and $\{Q\}$ be cogredient rays.

UNIVERSITY OF CHICAGO.

Reprinted from DUKE MATHEMATICAL JOURNAL
Vol. 4, No. 3, September, 1938

SIMPLE LIE ALGEBRAS OVER A FIELD OF CHARACTERISTIC ZERO

BY N. JACOBSON

The present paper gives a resumé and extension of the theory of simple Lie algebras over an arbitrary field Φ of characteristic 0 developed in recent papers by Landherr and by the author.[1] The extension consists in part in dropping the restriction of normality. Isomorphisms and automorphisms of simple but not necessarily normal simple Lie algebras are considered. We have also discussed the problem of cogredience of matrices arising in this connection and in the last sections have considered in detail the theory for a real closed field and sketched the theory also for $\mathfrak{p}$-adic fields. In the latter discussion we have confined ourselves to referring to results in the literature that are applicable here and have merely supplemented these with the theory of Hermitian and skew-Hermitian matrices having elements in a quaternion algebra.

1. **Preliminaries.** If $\mathfrak{A}$ is an associative algebra over a field Φ, it is readily seen that the elements of $\mathfrak{A}$ constitute a Lie algebra $\mathfrak{A}_l$ relative to the composition $[a, b] = ab - ba$. Evidently if $\mathfrak{A} \cong \mathfrak{B}$, $\mathfrak{A}_l \cong \mathfrak{B}_l$ and if $a \rightarrow b$ is an anti-isomorphism between $\mathfrak{A}$ and $\mathfrak{B}$, then $a \rightarrow -b$ is an isomorphism between $\mathfrak{A}_l$ and $\mathfrak{B}_l$. In particular if S is an automorphism (anti-automorphism) in $\mathfrak{A}$, then S $(-S\colon a \rightarrow -a^S)$ is an automorphism in $\mathfrak{A}_l$. It is clear that the elements left invariant by an automorphism in a Lie algebra form a subalgebra. Hence if S is an anti-automorphism in $\mathfrak{A}$, the set $\mathfrak{S}_S$ of S-skew elements b $(b^S = -b)$ form a Lie subalgebra of $\mathfrak{A}_l$.

The anti-automorphisms S and T of $\mathfrak{A}$ over Φ are *cogredient* if there exists an automorphism G of $\mathfrak{A}$ over Φ, such that $T = G^{-1}SG$. In this case G maps $\mathfrak{S}_S$ on $\mathfrak{S}_T$ and hence G is an isomorphism between these Lie algebras. If $S = T$, i.e., $SG = GS$, then G induces an automorphism in $\mathfrak{S}_S$. If G is inner, say $a^G = g^{-1}ag$, the condition $GS = SG$ is equivalent to $gg^S \in$ the centrum of $\mathfrak{A}$.

If $\mathfrak{L}$ is any (Lie) subalgebra of $\mathfrak{A}_l$, we define the *enveloping ring* $\mathfrak{R}$ of $\mathfrak{L}$ in $\mathfrak{A}$ to be the smallest subring of $\mathfrak{A}$ containing all the elements of $\mathfrak{L}$. $\mathfrak{R}$ is clearly the totality of elements of the form $\sum l_1 l_2 \cdots l_k$, where $l_i \in \mathfrak{L}$. Since $l\alpha \in \mathfrak{L}$ for any l in $\mathfrak{L}$ and any α in Φ, it is evident that $\mathfrak{R}$ is an algebra over Φ.

If $\mathfrak{L}$ is an arbitrary Lie algebra over Φ, it is well known that the correspondence between the element a in $\mathfrak{L}$ and the linear transformation $\mathbf{A}$ defined by $x\mathbf{A} \equiv [x, a]$ for x variable in $\mathfrak{L}$ is a representation, called the *adjoint representation*, of $\mathfrak{L}$:

Received March 11, 1938; presented to the American Mathematical Society, April 15, 1938.

[1] Landherr [1], Jacobson [1] and [2]. Numbers in brackets refer to the bibliography at the end of the paper.

if $a \to \mathbf{A}$, $b \to \mathbf{B}$, then $a + b \to \mathbf{A} + \mathbf{B}$, $a\alpha \to \mathbf{A}\alpha$ and $[a, b] \to [\mathbf{A}, \mathbf{B}] = \mathbf{AB} - \mathbf{BA}$. Let $\mathfrak{R}$ be the enveloping ring of the linear transformation $\mathbf{A}$ in the ring of all linear transformations. If $\mathfrak{L}$ is simple, it can be shown that the centrum of $\mathfrak{R}$ is an algebraic field Σ containing Φ and $\mathfrak{R} \cong \Sigma_m$ the m-rowed matrix ring[2] with elements in Σ.[3] Σ is called the *extended centrum* of $\mathfrak{L}$. $\mathfrak{L}$ may be regarded as an algebra over Σ and when this is done, it becomes *normal simple*, i.e., the extension $(\mathfrak{L} \text{ over } \Sigma)_\Omega$ is simple for Ω the algebraic closure of Σ. Moreover, Σ is the only field of operators $\supset \Phi$ relative to which $\mathfrak{L}$ is closed and normal simple.

If G is an isomorphism between the simple Lie algebras $\mathfrak{L}_1$ and $\mathfrak{L}_2$ over Φ, it induces an isomorphism between their extended centrums Σ_1 and Σ_2. Hence if we identify Σ_1 and Σ_2 as Σ by means of some fixed isomorphism (not necessarily G), $\mathfrak{L}_1$ and $\mathfrak{L}_2$ may be regarded as normal simple algebras over Σ. Then G induces an automorphism in Σ such that $(a_1\xi)^G = a_1^G\xi^G$ for a_1 in $\mathfrak{L}_1$ and ξ in Σ. If Ω is any over-field of Σ, $(\mathfrak{L}_1 \text{ over } \Sigma)_\Omega \cong (\mathfrak{L}_2 \text{ over } \Sigma)_\Omega$. By identifying $\mathfrak{L}_1$ and $\mathfrak{L}_2$ as $\mathfrak{L}$ we note that an automorphism G of $\mathfrak{L}$ over Φ defines an automorphism in Σ such that $(a\xi)^G = a^G\xi^G$. Hence if $\mathfrak{G}$ is the group of automorphisms of $\mathfrak{L}$ over Σ, $\mathfrak{G}_0$ the group of automorphisms of $\mathfrak{L}$ over Σ and $\mathfrak{X}$ the Galois group of Σ over Φ, then $\mathfrak{G}_0$ is invariant in $\mathfrak{G}$ and $\mathfrak{G}/\mathfrak{G}_0 \cong \mathfrak{X}_0$ a subgroup of $\mathfrak{X}$.

2. **Simple Lie algebras over an algebraically closed field.** From now on we suppose that Φ has characteristic 0 and in this section that Φ is also algebraically closed.

It is well known that the associative algebra Φ_n of n-rowed matrices ($n > 1$) with elements in Φ is simple and the derived algebra[4] Φ'_{nl} of Φ_{nl} is a simple Lie algebra of order $n^2 - 1$ over Φ. Φ'_{nl} may be defined also as the totality of matrices of trace 0 in Φ_{nl}. $1 \epsilon' \Phi'_{nl}$, where ϵ' means "is not an element of", and hence if $A_1, \cdots, A_m$, $m = n^2 - 1$, is a basis for Φ'_{nl}, $A_0, \cdots, A_m$ where $A_0 = 1$ is a basis for Φ_{nl} (or Φ_n). The enveloping ring of Φ'_{nl} in Φ_n is Φ_n. The algebras Φ'_{nl} for $n = 2, 3, \cdots$ constitute Cartan's class A. It has been shown by A. Weinstein[5] that if G is an automorphism of Φ'_{nl} over Φ, then there is a non-singular matrix G such that either $A^G = G^{-1}AG$ or $A^G = -G^{-1}A'G$ for all A in Φ'_{nl} and A' the transpose of the matrix A.

If n is odd it can be shown[6] that any involutorial anti-automorphism (i.a.a.) in Φ_n over Φ is cogredient to the i.a.a. $A \to A'$. The skew elements relative to this i.a.a. are the ordinary skew symmetric matrices, and if $n = 3, 5, 7, \cdots$ these form a simple Lie algebra $\mathfrak{S}$ of order $\frac{1}{2}n(n - 1)$ (Cartan's class B). The

[2] In general, if $\mathfrak{A}$ is a ring (algebra), we denote the r-rowed matrix ring (algebra) with elements in $\mathfrak{A}$ by $\mathfrak{A}_r$.

[3] See Jacobson [3] for the results of this paragraph and the next.

[4] The derived algebra $\mathfrak{L}' = [\mathfrak{L}, \mathfrak{L}]$ of a Lie algebra $\mathfrak{L}$ is the totality of sums of commutators $[x, y]$, x, y in $\mathfrak{L}$. The algebras Φ'_{nl} belong to one of Cartan's classes of simple algebras (cf. Cartan [1], Chapter 5).

[5] Weinstein [1].

[6] The results quoted in this paragraph and the next may be found in Jacobson [1].

enveloping ring of $\mathfrak{S}$ in Φ_n is Φ_n. If G is an automorphism of $\mathfrak{S}$ over Φ and $n > 5$, there exists a matrix G such that $GG' = \gamma \neq 0$ in Φ and $A^G = G^{-1}AG$ for all A in $\mathfrak{S}$.

If n is even $(= 2\nu)$, any i.a.a. in Φ_n is cogredient either to $A \to A'$ or to $A \to Q^{-1}A'Q$, where

$$Q = \begin{pmatrix} 0 & 1_\nu \\ -1_\nu & 0 \end{pmatrix}. \tag{1}$$

In the first case the skew elements are ordinary skew matrices and for $n = 6, 8, \cdots$ these form a simple Lie algebra of order $\frac{1}{2}n(n-1)$ (Cartan's class D). The automorphisms of $\mathfrak{S}$ if $n \neq 8$ have the same form as for the algebras of class B. In the second case for $n = 2, 4, \cdots$ we obtain a simple Lie algebra $\mathfrak{S}$ of order $\frac{1}{2}n(n+1)$ consisting of the matrices A such that $Q^{-1}A'Q = -A$ (Cartan's class C). The automorphisms of $\mathfrak{S}$ over Φ have the form $A^G = G^{-1}AG$, where G is a matrix such that $Q^{-1}G'QG = \gamma \neq 0$ or $G'QG = Q\gamma$. In either of the two cases the enveloping ring of $\mathfrak{S}$ is Φ_n.

Suppose now that $n > 5$ for class B, > 2 for class C and > 6 for class D. By a fundamental result of Cartan's[7] the algebras enumerated in this section and subject to these restrictions are not isomorphic and any simple Lie algebra over the algebraically closed field Φ is isomorphic to one of these algebras or to one of five other Lie algebras. The orders of these exceptional Lie algebras are 14, 52, 78, 133 and 248.

3. **Construction of simple Lie algebras.** Let $\mathfrak{A}$ denote a simple associative algebra over Φ, an arbitrary field of characteristic 0, and n the degree of $\mathfrak{A}$ over its centrum.

THEOREM 1. *If $\mathfrak{A}$ has centrum Σ and $n > 1$, then the derived algebra $\mathfrak{A}_l'$ is simple with Σ as its extended centrum.*[8]

$\mathfrak{A}_l'$ consists of sums of elements of the form $[a, b]$. For ξ in Σ, $[a, b]\xi = [a\xi, b] = [a, b\xi]$ and hence $\mathfrak{A}_l'$ as well as $\mathfrak{A}$ may be regarded as an algebra over Σ. $\mathfrak{A}$ is normal over Σ, $(\mathfrak{A} \text{ over } \Sigma)_\Omega \cong \Omega_n$ if Ω is the algebraic closure of Σ. Hence $(\mathfrak{A}_l \text{ over } \Sigma)_\Omega \cong \Omega_{nl}$ and $(\mathfrak{A}_l' \text{ over } \Sigma)_\Omega \cong \Omega_{nl}'$ a simple Lie algebra. Thus $\mathfrak{A}_l'$ is normal simple over Σ and the theorem follows from the remarks of §1.

The isomorphism $(\mathfrak{A} \text{ over } \Sigma)_\Omega \cong \Omega_n$ defines a representation of $\mathfrak{A}$ by matrices in Ω_n. In this representation the matrices of $\mathfrak{A}_l'$ have trace 0 and there are $m = n^2 - 1$ of them linearly independent over Ω since $(\mathfrak{A}_l' \text{ over } \Sigma)_\Omega \cong \Omega_{nl}'$. Hence the order of $\mathfrak{A}_l'$ over Φ is km if the order of Σ over Φ is k. Since $\mathfrak{A}_l'$ is not a subring of $\mathfrak{A}$, the order of the enveloping ring $\mathfrak{R}$ of $\mathfrak{A}_l'$ in $\mathfrak{A}$ exceeds m (over Σ) and so $\mathfrak{R} = \mathfrak{A}$.[9] The characteristic polynomials of the matrices of $\mathfrak{A}_l'$ all belong to the ring $\Sigma[\lambda]$.[10]

[7] Cartan [1], Chapter 5. We have substituted the algebra of class C with $n = 4$ for the isomorphic algebra of class B with $n = 5$.

[8] Landherr [1].

[9] If $\mathfrak{A}_l'$ is a subring of $\mathfrak{A}$, then the matrices of trace 0 form a subring of Ω_n and this is clearly impossible.

[10] Deuring [1], pp. 50–52.

THEOREM 2. *Suppose $\mathfrak{A}$ has centrum Σ of order k over Φ and the i.a.a. S of first kind. If n is odd, the order of $\mathfrak{S}_S$ over Φ is $\frac{1}{2}kn(n-1)$ and $\mathfrak{S}_S$ is simple if $n > 1$. If n is even, this order is $\frac{1}{2}kn(n-1)$ or $\frac{1}{2}kn(n+1)$. In the former case $\mathfrak{S}_S$ is simple if $n > 4$ and in the latter, if $n > 1$. The extended centrum of $\mathfrak{S}_S$ is Σ.*

By definition $\xi^S = \xi$ for every ξ in Σ. Hence if $b \in \mathfrak{S}_S$, $b\xi \in \mathfrak{S}_S$ also and $\mathfrak{S}_S$ may be regarded as an algebra over Σ. S is an i.a.a. in $\mathfrak{A}$ over Σ and has a unique extension to $(\mathfrak{A} \text{ over } \Sigma)_\Omega \cong \Omega_n$. The set of S-skew elements of $(\mathfrak{A}$ over $\Sigma)_\Omega$ is $(\mathfrak{S}_S \text{ over } \Sigma)_\Omega$. On the other hand, if n is odd, this set may be represented by the isomorphism $(\mathfrak{A} \text{ over } \Sigma)_\Omega \cong \Omega_n$ as the set of skew symmetric matrices, and if n is even, this set is representable as the set of skew symmetric matrices or as the set of matrices A such that $Q^{-1}A'Q = -A$, Q as in (1). Thus if n is restricted as in the hypothesis, $(\mathfrak{S}_S \text{ over } \Sigma)_\Omega$ is in one of Cartan's classes B, C or D. Hence $\mathfrak{S}_S$ is simple with Σ as its extended centrum.

By the above proof we obtain a representation of $\mathfrak{A}$ by matrices in Ω_n such that the matrices of $\mathfrak{S}_S$ form a basis for the matrices A such that $A' = -A$ or the matrices A such that $Q^{-1}A'Q = -A$. In either case, since the enveloping ring of $(\mathfrak{S}_S \text{ over } \Sigma)_\Omega$ is Ω_n, the enveloping ring of $\mathfrak{S}_S$ in $\mathfrak{A}$ is $\mathfrak{A}$ itself.

THEOREM 3. *If $\mathfrak{A}$ has centrum $\Sigma(q) \neq \Sigma$, $q^2 = \mu$ in Σ, $n > 1$ and S is an i.a.a. of second kind such that Σ is the set of S-symmetric elements of $\Sigma(q)$, then $\mathfrak{S}_S'$ is simple with Σ as its extended centrum.*

The above argument shows that $\mathfrak{S}_S$ and $\mathfrak{S}_S'$ may be regarded as algebras over Σ. If $a_0, \cdots, a_m$ is a basis for $\mathfrak{S}_S$ over Σ, it is easily seen that a_i is a basis for $\mathfrak{A}$ over $\Sigma(q)$. Hence $m = n^2 - 1$. If Ω is the algebraic closure of Σ and contains $\Sigma(q)$, $(\mathfrak{S}_S \text{ over } \Sigma)_\Omega \cong \Omega_{nl}$, $(\mathfrak{S}_S' \text{ over } \Sigma)_\Omega \cong \Omega_{nl}'$. Thus $\mathfrak{S}_S'$ is simple and has Σ as its extended centrum.

The order of $\mathfrak{S}_S'$ over Σ is m (km over Φ). Since the enveloping ring $\mathfrak{R}$ of $\mathfrak{S}_S'$ in $\mathfrak{A}$ is an algebra over Σ, as the author has shown,[11] $\mathfrak{R} = \mathfrak{A}$ if $n > 2$. Since $(\mathfrak{A} \text{ over } \Sigma(q))_\Omega \cong \Omega_n$, $\mathfrak{A}$ has a representation by matrices in Ω_n such that the matrices of $\mathfrak{S}_S'$ form a basis over Ω of all the matrices of trace 0 in Ω_n. The automorphism G induced by S in $\Sigma(q)$ can be extended to an automorphism $\bar{G}$ in Ω and $\bar{G}$ may be extended to an automorphism $\tilde{G}$ in Ω_n by the definition $A^{\tilde{G}} = (\alpha_{ij})^{\tilde{G}} = (\alpha_{ij}^{\bar{G}})$.[12] It is readily seen that the correspondence $\sum a_i\omega_i \to -\sum a_i\omega_i^{\bar{G}}$ is an anti-automorphism $\tilde{S}$ in $(\mathfrak{A} \text{ over } \Sigma(q))_\Omega$, i.e., in Ω_n extending S and the correspondence $A \to (A^{S\tilde{G}^{-1}})'$ is an automorphism in Ω_n leaving the elements of Ω invariant. It follows that there is a matrix S in Ω_n such that $A^{\tilde{S}} = S^{-1}(A')^{\tilde{G}}S = S^{-1}(A^{\tilde{G}})'S$.[13] The characteristic polynomials of the matrices of $\mathfrak{A}$ all belong to $\Sigma(q)[\lambda]$. Now suppose A is a matrix of $\mathfrak{S}_S'$ and its characteristic polynomial $\lambda^n - \alpha_1\lambda^{n-1} + \cdots + (-1)^n\alpha_n$ is in $\Sigma[\lambda]$. Since $A = -S^{-1}(A')^{\tilde{G}}S$, $\alpha_{2i+1}^{G} = -\alpha_{2i+1} = 0$. If this holds for all A in $\mathfrak{S}_S'$, it will hold for all matrices of trace 0 in Ω_n, and this is impossible if $n > 2$.

[11] Jacobson [2], p. 182.

[12] See, for example, the Anhang by Baer and Hasse to Steinitz, *Theorie der algebraischen Körpern*.

[13] Deuring [1].

4. **Simple Lie algebras of types** A, B, C, **and** D. If $\mathfrak{L}$ is a simple Lie algebra, Σ its extended centrum and $(\mathfrak{L} \text{ over } \Sigma)_\Omega$ is an algebra in Cartan's class A, B, C or D, then we say that $\mathfrak{L}$ has respectively *type* A, B, C or D. The order of $\mathfrak{L}$ over Σ is $n^2 - 1$, $\frac{1}{2}n(n-1)$, or $\frac{1}{2}n(n+1)$, and we suppose from now on that for type A, $n > 1$; for type B, $n > 5$; for type C, $n > 2$; and for type D, $n > 6$. The algebras obtained by Theorems 1 and 3 have type A, those of Theorem 2 have type B, C or D. If n is restricted as indicated, none of these algebras of different type are isomorphic since the extensions $(\mathfrak{L} \text{ over } \Sigma)_\Omega$ are not isomorphic. If an algebra of type A has a representation in Ω_n such that the representing matrices have characteristic polynomials all in $\Sigma[\lambda]$, then $\mathfrak{L}$ has *type* A_I, otherwise it has *type* A_{II}. Thus the Lie algebras of Theorem 1 have type A_I and it can be shown[14] that those of Theorem 3 have type A_{II} if $n > 2$. Hence an algebra of Theorem 1 is isomorphic to no algebra of Theorem 3 if $n > 2$.

We suppose in the sequel that if $\mathfrak{L}$ has type D, $n > 8$ (in addition to the conditions on n noted above for the other types). With this restriction we have

THEOREM 4. *If $\mathfrak{L}$ is a simple Lie algebra of type* A, B, C *or* D, *and Σ is its extended centrum, $\mathfrak{L}$ is isomorphic to one of the Lie algebras of Theorem* 1, 2 *or* 3.

For the algebras of type A_I this has been proved by Landherr [1] and for types A_{II}, B, C and D by the author [1] and [2]. The condition $(\mathfrak{L} \text{ over } \Sigma)_\Omega \cong$ to an algebra in Cartan's class A, B, C or D defines a representation of $\mathfrak{L}$ by matrices in Ω_n. The algebra $\mathfrak{A}$ may be taken as the enveloping ring of the matrices in this representation. If $\mathfrak{L}$ has type A_{II}, B, C or D, the i.a.a. S may be defined as the correspondence $\sum A_{i_1} \cdots A_{i_\nu} \to \sum (-1)^\nu A_{i_\nu} \cdots A_{i_1}$, where the A's are matrices of $\mathfrak{L}$.

5. **Isomorphisms and automorphisms.** The following theorems give conditions for the isomorphism of Lie algebras obtained by a construction of Theorems 1, 2 or 3 (n restricted as in the last section).

THEOREM 5. *If $\mathfrak{A}$ and $\mathfrak{B}$ are as in Theorem* 1 *and G is an isomorphism between $\mathfrak{A}_l'$ and $\mathfrak{B}_l'$ over Φ, then G may be realized as either an isomorphism or the negative of an anti-isomorphism between $\mathfrak{A}$ and $\mathfrak{B}$ over Φ.*[15]

By Theorem 1, $\mathfrak{A}$ and $\mathfrak{B}$ have isomorphic centrums and may be regarded as normal simple associative algebras over the same field $\Sigma \supset \Phi$. If Ω is the algebraic closure of Σ, we have seen that $\mathfrak{A}$ has a representation in Ω_n such that the matrices of $\mathfrak{A}_l'$ are $\sum_1^m A_i\xi_i$, $m = n^2 - 1$, $\xi \in \Sigma$ and the matrices of Ω_{nl}' are $\sum_1^m A_i\omega_i$, $\omega \in \Omega$. Similarly $\mathfrak{B}$ is representable in Ω_n such that the matrices of $\mathfrak{B}_l'$ are $\sum_1^m B_i\xi_i$, B_i linearly independent over Ω, and we may suppose that $B_i = A_i^G$. The automorphism G induced in Σ by the isomorphism G may be extended to an automorphism $\bar{G}$ in Ω and $\bar{G}$ may be extended to an automorphism $\tilde{G}$ of

[14] Jacobson [2], p. 183.
[15] Cf. Landherr [1].

Ω_n by the definition $A^{\tilde{G}} = (\alpha_{ij})^{\tilde{G}} = (\alpha_{ij}^{G})$. Suppose $[A_i, A_j] = \sum A_k \gamma_{kij}, \gamma \in \Sigma$. Then $[B_i, B_j] = \sum B_k \gamma_{kij}^{G}$ and $[A_i^{\tilde{G}}, A_j^{\tilde{G}}] = \sum A_k^{\tilde{G}} \gamma_{kij}^{G}$. Clearly $A_i^{G} \in \Omega'_{nl}$ and are linearly independent over Ω. Hence the correspondence $\sum A_i^{\tilde{G}} \omega_i \to \sum B_i \omega_i$ is an automorphism of Ω'_{nl} over Ω. It follows by Weinstein's theorem that there is a matrix G such that either $B_i = G^{-1} A_i^{\tilde{G}} G$ or $B_i = -G^{-1}(A_i^{\tilde{G}})' G = -G^{-1}(A'_i)^{\tilde{G}} G$. In the former case the correspondence $A \to G^{-1} A^{\tilde{G}} G$ is an automorphism in Ω_n which reduces to the isomorphism G for the matrices of $\mathfrak{A}'_l$ and in the latter $A \to G^{-1}(A')^{\tilde{G}} G = A^{U}$ is an anti-automorphism in Ω_n whose negative is G for the matrices of $\mathfrak{A}'_l$. Since the enveloping ring of $\mathfrak{A}'_l$ is $\mathfrak{A}$, the matrices of $\mathfrak{A}$ have the form $\sum A_{i_1} \cdots A_{i_p} \xi_{i_1} \cdots \xi_{i_p}$, $\xi \in \Sigma$, and hence G or U maps the matrices of $\mathfrak{A}$ on those of $\mathfrak{B}$ and leaves the elements of Φ unaltered.

COROLLARY. *If G is an automorphism of $\mathfrak{A}'_l$ over Φ, it may be realized either as a unique automorphism or the negative of a unique anti-automorphism of $\mathfrak{A}$ over Φ.*

We identify $\mathfrak{A}$ and $\mathfrak{B}$ in Theorem 5. The uniqueness of the automorphism or anti-automorphism is an immediate consequence of the fact that the enveloping ring of $\mathfrak{A}'_l$ is $\mathfrak{A}$.

If S and T are two anti-automorphisms of $\mathfrak{A}$ over Φ, then ST is an automorphism of $\mathfrak{A}$ over Φ. Hence if $\mathfrak{G}$ is the group of automorphisms of $\mathfrak{A}'_l$ over Φ and $\mathfrak{G}_0$ the subgroup of the automorphism induced by the automorphisms of $\mathfrak{A}$ over Φ, then $\mathfrak{G}_0$ is an invariant subgroup of index 1 or 2 in $\mathfrak{G}$.

If $n = 2$, $\mathfrak{G}_0 = \mathfrak{G}$. For in this case $\mathfrak{A}$ is $\cong$ either to Σ_2 or to $\mathfrak{D}$ the generalized quaternion algebra whose basis is $1, i, j, k$, where

$$
\begin{aligned}
i^2 &= \alpha, & j^2 &= \beta, & k^2 &= -\alpha\beta; \\
ij &= -ji = k, & jk &= -kj = -i\beta, & ki &= -ik = -j\alpha.
\end{aligned}
\tag{2}
$$

As we noted above Σ'_2 consists of the matrices of trace 0, and it is easily seen that its elements may also be characterized as the S_0-skew elements of Σ_2, where S_0 is the anti-automorphism $A \to Q^{-1} A' Q$, Q as in (1). In $\mathfrak{D}$ the correspondence $a = \alpha_0 + i\alpha_1 + j\alpha_2 + k\alpha_3 \to \bar{a} = \alpha_0 - i\alpha_1 - j\alpha_2 - k\alpha_3 \equiv a^{S_0}$ is an anti-automorphism whose skew elements are $i\alpha_1 + j\alpha_2 + k\alpha_3$, i.e., the elements of $\mathfrak{D}'_l$. Thus in either case $a^{S_0} = -a$ for the elements of $\mathfrak{A}'_l$. Hence the automorphism determined by $-S_0$ and by the identity automorphism are identical for the elements of $\mathfrak{A}'_l$ and $\mathfrak{G} = \mathfrak{G}_0$.

Suppose conversely that $\mathfrak{G} = \mathfrak{G}_0$. Then either there exists no anti-automorphisms of $\mathfrak{A}$ over Φ or there is an anti-automorphism S_0 such that $a^{S_0} = -a$ for all a in $\mathfrak{A}'_l$. For any ξ in Σ, $(a\xi)^{S_0} = a^{S_0} \xi^{S_0} = -a\xi^{S_0} = -a\xi$. Hence $\xi^{S_0} = \xi$ and $\mathfrak{S}_{S_0}$ the set of S_0-skew elements is a vector space over Σ. If $\mathfrak{S}_{S_0} > \mathfrak{A}'_l$ we must have $\mathfrak{S}_{S_0} = \mathfrak{A}$ which is impossible since $1 \notin \mathfrak{S}_{S_0}$. Thus $\mathfrak{S}_{S_0} = \mathfrak{A}'_l$ and if $a \in \mathfrak{A}'_l$ so does $a^3, a^5, \cdots$. It follows that $\operatorname{tr}(a^3) = 0$ in the representation of $\mathfrak{A}$ in Ω_n, i.e., $\operatorname{tr}(\Sigma A_i \xi_i)^3 = 0$ for all ξ in Σ. Hence $\operatorname{tr}(\sum A_i \omega_i)^3 = 0$ for all ω in Ω. This is impossible if $n > 2$, for then there exist matrices B in Ω'_{nl} such that $\operatorname{tr} B^3 \neq 0$. We have therefore shown that if $n > 2$ and $\mathfrak{A}$ over Φ has anti-automorphisms, then $\mathfrak{G} \neq \mathfrak{G}_0$.

Let $\mathfrak{F}$ denote the subgroup of $\mathfrak{G}_0$ consisting of the automorphisms induced by inner automorphisms of $\mathfrak{A}$. It is readily seen that $\mathfrak{F}$ is invariant in $\mathfrak{G}$ and is isomorphic to $\mathfrak{U}/\Sigma^*$, where $\mathfrak{U}$ is the group of units (non-singular elements) in $\mathfrak{A}$ and Σ^* is the group of multiple 1ξ, where $\xi \neq 0$ in the centrum.

THEOREM 6. *If $\mathfrak{A}$ and $\mathfrak{B}$ are as in Theorem 2, S and T the corresponding i.a.a.'s of first kind and G is an isomorphism between $\mathfrak{S}_S$ and $\mathfrak{S}_T$ over Φ, then G may be realized as an isomorphism between $\mathfrak{A}$ and $\mathfrak{B}$ over Φ.*

By Theorem 2, $\mathfrak{A}$ and $\mathfrak{B}$ may be regarded as normal simple over the same field Σ and by §4, $\mathfrak{S}_S$ and $\mathfrak{S}_T$ have the same type. We have seen that $\mathfrak{A}$ has a representation in Ω_n such that the matrices of $\mathfrak{S}_S$ are $\sum_1^m A_i\xi_i$, $m = \frac{1}{2}n(n-1)$ or $\frac{1}{2}n(n+1)$ and accordingly the A_i form a basis for either all skew-symmetric matrices (types B, D) or the matrices A such that $Q^{-1}A'Q = -A$, Q as in (1) (type C). The same result holds for $\mathfrak{B}$ and $\mathfrak{S}_T$ and we may suppose that $B_i = A_i^G$. As in the proof of Theorem 5 we obtain automorphisms $\bar{G}$ and $\tilde{G}$ such that $A^{\tilde{G}} = (\alpha_{ij}^{\bar{G}})$ and these reduce to G for the elements of Σ. Since $A_i^{\tilde{G}}$ are skew-symmetric or satisfy $Q^{-1}(A_i^{\tilde{G}})'Q = -A_i^{\tilde{G}}$ if the A_i do, the correspondence $\sum A_i^{\tilde{G}}\omega_i \rightarrow \sum B_i\omega_i$ is an automorphism in a Lie algebra of class B, C or D. Hence by the result quoted in §2, there is a matrix G such that $B_i = G^{-1}A_i^{\tilde{G}}G$ and the correspondence $A \rightarrow G^{-1}A^{\tilde{G}}G$ is an automorphism in Ω_n reducing to G for the matrices of $\mathfrak{S}_S$. Since the enveloping ring of $\mathfrak{S}_S$ is $\mathfrak{A}$, $A \rightarrow G^{-1}A^{\tilde{G}}G$ is an isomorphism between the matrices of $\mathfrak{A}$ and $\mathfrak{B}$ over Φ.

COROLLARY 1. *If $\mathfrak{A}$ is as in Theorem 2 and S and T are i.a.a.'s such that $\mathfrak{S}_S$ and $\mathfrak{S}_T$ are isomorphic over Φ, then S and T are cogredient.*

If we identify $\mathfrak{A}$ and $\mathfrak{B}$ in Theorem 6 we obtain an automorphism G in $\mathfrak{A}$ over Φ mapping $\mathfrak{S}_S$ into $\mathfrak{S}_T$. Thus T and $G^{-1}SG$ have the same effect on the elements of $\mathfrak{S}_T$. Since the enveloping ring of $\mathfrak{S}_T$ is $\mathfrak{A}$, T and $G^{-1}SG$ have the same effect in $\mathfrak{A}$, i.e., $T = G^{-1}SG$.

COROLLARY 2. *Any automorphism G of $\mathfrak{S}_S$ over Φ may be realized by a unique automorphism of $\mathfrak{A}$ over Φ commutative with S.*

By Theorem 6 we have an automorphism G of $\mathfrak{A}$ over Φ which reduces to G for the elements of $\mathfrak{S}_S$. If $a \in \mathfrak{S}_S$, $a^S = -a$ and hence G and S commute for the elements of $\mathfrak{S}_S$. Since the enveloping ring of $\mathfrak{S}_S$ is $\mathfrak{A}$, $GS = SG$ for all elements of $\mathfrak{A}$. The uniqueness of the extension G follows for the same reason.

This corollary shows that the group $\mathfrak{G}$ of automorphisms of $\mathfrak{S}_S$ over Φ is isomorphic to the group of the automorphisms of $\mathfrak{A}$ over Φ commutative with S. Let $\mathfrak{F}$ denote the set of automorphisms of $\mathfrak{S}_S$ determined by inner automorphisms of $\mathfrak{A}$. Then $\mathfrak{F}$ is invariant in $\mathfrak{G}$ and $\cong \mathfrak{U}_S/\Sigma^*$, where $\mathfrak{U}_S$ is the totality of *S-orthogonal* elements g, i.e., $gg^S = 1\gamma \neq 0$ in Σ and Σ^* is the set 1ξ, $\xi \neq 0$ in Σ.

THEOREM 7. *If $\mathfrak{A}$ and $\mathfrak{B}$ are as in Theorem 3 with $n > 2$, S and T the corresponding i.a.a.'s of second kind and G is an isomorphism between $\mathfrak{S}_S'$ and $\mathfrak{S}_T'$ over Φ, then $\mathfrak{G}$ may be realized as an isomorphism between $\mathfrak{A}$ and $\mathfrak{B}$ over Φ.*

By Theorem 3 the subfields of symmetric elements of the centrums of $\mathfrak{A}$ and $\mathfrak{B}$ are isomorphic, $\cong \Sigma$, and $\mathfrak{A}$ and $\mathfrak{B}$ may be regarded as simple algebras over Σ

having the centrums $P_1 = \Sigma(q_1)$, $q_1^2 = \mu_1 \in \Sigma$ and $P_2 = \Sigma(q_2)$, $q_2^2 = \mu_2 \in \Sigma$, respectively. We may embed P_1 and P_2 in the same algebraic closure Ω. Then $\mathfrak{A}$ is representable by matrices in Ω_n such that the matrices of $\mathfrak{S}_S'$ are $\sum_1^m A_i \xi_i$, $m = n^2 - 1$, $\xi \in \Sigma$ and the matrices of Ω_{nl}' are $\sum A_i \omega_i$. Similarly, we represent $\mathfrak{B}$ by matrices in Ω_n such that the matrices of $\mathfrak{S}_T'$ are $\sum B_i \xi_i$, where $B_i = A_i^G$. By the proof of Theorem 5 we obtain either an isomorphism G mapping $\mathfrak{A}$ over Φ into $\mathfrak{B}$ over Φ and reducing to G for the elements of $\mathfrak{S}_S'$, or we obtain an anti-isomorphism U such that $a^U = -a^G$ for the elements of $\mathfrak{S}_S'$. But in this case $UT = G$ is an isomorphism G of $\mathfrak{A}$ over Φ into $\mathfrak{B}$ over Φ reducing to G for the elements of $\mathfrak{S}_S'$. It follows, of course, that P_1 and P_2 are isomorphic.

COROLLARY 1. *If $\mathfrak{A}$ is as in Theorem 3 and S and T are i.a.a.'s such that $\mathfrak{S}_S'$ and $\mathfrak{S}_T'$ are isomorphic over Φ, then S and T are cogredient.*

COROLLARY 2. *Any automorphism G of $\mathfrak{S}_S'$ over Φ may be realized by a unique automorphism of $\mathfrak{A}$ over Φ commutative with S.*

The proofs are identical with those of Corollaries 1 and 2 to Theorem 6.

By the second of the present corollaries the group of automorphisms G of $\mathfrak{S}_S'$ over Φ is isomorphic to the subgroup of the automorphisms of $\mathfrak{A}$ over Φ commutative with S. The elements of $\mathfrak{G}$ determined by inner automorphisms for an invariant subgroup $\mathfrak{F} \cong \mathfrak{U}_S/P^*$, where $\mathfrak{U}_S$ is the totality of S-orthogonal elements and P^* is the set 1ξ, $\xi \neq 0$ in $P = \Sigma(q)$.

6. **Cogredience of i.a.a.'s and cogredience of matrices.** The above discussion reduces the problems of simple Lie algebras (with the exception of those having orders 14, 28, 52, 78, 133 and 248 over the extended centrum) to problems in simple associative algebras. In this section we shall see how the latter may be formulated as specific questions about matrices with elements in a division algebra.

It is well known that any simple associative algebra $\mathfrak{A}$ is isomorphic to a $\mathfrak{D}_r$, an r-rowed matrix algebra with elements in a division algebra $\mathfrak{D}$. $\mathfrak{D}$ is determined in the sense of isomorphism by $\mathfrak{A}$. We recall also that any automorphism leaving the elements of the centrum P invariant is inner.[16]

Suppose $\mathfrak{D}_r$ is involutorial anti-automorphic and S is an i.a.a. in $\mathfrak{D}_r$ over Φ. As has been shown by Albert[17] there exists an i.a.a. $a \to a^U$ in $\mathfrak{D}$ whose extension defined by $A^U = (a_{ij})^U = (a_{ji}^U)$ has the same effect in P as S and there exists a matrix S in $\mathfrak{D}_r$ such that $A^S = S^{-1}A^U S$. The matrix S is determined to within multiplication by elements of P. Let Σ denote the symmetric part of P relative to U or S. If S is of first kind ($P = \Sigma$), then the matrix S is either U-symmetric ($S^U = S$) or U-skew ($S^U = -S$), and if S is of second kind, its matrix may be normalized to be U-symmetric and will then be determined to within multiplication by elements of Σ. Thus in any case we may associate with S a *ray* $\{S\}$ of non-singular U-symmetric or U-skew matrices consisting

[16] These results may be found in Deuring [1].

[17] Albert [1], p. 909.

of the multiples of the fixed U-symmetric or U-skew matrix S by the elements $\neq 0$ of Σ.

Let E_{ij} denote the matrix basis of $\mathfrak{D}_r$ such that $A = (a_{ij}) \equiv \sum a_{ij}E_{ij}$ and suppose G is an automorphism of $\mathfrak{D}_r$ over Φ. Let $E_{ij}^G = F_{ij}$. Then $F_{ij}F_{kl} = \delta_{jk}F_{il}$ and hence there is a matrix K such that $F_{ij} = K^{-1}E_{ij}K$. Hence the automorphism G' such that $A^{G'} = KA^GK^{-1}$ leaves the E_{ij} invariant. Since $\mathfrak{D}$ may be characterized as the totality of elements of $\mathfrak{D}_r$ commutative with the E_{ij}, G' maps $\mathfrak{D}$ into itself. Consider the i.a.a. $G'^{-1}UG'$. It maps E_{ij} into E_{ji} and $\mathfrak{D}$ into itself.

The inner automorphisms of $\mathfrak{D}_r$ form an invariant subgroup in the group of automorphisms of $\mathfrak{D}_r$ over Φ. The factor group determined is isomorphic to a subgroup of the Galois group of P over Φ. Let $G_1 = 1, G_2, \cdots, G_l$ be representatives of the cosets of this factor group. By the above we may suppose that the G_i leave the E_{ij} unaltered and map $\mathfrak{D}$ into itself. Hence the i.a.a. $U_q = G_q^{-1}UG_q$ map E_{ij} into E_{ji} and $\mathfrak{D}$ into itself. Now suppose $T = G^{-1}SG$ is cogredient to S. The automorphism G has the form G_qH, where $A^H = H^{-1}AH$. Then a simple computation yields $A^T = T^{-1}A^{U_q}T$, where $T = H^{U_q}S^{G_q}H$. We say that the rays $\{S_1\}$ and $\{S_2\}$ are *U-cogredient* if there exists a non-singular matrix H and an element $\rho \neq 0$ in Σ such that $S_2 = H^US_1H\rho$. Thus the i.a.a.'s cogredient to S have the form $A^T = T^{-1}A^{U_q}T$, where the rays $\{T\}$ and $\{S^{G_q}\}$ are U_q-cogredient for some q, and conversely. Similarly the condition that the inner automorphism G where $A^G = G^{-1}AG$ commute with S is that $G^USG = S\rho$, $\rho \neq 0$ in Σ.

7. **Bilinear forms and cogredience of matrices.** In this section we suppose $\mathfrak{D}$ is an arbitrary quasi-field[18] of characteristic $\neq 2$ and $a \to \bar{a}$ is a fixed i.a.a. in $\mathfrak{D}$. In particular if $\mathfrak{D}$ is commutative, $\mathfrak{D}$ may be the identity mapping. Let $\mathfrak{R}$ be a vector space of r $(< \infty)$ dimensions over $\mathfrak{D}$. A *bilinear form* f in $\mathfrak{R}$ is a function of pairs of vectors $\mathbf{x}$, $\mathbf{y}$ in $\mathfrak{R}$ having values in $\mathfrak{D}$ such that

$$(\mathbf{x}_1 + \mathbf{x}_2, \mathbf{y}) = (\mathbf{x}_1, \mathbf{y}) + (\mathbf{x}_2, \mathbf{y}), \qquad (\mathbf{y}, \mathbf{x}_1 + \mathbf{x}_2) = (\mathbf{y}, \mathbf{x}_1) + (\mathbf{y}, \mathbf{x}_2),$$

$$(\mathbf{x}, \mathbf{y}a) = (\mathbf{x}, \mathbf{y})a, \qquad (\mathbf{x}a, \mathbf{y}) = \bar{a}(\mathbf{x}, \mathbf{y}), \qquad a \in \mathfrak{D}.$$

If $\mathbf{x}_1, \cdots, \mathbf{x}_r$ is a basis for $\mathfrak{R}$ and $(\mathbf{x}_i, \mathbf{x}_j) = s_{ij}$, then for $\mathbf{x} = \sum \mathbf{x}_i x_i$, $\mathbf{y} = \sum \mathbf{x}_i y_i$ we have $(\mathbf{x}, \mathbf{y}) = \sum \bar{x}_i s_{ij} y_j$, and conversely any matrix $S = (s_{ij})$ defines a bilinear form by this equation. S is called *the matrix* of f relative to the basis $\mathbf{x}_1, \cdots, \mathbf{x}_r$. A change to the basis $\mathbf{y}_1, \cdots, \mathbf{y}_r$, where $(\mathbf{y}_1, \cdots, \mathbf{y}_r) = (\mathbf{x}_1, \cdots, \mathbf{x}_r)H$, transforms S into the cogredient matrix $\bar{H}'SH$.

f is *Hermitian* if $\overline{(\mathbf{x}, \mathbf{y})} = (\mathbf{y}, \mathbf{x})$, *skew-Hermitian* if $\overline{(\mathbf{x}, \mathbf{y})} = -(\mathbf{y}, \mathbf{x})$. The conditions on S are respectively $\bar{S}' = S$ and $\bar{S}' = -S$. An element b of $\mathfrak{D}$ is represented by f if there exists a vector $\mathbf{u} \neq 0$ such that $(\mathbf{u}, \mathbf{u}) = b$. If f is Hermitian (skew-Hermitian), the elements represented by it are Hermitian (skew-Hermitian).

[18] I.e., not necessarily commutative field.

We suppose from now on that f is either Hermitian or skew-Hermitian. The vectors $\mathbf{u}$, $\mathbf{v}$ are *orthogonal* if $(\mathbf{u}, \mathbf{v}) = 0$ (or $(\mathbf{v}, \mathbf{u}) = 0$). If $\mathfrak{S}$ is a subspace of $\mathfrak{R}$, the totality of vectors orthogonal to all of the vectors of $\mathfrak{S}$ form a subspace $\mathfrak{S}'$ called the *orthogonal complement* of $\mathfrak{S}$. If $\mathfrak{R}' = 0$, f is said to be *degenerate*, otherwise *non-degenerate*. The conditions that $\mathbf{z} = \sum \mathbf{x}_i z_i$ belong to $\mathfrak{R}'$ are

$$(\mathbf{x}_i, \mathbf{z}) = \sum_{j=1}^{r} s_{ij} z_j = 0 \qquad (i = 1, \cdots, r).$$

Hence if the rank of S is s,[19] the dimensionality of $\mathfrak{R}'$ is $r - s$. In particular f is non-degenerate if and only if S is non-singular.

If $\mathfrak{D}$ is commutative, $\bar{d} = d$ and f is skew, then it is well known that f has a matrix of the form

$$\left(\begin{array}{cc|c} 0 & 1_\nu & \\ -1_\nu & 0 & \\ \hline & & 0 \end{array}\right), \qquad s = 2\nu. \tag{3}$$

Hence the rank of any skew matrix is even and the matrix is cogredient to (3). Any two skew matrices of the same rank are cogredient. In all other cases we proceed to show that f has a matrix of the form

$$\begin{pmatrix} b_1 & & & & & \\ & \ddots & & & & \\ & & b_s & & & \\ & & & 0 & & \\ & & & & \ddots & \\ & & & & & 0 \end{pmatrix}. \tag{4}$$

LEMMA. *If $f \neq 0$, i.e., there exist vectors $\mathbf{u}$, $\mathbf{v}$ such that $(\mathbf{u}, \mathbf{v}) \neq 0$, then there exists a vector $\mathbf{u}$ such that $(\mathbf{u}, \mathbf{u}) \neq 0$.*

If $(\mathbf{u}, \mathbf{u}) = 0$ for all $\mathbf{u}$ in $\mathfrak{R}$ we have

$$(\mathbf{u}, \mathbf{v}) + (\mathbf{v}, \mathbf{u}) = (\mathbf{u} + \mathbf{v}, \mathbf{u} + \mathbf{v}) - (\mathbf{u}, \mathbf{u}) - (\mathbf{v}, \mathbf{v}) = 0.$$

Hence for any a in $\mathfrak{D}$,

$$(\mathbf{u}, \mathbf{v}a) + (\mathbf{v}a, \mathbf{u}) = (\mathbf{u}, \mathbf{v})a + \bar{a}(\mathbf{v}, \mathbf{u}) = (\mathbf{u}, \mathbf{v})a - \bar{a}(\mathbf{u}, \mathbf{v}) = 0.$$

If f is Hermitian, set $a = \overline{(\mathbf{u}, \mathbf{v})} = (\mathbf{v}, \mathbf{u})$, $\bar{a} = (\mathbf{u}, \mathbf{v})$ in the second member of this equation, and we obtain $2(\mathbf{u}, \mathbf{v})(\mathbf{v}, \mathbf{u}) = 0$ contrary to the assumptions $f \neq 0$, $\mathfrak{D}$ of characteristic $\neq 2$. If f is skew-Hermitian, we obtain from the third member of the same equation that $\bar{a}(\mathbf{u}, \mathbf{v}) = (\mathbf{u}, \mathbf{v})a$. Choose $\mathbf{u}$, $\mathbf{v}$ such that $(\mathbf{u}, \mathbf{v}) \neq 0$; then $\bar{a} = (\mathbf{u}, \mathbf{v})a(\mathbf{u}, \mathbf{v})^{-1}$. Thus the correspondence $a \to \bar{a}$ is an automorphism as well as an anti-automorphism. It follows that $\mathfrak{D}$ is commutative, $\bar{a} = a$, and we have the case previously treated and outside of the present consideration.

[19] Cf. van der Waerden, *Moderne Algebra*, I, 2d ed., p. 109, or II, 1st ed., p. 116.

If $f \neq 0$, let $\mathbf{u}_1$ be any vector such that $(\mathbf{u}_1, \mathbf{u}_1) = b_1 \neq 0$, and suppose we have already found k vectors $\mathbf{u}_1, \cdots, \mathbf{u}_k$ such that $(\mathbf{u}_i, \mathbf{u}_j) = 0$ if $i \neq j$ and $(\mathbf{u}_i, \mathbf{u}_i) = b_i \neq 0$. Let $\mathfrak{R}_k$ denote the space generated by $\mathbf{u}_1, \cdots, \mathbf{u}_k$ and $\mathbf{E}_k$ the transformation in $\mathfrak{R}$ defined by

$$\mathbf{x}\mathbf{E}_k = \sum_{i=1}^{k} \mathbf{u}_i(\mathbf{u}_i, \mathbf{u}_i)^{-1}(\mathbf{u}_i, \mathbf{x}).$$

$\mathbf{E}_k$ is a *projection* of $\mathfrak{R}$ on $\mathfrak{R}_k$, i.e., a linear transformation mapping $\mathfrak{R}$ on $\mathfrak{R}_k$, leaving the elements of $\mathfrak{R}_k$ invariant and such that $(\mathbf{x}, \mathbf{y}\mathbf{E}_k) = (\mathbf{x}\mathbf{E}_k, \mathbf{y})$ for any $\mathbf{x}, \mathbf{y}$.[20] Hence if we set $\mathbf{x} = \mathbf{x}\mathbf{E}_k + \mathbf{x}(\mathbf{1} - \mathbf{E}_k) = \mathbf{x}_1 + \mathbf{x}_2$, where $\mathbf{x}_1 = \mathbf{x}\mathbf{E}_k$, $\mathbf{x}_2 = \mathbf{x}(\mathbf{1} - \mathbf{E}_k)$, $\mathbf{x}_1$ will belong to $\mathfrak{R}_k$, $\mathbf{x}_2$ to $\mathfrak{R}_k'$ and $\mathfrak{R} = \mathfrak{R}_k \oplus \mathfrak{R}_k'$. If $f \neq 0$ in $\mathfrak{R}_k'$, there exists a vector $\mathbf{v}_{k+1}$ such that for $\mathbf{u}_{k+1} = \mathbf{v}_{k+1}(\mathbf{1} - \mathbf{E}_k)$ we have $(\mathbf{u}_{k+1}, \mathbf{u}_{k+1}) = b_{k+1} \neq 0$. We repeat this process with $\mathfrak{R}_{k+1}$, the space of $\mathbf{u}_1, \cdots, \mathbf{u}_{k+1}$ and continue until, say for $\mathfrak{R}_s$, we obtain $f = 0$ in $\mathfrak{R}_s'$. We then choose $\mathbf{v}_{s+1}, \cdots, \mathbf{v}_r$ in $\mathfrak{R}$ such that $\mathbf{u}_{s+1} = \mathbf{v}_{s+1}(\mathbf{1} - \mathbf{E}_s), \cdots, \mathbf{u}_r = \mathbf{v}_r(\mathbf{1} - \mathbf{E}_s)$ form a basis for $\mathfrak{R}_s'$. The matrix of f relative to the basis $\mathbf{u}_1, \cdots, \mathbf{u}_r$ has the required diagonal form (4).

THEOREM 8. *Unless $\mathfrak{D}$ is commutative, $\bar{a} \equiv a$ and the matrix is skew, any Hermitian or skew-Hermitian matrix S is cogredient to a diagonal matrix.*

We note that b_i is any element represented by the form associated with S and any b_i may be replaced by $\bar{a}b_ia$, $a \neq 0$, $\bar{b}_i = \pm b_i$ according as S is Hermitian or skew-Hermitian. s in (4) is evidently the rank of S.

Let P denote the centrum of $\mathfrak{D}$ and Σ the subfield of symmetric elements of P. Then either $\mathrm{P} = \Sigma$ (first kind) or $\mathrm{P} = \Sigma(q)$, $q^2 = \mu$ in Σ (second kind). If $\mathfrak{D}$ has a finite basis over P, so has $\mathfrak{D}_r$, and we have seen that this algebra has a representation $A \rightarrow A_1$ by matrices in Ω_n, Ω the algebraic closure of P and n^2 is the order of $\mathfrak{D}_r$ over P. We write $N(A) = \det A_1$ and recall that $N(A) \in \mathrm{P}$. Suppose the i.a.a. is of first kind, $\mathrm{P} = \Sigma$. We have seen that $\bar{A}' \rightarrow S^{-1}A_1'S$, where S is either symmetric or skew in Ω_n. Hence if $B = \bar{G}'AG$, $N(B) = N(A)\gamma^2$, where $\gamma = N(G)$. Thus if Σ^* is defined as above, Σ_2^* is the subgroup of squares in Σ^* and $\Lambda = \Sigma^*/\Sigma_2^*$, then with every non-degenerate bilinear form f (or class of non-singular cogredient matrices) there is associated an element δ of Λ, namely, the element of Λ determined by $N(A)$ if A is a matrix of f. δ is called the *discriminant* of f (or of the class of matrices). If $\mathfrak{D}$ is of second kind, we have seen that if $A \rightarrow A_1$, then $\bar{A}' \rightarrow S^{-1}\bar{A}_1'S$ if $\omega \rightarrow \bar{\omega}$ is an extension in Ω of the automorphism induced by the i.a.a. in P. Thus $N(\bar{A}') = \overline{N(A)}$ and if A is Hermitian,[21] $N(A)$ is in Σ and if $B = \bar{G}'AG$, $N(B) = N(A)\gamma\bar{\gamma}$ and $\gamma = N(G)$. In this case we let Σ_N^* denote the elements of Σ^* which are norms $(\gamma\bar{\gamma})$ of elements of P and let $\Lambda = \Sigma^*/\Sigma_N^*$. Hence if f is a non-degenerate Hermitian form, we define its discriminant δ to be the element of Λ determined by $N(A)$ for A a matrix of f.

[20] $\mathbf{E}_k$ is uniquely determined by these properties. $\mathbf{1} - \mathbf{E}_k$ is a projection on $\mathfrak{R}_k'$.

[21] It is not necessary to consider skew-Hermitian matrices in this case. For if q is skew in P and A is skew-Hermitian, Aq is Hermitian.

Consider the special case where $\mathfrak{D}$ is the quaternion algebra with basis $1, i, j, k$ over Σ and products as in (2), and let the i.a.a. be defined by $\bar{a} = \alpha_0 - i\alpha_1 - j\alpha_2 - k\alpha_3$ if $a = \alpha_0 + i\alpha_1 + j\alpha_2 + k\alpha_3$. We may represent $\mathfrak{D}$ in Ω_2 by

$$(5)\quad 1 \rightarrow \begin{pmatrix} 1 & \\ & 1 \end{pmatrix}, \quad i \rightarrow \begin{pmatrix} \sqrt{\alpha} & \\ & -\sqrt{\alpha} \end{pmatrix}, \quad j \rightarrow \begin{pmatrix} 0 & 1 \\ \beta & 0 \end{pmatrix}, \quad k \rightarrow \begin{pmatrix} 0 & \sqrt{\alpha} \\ -\beta\sqrt{\alpha} & 0 \end{pmatrix},$$

and $\mathfrak{D}_r$ is represented in Ω_{2r} by replacing the element a_{ij} in A by its corresponding two-rowed matrix in (5). The condition $\bar{a} = a$ is equivalent to $a \in \Phi$ and $\bar{a} = -a$ is equivalent to $\operatorname{tr} a = a + \bar{a} = 0$. Then $a^2 + N(a) = 0$, where $N(a)$ is defined as above and $= a\bar{a} = \bar{a}a = -\alpha\alpha_1^2 - \beta\alpha_2^2 + \alpha\beta\alpha_3^2$. If $\bar{A}' = A$, we have seen that A is cogredient to B given by (4) and $\bar{b}_i \in \Sigma$. If A is non-singular, we have $N(B) \in \Sigma_2^*$ and hence $\delta(A) = 1$.[22] If $\bar{A}' = -A$ is non-singular, then $\delta(A)$ is the element of Λ determined by $N(b_1 \cdots b_r) = N(b_1) \cdots N(b_r)$.

8. **Real closed case.** Suppose Φ is real closed.[23] The division algebras over Φ are Φ, $\mathrm{P} = \Phi(\sqrt{-1})$ and $\mathfrak{D}$, Hamilton's quaternion algebra. Hence the simple associative algebras over Φ are Φ_n, P_n and $\mathfrak{D}_r$. Φ_n has the i.a.a. $A \rightarrow A'$ of first kind; P_n has the i.a.a. $A \rightarrow A'$ of the first kind and the i.a.a. $A \rightarrow \bar{A}'$, where $\bar{a}$ is the conjugate of a in P; $\mathfrak{D}_r$ has the i.a.a. $A \rightarrow \bar{A}'$, where $\bar{a}$ is the conjugate quaternion of a. The automorphisms of Φ_n and $\mathfrak{D}_r$ are all inner, but P_n has the outer automorphism $A \rightarrow \bar{A}$.

(a) Suppose $S \in \Phi_n$ and $S' = -S$ is non-singular. Then S is cogredient to Q given by (1). If $S' = S$ is non-singular, by Theorem 8 S is cogredient to one of the matrices

$$(6)\qquad S_p = \begin{pmatrix} 1 & & & & & \\ & \ddots & & & & \\ & & 1 & & & \\ & & & -1 & & \\ & & & & \ddots & \\ & & & & & -1 \end{pmatrix} \quad (p \text{ entries } 1) \qquad (p = 0, 1, \cdots, n).$$

By Sylvester's theorem on the invariance of the signature $2p - n$, distinct S_p's are not cogredient. Clearly no H exists such that $Q = H'S_pH\rho$ since S_p is symmetric and Q is skew. Suppose now that $S_q = H'S_pH\rho$. If $\rho > 0$, we have $S_q = K'S_pK$ for $K = H\sqrt{\rho}$, and if $\rho < 0$, then $S_q = -K'S_pK$ for $K = H\sqrt{-\rho}$. Hence either $q = p$ or $q = n - p$. It follows from the above theory that any i.a.a. in Φ_n is cogredient either to Q: $A \rightarrow Q^{-1}A'Q$ or to one of the

[22] In this case the discriminant as defined above does not distinguish between any non-singular Hermitian matrices. However, a determinant can be defined for these matrices having the desired properties. See Moore [1], Chapter II.

[23] For the classification of Lie algebras over Φ, cf. Cartan [2].

$[\frac{1}{2}n] + 1$ i.a.a.'s S_p: $A \to S^{-1}A'S_p$, where $p = 0, 1, \cdots, [\frac{1}{2}n]$ and none of these i.a.a.'s is cogredient.

(b) Let $S \in \mathrm{P}_n$ and $S' = -S$ be non-singular. S is cogredient to Q in (1). Any non-singular $S' = S$ is cogredient to 1, the identity matrix. It follows that any i.a.a. in P_n is cogredient either to Q: $A \to Q^{-1}A'Q$ or to U: $A \to A'$ and these two are not cogredient.

Suppose $S \in \mathrm{P}_n$ is non-singular and $\bar{S}' = S$. S is cogredient to S_p in (6) and by Sylvester's theorem distinct S_p's are not cogredient. Hence any i.a.a. of second kind in P_n is cogredient to S_p: $A \to S_p^{-1}\bar{A}'S_p$, where $p = 0, 1, \cdots, [\frac{1}{2}n]$. The condition that S_p and S_q be cogredient i.a.a.'s is that $S_q = \bar{H}'S_pH\rho$ or $S_q = \bar{H}'\bar{S}_pH\rho = \bar{H}'S_pH\rho$. Hence as in (a) no two of the S_p are cogredient for $p = 0, \cdots, [\frac{1}{2}n]$.

(c) Let $S \in \mathfrak{D}_r$ and $\bar{S}' = -S$ be non-singular. Then S is cogredient to

$$V = \begin{pmatrix} v & & & \\ & v & & \\ & & \ddots & \\ & & & v \end{pmatrix}, \tag{7}$$

where v is any element in $\mathfrak{D}$ such that $\bar{v} = -v$. By Theorem 8 it suffices to prove the

Lemma. *If $\bar{v} = -v$, $\bar{u} = -u \in \mathfrak{D}$, then v and u are cogredient in $\mathfrak{D}$.*

Note first that v and $v\alpha$ are cogredient if $\alpha > 0$ in Φ. For $\Phi(v) \cong \Phi(\sqrt{-1})$ and hence $\alpha = \bar{a}a$, a in $\Phi(v)$, and $v\alpha = \bar{a}va$. The condition $\bar{v} = -v$ is equivalent to $\mathrm{tr}\,(v) = v + \bar{v} = 0$. If $\alpha = [N(u)/N(v)]^{\frac{1}{2}}$ $(N(u) = u\bar{u} = \bar{u}u)$, then $N(v\alpha) = N(u)$. Since $\mathrm{tr}\,(v\alpha) = \mathrm{tr}\,(u) = 0$, $v\alpha$ and u are similar, say $v\alpha = g^{-1}ug = \gamma^{-1}\bar{g}ug$, $\gamma = N(g)$. Hence $v\alpha\gamma$ and u are cogredient and v and u are cogredient also.

If $\bar{S}' = S$ is non-singular in $\mathfrak{D}_r$, it is cogredient to S_p given by (5). Sylvester's theorem holds in this case also[24] and the argument above shows that any i.a.a. in $\mathfrak{D}_r$ is cogredient to V: $A \to V^{-1}\bar{A}'V$, V as in (7), or to S_p: $A \to S_p^{-1}\bar{A}'S_p$, $p = 0, 1, \cdots, [\frac{1}{2}r]$, and no two of these are cogredient.

If Φ is extended to its algebraic closure P, Φ_n extends to P_n and the i.a.a. Q extends to Q in P_n. Hence the Lie algebra $\mathfrak{S}_Q$ in Φ_n is of type C if $n > 1$. The i.a.a. S_p in Φ_n is cogredient in P_n to U and hence the Lie algebras $\mathfrak{S}_{S_p}$ in Φ_n have type B or D according as n is odd and > 1 or even and > 4.

The extension $\mathfrak{D}_{r\mathrm{P}} \cong \mathrm{P}_{2r}$. The matrices of $\mathfrak{S}_V$ in $\mathfrak{D}_r$ are $A = (a_{ij})$, where $a_{ji} = -v^{-1}\bar{a}_{ij}v$. In particular for $i = j$ we have $a_{ii} = -v^{-1}\bar{a}_{ii}v$ and hence $\overline{va_{ii}} = va_{ii} \in \Phi$ and $a_{ii} = v\alpha_{ii}$, $\alpha_{ii} \in \Phi$. It follows that the order of S_V over Φ is $r + 4r(r-1)/2 = 2r(2r-1)/2$.[25] Thus $\mathfrak{S}_{V\mathrm{P}} \cong \mathfrak{S}_U$ in P_{2r} and $\mathfrak{S}_V$ has type D if $r > 2$. Similarly we obtain the order of $\mathfrak{S}_{S_p}$ in $\mathfrak{D}_r$ as $3r + 4r(r-1)/2 = 2r(2r+1)/2$ and hence if $r > 0$, $\mathfrak{S}_{S_p}$ has type C. These results give the follow-

[24] Moore [1], p. 193.

[25] This can be seen more directly by examining the representation of $\mathfrak{D}_r$ given in §7.

ing list of non-isomorphic simple Lie algebras over Φ together with their automorphisms.

A_{I}. The algebras Φ_n', P_n' and $\mathfrak{D}_r'$ for $n, r > 1$. The orders over Φ are respectively $n^2 - 1$, $2(n^2 - 1)$ and $(2r)^2 - 1$. As before let $\mathfrak{G}$ be the group of automorphisms, $\mathfrak{G}_0$ the subgroup of the automorphisms determined by automorphisms of the associative algebra and $\mathfrak{F}$ the subgroup of $\mathfrak{G}_0$ of elements given by inner automorphisms of the associative algebra. For Φ_2' and $\mathfrak{D}_2'$ we have $\mathfrak{G} = \mathfrak{G}_0 = \mathfrak{F} \cong \mathfrak{U}/\Phi^*$, $\mathfrak{U}$ the group of units and Φ^* the elements $\neq 0$ of Φ. For P_2' we have $\mathfrak{G} = \mathfrak{G}_0 > \mathfrak{F} \cong \mathfrak{U}/P^*$ and $\mathfrak{F}$ has index 2 in $\mathfrak{G}$. An element of $\mathfrak{G}$ not in $\mathfrak{F}$ is $A \to \bar{A}$. If $r, n > 2$ we see for Φ_n' and $\mathfrak{D}_r'$ that $\mathfrak{G} > \mathfrak{G}_0 = \mathfrak{F} \cong \mathfrak{U}/\Phi^*$. $\mathfrak{G}_0$ has index 2 in $\mathfrak{G}$ and an element of $\mathfrak{G}$ not in $\mathfrak{G}_0$ is $A \to A'$ (in Φ_n') or $A \to \bar{A}'$ (in $\mathfrak{D}_r'$). For P_n', $n > 2$, we have $\mathfrak{G} > \mathfrak{G}_0 > \mathfrak{F} \cong \mathfrak{U}/P^*$. An element of $\mathfrak{G}$ not in $\mathfrak{G}_0$ is $A \to A'$ and one in $\mathfrak{G}_0$ not in $\mathfrak{F}$ is $A \to \bar{A}$. These automorphisms commute and hence $\mathfrak{G}/\mathfrak{F}$ is isomorphic to the Vierergruppe.

A_{II}. These are the algebras $\mathfrak{S}_{S_p}$, $p = 0, \cdots, [\frac{1}{2}n]$, in P_n, $n > 2$. The orders are $n^2 - 1$ over Φ. The automorphism $A \to \bar{A}$ commutes with S_p. Hence $\mathfrak{F}$ has index 2 in $\mathfrak{G}$. $\mathfrak{F} \cong \mathfrak{U}_{S_p}/P^*$, where $\mathfrak{U}_{S_p}$ consists of the matrices U such that $\bar{U}'S_pU = S_p\rho$, where S_p is the matrix (6) associated with the i.a.a. S_p. Comparing signature we obtain that $\rho > 0$ unless n is even and $p = \frac{1}{2}n$. In the latter case S_p and $-S_p$ are cogredient matrices and hence there are U's in $\mathfrak{U}_{S_p}$ for which $\rho < 0$. The totality of U's with $\rho > 0$ form an invariant subgroup $\mathfrak{U}_{S_p}^+$ of index 2 in $\mathfrak{U}_{S_p}$ containing P^*. Hence $\mathfrak{F} \cong \mathfrak{U}_{S_p}/P^*$ has an invariant subgroup $\mathfrak{F}^+$ of index 2. $\mathfrak{F}^+$ is invariant in $\mathfrak{G}$ also and $\mathfrak{G}/\mathfrak{F}^+ \cong$ the Vierergruppe.

B. The algebras $\mathfrak{S}_{S_p}$ in Φ_n and $\mathfrak{S}_U$ in P_n for n odd and > 5 and $p = 0, \cdots, [\frac{1}{2}n]$. The orders are respectively $n(n - 1)/2$ and $2n(n - 1)/2$. For $\mathfrak{S}_{S_p}$ we have $\mathfrak{G} = \mathfrak{F} \cong \mathfrak{U}_{S_p}/\Phi^*$, where $\mathfrak{U}_{S_p}$ is the set of matrices U such that $U'S_pU = S_p\rho$. ρ is necessarily > 0. For $\mathfrak{S}_U$, $\mathfrak{F}$ has index 2 in $\mathfrak{G}$ since $A \to \bar{A}$ is in $\mathfrak{G}$ but not in $\mathfrak{F}$. In this case $\mathfrak{F} \cong \mathfrak{U}_U/P^*$, where $\mathfrak{U}_U$ is the group of matrices U such that $U'U = 1\rho$, ρ in P.

C. The algebras $\mathfrak{S}_Q$ in Φ_n and $\mathfrak{S}_Q$ in P_n for n even and > 2 and the algebras $\mathfrak{S}_{S_p}$ in $\mathfrak{D}_r$ for $r > 1$. The orders are respectively $n(n + 1)/2$, $2n(n + 1)/2$ and $2r(2r + 1)/2$. For $\mathfrak{S}_Q$ in Φ_n, $\mathfrak{G} = \mathfrak{F} \cong \mathfrak{U}_Q/\Phi^*$, $\mathfrak{U}_Q$ the set of matrices U such that $U'QU = Q\rho$. The matrices U such that $\rho > 0$ form an invariant subgroup $\mathfrak{U}_Q^+$ of index 2 in $\mathfrak{U}_Q$ and $\mathfrak{U}_Q^+$ contains Φ^*. Hence $\mathfrak{F}$ has an invariant subgroup $\mathfrak{F}^+$ of index 2. For $\mathfrak{S}_Q$ in P_n, $\mathfrak{F}$ has index 2 in $\mathfrak{G}$ and is isomorphic to $\mathfrak{U}_Q/P^*$, $\mathfrak{U}_Q$ the matrices U such that $U'QU = Q\rho$, ρ in P. For $\mathfrak{S}_{S_p}$ in $\mathfrak{D}_r$, $\mathfrak{G} = \mathfrak{F} \cong \mathfrak{U}_{S_p}/\Phi^*$. If $U \in \mathfrak{U}_{S_p}$, i.e., $\bar{U}'S_pU = S_p\rho$, then $\rho > 0$ unless r is even and $p = \frac{1}{2}r$ and in this case the U's with $\rho > 0$ form an invariant subgroup $\mathfrak{U}_{S_p}^+$ of index 2 in $\mathfrak{U}_{S_p}$ and containing Φ^*. Hence for $p = \frac{1}{2}r$, $\mathfrak{F}$ has an invariant subgroup of index 2.

D. The algebras $\mathfrak{S}_{S_p}$ in Φ_n and $\mathfrak{S}_U$ in P_n for n even and > 8 and $p = 0, \cdots, \frac{1}{2}n$ and the algebras $\mathfrak{S}_V$ in $\mathfrak{D}_r$ for $r > 4$. The orders are respectively $n(n - 1)/2$, $2n(n - 1)/2$ and $2r(2r - 1)/2$. For $\mathfrak{S}_{S_p}$, $\mathfrak{G} = \mathfrak{F} \cong \mathfrak{U}_{S_p}/\Phi^*$. If $p = \frac{1}{2}n$, $\mathfrak{F}$ has an invariant subgroup $\mathfrak{F}^+$ of index 2 determined as above. For $\mathfrak{S}_U$ in P_n, $\mathfrak{F}$ has index 2 in $\mathfrak{G}$. The automorphism $A \to \bar{A}$ is in $\mathfrak{G}$ but not in $\mathfrak{F}$. $\mathfrak{F} \cong \mathfrak{S}_U/\mathrm{P}^*$. For $\mathfrak{S}_V$ in $\mathfrak{D}_r$, $\mathfrak{G} = \mathfrak{F} \cong \mathfrak{U}_V/\Phi^*$. $\mathfrak{U}_V$ is the group of matrices U such that $\bar{U}'VU = V\rho$. As before $\mathfrak{F}$ has an invariant subgroup $\mathfrak{F}^+$ of index 2.

9. **$\mathfrak{p}$-adic fields.** The division algebras over a $\mathfrak{p}$-adic field Φ have been determined by Hasse.[26] However, their automorphisms seem not to have been treated in the general case of non-normal algebras. We shall therefore restrict our attention to normal simple Lie algebras $\mathfrak{L}$ and their automorphisms. Since any simple Lie algebra is normal simple over its extended centrum, the only loss of generality here is in connection with the automorphisms of $\mathfrak{L}$.

There are $\phi(m)$ (Euler function) distinct normal division algebras over m^2 over Φ. These are anti-automorphic if and only if $m = 1$ or 2. If $m > 2$ the $\phi(m)$ algebras may be paired into $\frac{1}{2}\phi(m)$ pairs consisting of an algebra and its anti-isomorphic algebra. Any normal simple algebra of order n^2 over Φ has the form $\mathfrak{D}_r$, where $\mathfrak{D}$ is a normal division algebra of order m^2 and $n = mr$. It follows that there are $\sum_{\substack{m/n \\ m>2}} \frac{1}{2}\phi(m) + 2 = [\frac{1}{2}(n + 3)]$ non-isomorphic normal simple Lie algebras of type $\mathrm{A_I}$ and order $n^2 - 1$ over Φ to which any Lie algebra of this type is isomorphic.

As the author has shown, the only division algebras of second kind over Φ are the commutative fields $\mathrm{P} = \Phi(q)$, $q^2 = \mu$ in Φ.[27] As above we denote the set of non-zero elements of Φ by Φ^* the set of squares in Φ^* by Φ_2^* and Φ^*/Φ_2^* by Λ. If $\mathfrak{p} \nmid 2$ the order of Λ is 4 and if $\mathfrak{p} \mid 2$ its order is 2^{l+2}, where l is the order of Φ over the field of 2-adic numbers.[28] Hence for $\mathfrak{p} \nmid 2$ we have three non-isomorphic quadratic extensions of Φ and for $\mathfrak{p} \mid 2$, $2^{l+2} - 1$ such extensions. To obtain the Lie algebras of type $\mathrm{A_{II}}$ we consider the algebras P_n for $n > 2$ and classify the i.a.a.'s of second kind in P_n and hence the rays of Hermitian matrices. It has been shown by Landherr[29] that if n is odd there are two cogredience classes, and if n is even one class. In the respective cases we have for $\mathfrak{p} \nmid 2$ three or six non-isomorphic Lie algebras and for $\mathfrak{p} \mid 2$, $2^{l+2} - 1$ or $2(2^{l+2} - 1)$ such algebras.

To obtain the Lie algebras of type B, C and D we consider the anti-automorphic associative algebras of first kind Φ_n and $\mathfrak{D}_r$, $\mathfrak{D}$ the quaternion division algebra over Φ. The cogredience problem for rays of symmetric matrices has been treated by Hasse. We refer to his papers ([2], [3], and [4]) recalling merely that the number of cogredience classes depends only on $\mathfrak{p}$ and on the residue of n mod 2. The corresponding Lie algebras are of type B or D according as n is odd and > 1 or even and > 4. If we consider the i.a.a.'s associated with skew matrices, we obtain one other class, the resulting Lie algebra having type C if $n > 1$.

[26] Hasse [1].
[27] Jacobson [4].
[28] Hasse [2], p. 115.
[29] Landherr [1]. Landherr obtains here the algebras of type $\mathrm{A_{II}}$ by a method different from ours.

It remains to discuss $\mathfrak{D}_r$, where $\mathfrak{D}$ and the i.a.a. are as in §7. For $a = \alpha_0 + i\alpha_1 + j\alpha_2 + k\alpha_3$ we have $N(a) = \alpha_0^2 - \alpha\alpha_1^2 - \beta\alpha_2^2 + \alpha\beta\alpha_3^2$. Since any non-degenerate quadratic form in four variables over Φ represents all elements of Φ^*, it follows that every element of Φ^* is a norm of an element in $\mathfrak{D}$. Hence the elements $b_i = \bar{b}_i = \beta_i \in \Phi^*$ in (4) may be replaced by 1 and any non-singular Hermitian matrix in $\mathfrak{D}_r$ is cogredient to 1. There is therefore a single cogredience class of i.a.a.'s given by Hermitian matrices. A simple computation shows that the order of the corresponding Lie algebra $\mathfrak{S}_U$ is $2r(2r + 1)/2$ and hence $\mathfrak{S}_U$ has type C if $r > 0$.

LEMMA 1. *The skew Hermitian elements u and v ($\neq 0$) in $\mathfrak{D}$ are cogredient if and only if $\Phi(u)$ and $\Phi(v)$ are isomorphic, i.e., $N(u)/N(v) \in \Phi_2^*$.*

Clearly if $u = \bar{g}vg$ and $\mu = N(u)$, $\nu = N(v)$, $\gamma = N(g)$, then $\mu = \gamma^2\nu$. Since $u^2 = -\mu$, $v^2 = -\nu$, $\Phi(u) \cong \Phi(v)$. Now suppose $\Phi(u) = \Phi(v)$. Then $v = u\rho$. If $\rho = N(q)$ for q in $\Phi(u)$, we obtain $v = \bar{q}uq$ is cogredient to u. If ρ is not a norm, $\rho = \sigma\tau$, where σ is any non-norm and τ is a norm in $\Phi(u)$. There is an element w in $\mathfrak{D}$ such that $w^{-1}uw = -u$ and $w^2 = \sigma$ is not a norm in $\Phi(u)$. Thus $\bar{w}uw = u\sigma$ and $v = u\rho = u\sigma\tau$ is cogredient to $u\sigma$ and to u. If $\Phi(u)$ is isomorphic to $\Phi(v)$, $\mu = \gamma^2\nu$ and hence $N(u\gamma) = N(v)$. There exists an element z in $\mathfrak{D}$ such that $u\gamma = z^{-1}vz$. Therefore v is cogredient to $u\gamma\zeta$ if $\zeta = N(z)$ and hence v is cogredient to u.

We recall that $\mathfrak{D}$ contains all quadratic extensions P of Φ.[30] It follows that if $\mathfrak{p} \nmid 2$, there are 3, and if $\mathfrak{p} \mid 2$, $2^{l+2} - 1$ cogredience classes of skew-Hermitian elements of $\mathfrak{D}$.

LEMMA 2. *If f is a skew-Hermitian form and* dim $\mathfrak{R}$ *over* $\mathfrak{D} = r > 3$, *then f represents* 0.

If f is degenerate, this is trivial. Hence suppose it non-degenerate and let $\mathbf{u}_1, \mathbf{u}_2, \cdots, \mathbf{u}_r$ be a basis for $\mathfrak{R}$ such that the matrix of f is (4) with $s = r$. Then for $\mathbf{w} = \sum \mathbf{u}_i w_i$ we have $(\mathbf{w}, \mathbf{w}) = \sum \bar{w}_i b_i w_i$. By the preceding lemma we may choose w_i so that $\bar{w}_i b_i w_i = b_i\beta_i$, where β_i is arbitrary in Φ. Since the space $\mathfrak{S}$ of skew-Hermitian elements of $\mathfrak{D}$ has three dimensions over Φ and $r > 3$, we may choose β_i not all 0 such that $\sum b_i\beta_i = \sum \bar{w}_i b_i w_i = 0$.

LEMMA 3. *If f is a non-degenerate skew-Hermitian form and represents* 0, *then f represents every skew-Hermitian element of $\mathfrak{D}$.*

Suppose $(\mathbf{u}, \mathbf{u}) = 0$, $\mathbf{u} \neq 0$. Since f is non-degenerate, there is a vector $\mathbf{v}$ such that $(\mathbf{u}, \mathbf{v}) = d \neq 0$. Then replacing $\mathbf{v}$ by $\mathbf{v}d^{-1}$, we may suppose $(\mathbf{u}, \mathbf{v}) = 1$. Let $(\mathbf{v}, \mathbf{v}) = e$. If u is any element of $\mathfrak{S}$, $u - e \in \mathfrak{S}$, and it is easily seen that if $\mathbf{w} = \mathbf{u}\frac{1}{2}(e - u) + \mathbf{v}$, $(\mathbf{w}, \mathbf{w}) = u$.

LEMMA 4. *If $r > 2$ and f is non-degenerate, then f represents every skew-Hermitian element of $\mathfrak{D}$.*

By Lemmas 2 and 3 we may suppose that $r = 3$ and that f does not represent 0. Then b_1, b_2, b_3 in Lemma 2 are linearly independent and hence form a basis for $\mathfrak{S}$. Hence if $u \in \mathfrak{S}$, $u = \sum b_i\beta_i = \sum \bar{w}_i b_i w_i$ is represented by f.

[30] Deuring [1], p. 113.

LEMMA 5. *If $r = 2$ and f and g are non-degenerate skew-Hermitian forms, then there is an element $u \neq 0$ represented by both f and g.*

By Lemma 3 we may suppose that neither f nor g represents 0. Relative to suitable bases of $\mathfrak{R}$ the matrices of f and g are respectively

$$(8) \qquad B_0 = \begin{pmatrix} b_1 & \\ & b_2 \end{pmatrix}, \qquad C_0 = \begin{pmatrix} c_1 & \\ & c_2 \end{pmatrix}.$$

Consider the form in four-space over $\mathfrak{D}$ having the matrix

$$\begin{pmatrix} b_1 & & & \\ & b_2 & & \\ & & -c_1 & \\ & & & -c_2 \end{pmatrix}.$$

By Lemma 2 there are elements w_1, w_2, w_3 and w_4 not all 0 such that $\bar{w}_1 b_1 w_1 + \bar{w}_2 b_2 w_2 = \bar{w}_3 c_1 w_3 + \bar{w}_4 c_2 w_4 = u$, $u \neq 0$, since f does not represent 0.

LEMMA 6. *The matrices B_0 and C_0 in* (8) *are cogredient if and only if they have the same discriminant.*

By Lemma 5 and the proof of Theorem 8 these matrices are cogredient respectively to

$$\begin{pmatrix} u & \\ & b_3 \end{pmatrix} \quad \text{and} \quad \begin{pmatrix} u & \\ & c_3 \end{pmatrix}.$$

Since the discriminants are equal, $N(b_3) = \gamma^2 N(c_3)$. Hence by Lemma 1, b_3 and c_3 are cogredient and the above matrices are cogredient also.

If f is any non-degenerate skew-Hermitian form in $\mathfrak{R}$ over $\mathfrak{D}$ and $u \neq 0$ is arbitrary in $\mathfrak{D}$, it follows by the proof of Theorem 8 that we may suppose $b_1 = b_2 = \cdots = b_{r-2} = u$ in (4), i.e., any non-singular skew-Hermitian matrix is cogredient to a matrix of the form

$$(9) \qquad \begin{bmatrix} u & & & & \\ & \ddots & & & \\ & & u & & \\ & & & b_{r-1} & \\ & & & & b_r \end{bmatrix},$$

where $u \neq 0$ is arbitrary in $\mathfrak{S}$. Hence we have the following criterion:

THEOREM 9. *Two non-singular skew-Hermitian matrices with elements in a $\mathfrak{p}$-adic quaternion algebra are cogredient if and only if they have the same discriminant.*

We may suppose that B is the matrix (9) and C is diagonal with elements $u, \cdots, u, c_{r-1}, c_r$. If the discriminants $\delta(B) = \delta(C)$, then $\delta(B_0) = \delta(C_0)$ for

$$B_0 = \begin{pmatrix} b_{r-1} & \\ & b_r \end{pmatrix} \quad \text{and} \quad C_0 = \begin{pmatrix} c_{r-1} & \\ & c_r \end{pmatrix},$$

and hence by Lemma 6 these matrices are cogredient. But then B and C are cogredient.

Let γ be any element of Φ^* and c an element of $\mathfrak{D}$ such that $N(c) = \gamma$. $\Phi(c)$ is a quadratic field and contains v a skew-Hermitian element. Then $\Phi(c) = \Phi(v)$. As we have seen, there is an element w_1 in $\mathfrak{D}$ such that $w_1^{-1}vw_1 = -v$, and since $w_1^2 = \sigma$, w_1 is in $\mathfrak{S}$. $w_2 = w_1^{-1}c$ satisfies these conditions also. Hence $\gamma = N(w_1)N(w_2)$ and $w_1, w_2 \in \mathfrak{S}$. If we replace w_2 by w_2v^{-1} and set $w_3 = v$, then $w_1, w_2, w_3 \in \mathfrak{S}$ and $\gamma = N(w_1)N(w_2)N(w_3)$. Thus if r is even, the diagonal matrix with elements $u, \cdots, u, v_1, v_2$ has norm $= \gamma\delta^2$, and if r is odd and > 1, the diagonal matrix $u, \cdots, u, v_1, v_2, v_3$ has norm $= \gamma\delta^2$. In either case if $r > 1$ every element of Λ is a discriminant of a skew-Hermitian matrix. If $\mathfrak{p} \nmid 2$, there are four cogredience classes, and if $\mathfrak{p} \mid 2$, there are 2^{l+2} such classes. The corresponding Lie algebras have type 0 if $r > 2$ (order $= 2r(2r - 1)/2$).

The automorphisms of the above Lie algebras may be discussed along the lines indicated in the last section.

Bibliography

A. A. ALBERT.

1. *Involutorial simple algebras and real Riemann matrices*, Annals of Math., vol. 36 (1935), pp. 886-964.

E. CARTAN.

1. *Thèse*, Paris, 1894.
2. *Les groupes réels simples et continus*, Annales de l'École Normale, vol. 31(1914), pp. 263-355.

M. DEURING.

1. *Algebren*, Berlin, 1935.

H. HASSE.

1. *Über p-adische Schiefkörper*, Math. Annalen, vol. 104(1931), pp. 495-534.
2. *Darstellbarkeit von Zahlen durch quadratische Formen in einem beliebigen algebraischen Zahlkörper*, Journal für Math., vol. 153(1924), pp. 113-130.
3. *Symmetrische Matrizen im Körper der rationalen Zahlen*, Journal für Math., vol. 153 (1924), pp. 12-43.
4. *Äquivalenz quadratischer Formen in einem beliebigen algebraischen Zahlkörper*, Journal für Math., vol. 153(1924), pp. 158-162.

N. JACOBSON.

1. *A class of normal simple Lie algebras of characteristic zero*, Annals of Math., vol. 38(1937), pp. 508-517.
2. *Simple Lie algebras of type* A, Annals of Math., vol. 39(1938), pp. 181-188.
3. *A note on non-associative algebras*, this Journal, vol. 3(1937), pp. 544-548.
4. *A note on topological fields*, Am. Jour. of Math., vol. 59(1937), pp. 889-894.

W. LANDHERR.

1. *Über einfache Liesche Ringe*, Hamb. Abhandlungen, vol. 11(1935), pp. 41-64.
2. *Liesche Ringe vom Typus A*, Abhandlungen der Hansischen Univ., vol. 12(1938), pp. 200-241.

E. H. MOORE.

1. *General Analysis*, I, Am. Philos. Soc. Publication, Philadelphia, 1935.

A. WEINSTEIN.

1. *Fundamentalsatz der Tensorrechnung*, Math. Zeitsch., vol. 16(1923), pp. 78-91.

E. WITT.

1. *Theorie der quadratischen Formen in beliebigen Körpern*, Journal für Math., vol. 176(1936-1937), pp. 31-44.

UNIVERSITY OF NORTH CAROLINA.

NORMAL SEMI-LINEAR TRANSFORMATIONS.* [1]

By N. JACOBSON.

The present paper is devoted to a discussion of vector spaces relative to a totally regular (definite) bilinear form. The coefficients are taken in any quasi-field $\mathfrak{F}$ having an involutorial anti-automorphism. We consider semi-linear transformations (s. l. t.'s) and define the adjoints of such transformations and normality, generalizing the well-known notion due to Toeplitz. It is shown that a normal s. l. t. is always completely reducible and in a number of instances is orthogonally completely reducible in the sense defined in § 2. These cases are (1) any unitary, self-adjoint or skew s. l. t., (2) certain cases where $\mathfrak{F}$ is similar to the real field, the complex field, or the quasi-field of real quaternions. Some of these results may be formulated in a simple fashion as theorems on matrices. In § 9 we re-state the theory in terms of projective geometry.

1. Let $\mathfrak{F}$ be an arbitrary quasi-field with an involutorial anti-automorphism (i. a. a.) $\alpha \to \bar{\alpha}$: [2]

$$\overline{\alpha + \beta} = \bar{\alpha} + \bar{\beta} \qquad \overline{\alpha\beta} = \bar{\beta}\bar{\alpha} \qquad \bar{\bar{\alpha}} = \alpha$$

and $\mathfrak{R}$ a vector space of n dimensions over $\mathfrak{F}$. In particular if $\mathfrak{F}$ is commutative we may have $\bar{\alpha} \equiv \alpha$. A bilinear form $f = (x, y)$ is a function of pairs of vectors x, y in $\mathfrak{R}$ with values in $\mathfrak{F}$ such that

$$(x_1 + x_2, y) = (x_1, y) + (x_2, y) \qquad (x, y_1 + y_2) = (x, y_1) + (x, y_2)$$
$$(x, y\alpha) = (x, y)\alpha \qquad (x\alpha, y) = \bar{\alpha}(x, y).$$

If $x_1, x_2, \cdots, x_n$ is a basis for $\mathfrak{R}$ over $\mathfrak{F}$ and $(x_i, x_j) = \alpha_{ij}$, we call $A = (\alpha_{ij})$ the matrix of f relative to $x_1, x_2, \cdots, x_n$. Then for $x = \Sigma x_i\xi_i$ and $y = \Sigma x_i\eta_i$ we have $(x, y) = \Sigma\bar{\xi}_i\alpha_{ij}\eta_j$. Thus f is determined by its matrix and conversely any matrix may be used to define a bilinear form. If we change the basis to

---

* Received May 12, 1938.

[1] Presented to the Society April 15, 1938.

[2] Besides the well-known instances of such quasi-fields we note the following. Let $\mathfrak{D}$ be a domain of integrity with an i. a. a. and suppose $\mathfrak{D}$ has a quotient field $\mathfrak{F}$, i. e. $\mathfrak{F}$ consists of the elements $\alpha\beta^{-1}$, α, β in $\mathfrak{D}$. (Cf. Ore, "Linear equations in non-commutative fields," *Annals of Mathematics*, vol. 32 (1931), p. 466). It is readily verified that $\alpha\beta^{-1} \to \bar{\beta}^{-1}\bar{\alpha}$ is an i. a. a. in $\mathfrak{F}$. For example we may take $\mathfrak{D}$ to be a ring of differential operators and $\bar{\alpha}$ the adjoint of the operator α.

$y_1, y_2, \cdots, y_n$ where $y_i = \Sigma x_j \mu_{ji}$, $M = (\mu_{ij})$ non-singular then the matrix of f relative to $y_1, y_2, \cdots, y_n$ is $\bar{M}'AM$, $M' = (\nu_{ij})$, $\nu_{ij} = \mu_{ji}$.

f is hermitian if $\overline{(x, y)} = (y, x)$, skew-hermitian if $\overline{(x, y)} = -(y, x)$. The conditions on the matrix of f are respectively $\bar{A}' = A$ and $\bar{A}' = -A$. We suppose from now on that one of these cases obtains and in addition that f is *totally-regular* [3] in the sense that $(u, u) = 0$ only if $u = 0$. x and y are said to be orthogonal if $(x, y) = 0$ (or $(y, x) = 0$). If $\mathfrak{S}$ is a subspace the set of vectors y orthogonal to all x in $\mathfrak{S}$ is a subspace $\mathfrak{S}'$ called the orthogonal complement of $\mathfrak{S}$.

Suppose $u_1, u_2, \cdots, u_k$ are vectors such that $(u_i, u_i) = \beta_i \neq 0$ and $(u_i, u_j) = 0$ if $i \neq j$ and let R_k denote the space spanned by the u's. Consider the transformation $\boldsymbol{E}_k$ such that

$$x\boldsymbol{E}_k = \sum_{i=1}^{k} u_i (u_i, u_i)^{-1} (u_i, x).$$

$\boldsymbol{E}_k$ is linear, maps $\mathfrak{R}$ into $\mathfrak{R}_k$, leaves the elements of $\mathfrak{R}_k$ invariant and sends $\mathfrak{R}'_k$ into 0. We call $\boldsymbol{E}_k$ an orthogonal projection of $\mathfrak{R}$ on $\mathfrak{R}_k$. Note also that $(x, y\boldsymbol{E}_k) = (x\boldsymbol{E}_k, y)$ and hence $\boldsymbol{1} - \boldsymbol{E}_k$ is an orthogonal projection on $\mathfrak{R}'_k$. For any x we have $x = x\boldsymbol{E}_k + x(\boldsymbol{1} - \boldsymbol{E}_k) = x_k + x'_k$ where $x_k \in \mathfrak{R}_k$, $x'_k \in \mathfrak{R}'_k$. Hence $\mathfrak{R} = \mathfrak{R}_k + \mathfrak{R}'_k$, $\mathfrak{R}_k \cap \mathfrak{R}'_k = 0$.[4] If $\mathfrak{R}_k \neq \mathfrak{R}$ we may choose a vector of the form $x(\boldsymbol{1} - \boldsymbol{E}_k) \neq 0$ and using it as u_{k+1} we obtain $u_1, u_2, \cdots, u_{k+1}$ such that $(u_i, u_i) = \beta_i \neq 0$ and $(u_i, u_j) = 0$ if $i \neq j$. If we begin with any vector $\neq 0$ as u_1 we obtain a basis of this type, called orthogonal, for the whole space. If $\mathfrak{S}$ is any subspace we may obtain an orthogonal basis $v_1, v_2, \cdots, v_r$ for $\mathfrak{S}$ and supplement it with $v_{r+1}, v_{r+2}, \cdots, v_n$ to obtain an orthogonal basis for $\mathfrak{R}$. $v_{r+1}, v_{r+2}, \cdots, v_n$ generate $\mathfrak{S}'$ and hence $\mathfrak{R} = \mathfrak{S} + \mathfrak{S}'$, $\mathfrak{S} \cap \mathfrak{S}' = 0$. We write $\mathfrak{R} = \mathfrak{S} \oplus \mathfrak{S}'$ in place of these two equations.

Let $\mathfrak{L}$ denote the lattice of subspaces of $\mathfrak{R}$. The correspondence $\mathfrak{S} \to \mathfrak{S}'$ is $(1-1)$ and involutorial, $\mathfrak{S}'' = \mathfrak{S}$. We note also that it is an anti-automorphism of $\mathfrak{L}$, i. e.,

$$(\mathfrak{S}_1 + \mathfrak{S}_2)' = \mathfrak{S}_1' \cap \mathfrak{S}_2' \qquad (\mathfrak{S}_1 \cap \mathfrak{S}_2)' = \mathfrak{S}'_1 + \mathfrak{S}'_2.$$ [5]

If $\mathfrak{R} = \mathfrak{R}_1 + \mathfrak{R}_2 + \cdots + \mathfrak{R}_l$ where $\mathfrak{R}_i \neq 0$ and $\mathfrak{R}_i \subset \mathfrak{R}'_j$ if $i \neq j$ then we write $\mathfrak{R} = \mathfrak{R}_1 \oplus \mathfrak{R}_2 \oplus \cdots \oplus \mathfrak{R}_l$. It follows that every vector is ex-

---

[3] Cf. Weyl, "On generalized Riemann matrices," *Annals of Mathematics*, vol. 35 (1934), p. 715 and Birkhoff and v. Neumann, "The logic of quantum mechanics," *Annals of Mathematics*, vol. 37 (1936), pp. 823-843. Evidently the assumption of total-regularity excludes the case $\mathfrak{F}$ commutative, $\bar{\alpha} \equiv \alpha$ and f skew.

[4] It follows that $\boldsymbol{E}_k$ is the only orthogonal projection of $\mathfrak{R}$ on $\mathfrak{R}_k$.

[5] Cf. Birkhoff and v. Neumann, *loc. cit.* in 3.

pressible uniquely in the form $x_1 + x_2 + \cdots + x_l$, $x_i \in \mathfrak{R}_i$. If $u_i^{(1)}$, $u_i^{(2)}, \cdots, u_i^{(n_i)}$ is an orthogonal basis for $\mathfrak{R}_i$ then the $u_i^{(j)}$, $i = 1, \cdots, l$; $j = 1, \cdots, n_i$ form an orthogonal basis for $\mathfrak{R}$.

2. We recall the definition of a semi-linear transformation (s. l. t.) $\boldsymbol{T}$ with automorphism S as a single valued mapping of $\mathfrak{R}$ on itself such that

$$(x + y)\boldsymbol{T} = x\boldsymbol{T} + y\boldsymbol{T} \qquad (x\alpha)\boldsymbol{T} = (x\boldsymbol{T})\alpha^S.$$

If $x_i\boldsymbol{T} = \Sigma x_j\tau_{ji}$ and $x = \Sigma x_i\xi_i$ then $x\boldsymbol{T} = \Sigma x_i\tau_{ij}\xi_j{}^S$. $\boldsymbol{T}$ is determined by the matrix $T = (\tau_{ij})$ and by the automorphism S. If $\mathfrak{S}$ is a subspace then $\mathfrak{S}\boldsymbol{T}$ the set of vectors of the form $y\boldsymbol{T}$, $y \in \mathfrak{S}$ is a subspace also. The sum of two s. l. t.'s with the same automorphism is another such s. l. t. If $\boldsymbol{T}_1$ and $\boldsymbol{T}_2$ have respectively the matrices T_1 and T_2 and the automorphisms S_1 and S_2 then $\boldsymbol{T}_1\boldsymbol{T}_2$ is an s. l. t. with matrix $T_2T_1{}^{S_2}$ and automorphism S_1S_2. The scalar multiplication $x \to x\mu = x\boldsymbol{M}$ is an s. l. t. with matrix 1μ and automorphism $\alpha \to \mu^{-1}\alpha\mu$. If $y_1, y_2, \cdots, y_n$ is a basis such that $y_i = \Sigma x_j\mu_{ji}$, the matrix of $\boldsymbol{T}$ relative to the y's is $M^{-1}TM^S$, $M = (\mu_{ij})$.[6]

If Ω is a set of s. l. t.'s then the system of subspaces invariant under all s. l. t.'s of Ω is a sublattice $\mathfrak{L}(\Omega)$ of $\mathfrak{L}$. $\mathfrak{S}$ in $\mathfrak{L}(\Omega)$ is irreducible if it contains no proper invariant subspace. We recall also that $\mathfrak{L}(\Omega)$ or Ω is completely reducible if for every $\mathfrak{S}_1$ in $\mathfrak{L}(\Omega)$ there is an $\mathfrak{S}_2$ in this lattice such that $\mathfrak{R} = \mathfrak{S}_1 + \mathfrak{S}_2$, $\mathfrak{S}_1 \cap \mathfrak{S}_2 = 0$. In view of the chain conditions this is equivalent, as is well-known, to the condition $\mathfrak{R} = \mathfrak{R}_1 + \mathfrak{R}_2 + \cdots + \mathfrak{R}_l$, where the $\mathfrak{R}_i$ are irreducible.[7] Ω is *orthogonally completely reducible* if $\mathfrak{L}(\Omega) \supset \mathfrak{S}'$ for every $\mathfrak{S}$ in $\mathfrak{L}(\Omega)$. If $\{\Omega_a\}$ is a set of sets of s. l. t.'s and Ω the logical sum of the Ω_a then $\mathfrak{L}(\Omega)$ is the intersection $\Delta\mathfrak{L}(\Omega_a)$. Hence it follows directly from the definition that if each Ω_a is orthogonally completely reducible, then so is Ω.

Now suppose $\mathfrak{L}(\Omega) \supset \mathfrak{S}'$ for every irreducible $\mathfrak{S}$ in this lattice and let $\mathfrak{U}$ be arbitrary in $\mathfrak{L}(\Omega)$. Suppose $\mathfrak{U} \supset \mathfrak{S}_1$ irreducible in $\mathfrak{L}(\Omega)$. Then $\mathfrak{U} \cap \mathfrak{S}'_1 = \mathfrak{U}_1 \in \mathfrak{L}(\Omega)$ and $\mathfrak{U} = \mathfrak{U} \cap \mathfrak{R} = \mathfrak{U} \cap (\mathfrak{S}_1 \oplus \mathfrak{S}'_1) = \mathfrak{S}_1 \oplus \mathfrak{U}_1$.[8] If $\mathfrak{U}_1$ is not reducible we repeat this process and obtain $\mathfrak{U}_1 = \mathfrak{S}_2 \oplus \mathfrak{U}_2$ where $\mathfrak{S}_2$, $\mathfrak{U}_2 \in \mathfrak{L}(\Omega)$ and $\mathfrak{S}_2$ is irreducible. Then $\mathfrak{U} = \mathfrak{S}_1 \oplus \mathfrak{S}_2 \oplus \mathfrak{U}_2$. Continuing in this way we obtain $\mathfrak{U} = \mathfrak{S}_1 \oplus \mathfrak{S}_2 \oplus \cdots \oplus \mathfrak{S}_k$, $\mathfrak{S}_i$ irreducible in

---

[6] Cf. Jacobson, "Pseudo-linear transformations," *Annals of Mathematics*, vol. 38 (1937), pp. 484-507 for the results of this paragraph.

[7] See for example van der Waerden's *Moderne Algebra I*, 1st ed., p. 143, or 2nd ed., p. 155.

[8] We are applying the rule that if $\mathfrak{S}_1 \supset \mathfrak{S}_2$ then

$$\mathfrak{S}_1 \cap (\mathfrak{S}_2 + \mathfrak{S}_3) = \mathfrak{S}_2 + (\mathfrak{S}_1 \cap \mathfrak{S}_3).$$

$\mathfrak{L}(\Omega)$. It follows that $\mathfrak{U}' = \mathfrak{S}'_1 \cap \mathfrak{S}'_2 \cap \cdots \cap \mathfrak{S}'_k \in \mathfrak{L}(\Omega)$ and hence Ω is orthogonally completely reducible. If we apply this process to $\mathfrak{R}$ we obtain $\mathfrak{R} = \mathfrak{R}_1 \oplus \mathfrak{R}_2 \oplus \cdots + \mathfrak{R}_l$ where the $\mathfrak{R}_i$ are irreducible in $\mathfrak{L}(\Omega)$.[9] Hence we obtain an orthogonal basis for $\mathfrak{R}$ relative to which all the s. l. t.'s in Ω have the form

$$\begin{pmatrix} T_1 & & & & \\ & T_2 & & & \\ & & \cdot & & \\ & & & \cdot & \\ & & & & \cdot \\ & & & & & T_l \end{pmatrix}$$

where the system T_i is irreducible.

3. If $\boldsymbol{T}$ is any single valued mapping of R into itself, S an automorphism, we define an *S-adjoint* $\boldsymbol{T}_S$ of $\boldsymbol{T}$ as a transformation such that

$$(x, y\boldsymbol{T})^S = (x\boldsymbol{T}_S, y) \tag{1}$$

for all x and y. If $\boldsymbol{T}_S^{(1)}$ and $\boldsymbol{T}_S^{(2)}$ are two such transformations then $(x\boldsymbol{T}_S^{(1)}, y) = (x\boldsymbol{T}_S^{(2)}, y)$ and hence $(x\boldsymbol{T}_S^{(1)} - x\boldsymbol{T}_S^{(2)}, y) = 0$ and $x\boldsymbol{T}_S^{(1)} = x\boldsymbol{T}_S^{(2)}$. Thus if $\boldsymbol{T}_S$ exists it is single valued and unique. Since

$$\begin{aligned} ((x_1 + x_2)\boldsymbol{T}_S, y) &= (x_1 + x_2, y\boldsymbol{T})^S = (x_1, y\boldsymbol{T})^S + (x_2, y\boldsymbol{T})^S \\ &= (x_1\boldsymbol{T}_S, y) + (x_2\boldsymbol{T}_S, y) = (x_1\boldsymbol{T}_S + x_2\boldsymbol{T}_S, y), \end{aligned}$$

$(x_1 + x_2)\boldsymbol{T}_S = x_1\boldsymbol{T}_S + x_2\boldsymbol{T}_S$. Similarly we obtain $\alpha\boldsymbol{T}_S = \boldsymbol{T}_S\alpha^{\bar{S}}$ where α is the scalar multiplication $x \to x\alpha$ and $\bar{S}$ is the automorphism $\alpha \to \overline{\bar{\alpha}^S} \equiv \alpha^{\bar{S}}$. Thus $\boldsymbol{T}_S$ is an s. l. t. with automorphism $\bar{S}$. From (1) we obtain $(x, y\boldsymbol{T}_S)^{\bar{S}^{-1}} = (x\boldsymbol{T}, y)$. Hence $\boldsymbol{T}$ is the $\bar{S}^{-1}$ adjoint of $\boldsymbol{T}_S$ if $\boldsymbol{T}_S$ exists and a necessary condition that $\boldsymbol{T}_S$ exist is therefore that $\boldsymbol{T}$ be a s. l. t. with automorphism $S^{-1} = \overline{\bar{S}^{-1}}$. This condition is also sufficient. For let $\boldsymbol{T}_S$ be the s. l. t. with matrix $T^* = A^{-1}(\bar{T}')^{\bar{S}}A^{\bar{S}}$ (T the matrix of $\boldsymbol{T}$ and A that of the bilinear form relative to the basis $x_1, x_2, \cdots, x_n$) and automorphism S. Then

$$\begin{aligned} (xy\boldsymbol{T})^S &= (\Sigma x_i\xi_i, \Sigma x_j\tau_{ja}\eta_a{}^{S^{-1}})^S = \Sigma\bar{\xi}_i{}^S\alpha_{ij}{}^S\tau_{ja}{}^S\eta_a \\ (x\boldsymbol{T}_S, y) &= (\Sigma x_a\tau^*{}_{ai}\xi_i{}^{\bar{S}}, \Sigma x_j\eta_j) = \Sigma\overline{\xi_i{}^{\bar{S}}\bar{\tau}^*{}_{ai}}\alpha_{aj}\eta_j \\ &= \Sigma\bar{\xi}_i{}^S\bar{\tau}^*{}_{ai}\alpha_{aj}\eta_j \end{aligned}$$

and $(x\boldsymbol{T}_S, y) = (x, y\boldsymbol{T})^S$ since $\overline{T^{*\prime}}A = A^S T^S$. We call $\boldsymbol{T}_S$ the *adjoint* of T

---

[9] The converse does not hold. There exist Ω's for which $\mathfrak{R} = \mathfrak{R}_1 \oplus \cdots \oplus \mathfrak{R}_l$ where the $\mathfrak{R}_i$ are irreducible but Ω is not orthogonally completely reducible.

and denote it more simply as $\boldsymbol{T}^*$. The relation is a symmetric one: $\boldsymbol{T}^{**} = \boldsymbol{T}$. If $\boldsymbol{T}_1$ and $\boldsymbol{T}_2$ are two s. l. t.'s with automorphisms S_1^{-1} and S_2^{-1}, then

$$(x, y\boldsymbol{T}_1\boldsymbol{T}_2)^{S_2S_1} = (x\boldsymbol{T}^*{}_2, y\boldsymbol{T}_1)^{S_1} = (x\boldsymbol{T}^*{}_2\boldsymbol{T}^*{}_1, y),$$

i. e., $(\boldsymbol{T}_1\boldsymbol{T}_2)^* = \boldsymbol{T}^*{}_2\boldsymbol{T}^*{}_1$. If $S_1 = S_2 = S$ then

$$(x, y)(\boldsymbol{T}_1 + \boldsymbol{T}_2))^S = (x, y\boldsymbol{T}_1)^S + (x, y\boldsymbol{T}_2)^S = (x(\boldsymbol{T}^*{}_1 + \boldsymbol{T}^*{}_2), y)$$

or $(\boldsymbol{T}_1 + \boldsymbol{T}_2)^* = \boldsymbol{T}^*{}_1 + \boldsymbol{T}^*{}_2$. We note finally that if M is the s. l. t. $x \rightarrow x\mu$ then M^* is the s. l. t. $x \rightarrow x\bar{\mu}$ since $\mu(x, y\mu)\mu^{-1} = (x\bar{\mu}, y)$.[10]

Let Ω be a set of s. l. t.'s and Ω^* the set consisting of the adjoints of the transformations in Ω. It is readily seen that if $\mathfrak{S} \in \mathfrak{L}(\Omega)$ then $\mathfrak{S}' \in \mathfrak{L}(\Omega^*)$. Hence $\mathfrak{L}(\Omega)$ and $\mathfrak{L}(\Omega^*)$ are anti-isomorphic. The condition that $\mathfrak{L}(\Omega)$ be orthogonally completely reducible is that $\mathfrak{L}(\Omega) = \mathfrak{L}(\Omega^*)$ and hence we have the theorem:

THEOREM 1. *If Ω is a set of semi-linear transformations containing the adjoint $\boldsymbol{T}^*$ of every $\boldsymbol{T}$ in Ω, then Ω is orthogonally completely reducible.*

4. $\boldsymbol{T}$ is a *normal* s. l. t. if $\boldsymbol{T}^*\boldsymbol{T} = \boldsymbol{T}\boldsymbol{T}^*$. Evidently if $\boldsymbol{T}$ is normal so is $\boldsymbol{T}^*$ and if $\boldsymbol{T}_1$ and $\boldsymbol{T}_2$ are normal and $\boldsymbol{T}_1\boldsymbol{T}_2 = \boldsymbol{T}_2\boldsymbol{T}_1$, $\boldsymbol{T}_1\boldsymbol{T}^*{}_2 = \boldsymbol{T}^*{}_2\boldsymbol{T}_1$ then $\boldsymbol{T}_1\boldsymbol{T}_2$ is normal. In particular the powers of a normal s. l. t. are normal. Special cases of normal s. l. t.'s are $\boldsymbol{T}$ *self-adjoint* where $\boldsymbol{T}^* = \boldsymbol{T}$, $\boldsymbol{T}$ *skew* where $\boldsymbol{T}^* = -\boldsymbol{T}$ and $\boldsymbol{T}$ *unitary* where $\boldsymbol{T}\boldsymbol{T}^* = \boldsymbol{T}^*\boldsymbol{T} = 1$.

Suppose $\boldsymbol{T}$ is normal and z is a vector such that $z\boldsymbol{T} = 0$. Then $(z\boldsymbol{T}^*, z\boldsymbol{T}^*) = (z, z\boldsymbol{T}^*\boldsymbol{T})^S = (z, z\boldsymbol{T}\boldsymbol{T}^*)^S = 0$ and hence $z\boldsymbol{T}^* = 0$. If $\mathfrak{S} \in \mathfrak{L}(\boldsymbol{T})$, $\mathfrak{S}\boldsymbol{T}^* \in \mathfrak{L}(\boldsymbol{T})$ also. Suppose $\mathfrak{S}$ is irreducible. Then either $\mathfrak{S}\boldsymbol{T} = \mathfrak{S}\boldsymbol{T}^* = 0$ or the mapping $y \rightarrow y\boldsymbol{T}^*$ is $(1-1)$ from $\mathfrak{S}$ to $\mathfrak{S}\boldsymbol{T}^*$. It follows that any invariant subspace of $\mathfrak{S}\boldsymbol{T}^*$ has the form $\mathfrak{U}\boldsymbol{T}^*$ where $\mathfrak{U}$ is invariant in $\mathfrak{L}(\boldsymbol{T})$. Hence $\mathfrak{U}\boldsymbol{T}^*$ can not be proper. Thus in any case if $\mathfrak{S}$ is irreducible either $\mathfrak{S}\boldsymbol{T}^* = 0$ or $\mathfrak{S}\boldsymbol{T}^*$ is irreducible. If $\boldsymbol{T}^*$ is linear (l. t.) $\mathfrak{S}\boldsymbol{T}^* \neq 0$ is similar to $\mathfrak{S}$.

Let $\mathfrak{R}_0$ be the join of all irreducible subspaces $\mathfrak{S}$ of $\mathfrak{L}(\boldsymbol{T})$. $\mathfrak{R}_0$ is invariant under $\boldsymbol{T}^*$ and hence $\mathfrak{R}'_0 \in \mathfrak{L}(\boldsymbol{T})$. If $\mathfrak{R}'_0 \neq 0$, it contains $\mathfrak{S}_0 \neq 0$ which is irreducible in $\mathfrak{L}(\boldsymbol{T})$. But this is impossible since $\mathfrak{R}_0 \cap \mathfrak{R}'_0 = 0$. Thus $\mathfrak{R}'_0 = 0$, i. e., $\mathfrak{R} = \mathfrak{R}_0$. $\mathfrak{R}$ is a join of irreducible invariant subspaces and hence is a join of a finite number of such subspaces. We therefore have

[10] This discussion is valid for any non-degenerate hermitian or skew-hermitian form. No use has been made of the total-regularity of f.

4

THEOREM 2. *Any normal s. l. t. is completely reducible.*

Remark. It is readily seen that the above proof holds also under the weaker assumption that $\boldsymbol{TT^*} = \boldsymbol{T^*T}\mu$, $\mu \neq 0$ in S.

5. In the remainder of the paper we discuss some special conditions that a normal s. l. t. be orthogonally completely reducible. If $\boldsymbol{T}^* = \pm \boldsymbol{T}$ then any subspace $\mathfrak{S}$ invariant under $\boldsymbol{T}$ is invariant under $\boldsymbol{T}^*$. Hence $\mathfrak{S}' \epsilon \mathfrak{L}(\boldsymbol{T})$. If $\boldsymbol{T^*T} = \boldsymbol{TT^*} = 1$ then $\mathfrak{S}\boldsymbol{T}$ is an invariant subspace of $\mathfrak{S}$ and since $\boldsymbol{T}$ is non-singular $\mathfrak{S}\boldsymbol{T} = \mathfrak{S}$ and $\mathfrak{S}\boldsymbol{T}^* = \mathfrak{S}\boldsymbol{T}^{-1} = \mathfrak{S}$. Thus $\mathfrak{S}' \epsilon \mathfrak{L}(\boldsymbol{T})$ in this case also and we have the theorem:

THEOREM 3. *Any symmetric, skew or unitary semi-linear transformation is orthogonally completely reducible.*

If $\boldsymbol{T}$ is any s. l. t. with automorphism S^{-1} we define the ring $\mathfrak{F}[t, S^{-1}]$ as the ring of polynomials in the indeterminate t with coefficients in $\mathfrak{F}$ such that $\alpha t = t\alpha^{S^{-1}}$ or $t\alpha = \alpha^S t$. Then any irreducible invariant subspace $\mathfrak{S}$ of $\boldsymbol{T}$ is generated by a single vector x such that

$$x\phi(\boldsymbol{T}) = x\boldsymbol{T}^r - x\boldsymbol{T}^{r-1}\beta_1 - \cdots - x\beta_r = 0$$

and $\phi(t) = t^r - t^{r-1}\beta_1 - \cdots - \beta_r$ is irreducible in $\mathfrak{F}[t, S^{-1}]$.[11] The condition that two such spaces be similar is that the corresponding polynomials be similar in the sense of Ore.

6. Suppose now that $\mathfrak{F}$ is a commutative real closed field[12] and $\bar{\alpha} \equiv \alpha$. If $x_1, x_2, \cdots, x_n$ is an orthogonal basis for $\mathfrak{R}$ then either $(x_i, x_i) > 0$ for all i or $(x_i, x_i) < 0$ (in the algebraic ordering of $\mathfrak{F}$). For if $(x_1, x_1) = \beta_1 > 0$ and $(x_2, x_2) = \beta_2 < 0$ then $(z, z) = 0$ if $z = x_1 + x_2\lambda$, $\lambda = (-\beta_1/\beta_2)^{\frac{1}{2}}$. If we replace f by $-f$ we do not change the definition of the adjoint. Hence we may suppose that f is positive definite: $(x_i, x_i) = \beta_i > 0$. By replacing x_i by $x_i\beta_i^{-\frac{1}{2}} = u_i$ we obtain a basis such that $(u_i, u_j) = \delta_{ij}$. A basis of this type will be called *Cartesian*. It is clear from the computation of § 1 that the passage from one Cartesian basis to another is given by an orthogonal matrix $M(MM' = M'M = 1)$.

If $\boldsymbol{T}$ is an l. t. with matrix T relative to the Cartesian basis $u_1, u_2, \cdots, u_n$, $\boldsymbol{T}^*$ has the matrix T' relative to this basis. $\boldsymbol{T}$ is normal if and only if $TT' = T'T$ and $\boldsymbol{T}$ is orthogonal (unitary), self-adjoint or skew according as

[11] Jacobson, *loc. cit.* in 6.

[12] In the sense of Artin-Schreier. Cf. van der Waerden, *Moderne Algebra I*, p. 227, 1st ed., or p. 235, 2nd ed.

T is orthogonal, symmetric or skew. Let $\mathfrak{R}_1$ be the join of all irreducible invariant subspaces similar to a given one and suppose $\mathfrak{R}_1 \boldsymbol{T} \neq 0$. As above $\mathfrak{R}_1$ is also in $\mathfrak{L}(\boldsymbol{T}^*)$ and hence in $\mathfrak{L}(\boldsymbol{L})$, where $\boldsymbol{L} = \boldsymbol{T}\boldsymbol{T}^* = \boldsymbol{T}^*\boldsymbol{T}$. We may decompose $\mathfrak{R}_1$ as $\mathfrak{S}^{(1)} + \cdots + \mathfrak{S}^{(k)}$ where $\mathfrak{S}^{(i)}$ consists of all vectors y_i such that $y_i \boldsymbol{L} = y_i \alpha_i$, $\alpha_i > 0$.[13] $\boldsymbol{T}$ acting in $\mathfrak{S}^{(i)}$ has the form $\beta_i \boldsymbol{O}_i$ where $\boldsymbol{O}_i$ is orthogonal and $\beta_i = \alpha_i^{\frac{1}{2}}$. If $\mathfrak{S}_1^{(i)}$ and $\mathfrak{S}_1^{(j)}$ are irreducible elements of $L(\boldsymbol{T})$ contained in $\mathfrak{S}^{(i)}$ and $\mathfrak{S}^{(j)}$ respectively and T_i and T_j the matrices of $\boldsymbol{T}$ relative to orthogonal bases in these spaces, $T_i = \beta_i O_i$, $T_j = \beta_j O_j$, O_i and O_j orthogonal. Since $S^{-1} T_i S = T_j$, $S^{-1} O_i S = O_j \beta_j \beta_i^{-1}$ and since the roots of an orthogonal matrix have absolute value 1 this implies that $\beta_i = \beta_j$. Thus $\boldsymbol{T}$ acting in $\mathfrak{R}_1$ is a positive multiple of an orthogonal l. t. If $\mathfrak{S}_1$ is any irreducible subspace in $\mathfrak{L}(\boldsymbol{T})$ either $\mathfrak{S}_1 \boldsymbol{T} = \mathfrak{S}_1 \boldsymbol{T}^* = 0$ or $\mathfrak{S}_1$ can be embedded in a space $\mathfrak{R}_1$. It follows that $\mathfrak{S}_1$ is also invariant under $\boldsymbol{T}^*$ and by the remarks of § 2 we have

THEOREM 4. *If $\boldsymbol{T}$ is a normal linear transformation in $\mathfrak{R}$ over a real closed field $\mathfrak{F}$, $\boldsymbol{T}$ is orthogonally completely reducible.*

Now by Theorem 4 we may write $\mathfrak{R} = \mathfrak{S}_1 \oplus \mathfrak{S}_2 \oplus \cdots \oplus \mathfrak{S}_l$ where the $\mathfrak{S}_i$ are irreducible and invariant with respect to T and hence are either 1 or 2-dimensional since the irreducible polynomials in $\mathfrak{F}[t]$ have degree 1 or 2. If we choose Cartesian bases in the spaces $\mathfrak{S}_i$ we obtain such a basis for $\mathfrak{R}$ and relative to this basis the matrix of $\boldsymbol{T}$ has the completely reduced form

$$(2) \qquad \begin{pmatrix} T_1 & & & & \\ & T_2 & & & \\ & & \cdot & & \\ & & & \cdot & \\ & & & & \cdot \\ & & & & & T_l \end{pmatrix}$$

We therefore have the following theorem due to Murnaghan and Wintner:[14]

THEOREM 5. *If $\boldsymbol{T}$ is a normal matrix with elements in a real closed field then there exists an orthogonal matrix M such that $M^{-1}TM = M'TM$ has the form (2) where the T_i are one and two dimensional.*

[13] If $\boldsymbol{L}$ is a self-adjoint l. t. a simple computation shows that its characteristic roots are real. If $\mathfrak{S}$ is the set of vectors u such that $u\boldsymbol{L} = u\alpha$ the space orthogonal to $\mathfrak{S}$ is also invariant under $\boldsymbol{L}$ and hence if we continue this process we obtain the decomposition. If $\boldsymbol{L} = \boldsymbol{T}\boldsymbol{T}^*$, $\alpha(u, u) = (u\boldsymbol{T}\boldsymbol{T}^*, u) = (u\boldsymbol{T}, u\boldsymbol{T}) \geqq 0$ and hence $\alpha \geqq 0$.

[14] "A canonical form for real matrices under orthogonal transformations." *Proceedings of the National Academy of Sciences*, vol. 17 (1931), pp. 417-420.

By changing the sign of one of the vectors if necessary we may suppose that M is proper, i. e., $\det M = 1$.

As has been shown by Murnaghan and Wintner the condition that a two dimensional matrix be normal is that it be symmetric or have the form

$$\begin{pmatrix} \alpha & \beta \\ -\beta & \alpha \end{pmatrix} \tag{3}$$

where $\beta \neq 0$. If T is symmetric it is well known that it can be transformed into diagonal form by an orthogonal matrix and conversely. Thus the matrices (3) constitute the set of irreducible 2-rowed normal matrices. Note that $TT' = 1\gamma$, $\gamma = \alpha^2 + \beta^2 \neq 0$ and hence T is a positive multiple of an orthogonal matrix O as was shown in the proof of Theorem 4.

THEOREM 6. *T is an irreducible normal matrix if and only if it is 1-rowed or a positive multiple of a 2-rowed orthogonal non-symmetric matrix O.*

If T is symmetric, skew orthogonal then O is respectively symmetric skew or orthogonal and Theorems 5 and 6 give the well-known results on the orthogonal reduction of these matrices.

7. Let $\mathfrak{F} = \mathfrak{F}_0(\sqrt{-1})$ where $\mathfrak{F}_0$ is real closed and $\bar{\alpha} = \alpha_0 - \sqrt{-1}\beta_0$ for $\alpha = \alpha_0 + \sqrt{-1}\beta_0$. $\mathfrak{F}$ is algebraically closed. If f is a skew hermitian form, $\sqrt{-1}f$ is hermitian. Hence as in § 6 we may suppose that f is hermitian and positive definite. Then $\mathfrak{R}$ has a *unitary* basis $u_1, u_2, \cdots, u_n$, i. e., $(u_i, u_j) = \delta_{ij}$. The change to a second unitary basis is given by a unitary matrix M ($\bar{M}'M = M\bar{M}' = 1$).

An l. t. $\boldsymbol{T}$ is normal if and only if $\bar{T}'T = T\bar{T}'$ for the matrix T of $\boldsymbol{T}$ relative to a unitary basis. Since $\mathfrak{F}$ is algebraically closed the irreducible invariant subspaces of $\boldsymbol{T}$ are one dimensional. Hence, as before, we obtain orthogonal complete reducibility and hence the following theorem of Toeplitz: [15]

THEOREM 7. *If T is a normal matrix with elements in $\mathfrak{F} = \mathfrak{F}_0(\sqrt{-1})$, $\mathfrak{F}_0$ real closed, then there exists a unitary matrix M such that $M^{-1}TM = \bar{M}'TM$ is diagonal.*

If T is hermitian, skew hermitian or unitary, the diagonal elements of the normal form are respectively real, pure imaginary or of absolute value $= 1$.

We consider next the anti-linear transformations (a. l. t.) in $\mathfrak{R}$. These are the s. l. t.'s $\boldsymbol{T}$ with automorphisms S such that $\alpha^S = \bar{\alpha}$. Let T be the matrix of $\boldsymbol{T}$ relative to a unitary basis. The computation on p. 48 shows that

[15] Toeplitz, "Das algebraische Analogon zu einen Satze von Fejér," *Mathematische Zeitschrift*, Bd. 2 (1918), pp. 187-197.

$\boldsymbol{T}^*$ has the matrix T' and automorphism S $(S^2 = E)$. We write $\boldsymbol{T} = (T, S)$ and $\boldsymbol{T}^* = (T', S)$. Then $\boldsymbol{T}^*\boldsymbol{T}$ and $\boldsymbol{T}\boldsymbol{T}^*$ are l. t.'s with matrices $T\bar{T}'$ and $T'\bar{T}$ respectively. The condition that $\boldsymbol{T}$ be normal is $T\bar{T}' = T'\bar{T} = \overline{T'T}$. $\boldsymbol{T}$ is self-adjoint, skew or unitary if and only if T is symmetric, skew or unitary.

$\boldsymbol{TT}^*$ is a self-adjoint l. t. In an invariant subspace of this l. t. we can choose a unitary basis of vectors u such that $u\boldsymbol{TT}^* = u\alpha$. Then $\bar{\alpha}(u, u) = (u\boldsymbol{TT}^*, u) = (u\boldsymbol{T}, u\boldsymbol{T}) = (u\boldsymbol{T}, u\boldsymbol{T}) \geq 0$ and $\bar{\alpha} = \alpha \geq 0$.

Now suppose that $\mathfrak{S}_1$ is irreducible and invariant relative to $\boldsymbol{T}$. If $\mathfrak{S}_1\boldsymbol{T} = 0$, $\mathfrak{S}_1\boldsymbol{T}^* = 0$ also. Hence suppose $\mathfrak{S}_1\boldsymbol{T} = \mathfrak{S}_1$. Then $\mathfrak{S}_1\boldsymbol{T}^2 = \mathfrak{S}_1$ and since $\boldsymbol{T}^2$ is a normal l. t. by the above, $\mathfrak{S}_1(\boldsymbol{T}^*)^2 = \mathfrak{S}_1$. Set $\boldsymbol{L} = \boldsymbol{TT}^* = \boldsymbol{T}^*\boldsymbol{T}$, $\boldsymbol{L}^2 = \boldsymbol{T}^2(\boldsymbol{T}^*)^2 = (\boldsymbol{TT}^*)^2$. The space $\mathfrak{R}_1 = \mathfrak{S}_1 + \mathfrak{S}_1\boldsymbol{T}^* = \mathfrak{S}_1 + \mathfrak{S}_1\boldsymbol{L}$ is invariant relative to $\boldsymbol{T}$ and $\boldsymbol{T}^*$. If $y_1 \neq 0$ is a vector in $\mathfrak{S}_1$ such that $y_1\boldsymbol{L}^2 = y_1\alpha$, $(\alpha > 0)$, then since every vector in $\mathfrak{S}_1$ has the form $y_1\mu(\boldsymbol{T})$ and $\boldsymbol{L}^2\mu(\boldsymbol{T}) = \mu(\boldsymbol{T})\boldsymbol{L}^2$, $\alpha\mu(\boldsymbol{T}) = \mu(\boldsymbol{T})\alpha$, we have $y_1\boldsymbol{L}^2 = y_1\alpha$ for every y_1 in $\mathfrak{S}_1$. Then $y_2\boldsymbol{L}^2 = y_2\alpha$ for every y_2 in $\mathfrak{S}_1\boldsymbol{T}^*$ and $x_1\boldsymbol{L}^2 = x_1\alpha$ for all x_1 in $\mathfrak{R}_1$. If $x_1\boldsymbol{L} = x_1\beta$, $\beta > 0$ and hence $\beta = \alpha^{\frac{1}{2}}$. It follows that $x_1\boldsymbol{L} = x_1\beta$ for all x_1 in $\mathfrak{R}_1$. Then $\mathfrak{S}_1\boldsymbol{T}^* = \mathfrak{S}_1\boldsymbol{L} = \mathfrak{S}_1 \epsilon \mathfrak{L}(\boldsymbol{T}^*)$. We therefore have the following theorem:

THEOREM 8. *If $\boldsymbol{T}$ is a normal anti-linear transformation in $\mathfrak{R}$ over $\mathfrak{F} = \mathfrak{F}_0(\sqrt{-1})$ and $\mathfrak{F}_0$ real closed, then $\boldsymbol{T}$ is orthogonally completely reducible.*

If $\boldsymbol{T}$ is irreducible $\mathfrak{R}$ is 1- or 2-dimensional.[16] Evidently any a. l. t. in 1-dimensions is normal. If $\boldsymbol{T} = \mathbf{1}\tau$, $\tau = \beta\epsilon$ where $\beta \geq 0$ and $\epsilon\bar{\epsilon} = 1$. Thus $\boldsymbol{T}$ is a non-negative multiple of a unitary a. l. t. Suppose next that $\mathfrak{R}$ is 2-dimensional. We have seen that a necessary condition for irreducibility is that $\boldsymbol{TT}^* = 1\beta = \boldsymbol{T}^*\boldsymbol{T}$, $\beta > 0$ and hence $\boldsymbol{T} = \gamma\boldsymbol{U}$, $\gamma = \beta^{\frac{1}{2}}$ and $\boldsymbol{U}$ is unitary. If $\boldsymbol{T}^* = \boldsymbol{T}$, $\boldsymbol{T}^2 = \beta$ and if x is any vector and $\gamma = \beta^{\frac{1}{2}}$, $y = x\boldsymbol{T} + x\gamma$ then $y\boldsymbol{T} = y\gamma$. Thus either y or x is an invariant vector $\neq 0$, contrary to the irreducibility of $\boldsymbol{T}$. Suppose conversely that $\boldsymbol{T} = \gamma\boldsymbol{U}$ and $\boldsymbol{T}$ is not self-adjoint. $\boldsymbol{T}$ is evidently normal. If $\boldsymbol{T}$ is reducible, then there is a unitary basis u, v such that $u\boldsymbol{T} = u\mu$, $v\boldsymbol{T} = v\nu$. It follows readily that $\boldsymbol{T}$ is self-adjoint. We have therefore proved

THEOREM 9. *If $\boldsymbol{T}$ is an irreducible normal anti-linear transformation, $\mathfrak{R}$ is 1- or 2-dimensional. In the former case $\boldsymbol{T}$ is arbitrary and in the latter, $\boldsymbol{T}$ is a positive multiple of a unitary a. l. t. and is not self-adjoint and conversely.*

---

[16] Asano and Nakayama, "Über halblineare Transformationen," *Mathematische Annalen*, Bd. 115 (1937), p. 110.

Note that $\boldsymbol{U}$ and γ are uniquely determined by $\boldsymbol{T}$ if $\gamma > 0$. For if $\boldsymbol{T} = \gamma \boldsymbol{U} = \gamma_1 \boldsymbol{U}_1$, $\gamma\gamma_1^{-1}$ is unitary and hence $\gamma = \gamma_1$, $\boldsymbol{U} = \boldsymbol{U}_1$.

If $\boldsymbol{T}$ is any normal a. l. t. and $\mathfrak{R} = \mathfrak{R}_1 \oplus \mathfrak{R}_2 \oplus \cdots \oplus \mathfrak{R}_l$ where the $\mathfrak{R}_i$ are irreducible we may choose a unitary basis in each $\mathfrak{R}_i$. The new matrix of $\boldsymbol{T}$ will be $M^{-1}T\bar{M} = N'TN$ where M and N are unitary and $N = \bar{M}$. $M^{-1}T\bar{M}$ will be completely reduced. Hence we have the following theorem:

THEOREM 10. *If T is a matrix with elements in $\mathfrak{F} = \mathfrak{F}_0(\sqrt{-1})$, $\mathfrak{F}_0$ real closed and $T\bar{T}' = T'\bar{T}$, then there exists a unitary matrix M such that*

$$M^{-1}T\bar{M} = N'TN = \begin{pmatrix} T_1 & & & & \\ & T_2 & & & \\ & & \cdot & & \\ & & & \cdot & \\ & & & & T_l \end{pmatrix}, \quad N = \bar{M} \tag{4}$$

where T_i is either 1-rowed or a positive multiple of a 2-rowed unitary matrix which is not symmetric.

We may, of course, re-arrange the T_i's in an arbitrary manner without destroying the validity of the result.

COROLLARY 1. *If T is a symmetric matrix with elements in $\mathfrak{F}_0(\sqrt{-1})$, there exists a unitary matrix M such that $D = M^{-1}T\bar{M} = N'TN$ is diagonal.*

If D has diagonal elements $\delta_i = \beta_i\epsilon_i$, $\beta_i \geq 0$, $\epsilon_i\bar{\epsilon}_i = 1$, we may replace M by the unitary matrix MQ where Q is the diagonal matrix with elements $\epsilon_i^{\frac{1}{2}}$. Then D is replaced by a diagonal matrix whose elements are all ≥ 0.

COROLLARY 2. *If T is a skew matrix with elements in $\mathfrak{F}_0(\sqrt{-1})$ there exists a unitary matrix such that $M^{-1}T\bar{M} = N'TN$ has the form (4) where $T_i = 0$ or is 2-rowed and skew.*

As in the symmetric case we may suppose that the matrices $T_i \neq 0$ are positive multiples of the matrix $\begin{pmatrix} 0 & 1 \\ -1 & 0 \end{pmatrix}$.

COROLLARY 3. *If T is a unitary matrix with elements in $\mathfrak{F}_0(\sqrt{-1})$, there exists a unitary matrix M such that $M^{-1}T\bar{M} = N'TN$ has the form (4) where the T_i are unitary.*

In this case it is readily seen that we may normalize the T_i to have determinant $= 1$.

Now suppose that T and R are any two matrices such that $T\bar{T}' = T'\bar{T}$, $R\bar{R}' = R'\bar{R}$ and $T = M^{-1}R\bar{M}$ where M is unitary. Then $R\bar{R}$ and $T\bar{T} = M^{-1}R\bar{R}M$ are similar. We proceed to prove the converse. It has been shown by Nakayama and by Haantjes [17] that if R and T are any two matrices with elements in $\mathfrak{F}_0(\sqrt{-1})$ such that rank $R\bar{R}\cdots\bar{R}^{(i)} = \text{rank } T\bar{T}\cdots\bar{T}^{(i)}$ and $R\bar{R}$ and $T\bar{T}$ are similar, then there exists a matrix A such that $A^{-1}R\bar{A} = T$. Moreover in the present case it is clear from (4) that rank $R = \text{rank } R\bar{R}\cdots\bar{R}^{(i)} = \text{rank } T\bar{T}\cdots\bar{T}^{(i)} = \text{rank } T$. Hence we may suppose that $A^{-1}R\bar{A} = T$ and also that R and T have the form (4). Thus R and T are matrices of the same normal a. l. t. $\boldsymbol{T}$. The irreducible matrices R_i and T_i correspond to two decompositions of R into irreducible invariant subspaces relative to T. Hence by the Jordan-Hölder theorem, we may suppose that $A_i^{-1}R_i\bar{A}_i = T_i$, $i = 1, 2, \cdots, l$. It therefore suffices to prove the following

LEMMA. *If R and T satisfy $R\bar{R}' = R'\bar{R}$, $T\bar{T}' = T'\bar{T}$ and these matrices belong to the same irreducible normal a. l. t., then there exists a unitary matrix M such that $M^{-1}R\bar{M} = T$.*

We have seen that $A^{-1}R\bar{A} = T$ and T and R have the form βU and γV where U and V are unitary and $\beta, \gamma \geqq 0$. The case $\beta = \gamma = 0$ is trivial. If $\beta \neq 0$ we obtain $A^{-1}V\bar{V}A = (\beta\gamma^{-1})^2 U\bar{U}$. Since the roots of the unitary matrices $U\bar{U}$ and $V\bar{V}$ have absolute value 1, this implies $\beta^2 = \gamma^2$ and $\beta = \gamma$. Thus we may suppose that R and T are unitary. Then $T = \bar{A}'R(A')^{-1}$ and $BR = R\bar{B}$ if $B = A\bar{A}'$. As the author has shown [18] the matrices X satisfying the equation $XR = R\bar{X}$ form a division algebra over $\mathfrak{F}_0$. On the other hand B is hermitian and positive definite and hence it satisfies a quadratic equation whose roots are real and positive in $\mathfrak{F}_0$. Hence $B = \delta 1$, $\delta > 0$ and $A = \delta^{\frac{1}{2}}M$ where M is unitary. Then $T = M^{-1}R\bar{M}$.

THEOREM 11. *A necessary and sufficient condition that two matrices R and T with elements in $\mathfrak{F}_0(\sqrt{-1})$ such that $R\bar{R}' = R'\bar{R}$, $T\bar{T}' = T'\bar{T}$ be related by a unitary matrix M in the manner $M^{-1}R\bar{M} = N'RN = T\ (N = \bar{M})$ is that $R\bar{R}$ and $T\bar{T}$ be similar.*

8. We suppose in this section that $\mathfrak{F}$ is Hamilton's quaterion algebra over a real closed field $\mathfrak{F}_0$. Besides the well-known i. a. a.

$$\alpha = \alpha_0 + i\alpha_1 + j\alpha_2 + k\alpha_3 \rightarrow \bar{\alpha} = \alpha_0 - i\alpha_1 - j\alpha_2 - k\alpha_3$$

we have the i. a. a.'s $\alpha \rightarrow \alpha^V = \nu^{-1}\bar{\alpha}\nu$ where $\nu = -\nu$.[19] However if $f = (x, y)$

[17] Nakayama, "Über die Klassifikation halblinearer Transformationen," *Proc. Phys.-Math. Soc. Japan*, vol. 19 (1937); Haantjes, "Halblineare transformationen," *Mathematische Annalen*, Bd. 114 (1937), pp. 292-304.

[18] *Loc. cit.* in 6, p. 503.

[19] There are the only i. a. a.'s in $\mathfrak{F}$ leaving the elements of $\mathfrak{F}_0$ unaltered. See, for example, Albert, "Involutorial simple algebras and real Riemann matrices," *Annals of Mathematics*, vol. 36 (1935), p. 897.

is a bilinear form with the i. a. a. V, then the condition $(x\alpha, y) = \nu^{-1}\bar{\alpha}\nu(x, y)$ implies that $\nu(x\alpha, y) = \bar{\alpha}\nu(x, y)$ and hence νf is a bilinear form with the usual i. a. a. $\alpha \rightarrow \bar{\alpha}$. It is readily seen that f is hermitian or skew if and only if νf is respectively skew or hermitian. Since the adjoints of an l. t. defined relative to f and νf are equal we shall consider only the i. a. a. $\alpha \rightarrow \bar{\alpha}$.

We shall show next that if f is totally regular and dim $R > 1$ then f can not be skew. For suppose u and v are orthogonal and $(u, u) = \gamma$, $(v, v) = \delta$ where $\bar{\gamma} = -\gamma$, $\bar{\delta} = -\delta$. It can be shown that $-\gamma$ and δ are cogredient in the sense that there is an α in $\mathfrak{F}$ such that $\delta = -\bar{\alpha}\gamma\alpha$.[20] Then $(u\alpha + v, u\alpha + v) = 0$ contrary to the total regularity. Thus f is hermitian and as in the complex case we may suppose that it is positive definite.

It follows that $\mathfrak{R}$ has a *unitary* basis of vectors $u_1, u_2, \cdots, u_n$ such that $(u_i, u_j) = \delta_{ij}$. Any other unitary basis is obtained from the u's by transformation with a unitary matrix. The l. t. $\boldsymbol{T}$ is normal if and only if its matrix T relative to the u's satisfies $T\bar{T}' = \bar{T}'T$.

If $\phi(t) \in \mathfrak{F}[t]$, $\phi^*(t) = \phi(t)\bar{\phi}(t) \in \mathfrak{F}_0[t]$ the centrum of $\mathfrak{F}[t]$. Evidently $\phi^*(t)$ can be decomposed into linear factors in $\mathfrak{F}[t]$ and hence by the factorization theorem[21] $\phi(t)$ has linear factors. Thus the only irreducible polynomials in $\mathfrak{F}[t]$ are linear and hence the irreducible subspaces of an l. t. are 1-dimensional.

Suppose $\boldsymbol{T}$ is normal and $\mathfrak{S}$ is irreducible in $\mathfrak{L}(\boldsymbol{T})$. If $u \neq 0$ is in $\mathfrak{S}$, $u\boldsymbol{T} = u\beta$ and

$$(u\bar{\beta}, u) = \beta(u, u) = (u, u)\beta = (u, u\beta) = (u, u\boldsymbol{T}) = (u\boldsymbol{T}^*, u)$$

since $(u, u) \in \mathfrak{F}_0$. Hence $(u(\bar{\beta} - \boldsymbol{T}^*), u) = 0$ and

$$\begin{aligned}(u(\bar{\beta} - \boldsymbol{T}^*), u(\bar{\beta} - \boldsymbol{T}^*)) &= (u(\bar{\beta} - \boldsymbol{T}^*), u\bar{\beta}) - (u(\bar{\beta} - \boldsymbol{T}^*), u\boldsymbol{T}^*) \\ &= -(u(\bar{\beta} - \boldsymbol{T}^*), u\boldsymbol{T}^*) \\ &= -(u\boldsymbol{T}(\bar{\beta} - \boldsymbol{T}^*), u) \\ &= -\bar{\beta}(u(\bar{\beta} - \boldsymbol{T}^*), u) = 0.\end{aligned}$$

Hence $u\boldsymbol{T}^* = u\bar{\beta}$, i. e., $\mathfrak{S}$ is invariant relative to $\boldsymbol{T}^*$.

THEOREM 12. *If $\boldsymbol{T}$ is a normal l. t. in $\mathfrak{R}$ over $\mathfrak{F}$ where $\mathfrak{F}$ is a quaternion algebra over a real closed field, then $\boldsymbol{T}$ is orthogonally completely reducible.*

As before, we obtain

---

[20] Jacobson, "Simple Lie algebras of characteristic zero," *Duke Mathematical Journal*, vol. 4 (1938), p. 546.

[21] Ore, "Non-commutative polynomials," *Annals of Mathematics*, vol. 34 (1933), p. 494.

THEOREM 13. *If T is normal matrix* ($T\bar{T}' = \bar{T}'T$) *with elements in* $\mathfrak{F}$, *there exists a unitary matrix M such that $M^{-1}TM = \bar{M}'TM$ is diagonal.*[22]

In the special cases $\boldsymbol{T}^* = \boldsymbol{T}$, $\boldsymbol{T}^* = -\boldsymbol{T}$ and $\boldsymbol{T}\boldsymbol{T}^* = 1$ the diagonal elements β_i of T satisfy the conditions $\bar{\beta}_i = \beta_i \in \mathfrak{F}_0$, $\bar{\beta}_i = -\beta_i$ and $\beta_i\bar{\beta}_i = 1$ respectively.

9. The above results may be interpreted also in projective geometry. We define the projective space $\mathfrak{P}$ of $(n-1)$-dimensions over $\mathfrak{F}$ as the set of subspace of the n-dimensional vector space $\mathfrak{R}$ over $\mathfrak{F}$. The points of $\mathfrak{P}$ are the rays $x\alpha$, $x \neq 0$ and fixed and α variable. The correspondence π: $\mathfrak{S} \to \mathfrak{S}' \equiv \mathfrak{S}\pi$ determined by a totally regular hermitian or skew-hermitian form f is called an *elliptic polarity*.[23]

It is known that any $(1-1)$ mapping τ of the elements of (the subspaces of $\mathfrak{R}$) which preserves incidences is determined by a non-singular s. l. t. $\boldsymbol{T}$.[24] τ will be called a semi-collination. Two s. l. t.'s $\boldsymbol{T}_1$ and $\boldsymbol{T}_2$ define the same τ if and only if $\boldsymbol{T}_1 = \boldsymbol{T}_2\rho$, $\rho \neq 0$ in $\mathfrak{F}$. τ is a collineation if the automorphism of $\boldsymbol{T}$ is inner; hence, if and only if $\boldsymbol{T}$ may be chosen as an l. t.

If τ is a semi-collineation, then so is $\tau' = \pi\tau\pi = \pi^{-1}\tau\pi$. Since $(x, y) = 0$ if and only if $(x(\boldsymbol{T}^*)^{-1}, y\boldsymbol{T}) = 0$, we see that if $\mathfrak{S}$ is a subspace and $\mathfrak{S}\pi$ its polar then $\mathfrak{S}(\boldsymbol{T}^*)^{-1}$ is the polar of $\mathfrak{S}\pi\tau$. Hence $\mathfrak{S}(\boldsymbol{T}^*)^{-1} = \mathfrak{S}\pi\tau\pi$. Thus the semi-collineation τ' is determined by $(\boldsymbol{T}^*)^{-1}$. We shall call τ normal if $\tau\tau' = \tau'\tau$. The condition on $\boldsymbol{T}$ is $\boldsymbol{T}\boldsymbol{T}^* = \boldsymbol{T}^*\boldsymbol{T}\mu$, $\mu \in \mathfrak{F}$. Notable special cases are given by the condition $\tau' = \tau$ and $\tau' = \tau^{-1}$.

It is known that if $\mathfrak{P}$ satisfies certain topological conditions, then $\mathfrak{F}$ must be either the field of real numbers, the field of complex numbers or the quasi-field of real quaternions.[25] We shall now discuss these cases separately.

(1). Suppose $\mathfrak{F}$ is the field of real numbers. The only automorphism or anti-automorphism of $\mathfrak{F}$ is the identity. It follows as in § 6 that if π is an elliptic polarity, the corresponding f may be assumed positive definite. Every semi-collineation τ is a collineation.

Suppose τ is normal. The condition $\tau'\tau = \tau\tau'$ is equivalent to $\boldsymbol{T}^*\boldsymbol{T}$

---

[22] Cf. Teichmüller, " Operatoren im Wachsschen Raum," *Crelle*, Bd. 174 (1935), p. 111.

[23] Veblen and Young, *Projective Geometry II*, p. 218. Cf. also Cartan, *La géometrie projective complexe*, p. 139. Cartan calls the correspondences determined by forms whose anti-automorphisms are different from the identity *anti-polarities*. From the present point of view it does not seem worth while to make this distinction.

[24] Cf. Brauer, " A characterization of null systems in projective space," *Bulletin of the American Mathematical Society*, vol. 42 (1936), pp. 247-254.

[25] Veblen and Young, *Projective Geometry II*, chap. 1, and Kolmogoroff, " Zur Begründung der projektiven Geometrie," *Annals of Mathematics*, vol. 33 (1932), pp. 175-176.

$= \boldsymbol{TT}^*\mu$, $\mu \neq 0$ in $\mathfrak{F}$. Hence $TT' = T'T\mu$ if T is the matrix of $\boldsymbol{T}$ relative to a suitable basis. If $T = (\tau_{ij})$, $\operatorname{tr} TT' = \Sigma\tau_{ij}{}^2 \neq 0$ and since $\operatorname{tr} TT' = \operatorname{tr} T'T$, we obtain $\mu = 1$ and hence $\boldsymbol{T}$ is a normal l. t. $\tau' = \tau$ if and only if $\boldsymbol{T}^*\boldsymbol{T} = 1\rho$, $\rho > 0$. If we replace $\boldsymbol{T}$ by $\boldsymbol{T}\rho^{-\frac{1}{2}}$, we see that $\boldsymbol{T}$ can be taken as a unitary (orthogonal) l. t. On the other hand $\tau' = \tau^{-1}$ if and only if $\boldsymbol{T}^* = \boldsymbol{T}\mu$ and since $\boldsymbol{T}^{**} = \boldsymbol{T}$, we obtain $\mu^2 = 1$ and $\boldsymbol{T}^* = \pm \boldsymbol{T}$.

(2). $\mathfrak{F}$ is the field of complex numbers. In this case we restrict ourselves to continuous polarities and semi-collineations. The corresponding anti-automorphism or automorphism will be continuous. Hence we have just two types, the identity and the mapping $\alpha \rightarrow \bar{\alpha}$. A bilinear form with the identity as its anti-automorphism can not be totally regular. It follows (§ 7) that any elliptic polarity is determined by a positive definite hermitian form.

There are two types of continuous semi-collineations: the collineations and the semi-collineations with the automorphism $\alpha \rightarrow \bar{\alpha}$. The latter are called *anti-collineations.* If τ is a normal collineation then $\boldsymbol{TT}^* = \boldsymbol{T}^*\boldsymbol{T}\mu$ and conversely. Hence $T\bar{T}' = \bar{T}'T\mu$ and since $\operatorname{tr} TT' = \Sigma|\tau_{ij}|^2 \neq 0$ we obtain $\mu = 1$ and hence $\boldsymbol{T}$ is normal. The condition $\tau' = \tau$ means that $\boldsymbol{TT}^* = 1\rho$ and as before $\boldsymbol{T}$ may be taken unitary. $\tau' = \tau^{-1}$ is equivalent to $\boldsymbol{T}^* = \boldsymbol{T}\rho$ and hence $\boldsymbol{T}^{**} = \boldsymbol{T}^*\bar{\rho}\rho = \boldsymbol{T}\bar{\rho}\rho = \boldsymbol{T}$, $\bar{\rho}\rho = 1$. If we replace $\boldsymbol{T}$ by $\boldsymbol{T}\sigma$ where $\rho = \bar{\sigma}^{-1}\sigma$ we see that τ is determined by a self-adjoint l. t.

If τ is a normal anti-collineation, $\boldsymbol{TT}^* = \boldsymbol{T}^*\boldsymbol{T}\mu$. The matrices of these l. t. are respectively $T\bar{T}'$ and $T'\bar{T}\mu$. It follows again that $\mu = 1$ and $\boldsymbol{T}$ is a normal a. l. t. and conversely. $\tau' = \tau$ if and only if $\boldsymbol{T}$ can be taken as a unitary a. l. t. and $\tau' = \tau^{-1}$ if and only if $\boldsymbol{T}$ is either self-adjoint or skew.

(3). $\mathfrak{F}$ is the field of real quaternions. Any anti-automorphism or automorphism of $\mathfrak{F}$ maps the center $\mathfrak{F}_0$ into itself and since $\mathfrak{F}_0$ is the field of real numbers, the elements of $\mathfrak{F}_0$ are left invariant. It follows that the anti-morphisms of $\mathfrak{F}_0$ are $\alpha \rightarrow \bar{\alpha}$ and $\alpha \rightarrow \nu^{-1}\bar{\alpha}\nu$, $\bar{\nu} = -\nu$ and the automorphisms are all inner. Hence as we saw in § 8, the elliptic polarities in $\mathfrak{P}$ are determined by positive definite hermitian forms. The semi-collineations in $\mathfrak{P}$ are all collineations. τ is normal if and only if the corresponding l. t. is normal; $\tau' = \tau$ if and only if T can be chosen unitary; $\tau' = \tau^{-1}$ if and only if $\boldsymbol{T}$ is self-adjoint or skew.

The results of § 6-8 may be translated to the present situation. For example we have the following theorem:

Let $\mathfrak{P}$ be a continuous projective space in the above sense, π a continuous elliptic polarity and τ a continuous normal collineation or anti-collineation in $\mathfrak{P}$. Then if $\mathfrak{S}$ is a hyperplane invariant under τ, its polar plane $\mathfrak{S}\pi$ is also invariant under τ.

UNIVERSITY OF NORTH CAROLINA.

Correction

The proof of Theorem 4 at the top of page 51 contains an error on line 10. The statement $\beta_i = \beta_j$ needs to be replaced by $\beta_i = \pm\beta_j$. This invalidates the rest of the argument. To obtain a correct proof we prove first Theorem 5 by the method given on pages 182–186 of the author's *Lectures in Abstract Algebra II*. Also by Theorem 8 of this book we have $\mathfrak{R} = \mathfrak{S}_1 \oplus \cdots \oplus \mathfrak{S}_l$ where the $\mathfrak{S}_i$ are irreducible $\mathfrak{L}(T)$ and either $\mathfrak{S}_i$ is one dimensional or it is two dimensional and has a Cartesian base such that the matrix of the restriction T_i of T to $\mathfrak{S}_i$ has the form (3). T_i is either self-adjoint or is a positive multiple of an orthogonal linear transformation.

We now arrange the $\mathfrak{S}_i$ so that the first n_1 are isomorphic as $\mathfrak{F}[T]$-modules, the next n_2 are isomorphic but not isomorphic to any $\mathfrak{S}_i$, $1 \leq i \leq n_1$, the next n_3 are isomorphic but not isomorphic to any $\mathfrak{S}_i$, $1 \leq i \leq n_1 + n_2$, etc. Put $\mathfrak{R}_1 = \mathfrak{S}_1 + \cdots + \mathfrak{S}_{n_1}$, $\mathfrak{R}_2 = \mathfrak{S}_{n_1+1} + \cdots + \mathfrak{S}_{n_1+n_2}$, $\mathfrak{R}_3 = \mathfrak{S}_{n_1+n_2+1} + \cdots + \mathfrak{S}_{n_1+n_2+n_3}, \ldots$. Then the $\mathfrak{R}_i$ are the homogeneous components of $\mathfrak{R}$ as $\mathfrak{F}[T]$-module and any irreducible submodule $\mathfrak{S}$ of $\mathfrak{R}$ is contained in some one of the $\mathfrak{R}_j$. If $\mathfrak{S}_i$ is two dimensional and we choose an orthogonal base for $\mathfrak{S}_i$ so that the matrix of the restriction T_i of T to $\mathfrak{S}_i$ is

$$\begin{pmatrix} \alpha_i & \beta_i \\ -\beta_i & \alpha_i \end{pmatrix}$$

then the characteristic polynomial of T_i is $\lambda^2 - 2\alpha_i\lambda + (\alpha_i^2 + \beta_i^2)$ and $T_i = \mu_i O_i$ where $\mu_i = (\alpha_i^2 + \beta_i^2)^{1/2}$ and O_i is orthogonal. If $\mathfrak{S}_k \cong \mathfrak{S}_i$ then T_i and T_k have the same characteristic polynomials and hence $\mu_i = \mu_k$. It follows that if $\mathfrak{R}_j$ is the homogeneous component containing $\mathfrak{S}_i$ and $\mathfrak{S}_k$ then the restriction of T to $\mathfrak{R}_j$ is a multiple of an orthogonal transformation. This implies that $\mathfrak{R}_j T^* \subset \mathfrak{R}_j$. Moreover, if $\mathfrak{S}$ is an irreducible $\mathfrak{F}[T]$-submodule of $\mathfrak{R}_j$ then the argument shows that $\mathfrak{S}T^* \subset \mathfrak{S}$. Similarly, we see that if $\mathfrak{S}_i$ is one dimensional and $\mathfrak{R}_j$ is the homogeneous component containing $\mathfrak{S}_i$ then the restriction of T to $\mathfrak{R}_j$ is a scalar. Then again $\mathfrak{S}T^* \subset \mathfrak{S}$ for any irreducible submodule $\mathfrak{S}$ of $\mathfrak{R}_j$. Since every irreducible submodule is contained in some $\mathfrak{R}_j$ we see that $\mathfrak{S}T^* \subset \mathfrak{S}$ for every irreducible $\mathfrak{S}$. By the complete reducibility property stated in Theorem 2 and the fact that any submodule of a completely reducible module is completely reducible, any submodule $\mathfrak{S}$ of $\mathfrak{R}$ is a sum of irreducibles. Hence $\mathfrak{S}T^* \subset \mathfrak{S}$. This implies Theorem 4.

Reprinted from
American Journal of Mathematics
January 1939.

AN APPLICATION OF E. H. MOORE'S DETERMINANT OF A HERMITIAN MATRIX*

N. JACOBSON

E. H. Moore defined a determinant for any hermitian matrix with elements in a "number system of type B."† In more descriptive language, a system Φ of this type may be characterized as a quasi-field of characteristic not equal to 2 in which there is defined an involutorial anti-automorphism or *involution* $a \rightarrow \bar{a}$:

$$\overline{a+b} = \bar{a} + \bar{b}, \qquad \overline{ab} = \bar{b}\bar{a}, \qquad \bar{\bar{a}} = a,$$

such that the symmetric elements ($\bar{a}=a$) are contained in the center. It follows readily that Φ is either commutative with $\bar{a} \equiv a$, a quadratic field over the field of symmetric elements, or a generalized quaternion algebra over this field. An examination of Moore's theory of determinants shows that it is entirely integral, and hence it is valid if Φ is any ring with an identity in which there is a unique element $1/2$ such that $2(1/2) \equiv 1/2+1/2=1$ and which has an involution $a \rightarrow \bar{a}$ whose symmetric elements are in the center Γ of Φ.

The uniqueness of $1/2$ implies its symmetry. If $2a \equiv a+a=0$, then $0=(a+a)/2=a/2+a/2=(1/2+1/2)a=a$. Let Σ and P respectively denote the sets of symmetric and of skew elements ($\bar{a}=-a$) of Φ. Then Σ and P are subgroups under the operation $+$. If $b \in \Sigma \cap \mathrm{P}$, $b=-b$, $2b=0$, and hence $b=0$. For any a we have

$$a = \tfrac{1}{2}(a + \bar{a}) + \tfrac{1}{2}(a - \bar{a}) = Sa + Va,$$

where $Sa \in \Sigma$, $Va \in \mathrm{P}$. Thus the additive group of Φ is a direct sum of Σ and P. We call Sa and Va respectively the scalar and the vector parts of a. Now Σ is a subring of Γ, and P is closed under multiplication by elements in Σ and under commutation $[u, v]=uv-vu$. Hence, for any two elements a, b, $[a, b]=[Va, Vb] \in \mathrm{P}$. Thus $S[a, b]=0$ and since, in general, $S(a+b)=Sa+Sb$, $Sab=Sba$. Moreover, $a\bar{a}$ and $\bar{a}a$ are symmetric, $a\bar{a}-\bar{a}a$ skew. Hence $a\bar{a}=\bar{a}a$. As usual we call this element, the norm of a, Na and note that $Nab=(Na)(Nb)$. Any element a satisfies a quadratic equation with coefficients in Σ, namely,

$$x^2 - (2Sa)x + Na = 0.$$

* Presented to the Society, February 25, 1939.

† See Moore and Barnard, *General Analysis* I, American Philosophical Society Publication, chap. 2. We refer to this volume as M-B.

Thus all the properties noted in M-B, page 103, which do not involve inverses hold in the present case.

We consider the ring Φ_n of n-rowed square matrices with elements in Φ. The correspondence $A=(a_{ij})\rightarrow\overline{A}'=(b_{ij})$, where $b_{ij}=\bar{a}_{ji}$ is an involution in Φ_n. As is customary, we call the symmetric elements $(\overline{A}'=A)$ under this involution hermitian and we denote their totality as H. If A, B are in H, X arbitrary, then $A\pm B$, $AB+BA$ and $\overline{X}'AX$ are in H.

Suppose $A=(a_{ij})$ is hermitian. If σ is a set of distinct indices in the range $1,\cdots,n$ and f is in σ, we define

$$s_\sigma = \sum (-1)^r a_{fh_1}a_{h_1h_2}\cdots a_{h_{r-1}h_r}a_{h_rf},$$

summed on all permutations of the indices h_i of σ which are not equal to f. It can be shown that s_σ is in Σ and does not depend on the particular index f selected in σ.* Moore's determinant is defined as

$$\det_M A = \sum s_{\sigma_1}s_{\sigma_2}\cdots s_{\sigma_k},$$

summed on all partitions of $1,\cdots,n$ into disjoint sets σ_i. The proofs in M-B, pages 110–123, are valid for the present case. We require in particular the existence of the adjoint, that is, a certain hermitian matrix $\text{adj}_M A$ whose elements are polynomials of $(n-1)$st degree in those of A, such that

$$A(\text{adj}_M A) = (\text{adj}_M A)A = (\det_M A)1,$$

where 1 is the identity matrix. It follows from this as usual that A satisfies its characteristic equation $\det_M(x1-A)=0$ of nth degree in x with coefficients in Σ that are polynomials in the elements of the matrix A.†

We now apply this result to ordinary matrix theory. Let Σ be any field of characteristic not equal to 2, Σ_{2n} the $2n$-rowed matrix ring over Σ and Q the matrix

$$Q = \begin{pmatrix} q & & & \\ & q & & \\ & & \ddots & \\ & & & q \end{pmatrix}, \qquad q = \begin{pmatrix} 0 & 1 \\ -1 & 0 \end{pmatrix}.$$

The correspondence $A\rightarrow Q^{-1}A'Q$ is an involution in Σ_{2n} and, if

* M-B, p. 114.

† See, for example, Wedderburn, *Lectures on Matrices*, p. 23, or Albert, *Modern Higher Algebra*, p. 78. Compare also M-B, p. 128.

$$A = \begin{pmatrix} a_{11} & a_{12} & \cdots & a_{1n} \\ \cdot & \cdot & \cdots & \cdot \\ \cdot & \cdot & \cdots & \cdot \\ a_{n1} & a_{n2} & \cdots & a_{nn} \end{pmatrix},$$

where $a_{ij} \varepsilon \Sigma_2$,

$$Q^{-1}A'Q = \begin{pmatrix} \bar{a}_{11} & \bar{a}_{21} & \cdots & \bar{a}_{n1} \\ \cdot & \cdot & \cdots & \cdot \\ \cdot & \cdot & \cdots & \cdot \\ \bar{a}_{1n} & \bar{a}_{2n} & \cdots & \bar{a}_{nn} \end{pmatrix},$$

where, in general,

$$\bar{a} = \begin{pmatrix} \delta & -\beta \\ -\gamma & \alpha \end{pmatrix} \quad \text{if} \quad a = \begin{pmatrix} \alpha & \beta \\ \gamma & \delta \end{pmatrix}.$$

The correspondence $a \rightarrow \bar{a}$ is an involution in $\Phi = \Sigma_2$ satisfying the above conditions, and $A \rightarrow Q^{-1}A'Q$ is equivalent to $A = (a_{ij}) \rightarrow \bar{A}'$. Hence, if A is symmetric in the sense that $Q^{-1}A'Q = A$ $(\bar{A}' = A)$, the above result shows that A satisfies an equation of the nth degree in x with coefficients which are polynomials in the elements $\alpha, \beta, \cdots$.

Now suppose R is any nonsingular matrix such that $R' = -R$. If B satisfies $R^{-1}B'R = B$ and V is any matrix,

$$(V'RV)^{-1}(V^{-1}BV)'(V'RV) = V^{-1}BV.$$

We may choose V so that $V'RV = Q$ as above.* Thus $A = V^{-1}BV$ is symmetric with respect to the involution $A \rightarrow \bar{A}'$. Hence B satisfies an equation $\phi(x; \alpha, \beta, \cdots)$ of the nth degree with coefficients polynomials in the elements of B. Let $f(x; \alpha, \beta, \cdots)$ be the ordinary characteristic polynomial and $\psi(x; \alpha, \beta, \cdots)$ the minimum polynomial obtained when we regard the elements $\alpha, \beta, \cdots$ in the general matrix satisfying our condition as indeterminates.† As is well known, $f(x)$ and $\phi(x)$ are divisible by $\psi(x)$ and $f(x)$ and $\psi(x)$ have the same irreducible factors.

We now choose W so that

$$W'RW = S = \begin{pmatrix} 0 & 1_n \\ -1_n & 0 \end{pmatrix},$$

where 1_n is the identity in Σ_n. Then $C = W^{-1}BW$ satisfies $S^{-1}C'S = C$. If

* Wedderburn, p. 96 or Albert, p. 108, loc. cit.

† The conditions on $A = V^{-1}BV$ are $a_{ii} = \alpha_{ii}1_2$ and $a_{ji} = \bar{a}_{ij}$, $i \neq j$. We regard α_{ii} and the elements of a_{ij}, $i < j$ as indeterminates. If we replace Σ by the polynomial ring obtained by adjoining these indeterminates to Σ, we obtain as before $\phi(B) = 0$.

$$C = \begin{pmatrix} L & M \\ N & P \end{pmatrix},$$

L, M, and so on, being in Σ_n, it follows that $M' = -M$, $N' = -N$, $P = L'$ and conversely any matrix of this form satisfies the condition. If we specialize the indeterminates so that $M = N = 0$, we obtain a matrix with irreducible minimum polynomial of degree n. It follows that $\psi(x)$ for the general matrix has degree not less than n and hence $\psi(x) = \phi(x)$ and $f(x) = [\phi(x)]^2$.

THEOREM. *If B is a matrix of $2n$ rows and columns with elements in a field of characteristic not equal to 2 such that $R^{-1}B'R = B$, where R is any non-singular skew symmetric matrix, then the characteristic polynomial $f(x)$ of B has the form $[\phi(x)]^2$, where the coefficients of $\phi(x)$ are polynomials in the elements of B and $\phi(B) = 0$. If the elements of the general matrix B are regarded as indeterminates, then $\phi(x)$ is irreducible.*

UNIVERSITY OF NORTH CAROLINA

Reprinted from
Bulletin of the American Mathematical Society
October 1939.

ANNALS OF MATHEMATICS
Vol. 40, No. 4, October, 1939

STRUCTURE AND AUTOMORPHISMS OF SEMI-SIMPLE LIE GROUPS IN THE LARGE[1]

BY N. JACOBSON

(Received May 3, 1939)

The present paper attempts to fill in several gaps in the literature relating Lie algebras to Lie groups. It is well known that there is a (1–1) correspondence between Lie subalgebras of the Lie algebra $\mathfrak{L}$ of $\mathfrak{G}$ and its closed local subgroups. However in general different closed local subgroups may generate the same closed subgroup of $\mathfrak{G}$. We can show nevertheless that $\mathfrak{G}$ is semi-simple (simple) if and only if $\mathfrak{L}$ is semi-simple (simple). We also give a new set of linear groups which represent the classes of locally isomorphic simple Lie groups and which is somewhat simpler than Cartan's original list.[2] Omitting a finite number of exceptions these are merely the important geometric linear groups with real complex and quaternionic elements (unimodular, orthogonal and symplectic[3] groups). We determine the group $\mathfrak{A}$ of bicontinuous automorphisms of these "unexceptional" groups and discuss the structure of $\mathfrak{A}/\mathfrak{J}$, $\mathfrak{J}$ the set of inner automorphisms.

1. We begin with a brief resumé of the local theory. A topological space U is a *local group*[4] (group germ, group nucleus) if it contains an open set V in which a composition xy in U is defined such that

1. If x, y, z, xy, yz are in V then $x(yz) = (xy)z$.
2. There is a point 1 in V such that $1x = x1 = x$ for any x in V.
3. For each x in V there is an x^{-1} in V such that $xx^{-1} = x^{-1}x = 1$.
4. xy and x^{-1} are continuous in x and y.

U is a *local Lie group* if it is an r-cell and in place of 4. we have

4′. xy and x^{-1} are analytic in x and y in the sense that the coördinates are analytic functions of those of x and y.

W is a local subgroup of U if it is a local group relative to the same composition and topology as defined in U. Two local subgroups are regarded as identical if their intersection is open in each of the local groups. W is invariant if for each neighborhood W_1 of 1 in W there exist symmetric[5] neighborhoods N and Z of the identity in U and W respectively such that $x^{-1}zx \in W_1$ if $x \in N$ and $z \in Z$. Two

---

[1] Presented to the National Academy of Sciences Oct. 24, 1938.

[2] Cartan [2].

[3] This term has been introduced by Prof. Weyl (The classical groups, Princeton 1939) in place of the earlier and undesirable terms *Abelian* or *complex* linear groups.

[4] Sometimes called *group germ* or *group nucleus*. The present term has been adopted in the forthcoming translation of Pontrjagin's book on continuous groups.

[5] W_1 is symmetric if W_1^{-1} the set of points w_1^{-1}, w_1 in W_1, $= W_1$.

local groups U and U' are isomorphic if there exists a homeomorphism $x \to x'$ between suitable symmetric neighborhoods N and N' of 1 and $1'$ such that if x, y, xy are in N then $x'y' = (xy)'$ is in N'. It follows that $1 \to 1'$, $x^{-1} \to (x')^{-1}$. We may also suppose that if $x', y', x'y'$ are in N' then xy is in N.[6]

Suppose now that U is a local Lie group and let M be a cell neighborhood of 1 in which canonical parameters[7] are defined. Then if $x = (\xi_1, \cdots, \xi_r) \in N$ a concentric sphere of half the radius of M then x^2 has coördinates $2\xi_i$ and x has a unique square root $x^{\frac{1}{2}}$ in N. We remark that x may not have a unique n^{th} root $(n > 2)$ in N but it has a unique *proper* n^{th} root y in the sense that $y, y^2, \cdots y^{n-1}$ are all in N. For our purpose the existence and uniqueness of square roots suffices. We define $x^{\frac{1}{4}} = (x^{\frac{1}{2}})^{\frac{1}{2}}, \cdots, x^{p/2^m} = x^{(p-1)/2^m} x^{1/2^m}$ if $0 \leqq p \leqq 2^m$ and $x^{-p/2^m} = (x^{p/2^m})^{-1}$. We define x^α for $-1 \leqq \alpha \leqq 1$ by a limiting process. Finally if $|\alpha| > 1$ but $(\alpha\xi_1, \cdots, \alpha\xi_r)$ is still in N we define x^α as the element y in N such that $y^{1/\alpha} = x$. The canonical coördinates of x^α are $\alpha\xi_i$. Since the usual rules for powers hold the elements x^α, x fixed form a 1-dimensional local subgroup and these local subgroups cover the neighborhood N without overlapping.

It has been shown by G. Birkhoff[8] that the Lie algebra associated with U may be introduced in the following way: If x and y are in a suitable neighborhood N' of 1 then $\lim_{\alpha \to 0} (x^\alpha y^\alpha)^{1/\alpha} = x + y$ and $\lim_{\alpha \to 0} (x^\alpha y^\alpha x^{-\alpha} y^{-\alpha})^{1/\alpha^2} = [x, y]$ exist. The coördinates of $x + y$ are $\xi_i + \eta_i$ if $x = (\xi_1, \cdots, \xi_r)$, $y = (\eta_1, \cdots, \eta_r)$. $[x, y]$ is bilinear in x and y and satisfies

$$[x, y] = -[y, x] \quad [x, [y, z]] + [y, [z, x]] + [z, [x, y]] = 0$$

($-x$ denotes the elements with coördinates $-\xi_i$, $x + (-x) = 0 \equiv 1$). We may regard N' as the interior of a sphere in r-dimensional Cartesian space $\mathfrak{L}$ and $x + y$ as vector addition in this space. If x and y are arbitrary in $\mathfrak{L}$ we may choose $\rho \neq 0$ so that $\rho x, \rho y \in N'$ and define $[x, y] = \rho^{-2}[\rho x, \rho y]$. This is independent of ρ and together with the operation $+$ defines $\mathfrak{L}$ as a Lie algebra over the field of real numbers. We call $\mathfrak{L}$ *the Lie algebra* of U. The operations in $\mathfrak{L}$ have been defined by algebraic and topological processes in U. Conversely we may express the group operation by

$$xy = x + y + \tfrac{1}{2}[x, y] + \cdots$$

where all the terms are pure commutators. It follows that any isomorphism between local Lie groups induces an isomorphism between their Lie algebras and conversely.

It has been shown by E. Cartan that if W is a closed local subgroup of U there is a neighborhood of 1 in W whose elements all belong to a subalgebra $\mathfrak{M}$ of $\mathfrak{L}$.[9]

[6] If we choose N_1 so that $N_1^2 \subset N$ and the corresponding N_1' then this will hold for N_1 in place of N.

[7] Cf. Eisenhart [1], p. 44 or Birkhoff [1], p. 73.

[8] Birkhoff [1]. Cf. also Pontrjagin's book. For the definitions needed from the theory of Lie algebras see Jacobson [1], p. 875.

[9] Cartan [3], p. 22.

$\mathfrak{M}$ may be defined as the set of all the limiting directions of elements of W converging to 1. Thus W is a local Lie group with $\mathfrak{M}$ as its Lie algebra. $\mathfrak{M} = 0$ if and only if W is discrete. It follows also that any totally disconnected closed local subgroup of U is discrete. W is invariant if and only if $\mathfrak{M}$ is an ideal.

2. If U is a neighborhood of 1 in a topological group $\mathfrak{G}$ it is a local group relative to the multiplication and topology defined in $\mathfrak{G}$. We recall that $\mathfrak{G}_1$ and $\mathfrak{G}_2$ are locally isomorphic if they have isomorphic local groups U_1 and U_2. $\mathfrak{G}$ is a *Lie group* if U can be taken as a local Lie group. It follows that the Lie groups $\mathfrak{G}_1$ and $\mathfrak{G}_2$ are locally isomorphic if and only if their Lie algebras (the Lie algebras of the local Lie groups U_1 and U_2) $\mathfrak{L}_1$ and $\mathfrak{L}_2$ are isomorphic. If $\mathfrak{G}_1$ is connected and simply connected Schreier has shown that if U_1 is any neighborhood of 1 in $\mathfrak{G}_1$ there exists a neighborhood $V_1 \subset U_1$ such that the elements of $\mathfrak{G}_1$ are representable as $v_1 v_2 \cdots v_m$, v_i in V_1 and $v_1 v_2 \cdots v_m = 1$ if and only if this relation can be reduced to $1 = 1$ by a sequence of identifications $uv = w$ where u, v, w are in U_1.[10] It follows that if $\mathfrak{G}_2$ is connected any local isomorphism between $\mathfrak{G}_1$ and $\mathfrak{G}_2$ is obtained from a unique continuous homomorphism of $\mathfrak{G}_1$ into $\mathfrak{G}_2$. The group $\mathfrak{D}$ mapped into 1 in this mapping is discrete. If $\mathfrak{G}_2$ is simply connected also, $\mathfrak{D} = 1$ and the mapping is an isomorphism. In particular if $\mathfrak{G}_1 = \mathfrak{G}_2 = \mathfrak{G}$ any local automorphism corresponds to a (bicontinuous) automorphism of $\mathfrak{G}$. If $\mathfrak{G}$ is not simply connected there may exist local automorphism which do not generate automorphisms in the large. On the other hand any automorphism determines a local automorphism of $\mathfrak{G}$ and if $\mathfrak{G}$ is connected distinct automorphisms will define distinct local automorphisms. Since the latter induce linear transformations in $\mathfrak{L}$ it is natural to topologize the group $\mathfrak{A}$ of automorphism using the usual Euclidean topology in the space of linear transformations.

Any closed subgroup $\mathfrak{H}$ of $\mathfrak{G}$ determines a closed local subgroup W of U and hence a subalgebra $\mathfrak{M}$ of $\mathfrak{L}$. Hence $\mathfrak{H}$ is a Lie group. If $\mathfrak{H}$ is invariant so is W and hence $\mathfrak{M}$ is an ideal. If $\mathfrak{M}$ is any subalgebra of $\mathfrak{L}$ the closed local group W determined by $\mathfrak{M}$ may not be a neighborhood of 1 of any closed subgroup of $\mathfrak{G}$.[11] If $\mathfrak{H}^*$ is the smallest closed subgroup containing W, W^* its local group and $\mathfrak{M}^*$ its Lie algebra then $W^* \geqq W$ and $\mathfrak{M}^* \geqq \mathfrak{M}$. If $\mathfrak{M}$ is an ideal W is an invariant local group and $\mathfrak{H}^*$ is invariant in $\mathfrak{G}_1$ the component of 1 in $\mathfrak{G}$. Hence $\mathfrak{M}^*$ is an ideal. Similarly if $\mathfrak{M}$ is commutative so is $\mathfrak{M}^*$. Since the elements $(g^\alpha h^\alpha g^{-\alpha} h^{-\alpha})^{1/\alpha^2}$ converge in the direction through $[g, h]$ it follows from Cartan's theorem that the Lie algebra of the derived group[12] $\mathfrak{G}'$ contains the derived algebra $\mathfrak{L}'$.

We shall call a topological group $\mathfrak{G}$ topologically semi-simple or briefly *t-semi-simple* if every closed commutative invariant subgroup of $\mathfrak{G}$ is discrete. If $\mathfrak{G}_1$ is

[10] Schreier [1], p. 25.

[11] A piece of a geodesic in an irrational direction on the torus is an example of this type.

[12] The smallest closed subgroup containing all the commutators $g_1^{-1} g_2^{-1} g_1 g_2$.

locally isomorphic to $\mathfrak{G}$, $\mathfrak{H}_1$ a closed commutative invariant subgroup of $\mathfrak{G}_1$ then $\mathfrak{H}_1$ is discrete. For, otherwise $\mathfrak{H}_1$ determines a local subgroup of this type and hence by the local isomorphism a commutative invariant local subgroup not discrete in $\mathfrak{G}$. If $\mathfrak{G}$ is connected and $\mathfrak{D}$ a discrete invariant subgroup, $\mathfrak{D} \leqq \mathfrak{C}$ the center of $\mathfrak{G}$.[13] Hence if $\mathfrak{D}$ is a commutative invariant subgroup its closure $\bar{\mathfrak{D}}$ is discrete and $\bar{\mathfrak{D}} = \mathfrak{D} \leqq \mathfrak{C}$ also. The natural homomorphism between $\mathfrak{G}$ and $\mathfrak{G}/\mathfrak{C}$ is a local isomorphism. Hence $\mathfrak{G}/\mathfrak{C}$ is t-semi-simple. The group $\mathfrak{C}_1 \geqq \mathfrak{C}$ which corresponds to the center of $\mathfrak{G}/\mathfrak{C}$ is discrete and invariant.[14] Thus $\mathfrak{C}_1 = \mathfrak{C}$, $\mathfrak{G}/\mathfrak{C}$ has no commutative invariant subgroup $\neq 1$ and hence no discrete invariant subgroup $\neq 1$.

$\mathfrak{G}$ is *t-simple* if every closed invariant subgroup of $\mathfrak{G}$ is either open or discrete. It follows that $\mathfrak{G}$ is t-semi-simple unless its component of 1 is commutative. If $\mathfrak{G}$ is connected its closed invariant subgroups $\neq \mathfrak{G}$ are discrete. Then either $\mathfrak{G} = \mathfrak{C}$ or the latter is discrete. Furthermore $\mathfrak{G}/\mathfrak{C}$ is t-simple and has no discrete invariant subgroup $\neq 1$. Hence $\mathfrak{G}/\mathfrak{C}$ is simple in the usual sense. We suppose henceforth that if $\mathfrak{G}$ is t-simple it is also t-semi-simple.

Now suppose that $\mathfrak{G}$ is a t-semi-simple Lie group, $\mathfrak{L}$ its Lie algebra. $\mathfrak{L}$ is semi-simple. For if $\mathfrak{M}$ is a commutative ideal, $\mathfrak{H}^*$ the smallest closed subgroup containing $\mathfrak{M}$ is commutative and invariant. The converse that $\mathfrak{G}$ is t-semi-simple if $\mathfrak{L}$ is semi-simple is trivial. If $\mathfrak{G}$ is t-simple (and t-semi-simple) $\mathfrak{L}$ is semi-simple and if not simple, $\mathfrak{L} = \mathfrak{L}_1 \oplus \mathfrak{L}_2$[15] where $\mathfrak{L}_1$ and $\mathfrak{L}_2$ are proper ideals. $\mathfrak{L}_i$ generates $\mathfrak{G}_1$ the component of 1 but since the elements of $\mathfrak{L}_1$ and $\mathfrak{L}_2$ commute, $\mathfrak{G}_1$ is commutative contrary to assumption. Conversely if $\mathfrak{L}$ is simple, $\mathfrak{G}$ is t-simple.

Suppose again that $\mathfrak{G}$ is t-semi-simple and $\mathfrak{L}_1$ is an ideal in $\mathfrak{L}$, $\mathfrak{G}_1$ the closed invariant subgroup determined by $\mathfrak{L}_1$ and $\mathfrak{L}_1^*$ the Lie algebra of $\mathfrak{G}_1$. Since $\mathfrak{L}_1^*$ is an ideal in $\mathfrak{L}$ it is semi-simple and hence $\mathfrak{L}_1^* = \mathfrak{L}_1 \oplus \mathfrak{M}$. The elements of $\mathfrak{M}$ commute with those of $\mathfrak{L}_1$ and hence they generate a subgroup of the center of $\mathfrak{G}_1$. It follows that $\mathfrak{M} = 0$, $\mathfrak{L}_1^* = \mathfrak{L}_1$. Hence in this case we have a (1–1) correspondence between closed connected invariant subgroups of $\mathfrak{G}$ and ideals of $\mathfrak{L}$. Suppose $\mathfrak{G}$ is connected and $\mathfrak{L} = \mathfrak{L}_1 \oplus \cdots \oplus \mathfrak{L}_s$, $\mathfrak{L}_i$ simple and $\mathfrak{G}_i$ the group generated by $\mathfrak{L}_i$. Then $\mathfrak{G} = \mathfrak{G}_1\mathfrak{G}_2 \cdots \mathfrak{G}_s$. The component of 1 of $\mathfrak{D}_i = \mathfrak{G}_i \cap (\mathfrak{G}_1 \cdots \mathfrak{G}_{i-1}\mathfrak{G}_{i+1} \cdots \mathfrak{G}_s)$ corresponds to $\mathfrak{L}_i \cap (\mathfrak{L}_1 + \cdots + \mathfrak{L}_{i+1} + \mathfrak{L}_{i+1} + \cdots + \mathfrak{L}_s) = 0$ and hence consists of the point 1. Thus $\mathfrak{D}_i$ is discrete and is contained in $\mathfrak{C}$ the center. If $\mathfrak{C} = 1$, $\mathfrak{G}$ is a direct product of simple groups. In the general case $\mathfrak{G}/\mathfrak{C}$ is a direct product of simple groups. It has been shown by Cartan[16] that the inner automorphisms $(\mathfrak{G}/\mathfrak{C})$ form the component of 1 in

[13] If $d \in \mathfrak{D}$ then the set of elements $g^{-1}dg$, $g \in \mathfrak{G}$ is connected since $g \to g^{-1}dg$ is a continuous mapping. Hence $g^{-1}dg$ consists of a single point. i.e. $g^{-1}dg = 1$.

[14] If $\mathfrak{H}/\mathfrak{C}$ and $\mathfrak{C}$ are discrete then $\mathfrak{H}$ is discrete. For $\mathfrak{C}$ the inverse image of the open set 1 is open in $\mathfrak{H}$. Since $\mathfrak{C}$ is discrete any point contained in $\mathfrak{C}$ is also open in $\mathfrak{H}$ and hence $\mathfrak{H}$ is discrete.

[15] Cartan [1], p. 52.

[16] Cartan [4], p. 8.

the group of automorphisms $\mathfrak{A}$. Thus this component has no commutative invariant subgroup and if $\mathfrak{G}$ is t-simple the component is simple.

3. Let $\mathfrak{R}$ be a complete valued associative ring with an identity[17] i.e. there is defined a real valued function $|x|$ for x in $\mathfrak{R}$ such that

1. $|x| \geqq 0, = 0$ if and only if $x = 0$,
2. $|x| = |-x|$,
3. $|x + y| \leqq |x| + |y|$,
4. $|xy| \leqq |x||y|$,
5. $\mathfrak{A}$ is complete in the topology defined by the metric $d(x, y) = |x - y|$.

The set of units u in $\mathfrak{R}$ form a group $\mathfrak{U}$. Since the product is continuous in $\mathfrak{R}$ by 3 and 4 it is continuous in the subspace $\mathfrak{U}$. If $|x| < 1$ it follows as usual that $y = 1 + x + x^2 + \cdots$ exists and $(1 - x)y = y(1 - x) = 1$. Also for a given $\epsilon > 0$ there exists $\delta > 0$ such that $|x| < \delta$ implies $|y - 1| < \epsilon$. Hence $\mathfrak{U}$ is open in $\mathfrak{R}$ and x^{-1} is continuous at $x = 1$. Now suppose N is any open set containing a^{-1} then aN is open and contains 1 since $x \rightarrow xa$ is a homeomorphism. There exists a neighborhood M of 1 such that $M^{-1} \leqq aN$. Hence $N' = Ma$ is an open set about a such that $(N')^{-1} \leqq N$. Thus x^{-1} is continuous at every point and $\mathfrak{U}$ is a topological group.

We suppose now that $\mathfrak{R}$ is a simple algebra over R the field of real numbers, i.e. $\mathfrak{R} = R_n$, C_n or Q_n the set of $n \times n$ matrices with elements in the real, complex or quaternionic fields. If $x = (\xi_{ij}) = \sum \xi_{ij}e_{ij}$, e_{ij} a matric basis set $|x| = (\sum \xi_{ij}\bar{\xi}_{ij})^{\frac{1}{2}}$. Then $\mathfrak{R}$ is a complete valued ring.[18] According as $\mathfrak{R} = R_n$, C_n or Q_n we write $\mathfrak{U} = L(R, n)$, $L(C, n)$ or $L(Q, n)$. Using the methods introduced by v. Neumann[19] we can define $\exp x = 1 + x + \dfrac{x^2}{2!} + \cdots$ for any x and $\log x = (x - 1) - \dfrac{(x-1)^2}{2} +$ for $|x - 1| < 1$. These functions are continuous and

$$\exp(\log x) = x \quad \text{if} \quad |x - 1| < 1$$

$$\log(\exp x) = x \quad \text{if} \quad |x| < \log 2.$$

Thus the mapping $x \rightarrow \exp x$ is a homeomorphism between the neighborhood $|x| < \log 2$ of 0 and a neighborhood M of 1. Since $\exp x \exp(-x) = 1$ the latter is in $\mathfrak{U}$. It is readily seen that $\mathfrak{U}$ is a Lie group. Since

$$(\exp x)\,(\exp y) = \exp\,(x + y)$$

if $xy = yx$ we have $(\exp x)^2 = \exp 2x$ and hence the real coördinates of ξ_{ij} in $\exp x$, $x = (\xi_{ij})$ are canonical. Hence if N is the image of $|x| < \delta_1$, $\delta_1 \leqq \frac{1}{2}\log 2$ under the mapping $x \rightarrow a = \exp x$ then a has the unique square root $a^{\frac{1}{2}} = \exp \frac{1}{2}x$.

[17] Deuring [1], p. 93.
[18] von Neumann [1], p. 6. The proof given there holds also for R_n and Q_n.
[19] von Neumann [1].

As in §2 we have a^α defined topologically-algebraically and $a^\alpha = \exp(\alpha x)$. Since

$$|\exp x - \exp y| = O|x - y| \qquad |\log x - \log y| = O|x - y|$$

in our neighborhoods we can prove for $b = \exp y$

$$(a^\alpha b^\alpha)^{1/\alpha} = \exp\left[\frac{1}{\alpha}\log(\exp \alpha x \exp \alpha y)\right] \rightarrow \exp(x + y)$$

and

$$(a^\alpha b^\alpha a^{-\alpha} b^{-\alpha})^{1/\alpha^2} \rightarrow \exp[x, y] \qquad [x, y] = xy - yx$$

if $|x|, |y| < \delta_2$.[20] Thus the correspondence $x \rightarrow \exp x = a$ sets up an isomorphism between the Lie algebra of $\mathfrak{U}$ and the Lie algebra $\mathfrak{R}$ in which $x + y$ is as in the associative algebra and $[x, y] = xy - yx$.

If $\mathfrak{G}$ is a closed subgroup of $\mathfrak{U}$ the result of Cartan's quoted above shows that $\mathfrak{G}$ has a neighborhood of 1 consisting of the matrices $\exp x$ where $|x| < \rho$ and $x \in \mathfrak{L}$ a Lie subalgebra of $\mathfrak{R}$.

4. If $\mathfrak{U} = L(R, n)$ or $L(C, n)$ it is well known[21] that the derived groups $L'(R, n) = L_1(R, n)$ the set of matrices of determinant 1 and $L'(C, n) = L_1(C, n)$. Since $\det(\exp x) = \exp(\operatorname{tr} x)$, $\operatorname{tr} x = \sum \xi_{ii}$ if $x = (\xi_{ij})$ the Lie algebras of $L'(R, n)$ and $L'(C, n)$ are respectively the derived algebras R_n' and C_n'. Q_n may be represented by $2n \times 2n$ complex matrices. We define tr a, det a as the trace and determinant in this representation. As is well-known tr a, det $a \in R$. It is readily seen that Q_n' consists of the elements of trace 0. We show below (§8) that $\det a > 0$ for a in $L(Q, n)$ and $L'(Q, n)$ consists of the elements of determinant 1. Hence we have, as before, that the Lie algebra of $L'(Q, n)$ is Q_n'. R_n', $C_n'(n > 1)$, Q_n' are simple and hence $L'(R, n)$, $L'(C, n)$, $L'(Q, n)$ are t-simple Lie groups.[22] We call these the real, complex and quaternionic unimodular groups.

If S is an automorphism in $\mathfrak{R}$[23] it induces an automorphism in $\mathfrak{U}'$ and in the Lie algebra $\mathfrak{R}'$. If $a = \exp x$ by the continuity of S we have $a^S = \exp x^S$ and hence S in $\mathfrak{R}'$ induces S in $\mathfrak{U}'$. If S is an anti-automorphism in $\mathfrak{R}$, $a \rightarrow (a^S)^{-1}$ and $x \rightarrow -x^S$ are automorphisms in $\mathfrak{U}'$ and $\mathfrak{R}'$ respectively. Since $(a^S)^{-1} = \exp(-x^S)$, $x \rightarrow -x^S$ generates the automorphism in $\mathfrak{U}'$. It has been shown that every automorphism in $\mathfrak{R}'$ is of one of these two types. Applying this we obtain the following groups $\mathfrak{A}$ of automorphisms for the unimodular groups.

1. $L'(R, n)$. The automorphism of R_n are $x \rightarrow s^{-1}xs$, the anti-automorphisms

[20] Cf. Birkhoff [1], p. 78.

[21] A proof of this theorem is given in §8.

[22] The results on Lie algebras required here and in the rest of this section may be found in Jacobson [2], pp. 545-548.

[23] We mean here an automorphism in the algebra $\mathfrak{R}$, i.e. $(x\alpha)^S = x^S\alpha$ for real α. This is equivalent to the condition that S be a continuous automorphism in the ring $\mathfrak{R}$.

are $x \to s^{-1}x's$ where x' is the transposed of x. These give the automorphisms $a \to s^{-1}as$ and $a \to s^{-1}(a')^{-1}s$ in $L'(R, n)$. The former set form a subgroup $\cong L(R, n)/R^*$, R^* the set of matrices $\alpha 1 \neq 0$ and this subgroup has index 1 or 2 according as $n = 2$ or $n > 2$. If n is even $R^* \leqq L^+(R, n)$ the set of matrices of positive determinant and $L^+(R, n)/R^*$ has index 2 in $L(R, n)/R^*$. Since any matrix in $L^+(R, n)$ has the form αa_1, $\alpha \in R^*$ and $a_1 \in L'(R, n)$ we may choose two elements a_1, $-a_1$ in each class mod R^*. Hence $L^+(R, n)/R^* \cong L'(R, n)/D \cong \mathfrak{J}$, $D = (1, -1)$, $\mathfrak{J}$ the group of inner automorphisms. If n is odd every class mod R^* contains matrices in $L^+(R, n)$ and hence $L(R, n)/R^* = L^+(R, n)/R^*_+$, $R^*_+ = R^* \cap L^+(R, n)$. In this case $L^+(R, n)/R^*_+ \cong L_1(R, n) \cong \mathfrak{J}$.

2. $L'(C, n)$. Similar considerations show that the automorphisms here are $a \to s^{-1}as$, $a \to s^{-1}(a')^{-1}s$, $a \to s^{-1}\bar{a}s$, $a \to s^{-1}(\bar{a}')^{-1}s$, $\bar{a} = (\bar{\alpha}_{ij})$ if $a = (\alpha_{ij})$. The first set has index 2 or 4 in $\mathfrak{A}$ according as $n = 2$ or $n > 2$ and is isomorphic to $L(C, n)/C^*$, C^* the set $\alpha 1$, $\alpha \neq 0$ in C. $L(C, n)/C^* \cong L'(C, n)/D \cong \mathfrak{J}$, D the set $\zeta 1$, where $\zeta^n = 1$.

3. $L'(Q, n)$. The automorphism are $a \to s^{-1}as$, $a \to s^{-1}(\bar{a}')^{-1}s$. If $n = 1$ the last set is superfluous and if $n > 1$ the first set has index 2 and is isomorphic to $L(Q, n)/R^* \cong L'(Q, n)/D \cong \mathfrak{J}$, $D = (1, -1)$.

5. Suppose S is an involutorial anti-automorphism in the associative algebra $\mathfrak{R}$, and $\mathfrak{U}'_S$ be the set of elements invariant under this automorphism $a \to (a^S)^{-1}$ of $\mathfrak{U}'$. Because of the continuity $\mathfrak{U}'_S$ is a closed subgroup of $\mathfrak{U}'$. If $a = \exp x$, $(a^S)^{-1} = \exp(-x^S)$ and $|x|$ is sufficiently small, $a = (a^S)^{-1}$ implies $x^S = -x$. Conversely if $x^S = -x$, x in R' then $\exp x$ is in $\mathfrak{U}'_S$. Thus the Lie algebra of $\mathfrak{U}'_S$ is $\mathfrak{S}'_S$ the derived algebra of $\mathfrak{S}_S$ the set of S-skew elements. Except for certain small values of n (cf. below) the algebras $\mathfrak{S}'_S$ are all simple and hence the $\mathfrak{U}'_S$ are t-simple. If S_1 and S_2 are cogredient anti-automorphisms in the same $\mathfrak{R}$ i.e. $S_2 = G^{-1}s_1G$, G an automorphism, it is evident that $\mathfrak{U}'_{S_1}$ and $\mathfrak{U}'_{S_2}$ are equivalent. Otherwise for n sufficiently large $\mathfrak{U}'_{S_1}$ and $\mathfrak{U}'_{S_2}$ are not locally isomorphic. If we identify $\mathfrak{U}'_{S_1}$ and $\mathfrak{U}'_{S_2}$ we find that the automorphisms of this group have the form $a \to a^G$ where G is an automorphism of $\mathfrak{R}$ commutative with S. Applying these principles we obtain the following t-simple groups and their automorphisms.

1. The real orthogonal groups $O(R, n, \nu)$ consisting of the real matrices a such that $a's_\nu a = s_\nu$, $\det a = 1$ where $s_\nu = e_{11} + \cdots + e_{pp} - e_{p+1,p+1} - \cdots - e_{nn}$, $\nu = 2p - n = n, n-2, \cdots, 0$ or 1 as n is even or odd and $n > 8$ if even, > 5 if odd. The group of automorphisms $\mathfrak{A}$ consists of the mappings $a \to b^{-1}ab$ where b satisfies $b's_\nu b = \rho s_\nu$, $\rho \neq 0$. Hence $\mathfrak{A}$ is the factor group of these matrices with respect to R^*. If n is odd $\rho > 0$ since the signatures of $b's_\nu b$ and s_ν are the same. We may choose in each class mod R^* a unique b_1 such that $b_1's_\nu b_1 = s_\nu$ and $\det b_1 = 1$. Thus $\mathfrak{A} = \mathfrak{J} \cong O(R, n, \nu)$. If n is even and $\nu \neq 0$, $\rho > 0$ and we may again choose a b_1 such that $b_1's_\nu b_1 = s_\nu$. However the elements of determinant 1 and -1 do not belong to the same class. The former form a subgroup of index $2 \cong O(R, n, \nu)/D \cong \mathfrak{J}$, $D = (1, -1)$. If n is even and $\nu = 0$ the representatives may be selected so that either $b_1's_0b_1 = s_0$ or $b_1's_0b_1 =$

$-s_0$. The former elements having det $= 1$ form an invariant subgroup of index $4 \cong O(R, n, \nu)/D, D = (1, -1)$.

2. The real symplectic group $S(R, 2n)$: a in R_n such that $a'qa = q, q' = -q$ and $n > 2$.[24] $\mathfrak{A}$ consists of $a \rightarrow b^{-1}ab$, $b'qb = \rho q$, $\rho \in R^*$. These elements for which $\rho > 0$ form a subgroup of index $2 = \mathfrak{J} \cong S(R, n)/D, D = (1, -1)$.

3. The complex orthogonal group $O(C, n)$: $a'a = 1$, a in $L'(C, n)$ and $n > 8$ if even, > 5 if odd. $\mathfrak{A}$ consists of $a \rightarrow b^{-1}ab$ and $a \rightarrow b^{-1}\bar{a}b$ where $b'b = \rho^1$, $\rho \in C^*$. The former form a subgroup of index $2 = \mathfrak{J} \cong O(C, n)/D, D = 1$, or $=$ $(1, -1)$ according as n is odd or even.

4. The complex symplectic group $S(C, 2n)$: $a'qa = q$ $q' = -q$, a in C_n and $n > 2$.[24] $\mathfrak{A}$ consists of $a \rightarrow b^{-1}ab$ and $a \rightarrow b^{-1}\bar{a}b$, $b'qb = \rho q$, $\rho \in C^*$. $\mathfrak{A} = \mathfrak{J} \cong S(C, 2n)/D, D = (1, -1)$.

5. The complex unitary, unimodular groups $U'(C, n, \nu)$: $\bar{a}'s_\nu a = s_\nu$, s_ν and ν as in 1, $a \in L'(C, n)$, $n > 2$. $\mathfrak{A}$ consists of $a \rightarrow b^{-1}ab$ and $a \rightarrow b^{-1}\bar{a}b$, $\bar{b}'s_\nu b = \rho s_\nu$, $\rho \in R^*$. The former forms a subgroup of index 2 and unless n is even and $\nu = 0$, this subgroup $= \mathfrak{J} \cong U'(C, n, \nu)/D$, D the set $\zeta 1$, $\zeta^n = 1$. If n is even and $\nu = 0$, $\mathfrak{J}$ has index 4 in $\mathfrak{A}$.

6. The quaternionic unitary groups $U(Q, n, \nu)$; a in Q_n such that $\bar{a}'s_\nu a = s_\nu$, s_ν and ν as in 1 $n > 1$.[24] $\mathfrak{A}$ consists of $a \rightarrow b^{-1}ab$, $\bar{b}'s_\nu b = \rho s_\nu$, $\rho \in R^*$. As before $\mathfrak{A} = \mathfrak{J} \cong U(Q, n, \nu)/D, D = (1, -1)$ unless n is even, $\nu = 0$ when $\mathfrak{J}$ has index 2 in $\mathfrak{A}$.

7. The quaternionic symplectic group $S(Q, n)$: a in Q_n such that $\bar{a}'qa = q$ where $\bar{q}' = -q$, $n > 4$.[24] $\mathfrak{A}$ consists of $a \rightarrow b^{-1}ab$, $\bar{b}'qb = \rho q$, $\rho \in R^*$. The elements with $\rho > 0$ form a subgroup of index $2 = \mathfrak{J} \cong S(Q, n)/D, D = (1, -1)$.

The groups enumerated here are not locally isomorphic to any of the unimodular groups since their Lie algebras are not isomorphic. Altogether these ten classes include all t-simple Lie groups (relative to local isomorphism) except those of dimension 14, 28, 52, 78, 133, 248, 56, 104, 156, 266 and 498. The groups $O(R, n, n)$, $U'(C, n, n)$ and $U(Q, n, n)$ in our list are compact. This may be seen by noting that the conditions imposed entail the boundedness of the coördinates of these matrices.

8. Suppose F is any quasi-field, F_n the ring of $n \times n$ matrices over F and $L(F, n)$ the group of units in F_n. The method given by Dickson[25] for the case F a finite field may be applied to prove that any element of $L(F, n)$ has the form bd_n where b is a product of matrices of the type $1 + \theta e_{rs}$, $r \neq s$ and $d_n = 1 + (\delta - 1)e_{nn}$, $\delta \neq 0$.

We have

$$(1 + \theta e_{rs})(1 + (\delta - 1)e_{rr})(1 + \theta e_{rs})^{-1}(1 + (\delta - 1)e_{rr})^{-1} = 1 + (1 - \delta)\theta e_{rs}.$$

It follows that if F has more than two elements, b is a product of commutators.

---

[24] These elements all have determinant 1. Cf. v. d. Waerden 1, p. 10 for the cases $S(R, 2n)$ and $S(C, 2n)$. For the other cases see §8.

[25] Dickson [1], p. 78.

d_n is a commutator if δ is a commutator in F. If F is commutative and det $a = 1$, det $d_n = \delta = 1$ and hence a is a product of commutators.

This result applied to $F = R$, $= C$ shows that $L'(R, n) = L_1(R, n)$, $L'(C, n) = L_1(C, n)$ and $L^+(R, n)$ consists of the elements bd_n with $\delta > 0$. If $F = Q$ we employ the representation by complex matrices and obtain that $L'(Q, n) \leqq L_1(Q, n)$. Since det $b = 1$, det $d_n = N(\delta) = \delta\bar{\delta} > 0$. $N(\delta) = 1$ for a in $L_1(Q, n)$. If $\delta = 1$ it is evidently a commutator and if $\delta = -1$, $iji^{-1}j^{-1} = \delta$ (1, i, j, k the quaternion units). If $\delta \neq 1, -1$, $R(\delta)$ is isomorphic to C and hence we have ξ in $R(\delta)$ so that $\xi^2 = \delta$, $\xi\bar{\xi} = 1$. These exists a μ such that $\mu\alpha = \bar{\alpha}\mu$ if $\alpha \in R(\delta)$. Hence $\delta = \xi\mu\xi^{-1}\mu^{-1}$ and in all cases δ is a commutator. It follows that $L'(Q, n) = L_1(Q, n)$.

Any element θ in R, C or Q can be joined to 0 by an arc in these fields. It follows that the matrices b can be joined by arcs in $L'(R, n)$, $L'(C, n)$, $L'(Q, n)$ to 1. If δ is any element $\neq 0$ in C or Q or $\delta > 0$ in R then it can be joined to 1 by an arc which avoids the point 0. Finally if $N(\delta) = 1$ in Q, may be joined by an arc consisting of points of norm 1 to 1. It follows that $L^+(R, n)$, $L(C, n)$, $L(Q, n)$, $L'(R, n)$, $L'(C, n)$ and $L'(Q, n)$ are connected groups.

BIBLIOGRAPHY

G. Birkhoff

[1] "Analytical groups," *Trans. Am. Math. Soc. 43* (1938), pp. 61–101.

E. Cartan

[1] *Thèse*, Paris, Nony 1894.

[2] "Les groupes réels simples et continus," *Annales de l'École Normale, 31* (1914), pp. 263–355.

[3] "La théorie des groupes finis et continus et l'analysis situs," *Mem. des Sci. Math.* 1930.

[4] "Groupes simples clos et ouverts et géometrie riemanniene," *Jour. Math. 2* (1929), pp. 1–33.

M. Deuring

[1] *Algebren, Ergebnisse der Math.*, 1935.

L. E. Dickson

[1] *Linear groups.*

L. P. Eisenhart

[1] *Continuous groups of transformations*, Princeton, 1933.

N. Jacobson

[1] "Rational methods in the theory of Lie algebras," *Annals of Math. 36* (1935), pp. 875–881.

[2] "Simple Lie algebras over a field of characteristic zero," *Duke Math. Jour. 4* (1938), pp. 534–551.

J. von Neumann

[1] "Gruppen von linearen Transformationen," *Math. Zeits. 30* (1929), pp. 3–42.

O. Schreier

[1] "Abstrakte kontinuierliche Gruppen," *Hamb. Abhand.* 4 (1926), pp. 15–32.

B. L. van der Waerden

[1] *Gruppen von linearen Transformationen, Ergebnisse der Math.*, 1935.

University of North Carolina.

Reprinted from DUKE MATHEMATICAL JOURNAL
Vol. 5, No. 4, December, 1939

CAYLEY NUMBERS AND NORMAL SIMPLE LIE ALGEBRAS OF TYPE G

BY N. JACOBSON

In an earlier paper[1] we discussed the set $\mathfrak{D}(\mathfrak{A})$ of derivations in an arbitrary algebra $\mathfrak{A}$ (not necessarily associative), i.e., the operators D in $\mathfrak{A}$ such that

$$(x + y)D = xD + yD, \qquad (x\alpha)D = (xD)\alpha, \qquad (xy)D = x(yD) + (xD)y,$$

α being in the underlying field Φ. We noted that $\mathfrak{D}$ is closed with respect to addition, scalar multiplication and commutation $[D, E] \equiv DE - ED$. Hence $\mathfrak{D}$ is a Lie algebra over Φ. We shall show here that if $\mathfrak{A}$ is a generalized Cayley algebra and Φ is of characteristic 0, then $\mathfrak{D}$ is normal simple of type G and all such Lie algebras may be obtained in this way. The derivation algebras are isomorphic if and only if the Cayley systems are. If Φ is algebraically closed, these results have been indicated by Cartan.[2] The extension to the general case given here depends essentially on the determination of the automorphisms of $\mathfrak{D}$ in the algebraically closed case. The structure of Cayley systems has been obtained by Zorn.[3] We give several extensions of his theory.

We require below the theorem that if $\mathfrak{A}_P$ is the algebra obtained by extending Φ to P, then $\mathfrak{D}(\mathfrak{A}_P) = \mathfrak{D}_P$.[4] We note also that if S is either an automorphism or anti-automorphism in $\mathfrak{A}$ such that $(x\alpha)S = (xS)\alpha^s$, where $\alpha \rightarrow \alpha^s$ is an automorphism in Φ, then $S^{-1}DS$ is a derivation for every D in $\mathfrak{D}$. If $\alpha^s \equiv \alpha$, $D \rightarrow S^{-1}DS$ is an automorphism of $\mathfrak{D}$ over Φ.

1. Let Q be a (generalized) quaternion algebra over a field of characteristic $\neq 2$. We do not exclude the possibility that $Q = \Phi_2$, the 2-rowed matrix algebra. A (generalized) Cayley algebra $\mathfrak{A}$ is a vector space of order 2 over Q, $\mathfrak{A} = Q1 + Qe_4$, in which

$$(a + be_4)(c + de_4) = (ac + \bar{d}b\alpha_4) + (da + b\bar{c})e_4, \tag{1}$$

where $\alpha_4 \neq 0$. If Q has basis $(1, e_1, e_2, e_3)$ such that $e_1^2 = \alpha_1$, $e_2^2 = \alpha_2$, $\alpha_i \neq 0$, $e_1e_2 = -e_2e_1 = e_3$ and $e_4^2 = \alpha_4$, then $1, e_1, \cdots, e_7$ is a basis for $\mathfrak{A}$ if $e_5 = e_1e_4$, $e_6 = e_4e_2$, $e_7 = e_3e_4$. $(1, e_1, e_4, e_5)$, $(1, e_4, e_2, e_6)$, $(1, e_3, e_4, e_7)$, $(1, e_1, e_6, e_7)$, $(1, e_2, e_5, e_7)$ and $(1, e_3, e_5, -e_6\alpha_1)$ are quaternion algebras. If e_i, e_j, e_k do not belong to one of these algebras, then $(e_ie_j)e_k = -e_i(e_je_k)$. It is sometimes

Received December 28, 1938; presented to the American Mathematical Society, December 28, 1938.

[1] *Abstract derivation and Lie algebras*, Trans. Amer. Math. Soc., vol. 42(1937), pp. 206–224.
[2] *Les groupes réels simples et continus*, Ann. de l'École Normale, vol. 31(1914), p. 298.
[3] *Alternativkörper und quadratische Systeme*, Hamb. Abhandl., vol. 9(1933), pp. 395–402.
[4] Loc. cit. (footnote 1), p. 213.

convenient to use

$$(2)\qquad (ae_4)b = (a\bar{b})e_4, \qquad a(be_4) = (ba)e_4, \qquad (ae_4)(be_4) = \bar{b}a\alpha_4$$

in place of (1).

Abstract characterizations of $\mathfrak{A}$ have been given by Zorn.[5] We recall the following properties. Any two elements of $\mathfrak{A}$ generate an associative algebra. If $x = a + be_4$, define $\bar{x} = \bar{a} - be_4$. Then the correspondence $x \to \bar{x}$ is an involutorial anti-automorphism in $\mathfrak{A}$. If $N(x) = x\bar{x} = \bar{x}x$, $\text{tr}\,(x) = x + \bar{x}$, these belong to Φ and $x^2 - \text{tr}\,(x)x + N(x) = 0$. We have $N(xy) = N(x)N(y)$, $\text{tr}\,(x)$ is linear and $\text{tr}\,xy = \text{tr}\,yx$. Hence the function $(x, y) = \frac{1}{2}\,\text{tr}\,x\bar{y}$ is a symmetric non-degenerate bilinear form whose corresponding quadratic form is $N(x)$.

We call the elements of trace 0 ($\bar{x} = -x$) vectors and let $\mathfrak{A}'$ denote their totality. Relative to (x, y) and $x \times y = \frac{1}{2}(xy - yx)$ we obtain a vector calculus whose fundamental properties are

$$(3)\qquad (x \times y, x \times y) = (x, x)(y, y) - (x, y)^2,$$

$$(4)\qquad (x \times y) \times z + x \times (y \times z) = 2y(x, z) - x(y, z) - z(x, y).$$

Equation (4) may be proved by noting the linearity and verifying for the basis e_i. $\mathfrak{A}$ is determined by the structure of $\mathfrak{A}'$ since $xy = -(x, y) + x \times y$ and hence

$$(\alpha + x)(\beta + y) = \alpha\beta + y\alpha + x\beta - (x, y) + x \times y.$$

Suppose f_1, f_2 are vectors such that $(f_1, f_1) \neq 0$, $(f_2, f_2) \neq 0$, $(f_1, f_2) = 0$. Let $f_3 = f_1 \times f_2$. Then by (3) and (4)

$$f_2 \times f_3 = f_1(f_2, f_2), \qquad f_3 \times f_1 = f_2(f_1, f_1),$$

$$(f_2, f_3) = (f_3, f_1) = 0, \qquad (f_3, f_3) = (f_1, f_1)(f_2, f_2).$$

We may choose an f_4 orthogonal ($(f_i, f_4) = 0$) to f_1, f_2, f_3. The vectors $f_5 = f_1 \times f_4$, $f_6 = f_4 \times f_2$, $f_7 = f_3 \times f_4$ together with f_1, f_2, f_3 are mutually orthogonal and their vector and scalar products are determined by (f_1, f_1), (f_2, f_2) and (f_4, f_4). $(1, f_1, \cdots, f_7)$ is a basis of the same type as $(1, e_1, \cdots, e_7)$ and $\mathfrak{A} = R + Rf_4$, $R = (1, f_1, f_2, f_3)$. R may be taken as any quaternion subalgebra of $\mathfrak{A}$. Hence in the original notation we may suppose that Q is any quaternion algebra in $\mathfrak{A}$.

In the generation (Q, e_4) we may replace e_4 by de_4. Then α_4 is replaced by $\alpha_4 N(d)$. Conversely, if $c + de_4$ satisfies the first two conditions in (2), we obtain by the linearity that c also satisfies them and hence $c = 0$. If $Q \cong \Phi_2$, any element is a norm in Φ_2 and hence we may suppose $\alpha_4 = 1$. If $(x, x) \neq 0$, x has an inverse. On the other hand, if there exists an $x \neq 0$ such that $(x, x) = 0$, either Q is matric or α_4 is a norm in Q, and we may then suppose $\alpha_4 = 1$. We readily obtain a matric subalgebra of $\mathfrak{A}$ and a normalized generation of the type described.

[5] Loc. cit. (footnote 3). Cf. also Zorn, *Theorie der alternativen Ringe*, Hamb. Abhandl., vol. 8(1931), pp. 123–147.

THEOREM. *Any two Cayley algebras which are not division algebras are isomorphic.*

If $\mathfrak{A}_1 = (e_0, \cdots, e_7)$ and $\mathfrak{A}_2 = (f_0, \cdots, f_7)$ are isomorphic Cayley algebras, the matrices (e_i, e_j), (f_i, f_j) are cogredient. Suppose conversely that these norm forms are equivalent; i.e., their matrices are cogredient. Choose e_1, e_2 in $\mathfrak{A}_1'$ such that $(e_1, e_2) = 0$, $(e_1, e_1) \neq 0$, $(e_2, e_2) \neq 0$ and f_1, f_2 in $\mathfrak{A}_2'$ such that $(f_i, f_j) = (e_i, e_j)$. By (3) and (4), $(1, e_1, e_2, e_3 = e_1 \times e_2)$ and $(1, f_1, f_2, f_3 = f_1 \times f_2)$ are isomorphic under the correspondence $e_i \to f_i$. If the other e's and f's are orthogonal to these, the matrices determined by them are cogredient.[6] Hence we may suppose that $(e_4, e_4) = (f_4, f_4)$ and then the correspondence $e_4 \to f_4$ will determine an extension of the isomorphism between $\mathfrak{A}_1$ and $\mathfrak{A}_2$.

THEOREM. *Any two Cayley algebras which have equivalent norm forms are isomorphic.*

THEOREM. *If Q and R are quaternion subalgebras of a Cayley algebra $\mathfrak{A}$, $e \to f = eS$ an isomorphism between them, then S may be extended to an automorphism in $\mathfrak{A}$.*

If Φ is extended to P, $\mathfrak{A}_P$ the extended algebra is evidently a Cayley algebra. Conversely, if $\mathfrak{A}_P$ is a Cayley algebra, $\mathfrak{A}$ has a non-degenerate form (x, y) defined in it. It follows that $\mathfrak{A}$ contains a quaternion subalgebra Q and an element e_4 orthogonal to Q, $(e_4, e_4) \neq 0$ and hence $\mathfrak{A}$ is a Cayley algebra.

2. Suppose D is a derivation in $\mathfrak{A}$. $1D = 0$ and hence (1) the set of multiples 1α is an invariant subspace relative to the derivation algebra $\mathfrak{D}$. Let $x \in \mathfrak{A}'$ and set $xD = x'$. Since $x^2 = -N(x)$, $xx' + x'x = 0$, $x(x')^2 = (x')^2x$. From $(x')^2 = x' \operatorname{tr}(x') - N(x')$, we obtain $xx' \operatorname{tr}(x') = 0$. If $xx' = 0$ and $x' \neq 0$, $x^2 = 0$. We may generate $\mathfrak{A}'$ by elements x such that $x^2 \neq 0$ and for these we obtain $\operatorname{tr} x' = 0$. It follows that this holds for every x in $\mathfrak{A}'$. Thus $\mathfrak{A}'$ is also invariant relative to $\mathfrak{D}$ and $\mathfrak{A}$ is mapped into $\mathfrak{A}'$ by $\mathfrak{D}$. If $x \in \mathfrak{A}'$, $\bar{x}D = -xD = \overline{xD}$. Since $\bar{1}D = \overline{1D}$, D commutes with the anti-automorphism $x \to \bar{x}$ in $\mathfrak{A}$.

If A is a linear transformation in the vector space $\mathfrak{A}'$, we define its adjoint as the linear transformation A^* such that $(x, yA) = (xA^*, y)$ for all x, y.[7] The correspondence $A \to A^*$ is an involutorial anti-automorphism in the algebra of linear transformations. Since $(x\bar{y})D \in \mathfrak{A}'$, $0 = \operatorname{tr}(x\bar{y})D = \operatorname{tr}(xD)\bar{y} + \operatorname{tr} x(\overline{yD}) = (xD, y) + (x, yD)$. Thus $D^* = -D$, i.e., every derivation is a skew-symmetric linear transformation.

If $\mathfrak{A} = (Q, e_4)$ and D is a derivation in Q,[8] it follows from (2) that we obtain all possible extensions of D to derivations in $\mathfrak{A}$ by setting $e_4D = ce_4$, c any

[6] This is proved by Witt, *Theorie der quadratischen Formen in beliebigen Körpern*, Jour. für Math., vol. 176(1936), p. 34.

[7] Cf. Jacobson, *Normal semi-linear transformations*, Amer. Jour. of Math., vol. 61(1939), p. 48.

[8] The derivations in Q are all inner, i.e., $xD = xd - dx$ for a fixed element d in Q. See Jacobson, loc. cit. (footnote 1), p. 215.

element in $Q' = \mathfrak{A}' \cap Q$. Any D is determined by its effect on e_1, e_2 and e_4. We shall show that D has the form

$$(5) \qquad e_1 D = \sum_1^7 e_i \lambda_i, \qquad e_2 D = \sum_1^7 e_i \mu_i, \qquad e_4 D = \sum_1^7 e_i \nu_i,$$

where

$$(6) \quad \lambda_1 = \mu_2 = \nu_4 = \mu_1\alpha_1 + \lambda_2\alpha_2 = \nu_1\alpha_1 + \lambda_4\alpha_4 = \nu_2\alpha_2 + \mu_4\alpha_4$$
$$= \lambda_6\alpha_4\alpha_2 + \mu_5\alpha_1\alpha_4 - \nu_3\alpha_1\alpha_2 = 0.$$

Equations (6) are obtained from $e_i(e_iD) + (e_iD)e_i = 0$ $(i = 1, 2, 4)$, $\mathrm{tr}\,(e_1e_2)D = \mathrm{tr}\,(e_1e_4)D = \mathrm{tr}\,(e_2e_4)D = \mathrm{tr}\,(e_3e_4)D = 0$. Now suppose D satisfies (5) and define $e_3D = (e_1D)e_2 + e_1(e_2D)$, $e_5D = (e_1D)e_4 + e_1(e_4D)$, etc. By subtracting a suitable derivation obtained by extending a derivation in $(1, e_1, e_2, e_3)$ by $e_4 \to ce_4$, we obtain an E of the type (5) for which $\lambda_2 = \lambda_3 = \mu_1 = \mu_3 = \nu_5 = \nu_6 = \nu_7 = 0$. Similarly, if we use $(1, e_1, e_6, e_7)$ with e_4, we may obtain an F for which $\lambda_6 = \lambda_7 = \nu_2 = \nu_3 = 0$ also. Then by (6) $\mu_4 = \mu_5 = 0$. If we use $(1, e_1, e_4, e_5)$ with e_2 we obtain $\lambda_4 = \lambda_5 = \mu_6 = \mu_7 = \nu_1 = 0$. Thus D is a sum of derivations and hence is itself a derivation. D has order 14 over Φ.

We wish to show next that $\mathfrak{D}$ is an irreducible set of linear transformations in $\mathfrak{A}'$. For this purpose we consider the enveloping algebra $\mathfrak{R}$, i.e., the smallest algebra of linear transformations containing the operators of $\mathfrak{D}$ in $\mathfrak{A}'$. The derivations defined by

$$e_1D = e_2\alpha_1\lambda + e_3\mu, \qquad e_2D = -e_1\alpha_2\lambda - e_3\nu, \qquad e_4D = 0$$

have the matrices

$$\begin{pmatrix} L_1 & \\ & L_2 \end{pmatrix}$$

relative to the basis $(e_1, e_2, \cdots, e_7)$,[9] where

$$L_1 = \begin{pmatrix} 0 & -\lambda\alpha_2 & \mu\alpha_2 \\ \lambda\alpha_1 & 0 & -\nu\alpha_1 \\ \mu & -\nu & 0 \end{pmatrix}$$

and L_2 is a 4-rowed matrix. It is readily verified that the enveloping algebra of these matrices includes the matrices

$$\begin{pmatrix} A & \\ & B \end{pmatrix},$$

where A is an arbitrary 3-rowed matrix. Furthermore, if E is the derivation defined by $e_1E = e_2E = 0$, $e_4E = e_5$, then $\alpha_1^{-1}E^2$ has the matrix

$$\begin{pmatrix} 0 & \\ & 1 \end{pmatrix}.$$

[9] The matrix M of a linear transformation A relative to the basis $(e_1, e_2, \cdots, e_n)$ is defined by $(e_1A, e_2A, \cdots, e_nA) = (e_1, e_2, \cdots, e_n)M$.

Thus R contains the linear transformations whose matrices are

$$\begin{pmatrix} A & \\ & 0 \end{pmatrix} = \begin{pmatrix} A & \\ & B \end{pmatrix} - \begin{pmatrix} A & \\ & B \end{pmatrix}\begin{pmatrix} 0 & \\ & 1 \end{pmatrix}.$$

In particular $\mathfrak{R}$ contains the linear transformation E_{11} such that $e_1E_{11} = e_1$, $e_iE_{11} = 0$ and E_{12} such that $e_1E_{12} = e_2$, $e_iE_{12} = 0$ for $i \neq 1$. If we replace e_1, e_2 by the pair e_i, e_j, we see that $\mathfrak{R}$ contains E_{ii} and E_{ij} and hence $\mathfrak{R}$ is the complete set of linear transformations in $\mathfrak{A}'$. Evidently this implies that $\mathfrak{A}'$ contains no proper subspace invariant under all the transformations of $\mathfrak{D}$.

3. Assume from now on that Φ has characteristic 0. The irreducibility of $\mathfrak{D}$ implies that $\mathfrak{D} = \mathfrak{D}_1 \oplus \mathfrak{D}_2 \oplus \cdots \oplus \mathfrak{D}_r$ a direct sum of simple Lie algebras.[10] If D is any element of $\mathfrak{D}$, $D = \sum D_i$, D_i in $\mathfrak{D}_i$. The condition that two such elements be commutative is that their components D_i be commutative. It follows that the subalgebra $\mathfrak{C}(D)$ of derivations commutative with a fixed D is a direct sum of r subalgebras $\mathfrak{C}(D) \cap \mathfrak{D}_i \neq 0$. Now let D be the derivation defined by $e_1D = e_2\alpha_1$, $e_2D = -e_1\alpha_2$, $e_4D = 0$. Then

$$(e_1D, e_2D, \cdots, e_7D) = (e_2\alpha_1, -e_1\alpha_2, 0, 0, -e_6\alpha_1, e_5\alpha_2, 0)$$

and the conditions $e_1[D, E] = e_2[D, E] = e_4[D, E] = 0$ imply that

$$e_1E = e_2\lambda_2 + e_5\lambda_5 + e_6\lambda_6, \qquad e_2E = e_1\mu_1 + e_5\mu_5 + e_6\mu_6, \qquad e_4E = e_3\nu_3 + e_7\nu_7,$$

where

$$\mu_1\alpha_1 + \lambda_2\alpha_2 = \mu_5\alpha_1 - \lambda_6\alpha_2 = \mu_6 + \lambda_5 = \nu_3\alpha_1 - 2\lambda_6\alpha_4 = 0.$$

It follows that $\mathfrak{C}(D)$ has the basis D, E_1, E_2, E_3, where

$$(e_1E_1, e_2E_1, \cdots, e_7E_1) = (e_5, -e_6, -2e_7, 0, e_1\alpha_4, -e_2\alpha_4, -2e_3\alpha_4),$$

$$(e_1E_2, e_2E_2, \cdots, e_7E_2) = \left(e_6, e_5\frac{\alpha_2}{\alpha_1}, 2e_4\alpha_2, 2e_3\frac{\alpha_4}{\alpha_1}, e_2\alpha_4, e_1\frac{\alpha_4\alpha_2}{\alpha_1}, 0\right),$$

$$(e_1E_3, e_2E_3, \cdots, e_7E_3) = (e_2\alpha_1, -e_1\alpha_2, 0, 2e_7, e_6\alpha_1, -e_5\alpha_2, -2e_4\alpha_1\alpha_2),$$

whence

$$[E_1, E_2] = 2E_3\frac{\alpha_4}{\alpha_1}, \qquad [E_2, E_3] = -2E_1\alpha_2, \qquad [E_3, E_1] = -2E_2\alpha_1.$$

This table shows that the only ideals in $\mathfrak{C}(D)$ are (D) and (E_1, E_2, E_3) and $\mathfrak{C}(D) = (D) \oplus (E_1, E_2, E_3)$. Hence $r \leqq 2$ and if $r = 2$, $D \in \mathfrak{D}_1$ and $(E_1, E_2, E_3) \subset \mathfrak{D}_2$. The derivation $D' = \frac{1}{2}(D - E_3)$ satisfies $e_4D' = -e_7$, $e_7D' = e_4\alpha_1\alpha_2$, $e_1D' = 0$. Hence if we replace e_1, e_2, e_4 by e_4, e_7, e_1, the above argument shows that $D' \in \mathfrak{D}_1$ or $D' \in \mathfrak{D}_2$. Either case contradicts $\mathfrak{D}_1 \cap \mathfrak{D}_2 = 0$. Hence $r = 1$ and we have proved the following theorem.

[10] The irreducibility implies that $\mathfrak{D} = \mathfrak{S} \oplus \mathfrak{D}'$, where $\mathfrak{S}$ is a commutative algebra and $\mathfrak{D}'$ is semi-simple (Jacobson, *Rational methods in the theory of Lie algebras*, Annals of Math., vol. 36(1935), p. 877). By Cartan's theorem $\mathfrak{D}'$ is a direct sum of simple algebras (Thèse, p. 53). Evidently any commutative algebra is a direct sum of simple algebras of order 1.

THEOREM. *$\mathfrak{D}$ is a simple Lie algebra.*

Thus if Φ is algebraically closed, $\mathfrak{D}$ is the simple Lie algebra of order 14 in the Killing-Cartan list.[11] We may now apply Cartan's theory of representations.[12] The weights of any representation of $\mathfrak{D}$ have the form $m_1\lambda_1 + m_2\lambda_2 + m_3\lambda_3$, where $m_1 + m_2 + m_3 = 0$, $3m_i$ and $m_i - m_j$ are integers. If Λ is a weight, so are all the forms $\Lambda - \omega_i$, $\Lambda - 2\omega_i$, $\cdots$, $\Lambda - 3m_i\omega_i = -m_i\lambda_i + (m_i + m_j)\lambda_j + (m_i + m_k)\lambda_k$, $\omega_i = \frac{2}{3}\lambda_i - \frac{1}{3}\lambda_j - \frac{1}{3}\lambda_k$ and $\Lambda - \omega_{ij}$, $\Lambda - 2\omega_{ij}$, $\cdots$, $\Lambda - (m_i - m_j)\omega_{ij} = m_j\lambda_i + m_i\lambda_j + m_k\lambda_k$, $\omega_{ij} = \lambda_i - \lambda_j$. The last equation shows that if Λ is a weight, so are all the forms obtained by permuting the m's in Λ. The highest weight of an irreducible representation satisfies $m_1 \geqq m_2 \geqq m_3$, $m_2 \leqq 0$.

LEMMA 1. *The representation in $\mathfrak{A}'$ is the only one (in the sense of similarity) of order $\leqq 7$.*

Suppose Λ is the highest weight of such a representation. We distinguish three cases:

(1) $m_2 < 0$, $m_2 > m_3$. Here $| m_i | \neq | m_j |$, $i \neq j$ and hence we obtain at least 12 distinct weights.

(2) $m_2 = 0$. Then $\Lambda = k(\lambda_1 - \lambda_2)$ and k is an integer. Again we obtain more than 7 distinct weights.

(3) $m_2 < 0$ and $m_2 = m_3$. Here $\Lambda = k(\frac{2}{3}\lambda_1 - \frac{1}{3}\lambda_2 - \frac{1}{3}\lambda_3)$ and we obtain more than 7 distinct weights unless $k = 1$. In this case we obtain 6 distinct weights $\neq 0$ and the highest is $\frac{2}{3}\lambda_1 - \frac{1}{3}\lambda_2 - \frac{1}{3}\lambda_3$.

Since the highest weights of any two representations under consideration are equal, these representations are similar.

Let $D \to D^{\bar{S}} = \tilde{D}$ be an automorphism in $\mathfrak{D}$. This correspondence defines a second irreducible representation of $\mathfrak{D}$ in $\mathfrak{A}'$, and hence by Lemma 1 there exists a linear transformation S in $\mathfrak{A}'$ such that $\tilde{D} = S^{-1}DS$ for all D restricted to $\mathfrak{A}'$. Since D, $\tilde{D}$ are skew, $\tilde{D} = S^*D(S^*)^{-1}$ and $SS^*D = DSS^*$. Since the enveloping algebra of $\mathfrak{D}$ is the complete set of linear transformations, $SS^* = 1\sigma$, $\sigma \neq 0$. We may replace S by $S_1 = S\tau$, $\tau^2 = \sigma^{-1}$ and obtain that S_1 is orthogonal, i.e., $(xS_1, yS_1) = (x, y)$. Define $1S_1 = 1$. Then $\tilde{D} = S_1^{-1}DS_1$ in $\mathfrak{A}$ and S_1 is orthogonal. Let $(1, e_1, \cdots, e_7)$ be a basis as in §1 and $f_i = e_iS$. Then $(e_i, e_j) = (f_i, f_j)$ and hence $e_i^2 = f_i^2$. Let D be the derivation such that $(e_1D, e_2D, \cdots, e_7D) = (e_2\alpha_1, -e_1\alpha_2, 0, -e_7, -2e_6\alpha_1, 2e_5\alpha_2, e_4\alpha_1\alpha_2)$. Evidently the multiples of e_3 are the only elements in $\mathfrak{A}'$ annihilated by D, and hence the multiples of f_3 are the only elements annihilated by $\tilde{D}$. Since

$$(f_1f_2)\tilde{D} = (f_1\tilde{D})f_2 + f_1(f_2\tilde{D}) = f_2^2\alpha_1 - f_1^2\alpha_2 = 0,$$

$f_1f_2 = f_3\rho$, and since

$$(f_3, f_3) = (e_3, e_3) = -(e_1, e_1)(e_2, e_2) = -(f_1, f_1)(f_2, f_2) = (f_1f_2, f_1f_2),$$

[11] Cartan, Thèse, Paris, 1894, p. 94.

[12] Cf. Cartan, *Les groupes projectifs qui ne laissent invariante aucune multiplicité plane*, Bull. Soc. Math. de France, vol. 41(1913), pp. 53–96.

we have $\rho = \pm 1$, or $f_1 f_2 = \pm f_3$. Similarly $f_i f_j = f_k \rho_{ij} \epsilon_{ij}$, $\epsilon_{ij} = \pm 1$ if $e_i e_j = e_k \rho_{ij}$. By changing the sign of τ if necessary, we may suppose $f_1 f_2 = f_3$. Then $f_2 f_3 = f_1 \alpha_2$, $f_3 f_1 = f_2 \alpha_1$ and $e_1 \to f_1$, $e_2 \to f_2$, $e_4 \to f_4$ leads to an automorphism S_2 in $\mathfrak{A}$ such that $T = S_1 S_2^{-1}$ satisfies

$$1T = 1, \qquad e_i T = e_i \epsilon_i, \qquad \epsilon_i = \begin{cases} 1 & \text{if } i = 1, 2, 3, 4, \\ \pm 1 & \text{if } i = 5, 6, 7, \end{cases}$$

and $D_1 = T^{-1} D T \in \mathfrak{D}$ if D does. Using our special D, we obtain

$$\begin{aligned} e_5 D_1 &= e_5 T^{-1} D T = -2 e_6 \epsilon_5 \epsilon_6 \alpha_1 \\ &= (e_1 e_4) D_1 = -e_6(\epsilon_7 + 1)\alpha_1, \end{aligned}$$

and hence $\epsilon_7 = 1$, $\epsilon_5 \epsilon_6 = 1$. If we use the derivation E such that $e_1 E = e_3$, $e_3 E = e_1 \alpha_2$, $e_4 E = e_6$ and compute $e_5 E_1$ in two ways, we obtain $\epsilon_6 = \epsilon_5 = 1$. Thus $T = 1$, $S_1 = S_2$.

LEMMA 2. *If $D \to D^{\bar{S}}$ is an automorphism of $\mathfrak{D}(\mathfrak{A})$, $\mathfrak{A}$ the Cayley algebra over an algebraically closed field Φ, then there is a unique automorphism S in $\mathfrak{A}$ such that $D^{\bar{S}} = S^{-1} D S$.*

If S' and S'' both satisfy this condition, $1S' = 1S'' = 1$ and S' and S'' send $\mathfrak{A}'$ into itself. $(S')^{-1} S''$ commutes with all D in $\mathfrak{A}'$. Hence $S'' = \rho S'$. It follows immediately that $\rho = 1$.

4. Suppose $\mathfrak{A}_1 = (1, e_1, \cdots, e_7)$ and $\mathfrak{A}_2 = (1, f_1, \cdots, f_7)$ are two Cayley algebras over Φ of characteristic 0 such that $\mathfrak{D}(\mathfrak{A}_1) \cong \mathfrak{D}(\mathfrak{A}_2)$, where, say, $D \to E$ gives the isomorphism. If Ω is the algebraic closure of Φ, there is just one Cayley algebra $\mathfrak{A}$ over Ω and we may regard $\mathfrak{A}_1$ and $\mathfrak{A}_2$ as subrings of $\mathfrak{A}$ and $1, e_1, \cdots, e_7$ and $1, f_1, \cdots, f_7$ as bases for $\mathfrak{A}$ over Ω. Let $D_1, \cdots, D_{14}$ and $E_1, \cdots, E_{14}$ be corresponding bases for $\mathfrak{D}(\mathfrak{A}_1)$ and $\mathfrak{D}(\mathfrak{A}_2)$. Either of these sets forms a basis for $\mathfrak{D}(\mathfrak{A})$ over Ω. If

$$e_i D_k = \sum e_j \mu_{ji}^{(k)}, \qquad f_i E_k = \sum f_j \nu_{ji}^{(k)}, \qquad (i, j = 1, \cdots, 7; k = 1, \cdots, 14),$$

the matrices $\sum M^{(k)} \omega_k$ and $\sum N^{(k)} \omega_k$, $M^{(k)} = (\mu_{ij}^{(k)})$, $N^{(k)} = (\nu_{ij}^{(k)})$ correspond in different representations of $\mathfrak{D}(\mathfrak{A})$. Hence there exists a matrix Q with elements in Ω such that $Q^{-1} M^{(k)} Q = N^{(k)}$. Since the M's and N's have elements in Φ, we may suppose that Q has elements in Φ also and by choosing the basis $f_1, \cdots, f_7$ in $\mathfrak{A}_1'$ suitably, we may assume $M^{(k)} = N^{(k)}$. Let T be the linear transformation defined by $1T = 1$, $e_i T = f_i$. Then $D_k \to T^{-1} D_k T$ defines an automorphism in $\mathfrak{D}(\mathfrak{A})$. By the proof of Lemma 2, there exists a $\tau \neq 0$ such that $1, f_1 \tau, \cdots, f_7 \tau$ satisfies the same multiplication table as $1, e_1, \cdots, e_7$. Since $e_1 e_2 = e_3$, $f_1 f_2 \tau = f_3$ and $\tau \in \Phi$. Thus

$$S = \begin{cases} 1 & \text{in } (1), \\ T\tau & \text{in } \mathfrak{A}' \end{cases}$$

induces an isomorphism between $\mathfrak{A}_1$ over Φ and $\mathfrak{A}_2$ over Φ such that $E = S^{-1} D S$. S is unique since its extension is an automorphism in $\mathfrak{A}$ over Ω.

THEOREM. *If $\mathfrak{A}_1$ and $\mathfrak{A}_2$ are Cayley algebras over Φ such that there exists an isomorphism $D \to E$ between their derivation algebras, then there exists a unique isomorphism S between $\mathfrak{A}_1$ and $\mathfrak{A}_2$ such that $E = S^{-1}DS$.*

THEOREM. *If $\mathfrak{A}$ is a Cayley algebra over Φ, $\mathfrak{D}$ its derivation algebra, then the group of automorphisms of $\mathfrak{D}$ is isomorphic to the group of automorphisms of $\mathfrak{A}$.*

If S is an automorphism in $\mathfrak{A}$, $D \to S^{-1}DS = D^{\bar{S}}$ is one in $\mathfrak{D}$. By the preceding theorem any $\bar{S}$ has this form and the corresponding S is unique.

5. A Lie algebra $\mathfrak{L}$ is of *type G* if $\mathfrak{L}_\Omega$ is isomorphic to the derivation algebra of the Cayley algebra over Ω. The latter is one of the five exceptional algebras in Cartan's list of the simple Lie algebras over Ω. If $\mathfrak{A}$ is any Cayley algebra over Φ, we have noted that $\mathfrak{D}_\Omega = \mathfrak{D}(\mathfrak{A}_\Omega)$ so that $\mathfrak{D}$ is of type G.

Now suppose $\mathfrak{L}$ is an arbitrary Lie algebra of type G. The elements of $\mathfrak{L}$ may be represented as derivations in $(1, e_1, \cdots, e_7)$ over Ω, where $\alpha_1 = \alpha_2 = \alpha_4 = 1$ in the notation of §2. Thus $\mathfrak{L} \cong$ the set $\sum_1^{14} D_i\gamma_i$, γ in Φ, and $\sum D_i\omega_i$, ω in Ω, give all the derivations in the algebra over Ω. The D's are determined as in (5) and the λ_i, μ_i, ν_i obtained by the derivations $\sum D_i\gamma_i$ generate a finite algebraic extension of Φ and may be taken as elements in a finite Galois extension P of Φ. Thus $\mathfrak{L}_P$ is isomorphic to the Cayley algebra of $(1, e_1, \cdots, e_7)$ over P.

If $s \in \mathfrak{G}$ the Galois group of P over Φ, the correspondence $x = \sum_0^7 e_i\rho_i \to \sum e_i\rho_i^s = xS$ $(e_0 = 1)$ is an automorphism of $(1, e_1, \cdots, e_7)$ over P regarded as an algebra over Φ and $(x\rho)S = (xS)\rho^s$. It follows that $E_i = S^{-1}D_iS$ are linearly independent (over P) derivations and have the same multiplication table as the D_i. Hence $\sum D_i\sigma_i \to \sum E_i\sigma_i$, σ in P, is an automorphism in the derivation algebra and there exists an automorphism $\bar{S}$ of $(1, e_1, \cdots, e_7)$ over P such that $\sum E_i\sigma_i = \bar{S}^{-1}(\sum D_i\sigma_i)\bar{S}$. In particular, if $S_1 = S\bar{S}^{-1}$, $S_1^{-1}(\sum D_i\gamma_i)S_1 = \sum D_i\gamma_i$. Let $\mathfrak{A}$ be the subset of elements y such that $yS_1 = y$ for all s in $\mathfrak{G}$. $\mathfrak{A}$ is an algebra over Φ containing 1.

If $st = u$ in $\mathfrak{G}$, $ST = U$ and

$$\begin{aligned} \bar{U}^{-1}D_i\bar{U} &= U^{-1}D_iU = T^{-1}S^{-1}D_iST = T^{-1}\bar{S}^{-1}D_i\bar{S}T \\ &= T^{-1}\bar{S}^{-1}T(T^{-1}D_iT)T^{-1}\bar{S}T = (T^{-1}\bar{S}^{-1}T)\bar{T}^{-1}D_i\bar{T}(T^{-1}\bar{S}T). \end{aligned}$$

$\bar{U}$ and $\bar{T}(T^{-1}\bar{S}T)$ are automorphisms in $(1, e_1, \cdots, e_7)$ over P and hence by the uniqueness noted above they are equal. If $U_1 = U\bar{U}^{-1}$, $T_1 = T\bar{T}^{-1}$, it follows that $S_1T_1 = U_1$. We require the

LEMMA. *Let $\mathfrak{R}$ be a vector space of order n over P, a separable Galois field over Φ. Suppose $s \to S_1$ is a (1-1) representation of the Galois group $\mathfrak{G}$ of P over Φ by semi-linear transformations such that $\mu S_1 = S_1\mu^s$. Then the order over Φ of $\mathfrak{R}_0$ the set of elements invariant under all S_1 is n and its extension $\mathfrak{R}_0\mathrm{P} = \mathfrak{R}$.*

The set of operators $\sum S_1\mu_s$, μ_s in P, is a cross product (P, S_1, 1) and hence is isomorphic to Φ_k the k-rowed matrix algebra over Φ. We may regard $\mathfrak{R}$ as a vector space over Φ. Then its order will be nk and the operators in (P, S_1, 1) are linear transformations and therefore determine matrices in Φ_{nk}. Thus we obtain a representation of Φ_k in Φ_{nk} such that the identities correspond. It is well known that this representation decomposes into n irreducible parts all equivalent to Φ_k. Hence we may write any x in $\mathfrak{R}$ uniquely as $x_1 + \cdots + x_n$, where the x_i form an invariant subspace $\mathfrak{R}^{(i)}$ relative to all $\sum S_1\mu_s$. The condition $xS_1 = x$ is equivalent to $x_iS_1 = x_i$ and therefore $\mathfrak{R}_0 = \mathfrak{R}_0^{(1)} + \cdots + \mathfrak{R}_0^{(n)}$, if $\mathfrak{R}_0^{(i)} = \mathfrak{R}^{(i)} \cap \mathfrak{R}_0$. Because of the similarity of the transformations induced by S_1 in $\mathfrak{R}^{(i)}$ and $\mathfrak{R}^{(j)}$ the order $(\mathfrak{R}_0^{(i)} : \Phi) = (\mathfrak{R}_0^{(j)} : \Phi)$ and hence $(\mathfrak{R}_0 : \Phi) = n(\mathfrak{R}_0^{(i)} : \Phi)$. Now if we consider the special case where $n = 1$ and the operators S_1 are the elements of the Galois group, we see that the elements left invariant by S_1 form a 1-dimensional subspace, namely, Φ itself.[13] It follows that $\mathfrak{R}_0$ has order n over Φ. If y_i is a vector in $\mathfrak{R}^{(i)}$ such that $y_iS_1 = y_i$, $\mathfrak{R}^{(i)}$ consists of the multiples $y_i\rho$ and hence $\mathfrak{R} = \mathfrak{R}_0\mathrm{P}$.

Returning to our special case, we see that $\mathfrak{A}$ has order 8 over Φ and $\mathfrak{A}\mathrm{P} = (1, e_1, \cdots, e_7)$ over P. Hence $\mathfrak{A}\mathrm{P} \cong \mathfrak{A}_\mathrm{P}$ and $\mathfrak{A}$ is a Cayley algebra. If $D = \sum D_i\gamma_i$, $(yD)S_1 = yD$, i.e., $yD \in \mathfrak{A}$ and hence $\sum D_i\gamma_i$ is a derivation in $\mathfrak{A}$. Since the D_i are independent, we obtain here all the derivations in $\mathfrak{A}$.

THEOREM. *A necessary and sufficient condition that a Lie algebra $\mathfrak{L}$ over Φ be of type G is that $\mathfrak{L} = \mathfrak{D}(\mathfrak{A})$, $\mathfrak{A}$ a Cayley algebra over Φ.*

6. The above results establish a complete equivalence between the problems of classifying and obtaining the automorphisms of Lie algebras of type G and the analogous problems for Cayley algebras. As was noted also the classification of Cayley algebras can be reduced to a question of equivalence of certain quadratic forms in eight variables. For certain special fields, this is readily accomplished. In particular, if Φ is real closed, there are two Cayley algebras and if Φ is an algebraic number field, there are 2^{r_1} such algebras, where r_1 is the number of real conjugate fields.

We remark finally that the arguments above are quite general. The special considerations are all contained in the study of the structure and automorphisms for the algebras over an algebraically closed field. We hope to apply this method to other types of Lie algebras in a later paper.

UNIVERSITY OF NORTH CAROLINA.

[13] This method may be used to prove that (P, S, 1) $\cong \Phi_k$. For we obtain here a realization of the cross product by linear transformations in P over Φ. Since the order of (P, S, 1) $= k^2$, this set includes all linear transformations and hence is isomorphic to Φ_k.

Annals of Mathematics
Vol. 41, No. 1, January, 1940

THE FUNDAMENTAL THEOREM OF THE GALOIS THEORY FOR QUASI-FIELDS[1]

By N. Jacobson

(Received February 13, 1939)

It has been noted recently by Artin and by Baer that one obtains the fundamental theorem of the Galois theory most readily by starting with a finite group $\mathfrak{G}$ of automorphisms in a field P and determining the structure of P over Φ the set of invariant elements. One proves that P is finite, separable and normal and $\mathfrak{G}$ is its Galois group over Φ. The theorem is completed by proving that any isomorphism between subfields over Φ in a finite, separable and normal extension P of Φ can be extended to an automorphism in P. It follows that the Galois group of P over Φ has Φ as its set of invariant elements. The correspondence between subfields and subgroups follows readily. The usual proofs of these theorems are obtained by strictly commutative methods (symmetric functions, unique factorization of polynomials).

In the present paper we begin with an arbitrary quasi-field P and a finite group of outer automorphisms $\mathfrak{G}$ acting in P and establish the correspondence between subgroups of $\mathfrak{G}$ and sub-quasi-fields Σ between P and Φ the set of invariant elements. The methods are necessarily those of non-commutative algebra. The particular tool used is the theory of simple rings. We obtain some applications to division algebras. The first section is introductory containing results that are for the most part well known. In the last section we give a generalization of Hilbert's theorem on the elements of norm 1 in a cyclic field.

1. By a transformation T we shall understand a single-valued correspondence between the elements of a set P and those of a subset. We denote its effect on α by αT. T and S are regarded as equal only when they have the same effect on all α's of the set. The product ST is defined by $\alpha(ST) = (\alpha S)T$. If P is a group written additively $S + T$ is the transformation such that $\alpha(S + T) = \alpha S + \alpha T$ and T is an endomorphism if $(\alpha + \beta)T = \alpha T + \beta T$. With these conventions it is readily verified that the set of endomorphisms of a commutative group form a ring with an identity, namely, the identity transformation. The zero element is O given by $\alpha O = 0$ and the negative of T is $\alpha(-T) = -\alpha T$. If T is $(1 - 1)$ its inverse is also an endomorphism. When we speak of an endomorphism of a ring P we shall mean an endomorphism of its additive group.

A commutative group P in which there is defined a set of endomorphisms

[1] Presented to the Society April 7, 1939.

forming a quasi-field $\mathbf{\Phi}$ which includes the identity transformation is called a vector space. P has finite dimensionality $(P:\mathbf{\Phi}) = n$ if it contains n elements $\alpha_1, \alpha_2, \cdots, \alpha_n$ such that every element is expressible uniquely in the form $\alpha_1\xi_1 + \alpha_2\xi_2 + \cdots + \alpha_n\xi_n$, $\xi_i \in \mathbf{\Phi}$. As is well known n is an invariant and may be characterized as the maximum number of linearly independent elements in P. We suppose hereafter that $(P:\mathbf{\Phi})$ is finite.

An endomoprhism $\boldsymbol{A}$ commutative with all the transformations in $\mathbf{\Phi}$ is called a linear transformation (l. t.) of P over $\mathbf{\Phi}$. Their totality is a ring $\mathbf{\Lambda}$. Now suppose $\alpha_1, \alpha_2, \cdots, \alpha_n$ is a basis for P over $\mathbf{\Phi}$. Clearly $\boldsymbol{A}$ is determined by $\alpha_r \boldsymbol{A}$ and we may choose $\alpha_r \boldsymbol{A}$ arbitrarily and define a l. t. by the condition $(\sum \alpha_i \xi_i)\boldsymbol{A} = \sum(\alpha_i \boldsymbol{A})\xi_i$. In particular we define $\boldsymbol{E}_{ij}$ by $\alpha_r \boldsymbol{E}_{ij} = \delta_{ir}\alpha_j$ and ξ' by $\alpha_r\xi' = \alpha_r\xi$ for any ξ in $\mathbf{\Phi}$.[1a] The totality of transformations ξ' forms a quasi-field $\mathbf{\Phi}'$ anti-isomorphic to $\mathbf{\Phi}$ since $\alpha_r(\xi\mathbf{n})' = \alpha_r\xi\mathbf{n} = \alpha_r\xi'\mathbf{n} = \alpha_r\mathbf{n}\xi' = \alpha_r\mathbf{n}'\xi'$. Evidently P is a vector space over $\mathbf{\Phi}'$ and since $\alpha_r\xi' = \alpha_r\xi$, $\alpha_1, \alpha_2, \cdots, \alpha_n$ form a basis for this space. It is readily seen that $\alpha_r\xi'\boldsymbol{E}_{ij} = \alpha_r\boldsymbol{E}_{ij}\xi'$ and hence $\xi'\boldsymbol{E}_{ij} = \boldsymbol{E}_{ij}\xi'$. If $\alpha_r\boldsymbol{A} = \sum \alpha_j\xi_{jr}$ then $\boldsymbol{A} = \sum \boldsymbol{E}_{ij}\xi'_{ji}$ and hence the set of l. t.'s is identical with the set of these transformations. If $\sum \boldsymbol{E}_{ij}\xi'_{ji} = \mathbf{0}$ then $\sum \alpha_j\xi_{jr} = 0$ and by the uniqueness we have $\xi_{jr} = \mathbf{0}$. If we regard P as a space over $\mathbf{\Phi}'$ and use the basis α_i, we obtain $\alpha_r\xi'' = \alpha_r\xi$ and hence the l. t.'s ξ and ξ'' are identical. The $\boldsymbol{E}_{ij}$ are l. t.'s in P over $\mathbf{\Phi}'$ and the general form of such a transformation is $\Sigma\, \boldsymbol{E}_{ij}\xi_{ji}$. We note also that

$$\boldsymbol{E}_{11} + \boldsymbol{E}_{22} + \cdots + \boldsymbol{E}_{nn} = \mathbf{1}, \qquad \boldsymbol{E}_{ij}\boldsymbol{E}_{kl} = \delta_{jk}\boldsymbol{E}_{il}.$$

Hence if $\boldsymbol{E}_{pq}(\Sigma\, \boldsymbol{E}_{ij}\xi'_{ji}) = (\Sigma\, \boldsymbol{E}_{ij}\xi'_{ji})\boldsymbol{E}_{pq}$,

$$\Sigma\, \boldsymbol{E}_{pj}\xi'_{jq} = \Sigma\, \boldsymbol{E}_{iq}\xi'_{pi}$$

and $\xi'_{ij} = \boldsymbol{O}$ if $i \neq j$, $\xi'_{ii} = \xi'$ and similarly the condition that $\Sigma\, \boldsymbol{E}_{ij}\xi_{ji}$ commute with all $\boldsymbol{E}_{pq}$ is that it belong to $\mathbf{\Phi}$. Now if $\boldsymbol{T}(\boldsymbol{T}')$ is any endomorphism commutative with all the l. t.'s $\Sigma\, \boldsymbol{E}_{ij}\xi'_{ji}(\Sigma\, \boldsymbol{E}_{ij}\xi_{ji})$ it is an l. t. in P over $\mathbf{\Phi}'(\mathbf{\Phi})$ and commutes with the $\boldsymbol{E}_{pq}$. Hence $\boldsymbol{T} \in \mathbf{\Phi}(\boldsymbol{T}' \in \mathbf{\Phi}')$.

If P is a ring with an identity the transformations $\alpha \to \alpha\xi = \alpha\boldsymbol{\xi}$ are endomorphisms and their totality is a ring $\mathbf{P}$ isomorphic to P. The transformations $\alpha \to \xi\alpha = \alpha\xi'$ are endomorphisms and form a ring $\mathbf{P}'$ anti-isomorphic to P. The elements of $\mathbf{P}$ commute with those of $\mathbf{P}'$. Furthermore if $\boldsymbol{T}'$ is any endomorphism commutative with all the $\boldsymbol{\xi}$ and $1\boldsymbol{T}' = \tau$, then $\alpha\boldsymbol{T}' = (1\alpha)\boldsymbol{T}' = (1\boldsymbol{\alpha})\boldsymbol{T}' = (1\boldsymbol{T}')\boldsymbol{\alpha} = \tau\boldsymbol{\alpha} = \tau\alpha = \alpha\boldsymbol{\tau}'$. Hence $\boldsymbol{T}' \in \mathbf{P}'$. Similarly $\mathbf{P}$ is the complete set of endomorphisms commutative with those of $\mathbf{P}'$.

If $P \geqq \Phi > 1$ and Φ is a quasi-field the transformations corresponding to the elements of Φ form a quasi-field $\mathbf{\Phi}$ and P is a vector space over $\mathbf{\Phi}$. If left multiplication is used in place of right multiplication we obtain a quasi-field $\mathbf{\Phi}'$ and P is a vector space over $\mathbf{\Phi}'$. Thus, for example, the set of l. t.'s of any vector space over $\mathbf{\Phi}$ contains $\mathbf{\Phi}'$ and may be regarded as a space over $\mathbf{\Phi}'$. Now suppose

[1a] Note that the ξ' as well as the $\boldsymbol{E}_{ij}$ depend on the basis $\alpha_1, \alpha_2, \cdots, \alpha_n$.

E is a space over $\mathbf{P}$ and $\mathbf{P} \geqq \mathbf{\Phi} > \mathbf{1}$ where the dimensionalities $(E:\mathbf{P})$ and $(\mathbf{P}:\mathbf{\Phi})$ are finite. If $\alpha_1, \alpha_2, \cdots, \alpha_n$ is a basis for E over $\mathbf{P}$, $\xi_1, \xi_2, \cdots, \xi_r$ one for $\mathbf{P}$ over $\mathbf{\Phi}$ then $\beta_{ij} = \alpha_i \xi_j$ form a basis for E over $\mathbf{\Phi}$ and hence $(E:\mathbf{\Phi}) = (E:\mathbf{P})(\mathbf{P}:\mathbf{\Phi})$.

We recall also the definition of a semi-linear transformation T with automorphism S (in $\mathbf{P}$) as an endomorphism in E such that $(\alpha\xi)T = (\alpha T)\xi^S$ for every ξ in $\mathbf{P}$.

2. Let P be an arbitrary quasi-field and $\mathfrak{G}$ a group of n automorphisms $\mathbf{1}$, $S, T, \cdots, U$ in P $((\alpha + \beta)^S = \alpha^S + \beta^S, (\alpha\beta)^S = \alpha^S\beta^S)$. We suppose that no $S \neq \mathbf{1}$ is inner, i.e. has the form $\alpha \rightarrow \xi^{-1}\alpha\xi$. Let $(\mathbf{P}, S)$ denote the set of endomorphisms of P generated by the transformations ξ of P, $\alpha \rightarrow \alpha\xi = \alpha\xi$ and the automorphisms S. $(\mathbf{P}, S)$ consists of the sums $\Sigma\, S\xi_S$ and $\xi S = S\xi^S$ where ξ^S corresponds to ξ^S. Suppose

$$S\xi_S + T\xi_T + \cdots + U\xi_U = O. \tag{1}$$

If $\xi_S \neq O \neq \xi_T$ we multiply (1) on the left by $\mathfrak{n}$ and on the right by $\xi_S^{-1}\mathfrak{n}^S\xi_S$ and subtract to obtain

$$T(\mathfrak{n}^T\xi_T - \xi_T\xi_S^{-1}\mathfrak{n}^S\xi_S) + \cdots = O$$

which has at least one term less than (1). Since TS^{-1} is not inner there exists an η such that

$$\eta^{TS-1} \neq (\xi_S^{S^{-1}}(\xi_T^{S^{-1}})^{-1})^{-1}\eta(\xi_S^{S^{-1}}(\xi_T^{S^{-1}})^{-1})$$

and hence we may suppose that the new coefficient of T is $\neq 0$. Continuing in this way we get finally a single terms $S\zeta_S = 0$ where $\zeta_S \neq 0$. Since S and ζ_S have inverses this is impossible. Thus the space $(\mathbf{P}, S)$ over $\mathbf{P}$ has n dimensions.

$(\mathbf{P}, S)$ is a simple ring with an identity. For if $\mathfrak{B}$ is a two-sided ideal and contains $\sum S\xi_S \neq O$ the above argument shows that $\mathfrak{B}$ contains some $S\zeta_S$, $\zeta_S \neq O$. Then $\mathfrak{B} > \mathbf{1}$ and $\mathfrak{B} = (\mathbf{P}, S)$. Any right ideal $\mathfrak{J}$ of $(\mathbf{P}, S)$ is a subspace over $\mathbf{P}$. It follows that $(\mathbf{P}, S)$ satisfies the chain conditions for right ideals. Hence the following theorems hold.[2]

Any two irreducible right ideals are similar, i.e. there exists a (1–1) correspondence $X \rightarrow \bar{X}$ between $\mathfrak{J}$ and $\bar{\mathfrak{J}}$ such that

$$\overline{X + Y} = \bar{X} + \bar{Y} \qquad \overline{XA} = \bar{X}A$$

for any A in $(\mathbf{P}, S)$. In particular, if we restrict A to be in $\mathbf{P}$ we see that $X \rightarrow \bar{X}$ is a (1–1) l. t. between $\mathfrak{J}$ and $\bar{\mathfrak{J}}$ and hence $(\mathfrak{J}:\mathbf{P}) = (\bar{\mathfrak{J}}:\mathbf{P})$. Since the elements $(\mathbf{1} + S + T + \cdots + U)\xi$, ξ variable, form a right ideal of dimension 1, we see that $(\mathfrak{J}:\mathbf{P}) = 1$ for every irreducible $\mathfrak{J}$.

$(\mathbf{P}, S) = \mathfrak{J}_1 + \mathfrak{J}_2 + \cdots + \mathfrak{J}_r$ (direct sum) where the $\mathfrak{J}_i$ are irreducible right ideals. Hence $n = ((\mathbf{P}, S):\mathbf{P}) = r(\mathfrak{J}_i:\mathbf{P}) = r$. It follows that $(\mathbf{P}, S) = \mathbf{\Psi}_n$ a

[2] See v. d. Waerden, *Moderne Algebra*, II, pp. 149–177 or Deuring, *Algebren* pp. 8–25.

matrix ring, i.e. the elements of $(\mathbf{P}, S)$ have the form $\Sigma E_{ij}\psi_{ij}$ where the ψ's form a quasi-field, $E_{ij} \neq 0$, commute with the ψ's and satisfy

$$(2) \qquad E_{11} + E_{22} + \cdots + E_{nn} = 1 \qquad E_{ij}E_{kl} = \delta_{jk}E_{il} .$$

Let α be any element $\neq 0$ in P. By (2) there is an E_{pp} such that $\alpha E_{pp} \neq 0$ and since $\alpha E_{pp} = \alpha E_{pi} E_{ip}$, this implies that no $\alpha_i = \alpha E_{pi}$ is 0. Suppose

$$\alpha_1\psi_1 + \alpha_2\psi_2 + \cdots + \alpha_n\psi_n = 0.$$

Then $(\alpha E_{p1}\psi_1 + \alpha E_{p2}\psi_2 + \cdots + \alpha E_{pn}\psi_n)E_{ii} = (\alpha E_{pi})\psi_i = 0$ which implies $\psi_i = 0$. On the other hand since Ψ_n includes $\mathbf{P}$, any element γ in P can be obtained from a fixed $\beta \neq 0$ by operating with a suitable $\sum E_{ij}\psi_{ij}$. Thus

$$\gamma = (\alpha E_{pp}) \sum E_{ij}\psi_{ij} = \alpha_1\psi_1 + \alpha_2\psi_2 + \cdots + \alpha_n\psi_n .$$

Hence the α_i form a basis for P over Ψ and from their definition we have $\alpha_r E_{ij} = \delta_{ir}\alpha_j$.

As in §1 we associate with each ψ the transformation ψ' defined by $(\sum \alpha_i\psi_i)\psi' = \sum \alpha_i\psi\psi_i$ and denote the totality of these transformations as Ψ'. P is a vector space over Ψ' with a basis $\alpha_1, \alpha_2, \cdots, \alpha_n$ as a basis. The transformations $\sum E_{ij}\psi_{ij}$ form the complete set of l. t.'s in P over Ψ' and the elements of Ψ' may be characterized as the endomorphisms commutative with all the $\sum E_{ij}\psi_{ij}$.

On the other hand $\Psi_n = (\mathbf{P}, S)$ and if an endomorphism commutes with all of $\mathbf{P}$ it is in $\mathbf{P}'$ the set of left multiplications $\alpha \to \xi\alpha = \alpha\xi'$ and if in addition it commutes with all the S's we must have ξ in Φ. We have therefore shown that $\Psi' = \mathbf{\Phi}'$ and $(\mathbf{P}, S)$ is the complete set of l. t.'s of P over $\mathbf{\Phi}'$. If we use $(\mathbf{P}', S)$ in place of $(\mathbf{P}, S)$ we obtain $(\mathrm{P}:\mathbf{\Phi}) = n$ and $(\mathbf{P}', S)$ is the complete set of l. t.'s in the vector space P over $\mathbf{\Phi}$. Hence we have the important theorem.

THEOREM 1. *Let* P *be an arbitrary quasi-field,* $\mathfrak{G}$ *a finite group of n outer automorphisms acting in* P *and* Φ *the sub-quasi-field of invariant elements. Then the order* $(\mathrm{P}:\mathbf{\Phi}) = (\mathrm{P}:\mathbf{\Phi}') = n$ *and* $(\mathbf{P}, S)$ *the set of transformations* $\Sigma S\xi_S$ *is the complete set of linear transformations of* P *over* $\mathbf{\Phi}'$, $(\mathbf{P}', S)$ *the complete set of linear transformations of* P *over* $\mathbf{\Phi}$.

Suppose V is any automorphism in P leaving the elements of Φ unaltered. Then V is a l. t. in P over $\mathbf{\Phi}'$ and hence $V = \sum S\xi_S$. For every η we have $\mathbf{n}V - V\mathbf{n}^V = O$. Hence $\sum S(\mathbf{n}^S\xi_S - \xi_S\mathbf{n}^V) = O$ and if $\xi_S \neq 0$, $\eta^V = \xi_S^{-1}\eta^S\xi_S$. Since no $S \neq 1$ is inner this holds for just one S and $V = SA$ where A is inner. A must leave the elements of Φ invariant. Hence if $\alpha^A = \xi^{-1}\alpha\xi$, ξ commutes with all the α's in Φ.

If $\mathfrak{H}$ is a subgroup of $\mathfrak{G}$ we denote the sub-quasi-field of elements invariant under the transformations of $\mathfrak{H}$ by $\mathrm{P}(\mathfrak{H})$ and if Σ is any sub-quasi-field between Φ and P we denote the subgroup of $\mathfrak{G}$ leaving the elements of Σ invariant by $\mathfrak{G}(\Sigma)$. Note that $\Phi \leqq \mathrm{P}(\mathfrak{H}) \leqq \mathrm{P}$, $\mathrm{P}(\mathfrak{G}) = \Phi$ $\mathrm{P}(1) = \mathrm{P}$. The following is the fundamental theorem of the Galois theory.

THEOREM 2. *The correspondences* $\mathfrak{H} \to \mathrm{P}(\mathfrak{H})$ *and* $\Sigma \to \mathfrak{G}(\Sigma)$ *are inverses of each other. Either one sets up a* (1–1) *correspondence between the subgroups of* $\mathfrak{G}$

and the quasi-fields Σ *between* Φ *and* P. *The order* $(P:\Sigma) = (P:\Sigma') =$ *order of* $\mathfrak{G}(\Sigma)$ *and* $(\Sigma:\Phi) = (\Sigma:\Phi') =$ *index of* $\mathfrak{G}(\Sigma)$.

Let $\mathfrak{H}$ be a subgroup of $\mathfrak{G}$ and $P(\mathfrak{H})$ the set of invariant elements. If T is an automorphism of $\mathfrak{G}$ leaving the elements of $P(\mathfrak{H})$ invariant, $T = SA$ where S is in $\mathfrak{H}$ and A is inner. Since $A = S^{-1}T$ is in $\mathfrak{G}$ this implies that $A = 1$, $S = T \in \mathfrak{H}$. Thus $\mathfrak{G}(P(\mathfrak{H})) = \mathfrak{H}$. Now suppose Σ is given where $\Phi \leqq \Sigma \leqq P$ and let $\mathbf{\Lambda}$ be the set of l. t.'s of P over $\mathbf{\Sigma}'$. Then $\mathbf{\Lambda} \leqq (\mathbf{P}, S)$. If $\sum S\, \xi_S \in \mathbf{\Lambda}$

$$\sum S\mathfrak{u}'^{S}\xi_S = \mathfrak{u}' \sum S\xi_S = (\sum S\xi_S)\mathfrak{u}' = \sum S\mathfrak{u}'\xi_S$$

Hence $\sum S(\mathfrak{u}'^{S} - \mathfrak{u}')\xi_S = O$. The process used above shows that $\xi_S = O$ unless $\mu^S = \mu$ for all μ in Σ, i. e. unless $S \in \mathfrak{G}(\Sigma) = \mathfrak{H}$. Hence $\mathbf{\Lambda}$ consists of the transformations $\sum_{S \in \mathfrak{H}} S\xi_S$. $\mathbf{\Sigma}'$ is the complete set of endomorphisms commutative with those of $\mathbf{\Lambda}$. On the other hand the form of the elements of $\mathbf{\Lambda}$ shows that these transformations are precisely the $\boldsymbol{\zeta}$'s such that $\zeta \in P(\mathfrak{H})$. Hence $P(\mathfrak{G}(\Sigma)) = \Sigma$. The order relations follow from Theorem 1.

If $\Sigma = P(\mathfrak{H})$, $\Sigma^S = P(S^{-1}\mathfrak{H}S)$. Hence $\mathfrak{H}$ is invariant if and only if Σ is transformed into itself by all the elements of $\mathfrak{G}$. If $\bar{1}, \bar{S}, \cdots$ are representatives of the cosets of $\mathfrak{H}$, the transformations induced in Σ by these elements are distinct and depend only on the cosets. Their totality is a group $\bar{\mathfrak{G}} \cong \mathfrak{G}/\mathfrak{H}$.

3. If P is commutative, ξ any element in P, let $\xi, \xi^S, \cdots, \xi^T$ be its distinct conjugates. The coefficients of

$$(x - \xi)(x - \xi^S) \cdots (x - \xi^T)$$

are invariant under $\mathfrak{G}$ and therefore belong to Φ. Hence every element of P satisfies an equation with coefficients in Φ and with distinct roots in P. It follows that P is a separable, normal extension of Φ. To complete the Galois theory for finite extensions along these lines one needs only the converse theorem that if P is a finite, separable, normal extension of Φ, then the Galois group leaves invariant only the elements of Φ.[3]

Let Γ be the center of P and $\Delta = \Gamma \cap \Phi$. The automorphisms S induce automorphisms S' in Γ and Δ is the set of invariant elements relative to $\mathfrak{G}'$ the set of S''s. Suppose first that the order of $\mathfrak{G}'$ = order of $\mathfrak{G} = n$. $\Phi\Gamma$ the join of Φ and Γ is invariant relative to **1** only. Hence $\Phi\Gamma = P$. If $\gamma_1, \gamma_2, \cdots, \gamma_n$ is a basis for Γ over Δ every element of P has the form $\gamma_1\varphi_1 + \gamma_2\varphi_2 + \cdots + \gamma_n\varphi_n$, φ_i in Φ and since $(P:\mathbf{\Phi}) = n$, these elements form a basis for P over $\mathbf{\Phi}$. Hence if we regard P, Φ and Γ as algebras over Δ we have $P = \Phi \times \Gamma$.[4] It is then evident that the center of Φ is Δ, i. e. Φ is a normal division algebra over $\mathbf{\Delta}$.[5]

---

[3] v. d. Waerden, *Moderne Algebra* I, p. 150, 1 st ed. or p. 161, 2nd ed.

[4] If P is an algebra over $\mathbf{\Delta}$ with an identity 1, Φ and Γ subalgebras containing 1, $P = \Phi \times \Gamma$ means that 1) $P = \Phi\Gamma$, 2) the elements of Φ commute with those of Γ and 3) if $\varphi_1, \varphi_2, \cdots, \varphi_m$ are linearly independent in Φ over $\mathbf{\Delta}$ and $\gamma_1, \gamma_2, \cdots, \gamma_r$ are independent in Γ over $\mathbf{\Delta}$ then $\varphi_i\gamma_j$ are independent in P over $\mathbf{\Delta}$.

[5] Since $P = \Phi \times \Gamma$ any element commutative with all $\varphi \in \Phi$ is in Γ. It follows that if V is any automorphism leaving the elements of Φ invariant then $V \in \mathfrak{G}$.

The converse of this is clear. If $\mathrm{P} = \Phi \times \Gamma$ where Φ is a normal division algebra over $\boldsymbol{\Delta}$ and Γ is a separable, normal field over $\boldsymbol{\Delta}$ then the Galois group of Γ has order $n = (\Gamma : \boldsymbol{\Delta})$. These automorphisms have a natural extension in $\mathrm{P} = \Phi \times \Gamma$ obtained by requiring that they leave the elements of Φ unaltered. The automorphisms thus obtained are outer.

THEOREM 3. *Let* P *be a quasi-field,* Γ *its center. A necessary and sufficient condition that* P *have a group of n outer automorphisms which induce distinct automorphisms in* Γ *is that* $\mathrm{P} = \Phi \times \Gamma$ *where* Φ *is a normal division algebra over* $\boldsymbol{\Delta}$ *a subfield of* $\boldsymbol{\Gamma}$ *and* Γ *is separable and normal of order n over* $\boldsymbol{\Delta}$.

If $(\mathrm{P} : \boldsymbol{\Gamma})$ is finite any automorphism leaving the elements of Γ invariant is inner.[6] Hence in this case the group $\mathfrak{G}'$ is necessarily of order n.

THEOREM 4. *If* P *is a quasi-field such that* $(\mathrm{P} : \boldsymbol{\Gamma})$ *is finite, a necessary and sufficient condition that* P *have a group of n outer automorphisms is that* $\mathrm{P} = \Phi \times \Gamma$ *where* Φ *is a normal division algebra over* $\boldsymbol{\Delta}$ *a subfield of* $\boldsymbol{\Gamma}$ *and* Γ *is separable and normal of order n over* $\boldsymbol{\Delta}$.

Now suppose Γ is a $\mathfrak{p}$-adic field[7] and $(\mathrm{P} : \boldsymbol{\Gamma}) = m^2$. Then $(\Phi : \boldsymbol{\Delta}) = m^2$ and by a known theorem on $\mathfrak{p}$-adic algebras $\Phi \times \Gamma$ is a division algebra if and only if $(m, n) = 1$.[8] Hence we have the theorem.

THEOREM 5. *If* P *is a quasi-field with center* Γ *a* $\mathfrak{p}$-adic *field and* $(\mathrm{P} : \boldsymbol{\Gamma}) = m^2$, *then any group of outer automorphisms has order prime to m.*

In particular P has no outer automorphism S of order l a prime divisor of m since $S, S^2, \cdots, S^l$ would then be outer automorphisms.

We return now to the general case. Let $\Delta = \Gamma \cap \Phi$, $\mathfrak{G}'$ the group of induced automorphisms and $\mathfrak{H}$ the subgroup of elements T such that $\gamma^T = \gamma$ for all γ in Γ. $\mathfrak{H}$ is an invariant subgroup of $\mathfrak{G}$ and $\mathfrak{G}' \cong \mathfrak{G}/\mathfrak{H}$. If $\Sigma = \mathrm{P}(\mathfrak{H})$, $\mathrm{P} \geqq \Sigma \geqq \Phi$ and the automorphisms S of $\mathfrak{G}$ induce automorphisms $\bar{S}$ in Σ which form a group $\bar{\mathfrak{G}} \cong \mathfrak{G}/\mathfrak{H}$. Since $\Sigma \geqq \Gamma$ the $\bar{S}$ are outer and induce distinct automorphisms S' in Γ. Theorem 3 may then be applied to Σ. We shall not attempt a more refined analysis of the structure of P here. We merely give a very simple example to show that P may be $> \Sigma$.

4. The following example due to Köthe shows that the automorphisms of $\mathfrak{G}$ need not be distinct in Γ.[9] Let Γ_0 be any field of characteristic $\neq 2$ and $\Gamma = \Gamma_0(x_1, y_1, x_2, y_2, \cdots)$ where the x_i and y_i are distinct indeterminates. Let $\mathrm{P}^{(m)} = (x_1, y_1) \times (x_2, y_2) \times \cdots \times (x_m, y_m)$ where (x_i, y_i) is the generalized quaternion algebra with basis $1, \xi_i, \eta_i, \xi_i\eta_i$ such that

$$\xi_i^2 = x_i, \qquad \eta_i^2 = y_i, \qquad \xi_i\eta_i = -\eta_i\xi_i.$$

$\mathrm{P}^{(m)}$ is a quasi-field and hence so is $\mathrm{P} = \sum_1^\infty \mathrm{P}^{(m)}$. If $\zeta_m = \xi_1\xi_2 \cdots \xi_m$ then $\alpha_m \to \zeta_m^{-1}\alpha_m\zeta_m$ is an automorphism S_m of period 2 in $\mathrm{P}^{(m)}$ leaving the elements

[6] Deuring p. 43.

[7] i.e. a finite extension of the field of ordinary p-adic numbers of a field of power series in one indeterminate with coefficients in a finite field.

[8] Deuring p. 113 and p. 137.

[9] "Schiefkörper unendlichen Ranges über den Zentrum," *Math. Annalen* 105, p. 24.

of Γ unaltered. S_m has the same effect as S_{m-1} in $P^{(m-1)}$. Hence $S = \lim S_m$ is defined and is an automorphism in P. S is not inner, for if $\alpha^S = \zeta^{-1}\alpha\zeta$, ζ is in $P^{(N)}$ for N sufficiently large and hence

$$\eta_{N+1}^S = (\xi_1 \cdots \xi_{N+1})^{-1}\eta_{N+1}(\xi_1 \cdots \xi_{N+1})^{-1} = -\eta_{N+1} = \zeta^{-1}\eta_{N+1}\zeta = \eta_{N+1}$$

which is impossible. $S^2 = 1$ and $S = 1$ in Γ. Since $\zeta_m^2 = x_1 x_2 \cdots x_m$ the dimensionality over Γ of $\Phi^{(m)} = P^{(m)} \cap \Phi$ is 2^{2m-1}.[10] It follows that 1, η_1 is a basis for $P^{(m)}$ over $\Phi^{(m)}$ and hence for P over Φ.

5. Let E be a vector space over **P**, $(E:\mathbf{P}) = r$ and $T_1 = 1, T_S, \cdots, T_U$ a finite group of n semi-linear transformations such that the automorphisms S in **P** associated with T_S ($\xi T_S = T_S \xi^S$) are distinct and outer. Let **Φ** denote the set of invariant elements of **P** and H those of E. H is a space over **Φ**. The transformations $\sum T_S \xi_S$ form a ring isomorphic to (**P**, S) defined above.[11] Hence these transformations may also be represented in the form $\sum E_{ij}\psi_{ij}$ as in §2. If $\zeta = \epsilon E_{pp} \neq 0$ then any vector in $E' = \zeta \sum T_S \xi_S$ has the form $\zeta_1\psi_1 + \zeta_2\psi_2 + \cdots + \zeta_n\psi_n$ where $\zeta_i = \zeta E_{pi}$ and $\psi_i \in \Psi$. It follows that if $\mu \neq 0$ is any vector in E′, $\mu\Psi_n = E'$. The transformations induced by the $\sum T_S\xi_S$ in E′ form a ring isomorphic to (**P**, S) also. Since $E = \sum T_S \neq 0$ there is a μ in E′ such that $\mu E = \eta \neq 0$. Then $\eta \in H$ and $\zeta = \eta \sum T_S \xi_S = \eta\xi$, ξ in **P**. Since any $\epsilon = \epsilon E_{11} + \epsilon E_{22} + \cdots + \epsilon E_{nn}$, ϵ is a linear combination of elements η_i in H with coefficients in **P**. Hence E has a basis $\eta_1, \eta_2, \cdots, \eta_r$ over **P**. If $\eta = \Sigma \eta_i \xi_i$ is in H it follows that $\xi_i \in \mathbf{\Phi}$ and $\eta_1, \eta_2, \cdots, \eta_r$ form a basis for H over **Φ**.

THEOREM 6. *Let* E *be a vector space over* **P**, $(E:\mathbf{P}) = r$ *and* $T_1 = 1, T_S, \cdots, T_U$ *a finite group of semi-linear transformations whose induced automorphisms* $1, S, \cdots, U$ *in* **P** *are distinct and outer. If* H *is the set of vectors invariant under the* T_S, **Φ** *the set of elements invariant under the* S, *then* H *is a space over* **Φ** *such that* $(H:\mathbf{\Phi}) = r$ *and the extension* $H\mathbf{P} = E$.

We recall that if T_S is the matrix of T_S relative to the basis $\epsilon_1, \epsilon_2, \cdots, \epsilon_r$, i. e. $(\epsilon_1 T_S, \epsilon_2 T_S, \cdots, \epsilon_r T_S) = (\epsilon_1, \epsilon_2, \cdots, \epsilon_r)T_S$, then the matrix relative to a second basis $(\epsilon_1, \epsilon_2, \cdots, \epsilon_r)A$ is $A^{-1}T_S A^S$.[12] The conditions $T_S T_T = T_U$ when $ST = U$ are equivalent to $T_T T_S^T = T_{ST}$. Since the matrices of the T_S are all 1 relative to $(\eta_1, \eta_2, \cdots, \eta_r)$ we have proved

THEOREM 7. *If* 𝔖 *is a finite group of outer automorphisms in a quasi-field* P *and* T_S *are matrices with elements in* P *such that* $T_1 = 1$, $T_S T_S^T = T_{ST}$, *then there exists a non-singular matrix* A *such that* $T_S = A^{-1}A^S$ *for all* S.

UNIVERSITY OF NORTH CAROLINA.

[10] If K is a subfield of a normal division algebra P and Λ is the set of elements commutative with those of K, then $(P:\Gamma) = (K:\Gamma)(\Lambda:\Gamma)$. See Deuring p. 44.

[11] Evidently $\sum S\xi_S \rightarrow \sum T_S\xi_S$ is a homomorphism. Since (**P**, S) is simple this correspondence is an isomorphism.

[12] Jacobson, "Pseudo-linear transformations," *Annals of Math.* 38 (1937) p. 486.

Reprinted from
Bulletin of the American Mathematical Society, April 1940.

A NOTE ON HERMITIAN FORMS[1]

N. JACOBSON

In this note we effect a reduction of the theory of hermitian forms of two particular types (coefficients in a quadratic field or in a quaternion algebra with the usual anti-automorphism) to that of quadratic forms. The main theorem (§2) enables us to apply directly the known results on quadratic forms. This is illustrated in the discussion in §3 of a number of special cases.

Let Φ be an arbitrary quasi-field of characteristic different from 2 in which an involutorial anti-automorphism $\alpha \to \bar{\alpha}$ is defined. For the present we do not exclude the cases where Φ is commutative and $\bar{\alpha} \equiv \alpha$ or Φ is a quadratic field with $\alpha \to \bar{\alpha}$ as its automorphism. Suppose $\mathfrak{R}$ is an n-dimensional vector space over Φ. We define a bilinear form (x, y) as a function of pairs of vectors with values in Φ, such that

$$(1)\quad \begin{aligned} (x_1 + x_2, y) &= (x_1, y) + (x_2, y), \qquad & (x, y_1 + y_2) &= (x, y_1) + (x, y_2), \\ (x, y\alpha) &= (x, y)\alpha, & (x\alpha, y) &= \bar{\alpha}(x, y), \end{aligned}$$

for all x, y in $\mathfrak{R}$ and α in Φ. If $x_1, x_2, \cdots, x_n$ is a basis for $\mathfrak{R}$ and $(x_i, x_j) = \alpha_{ij}$, the matrix $A = (\alpha_{ij})$ is called the matrix of (x, y) relative to this basis. By (1) it determines (x, y) as $\sum \bar{\xi}_i \alpha_{ij} \eta_j$, if $x = \sum x_i \xi_i$ and $y = \sum x_i \eta_i$. If $y_1, y_2, \cdots, y_n$ where $y_i = \sum x_j \rho_{ji}$ is a second basis for $\mathfrak{R}$ where $R = (\rho_{ij})$ is nonsingular, the matrix of (x, y) relative to this basis is $\bar{R}'AR$. We call A and $\bar{R}'AR$ cogredient. The form (x, y) is hermitian (skew-hermitian), if $(y, x) = \overline{(x, y)}$ ($(y, x) = -\overline{(x, y)}$). This is equivalent to the condition $\bar{A}' = A$ ($\bar{A}' = -A$).

It is readily seen that we may pass from the basis y_i to the x's by a sequence of substitutions of the following two types:

I. $y_i \to y_i$, $(i \neq r)$, $y_r \to y_r + y_s\theta$, $(s \neq r)$.

II. $y_i \to y_i$, $(i \neq r)$, $y_r \to y_r\theta$, $(\theta \neq 0)$.

It follows that we may pass from a matrix to any other matrix cogredient to it by a sequence of transformations of the corresponding types:

I. Addition of the sth column multiplied on the right by θ to the rth together with addition of the sth row multiplied on the left by $\bar{\theta}$ to the rth.

II. Multiplication of the rth column on the right by $\theta \neq 0$ together with multiplication of the rth row on the left by $\bar{\theta}$.

We showed in an earlier paper that any hermitian form or skew-

[1] Presented to the Society, October 28, 1939.

hermitian form with $\bar{\alpha}\neq\alpha$ has a matrix in diagonal form; that is, there is a basis $u_1, u_2, \cdots, u_n$ for $\mathfrak{R}$ such that[2] $(u_i, u_j)=0$, if $i\neq j$. We call a basis of this type orthogonal and u, v orthogonal, if $(u, v)=0$. If $(u_i, u_i)=\beta_i\neq 0$ for $i\leqq r$ and $(u_i, u_i)=0$ for $i>r$, we obtain the diagonal matrix

$$[\beta_1, \beta_2, \cdots, \beta_r, 0, \cdots, 0] \tag{2}$$

for our form.[3] The element β_1 may be taken to be any nonzero element represented by the form, that is, any element for which a u_1 exists such that $(u_1, u_1)=\beta_1$, β_2 is any element represented by (x, y) restricted to the space of vectors orthogonal to u_1, and so on. We note also that β_i may be replaced by $\bar{\gamma}_i\beta_i\gamma_i$, $(\gamma_i\neq 0)$.

The space $\mathfrak{R}_0$ generated by $u_{r+1}, u_{r+2}, \cdots, u_n$ may be characterized as the totality of vectors z, such that $(x, z)=0$ for all x. The space $\mathfrak{R}_1$ generated by $u_1, u_2, \cdots, u_r$ satisfies the condition $\mathfrak{R}=\mathfrak{R}_0+\mathfrak{R}_1$, $\mathfrak{R}_0 \cap \mathfrak{R}_1=0$. If $\mathfrak{R}_2$ is a second space of this sort, it has a basis of the form u_i+z_i, $(i=1, \cdots, r)$, and hence the matrices of (x, y) in $\mathfrak{R}_1$ and in $\mathfrak{R}_2$ are cogredient. We may therefore restrict our attention to nondegenerate forms $(\mathfrak{R}_0=0)$ and shall do so in the remainder of this note.

Two nondegenerate forms $(x, y)_1$ and $(x, y)_2$ in $\mathfrak{R}$ and $\mathfrak{R}'$ respectively are *cogredient* if there is a (1-1) correspondence $x\rightarrow x'$ between $\mathfrak{R}$ and $\mathfrak{R}'$ such that[4] $(x, y)_1=(x', y')_2$. It follows that

$$(x', (y_1 + y_2)')_2 = (x', y_1' + y_2')_2$$

and hence that $(y_1+y_2)'=y_1'+y_2'$. Similarly $(y\alpha)'=y'\alpha$ and so $x\rightarrow x'$ is a linear transformation and $\mathfrak{R}$ and $\mathfrak{R}'$ have the same dimensionality. If $x_1, x_2, \cdots, x_n$ is a basis for $\mathfrak{R}$, then $x_1', x_2', \cdots, x_n'$ is one for $\mathfrak{R}'$. The matrix of $(x, y)_1$ relative to the first basis is the same as that of $(x', y')_2$ relative to the second. Hence the matrices of $(x, y)_1$ and $(x', y')_2$ relative to any bases are cogredient and conversely cogredience of the matrices implies that of the forms.

We shall suppose from now on that Φ is either a quadratic field $\Phi_0(i)$, $i^2=-\lambda$ and $\bar{\alpha}=\alpha_0-i\alpha_1$ for $\alpha=\alpha_0+i\alpha_1$ or that $\Phi=\Phi_0(i, j)$ is a quaternion algebra in which $i^2=-\lambda$, $j^2=-\mu$, $k=ij=-ji$ and $\bar{\alpha}=\alpha_0-i\alpha_1-j\alpha_2-k\alpha_3$ for $\alpha=\alpha_0+i\alpha_1+j\alpha_2+k\alpha_3$. We suppose also that (x, y) is hermitian. Then $(x, x)\in\Phi_0$ and any β in (2) may be replaced

[2] *Simple Lie algebras over a field of characteristic zero*, Duke Mathematical Journal, vol. 4 (1938), p. 542.

[3] The above notation for diagonal matrices will be used throughout this note.

[4] $\mathfrak{R}$ and $\mathfrak{R}'$ have the same quasi-field and anti-automorphism.

by $\beta N(\gamma)$, $N(\gamma)=\bar{\gamma}\gamma$. Let Φ_0' be the multiplicative group of nonzero elements in Φ_0, Φ_0^* the subgroup of norms, and $\Gamma=\Phi_0'/\Phi_0^*$. A determinant for any hermitian matrix A has been defined by E. H. Moore.[5] We recall that, if a matrix B has the form (2) with $r=n$, then $\det B=\beta_1\beta_2\cdots\beta_n$ and, if $A=\overline{R}'BR$, $\det A=N(\rho)\det B$. Thus the coset of $\det A$ in Γ is an invariant of the class of matrices cogredient to A (or an invariant of the form). We shall call this coset the discriminant of A (or of the form).

$\mathfrak{R}$ may be regarded as a vector space of $2n$ or $4n$ dimensions over Φ_0 and

$$\{x, y\} = (1/2)[(x, y) + (y, x)] = (1/2)\operatorname{tr}(x, y) \tag{3}$$

is a symmetric form in $\mathfrak{R}$ over Φ_0. The symmetric form $\{x, y\}$ satisfies the special condition

$$\{x\alpha, y\alpha\} = \{x, y\}N(\alpha), \tag{4}$$

whence

$$\{x\bar{\alpha}, y\} = \{x\bar{\alpha}, y\bar{\alpha}^{-1}\bar{\alpha}\} = \{x, y\bar{\alpha}^{-1}\}N(\alpha) = \{x, y\alpha\}.$$

Hence, if $\bar{\alpha}=-\alpha$, $\{x, x\alpha\}=-\{x\alpha, x\}=0$. Conversely, if $\{x, y\}$ is any symmetric bilinear form in $\mathfrak{R}$ over Φ_0 such that (4) holds, (x, y) defined by

$$(x, y) = \begin{cases} \{x, y\} - (i/\lambda)\{x, yi\}, & \text{if } \Phi = \Phi_0(i), \\ \{x, y\} - (i/\lambda)\{x, yi\} - (j/\mu)\{x, yj\} \\ \qquad - (k/\lambda\mu)\{x, yk\}, & \text{if } \Phi = \Phi_0(i, j), \end{cases} \tag{5}$$

is hermitian in $\mathfrak{R}$ over Φ. The relation between (x, y) and $\{x, y\}$ is a reciprocal one and $\{x, y\}$ is nondegenerate if (x, y) is.[6]

Evidently, if $(x, y)_1$ in $\mathfrak{R}$ over Φ and $(x', y')_2$ in $\mathfrak{R}'$ over Φ are cogredient, then $\{x, y\}_1$ and $\{x', y'\}_2$ are cogredient also. Suppose now that $\{x, y\}_1$ and $\{x', y'\}_2$ are cogredient. Then we have u_1 and u_1', such that[7] $(u_1, u_1)_1=\{u_1, u_1\}_1=\{u_1', u_1'\}_2=(u_1', u_1')_2=\beta_1\neq 0$. Let $\mathfrak{R}_1$ and $\mathfrak{R}_1'$ respectively denote the spaces of vectors orthogonal to u_1 and u_1' relative to $(x, y)_1$ and $(x', y')_2$. The space $\mathfrak{R}_1$ may also be characterized as the set of vectors orthogonal to u_1, u_1i if $\Phi=\Phi_0(i)$, or to u_1, u_1i, u_1j, u_1k, if $\Phi=\Phi_0(i, j)$, with respect to $\{x, y\}_1$. A similar

[5] *General Analysis*, I, American Philosophical Society Publication, Philadelphia, 1935.

[6] We make use of the relation $a=\alpha_0+i\alpha_1+j\alpha_2+k\alpha_3=(1/2)[\operatorname{tr} a-(i/\lambda)\operatorname{tr} ai-(j/\mu)\operatorname{tr} aj-(k/\lambda\mu)\operatorname{tr} ak]$.

[7] There exists a vector u_1 such that $(u_1, u_1)\neq 0$. Cf. Jacobson, loc. cit.

characterization holds for $\mathfrak{R}_1'$. The matrix of $\{x, y\}_1$ relative to u_1, $u_1 i$ or $u_1, u_1 i, u_1 j, u_1 k$ and of $\{x', y'\}_2$ relative to $u_1', u_1' i$ or $u_1', u_1' i, u_1' j, u_1' k$ is

$$(6) \qquad [\beta_1, \lambda\beta_1] \quad \text{or} \quad [\beta_1, \lambda\beta_1, \mu\beta_1, \lambda\mu\beta_1].$$

Hence it follows from a theorem of Witt[8] that $\{x, y\}_1$ and $\{x', y'\}_2$ are cogredient when restricted to $\mathfrak{R}_1$ and $\mathfrak{R}_1'$. By induction $(x, y)_1$ and $(x', y')_2$ are cogredient. Thus we have proved the following theorem:

THEOREM. *A necessary and sufficient condition that two hermitian forms $(x, y)_1$ and $(x, y)_2$ be cogredient is that the corresponding symmetric forms $\{x, y\}_1$ and $\{x, y\}_2$ be cogredient.*

If $u_1, u_2, \cdots, u_n$ is an orthogonal basis, $(u_i, u_i) = \beta_i$, then $u_1, u_1 i, u_2, u_2 i, \cdots, u_n, u_n i$ or $u_1, u_1 i, u_1 j, u_1 k, \cdots, u_n, u_n i, u_n j, u_n k$ is an orthogonal basis for $\mathfrak{R}$ over Φ relative to $\{x, y\}$ and the corresponding matrix, where B_i is as in (6), is

$$(7) \qquad [B_1, B_2, \cdots, B_n].$$

We consider now some special cases:

(1) $\Phi_0(i)$, *where Φ_0 is a field in which every nondegenerate symmetric form in 5 or more variables is a null-form.* Examples of such fields are (a) any p-adic field, (b) an algebraic function field of one variable over a finite constant field.[9] In these cases any nondegenerate symmetric form in 4 or more variables represents every $\alpha \neq 0$ in Φ_0. For, if $\{x, y\}$ represents 0, say $\{u, u\} = 0$, we choose v such that $\{u, v\} = \beta \neq 0$. Then $\{u\xi + v\eta, u\xi + v\eta\} = \eta(2\beta\xi + \gamma\eta)$, $\gamma = \{v, v\}$ and the equation $\eta(2\beta\xi + \gamma\eta) = \alpha$ can be solved for ξ, η in Φ_0. If $\{x, y\}$ does not represent 0, we form the vector space of $(n+1)$ dimensions by adjoining z to $\mathfrak{R}$, and define $\{x\xi + z\eta, x\rho + z\sigma\} = \{x, x\}\xi\rho - \alpha\eta\sigma$. Since this form represents 0, we have $\{x, x\}\xi^2 - \alpha\eta^2 = 0$ for $\eta \neq 0$ since $\{x, x\} \neq 0$. Thus $\{x\xi\eta^{-1}, x\xi\eta^{-1}\} = \alpha$. It follows that any hermitian form in a space of 2 or more dimensions represents any α in Φ_0. Hence we may choose $\beta_1 = \beta_2 = \cdots = \beta_{n-1} = 1$ in (2). Thus two forms are cogredient, if, and only if, they have the same discriminant.

(2) $\Phi_0(i, j)$, Φ_0 *of the same type as in case* (1). Here we may take $\beta_1 = \cdots = \beta_n = 1$ and hence all nondegenerate forms are cogredient.

(3) $\Phi_0(i)$, Φ_0 *a real closed field.* Here we may suppose $\lambda = 1$ and we may suppose $\beta_1 = \cdots = \beta_p = 1$, $\beta_{p+1} = \cdots = \beta_n = -1$. For $\{x, y\}$ we

---

[8] *Theorie der quadratischen Formen in beliebigen Körpern*, Journal für die reine und angewandte Mathematik, vol. 176 (1936–1937), p. 34.

[9] Witt, loc. cit., p. 40, and Albert, *Quadratic null forms over a function field*, Annals of Mathematics, (2), vol. 39 (1938), pp. 494–505.

obtain $2p$ values $+1$ and $(2n-2p)$ values -1 in the diagonal form. Since the signature is an invariant for bilinear forms it is invariant also for the hermitian form (x, y).

(4) $\Phi_0(i, j)$, *Φ_0 a real closed field.* The considerations are similar to case (3). We find that two nondegenerate hermitian forms are cogredient if and only if they have the same signatures.[10]

(5) $\Phi_0(i)$, *Φ_0 an algebraic number field.* As is well known, the symmetric forms $\{x, y\}_1$ and $\{x, y\}_2$ in $\mathfrak{R}$ over Φ_0 are cogredient, if, and only if, they are cogredient in every p-adic extension of Φ_0. Suppose first that p is a prime spot such that $(-\lambda/p)=1$, that is, $-\lambda$ is a square in the p-adic field[11] Φ_{0p}. Then the matrix B_i in (7) is cogredient in Φ_{0p} to $[\beta_i, -\beta_i]$ and hence also[12] to $[1, -1]$. Thus any two matrices of the form (7) are cogredient. If $(-\lambda/p)=-1$, $\Phi_{0p}(i)$ is a quadratic field over Φ_{0p}. Hence $\{x, y\}_1$ and $\{x, y\}_2$ are cogredient, if, and only if, $(x, y)_1$ and $(x, y)_2$ are cogredient in $\mathfrak{R}$ over $\Phi_{0p}(i)$. The condition for this is that the discriminants be the same when p is finite and the signatures be the same when p is infinite. Combining these results, we see that a necessary and sufficient condition that two nondegenerate hermitian forms in $\mathfrak{R}$ over Φ be cogredient is that they have the same discriminant and the same signature at the infinite prime spots for which λ is positive.[13]

(6) $\Phi_0(i, j)$, *Φ_0 an algebraic number field.* To obtain conditions for cogredience of $(x, y)_1$ and $(x, y)_2$ we again consider $\{x, y\}_1$ and $\{x, y\}_2$. Let p be a prime spot at which $((-\lambda, -\mu)/p)=1$, that is, $\Phi_{0p}(i, j)$ is a matrix algebra. Then either $-\lambda$ is a square in Φ_{0p} or $-\mu$ is a norm in $\Phi_{0p}(i)$. In the first case B_i is cogredient to $[\beta_i, -\beta_i, \mu\beta_i, -\mu\beta_i]$ and hence to $[1, -1, 1, -1]$. If $-\mu$ is a norm in $\Phi_{0p}(i)$, the bilinear form with matrix $[\beta_i, \lambda\beta_i, \mu\beta_i]$ represents 0 and hence is cogredient to $[1, -1, -\lambda\mu\beta_i]$, and again (6) is cogredient to (8). If p is a prime spot for which $\Phi_{0p}(i, j)$ is a division algebra, $(x, y)_1$ and $(x, y)_2$ are always cogredient, if p is finite, and these forms are cogredient for p infinite, if, and only if, they have the same signatures. Thus a necessary and sufficient condition that $(x, y)_1$ and $(x, y)_2$ in $\mathfrak{R}$ over Φ be cogredient is that these forms have the same signatures for all infinite prime spots for which $\Phi_{0p}(i, j)$ is a division algebra.

UNIVERSITY OF NORTH CAROLINA

[10] E. H. Moore, loc. cit., p. 193.

[11] See Witt and the references cited there to Hasse's papers.

[12] This is a consequence of Witt's theorem that any two symmetric forms in two variables which are nonsingular and represent 0 are cogredient (Witt, p. 34).

[13] Cf. Landherr, *Äquivalenz Hermitescher Formen über einem beliebigen algebraischen Zahlkörper*, Abhandlungen aus dem mathematischen Seminar der Hansischen Universität, vol. 11 (1936), pp. 245–248.

Reprinted from
Transactions of the American Mathematical Society, July 1941.

RESTRICTED LIE ALGEBRAS OF CHARACTERISTIC p

BY

N. JACOBSON

In an earlier paper(1) [3] we noted certain identities which connect addition, scalar multiplication, commutation ($[ab]=ab-ba$), and pth powers in an arbitrary associative algebra of characteristic p ($\neq 0$). These lead naturally to the definition of a class of abstract algebras called restricted Lie algebras which in many respects bear a closer relation to Lie algebras of characteristic 0 than ordinary Lie algebras of characteristic p.

As is shown in the present paper any restricted Lie algebra $\mathfrak{L}$ may be obtained from an associative algebra by using the operations mentioned above. In fact $\mathfrak{L}$ determines a certain associative algebra $\mathfrak{U}$, called its u-algebra, such that $\mathfrak{L}$ is isomorphic to a subalgebra of $\mathfrak{U}_l$, the restricted Lie algebra defined by $\mathfrak{L}$; and if $\mathfrak{B}$ is any associative algebra such that $\mathfrak{B}_l$ contains a subalgebra homomorphic to $\mathfrak{L}$ and $\mathfrak{B}$ is the enveloping algebra of this subset then $\mathfrak{U}$ is homomorphic to $\mathfrak{B}$. The algebra $\mathfrak{U}$ has an anti-automorphism relative to which the elements corresponding to those in $\mathfrak{L}$ are skew. For ordinary Lie algebras an algebra having these properties has been defined by G. Birkhoff [2] and by Witt [5]. In their case however, the associative algebra has an infinite basis even when the Lie algebra has a finite basis whereas here $\mathfrak{U}$ has a finite basis if and only if $\mathfrak{L}$ has. Consequently every restricted Lie algebra $\mathfrak{L}$ with a finite basis has a (1-1) representation by finite matrices. The theory of representations of $\mathfrak{L}$ can be reduced to that of the associative algebra $\mathfrak{U}$. Thus, for example, there are only a finite number of inequivalent irreducible representations.

The most natural way to obtain a restricted Lie algebra is as a derivation algebra of an arbitrary algebra $\mathfrak{A}$, i.e., as the set of transformations $D: a\rightarrow aD$ in $\mathfrak{A}$ such that

$$(a+b)D = aD + bD, \qquad (a\alpha)D = (aD)\alpha, \qquad (ab)D = (aD)b + a(bD).$$

If $\mathfrak{A}=\mathfrak{L}$ is itself restricted ($ab=[ab]$) the derivations which satisfy

$$a^pD = [[aD, \overbrace{a]\, a] \cdots a}^{p-1}]$$

are called restricted. They are precisely the derivations of $\mathfrak{L}$ which can be extended to derivations of the u-algebra $\mathfrak{U}$. Hence their totality is a restricted Lie algebra $\mathfrak{D}_0$. Using $\mathfrak{D}_0$ and $\mathfrak{L}$ we may define a restricted holomorph $\mathfrak{H}_0$ of $\mathfrak{L}$. $\mathfrak{H}_0$ is a restricted Lie algebra.

Presented to the Society, September 12, 1940; received by the editors June 30, 1940.

(1) Numbers in brackets refer to the bibliography at the end of the paper.

The considerations in the present paper apply for the most part to restricted Lie algebras with an infinite basis as well as to those with a finite basis. A special result, however, for Lie algebras with a finite basis is that the nilpotency of $\mathfrak{L}$ implies that of the u-algebra (§6). In a later paper we hope to discuss certain classes of simple restricted Lie algebras with a finite basis.

1. **Restricted Lie algebras. Definitions.** If $\mathfrak{A}$ is an associative algebra and the commutator $[ab]=ab-ba$, it is well known that

$$[ab] = -[ba], \qquad [a[bc]] + [b[ca]] + [c[ab]] = 0.$$

If, in addition $\mathfrak{A}$ has characteristic p ($\neq 0$) then the following identities hold:

$$(a+b)^p = a^p + b^p + s(a, b), \qquad [[\overbrace{ab]b]\cdots b}^{p}] = [ab^p],$$

where $s(a, b) = s_1(a, b) + s_2(a, b) + \cdots + s_{p-1}(a, b)$ and the $(p-i)s_i(a, b)$ is the coefficient of λ^{p-i-1} in

$$[\cdots[a, \lambda a + b], \lambda a + b], \cdots, \lambda a + b]$$

([3], and [6]). We are thus led to define a *restricted Lie algebra* $\mathfrak{L}$ over a field Φ of characteristic p as a vector space over Φ in which operations $[ab]$ and $a^{[p]}$ are defined such that

$$(1) \qquad [ab] = -[ba], \qquad [a[bc]] + [b[ca]] + [c[ab]] = 0,$$

$$(2) \qquad [a, b_1 + b_2] = [ab_1] + [ab_2],$$

$$(3) \qquad [ab]\alpha = [a, b\alpha] = [a\alpha, b], \qquad \alpha \text{ in } \Phi,$$

$$(4) \qquad (a+b)^{[p]} = a^{[p]} + b^{[p]} + s(a, b),$$

$$(5) \qquad (a\alpha)^{[p]} = a^{[p]}\alpha^p,$$

$$(6) \qquad [\cdots[\overbrace{ab]b]\cdots b}^{p}] = [ab^{[p]}].$$

A subspace $\mathfrak{B}$ of $\mathfrak{L}$ is a *subalgebra* if $\mathfrak{B} \supset b^{[p]}$ and $[b_1b_2]$ for all b, b_1, b_2 in $\mathfrak{B}$. $\mathfrak{B}$ is an *ideal* if it contains also $[ba]$ for all b in $\mathfrak{B}$, a in $\mathfrak{L}$. A correspondence $a \to a^S$ between two restricted Lie algebras is a *homomorphism* if

$$(7) \qquad \begin{gathered} (a+b)^S = a^S + b^S, \qquad (a\alpha)^S = a^S\alpha, \qquad [ab]^S = [a^Sb^S], \\ (a^{[p]})^S = (a^S)^{[p]}. \end{gathered}$$

If S is (1-1) it is an *isomorphism* and if besides S is a correspondence within $\mathfrak{L}$ it is an *automorphism*.

If $x_1, x_2, \cdots$ (possibly infinite) is a basis for $\mathfrak{L}$, then $[x_ix_j] = \sum x_q\gamma_{qij}$, $x_i^{[p]} = \sum x_r\mu_{ri}$ (finite sums) where

$$(8) \qquad \gamma_{qij} = -\gamma_{qji}, \qquad \sum_q \gamma_{rqk}\gamma_{qij} + \sum_q \gamma_{rqi}\gamma_{qjk} + \sum_q \gamma_{rqj}\gamma_{qki} = 0,$$

$$(9) \qquad \sum_r \gamma_{sir}\mu_{rj} = \sum_q \gamma_{sq_{p-1}j}\gamma_{q_{p-1}q_{p-2}j}\cdots\gamma_{q_1ij}.$$

These equations are equivalent to

$$[x_ix_j] = -[x_jx_i],$$

$$[[x_ix_j]x_k] + [[x_jx_k]x_i] + [[x_kx_i]x_j] = 0, \quad [x_ix_j^{[p]}] = [\cdots[x_ix_j]\cdots x_j],$$

respectively. The γ's and μ's are the *constants of multiplication* of $\mathfrak{L}$. If $a \to a^S$ is an isomorphism between $\mathfrak{L}$ and $\mathfrak{L}^S$, $x_1^S, x_2^S, \cdots$ form a basis for $\mathfrak{L}^S$ and the x_i^S have the same constants of multiplication as the x_i's. On the other hand if $\mathfrak{M}$ is any restricted Lie algebra with basis $y_1, y_2, \cdots$ in (1-1) correspondence with the x_i such that $[y_iy_j] = \sum y_q\gamma_{qij}$, $y_i^{[p]} = \sum y_r\mu_{ri}$ then it is readily seen that the correspondence $\sum x_i\alpha_i \to \sum y_i\alpha_i$ is an isomorphism.

We have noted above that any associative algebra $\mathfrak{A}$ of characteristic p becomes a restricted Lie algebra $\mathfrak{A}_l$ when $[ab]$ is defined as $ab-ba$ and $a^{[p]} = a^p$. A homomorphism between $\mathfrak{L}$ and a subalgebra of Φ_{nl}, Φ_n the algebra of $n \times n$ matrices, is called a *representation*. Irreducibility, decomposability, equivalence, etc., of representations are defined as usual. As is well known these depend on the irreducibility, etc., of the enveloping algebra of the representing matrices.

2. **The u-algebra of a restricted Lie algebra.** Let $x_1, x_2, \cdots$ (possibly infinite) be a basis over Φ of a vector space $\mathfrak{L}$. We set $[x_ix_j] = \sum x_q\gamma_{qij}$ and $x_i^{[p]} = \sum x_r\mu_{ri}$ where the γ's are μ's satisfy (8) and (9). Then for $a = \sum x_i\alpha_i$, $b = \sum x_i\beta_i$ (finite sums) we set $[ab] = \sum x_k\gamma_{kij}\alpha_i\beta_j$. Evidently (8) implies that $\mathfrak{L}$ is a Lie algebra relative to $[ab]$. Equation (9) is equivalent to

$$[x_ix_j^{[p]}] = [\cdots \overbrace{[x_ix_j]\cdots x_j}^{p}].$$

We shall show that $\mathfrak{L}$ is a restricted Lie algebra relative to a suitable definition of $a^{[p]}$.

Let $\mathfrak{A}$ be the vector space with the basis $x_1^{\kappa_1}x_2^{\kappa_2}\cdots x_n^{\kappa_n}$, $\kappa_i \geqq 0$ integers and at least one $\kappa_i > 0$, $n = 1, 2, \cdots$. If only a finite number of monomials are being considered we may write them in terms of the same x's. A product

$$(x_1^{\kappa_1}x_2^{\kappa_2}\cdots x_n^{\kappa_n})(x_1^{\lambda_1}x_2^{\lambda_2}\cdots x_n^{\lambda_n})$$

is defined by repeated "straightenings," i.e., substitutions for x_ix_j when $i > j$ of the expression $x_jx_i + \sum x_k\gamma_{kij}$. It has been shown by G. Birkhoff [2] and by Witt [5] that this product is uniquely defined in $\mathfrak{A}$ and is associative.

Let $\mathfrak{B}$ be the ideal in $\mathfrak{A}$ having the basis $y_i = x_i^p - x_i^{[p]}$. Since

$$[bx_i^p] = [\cdots[bx_i]x_i]\cdots x_i] = [bx_i^{[p]}], \tag{10}$$

y_i commutes with every linear b and hence with every element of $\mathfrak{A}$.

If the term $x_1^{\kappa_1}\cdots x_n^{\kappa_n}$ has degree $\geqq p$ in x_i we may replace x_i^p by $y_i + x_i^{[p]}$. After a finite number of such substitutions we may write any $a = \sum x_1^{\kappa_1}\cdots x_n^{\kappa_n}$

$\cdot\rho_{\kappa_1\cdots\kappa_n}$ in the form $\sum x_1^{\lambda_1}\cdots x_n^{\lambda_n}u_{\lambda_1\cdots\lambda_n}$ where $\lambda_i<p$ and the u's are polynomials in $y_1, y_2, \cdots, y_n$. Thus any $b=\sum a_iy_i+\sum y_i\alpha_i$ in $\mathfrak{B}$ has the form $\sum x_1^{\lambda_1}\cdots x_n^{\lambda_n}\cdot v_{\lambda_1\cdots\lambda_n}$, v a polynomial in $y_1, y_2, \cdots, y_n$ with no constant term. Now

$$\begin{aligned} x_1^{\lambda_1}\cdots x_n^{\lambda_n}y_1^{m_1}\cdots y_n^{m_n} &= x_1^{\lambda_1}(x_1^p-x_1^{[p]})^{m_1}\cdots x_n^{\lambda_n}(x_1^p-x_1^{[p]})^{m_n}\\ &= x_1^{\lambda_1+pm_1}\cdots x_n^{\lambda_n+pm_n}+\cdots \end{aligned}$$

where the terms not indicated have degree $<\sum\lambda_i+p\sum m_i$. Consider the terms of maximum degree $N=\sum\lambda_i+p\sum m_i$ in b where we suppose $b\neq 0$ and hence one of the v's is $\neq 0$. Since at least one $m>0$, $N>p$. Two terms of maximum degree are different if $(\lambda_1, \cdots, \lambda_n; m_1, \cdots, m_n)\neq(\lambda_1', \cdots, \lambda_n'; m_1', \cdots, m_n')$. Hence these terms occur only once and can not cancel off. It follows that when b is written in its normal form $\sum x_1^{\kappa_1}\cdots x_n^{\kappa_n}\rho_{\kappa_1\cdots\kappa_n}$ at least one of the x's has degree $>p$. Thus the classes $\{x_1^{\lambda_1}\cdots x_n^{\lambda_n}\}$, $\lambda_i<p$, omitting $\{x_1^0\cdots x_n^0\}$, determined by the elements $x_1^{\lambda_1}\cdots x_n^{\lambda_n}$ modulo $\mathfrak{B}$ form a basis for the difference algebra $\mathfrak{U}=\mathfrak{A}/\mathfrak{B}$.

Since the classes $\{x_1\}, \{x_2\}, \cdots$ are linearly independent and $[\{x_i\}\{x_j\}]=\sum\{x_q\}\gamma_{qij}$ the correspondence $\sum x_i\alpha_i\rightarrow\sum\{x_i\}\alpha_i$ is an isomorphism between $\mathfrak{L}$ and an ordinary Lie subalgebra $\{\mathfrak{L}\}$ of $\mathfrak{U}_l$. Since $\{x_i\}^p=\{x_i^p\}=\{x_i^{[p]}\}=\sum\{x_r\}\mu_{ri}$, $\{\mathfrak{L}\}$ is a (restricted) subalgebra of $\mathfrak{U}_l$. Hence if we define $(\sum x_i\alpha_i)^{[p]}$ as the element corresponding to $\{\sum x_i\alpha_i\}^p$, $\mathfrak{L}$ becomes a restricted Lie algebra in which $x_i^{[p]}$ is as originally given.

If $\mathfrak{L}$ is a restricted Lie algebra to begin with, then the correspondence $\sum x_i\alpha_i\rightarrow\{\sum x_i\alpha_i\}$ is an isomorphism. Now suppose $\sum x_i\alpha_i\rightarrow\sum\bar{x}_i\alpha_i$ is a homomorphism between $\mathfrak{L}$ and $\bar{\mathfrak{L}}$, a subalgebra of $\mathfrak{B}_l$, where we suppose that any element of $\mathfrak{B}$ is a polynomial in the elements of $\bar{\mathfrak{L}}$. We have

$$\bar{x}_i\bar{x}_j=\bar{x}_j\bar{x}_i+\sum\bar{x}_q\gamma_{qij},\qquad \bar{x}_i^p=\sum\bar{x}_r\mu_{ri}.$$

The first set of equations implies that $\sum x_1^{\kappa_1}\cdots x_n^{\kappa_n}\rho_{\kappa_1\cdots\kappa_n}\rightarrow\sum\bar{x}_1^{\kappa_1}\cdots\bar{x}_n^{\kappa_n}\rho_{\kappa_1\cdots\kappa_n}$ is a homomorphism between $\mathfrak{A}$ and $\mathfrak{B}$(2). Because of the second set of equations $x_i^p-x_i^{[p]}$ are mapped into 0 and so our correspondence induces a homomorphism between $\mathfrak{U}=\mathfrak{A}/\mathfrak{B}$ and $\mathfrak{B}$. This mapping is an extension of the homomorphism between $\mathfrak{L}$ and $\bar{\mathfrak{L}}$. Thus we have proved the following theorem.

THEOREM 1. *If $\mathfrak{L}$ is a restricted Lie algebra there exists an associative algebra $\mathfrak{U}$ having the following properties*: 1. *$\mathfrak{L}$ is isomorphic to a subalgebra $\{\mathfrak{L}\}$ of $\mathfrak{U}_l$.* 2. *$\mathfrak{U}$ is the enveloping algebra of $\{\mathfrak{L}\}$.* 3. *If $\mathfrak{L}$ is homomorphic to a subalgebra $\bar{\mathfrak{L}}$ of any $\mathfrak{B}_l$ where $\mathfrak{B}$ is the enveloping algebra of $\bar{\mathfrak{L}}$, then $\mathfrak{U}$ is homomorphic to $\mathfrak{B}$.*

This theorem shows that conditions (1) to (6) are characteristic of the functions $a+b$, $a\alpha$, $[ab]$ and a^p in an associative algebra of characteristic p. The above considerations show also that equations (8) and (9) on the con-

(2) See [2] or [5].

stants γ and μ insure that the vector space $\mathfrak{L}$ be a restricted Lie algebra. If P is an extension of the field Φ the extended vector space of elements of the form $\sum x_i\xi_i$, ξ in P, is a Lie algebra over P. We denote this extended algebra as $\mathfrak{L}P$ since, as is easily shown, it does not depend on the particular choice of basis. In the remainder of the paper we shall denote $a^{[p]}$ by a^p when there is no risk of confusion and shall call $\mathfrak{U}$ the u-algebra of $\mathfrak{L}$.

Suppose $x_{i_1} \cdots x_{i_m}$ is a monomial in $\mathfrak{A}$ and the first x_1 in this product occurs in the r_1th place. Then we may straighten this term by interchanging x_1 successively with the r_1-1 terms in front of it and obtain a polynomial in the x's having one term $x_1x_{j_2} \cdots x_{j_m}$ of degree m. We define the rank of $x_{i_1} \cdots x_{i_m}$ inductively as (r_1-1) plus rank of $x_{j_2} \cdots x_{j_m}$. A monomial of rank 0 is said to be in canonical form. Then $i_1 \geqq i_2 \geqq \cdots \geqq i_m$. Consider the correspondence $a=\sum x_1^{\kappa_1} \cdots x_n^{\kappa_n}\rho_{\kappa_1\cdots\kappa_n} \to a^J=\sum(-1)^{\kappa_1+\cdots+\kappa_n}x_n^{\kappa_n} \cdots x_1^{\kappa_1}\rho_{\kappa_1\cdots\kappa_n}$. Evidently J is linear. We wish to show that it is an anti-automorphism. For this purpose it suffices to prove that

$$(x_{i_1} \cdots x_{i_m})^J = (-1)^m x_{i_m} \cdots x_{i_1}.$$

Suppose this holds for all products of $(m-1)$ or less x's and also for products of m x's whose ranks are less than those of the given monomial. Then

$$(x_{i_1} \cdots x_{i_m}) = (x_{i_1} \cdots x_{i_{j+1}}x_{i_j} \cdots x_{i_m}) + (x_{i_1} \cdots [x_{i_j}x_{i_{j+1}}] \cdots x_{i_m}),$$

where we may suppose that if the rank r of the original term is >0 that of $x_{j_1} \cdots x_{i_{j+1}}x_{i_j} \cdots x_{i_m}$ is $r-1$. We have

$$\begin{aligned}(x_{i_1} \cdots x_{i_m})^J &= (x_{i_1} \cdots x_{i_{j+1}}x_{i_j} \cdots x_{i_m})^J + (x_{i_1} \cdots [x_{i_j}x_{i_{j+1}}] \cdots x_{i_m})^J \\ &= (-1)^m(x_{i_m} \cdots x_{i_j}x_{i_{j+1}} \cdots x_{i_1}) \\ &\quad + (-1)^{m-1}(x_{i_m} \cdots [x_{i_j}x_{i_{j+1}}] \cdots x_{i_1}) \\ &= (-1)^m(x_{i_m} \cdots x_{i_1}).\end{aligned}$$

Since $(x_i^{[p]}-x_i^p)^J = -(x_i^{[p]}-x_i^p)$, J sends the ideal $\mathfrak{B}$ into itself and therefore induces an anti-automorphism in $\mathfrak{U}=\mathfrak{A}/\mathfrak{B}$. The elements $\sum\{x_i\}\alpha_i$ of $\{\mathfrak{L}\}$ are skew relative to the anti-automorphism.

By property (3) of $\mathfrak{U}$ any representation of $\mathfrak{L}$ determines a representation of $\mathfrak{U}$ and conversely. Questions of irreducibility, equivalence, etc., for $\mathfrak{L}$ are reducible to the corresponding questions for $\mathfrak{U}$. If $\mathfrak{L}$ has a finite basis $x_1, x_2, \cdots, x_n$, $\mathfrak{U}$ has the basis $\{x_1^{\lambda_1} \cdots x_n^{\lambda_n}\}$ $(\lambda_i<p)$ of p^n-1 elements. Since $\mathfrak{U}$ has a (1-1) representation in some Φ_m, $m \leqq p^n$, the same is true for $\mathfrak{L}$.

THEOREM 2. *Every restricted Lie algebra with a finite basis has a* (1-1) *representation.*

The number of inequivalent irreducible representations $(\neq 0)$ of $\mathfrak{U}$ is equal to the number of simple components of $\mathfrak{U}/\mathfrak{N}$, $\mathfrak{N}$ the radical of $\mathfrak{U}$. This implies

THEOREM 3. *There are only a finite number of inequivalent irreducible representations of a restricted Lie algebra with a finite basis.*

Example. $\mathfrak{L}$, the restricted Lie algebra with the basis x, y, z such that

$$[xy] = z, \qquad [xz] = [yz] = 0, \qquad x^p = y^p = z^p = z.$$

$\mathfrak{U}$ has the basis $x^i y^j z^k$, $i, j, k < p$ such that the above relations hold, it being understood that $[xy] = xy - yx$, etc. It is readily proved that $\mathfrak{U}$ is a direct sum of p algebras with bases $x^i y^j$ such that

$$xy - yx = \zeta, \qquad x^p = y^p = \zeta,$$

where $\zeta = 0, 1, \cdots, p-1$ in turn. If $\zeta = 0$ this algebra is nilpotent. Otherwise it is isomorphic to Φ_p. Hence there are $(p-1)$ inequivalent irreducible representations of $\mathfrak{L}$.

Theorem 2 is valid for Lie algebras of characteristic 0 though its proof given by Ado [1] is considerably more complicated than the present one. Theorem 3 is not true for algebras of characteristic 0. It is not known whether either of these results holds for ordinary Lie algebras of characteristic p.

3. **Ideals. Nilpotency.** If $\mathfrak{B}$ is an ideal in $\mathfrak{L}$ we define the sum, scalar product and commutator of the classes $\bar{a}$ modulo $\mathfrak{B}$ as usual by

$$\bar{a}_1 + \bar{a}_2 = \overline{a_1 + a_2}, \qquad \bar{a}\alpha = \overline{a\alpha}, \qquad [\bar{a}_1\bar{a}_2] = [\overline{a_1 a_2}].$$

If $a_1 - a_2 = b \in \mathfrak{B}$ then $a_1^p = a_2^p + b^p + s(a_2, b)$. Since b^p and $s(a_2, b) \in \mathfrak{B}$ we see that $a_1 \equiv a_2(\mathfrak{B})$ implies $a_1^p \equiv a_2^p(\mathfrak{B})$. Hence the definition $(\bar{a})^p = \overline{a^p}$ is unambiguous and together with the above operations it defines $\bar{\mathfrak{L}} = \mathfrak{L}/\mathfrak{B}$, the *difference algebra* of $\mathfrak{L}$ relative to $\mathfrak{B}$, as a restricted Lie algebra. The correspondence $a \rightarrow \bar{a}$ is a homomorphism between $\mathfrak{L}$ and $\bar{\mathfrak{L}}$. Conversely we may show in the usual manner that if $\mathfrak{L}$ is homomorphic to the restricted Lie algebra $\bar{\mathfrak{L}}$, $\bar{\mathfrak{L}} = \mathfrak{L}/\mathfrak{B}$ where $\mathfrak{B}$ is the set of elements mapped into 0 by the homomorphism.

If $\mathfrak{B}_1$ and $\mathfrak{B}_2$ are subspaces of $\mathfrak{L}$ we denote their sum as $\mathfrak{B}_1 + \mathfrak{B}_2$ and their commutator, i.e., the smallest space containing all $[b_1 b_2]$, $b_1 \in \mathfrak{B}_1$, $b_2 \in \mathfrak{B}_2$ by $[\mathfrak{B}_1\mathfrak{B}_2]$. Conditions (1), (2) and (3) imply

$$[\mathfrak{B}_1\mathfrak{B}_2] = [\mathfrak{B}_2\mathfrak{B}_1], \qquad [\mathfrak{B}_1[\mathfrak{B}_2\mathfrak{B}_3]] \leqq [\mathfrak{B}_2[\mathfrak{B}_3\mathfrak{B}_1]] + [\mathfrak{B}_3[\mathfrak{B}_1\mathfrak{B}_2]]. \tag{11}$$

Set $\mathfrak{L}^{[2]} = [\mathfrak{L}\mathfrak{L}], \cdots, \mathfrak{L}^{[i]} = [\mathfrak{L}^{[i-1]}\mathfrak{L}]$ and define $\mathfrak{L}^{p^k}$ to be the smallest subspace containing all a^{p^k} where $a^{p^k} = (a^{p^{k-1}})^p$. Thus $\mathfrak{L} \geqq \mathfrak{L}^{[2]} \geqq \mathfrak{L}^{[3]} \geqq \cdots$ and $\mathfrak{L} \geqq \mathfrak{L}^p \geqq \mathfrak{L}^{p^2} \geqq \cdots$, $\mathfrak{L}^{p^2} \leqq (\mathfrak{L}^p)^p$, etc. Hence if we define

$$\mathfrak{L}_i = \mathfrak{L}^{[i]} + (\mathfrak{L}^{[i-1]})^p + (\mathfrak{L}^{[i-2]})^{p^2} + \cdots + \mathfrak{L}^{p^{i-1}}$$

we have $\mathfrak{L} = \mathfrak{L}_1 \geqq \mathfrak{L}_2 \geqq \mathfrak{L}_3 \geqq \cdots$. By induction on j one readily establishes

$$[\mathfrak{L}^{[i]}\mathfrak{L}^{[j]}] \leqq \mathfrak{L}^{[i+j]}.$$

Since

$$[b_1^{p^k}, b_2^{p^l}] = [\cdots [b_1^{p^k} \overbrace{b_2]b_2] \cdots b_2}^{p^l}]$$

$$= [[\cdots \overbrace{[b_1[b_1 \cdots [b_1}^{p^k} b_2] \cdots], \overbrace{b_2]b_2] \cdots b_2}^{p^l-1}]$$

we have

$$[\mathfrak{L}^{[i-k]p^k}\mathfrak{L}^{[j-l]p^l}] \leqq \mathfrak{L}^{[(i-k)p^k+(j-l)p^l]} \leqq \mathfrak{L}^{[i+j]}. \tag{12}$$

This leads readily to

$$[\mathfrak{L}_i\mathfrak{L}_j] \leqq \mathfrak{L}_{i+j}, \qquad \mathfrak{L}_i^p \leqq \mathfrak{L}_{i+1}. \tag{13}$$

In particular $\mathfrak{L}_i$ is an ideal in $\mathfrak{L}$ and $\mathfrak{L}_i \geqq [\mathfrak{L}_{i-1}\mathfrak{L}] + \mathfrak{L}_{i-1}^p$. On the other hand from the definition of $\mathfrak{L}_i$ we have $\mathfrak{L}_i \leqq [\mathfrak{L}_{i-1}\mathfrak{L}] + \mathfrak{L}_{i-1}^p$. Hence $\mathfrak{L}_i = [\mathfrak{L}_{i-1}\mathfrak{L}] + \mathfrak{L}_{i-1}^p$. It follows that if $\mathfrak{L}_{i-1} = \mathfrak{L}_i$, $\mathfrak{L}_{i-1} = \mathfrak{L}_i = \mathfrak{L}_{i+1} = \cdots$. If $\mathfrak{L}_N = 0$ for N sufficiently large, $\mathfrak{L}$ is *nilpotent*. The smallest N for which this holds is called the *index* of nilpotency. If $\mathfrak{L}_N = 0$, $\mathfrak{L}^{p^{N-1}} = 0$ and $\mathfrak{L}^{[N]} = 0$. Conversely if $\mathfrak{L}^{[r]} = 0$ and $\mathfrak{L}^{p^s} = 0$ it is readily seen that $\mathfrak{L}_t = 0$ for $t = r + s - 1$.

4. **Restricted derivations.** The most natural instances of restricted Lie algebras are the derivation algebras(3). We recall that if $\mathfrak{A}$ is an arbitrary algebra (not necessarily associative) then a derivation D is defined to be a transformation $a \to aD$ in $\mathfrak{A}$ such that

$$(a + b)D = aD + bD, \quad (a\alpha)D = (aD)\alpha, \quad (ab)D = (aD)b + a(bD). \tag{14}$$

If $\mathfrak{A}$ has characteristic p the set $\mathfrak{D}$ of these transformations is closed under addition, scalar multiplication, commutation and pth powers. Thus $\mathfrak{D}$ is a restricted Lie algebra. If $\mathfrak{A}$ is associative

$$a^pD = (aD)a^{p-1} + a(aD)a^{p-2} + \cdots + a^{p-1}(aD) = [\cdots [aD, \overbrace{a]a] \cdots a}^{p-1}](^4).$$

It is therefore natural to confine our attention in the case that $\mathfrak{A} = \mathfrak{L}$ is a restricted Lie algebra to the derivations called *restricted* such that

$$a^pD = [\cdots [aD\overbrace{a]a] \cdots a}^{p-1}]. \tag{15}$$

Suppose D is a linear transformation in $\mathfrak{L}$ such that $[x_ix_j]D = [x_iD, x_j] + [x_i, x_jD]$, $x_1, x_2, \cdots$ a basis for $\mathfrak{L}$. Then D is a derivation. If $x_1^{\kappa_1} \cdots x_n^{\kappa_n}$, $\kappa_i = 1, 2, \cdots$, is a basis for the Birkhoff-Witt algebra $\mathfrak{A}$ we define the linear transformation D in $\mathfrak{A}$ by setting

(3) Cf. Jacobson [3].

(4) In general

$$ba^{p-1} + aba^{p-2} + \cdots + a^{p-1}b = [\cdots [b\overbrace{a]a] \cdots a}^{p-1}].$$

See [3, p. 209].

$$(16)\qquad \begin{aligned}(x_1^{\kappa_1}\cdots x_n^{\kappa_n})D = (x_1D)x_1^{\kappa_1-1}\cdots x_n^{\kappa_n} + \cdots + x_1^{\kappa_1-1}(x_1D)\cdots x_n^{\kappa_n}\\ + \cdots + x_1^{\kappa}\cdots x_n^{\kappa_n-1}(x_nD).\end{aligned}$$

By an induction similar to that of §2 we can show that

$$(17)\qquad (x_{i_1}\cdots x_{i_m})D = (x_{i_1}D)x_{i_2}\cdots x_{i_m} + \cdots + x_{i_1}\cdots(x_{i_m}D),$$

holds. Hence D is a derivation.

Now suppose

$$x_i^{[p]}D = [\cdots[x_iD, \overbrace{x_i]\cdots x_i}^{p-1}].$$

Then $(x_i^{[p]}-x_i^p)D=0$ and D maps the ideal $\mathfrak{B}$ whose basis is $x_i^{[p]}-x_i^p$ into itself. It follows that D induces a derivation in $\mathfrak{U}=\mathfrak{A}/\mathfrak{B}$ and hence is a restricted derivation in $\mathfrak{L}$. We have shown also that any restricted derivation is determined by a derivation of the u-algebra. The converse of this is clear. Hence we have proved

THEOREM 4. *A linear transformation D in $\mathfrak{L}$ is a restricted derivation if and only if $[x_ix_j]D=[x_iD, x_j]+[x_i, x_jD]$ and $x_i^pD=[\cdots[x_iD, x_i]\cdots x_i]$ for any basis $x_1, x_2, \cdots$. Every restricted derivation is induced by a derivation of the associative u-algebra $\mathfrak{U}$ of $\mathfrak{L}$. The set of restricted derivations forms a restricted Lie algebra.*

The last statement follows immediately from the second. The set of restricted derivations will be denoted by $\mathfrak{D}_0$. A consequence of the first part of Theorem 4 is that the restricted derivation algebra of $\mathfrak{L}_P$ is $\mathfrak{D}_{0P}$.

For any three elements a, b, l in $\mathfrak{L}$ we have

$$[a+b, l] = [al]+[bl],\ [a\alpha, l] = [al]\alpha,\ [[ab]l] = [[al]b]+[a[bl]],$$

$$[a^pl] = [\cdots[al]\overbrace{a]a]\cdots a}^{p-1}].$$

Thus the transformations $L: a\rightarrow[al]$ are restricted derivations which we call *inner*. The other parts of (1) to (6) show that the derivations corresponding to l_1+l_2, $l\alpha$, $[l_1l_2]$, l^p are respectively L_1+L_2, $L\alpha$, $[L_1L_2]$, L^p. Hence the inner derivations form a subalgebra $\mathfrak{J}$ of $\mathfrak{D}_0$ and $\mathfrak{L}$ is homomorphic to $\mathfrak{J}$ under the correspondence $l\rightarrow L$. The elements c mapped into 0 are those which satisfy $[ac]=0$ for all a and form the *center* $\mathfrak{C}$ of $\mathfrak{L}$. Hence $\mathfrak{J}\cong\mathfrak{L}/\mathfrak{C}$. Since

$$[al]D - [aD, l] = [a, lD],$$

$[LD]\in\mathfrak{J}$ for every L in $\mathfrak{J}$ and D in $\mathfrak{D}_0$, i.e., $\mathfrak{J}$ is an ideal.

Let $\mathfrak{H}_0$ be the vector space which is a direct sum of $\mathfrak{L}$ and $\mathfrak{D}_0$. The elements U of $\mathfrak{H}_0$ are uniquely representable in the form $a+D$, a in $\mathfrak{L}$, D in $\mathfrak{D}_0$. Hence if $x_1, x_2, \cdots$ is a basis for $\mathfrak{L}$ and $D_1, D_2, \cdots$ one for $\mathfrak{D}_0$, $x_1, x_2, \cdots; D_1, D_2, \cdots$ is a basis for $\mathfrak{H}_0$.

We define commutation in $\mathfrak{H}_0$ by

$$(18) \qquad [a + D, b + E] = [ab] + aE - bD + [DE].$$

It is readily seen that this satisfies conditions (1), (2), (3)[5]. We also have

$$[\cdots[\overbrace{ax_i]\cdots x_i}^{p}] = [ax_i^p], \qquad [\cdots[\overbrace{Dx_i]\cdots x_i}^{p}] = [Dx_i^p],$$

$$[\cdots[aD_j]\cdots D_j] = [aD_j^p], \qquad [\cdots[DD_j]\cdots D_j] = [DD_j^p],$$

or

$$[\cdots[Ux_i]\cdots x_i] = [Ux_i^p], \qquad [\cdots[UD_j]\cdots D_j] = [UD_j^p],$$

for all U in $\mathfrak{H}_0$. Hence, by §2, the definition

$$(19) \qquad \left(\sum x_i\alpha_i + \sum D_j\beta_j\right)^p = \sum x_i^p\alpha_i^p + \sum D_j^p\beta_j^p + s(x, D)$$

turns $\mathfrak{H}_0$ into a restricted Lie algebra called the *restricted holomorph* of $\mathfrak{L}$. Thus $(a+D)^p = a^p + D^p + s(a, D)$ in $\mathfrak{H}_0$.

5. **Relations to ordinary Lie algebras.** Suppose $\mathfrak{L}_1$ and $\mathfrak{L}_2$ are restricted Lie algebras and $a \to a^S$ is a mapping of $\mathfrak{L}_1$ into $\mathfrak{L}_2$ such that

$$(a + b)^S = a^S + b^S, \qquad (a\alpha)^S = a^S\alpha, \qquad [ab]^S = [a^Sb^S].$$

Then

$$[\cdots[xa]\cdots a]^S = [\cdots[x^Sa^S]\cdots a^S] = [x^S(a^S)^p],$$
$$[x, a^p]^S = [x^S(a^p)^S].$$

Hence $(a^p)^S - (a^S)^p \in \mathfrak{C}_2$ the center of $\mathfrak{L}_2$. If $\mathfrak{C}_2 = 0$, S is a homomorphism. Next we suppose D is a derivation in $\mathfrak{L} = \mathfrak{L}_1$. This implies that

$$\begin{aligned}[\cdots[xa]\cdots a]D &= [\cdots[xDa]\cdots a] + [\cdots[[x, aD]a]\cdots a] \\ &\quad + \cdots + [\cdots[[xa]a]\cdots aD] \\ &= [xD, a^p] + [x, [\cdots[aD, a]\cdots a]],\end{aligned}$$

since if A is the transformation $x \to [xa]$ and B is the transformation $x \to [x, aD]$,

$$BA^{p-1} + ABA^{p-2} + \cdots + A^{p-1}B = [\cdots[[BA]A]\cdots A]\text{[6]}.$$

On the other hand $[xa^p]D = [xD, a^p] + [x, a^pD]$. Hence $a^pD - [\cdots[aD, a]\cdots a]$ is in the center $\mathfrak{C}$ of $\mathfrak{L}$. If $\mathfrak{C} = 0$ every derivation is restricted.

If $\mathfrak{B}$ is a subset of $\mathfrak{L}$ closed with respect to addition, scalar multiplication and commutation, then $\mathfrak{B}^* = \mathfrak{B} + \mathfrak{B}^p + \mathfrak{B}^{p^2} + \cdots$ is the smallest subalgebra of $\mathfrak{L}$ containing $\mathfrak{B}$. If $[\mathfrak{B}\mathfrak{L}] \leqq \mathfrak{B}$, $\mathfrak{B}^*$ is an ideal.

---

[5] Cf. Zassenhaus [6, p. 57].

[6] See Footnote 4.

An ordinary Lie algebra $\mathfrak{A}$ whose center is 0 may always be imbedded in a restricted Lie algebra. For let $\mathfrak{D}$ be the derivation algebra of $\mathfrak{A}$. Since the center of $\mathfrak{A}$ is 0 the set $\mathfrak{J}$ of inner derivations forms an ideal of $\mathfrak{D}$ (regarded as an ordinary Lie algebra) isomorphic to $\mathfrak{A}$. If $\mathfrak{J}^* = \mathfrak{J} + \mathfrak{J}^p + \cdots$, $\mathfrak{J}^*$ is a restricted Lie algebra containing $\mathfrak{J}$. Since $[u^{p^k} v^{p^l}] = [\cdots [[u[\cdots [uv]\cdots],$ $v]\cdots v] \in \mathfrak{J}$ for any u, v in $\mathfrak{J}$ the ordinary Lie algebra $\mathfrak{J}^*/\mathfrak{J}$ is commutative.

Suppose $\mathfrak{L}$ is a restricted Lie algebra and $a \rightarrow A$ an absolutely irreducible representation of the ordinary Lie algebra determined by $\mathfrak{L}$, i.e., $a\alpha \rightarrow A\alpha$, $a+b \rightarrow A+B$, $[ab] \rightarrow [AB]$. We assume also that the representing matrices all have trace 0 and $p \nmid m$ where $m \times m$ are the dimensions of the matrices. We assert that our representation is one of the restricted Lie algebra. For if $a^p \rightarrow B$ we have $[XB] = [\cdots [XA]\cdots A] = [XA^p]$. Hence $A^p - B = 1\rho$ and since tr $A^p =$ tr $B = 0$, $A^p = B$.

6. **Algebras with a finite basis.** In this section we suppose $\mathfrak{L}$ has a finite basis. For any element a there is a least integer m such that $a, a^p, \cdots, a^{p^{m-1}}$ are linearly independent but a^{p^m} depends on $a, \cdots, a^{p^{m-1}}$. Then $a^{p^m} + a^{p^{m-1}}\alpha_1 + \cdots + a\alpha_m = 0$ or $f(a) = 0$ where $f(\lambda) = \lambda^{p^m} + \lambda^{p^{m-1}}\alpha_1 + \cdots + \lambda\alpha_m$. It follows that $a^{p^{m+1}} = (a^{p^m})^p$, $a^{p^{m+1}}, \cdots$ are linear combinations of $a, \cdots, a^{p^{m-1}}$ and hence these elements form a basis for the subalgebra generated by a.

A polynomial having the form of $f(\lambda)$ has been called a p-polynomial by Ore. The following facts were established by Ore: (1) A necessary and sufficient condition that $f(\lambda)$ be a p-polynomial is that its roots form a modulus (group under addition) with each root having multiplicity p^k, k fixed. (2) Any polynomial $\phi(\lambda)$ is a factor of a p-polynomial $f(\lambda)$. The $f(\lambda)$ of least degree with leading coefficient 1 is unique and is a divisor of any other p polynomial divisible by $\phi(\lambda)$(7). Thus suppose $a \rightarrow A$ is a (1-1) representation of $\mathfrak{L}$ and $\phi(\lambda)$ is the minimum polynomial of the linear transformation (or matrix) A. Then the p-polynomial $f(\lambda)$ associated with a is the one of least degree divisible by $\phi(\lambda)$.

An element a is *nilpotent* if $a^{p^m} = 0$ for some p^m. The least integer p^m for which this holds is the *index* of a.

THEOREM 5. *If $\mathfrak{L}$ is a Lie algebra with a finite basis and contains only nilpotent elements, then $\mathfrak{L}$ is nilpotent.*

Since $\mathfrak{L}$ has a finite basis the index of any a is $\leqq p^n$ where n is the dimensionality of $\mathfrak{L}$. Hence $a^{p^n} = 0$ and if A is the linear transformation $x \rightarrow [xa]$, $A^{p^n} = 0$. Thus $\mathfrak{L}^{p^n} = 0$ and as has been shown by Zorn [8], $\mathfrak{L}$ is nilpotent when regarded as an ordinary Lie algebra, i.e., $\mathfrak{L}^{[m]} = 0$. It follows as in §4 that $\mathfrak{L}$ is nilpotent.

The algebra $\mathfrak{L} > \mathfrak{L}_2 = \mathfrak{L}^{[2]} + \mathfrak{L}^p$. If $\mathfrak{M}$ is a subspace such that $\mathfrak{L} \geqq \mathfrak{M} \geqq \mathfrak{L}_2$, $\mathfrak{M}$ is a nilpotent ideal and $\mathfrak{L}/\mathfrak{M}$ is commutative and has all of its elements $\neq 0$ nil-

(7) Ore [4, p. 581].

potent of index p. We choose $\mathfrak{M}$ so that dim $\mathfrak{M}=n-1$ and let $x_2, x_3, \cdots, x_n$ be a basis for $\mathfrak{M}$ with $d, x_2, \cdots, x_n$ a basis for $\mathfrak{L}$.

THEOREM 6. *The u-algebra of a nilpotent Lie algebra with a finite basis is a nilpotent associative algebra.*

The theorem is trivial if $\mathfrak{L}$ has 1 dimension. Suppose it true for algebras of order $n-1$. Choose $\mathfrak{M}$ and $d, x_2, \cdots, x_n$ as indicated. Then the u-algebra $\mathfrak{U}$ is generated by $d, x_2, \cdots, x_n$ and the u-algebra $\mathfrak{V}$ of $\mathfrak{M}$ is generated by $x_2, \cdots, x_n$. $\mathfrak{V}$ is nilpotent. The elements of $\mathfrak{U}$ have the form

$$u = v_0 + dv_1 + \cdots + d^{p-1}v_{p-1} + d\beta_1 + d^2\beta_2 + \cdots + d^{p-1}\beta_{p-1}$$

where $v_i \in \mathfrak{V}$ and $\beta_i \in \Phi$. The *weight* of $d^i v_1^{(\kappa_1)} v_2^{(\kappa_2)} \cdots v_s^{(\kappa_s)}$ where

$$v^{(\kappa)} = [\cdots \overbrace{[vd]d \cdots d]}^{\kappa} \in \mathfrak{V}$$

is defined to be $\geqq i+s+\kappa_1+\cdots+\kappa_s$. Hence the weight of each term of u is $\geqq 1$. Since

$$v_1^{(\kappa_1)} v_2^{(\kappa_2)} \cdots v_s^{(\kappa_s)} d = d v_1^{(\kappa_1)} v_2^{(\kappa_2)} \cdots v_s^{(\kappa_s)} + v_1^{(\kappa_1+1)} v_2^{(\kappa_2)} \cdots v_s^{(\kappa_s)} + \cdots + v_1^{(\kappa_1)} v_2^{(\kappa_2)} \cdots v_s^{(\kappa_s+1)}$$

the weight of a product $u_1u_2 \cdots u_k$ is $\geqq k$. If the index of nilpotency of $\mathfrak{V}$ is m and $d^{p^t}=0$, then every term of weight $\geqq mp^t$ is 0. Thus $\mathfrak{U}$ is nilpotent of index $\leqq mp^t$.

A consequence of this theorem is that the irreducible representations of a nilpotent restricted Lie algebra are all 0. Hence any representation has matrices in triangular form with diagonal elements 0 if the basis is properly chosen.

BIBLIOGRAPHY

1. I. Ado, Bulletin de la Société Physico-mathématique de Kazan, vol. 6 (1935).

2. G. Birkhoff, *Representability of Lie algebras* $\cdots$, Annals of Mathematics, (2), vol. 38 (1937), pp. 326–332.

3. N. Jacobson, *Abstract derivation and Lie algebras*, these Transactions, vol. 42 (1937), pp. 206–224.

4. O. Ore, *On a special class of polynomials*, these Transactions, vol. 35 (1933), pp. 559–584.

5. E. Witt, *Treue Darstellung Liescher Ringe*, Journal für die reine und angewandte Mathematik, vol. 177 (1937), pp. 152–160.

6. H. Zassenhaus, *Über Liesche Ringe mit Primzahlcharakteristik*, Abhandlungen aus dem mathematischen Seminar der Hansischen Universität, 1939, pp. 1–100.

7. ———, *Endliche p-Gruppe und Lie-Ring mit der Charakteristic p*, ibid., pp. 200–207.

8. M. Zorn, *On a theorem of Lie*, Bulletin of the American Mathematical Society, vol. 42 (1936), p. 485.

UNIVERSITY OF NORTH CAROLINA,
CHAPEL HILL, N. C.

Reprinted from *American Journal of Mathematics*, July 1941.

CLASSES OF RESTRICTED LIE ALGEBRAS OF CHARACTERISTIC p. I *

By N. JACOBSON.

0. 1. In a previous paper [1] we discussed the Lie algebras, obtained from a simple associative algebra of characteristic 0 by defining the Lie algebra operation $[a, b]$ as $ab - ba$, and the subalgebras of skew elements of involutorial associative algebras of this type. The present paper develops a similar theory for algebras of characteristic p. The fundamental concept that enables us to obtain our extension is that of a restricted Lie algebra. The operations in these systems are $a + b, a\alpha, \alpha$ in the field, $[a, b]$ and a^p. Conditions for isomorphisms, the derivation algebras and the automorphism groups for the restricted Lie algebras under consideration are derived. The automorphism groups turn out to be the well-known classes of simple linear groups. As an application we obtain by a uniform method a generalization of the known isomorphism theorems between certain of these groups.

From the point of view of abstract Lie algebra theory the present paper contributes only examples of simple restricted Lie algebras of characteristic p. We can not prove, as in the characteristic 0 case, that these are "almost all" (all but the algebras of a finite number of exceptional orders) of the simple restricted Lie algebras with finite bases. In a continuation of this paper we shall consider another infinite class of algebras of this type. It is not known whether or not these together constitute "almost all" of the set of simple restricted Lie algebras with finite bases.

The main part of the discussion is restricted to normal algebras. In the last sections we indicate how the results may be extended to non-normal algebras. The derivation algebras of certain of these simple Lie algebras have the property that the difference algebras $\mathfrak{D}/\mathfrak{J}$, $\mathfrak{J}$ the algebra of inner derivation, are simple. This disproves a conjecture recently made by Zassenhaus.[2] It should be noted, however, that the underlying field for these examples is imperfect so that the question of the validity of Zassenhaus' conjecture for algebras with a finite basis over a perfect field remains open.

* Received January 25, 1941.

[1] "Simple Lie algebras over a field of characteristic zero," *Duke Mathematical Journal*, vol. 3 (1938), pp. 534-551. See also the references given there.

[2] "Über Liesche Ringe mit Primzahlcharakteristik," *Abhandl. Math. Sem. Hansischen Univ.*, vol. 13 (1939), p. 80.

0.2. A vector space $\mathfrak{L}$ over a field Φ of characteristic p $(\neq 0)$ is a *restricted Lie algebra* if there are defined in addition to the vector operations two functions $[a, b]$ and $a^{[p]}$ in $\mathfrak{L}$ such that the first is bilinear and satisfies

$$[a, b] = -[b, a] \qquad [a, [b, c]] + [b, [c, a]] + [c, [a, b]] = 0$$

and the second satisfies

$$(a\alpha)^{[p]} = a^{[p]}\alpha^p, \qquad [a, b^{[p]}] = [\cdots [a, \overbrace{b] \cdots b}^{p}]$$
$$(a + b)^{[p]} = a^{[p]} + b^{[p]} + s(a, b)$$

where $s(a, b) = s_1(a, b) + \cdots + s_{p-1}(a, b)$ and $(p - i)s_i(a, b)$ is the coefficient of λ^{p-i-1} in

$$[\cdots [a, \overbrace{\lambda a + b], \lambda a + b], \cdots, \lambda a}^{p-1} + b].$$

Subalgebras, ideals, difference algebras, homomorphisms, isomorphisms and automorphisms are defined in the obvious way.[3]

If $\mathfrak{A}$ is an associative algebra over Φ we obtain a restricted Lie algebra $\mathfrak{A}_l$ by defining $[a, b]$ as $ab - ba$ and $a^{[p]}$ as a^p. For example if Φ_n denotes the algebra of $n \times n$ matrices over Φ, Φ_{nl} is the Lie algebra determined by Φ_n. A homomorphism between a restricted Lie algebra $\mathfrak{L}$ and a subalgebra of Φ_{nl} is a representation of $\mathfrak{L}$. Of particular importance is the adjoint representation which associates to a the linear transformation or matrix $x \to [x, a]$, x variable in $\mathfrak{L}$.

If $\mathfrak{A}$ and $\mathfrak{B}$ are isomorphic it is evident that so also are $\mathfrak{A}_l$ and $\mathfrak{B}_l$ and if $\mathfrak{A}$ and $\mathfrak{B}$ are anti-isomorphic with $a \to b$ an anti-isomorphism between them then $a \to -b$ is an isomorphism between $\mathfrak{A}_l$ and $\mathfrak{B}_l$. For if $a_1 \to -b_1$, $a_2 \to -b_2$, $[a_1, a_2] = a_1a_2 - a_2a_1 \to -(b_2b_1 - b_1b_2) = [-b_1, -b_2]$ and $a_1^p \to -(b_1^p) = (-b_1)^p$. If $\mathfrak{A}$ has an anti-automorphism J the set $\mathfrak{S}(\mathfrak{A}, J)$ of skew elements $a^J = -a$ forms a subalgebra of the restricted Lie algebra $\mathfrak{A}_l$. A second anti-automorphism K is *cogredient* to J if $K = S^{-1}JS$ where S is an automorphism in $\mathfrak{A}$. When this holds $a \to a^S$ induces an isomorphism between $\mathfrak{S}(\mathfrak{A}, J)$ and $\mathfrak{S}(\mathfrak{A}, K)$. In particular if $\mathfrak{A} = \mathfrak{F}_r$ the algebra of $r \times r$ matrices with elements in an involutorial division algebra[4] $\mathfrak{F}$ and if $f \to \bar{f}$ is an involution in $\mathfrak{F}$ then J defined by $(f_{ij}) \to u^{-1}(\bar{f}_{ij})'u$ and K such that $(f_{ij})^K = v^{-1}(\bar{f}_{ij})'v$ are cogredient if $v = (\bar{t}'ut)_\rho$, where a' denotes the transposed

[3] See the author's paper "Restricted Lie algebras of characteristic p," *Transactions of the American Mathematical Society*, vol. 50 (1941), pp. 15-25.

[4] We are using Albert's terminology: an involution is an anti-automorphism of period 2 and an algebra is involutorial if it has an involution; A. A. Albert, *Structure of Algebras*, New York (1939), p. 151.

of a and ρ is in the center of $\mathfrak{A}$. For, we define $a^S = t^{-1}at$ and may verify that $K = S^{-1}JS$. If $J = S^{-1}JS$, $a \to a^S$ induces an automorphism in $\mathfrak{S}(\mathfrak{A}, J)$. Hence if g is a *J-orthogonal element* in the sense that $g^J g = g g^J = 1\gamma \neq 0$, γ in Φ then $a \to g^{-1}ag$ is an automorphism in $\mathfrak{S}(\mathfrak{A}, J)$. The set of J-orthogonal elements forms a group $O(\mathfrak{A}, J)$ under multiplication. We shall call the factor group $O(\mathfrak{A}, J)/\Phi^*$, Φ^* the set of elements $\neq 0$ in Φ, the projective orthogonal group $PO(\mathfrak{A}, J)$.

A *derivation* D in an arbitrary algebra $\mathfrak{A}$ (not necessarily associative) is a linear transformation in $\mathfrak{A}$ such that $(ab)D = (aD)b + a(bD)$. Their totality is a restricted Lie algebra $\mathfrak{D}$. If $\mathfrak{A}$ is itself a restricted Lie algebra we define a *restricted derivation* to be one which satisfies the condition $a^{[p]}D = [\ldots[aD, \overbrace{a]a]\cdots a}^{p-1}]$. These form a sub-algebra $\mathfrak{D}_0$ of $\mathfrak{D}$ which contains as ideal the set of inner derivations $a \to [a, b]$.[5]

If $\mathfrak{L}$ is a restricted Lie algebra with a finite basis $x_1, \cdots, x_n$ over Φ such that $[x_i, x_j] = \sum x_k\gamma_{kij}$, $x_i^{[p]} = \sum x_k\mu_{ki}$ the γ's and μ's are constants of multiplication. For any extension P of Φ we may define an algebra $\mathfrak{L}_\mathrm{P}$ using the same constants as for $\mathfrak{L}$ but embracing all $\sum x_i\rho_i$, ρ in P. We recall that if $\mathfrak{D}(\mathfrak{D}_0)$ is the derivation (restricted derivation) algebra of $\mathfrak{L}$, $\mathfrak{D}_\mathrm{P}(\mathfrak{D}_{0\mathrm{P}})$ is the corresponding algebra for $\mathfrak{L}_\mathrm{P}$.

$\mathfrak{L}$ is *simple* if it has no proper ideals. If $\mathfrak{L}_\mathrm{P}$ is simple it is evident that $\mathfrak{L}$ is simple. The elements a, b *commute* if $[a, b] = -[b, a] = 0$. The *center* of $\mathfrak{L}$ is the set of elements c such that $[a, c] = 0$ for all a in $\mathfrak{L}$.

I. ALGEBRAS OF TYPE A.

1.1. A lemma on representations. Let $\mathfrak{L}$ be a restricted Lie algebra with a finite basis over Φ and $a \to A$ a representation of $\mathfrak{L}$ in Φ_{Nl}. Suppose that $\mathfrak{L}$ contains a commutative subalgebra $\mathfrak{H}$ with a basis h_i, $i = 1, \cdots, m$ such that $h_i^p = h_i$ and that $\mathfrak{H}$ is maximal in the sense that the only elements a such that $[h, a] = 0$ for all h in $\mathfrak{H}$ are the elements of $\mathfrak{H}$. If $h_i \to H_i$ we have $H_i^p = H_i$ and hence we may assume that all of the H_i have diagonal form $\{m_{1i}, m_{2i}, \cdots, m_{Ni}\}$ where $m_{ji} = 0, 1, \cdots, p-1$. Then the general element $h = \sum h_i\lambda_i$ of $\mathfrak{H}$ is represented by $H = \{\Lambda_1, \Lambda_2, \cdots, \Lambda_N\}$ where the Λ's are linear forms in the λ's with coefficients $0, 1, \cdots, p-1$ and are called the *weights* of h in the representation. If we apply this to the adjoint representation we obtain a basis $h_1, h_2, \cdots, h_m, e_\alpha, e_\beta \cdots$ where $[e_\alpha, h] = e_\alpha\alpha$, α a linear form $\neq 0$ in the λ's. The α's are the *roots* of h.

[5] See [3] for the proof of this and the results of the next paragraph.

Now suppose that e_a and e_{-a} are contained in the normalized basis. $[e_{-a}, e_a]$ commutes with h since

$$[[e_{-a}, e_a], h] = [[e_{-a}, h], e_a] + [e_{-a}, [e_a, h]] = - [e_{-a}, e_a]\alpha + [e_{-a}, e_a]\alpha = 0.$$

Hence $[e_{-a}, e_a] = h_a \in \mathfrak{H}$. We consider again an arbitrary representation where $h \to H$, $e_a \to E_a$ etc. Suppose that $y_0 \neq 0$ is a vector in the representation space such that $y_0 H = y_0 \Lambda$, Λ one of the Λ_i and $y_0 E_a = 0$. The vector $y_i = y_0 E_{-a}{}^i$ satisfies $y_i H = y_i(\Lambda - i\alpha)$. Hence either all of the forms $\Lambda - i\alpha$, $i = 0, 1, \cdots, p-1$ are weights or there is a k, $0 \leqq k < p-1$ such that $y_k \neq 0$, $y_{k+1} = 0$. By induction we may prove that

$$y_i E_a = y_{i-1}\mu_i; \qquad \mu_i = i\Lambda_a - \frac{i(i-1)}{2}\alpha_a$$

where $\Lambda_a = \Lambda(h_a)$ the value of Λ for h_a and $\alpha_a = \alpha(h_a)$. Since $y_{k+1} = 0$, $y_k \neq 0$ this implies $\mu_{k+1} = 0$.

Lemma 1. *Let $\mathfrak{L}$ be a restricted Lie algebra with a finite basis and $\mathfrak{H}$ a maximal commutative subalgebra with a basis h_i such that $h_i{}^p = h_i$. Suppose that e_a, e_{-a} are elements of $\mathfrak{L}$ such that $[e_a, h] = e_a\alpha$ for $h = \sum h_i\lambda_i$ where α is a linear form $\neq 0$ in the λ's and that $y_0 \neq 0$ is a vector of a representation space such that $y_0 H = y_0\Lambda$, $y_0 E_a = 0$. Then either all the forms $\Lambda - i\alpha$ are weights of h in this representation or there is a k, $0 \leqq k < p-1$ such that $(k+1)\Lambda_a - \dfrac{k(k+1)}{2}\alpha_a = 0$, $\Lambda_a = \Lambda(h_a)$, $h_a = [e_{-a}, e_a]$ and $\Lambda - k\alpha$ is a weight.*

1. 2. Ideals in Φ_{nl}. We begin with the restricted Lie algebra Φ_{nl}, Φ an infinite field of characteristic p. As usual let e_{ij} denote a matrix basis for Φ_n, i. e. $e_{ij}e_{kl} = \delta_{jk}e_{il}$. Set $h_i = e_{ii}$, $h = \sum h_i\lambda_i$ and if $p \neq 2$, $e_{\lambda_i - \lambda_j} = e_{ji}$, $i \neq j$. For $p = 2$ we denote instead e_{ij}, $i < j$, by $e^{(1)}_{\lambda_i+\lambda_j}$ and e_{ji} as $e^{(2)}_{\lambda_i+\lambda_j}$. For $p \neq 2$ we obtain the following multiplication table

$$(1) \qquad \begin{aligned} &[h_i, h_j] = 0 \\ &[e_a, h] = e_a\alpha, \qquad \alpha = \lambda_i - \lambda_j \\ &[e_a, e_\beta] = \begin{cases} 0 \text{ if } \alpha+\beta \text{ is not a root, } \alpha \neq -\beta \\ \pm e_{a+\beta} \text{ if } \alpha+\beta \text{ is a root, } \alpha \neq -\beta \end{cases} \\ &[e_{-a}, e_a] = h_a = h_i - h_j \end{aligned}$$

and

$$(2) \qquad e_a{}^p = 0, \quad h_i{}^p = h_i.$$

We note that $\alpha_a = \alpha(h_a) = 2$. If $p = 2$ we have to replace e_a and e_{-a} by $e_a{}^{(1)}$ and $e_a{}^{(2)}$ and obtain a similar table.

We have shown elsewhere [6] that the only ideals in Φ_{nl} are (1) (the multiples of $1 = \sum h_i$) and Φ'_{nl}, the algebra generated by the elements $[a, b]$. The former is the center of Φ_{nl} and the latter may be characterized as the set of elements of trace 0. If $p \nmid n$, $\Phi'_{nl} \wedge (1) = 0$ and $\Phi_{nl} = \Phi'_{nl} \oplus (1)$.

1.3. Automorphisms of Φ_{nl}. We note first that the following mappings are automorphisms: $a \to a + i \operatorname{tr} a$, $i = 0, 1, 2, \cdots, p-1$, $ni + 1 \not\equiv 0 \pmod p$, $a \to t^{-1}at$, $a \to -a'$. The last two are clear. The first follows since $\operatorname{tr} a^p = (\operatorname{tr} a)^p$ and $ni + 1 \not\equiv 0$ insures that $a + i \operatorname{tr} a \neq 0$ if $a \neq 0$. We shall show that these generate the group of automorphisms of Φ_{nl} over Φ. Suppose that $h \to h^S$, $e_a \to e_a{}^S$ is an arbitrary automorphism. Then this correspondence is a second representation of Φ_{nl} in Φ_{nl}. Let $\Lambda_1, \Lambda_2, \cdots, \Lambda_n$ be the weights of h in the second representation. If T is the automorphism $a \to t^{-1}at$ which transforms h^S into the diagonal form $\{\Lambda_1, \Lambda_2, \cdots, \Lambda_n\}$ we have $h^{ST} = \{\Lambda_1, \Lambda_2, \cdots, \Lambda_n\}$.

Since $h_1{}^{ST}, \cdots, h_n{}^{ST}$ are linearly independent the linear forms $\Lambda_1, \Lambda_2, \cdots, \Lambda_n$ are linearly independent mod p. Hence if $\Lambda = \sum m_i\lambda_i$ is a weight at most one of the forms $\Lambda + \alpha$, $\Lambda - \alpha$, $\alpha = \lambda_i - \lambda_j$, is a weight and we may suppose first that $\Lambda + \alpha$ is not. Thus if $y_0 \neq 0$, $y_0H = y_0\Lambda$ then $y_0E_a = 0$ and by the above lemma if $p \neq 2$, $\Lambda_a = m_i - m_j = 0, 1$ according as $y_0E_{-a} = 0$ or $\neq 0$. If $\Lambda - \alpha$ is not a weight but $\Lambda + \alpha$ is, $\Lambda_{-a} = m_j - m_i = 0, 1$. Since $\Lambda - (m_i - m_j)\alpha$ is a weight the weights are invariant under permutation of the λ's. If $p = 2$ and Λ is a weight we have that either $\Lambda + \alpha$ is a weight or for any y_0 such that $y_0H = y_0\Lambda$, $y_0E_a{}^{(i)} = 0$. Then $y_0H_a = 0$ and $\Lambda_a = m_i + m_j = 0$. Thus in this case also the weights are invariant under permutation of the λ's.

If $\Lambda = m_1(\lambda_1 + \cdots + \lambda_{n_1}) + m_2(\lambda_{n_1+1} + \cdots + \lambda_{n_1+n_2}) + \cdots + m_k(\cdots + \lambda_{n_1 \ldots +n_k})$ where $m_i \neq m_j$ and $k \leq p$ is a weight, we obtain by permuting the λ's $\dfrac{n!}{n_1! \cdots n_k!}$ $(\sum n_i = n)$ distinct weights. This number is $> n$ unless $k = 1$ or $k = 2$ and $n_1 = n - 1$, $n_2 = 1$ or $n_1 = 1$, $n_2 = n - 1$. Since there is a weight not of the form $m(\lambda_1 + \cdots + \lambda_n)$ we may suppose $k = 2$, $\Lambda = m_1(\lambda_1 + \cdots + \lambda_{n-1}) + m_2\lambda_n$ and $m_1 - m_2 = \pm 1$. It follows that $h_1{}^{ST} + \cdots + h_n{}^{ST}$ is the scalar matrix $q = (n-1)m_1 + m_2 = nm_1 \pm 1 \neq 0$.

Consider the automorphism $a \to a^U = a + i \operatorname{tr} a$. The element h has weights $\Lambda + iq(\lambda_1 + \lambda_2 + \cdots + \lambda_n)$ in the representation $a \to a^{STU}$. Since $q \neq 0$ we may choose i so that $iq = -m_1$. Then $ni + 1 \not\equiv 0$ and h^{STU} will

[6] "Abstract derivation and Lie algebras," *Transactions of the American Mathematical Society*, vol. 42 (1937), pp. 216-217.

be the diagonal matrix $\{-\lambda_1,\cdots,-\lambda_n\}$ or $\{\lambda_1,\cdots,\lambda_n\}$. In the former case we combine with the automorphism $a\to -a' = a^V$ and obtain $h^{STUV}=h$.

Thus in any case we may suppose now that $\bar{S}=STU$ or $=STUV$ has the property $h^{\bar{S}}=h$. Since $[e_\alpha,h]=e_\alpha\alpha$, $[e_\alpha{}^{\bar{S}},h]=e_\alpha{}^{\bar{S}}\alpha$. If $p\neq 2$ we may suppose that the λ's are chosen so that the roots α are distinct and then $e_\alpha{}^{\bar{S}}=e_\alpha\xi_\alpha$. From $[e_{-\alpha},e_\alpha]=h_\alpha$, $[e_{-\alpha}{}^{\bar{S}},e_\alpha{}^{\bar{S}}]=[e_{-\alpha},e_\alpha]$ and $\xi_{-\alpha}\xi_\alpha=1$. From $[e_\alpha,e_\beta]=\pm e_{\alpha+\beta}$, $\xi_{\alpha+\beta}=\xi_\alpha+\xi_\beta$. Hence if $\xi_{\lambda_1-\lambda_2}=\cdots=\xi_{\lambda_1-\lambda_n}=1$, $\xi_{\lambda_2-\lambda_1}=\cdots=\xi_{\lambda_n-\lambda_1}=1$ and $\xi_\alpha=1$ for all α. $\bar{S}$ is then the identity automorphism. The automorphism $a\to\bar{h}^{-1}a\bar{h}=a^W$ where $\bar{h}=\Sigma h_i\bar{\lambda}_i$ sends h into itself and $e_{\lambda_i-\lambda_j}$ into $e_{\lambda_i-\lambda_j}\bar{\lambda}_j{}^{-1}\bar{\lambda}_i$. We may suppose that $\bar{\lambda}_j{}^{-1}\bar{\lambda}_1=\xi^{-1}_{\lambda_1-\lambda_j}$. Then $\bar{S}W$ is the identity and hence $\bar{S}=W^{-1}$ has the form $a\to\bar{h}a\bar{h}^{-1}$.

If $p=2$ we have for $\alpha=\lambda_1-\lambda_i$

$$e_\alpha{}^{(1)\bar{S}}=e_\alpha{}^{(1)}\lambda_{11}{}^{(i)}+e_\alpha{}^{(2)}\lambda_{12}{}^{(i)};\qquad e_\alpha{}^{(2)\bar{S}}=e_\alpha{}^{(1)}\lambda_{21}{}^{(i)}+e_\alpha{}^{(2)}\lambda_{22}{}^{(i)}. \tag{3}$$

Since $(e_\alpha{}^{(1)\bar{S}})^2=(e_\alpha{}^{(2)\bar{S}})^2=0$,

$$\lambda_{11}{}^{(i)}\lambda_{12}{}^{(i)}=\lambda_{21}{}^{(i)}\lambda_{22}{}^{(i)}=0. \tag{4}$$

Since $[e_\alpha{}^{(1)},e_\alpha{}^{(2)}]^{\bar{S}}=[e_\alpha{}^{(1)},e_\alpha{}^{(2)}]$,

$$\lambda_{11}{}^{(i)}\lambda_{22}{}^{(i)}+\lambda_{12}{}^{(i)}\lambda_{21}{}^{(i)}=1. \tag{5}$$

Hence either $\lambda_{11}{}^{(i)}=\lambda_{22}{}^{(i)}=0$ and $\lambda_{12}{}^{(i)}\lambda_{21}{}^{(i)}=1$ or $\lambda_{12}{}^{(i)}=\lambda_{21}{}^{(i)}=0$ and $\lambda_{11}{}^{(i)}\lambda_{22}{}^{(i)}=1$. From (1)

$$e_{ij}{}^{\bar{S}}=[e^{(2)\bar{S}}_{\lambda_1-\lambda_i},e^{(1)\bar{S}}_{\lambda_1-\lambda_j}]=e_{ij}\lambda_{22}{}^{(i)}\lambda_{11}{}^{(j)}+e_{ji}\lambda_{21}{}^{(i)}\lambda_{12}{}^{(j)}$$

if 1, i, j are unequal and since $[e_{ij},e_{ji}]^{\bar{S}}=[e_{ij},e_{ji}]$,

$$\lambda_{22}{}^{(i)}\lambda_{22}{}^{(j)}\lambda_{11}{}^{(i)}\lambda_{11}{}^{(j)}+\lambda_{21}{}^{(i)}\lambda_{21}{}^{(j)}\lambda_{12}{}^{(i)}\lambda_{12}{}^{(j)}=1 \tag{6}$$

and, hence, if $\lambda_{12}{}^{(i)}=\lambda_{21}{}^{(i)}=0$, $\lambda_{12}{}^{(j)}=\lambda_{21}{}^{(j)}=0$. If this holds we may show, as in the case $p\neq 2$, that $\bar{S}$ has the form $a\to\bar{h}a\bar{h}^{-1}$. Otherwise $\lambda_{11}{}^{(i)}=\lambda_{22}{}^{(i)}=1$, $i=2,\cdots,n$, and it follows readily that $a^{\bar{S}}=\bar{h}a'\bar{h}^{-1}$. We have therefore proved

LEMMA 2. *Any automorphism of Φ_{nl} is a product of automorphisms of the following types:*

I.	$a\to a+i\operatorname{tr}a$	$(i=0,1,\cdots,p-1,\ ni+1\neq 0)$.
II.	$a\to t^{-1}at$	
III.	$a\to -a'$.	

It is interesting to note that if $n \equiv 0 \pmod p$, $\operatorname{tr} 1 = 0$ and hence for any automorphism S, $1^S = \pm 1$.

The automorphisms of the form $a \to a + i \operatorname{tr} a$ form a group. The product of $a \to a + i \operatorname{tr} a$, $a \to a + j \operatorname{tr} a$ is $a \to a + (i + j + nij) \operatorname{tr} a$. Hence if $n \equiv 0 \pmod p$ this group is isomorphic to the additive group mod p and if $n \not\equiv 0 \pmod p$ the correspondence between $a \to a + i \operatorname{tr} a$ and $ni + 1$ is an isomorphism with the multiplicative group mod p. The automorphisms I and II form invariant subgroups. Any element in I commutes with any in II. If $n = 2$, $a' = - q^{-1}aq + \operatorname{tr} a$ if $q = \begin{pmatrix} 0 & 1 \\ -1 & 0 \end{pmatrix}$ so that I and II generate the whole group of automorphisms. If $n > 2$ the automorphisms I and II form an invariant subgroup of index 2. For otherwise we should have a matrix s such that $a' = - s^{-1}as - i \operatorname{tr} a$. For a of trace 0 we obtain $a = s's^{-1}as(s')^{-1}$. Hence $s's^{-1}$ commutes with all matrices of trace 0 and therefore with all matrices in Φ_n. It follows that $s's^{-1} = 1\rho$, $s = s\rho$ and since $s'' = s$, $s' = \pm s$. As we shall show in part II, if s is symmetric or skew the dimensionality of the space of matrices a such that $s^{-1}a's = -a$ is $n(n-1)/2$ or $n(n+1)/2$ (n even) and these numbers are $< n^2 - 1$ the dimensionality of the space of matrices of trace 0 if $n > 2$. Hence the equation $a' = - s^{-1}as - i \operatorname{tr} a$ can not hold for a fixed s and all a.

1.4. Restricted derivations of Φ_{nl}. Let D be a restricted derivation. We choose the $\lambda_i = \bar{\lambda}_i$ in $h = \Sigma h_i\lambda_i$ such that all of the p^n linear forms $\Sigma m_i\bar{\lambda}_i$, $m_i = 0, \cdots, p-1$ are distinct. Then the minimum p-polynomial satisfied by $\bar{h} = \Sigma h_i\bar{\lambda}_i$ is $\Pi(\lambda - \Sigma m_i\bar{\lambda}_i) = \lambda^{p^n} + \lambda^{p^{n-1}}\beta_{n-1} + \cdots + \lambda\beta_0$ where $\beta_0 = \prod_{\neq 0} (\Sigma m_i\bar{\lambda}_i) \neq 0$.[7] Thus

$$\bar{h}^{p^n} + \bar{h}^{p^{n-1}}\beta_{n-1} + \cdots + \bar{h}\beta_0 = 0.$$

Since D is restricted

$$(\bar{h}D)(\bar{H}^{p^n-1} + \bar{H}^{p^{n-1}-1}\beta_{n-1} + \cdots + \beta_0) = 0$$

where $\bar{H}$ denotes the operation $a \to [a, \bar{h}]$.

$$\bar{H}^{p^n-1} + \cdots + \beta_0 = \prod_{\neq 0} (\bar{H} - \Sigma m_i\bar{\lambda}_i).$$

Hence if we set $\bar{h}D = h_0 + \Sigma e_\alpha\sigma_\alpha$ we obtain $h_0 = 0$, $\bar{h}D = \Sigma e_\alpha\sigma_\alpha$. Set $\tau_\alpha = - \bar{\alpha}^{-1}\sigma_\alpha$ and subtract from D the derivation $a \to [a, \Sigma e_\alpha\tau_\alpha]$. The resulting derivation E has the property $\bar{h}E = 0$. Since $\bar{h}, \bar{h}^p, \cdots, \bar{h}^{p^{n-1}}$ are

[7] Cf. Ore "On a special class of polynomials," *Transactions of the American Mathematical Society*, vol. 35 (1933), pp. 564-565.

linearly independent they form a basis for the subalgebra $\mathfrak{H} = (h_1, h_2, \cdots, h_n)$. Since $(\bar{h}^{p^t})E = 0$, $h_i E = 0$. Now $[e_a, \bar{h}]E = [e_a E, \bar{h}] = (e_a E)\bar{\alpha}$ implies that for $p \neq 2$, $e_a E = e_a \xi_a$ and for $p = 2$, $e_a^{(1)}E = e_a^{(1)}\lambda_{11} + e_a^{(2)}\lambda_{12}$, $e_a^{(2)}E = e_a^{(1)}\lambda_{21} + e_a^{(2)}\lambda_{22}$. In the former case it follows readily that E, and hence D, is inner. In case $p = 2$ we use $(e_a^{(k)})^2 = 0$, $[e_a^{(k)}E, e_a^{(k)}] = 0$ and hence $e_a^{(k)}E = e_a^{(k)}\xi_a^{(k)}$ and we can prove that E is inner as in the case $p \neq 2$.

LEMMA 3. *Any restricted derivation of Φ_{nl} is inner.*

1.5. Restricted Lie algebras of type A. Let $\mathfrak{A}$ be a normal simple associative algebra of order n^2 over an arbitrary Φ of characteristic p and $\mathfrak{A}_l$ the restricted Lie algebra determined by $\mathfrak{A}$. $\mathfrak{A}_l$ will be called a restricted Lie algebra of *type* A_1. If Ω is the algebraic closure of Φ, $\mathfrak{A}_\Omega = \Omega_n$ and hence $\mathfrak{A}_{l\Omega} = \Omega_{nl}$. The algebra $\mathfrak{A}'_l$ generated by elements of the form $[a, b]$ is an ideal of order $n^2 - 1$ in $\mathfrak{A}_l$ since $(\mathfrak{A}'_l)_\Omega = \Omega'_{nl}$ is an ideal of this order in Ω_{nl}. We may choose a basis $a_1, \cdots, a_{n^2}$ for $\mathfrak{A}$ over Φ such that $a_1, \cdots, a_{n^2-1}$ is a basis for $\mathfrak{A}'_l$. Then these elements will constitute bases for Ω_n and Ω'_{nl} also, i. e. every element of Ω_n has the form $\sum_{j=1}^{n^2} a_i\omega_i$, $\omega \in \Omega$ and every element of Ω'_{nl} has the form $\sum_{j=1}^{n^2-1} a_j\omega_j$. It is well known that the characteristic polynomial of any element $a = \sum a_i\phi_i$ of $\mathfrak{A}$ has coefficients in Φ. In particular $\operatorname{tr} a \in \Phi$ and the correspondence $a \to a + i \operatorname{tr} a$ is an automorphsim of $\mathfrak{A}_l$ if $i = 0, 1, \cdots, p-1$ and $ni + 1 \not\equiv 0 \pmod{p}$.

Now suppose that $\mathfrak{A}$ is a simple associative algebra of order n^2 over its center $\mathrm{P} = \Phi(q)$ a quadratic extension of Φ and suppose that $\mathfrak{A}$ has an involution J of second kind.[8] Let $\mathfrak{S}(\mathfrak{A}, J)$ be the set of J-skew elements. It is known that there are n^2 elements $a_1, a_2, \cdots, a_{n^2}$ such that the elements of $\mathfrak{S}(\mathfrak{A}, J)$ are the sums $\sum a_i\phi_i$, $\phi \in \Phi$ and those of $\mathfrak{A}$ are $\sum a_i\rho_i$, ρ_i in P.[9] $\mathfrak{S}(\mathfrak{A}, J)$ is a restricted Lie algebra over Φ. Since $\mathfrak{S}(\mathfrak{A}, J)_\mathrm{P} = \mathfrak{A}_l$, $\mathfrak{S}(\mathfrak{A}, J)_\Omega = \Omega_{nl}$ if Ω is the algebraic closure of P. It follows as above that if $\mathfrak{S}(\mathfrak{A}, J)'$ denotes the space generated by the $[a, b]$ a, b in $\mathfrak{S}$, $\mathfrak{S}(\mathfrak{A}, J)'$ is a restricted Lie algebra such that $\mathfrak{S}'_\Omega = \Omega'_{nl}$. Hence we may suppose that $a_1, \cdots, a_{n^2-1}$ form a basis for $\mathfrak{S}'$ over Φ, for $\mathfrak{A}'_l$ over P and for Ω'_{nl} over Ω. Let $\Gamma = \Phi(\xi_1, \cdots, \xi_{n^2})$ be the field obtained from Φ by adjoining the indeterminates ξ_i and consider the algebra $(\mathfrak{A}$ over $\Phi)_\Gamma$. The involution J has a unique extension J to $(\mathfrak{A}$ over $\Phi)_\Gamma$. The element $a_\xi = \sum a_i\xi_i$ is skew relative to this involution.

[8] An involution J is of first kind or second kind according as it induces the identity automorphism or not in the center P. If J is of second kind then P is separable over Φ the field of invariant elements of P. Cf. Albert, *loc. cit.* [4], p. 153.

[9] Albert [4], p. 153.

We have

(7) $$(a_\xi)^n - (a_\xi)^{n-1} \operatorname{tr} a_\xi + \cdots = 0$$

and if we operate with J we obtain

(8) $$(-a_\xi)^n - (-a_\xi)^{n-1}(\operatorname{tr} a_\xi)^J + \cdots = 0.$$

Since (7) is the minimum equation of $\mathfrak{A}$ over P the coefficients of (7) and (8) differ by the sign $(-1)^n$.[10] Hence $(\operatorname{tr} a_\xi)^J = -\operatorname{tr} a_\xi$ and if $a = \Sigma a_i\phi_i$ is any element of $\mathfrak{S}(\mathfrak{A}, J)$, $\operatorname{tr} a \in \mathfrak{S}(\mathfrak{A}, J)$. It follows that $a \to a + i \operatorname{tr} a$ is an automorphism of $\mathfrak{S}$ over Φ if $i = 0, \cdots, p-1$ and $ni + 1 \not\equiv 0 \pmod{p}$. We shall call $\mathfrak{S}$ a restricted Lie algebra of *type* A_{II}.

1.6. The enveloping algebras. Consider an algebra $\mathfrak{S}$ of type A_{II} with $p \neq 2$. We may suppose that one of the $a_i = q$, $q^J = -q$, $q^2 = \mu$. Then $a_1 q, a_2 q, \cdots, a_{n^2} q$ are symmetric and together with $a_1, \cdots, a_{n^2}$ form a basis for $\mathfrak{A}$ over Φ. Thus the enveloping algebra over Φ of $\mathfrak{S}(\mathfrak{A}, J)$ is $\mathfrak{A}$. We wish to show that if $n > 2$, $\mathfrak{A}$ is the enveloping algebra of $\mathfrak{S}(\mathfrak{A}, J)'$. Suppose first that Φ is finite. There are no non-commutative division algebras over Φ and hence $\mathfrak{A} = \mathrm{P}_n$, $\mathrm{P} = \Phi(q)$. If $\alpha \to \bar{\alpha}$ is the automorphism in P we may suppose that J is the correspondence $a = (\alpha_{ij}) \to d^{-1}\bar{a}'d$, $\bar{a}' = (\bar{\alpha}_{ji})$, $d = \Sigma e_{ii}\delta_i$, δ_i in Φ.[11] Then the condition that $a \in \mathfrak{S}(\mathfrak{A}, J)$ is that $\bar{\alpha}_{ji} = -(\delta_i/\delta_j)\alpha_{ij}$. Hence $e_{ii}q$, $e_{ij}\delta_j - e_{ji}\delta_i$, $(e_{ij}\delta_j + e_{ji}\delta_i)q$, $i < j$, is a basis for $\mathfrak{S}(\mathfrak{A}, J)$. The elements $(e_{ii} - e_{jj})q$, $e_{ij}\delta_j - e_{ji}\delta_i$, $(e_{ij}\delta_j + e_{ji}\delta_i)q \in \mathfrak{S}(\mathfrak{A}, J)'$. If $n > 2$ choose $k \neq i, j$. Then $(e_{ii} - e_{kk})q(e_{ij}\delta_j - e_{ji}\delta_i) = e_{ij}q\delta_j$ is in the enveloping algebra $\mathfrak{B}$. Similarly e_{ij}, e_{ji}, $e_{ji}q \in \mathfrak{B}$. Hence also e_{ii}, $e_{ii}q \in \mathfrak{B}$ and $\mathfrak{B} = \mathfrak{A}$. Now suppose that Φ is infinite. It suffices to show that the enveloping algebra $\mathfrak{B} \geqq \mathfrak{S}(\mathfrak{A}, J)$. Since if $a \in \mathfrak{S}(\mathfrak{A}, J)$, $a^3, a^5, \cdots \in \mathfrak{S}(\mathfrak{A}, J)$ we must show that there is an a in $\mathfrak{S}(\mathfrak{A}, J)'$ such that a suitable $a^{2m+1} \notin \mathfrak{S}(\mathfrak{A}, J)'$. If this is not the case we should have for all m and ϕ_j in Φ, $\operatorname{tr}\left(\sum_1^{n^2-1} a_j\phi_j\right)^{2m-1} = 0$. Since Φ is infinite this implies that $\operatorname{tr}(\Sigma a_j\omega_j)^{2m-1} = 0$ and this is impossible if $n > 2$ since there are matrices of trace 0, say $\Sigma e_{ii}\omega_i$, $\Sigma \omega_i = 0$, such that $\Sigma \omega_i^{2m-1} \neq 0$. We need merely choose $2m - 1 \neq p^e$ and then satisfy the condition since $-(\omega_2 + \cdots + \omega_n)^{2m-1} + \omega_2^{2m-1} + \cdots + \omega_n^{2m-1} \not\equiv 0$.

Now let $p = 2$. We may suppose that $q^J = q + 1$. If $a_1, \cdots, a_{n^2}$ is a basis for $\mathfrak{S}(\mathfrak{A}, J)$, $a_1 q, \cdots, a_{n^2} q$ is a basis for $\mathfrak{S}q$ and $\mathfrak{A} = \mathfrak{S} + \mathfrak{S}q$

[10] Albert [4], p. 123.

[11] The involutions of second kind in P_n are of the form $a \to s^{-1}\bar{a}'s$ where $\bar{s}' = s$. Hence if the s associated with J is not diagonal we replace J by the cogredient involution $a \to d^{-1}\bar{a}'d$ where $d = \bar{g}'sg$ is diagonal.

and $\mathfrak{S} \triangle \mathfrak{S}q = 0$. Hence if the enveloping algebra $\mathfrak{B}$ over Φ of $\mathfrak{S}$ contains an element not in $\mathfrak{S}$ it will contain an element b in $\mathfrak{S}q$. If $\mathfrak{A}$ is a division algebra choose r, s in $\mathfrak{S}$ such that $[r, s] \neq 0$. Then $(rs)^J = sr \neq rs$ and hence $\mathfrak{B}$ contains an element aq, $a \neq 0$, in $\mathfrak{S}$. Then $a^{-1} \in \mathfrak{S}$ and $\mathfrak{B}$ contains q and therefore $\mathfrak{B} = \mathfrak{A}$.

If $\mathfrak{A}$ is not a division algebra $\mathfrak{A} = \mathfrak{F}_r$ where $r > 1$ and $\mathfrak{F}$ is a division algebra (commutative or not). We may suppose that the involution in $a \rightarrow d^{-1}\bar{a}'d$ where $a = (a_{ij})$, $\bar{a}' = (\bar{a}_{ji})$ and $a_{ji} \rightarrow \bar{a}_{ji}$ is an involution of the second kind in $\mathfrak{F}$ and $d = \Sigma\, e_{ii}d_i$, $\bar{d}_i = d_i$.[12] The condition that $a \in \mathfrak{S}(\mathfrak{A}, J)$ is $d_i a_{ij} = \bar{a}_{ji} d_j$. Hence if $\mathfrak{F} = \mathrm{P}$ the elements $e_{ii}, e_{ij}d_j + e_{ji}d_i$ and $e_{ij}d_jq + e_{ji}d_i(q+1) \in \mathfrak{S}$. Thus $e_{ii}(e_{ij}d_j + e_{ji}d_i) = e_{ij}d_j \in \mathfrak{B}$ and since $d_j \in \Phi$, $e_{ij} \in \mathfrak{B}$. Similarly $e_{ij}q \in \mathfrak{B}$ if $i \neq j$ and hence $e_{ii}, e_{ii}q \in \mathfrak{B}$ so that $\mathfrak{B} = \mathfrak{A}$. Finally suppose $\mathfrak{A} = \mathfrak{F}_r$ where $\mathfrak{F}$ is a non-commutative division algebra. The elements $e_{ii}a_{ii}$ where $d_i^{-1}\bar{a}_{ii}d_i = a_{ii}$ are in $\mathfrak{S}$. Since $a \rightarrow d_i^{-1}\bar{a}d_i$ is an involution of the second kind in $\mathfrak{F}$, $\mathfrak{B}$ includes every $e_{ii}b_{ii}$, b_{ii} arbitrary in $\mathfrak{F}$, and hence every b in $\mathfrak{F}$. Since $\mathfrak{B}$ contains $e_{ij}d_j + e_{ji}d_i$ it follows that $\mathfrak{B}$ contains all $e_{ij}b$ and $\mathfrak{B} = \mathfrak{A}$. As in the case $p \neq 2$ we may prove that $\mathfrak{A}$ is also the enveloping algebra of $\mathfrak{S}(\mathfrak{A}, J)'$ if $n > 2$.

LEMMA 4. *If $\mathfrak{A}$ is a simple associative algebra with center $\mathrm{P} = \Phi(q)$ a quadratic field and $\mathfrak{A}$ has an involution J of second kind, then the enveloping algebra over Φ of $\mathfrak{S}(\mathfrak{A}, J)$ is $\mathfrak{A}$. The same result holds for $\mathfrak{S}(\mathfrak{A}, J)'$ if $n > 2$.*

From now on we suppose that $n > 2$ for all Lie algebras of type A_{II}.

1.7. Isomorphism and automorphisms. Suppose $\mathfrak{L}_1$ and $\mathfrak{L}_2$ are isomorphic restricted Lie algebras of type A. We may suppose that $\mathfrak{L}_1$ is the set $\Sigma\, a_i\phi_i$, $\mathfrak{L}_2$ the set $\Sigma\, a_i^S\phi_i$ where S is the isomorphism and Ω_{nl} is the set $\Sigma\, a_i\omega_i$ or the set $\Sigma\, a_i^S\omega_i$. Thus the correspondence $\Sigma\, a_i\omega_i \rightarrow \Sigma\, a_i^S\omega_i$ is an automorphism in Ω_{nl} which extends the isomorphism between $\mathfrak{L}_1$ and $\mathfrak{L}_2$. Hence there exists an element g in Ω_n such that either $a_i^S = g^{-1}a_ig + j\,\mathrm{tr}\, a_i$, $j = 0, 1, \cdots, p-1$, $jn + 1 \neq 0$ or $a_i^S = -g^{-1}a'_ig + j\,\mathrm{tr}\, a_i$, $jn - 1 \neq 0$. If we replace a_i by $a_i \pm j\,\mathrm{tr}\, a_i = a_i^U$ where U is an automorphism in $\mathfrak{L}_1$ we may suppose $a_i^S = g^{-1}a_ig$ or $a_i^S = -g^{-1}a'_ig$. In the first case the correspondence S determines an isomorphism between the enveloping algebras over

[12] This is a consequence of the theorem that any hermitian matrix in $\mathfrak{F}_r$ is cogredient to a diagonal matrix. The proof is exactly as in [1] pp. 542-544 except for the proof of the Lemma that for any hermitian form $f = (x, y) \neq 0$ there is a vector u such that $(u, u) \neq 0$. This is seen as follows: If $(u, u) = 0$ for all u, $(u + v, u + v) = (u, u) + (v, v) + (u, v) + (v, u) = 0$. Hence $(u, v) = (v, u) = \overline{(u, v)}$ and this implies that $d \rightarrow \bar{d}$ is the identity transformation.

Φ of $\mathfrak{L}_1$ and of $\mathfrak{L}_2$ and in the second case $-S$ is an anti-isomorphism. If $\mathfrak{L}_1$ has type A_I the center of the enveloping algebra is Φ and if it has type A_{II} the center is $\mathrm{P} = \Phi(q)$. Hence we have

THEOREM 1. *A restricted Lie algebra of type A_I is not isomorphic to one of type A_{II}.*

By the same method we may prove as in the case of Φ of characteristic 0 the following theorems.

THEOREM 2. *The restricted Lie algebras $\mathfrak{A}_l$ and $\mathfrak{B}_l$ of type A_I are isomorphic if and only if the associative algebras $\mathfrak{A}$ and $\mathfrak{B}$ are either isomorphic or anti-isomorphic.*

THEOREM 3. *The restricted Lie algebras $\mathfrak{S}(\mathfrak{A}, J)$ and $\mathfrak{S}(\mathfrak{B}, K)$ of type A_{II} are isomorphic if and only if $\mathfrak{A}$ and $\mathfrak{B}$ are isomorphic and J and K are cogredient.*

THEOREM 4. *Any automorphism of $\mathfrak{A}_l$ has either the form $a \rightarrow t^{-1}at + i \operatorname{tr} a$, $i = 0, \cdots, p-1$, $ni + 1 \neq 0$, or $a \rightarrow -t^{-1}a't + i \operatorname{tr} a$, $ni - 1 \neq 0$ where $a \rightarrow a'$ is an anti-automorphism of $\mathfrak{A}$ over Φ.*

Of course the latter set need not exist.

THEOREM 5. *Any automorphism of $\mathfrak{S}(\mathfrak{A}, J)$ has the form $a \rightarrow a^S + i \operatorname{tr} a$ $i = 0, \cdots, p-1$, $ni + 1 \neq 0$, where S is an automorphism of $\mathfrak{A}$ over Φ which commutes with J.*

There are two types of S's: those which alter elements of P and those which are automorphisms of the normal simple algebra $\mathfrak{A}$ over P. The latter are inner and hence have the form $a \rightarrow s^{-1}as$ where $ss^J = 1\sigma$, σ in Φ.

By Lemma 3 and the property $\mathfrak{A}_{l\Omega} = \Omega_{nl}$, $\mathfrak{S}_{\Omega} = \Omega_{nl}$ we have

THEOREM 6. *Any restricted derivation of a restricted Lie algebra of type A is inner.*

1.8. The derived algebras. If $\mathfrak{L}$ has type A, $\mathfrak{L}_{\Omega} = \Omega_{nl}$ and $\mathfrak{L}'_{\Omega} = \Omega'_{nl}$. Hence by a previous result[13] we have

THEOREM 7. *If $\mathfrak{L}$ is a restricted Lie algebra of type A, $\mathfrak{L}'$ is simple if $p \nmid n$ and $\mathfrak{L}'/(1)$, (1) the set 1α, is simple if $p \mid n$ except when $p = n = 2$.*

By simplicity here we mean that there are no (restricted) ideals. As a

[13] *Loc. cit.*[6]

matter of fact this theorem holds when $\mathfrak{L}'$, $\mathfrak{L}'/(1)$ are regarded as ordinary Lie algebras, i. e. the algebras have no ordinary ideals other than (0) and the whole algebra.

In the remainder of this section we suppose $p \nmid n$. Then $\Omega_{ln} = \Omega'_{nl} \oplus (1)$. If $a \to a^S$ is an automorphism in Ω'_{nl} we define $1^S = 1$ and obtain an automorphism in Ω_{nl}. By Lemma 2 we have either $a^S = t^{-1}at$ or $a^S = -t^{-1}a't$ where $t \in \Omega_{nl}$. If we use the second part of Lemma 4 we obtain

THEOREM 1′. *The derived algebra of a restricted Lie algebra of type A_I is not isomorphic to the derived algebra of one of type A_{II} if $n > 2$ and $\not\equiv 0 \pmod p$.*

Similarly Theorems 2 and 3 hold with $\mathfrak{A}_l$, $\mathfrak{B}_l$, $\mathfrak{S}(\mathfrak{A}, J)$, $\mathfrak{S}(\mathfrak{A}, K)$ replaced by their derived algebras. Theorems 4 and 5 hold when this substitution is made and $i = 0$. If we note that any restricted derivation in Ω'_{nl} may be extended to Ω_{nl} by defining $1D = 0$ we obtain the fact that the restricted derivations of any $\mathfrak{A}'_l$, $\mathfrak{S}'$ with $p \nmid n$, are all inner. Finally we remark that since the algebras $\mathfrak{A}'_l$, $\mathfrak{S}'$ have no centers all of these results hold when we regard these algebras as ordinary Lie algebras. Thus for example if $a \to a^S$ is a correspondence in $\mathfrak{A}'_l$ such that $(a + b)^S = a^S + b^S$, $(a\alpha)^S = a^S\alpha$, $[a, b]^S = [a^S, b^S]$ then there is a t in $\mathfrak{A}$ such that either $a^S = t^{-1}at$ or $a^S = -t^{-1}a't$, $a \to a'$ an anti-automorphism. In the same way we may drop the word restricted in the statement of Theorem 6 for the derived algebra.

1. 9. The completeness theorem. We suppose that $\mathfrak{L}'$ is a restricted Lie algebra such that $\mathfrak{L}'_\Omega = \Omega'_{nl}$ and $p \nmid n$. Thus $\mathfrak{L}'$ is a subset of Ω'_{nl} and contains $n^2 - 1$ elements $a_1, a_2, \cdots, a_{n^2-1}$ such that every element of $\mathfrak{L}$ has the form $\Sigma\, a_j\phi_j$, ϕ in Φ and every element of Ω'_{nl} may be written in one and only one way as $\Sigma\, a_j\omega_j$, ω in Ω. The coefficients of the matrices a_j generate a finite algebraic extension Γ of Φ. Thus the a_j are linear combinations of $e_{ii} - e_{jj}$ and e_{ij}, $i \neq j$, with coefficients in Γ and hence the $e_{ii} - e_{ji}$, e_{ij} are linear combinations of the a_j with coefficients in Γ also. It follows that the set $\Sigma\, a_j\gamma_j$, γ in Γ, is Γ'_{nl} and so we may replace the algebraically closed field Ω by the finite extension Γ in our discussion.

LEMMA 5. *If $\mathfrak{L}$ is a restricted Lie algebra of type A over Φ there exists a separable extension Σ of Φ such that $\mathfrak{L}_\Sigma = \Sigma_{nl}$ and $\mathfrak{L}'_\Sigma = \Sigma'_{nl}$.*

From the theory of associative algebras we know that every normal simple algebra $\mathfrak{A}$ has a separable splitting field Σ, i. e., $\mathfrak{A}_\Sigma = \Sigma_n$. Hence if $\mathfrak{L} = \mathfrak{A}_l$, $\mathfrak{L}_\Sigma = \mathfrak{A}_{l\Sigma} = \Sigma_{nl}$. If $\mathfrak{L} = \mathfrak{S}(\mathfrak{A}, J)$, $\mathfrak{A}$ with center $\mathrm{P} = \Phi(q)$, $\mathfrak{L}_\mathrm{P} = \mathfrak{A}_{l\mathrm{P}}$.

There is a field Σ separable over P such that $\mathfrak{L}_\Sigma = \mathfrak{A}_{l\Sigma} = \Sigma_{nl}$. Since Σ is separable over Φ it has the required properties.

We may now prove the following "completeness" theorem.

THEOREM 8. *If $\mathfrak{L}'$ is a restricted Lie algebra over Φ such that $\mathfrak{L}'_\Omega = \Omega'_{nl}$ and $p \nmid n$ then either there exists a normal simple associative algebra $\mathfrak{A}$ such that $\mathfrak{L}' \cong \mathfrak{A}'_l$ or there exists a simple associative algebra with center $P = \Phi(q)$, a quadratic field over Φ, and with an involution J of the second kind such that $\mathfrak{L}' \cong \mathfrak{S}(\mathfrak{A}, J)'$.*

We suppose that Γ is chosen as above and let Δ be the maximal separable subfield of Γ. Then there is a chain of subfields $\Delta_0 = \Delta$, $\Delta_i = \Delta_{i-1}(x_i)$, $x_i^p = \xi_i \epsilon \Delta_{i-1}$ and $\Delta_u = \Gamma$. Suppose first that $u = 0$, i. e. that Γ is separable. Then we may extend Γ to $\bar{\Gamma}$ which is normal and separable over Φ and we have $\mathfrak{L}'_{\bar{\Gamma}} = \bar{\Gamma}'_{nl}$. Let $\mathfrak{G} = (1, s, \cdots, v)$ be its Galois group. If $a = (\alpha_{ij}) \epsilon \bar{\Gamma}'_{nl}$ we define $a^s = (\alpha_{ij}{}^s)$ and if $a = \sum a_j \bar{\gamma}_j$ define $a^{S_1} = \sum a_j{}^s \gamma_j$. Since $[a_j, a_k] = \sum a_p \alpha_{pjk}$, α_{pjk} in Φ, $a_j{}^p = \sum a_q \beta_{qj}$, β_{qj} in Φ, S_1 is an automorphism of $\bar{\Gamma}'_{nl}$ over $\bar{\Gamma}$. Hence there exists a t_s in $\bar{\Gamma}_n$ such that either $a^{S_1} = t_s{}^{-1} a t_s$ or $a^{S_1} = - t_s{}^{-1} a' t_s$. The s for which S_1 is of the first type form an invariant subgroup $\mathfrak{h}$ of index 1 or 2 in $\mathfrak{g}$ since the automorphisms of the form II form such a subgroup in the group of all automorphisms of $\bar{\Gamma}'_{nl}$ over $\bar{\Gamma}$.[14] Suppose first that $\mathfrak{h} = \mathfrak{g}$. Then the mappings S defined by $a^S = a^{sS_1^{-1}}$ generate automorphisms of the associative algebra $\bar{\Gamma}_n$ over Φ. The elements left invariant by the S form, therefore, an algebra $\mathfrak{A}$ over Φ. Since $a_1, \cdots, a_{n^2-1}$, $a_{n^2} = 1$, is a basis for $\bar{\Gamma}_n$ and $(\sum a_i \bar{\gamma}_i)^S = \sum a_i \bar{\gamma}_i{}^s$, $\mathfrak{A}$ consists of the elements $\sum a_i \phi_i$, ϕ_i in Φ. The elements $a_1, \cdots, a_{n^2-1}$ are linear combinations of commutators $[a_j, a_k]$ and therefore the set $\sum_1^{n^2-1} a_j \phi_j$ is $\mathfrak{A}'_l$. If $\mathfrak{h}$ has index 2 there is a quadratic subfield $P = \Phi(q)$ such that the Galois group of $\bar{\Gamma}$ over P is $\mathfrak{h}$. Then the set $\sum a_i \rho_i$, ρ_i in P, $a_{n^2} = 1$ is a simple algebra with P as center. Let v be an element in $\mathfrak{g}$ not in $\mathfrak{h}$ and $a_j{}^{V_1} = - t_v{}^{-1} a'_j t_v$. v induces the fundamental automorphism $\rho \rightarrow \bar{\rho}$ in P. Consider the correspondence J defined as $a^J = (\sum a_i \rho_i)^J = - \sum a_j \bar{\rho}_j + a_{n^2} \bar{\rho}_{n^2}$. Since $a^J = (t_v{}^{v^{-1}})^{-1} (a')^{v^{-1}} t_v{}^{v^{-1}}$, J is an anti-automorphism in $\mathfrak{A}$. By its form it is involutorial and its skew elements are $\sum a_j \phi_j + (a_{n^2} q)\phi$, $\bar{q} = -q$ in P.[15] Hence $\mathfrak{S}(\mathfrak{A}, J)'$ is the set $\sum a_j \phi_j = \mathfrak{L}'$.

Now suppose that the theorem holds for $u - 1$ and consider $\mathfrak{L}'_{\Delta_{u-1}}$. This

---

[14] Cf. the author's paper "Simple Lie algebras of type A," *Annals of Mathematics*, vol. 39 (1938), p. 186.

[15] If $p = 2$ we may take $q = 1$.

algebra becomes Γ'_{nl} when Δ_{u-1} is extended to $\Gamma = \Delta_{u-1}(x)$, $x = x_u$, $x^p = \xi \in \Delta_{u-1}$. Let d be a derivation in Γ over Δ_{u-1} such that the elements of Δ_{u-1} are the only d-constants.[16] Then $a = (\alpha_{ij}) \to a^d = (\alpha_{ij}{}^d)$ and $(\Sigma a_j\gamma_j)D_1 = \Sigma a_j{}^d\gamma_j$ are derivations in Γ_n over Δ_{u-1} and in Γ_{nl} over Γ, respectively. The last statement holds since the multiplication table $[a_j, a_k]$, $a_j{}^p$ has coefficients in Δ_{u-1}. D_1 is inner and, hence, is induced by a derivation of the associative algebra Γ_n over Γ. Hence $D = D_1 - d$ is a derivation in Γ_n over Δ_{u-1} and the set of D-constants is an algebra $\mathfrak{B}$ over Δ_{u-1}. Since $(\Sigma a_i\gamma_i)(D_1 - d) = (\Sigma a_i\gamma_i)D_1 - (\Sigma a_i{}^d\gamma_i + \Sigma a_i\gamma_i{}^d) = -\Sigma a_i\gamma_i{}^d$ the D-constants are the elements of the form $\Sigma a_i\delta_i$, δ_i in Δ_{u-1}. Hence $\mathfrak{L}'_{\Delta_{u-1}} = \mathfrak{B}'_l$. There is a separable extension Λ of Δ_{u-1} such that $\mathfrak{L}'_\Lambda = \mathfrak{B}'_{l\Lambda} = \Lambda'_{nl}$. As is well known $\Lambda = \bar{\Delta}_{u-1}$ where $\bar{\Delta}$ is the maximal separable subfield of Λ over Φ.[17] Hence by the hypothesis of the induction $\mathfrak{L}'$ has the form $\mathfrak{A}'_l$ or the form $\mathfrak{S}(\mathfrak{A}, J)'$.

II. RESTRICTED LIE ALGEBRAS OF TYPES *B*, *C*, *D*.

2.1. Involutions in Ω_n. Since the automorphisms of the associative algebra Ω_n are all inner, the involutions in Ω_n have the form $a^J = s^{-1}a's$ where $s' = \pm s$. We assume that Ω is algebraically closed. Then by replacing J, if necessary, by a cogredient involution we may suppose when $p \neq 2$ that s has one of the following forms

$$(9) \qquad s_1 = \begin{pmatrix} 0 & 1_\nu \\ -1_\nu & 0 \end{pmatrix}; \qquad s_2 = \begin{pmatrix} 1 & 0 & 0 \\ 0 & 0 & 1_\nu \\ 0 & 1_\nu & 0 \end{pmatrix}; \qquad s_3 = \begin{pmatrix} 0 & 1_\nu \\ 1_\nu & 0 \end{pmatrix}$$

where 1_ν is the identity matrix of ν rows and columns. The corresponding involutions will be denoted as J_1, J_2, J_3. A computation shows that $\mathfrak{S}(\Omega_n, J_i)$, $i = 1, 2, 3$, consists respectively of the following sets of matrices

$$\begin{pmatrix} a_{11} & a_{12} \\ a_{21} & a_{22} \end{pmatrix}, \quad a_{ij} \in \Omega_\nu, \quad a'_{11} = -a_{22}, \quad a'_{12} = a_{12}, \quad a'_{21} = a_{21}$$

$$\begin{pmatrix} 0 & v_1 & v_2 \\ -v'_2 & a_{11} & a_{12} \\ -v'_1 & a_{21} & a_{22} \end{pmatrix}, \quad a_{ij} \in \Omega_\nu, \quad a'_{11} = -a_{22}, \quad a'_{12} = -a_{12}, \quad a'_{21} = -a_{21}$$

and where $v_1 = (b_1, b_2, \cdots, b_\nu)$ and $v_2 = (b_{\nu+1}, \cdots, b_{2\nu})$

$$\begin{pmatrix} a_{11} & a_{12} \\ a_{21} & a_{22} \end{pmatrix}, \quad a'_{11} = -a_{22}, \quad a'_{12} = -a_{12}, \quad a'_{21} = -a_{21}.$$

---

[16] The existence of such derivations has been proved by Baer, "Algebraische Theorie der differentiierbaren Funktionenkörper. I," *Sitzungsberichte Heidelberger Akademie*, 1927, pp. 15-32.

[17] See Albert, *loc. cit.*[4], p. 102.

Thus for $\mathfrak{S}(\Omega_n, J_1)$ we have the basis $h_i = e_{ii} - e_{\nu+i,\,\nu+i}$; $e_{-\lambda_i+\lambda_j} = e_{ij} - e_{j+\nu,\,i+\nu}$ for $i \neq j$; $e_{-\lambda_i-\lambda_j} = e_{i,j+\nu} + e_{j,i+\nu}$, $i < j$; $e_{\lambda_i+\lambda_j} = e_{i+\nu,j} + e_{j+\nu,i}$, $i < j$; $e_{-2\lambda_i} = e_{i,i+\nu}$; $e_{2\lambda_i} = e_{i+\nu,i}$ where $i, j = 1, \cdots, \nu$. If we set $h = \Sigma\, h_i\lambda_i$ we obtain the following multiplication table

$$
(10)\quad
\begin{aligned}
&[h_i, h_j] = 0\\
&[e_\alpha, h] = e_\alpha\alpha, \qquad\qquad \alpha = \pm\lambda_i \pm \lambda_j\\
&[e_\alpha, e_\beta] = \begin{cases} 0 & \text{if } \alpha+\beta \text{ is not a root}\\ e_{\alpha+\beta}N_{\alpha\beta},\ N_{\alpha\beta} = \pm 1, \pm 2, & \text{if } \alpha+\beta \neq 0 \text{ is a root}\end{cases}\\
&[e_{-\alpha}, e_\alpha] = h_\alpha \neq 0
\end{aligned}
$$

where $h_{\lambda_i-\lambda_j} = h_i - h_j$, $h_{\lambda_i+\lambda_j} = h_i + h_j$, $h_{2\lambda_i} = h_i$ and

$$(11)\qquad h_i^p = h_i, \quad e_\alpha^p = 0.$$

For $\mathfrak{S}(\Omega_n, J_2)$ the following is a basis:

$$
\begin{array}{lll}
h_i = e_{i+1,\,i+1} - e_{i+\nu+1,\,i+\nu+1}; & e_{\lambda_i-\lambda_j} = e_{j+1\ i+1} - e_{i+\nu+1,\,j+\nu+1}, & i \neq j;\\
e_{\lambda_i+\lambda_j} = e_{i+\nu+1,\,j+1} - e_{j+\nu+1,\,i+1}, & & i < j;\\
e_{-\lambda_i-\lambda_j} = e_{j+1,\,i+\nu+1} - e_{i+1,\,j+\nu+1}, & & i < j;\\
e_{\lambda_i} = e_{1,\,i+1} - e_{i+\nu+1,\,1}; & e_{-\lambda_i} = e_{i+1,\,1} - e_{1,\,i+\nu+1} &
\end{array}
$$

where $i, j = 1, \cdots, \nu$. The multiplication table has the same form as (10) and (11) where $h_{\lambda_i-\lambda_j} = h_i - h_j$, $h_{\lambda_i+\lambda_j} = h_i + h_j$, $h_{\lambda_i} = h_j$. For $\mathfrak{S}(\Omega_n, J_3)$ the following is a basis:

$$
\begin{array}{lll}
h_i = e_{ii} - e_{\nu+i,\,\nu+i}; & e_{\lambda_i-\lambda_j} = e_{ji} - e_{i+\nu,\,j+\nu}, & i \neq j;\\
e_{\lambda_i+\lambda_j} = e_{i+\nu,\,j} - e_{j+\nu,\,i}, & & i < j;\\
e_{-\lambda_i-\lambda_j} = e_{j,\,i+\nu} - e_{i,\,j+\nu}, & & i < j
\end{array}
$$

where $i, j = 1, \cdots, \nu$. The multiplication table is the same as before with $N_{\alpha\beta} = \pm 1$, $h_{\lambda_i-\lambda_j} = h_i - h_j$, $h_{\lambda_i+\lambda_j} = h_i + h_j$.

Now suppose $p = 2$. In this case if s is alternate in the sense that $s' = s$ and the diagonal elements of s are all 0, it is cogredient to $q = \begin{pmatrix} 0 & 1_\nu \\ 1_\nu & 0 \end{pmatrix}$. If s is symmetric and not alternate it is cogredient to 1.[18] Accordingly we have to consider two Lie algebras $\mathfrak{S}(\Omega_n, J_1)$ and $\mathfrak{S}(\Omega_n, J_2)$. If we compare with the case $p \neq 2$ we see that the present $\mathfrak{S}(\Omega_n, J_1) = \bar{\mathfrak{S}}(\Omega_n, J_1) + \mathfrak{K}$ where $\mathfrak{S}$ consists of the matrices $\begin{pmatrix} a_{11} & a_{12} \\ a_{21} & a_{22} \end{pmatrix}$ with $a_{22} = a'_{11}$, a_{12} and a_{21} alternate

[18] See Albert, "Symmetric and alternate matrices in an arbitrary field. I," *Transactions of the American Mathematical Society*, vol. 43 (1938), p. 392.

and $\mathfrak{K}$ has the basis $e_{i+\nu, i}$, $e_{i, i+\nu}$, $i = 1, \cdots, \nu$. If $a \in \bar{\bar{\mathfrak{S}}}$, $b \in \mathfrak{K}$ then $[a, b] \in \dot{\mathfrak{S}}$ and $b^2 \in \bar{\bar{\mathfrak{S}}}$. $\bar{\bar{\mathfrak{S}}}$ has the following basis:

$$h_i = e_{ii} + e_{\nu+i, \nu+i}; \qquad f_{ij} = e_{ij} + e_{j+\nu, i+\nu}; \qquad i \neq j;$$
$$g_{ij} = g_{ji} = e_{i, j+\nu} + e_{j, i+\nu}, \qquad i \neq j;$$
$$k_{ij} = k_{ji} = e_{i+\nu, j} + e_{j+\nu, i}, \qquad i \neq j$$

and the multiplication table

$$(12) \quad \begin{aligned} &[f_{ij}, h] = f_{ij}(\lambda_i + \lambda_j), \ [g_{ij}, h] = g_{ij}(\lambda_i + \lambda_j), \ [k_{ij}, h] = k_{ij}(\lambda_i + \lambda_j) \\ &[f_{ij}, f_{kl}] = \delta_{jk} f_{il} + \delta_{il} f_{kj}, \ (k, l) \neq (i, j) \text{ or } (j, i) \\ &[f_{ij}, f_{ji}] = h_i + h_j \\ &[g_{ij}, k_{jk}] = f_{ik}, \ [g_{ij}, k_{ij}] = h_i + h_j \\ &[g_{ij}, f_{kj}] = g_{ik} \\ &[k_{ij}, f_{jk}] = k_{ik} \end{aligned}$$

where i, j, k are unequal. All other products are 0 and

$$(13) \qquad h_i^2 = h_i, \quad f_{ij}^2 = 0, \quad g_{ij}^2 = 0, \quad k_{ij}^2 = 0.$$

It follows that $\bar{\bar{\mathfrak{S}}}$ is the ideal generated by the elements $[c, d]$ and c^2, c, d in $\mathfrak{S}$.

The condition that $q^{-1}a'q = a$ is equivalent to $(qa)' = (qa)$ and $a \in \mathfrak{S}$ if and only if (qa) is alternate. Consider an arbitrary alternate matrix

$$b = \begin{pmatrix} b_{11} & b_{12} & \cdots & b_{1n} \\ b_{21} & b_{22} & \cdots & b_{2n} \\ \cdot & \cdot & \cdots & \cdot \\ \cdot & \cdot & \cdots & \cdot \\ \cdot & \cdot & \cdots & \cdot \\ b_{n1} & b_{n2} & \cdots & b_{nn} \end{pmatrix} \qquad b_{ij} = b_{ji}, \quad b_{ii} = 0, \quad n = 2\nu$$

Let $\Delta = \det b$ and $B_{ij} =$ cofactor of b_{ji} in b. Then $B_{ij} = B_{ji}$ and $B_{ii} = 0$. Hence we have the equations

$$b_{12}B_{21} + b_{13}B_{31} + \cdots + b_{1n}B_{n1} = \Delta$$
$$b_{j2}B_{21} + b_{j3}B_{31} + \cdots + b_{jn}B_{n1} = 0, \qquad (j = 3, \cdots, n)$$

and solving for B_{21} we obtain $B_{21}^2 = \Delta\Delta_2$ where Δ_2 is the determinant of an alternate matrix of $n - 2$ rows and columns. We assume that it is the square of a polynomial in the b_{ij} and obtain the result that Δ has this property, say $\Delta = B^2$. The equation $B_{21}^2 = (BB_2)^2$ then shows that B is a factor of B_{21}.

Similarly every B_{ij} is divisible by B. If we apply this result to $b = q\lambda + qa$, λ an indeterminate, we obtain that det $(\lambda + a)$ is a square $\phi(\lambda)^2$ in $\Omega[\lambda]$ and that adj $(\lambda + a)$ is divisible by $\phi(\lambda)$. It follows in the usual manner that $\phi(a) = 0$.[19] If we use the form of $a = \begin{pmatrix} a_{11} & a_{12} \\ a_{21} & a_{22} \end{pmatrix}$ and set $a_{11} = (\alpha_{ij}) = a'_{22}$ a simple computation shows that $\phi(\lambda)^2 = \lambda^n + (\alpha_{11} + \cdots + \alpha_{\nu\nu})^2\lambda^{n-2} + \cdots$. Thus $\phi(\lambda) = \lambda^\nu + (\mathrm{tr}' a)\lambda^{\nu-1} + \cdots$ where $\mathrm{tr}' a = \alpha_{11} + \cdots + \alpha_{\nu\nu}$.

LEMMA 6. *If* $a = \begin{pmatrix} a_{11} & a_{12} \\ a_{21} & a_{22} \end{pmatrix} \epsilon \bar{\mathfrak{S}}$ *where* $a_{11} = (\alpha_{ij})$ *then the characteristic polynomial of* a *has the form* $[\phi(\lambda)]^2$ *where* $\phi(\lambda) = \lambda^\nu + (\alpha_{11} + \cdots + \alpha_{\nu\nu})\lambda^{\nu-1} + \cdots$ *and* $\phi(a) = 0$. *This is true also when the elements of* a *are regarded as indeterminates.*

The second type of $\mathfrak{S}$ for $p = 2$, namely, $\mathfrak{S}(\Omega_n, J_2)$ has the basis $h_i = e_{ii}$, $e_{\lambda_i + \lambda_j} = e_{ij} + e_{ji}$ with $i < j$ and $i, j = 1, \cdots, n$. The multiplication table is

$$(14) \qquad \begin{aligned} &[h_i, h_j] = 0 \\ &[e_\alpha, h] = e_\alpha \alpha \\ &[e_\alpha, e_\beta] = \begin{cases} 0 \text{ if } \alpha + \beta \text{ is not a root} \\ e_{\alpha+\beta} \text{ if } \alpha + \beta \text{ is a root} \neq 0 \end{cases} \end{aligned}$$

and

$$(15) \qquad h_i^2 = h_i, \; e^2_{\lambda_i + \lambda_j} = h_i + h_j.$$

2. 2. Simplicity proofs. The enveloping algebras.

LEMMA 7. *If* $p \neq 2$, $\mathfrak{S}(\Omega_n, J_i)$ *is simple except when* $n = 4$, $i = 3$. *If* $p = 2$, $\bar{\mathfrak{S}}(\Omega_n, J_1)$ *with* $\nu = n/2 > 2$ *has only the proper ideals* (1) *of multiples of* 1 *and* $\bar{\mathfrak{S}}'$ *of elements* a *for which* $\mathrm{tr}' a = 0$. $\mathfrak{S}(\Omega_n, J_2)$, $p = 2$, $n > 2$ *has only the proper ideals* (1) *and* $\mathfrak{S}'$ *the elements of trace* 0.

Suppose $p \neq 2$ and let $\mathfrak{B}$ be an ideal $\neq 0$ in $\mathfrak{S}(\Omega_n, J_i)$ and $b = h_0 + \Sigma\, a_\alpha \beta_\alpha$ be an element $\neq 0$ in $\mathfrak{B}$. Choose $\bar{\lambda}_i$ in Ω so that the values $\bar{\alpha}$ of the roots α for these $\bar{\lambda}_i$ are all distinct and $\neq 0$. Then

$$[\cdots[[b, \bar{h}], \bar{h}] \cdots \bar{h}] = 0 + \Sigma\, e_\alpha \beta_\alpha \bar{\alpha}^k \quad (k \text{ brackets})$$

is in $\mathfrak{B}$ and since the Vandermonde determinant

[19] I am indebted to Dr. J. Williamson for the present proof.

$$\begin{vmatrix} 1 & 0 & 0 & \cdots \\ 1 & \bar{\alpha} & \bar{\alpha}^2 & \cdots \\ \cdot & & & \\ \cdot & & & \\ \cdot & & & \\ 1 & \bar{\epsilon} & \bar{\epsilon}^2 & \cdots \end{vmatrix} \neq 0$$

(number of rows = number of roots plus 1), h_0 and each $e_\alpha\beta_\alpha \in \mathfrak{B}$. If $\beta_\alpha \neq 0$, $\mathfrak{B}$ contains e_α and if $h_0 \neq 0$ at least one $[e_\alpha, h_0] = e_\alpha \alpha_0 \neq 0$ so that we may suppose that $\mathfrak{B}$ contains an e_α. Now except in the cases $i = 1$, $n = 2$; $i = 2$, $n = 1, 3$; $i = 3$, $n = 2, 4$ it is readily verified that we may obtain any root ρ in the form $(((\alpha + \beta) + \gamma) + \delta \cdots)$ where each parenthesis $(\alpha + \beta)$, $((\alpha + \beta) + \gamma) \cdots$ is a root $\neq 0$. It follows that $[\cdots[[e_\alpha, e_\beta], e_\gamma]\cdots] = e_\rho N_{\alpha\beta} N_{\beta\gamma} \cdots \neq 0$ and that $e_\rho \in \mathfrak{B}$ for every ρ. Since the elements $[e_{-\alpha}, e_\alpha] = h_\alpha$ generate the subalgebra $\mathfrak{H}$ of elements h_i we obtain $\mathfrak{B} = \mathfrak{S}$. If $i = 1$, $n = 2$, $\mathfrak{B}$ contains e_α and $[-e_\alpha, e_\alpha] = h_\alpha \neq 0$ and hence also $[e_{-\alpha}, h_\alpha] = e_{-\alpha}\beta \neq 0$ so that $\mathfrak{B} = \mathfrak{S}$. If $i = 2$, $n = 1$, $\mathfrak{S} = 0$ and if $i = 2$, $n = 2$ the argument is the same as for $\mathfrak{S}(\Omega_n, J_1)$. If $i = 3$, $n = 2$, $\mathfrak{S}$ has order 1 and if $i = 3$, $n = 4$ the result does not hold.

$p = 2$. Let $\mathfrak{B}$ be an ideal in $\bar{\mathfrak{S}}$ and containing b an element $\neq 1\alpha$. If $b = h_0 + \sum_{i \neq j} f_{ij}\rho_{ij} + \sum_{i<j} g_{ij}\sigma_{ij} + \sum_{i<j} k_{ij}\tau_{ij}$ then by the above argument h_0 and $u_{ij} = f_{ij}\rho_{ij} + f_{ji}\rho_{ji} + g_{ij}\sigma_{ij} + k_{ij}\tau_{ij}$ are in $\mathfrak{B}$. If $\sigma_{ij} \neq 0$ and $k \neq i, j$ then $[[u_{ij}, k_{jk}], f_{ji}]\sigma_{ij}^{-1} = f_{jk} \in \mathfrak{B}$ and similarly if $\tau_{ij} \neq 0$, $\mathfrak{B}$ contains an f_{jk}. If $\sigma_{ij} = \tau_{ij} = 0$, $\mathfrak{B}$ contains either $h_0 \neq 1\alpha$ or some $f_{ij}\rho_{ij} + f_{ji}\rho_{ji} \neq 0$. Now the Lie algebra generated by the h_i and f_{ij} is isomorphic to $\Omega_{\nu l}$. Hence by the structure of $\Omega_{\nu l}$ we obtain that $\mathfrak{B}$ contains all $\begin{pmatrix} a_{11} & 0 \\ 0 & a'_{11} \end{pmatrix}$ with $\operatorname{tr} a_{11} = 0$ and also $[g_{ij}, h_j + h_k] = g_{ij}$ and $[k_{ij}, h_j + h_k] = k_{ij}$. Thus $\mathfrak{B}$ contains all elements a of $\bar{\mathfrak{S}}$ with $\operatorname{tr}' a = 0$. If in addition $\mathfrak{B}$ contains an element not in $\bar{\mathfrak{S}}'$, $\mathfrak{B} = \bar{\mathfrak{S}}$.

In the case $\mathfrak{S}(\Omega_n, J_2)$, $p = 2$, $n > 2$ any $\mathfrak{B}$ which contains an element $b \neq 1\alpha$ contains an $e_\alpha \neq 0$ and hence all e_α and all $e_\alpha^2 = h_i + h_j$. It follows that $\mathfrak{B} = \mathfrak{S}(\Omega_n, J_2)'$ or $\mathfrak{B} = \mathfrak{S}(\Omega_n, J_2)$.

COROLLARY. *If* $p = 2$ *and* $\nu \not\equiv 0 \pmod 2$, $\bar{\mathfrak{S}}(\Omega_n, J_1) = \bar{\mathfrak{S}}(\Omega_n, J_1)' \oplus (1)$ *and* $\bar{\mathfrak{S}}'$ *is simple. If* $\nu \equiv 0 \pmod 2$ *the difference algebra* $\bar{\mathfrak{S}}'/(1)$ *is simple. Similarly* $\mathfrak{S}(\Omega_n, J_2) = \mathfrak{S}(\Omega_n, J_2)' \oplus (1)$ *when* $n \not\equiv 0 \pmod 2$ *and* $\mathfrak{S}'$ *is simple;* $\mathfrak{S}'/(1)$ *is simple when* $n \equiv 0 \pmod 2$.

If $\nu \equiv 0 \pmod 2$ $\bar{\mathfrak{S}}'$ does not contain (1), otherwise it does.

Remark. It should be noted that the proof of the above lemma is valid when the algebras are regarded as ordinary Lie algebras except when $p = 2$, $i = 2$.

In the remainder of the paper we suppose that for $p \neq 2$, $i = 2$, $n \geq 3$; if $i = 3$, $n \geq 6$ and if $p = 2$, $i = 1$, $\nu \geq 3$; if $i = 2$, $n \geq 3$. With these restrictions on n we have

LEMMA 8. *If $p \neq 2$ the enveloping algebra of $\mathfrak{S}(\Omega_n, J_i)$ is Ω_n. If $p = 2$ the enveloping algebra of $\bar{\mathfrak{S}}(\Omega_n, J_1)'$ or of $\mathfrak{S}(\Omega_n, J_2)'$ is Ω_n.*

$p \neq 2, i = 1$. The enveloping algebra $\mathfrak{B}$ contains all matrices of the form

$$\begin{pmatrix} a & \\ & -a' \end{pmatrix}\begin{pmatrix} 0 & b \\ 0 & 0 \end{pmatrix} = \begin{pmatrix} 0 & ab \\ 0 & 0 \end{pmatrix} \quad \text{if} \begin{pmatrix} 0 & b \\ 0 & 0 \end{pmatrix} \epsilon\, \mathfrak{B}.$$

and

$$\begin{pmatrix} 0 & b \\ 0 & 0 \end{pmatrix}\begin{pmatrix} -a' & \\ & a \end{pmatrix} = \begin{pmatrix} 0 & ba \\ 0 & 0 \end{pmatrix}.$$

Thus for the matrices $\begin{pmatrix} 0 & b \\ 0 & 0 \end{pmatrix}$ in $\mathfrak{B}$ the set $\{b\}$ forms a two sided ideal in Ω_ν (since a is arbitrary). Since Ω_ν is a simple associative algebra every $\begin{pmatrix} 0 & a \\ 0 & 0 \end{pmatrix}$ is in $\mathfrak{B}$ and similarly every $\begin{pmatrix} 0 & 0 \\ 0 & a \end{pmatrix}$ is in $\mathfrak{B}$. Then $\mathfrak{B}$ contains all

$$\begin{pmatrix} 0 & a \\ 0 & 0 \end{pmatrix}\begin{pmatrix} 0 & 0 \\ 1 & 0 \end{pmatrix} = \begin{pmatrix} a & 0 \\ 0 & 0 \end{pmatrix} \qquad \begin{pmatrix} 0 & 0 \\ 1 & 0 \end{pmatrix}\begin{pmatrix} 0 & a \\ 0 & 0 \end{pmatrix} = \begin{pmatrix} 0 & 0 \\ 0 & a \end{pmatrix}$$

and $\mathfrak{B} = \Omega_n$.

$p \neq 2$, $i = 3$. The above proof is valid.

$p \neq 2$, $i = 2$. The argument shows that $\mathfrak{B}$ contains all matrices of the form

$$\begin{pmatrix} 0 & 0 & 0 \\ 0 & a_{11} & a_{12} \\ 0 & a_{21} & a_{22} \end{pmatrix}$$

where a_{ij} are arbitrary in Ω_ν. $\mathfrak{B}$ contains also

$$(e_{12} - e_{\nu+1,1})(e_{1,\nu+1} - e_{21}) = -e_{11} - e_{\nu+1,\nu+1}$$

and hence

$$e_{11} \text{ and } e_{1,i+1} = e_{11}(e_{1,i+1} - e_{i+\nu+1,1}), \quad e_{i+\nu+1,1} = -(e_{1,i+1} - e_{i+\nu+1,1})e_{11}$$

and similarly $e_{i+1,1}$, $e_{1,i+\nu+1}$. Hence $\mathfrak{B} = \Omega_n$.

$p=2,\ i=1$. The enveloping algebra of matrices of the form $\begin{pmatrix} a & 0 \\ 0 & a' \end{pmatrix}$ with $\operatorname{tr} a = 0$ includes the matrices

$$\begin{pmatrix} a_1 & \\ & a_2 \end{pmatrix}, \quad \begin{pmatrix} a_2 & \\ & a_1 \end{pmatrix}$$

where a_1 is arbitrary and a_2 depends on a_1. Then by the argument in the first case $\mathfrak{B} = \Omega_n$.

$p=2,\ i=2$. $\mathfrak{B}$ contains $(e_{ii}+e_{jj})(e_{ik}+e_{ki}) = e_{ik}$ if i, j, k are unequal and $e_{ii} = e_{ik}e_{ki}$. Hence $\mathfrak{B} = \Omega_n$.

COROLLARY. *The sets of matrices of Lemma 8 are irreducible sets.*

2.3. Restricted derivations.

LEMMA 9. *If $p \neq 2$ the restricted derivations of $\mathfrak{S}(\Omega_n, J_i)$ are all inner. The same holds for $p=2$ and $i=2$. The restricted derivations of $\bar{\mathfrak{S}}(\Omega_n, J_1)$ are induced by inner derivations of Ω_{nl}.*

$p \neq 2$ or $p=2,\ i=2$. Choose $\bar{\lambda}_i$ in Ω so that the values $\Sigma\, m_i\bar{\lambda}_i$ of the p^n linear forms $\Sigma\, m_i\lambda_i$ are distinct. Then the minimum p-polynomial of

$$\bar{h} = \Sigma\, h_i\bar{\lambda}_i \text{ is } \Pi(\lambda - \Sigma\, m_i\bar{\lambda}_i) = \lambda^{p^\nu} + \cdots + \lambda\beta_0,\ \beta_0 = \prod_{\neq 0}(\Sigma\, m_i\bar{\lambda}_i) \neq 0.$$

Thus

$$\bar{h}^{p^\nu} + \cdots + \bar{h}\beta_0 = 0.$$

If $p=2,\ i=2$ the same result holds with n in place of ν. Suppose D is a restricted derivation. Then

$$(\bar{h}D)(\bar{H}^{p^\nu-1} + \cdots + \beta_0) = (\bar{h}D)\prod_{\neq 0}(\bar{H} - \Sigma\, m_i\lambda_i) = 0$$

where $\bar{H}$ is the mapping $x \rightarrow [x, \bar{h}]$. Hence $\bar{h}D = \Sigma\, e_\alpha\rho_\alpha$. If $\sigma_\alpha = -\bar{\alpha}^{-1}\rho_\alpha$, $[\bar{h}, \Sigma\, e_\alpha\sigma_\alpha] = \bar{h}D$. By subtracting the inner derivation $x \rightarrow [x, \Sigma\, e_\alpha\sigma_\alpha]$ from D we obtain the restricted derivation E such that $\bar{h}E = 0$. It follows that $\bar{h}^pE = \bar{h}^{p^2}E = \cdots = 0$ and since $\bar{h}, \bar{h}^p, \cdots$ form a basis for $\mathfrak{H}$, $hE = 0$ for all h in $\mathfrak{H}$. Since $[e_\alpha, \bar{h}] = e_\alpha\bar{\alpha}$, $[e_\alpha E, h] = e_\alpha E$ and $e_\alpha E = e_\alpha\xi_\alpha$. Since $[e_{-\alpha}, e_\alpha] = h_\alpha \neq 0$, $[e_{-\alpha}\xi_{-\alpha}, e_\alpha] + [e_{-\alpha}, e_\alpha\xi_\alpha] = 0$ and $\xi_{-\alpha} = -\xi_\alpha$. Similarly $[e_\alpha, e_\beta] = e_{\alpha+\beta}N_{\alpha+\beta}$ yields $\xi_{\alpha+\beta} = \xi_\alpha + \xi_\beta$ if $\alpha+\beta \neq 0$ is a root. By a fundamental set of roots (f. s.) we shall mean a set of roots $\alpha_1, \alpha_2, \cdots, \alpha_\nu$ (or $\alpha_1, \alpha_2, \cdots, \alpha_{n-1}$ if $p = i = 2$) which are linearly independent and such that any root ρ may be obtained as $((\pm\alpha_{i_1} \pm \alpha_{i_2}) \pm \alpha_{i_3}) \pm \cdots$ where each parenthesis is a root $\neq 0$. For $p \neq 2$, $i = 1$, $\lambda_1 - \lambda_2, \cdots, \lambda_1 - \lambda_\nu, \lambda_1 + \lambda_2$ is a f. s.; $p \neq 2$, $i = 2$, $\lambda_1, \lambda_2 \cdots \lambda_\nu$ is a f. s.; $p \neq 2$, $i = 3$, $\lambda_1 - \lambda_2, \cdots, \lambda_1 - \lambda_\nu$, $\lambda_1 + \lambda_2$ is a f. s.; $p = 2$, $i = 1$, $\lambda_1 + \lambda_2, \cdots, \lambda_1 + \lambda_\nu$; $p = i = 2$, $\lambda_1 + \lambda_2, \cdots$, $\lambda_1 + \lambda_n$ is a f. s. Now let $\alpha_1, \alpha_2, \cdots, \alpha_\nu$ $(\alpha_1, \cdots, \alpha_{n-1})$ be a f. s. By the linear independence we may solve the equation $\alpha_i = \xi_{\alpha_i}$ for $\lambda_i = \lambda_{i0}$ in Ω. Then if $h_0 = \Sigma h_i \lambda_{i0}$ and H_0 is the inner derivation $x \to [x, h_0]$ we have $h(E - H_0) = 0$, $e_{\alpha_i}(E - H_0) = 0$. Hence by the above $e_{-\alpha_i}(E - H_0) = 0$ and if $e_\rho = [\cdots[e_\alpha, e_\beta], e_\gamma \cdots] N_{\alpha\beta}^{-1} N_{\beta\gamma}^{-1} \cdots$ where $\alpha, \beta, \cdots = \pm \alpha_i$ we have $e_\rho(E - H_0) = 0$. Thus $E = H_0$ and D is inner.

$p = 2$, $i = 1$. If D is a restricted derivation the above argument shows that we may subtract an inner derivation from D to obtain E such that $hE = 0$, $h \in \mathfrak{H}$. Then

$$f_{ij}E = f_{ij}\alpha_{ij} + f_{ji}\beta_{ji} + g_{ij}\gamma_{ij} + k_{ij}\epsilon_{ij}.$$

Since $f_{ij}^2 E = [f_{ij}E, f_{ij}] = 0$, $\beta_{ji} = 0$. Now $k_i = e_{i,i+\nu}$ and $k_{i+\nu} = e_{i+\nu,i}$ are in $\mathfrak{S}$ and $[f_{i1}, k_{i+\nu}] = k_{1i}$, $[f_{1i}, k_i] = g_{1i}$ if $i > 1$ while all other products with f_{j1}, f_{1j} are 0. We note also that $[h, k_i] = [h, k_{i+\nu}] = 0$. Hence we may subtract the derivation $x \to [x, k]$ where $k = \Sigma k_i \gamma_{1i} + \Sigma k_{\nu+i}\epsilon_{i1}$ from E and obtain a restricted derivation F such that $hF = 0$, $f_{1i}F = f_{1i}\alpha_{1i} + k_{1i}\epsilon_{1i}$, $f_{i1}F = f_{i1}\alpha_{i1} + g_{1i}\gamma_{i1}$. Next we subtract a suitable derivation of the form $x \to [x, h_0]$, h_0 in $\mathfrak{H}$, to obtain G such that $hG = 0$, $f_{1i}G = k_{1i}\epsilon_{1i}$. Since $[f_{i1}, f_{1i}] = h_1 + h_i$ it follows that $f_{i1}G = g_{1i}\gamma_{i1}$. Since $[f_{i1}, k_1] = g_{1i}$, $[f_{1i}, k_{1+\nu}] = k_{1i}$ we can obtain by subtraction a K such that $hK = f_{21}K = f_{12}K = 0$, $f_{i1}K = g_{1i}\gamma_{i1}$, $f_{1i}K = k_{1i}\epsilon_{1i}$ for $i > 2$. Operating with K on $f_{21} = [[f_{21}, f_{1i}], f_{i1}]$ gives $\gamma_{i1} = \epsilon_{1i} = 0$ so that $f_{i1}K = f_{1i}K = 0$ and hence $f_{ij}K = 0$. Then

$$g_{ij}K = g_{ij}\rho_{ij} + f_{ij}\sigma_{ij} + f_{ji}\sigma_{ji} + k_{ij}\tau_{ij}, \quad \rho_{ij} = \rho_{ji}, \quad \tau_{ij} = \tau_{ji}$$

and since $[g_{ij}K, g_{ij}] = 0$, $\tau_{ij} = 0$. Let $k \neq i, j$. Then $[g_{ij}, f_{kj}] = g_{ik}$ and hence

$$g_{ik}\rho_{ik} + f_{ik}\sigma_{ik} + f_{ki}\sigma_{ki} = g_{ik}\rho_{ij} + f_{ki}\sigma_{ji}$$

and $\sigma_{ik} = 0$, $\rho_{ik} = \rho_{ij}$ or $\rho_{ij} = \rho$. In the same way we find that $k_{ij}K = k_{ij}\rho'$ and since $[g_{ij}, k_{ij}] = h_i + h_j$, $\rho = \rho'$. Choose α and β so that $\alpha + \beta = \rho$ and define $g = (\Sigma e_{ii})\alpha + (\Sigma e_{i+\nu,i+\nu})\beta$. Then $xK = [x, g]$.

2.4. Automorphisms. Let S be an automorphism in $\mathfrak{S}(\Omega_n, J_i)$ with $p \neq 2$ and $\{\Lambda\}$ the weights of h in the representation $x \to x^S$. Since this representation is $(1-1)$ there are ν independent weights. Since $(h^S)^{J_i} = -h^S$, if Λ is a weight so is $-\Lambda$. Hence if $i = 1, 3$ the weights are $\Lambda_1, \cdots, \Lambda_\nu, -\Lambda_1, \cdots, -\Lambda_\nu$ and if $i = 2$, $\Lambda_1, \cdots, \Lambda_\nu, 0, -\Lambda_1, \cdots, -\Lambda_\nu$ where $\Lambda_1, \cdots, \Lambda_\nu$ are linearly independent. Thus any subset of unequal weights not containing 0 is linearly independent if and only if it does not contain a weight together with its negative.

Evidently if α is a root, $\Lambda + \alpha, \Lambda, \Lambda - \alpha$ are linearly dependent. Hence if these are weights and $\Lambda \neq 0$ we have either $\Lambda + \alpha = -\Lambda$, $-\Lambda = \Lambda - \alpha$ or $\Lambda + \alpha = 0$, $\Lambda - \alpha = 0$. In the first two cases we have $\Lambda = \pm \alpha/2$ and $3\alpha/2$, $\alpha/2$, $-(\alpha/2)$ or $\alpha/2$, $-(\alpha/2)$, $-(3\alpha/2)$ are weights.[20] Then $-(3\alpha/2)$ or $3\alpha/2$ is a weight and since $\alpha/2 \neq 3\alpha/2, \neq -(3\alpha/2)$ this is impossible. The last two possibilities $\Lambda = \alpha, -\alpha$ can occur only when $i = 2$, for otherwise no weight is 0.

Suppose Λ is a weight $\neq \pm \alpha$, $\neq 0$ and $\Lambda + \alpha$ is not a weight but $\Lambda - \alpha$ is. If $\Lambda - 2\alpha$ is a weight, $\Lambda, \Lambda - \alpha, \Lambda - 2\alpha$ are dependent. Hence either $-\Lambda = \Lambda - \alpha$, $-\Lambda + \alpha = \Lambda - 2\alpha$, $-\Lambda = \Lambda - 2\alpha$ or $\Lambda = 2\alpha$. The first two may be excluded as before, the third since $\Lambda \neq \alpha$. If $\Lambda = 2\alpha$ then the weights are $2\alpha, \alpha, 0$. Then $-2\alpha, -\alpha$ are weights and this is impossible unless $p = 3$ and then we have a contradiction with $\Lambda \neq -\alpha$. Let x_0 be a vector such that $x_0 h^S = x_0 \Lambda$. Since $\Lambda + \alpha$ is not a weight $x_0 e_\alpha{}^S = 0$ and since one of $\Lambda - \alpha$, $\Lambda - 2\alpha$ is not a weight, either $x_0 e_{-\alpha}{}^S = 0$ or $x_0 e_{-\alpha}{}^S \neq 0$ and $x_0 (e_{-\alpha}{}^S)^2 = 0$. By Lemma 1 this implies $\Lambda_\alpha = 0$ or $\Lambda_\alpha = \alpha_\alpha/2$.

$i = 1, 3$. Let $\Lambda = \Sigma\, m_i \lambda_i$ be a weight, α a root. Either $\Lambda + \alpha$ or $\Lambda - \alpha$ is not a weight and we may suppose we have the first case. If $\alpha = \lambda_i - \lambda_j$, $h_\alpha = h_i - h_j$ and $\alpha_\alpha = 2$. Hence $\Lambda_\alpha = m_i - m_j = 0, 1$. If $\Lambda - \alpha$ were a weight we would obtain $m_i - m_j = 0, -1$. If $\Lambda_\alpha = 1$,

$$\Lambda - \alpha = \Lambda - (m_i - m_j)\alpha = \sum_{k \neq i,j} m_k \lambda_k + m_j \lambda_i + m_i \lambda_j$$

is a weight and similarly if $m_i - m_j = -1$, $\Lambda + \alpha = \Lambda - (m_i - m_j)\alpha$ is a weight. If $\alpha = \lambda_i + \lambda_j$, $h_\alpha = h_i + h_j$, $\alpha_\alpha = 2$. Here we obtain $m_i + m_j = 0$, ± 1 and if $m_i + m_j = 1$, $\Lambda - (m_i + m_j)\alpha = \Sigma\, m_k \lambda_k - m_j \lambda_i - m_i \lambda_j$. Then by the above $\Sigma\, m_k \lambda_k - m_i \lambda_i - m_j \lambda_j$ is a weight. Combining these results we obtain the following condition on the weights: The set of weights is invariant under permutation of the λ's and under change of sign of an even number of λ's. Either all $m_i = \pm \frac{1}{2}$ or one $m_i = \pm 1$ and all others are 0. If $i = 1$ we may use the root $\alpha = 2\lambda_i$ and obtain $m_i = 0$, or ± 1. Hence in this case

---

[20] Since we are assuming that $\Lambda + \alpha, \Lambda, \Lambda - \alpha$ are $\neq 0$, $p \neq 3$.

the weights are $\pm \lambda_i$. If $i = 3$ we must consider the possibility $m_i = \pm \frac{1}{2}$. By changing signs and permuting we obtain the weight

$$\tfrac{1}{2}(\lambda_1 + \cdots + \lambda_{\nu-2}) - \tfrac{1}{2}(\lambda_{\nu-1} + \lambda_\nu) \quad \text{or} \quad \tfrac{1}{2}(\lambda_1 + \cdots + \lambda_{\nu-1}) - \tfrac{1}{2}\lambda_\nu.$$

The first case is excluded if $\nu \geqq 5$ for then we obtain by permuting the λ's more than ν linearly independent weights. In the second case if $\nu \geqq 5$ we obtain ν independent weights by permuting and the additional weight $\frac{1}{2}(\lambda_1 + \cdots + \lambda_{\nu-3}) - \frac{1}{2}(\lambda_{\nu-2} + \lambda_{\nu-1} + \lambda_\nu)$ independent of these. If $\nu = 3$ we would obtain the weights $\pm \frac{1}{2}\lambda_1 \pm \frac{1}{2}\lambda_2 \pm \frac{1}{2}\lambda_3$ and this is impossible since there are only six weights. If $\nu = 4$ we can not exclude the following two additional possibilities $\Lambda_j = \frac{1}{2}(\sum_1^4 \lambda_i) - \lambda_j$ for $j = 1, \cdots, 4$ and

$$\Lambda_1 = \tfrac{1}{2}\lambda_1 + \tfrac{1}{2}\lambda_2 + \tfrac{1}{2}\lambda_3 + \tfrac{1}{2}\lambda_4, \qquad \Lambda_2 = \tfrac{1}{2}\lambda_1 + \tfrac{1}{2}\lambda_2 - \tfrac{1}{2}\lambda_3 - \tfrac{1}{2}\lambda_4,$$
$$\Lambda_3 = \tfrac{1}{2}\lambda_1 - \tfrac{1}{2}\lambda_2 + \tfrac{1}{2}\lambda_3 - \tfrac{1}{2}\lambda_4, \qquad \Lambda_4 = \tfrac{1}{2}\lambda_1 - \tfrac{1}{2}\lambda_2 - \tfrac{1}{2}\lambda_3 + \tfrac{1}{2}\lambda_4.$$

Thus unless $i = 3$, $\nu = 4$ the weights of h in the representation $x \to x^S$ are $\pm \lambda_i$.

$i = 2$. If Λ is a weight $\neq 0$, and $\neq$ any root the argument is similar to that for $i = 3$ and shows that $m_i - m_j = 0, \pm 1$, $m_i + m_j = 0, \pm 1$ and that the forms obtained from Λ by permuting the m's are weights. Since $\alpha = \lambda_i$ is a root, $h_{\lambda_i} = h_i$, and $\Lambda_\alpha = m_i$, $\alpha_\alpha = 1$. Hence we obtain $m_i = 0, \pm \frac{1}{2}$ and in the latter case $\Lambda - \alpha$ is a weight and is obtained from Λ by changing the sign of one m. It follows that all $\frac{1}{2}(\Sigma \pm \lambda_j)$ are weights and this is impossible if $\nu > 2$. Thus every $\Lambda \neq 0$ is a root. If $\Lambda = \lambda_i + \lambda_j$, $\alpha = \lambda_j$ then $\Lambda_\alpha = 1$ and $\Lambda - \alpha = \lambda_i = \Lambda'$ is a weight. Then if $\beta = \lambda_i - \lambda_j$, $\Lambda' - \beta = \lambda_j$ is a weight. This is impossible as is $\Lambda = \pm \lambda_i \pm \lambda_j$. Hence the weights here are $\pm \lambda_i$ also.

LEMMA 10. *If $p \neq 2$ and S is an automorphism of $\mathfrak{S}(\Omega_n, J_i)$ where if $i = 3$, $\nu \neq 4$ then there is a matrix q independent of λ such that $h^S = q^{-1}hq$.*

This is equivalent to the result established that the weights of h^S are $\pm \lambda_i$ if $i = 1, 3$ and $0, \pm \lambda_i$ if $i = 2$.

Now suppose $p = 2$, $i = 1$. If $a = \begin{pmatrix} a_{11} & a_{12} \\ a_{21} & a'_{11} \end{pmatrix}$ where $a'_{12} = a_{12}$, $a'_{21} = a_{21}$ with diagonal elements all 0 is in $\bar{\mathfrak{S}}(\Omega_n, J_1)$, we defined $\operatorname{tr}' a = \operatorname{tr} a_{11}$. Then

$$a^2 = \begin{pmatrix} a_{11}^2 + a_{12}a_{21} & a_{11}a_{12} + a_{12}a'_{11} \\ a_{21}a_{11} + a'_{11}a_{21} & a'_{11}{}^2 + a_{21}a_{12} \end{pmatrix}$$

and if $a_{12} = (\alpha_{ij})$, $a_{21} = (\beta_{ij})$ we have tr $a_{12}a_{21} = \Sigma\, \alpha_{ij}\beta_{ji} = 0$ by the conditions on a_{12}, a_{21}. Hence $\mathrm{tr}'\, a^2 = (\mathrm{tr}'\, a)^2$ and if and only if $\nu \equiv 0 \pmod 2$ the correspondence $a \to a + \mathrm{tr}'\, a$ is $(1-1)$ and hence an automorphism in $\bar{\mathfrak{S}}$.

Let S be any automorphism in $\bar{\mathfrak{S}}$. Let Λ be a weight in the representation $x \to x^S$ and x_0 a vector such that $x_0 h^S = x_0\Lambda$. If either $x_0 f_{ij}{}^S$ or $x_0 f_{ji}{}^S \neq 0$ we obtain $\Lambda + (\lambda_i + \lambda_j)$ as a weight also and if $x_0 f_{ij}{}^S = x_0 f_{ji}{}^S = 0$, $x_0(h_i{}^S + h_j{}^S) = 0$ and hence $m_i = m_j$ in $\Lambda = \Sigma\, m_i\lambda_i$. Thus the weights are invariant under permutation of the λ's. Since the characteristic polynomial of h^S is a perfect square the multiplicity of each Λ is even. If $\Lambda = \lambda_{k_1} + \cdots + \lambda_{k_\mu}$ is a weight we obtain, by permuting, $\binom{\nu}{\mu}$ distinct weights. Since there are ν linearly independent weights we may suppose that $0 < \mu < \nu$. By the assumption $\nu \geqq 3$ we obtain either $\mu = 1$ or $\mu = \nu - 1$. In the former case the weights are λ_i and each has multiplicity two. If $\mu = \nu - 1$ the weights are $\Sigma\, \lambda_j + \lambda_i$, $i = 1, \cdots, \nu$. Then $1^S = \Sigma\, h_i{}^S = 1(\nu - 1) \neq 0$ and if T denotes the automorphism $a \to a + \mathrm{tr}'\, a$ we see that ST has the weights λ_i each with multiplicity 2.

$p = 2$, $i = 2$. In this case $a \to a + \mathrm{tr}\, a$ is an automorphism if $n \equiv 0 \pmod 2$. Let S be an automorphism and x_0 a vector $\neq 0$ such that $x_0 h^S = x_0\Lambda$. Either $x_0 e_\alpha{}^S = 0$, $\alpha = \lambda_i + \lambda_j$ in which case $x_0(e_\alpha{}^S)^2 = x_0(h_i + h_j) = 0$, $m_i = m_j$ in $\Lambda = \Sigma\, m_i\lambda_i$ or $\Lambda + \alpha$ is a weight. Thus the weights are invariant under permutation and as above they are either λ_i, $i = 1, \cdots, n$ or $\Sigma\, \lambda_j + \lambda_i$ and in the latter case $n - 1 \not\equiv 0 \pmod 2$. We may then form ST, $a^T = a + \mathrm{tr}\, a$ and obtain an automorphism with weights λ_i.

LEMMA 11. *If $p = 2$ and S is an automorphism of $\bar{\mathfrak{S}}(\Omega_n, J_1)$ ($\mathfrak{S}(\Omega_n, J_2)$) and $\nu \not\equiv 0 \pmod 2$ ($n \not\equiv 0 \pmod 2$) then there exists a matrix q independent of λ such that $h^S = q^{-1}hq$. If $\nu \equiv 0 \pmod 2$ ($n \equiv 0 \pmod 2$) either this holds or the automorphism ST where $a^T = a + \mathrm{tr}'\, a$ ($a^T = a + \mathrm{tr}\, a$) has this property.*

LEMMA 12. *The matrix q in Lemma 10 or 11 may be normalized so that $qq^{J_i} = 1 = q^{J_i}q$.*

$p \neq 2$. Since $h^S = q^{-1}hq$, $-h^S = h^{SJ_i} = -q^{J_i}h(q^{J_i})^{-1}$. Hence $(qq^{J_i})h = h(qq^{J_i})$. It follows that $r = qq^{J_i}$ is a diagonal matrix. Since $r^{J_i} = r$ a simple computation shows that

$$r = \{\rho_1, \cdots, \rho_\nu, \rho_1, \cdots, \rho_\nu\}, \quad = \{\rho, \rho_1, \cdots, \rho_\nu, \rho_1, \cdots, \rho_\nu\},$$
$$= \{\rho_1, \cdots, \rho_\nu, \rho_1, \cdots, \rho_\nu\}$$

according as $i = 1, 2, 3$. It follows that we can find a diagonal matrix k such that $kk^{J_i} = r^{-1}$. The matrix $u = kq$ has the desired property.

$p = 2$, $i = 2$. In this case $r = qq^{J_2}$ is diagonal. Hence as above we may solve $h^2 = -r^{-1}$ and obtain $u = hq$ of the desired type.

$p = 2$, $i = 1$. Here $r = qq^{J_1}$ has the form

$$\begin{pmatrix} M_{11} & M_{12} \\ M_{21} & M_{22} \end{pmatrix}$$

where the M_{ij} are diagonal. If

$$q = \begin{pmatrix} q_{11} & q_{12} \\ q_{21} & q_{22} \end{pmatrix} \qquad r = qq^{J_1} = \begin{pmatrix} q_{11}q'_{22} + q_{12}q'_{21} & q_{11}q'_{12} + q_{12}q'_{11} \\ q_{21}q'_{22} + q_{22}q'_{21} & q_{21}q'_{12} + q_{22}q'_{11} \end{pmatrix}$$

and hence $M_{11} = M'_{22}$ and $M_{12} = b + b' = 0$, $M_{21} = c + c' = 0$. In this case too we may solve the equation $kk^{J_1} = r^{-1}$ to obtain a k commutative with h. The matrix kq satisfies the condition.

If U denotes the inverse $a \to uau^{-1}$ of the mapping determined by Lemma 10 or 11, the correspondence $\bar{S} = SU$ (or STU) is an automorphism in $\mathfrak{S}(\Omega_n, J_i)$ such that $h^{\bar{S}} = h$. From $[e_a, h] = e_a\alpha$ we obtain when $p \neq 2$ and when $p = i = 2$ $e_a{}^{\bar{S}} = e_a\xi_a$. If $p \neq 2$, $[e_{-a}, e_a] = h_a$ and hence $\xi_{-a} = \xi_a{}^{-1}$. Let $\alpha_1, \alpha_2, \cdots$ be the fundamental set of roots determined above. These are for $p \neq 2$ and $i = 1$, $\lambda_1 - \lambda_2, \cdots, \lambda_1 - \lambda_\nu, \lambda_1 + \lambda_2$; for $p \neq 2$, $i = 2$, λ_j and for $p \neq 2$, $i = 3$, $\lambda_1 - \lambda_2, \cdots, \lambda_1 - \lambda_\nu, \lambda_1 + \lambda_2$. If $i = 1, 3$ we solve the equations $\bar{\lambda}_1\bar{\lambda}_j{}^{-1} = \xi_{\lambda_1 - \lambda_j}$, $\bar{\lambda}_1\bar{\lambda}_2 = \xi_{\lambda_1+\lambda_2}$; if $i = 2$ solve $\bar{\lambda}_i = \xi_{\lambda_i}$. Then for $i = 1, 2, 3$, respectively, the matrices

$$g = \{\bar{\lambda}_1, \cdots, \bar{\lambda}_\nu, \bar{\lambda}_1{}^{-1}, \cdots, \bar{\lambda}_\nu{}^{-1}\}, \quad = \{1, \bar{\lambda}_1, \cdots, \bar{\lambda}_\nu, \bar{\lambda}_1{}^{-1}, \cdots, \bar{\lambda}_\nu{}^{-1}\}$$
$$= \{\bar{\lambda}_1, \cdots, \bar{\lambda}_\nu, \bar{\lambda}_1{}^{-1}, \cdots, \bar{\lambda}_\nu{}^{-1}\}$$

satisfy $g^{J_i}g = 1 = gg^{J_i}$ and $a^{\bar{S}} = g^{-1}ag$ for $a = h$, e_{a_i}. Hence the latter equation holds for e_{-a_i} and hence for all e_a. Thus $a^{\bar{S}} = g^{-1}ag$ for all a in $\mathfrak{S}$.

$p = 2$, $i = 2$. Here $e_a{}^2 = h_i + h_j$ and hence $\xi_a{}^2 = 1$, $\xi_a = 1$ and $\bar{S} = 1$. We consider finally the case $p = 2$, $i = 1$. Here

$$f_{ji}{}^{\bar{S}} = f_{ji}\alpha_{ji} + f_{ij}\beta_{ij} + g_{ij}\gamma_{ij} + k_{ij}\epsilon_{ij} \tag{16}$$

where the condition $(f_{j1}{}^{\bar{S}})^2 = 0$ yields $\alpha_{j1}\beta_{1j} = \gamma_{1j}\epsilon_{1j}$. If g is J_1-orthogonal we have seen that $x \to g^{-1}xg$ is an automorphism in $\mathfrak{S}$. Since $\bar{\mathfrak{S}}$ is a characteristic subalgebra $g^{-1}xg \in \bar{\mathfrak{S}}$ if x does. Now the matrix

$$d = \begin{pmatrix} d_{11} & d_{12} \\ d_{21} & d_{22} \end{pmatrix}$$

where

$d_{11}=\{\mu_1,\cdots,\mu_v\},\ d_{12}=\{\rho_1,\cdots,\rho_v\},\ d_{21}=\{\sigma_1,\cdots,\sigma_v\},\ d_{22}=\{\tau_1,\cdots,\tau_v\}$

is J_1-orthogonal if and only if $d_{11}d_{22}+d_{12}d_{21}=1$. The automorphism $a\rightarrow\bar{a}=d^{-1}ad=d^{J_1}ad$ sends the elements of $\mathfrak{H}$ into themselves and

$$\begin{aligned}\bar{f}_{j1}&=f_{j1}\mu_1\tau_j+f_{1j}\rho_1\sigma_j+g_{1j}\rho_1\tau_j+k_{1j}\mu_1\sigma_j\\ \bar{f}_{1j}&=f_{1j}\tau_1\mu_j+f_{j1}\rho_j\sigma_1+g_{1j}\tau_1\rho_j+k_{1j}\sigma_i\mu_j\\ \bar{g}_{1j}&=f_{ij}\tau_1\sigma_j+f_{j1}\tau_j\sigma_1+g_{1j}\tau_1\tau_j+k_{1j}\sigma_i\sigma_j\\ \bar{k}_{1j}&=f_{1j}\rho_1\mu_j+f_{ji}\mu_1\rho_j+g_{1j}\rho_1\rho_j+k_{1j}\mu_1\mu_j.\end{aligned}$$

Since $\alpha_{21}\beta_{12}=\gamma_{12}\epsilon_{12}$ we may solve the equations $\mu_1\tau_2=\alpha_{21}$, $\rho_1\sigma_2=\beta_{12}$, $\rho_1\tau_2=\gamma_{12}$, $\mu_1\sigma_2=\epsilon_{12}$. For this solution we can not have simultaneously $\mu_1=0$, $\rho_1=0$ or $\sigma_2=0$, $\tau_2=0$ since $f_{21}{}^{\bar{S}}\neq 0$. Hence we may determine the other parameters so that $d^{J_1}d=1$. The product of $\bar{S}$ and the inverse of $a\rightarrow\bar{a}$ is an automorphism $\bar{S}_1$ such that $f_{21}{}^{\bar{S}_1}=f_{21}$. Since $[f_{j1},f_{21}]=0$, $[f_{j1}{}^{\bar{S}_1},f_{21}]=0$ and if we use the notation (16) we obtain $\beta_{1j}=\gamma_{1j}=0$, $j=3,\cdots$. Since we do not have $\alpha_{j1}=\epsilon_{1j}=0$ we may find a d of the above type for which $\mu_1=\tau_1=\mu_2=\tau_2=1$, $\rho_1=\sigma_1=\rho_2=\sigma_2=0$ and $\tau_j=\alpha_{j1}$, $j=3,\cdots,\sigma_j=\epsilon_{1j}$. For the corresponding automorphism we have $f_{j1}=f_{j1}{}^{\bar{S}_1}$, $j=2,\cdots$. Then by multiplying $\bar{S}_1$ by the inverse of a suitable transformation $a\rightarrow\bar{a}$ we obtain $\bar{S}_2$ such that $f_{j1}{}^{\bar{S}_2}=f_{j1}$, $j=2,\cdots,\nu$. Now

$$f_{1j}{}^{\bar{S}_2}=f_{1j}\alpha_{1j}+f_{j1}\beta_{j1}+g_{j1}\gamma_{j1}+k_{j1}\epsilon_{j1}$$

and if $k\neq j$ the condition $[[f_{k1},f_{1j}],f_{j1}]=f_{k1}$ implies $\alpha_{1j}=1$ and $(f_{1j}{}^{\bar{S}_2})^2=0$, implies $\beta_{j1}=\gamma_{j1}\epsilon_{j1}$. Choose $\sigma_1=\epsilon_{21}$, $\rho_2=\gamma_{21}$, the other σ's and ρ's $=0$, $\mu_i=\tau_i=1$. The automorphism determined satisfies $\bar{h}=h$, $\bar{f}_{j1}=f_{j1}$, $\bar{f}_{12}=f_{12}{}^{\bar{S}_2}$. We then obtain an $\bar{S}_3$ such that $h^{\bar{S}_3}=h$, $f_{j1}{}^{\bar{S}_3}=f_{j1}$, $f_{12}{}^{\bar{S}_3}=f_{12}$. Then by $[f_{1j},f_{12}]=0$ we obtain $\beta_{j1}=\epsilon_{j1}=0$, $f_{1j}{}^{\bar{S}_3}=f_{1j}+g_{j1}\gamma_{j1}$, $j=3,\cdots,\nu$. Let $\rho_j=\gamma_{j1}$, $j=3,\cdots$ all the other ρ's $=0$, $\sigma_i=0$, $\mu_i=\tau_i=1$. We then obtain $\bar{h}=h$, $\bar{f}_{j1}=f_{j1}$, $\bar{f}_{1j}=f_{1j}{}^{\bar{S}_3}$ and therefore the product $\bar{S}_4$ of $\bar{S}_3$ and the inverse of a suitable transformation $a\rightarrow\bar{a}$ satisfies $h^{\bar{S}_4}=h$, $f_{j1}{}^{\bar{S}_4}=f_{j1}$, $f_{1j}{}^{\bar{S}_4}=f_{1j}$. Then $f_{ij}{}^{\bar{S}_4}=f_{ij}$ and

$$g_{1j}{}^{\bar{S}_4}=f_{1j}\mu_{1j}+f_{j1}\rho_{j1}+g_{1j}\sigma_{j1}+k_{1j}\tau_{j1}.$$

Using $g_{1k}=[g_{1j},f_{kj}]$ we obtain $\mu_{1j}=\tau_{j1}=0$, $\rho_{j1}=\rho_{k1}=\rho$, $\sigma_{j1}=\sigma_{k1}=\sigma$ and from $g_{1k}=[[g_{1j},f_{k1}],f_{1j}]$ we obtain $\rho=0$. Similarly we obtain $k_{1j}{}^{\bar{S}_4}=k_{1j}\zeta$. Since $[g_{1j},k_{1j}]=h_1+h_j$, $\sigma\zeta=1$. Finally we set $\rho_i=\sigma_i=0$, $\mu_i{}^2=\zeta$, $\tau_i{}^2=\sigma$ and obtain $\bar{h}=h^{\bar{S}_4}$, $\bar{f}_{ij}=f_{ij}{}^{\bar{S}_4}$, $\bar{g}_{1j}=g_{1j}{}^{\bar{S}_4}$, $\bar{k}_{1j}=k_{1j}{}^{\bar{S}_4}$. Since

$g_{jk} = [g_{1j}, f_{k1}]$, $g_{jk}{}^{\bar{S}_4} = \bar{g}_{jk}$ and similarly $\bar{k}_{jk} = k_{jk}{}^{\bar{S}_4}$. Hence $\bar{a} = a^{\bar{S}_4}$. We have therefore proved the following

LEMMA 13. *If $p \neq 2$ the automorphisms of $\mathfrak{S}(\Omega_n, J_i)$ all have the form $a \rightarrow g^{-1}ag$ where $g^{J_i}g = 1\gamma$ if $i = 1$, ν arbitrary; $i = 2$, $\nu \geq 3$; $i = 3$, $\nu \geq 3$, $\neq 4$. The same result holds when $p = 2$ for $\bar{\mathfrak{S}}(\Omega_n, J_1)$ with $\nu \not\equiv 0 \pmod 2$ $\nu \geq 3$ and for $\mathfrak{S}(\Omega_n, J_2)$ with $n \not\equiv 0 \pmod 2$ $n \geq 3$. In the case $\bar{\mathfrak{S}}(\Omega_n, J_1)$, $\nu \equiv 0 \pmod 2$ $\nu \geq 3$ we have in addition the automorphisms $a \rightarrow g^{-1}ag + \mathrm{tr}'\, a$ and for $\mathfrak{S}(\Omega_n, J_2)$, $n \equiv 0 \pmod 2$ $n \geq 3$ the remaining automorphisms are $a \rightarrow g^{-1}ag + \mathrm{tr}\, a$.*

2.5. Isomorphisms. The dimensionality of $\mathfrak{S}(\Omega_n, J_i)$, $i = 1, 2, 3$, is respectively $\nu(2\nu + 1)$, $\nu(2\nu + 1)$ and $\nu(2\nu - 1)$ for $p \neq 2$. Hence the only isomorphisms that can occur are between $\mathfrak{S}(\Omega_{2\nu}, J_1)$ and $\mathfrak{S}(\Omega_{2\nu+1}, J_2)$. Now suppose S is an isomorphism between $\mathfrak{S}(\Omega_{2\nu+1}, J_2)$ and $\mathfrak{S}(\Omega_{2\nu}, J_1)$. Consider the weights of h in the representation $a \rightarrow a^S$ of $\mathfrak{S}(\Omega_{2\nu+1}, J_2)$. Since h^S is skew, if Λ is a weight so is $-\Lambda$ and since $h_i{}^S$ are linearly independent there are ν linearly independent weights and hence these are $\Lambda_1, \cdots, \Lambda_\nu$ with the remaining weights $-\Lambda_1, \cdots, -\Lambda_\nu$. If α is a root of $\mathfrak{S}(\Omega_{2\nu+1}, J_2)$ and Λ is a weight either $\Lambda + \alpha$ or $\Lambda - \alpha$ is not a weight. It follows that $\Lambda_\alpha = 0$, ± 1 for $\alpha = \lambda_i - \lambda_j$ or $\lambda_i + \lambda_j$ and hence if $\Lambda = \Sigma\, m_i \lambda_i$, $m_i \pm m_j = 0, \pm 1$. We have also as in the last section that the weights are invariant under permutation of the λ's and arbitrary changes of sign of the m's. Hence if $\nu \geq 3$ the weights are $\pm \lambda_i$. Thus h^S is similar to the h given by our particular basis for $\mathfrak{S}(\Omega_{2\nu}, J_1)$. It follows by the proof of Lemma 12 that we may suppose that the similarity is given by a J_1-orthogonal element. Hence by combining S with a suitable automorphism we obtain an isomorphism T such that $h^T = h$ in $\mathfrak{S}(\Omega_{2\nu}, J_1)$. Then $[e_{\lambda_i}{}^T, h^T] = e_{\lambda_i}\lambda_i$ and this is impossible since λ_i is not a root in $\mathfrak{S}(\Omega_{2\nu}, J_1)$.

If $\nu = 1$, the basis of $\mathfrak{S}(\Omega_2, J_1)$ is e_+, e_-, h such that $e_+{}^p = 0$, $e_-{}^p = 0$, $h^p = h$ and if we replace these respectively by $e_+/\sqrt{2}$, $e_-/\sqrt{2}$, $h/2$ we obtain a basis whose multiplication table is that of the e's and h of $\mathfrak{S}(\Omega_3, J_2)$. If $\nu = 2$ it is readily verified that the following is an isomorphism between $\mathfrak{S}(\Omega_4, J_1)$ and $\mathfrak{S}(\Omega_5, J_2)$

$$\begin{array}{ll} h_1 \rightarrow h_1 + h_2 & h_2 \rightarrow h_1 - h_2 \\ e_{\lambda_1-\lambda_2} \rightarrow -e_{\lambda_2}\sqrt{2} & e_{\lambda_2-\lambda_1} \rightarrow -e_{-\lambda_2}\sqrt{2} \\ e_{\lambda_1+\lambda_2} \rightarrow e_{\lambda_1}\sqrt{2} & e_{-\lambda_1-\lambda_2} \rightarrow e_{-\lambda_1}\sqrt{2} \\ e_{2\lambda_1} \rightarrow e_{\lambda_1+\lambda_2} & e_{-2\lambda_1} \rightarrow e_{-\lambda_1-\lambda_2} \\ e_{2\lambda_2} \rightarrow e_{\lambda_1-\lambda_2} & e_{-2\lambda_2} \rightarrow e_{\lambda_2-\lambda_1} \end{array}$$

If $p = 2$ the algebras $\bar{\mathfrak{S}}(\Omega_{2\nu}, J_1)$ are not isomorphic to any $\mathfrak{S}(\Omega_n, J_2)$ since the former have outer derivations while the derivations of the latter are all inner.

LEMMA 13. *The only isomorphic pairs of algebras in the set* $\mathfrak{S}(\Omega_n, J_i)$, $\bar{\mathfrak{S}}(\Omega_n, J_1)$ *are* $\mathfrak{S}(\Omega_{n2}, J_1)$, $\mathfrak{S}(\Omega_3, J_2)$ *and* $\mathfrak{S}(\Omega_4, J_1)$, $\mathfrak{S}(\Omega_5, J_2)$.

2.6. Restricted Lie algebras of types *B, C, D*. Let $\mathfrak{A}$ be a normal simple associative algebra over Φ and suppose that $\mathfrak{A}$ has an involution J of the first kind.[9] The set $\mathfrak{S}(\mathfrak{A}, J)$ of J-skew elements is a restricted Lie algebra relative to $a + b$, $a\alpha$, $[a, b] = ab - ba$ and a^p. If Ω is the algebraic closure of Φ, $a_1, a_2, \cdots, a_{n^2}$ a basis for $\mathfrak{A}$ over Φ then $\mathfrak{A}_\Omega = \Omega_n$ and the correspondence $\Sigma\, a_i\omega_i \rightarrow \Sigma\, a_i^J\omega_i$ is an involution J in Ω_n. The skew elements relative to J form the extension $\mathfrak{S}(\mathfrak{A}, J)_\Omega$. On the other hand we may take this set to be one of the sets $\mathfrak{S}(\Omega_n, J_i)$ considered above. Following the notation introduced by Cartan in the characteristic 0 case we shall say that $\mathfrak{S}(\mathfrak{A}, J)$ is of type C, B, D according as $i = 1, 2, 3$. Thus if $p \neq 2$, $\mathfrak{S}(\mathfrak{A}, J)$ is a simple restricted Lie algebra except when its type is D and $n = 4$. If $p = 2$ and $\mathfrak{S}(\mathfrak{A}, J)_\Omega = \mathfrak{S}(\Omega_n, J_1)$ the algebra $\bar{\mathfrak{S}}(\mathfrak{A}, J)$ generated by the elements a^2 and $[a, b]$ extends to $\bar{\mathfrak{S}}(\mathfrak{A}, J)_\Omega = \bar{\mathfrak{S}}(\Omega_n, J_1)$ and hence has order $\nu(2\nu - 1)$, $\nu = n/2$.[21] Similarly $\bar{\mathfrak{S}}(\mathfrak{A}, J)'$, the subalgebra of $\bar{\mathfrak{S}}$ generated by $[a, b], a, b$ in $\bar{\mathfrak{S}}$, extends to $\bar{\mathfrak{S}}(\Omega_n, J_1)'$. It follows that if ν is odd and ≥ 3, $\bar{\mathfrak{S}}(\mathfrak{A}, J)'$ is simple and if ν is even and ≥ 4, $\bar{\mathfrak{S}}(\mathfrak{A}, J)'/(1)$ is simple. For $p = 2$ and $\mathfrak{S}(\mathfrak{A}, J)_\Omega = \mathfrak{S}(\Omega_n, J_2)$ and $\mathfrak{S}(\mathfrak{A}, J)'$ the algebra generated by $[a, b]$ extends to $\mathfrak{S}(\Omega_n, J_2)'$. Hence $\mathfrak{S}(\mathfrak{A}, J)'$ is simple if n is odd and ≥ 3 and $\mathfrak{S}(\mathfrak{A}, J)'/(1)$ is simple when n is even and ≥ 4. We note also that $\bar{\mathfrak{S}}(\Omega_n, J_2) = \mathfrak{S}(\Omega_n, J_2)$ and hence that $\bar{\mathfrak{S}}(\mathfrak{A}, J) = \mathfrak{S}(\mathfrak{A}, J)$.

THEOREM 9. *If $\mathfrak{A}$ is a normal simple associative algebra of order n^2 with an involution J of the first kind, then $\mathfrak{S}(\mathfrak{A}, J)$, the set of J-skew elements, is a simple restricted Lie algebra when $p \neq 2$ except when $\mathfrak{S}$ is of type D and $n = 4$. If $p = 2$ either $\bar{\mathfrak{S}}(\mathfrak{A}, J)'$ or $\bar{\mathfrak{S}}(\mathfrak{A}, J)'/(1)$ is simple when $\mathfrak{S}(\mathfrak{A}, J)_\Omega = \mathfrak{S}(\Omega_n, J_1)$ and $\nu = n/2 \geq 3$ and either $\mathfrak{S}(\mathfrak{A}, J)$ or $\mathfrak{S}(\mathfrak{A}, J)'/(1)$ is simple when $\mathfrak{S}(\mathfrak{A}, J)_\Omega = \mathfrak{S}(\Omega_n, J_2)$ and $n \geq 3$.*

Since the coefficients of the characteristic polynomials of the elements of $\mathfrak{A}$ in the representation in Ω_n all belong to Φ it follows that if $p = 2$ and

[21] In general, for any restricted Lie algebra $\mathfrak{L}$ we denote the sub-algebra generated by the elements $[a, b]$ and a by $\bar{\mathfrak{L}}$. It follows that $\bar{\mathfrak{L}}_P = \overline{(\mathfrak{L}_P)}$.

$a \epsilon \mathfrak{S}(\mathfrak{A}, J)'$ where $\mathfrak{S}(\mathfrak{A}, J)_\Omega = \mathfrak{S}(\Omega_n, J_1)$ then $\mathrm{tr}' a \epsilon \Phi$. Hence the correspondence $a \to a + \mathrm{tr}' a$ is an automorphism if $\nu \equiv 0 \pmod 2$. Similarly for the other type of $\mathfrak{S}(\mathfrak{A}, J)$ with $p = 2$ we have the automorphisms $a \to a + \mathrm{tr}\, a$ for $n \equiv 0 \pmod 2$.

In the remainder of Part II we make the following restrictions on the orders: In Theorems 10 and 13, $\nu \geq 1$, ≥ 3, ≥ 3 but $\neq 4$ according as the type is C, B or D. Otherwise the restrictions are as in Lemma 13. Now if $\mathfrak{S}(\mathfrak{A}, J)$ and $\mathfrak{S}(\mathfrak{B}, K)$ are isomorphic they are of the same type. If $p \neq 2$ and $a \to a^G$ is an isomorphism between two restricted Lie algebras $\mathfrak{S}(\mathfrak{A}, J)$ and $\mathfrak{S}(\mathfrak{B}, K)$ then we may regard these as subsets of the same $\mathfrak{S}(\Omega_n, J_i)$ such that the elements of $\mathfrak{S}(\mathfrak{A}, J)$ are $\Sigma\, a_i \phi_i$ uniquely, those of $\mathfrak{S}(\mathfrak{B}, K)$ are $\Sigma\, a_i{}^G \phi_i$ and those of $\mathfrak{S}(\Omega_n, J_i)$ are representable uniquely either as $\Sigma\, a_i \omega_i$ or $\Sigma\, a_i{}^G \omega_i$. Thus the correspondence $\Sigma\, a_i \omega_i \to \Sigma\, a_i{}^G \omega_i$ is an automorphism in $\mathfrak{S}(\Omega_n, J_i)$. Hence by Lemma 13 we obtain

THEOREM 10. *A necessary and sufficient condition that the restricted Lie algebras $\mathfrak{S}(\mathfrak{A}, J)$ and $\mathfrak{S}(\mathfrak{B}, K)$ be isomorphic is that $\mathfrak{A}$ and $\mathfrak{B}$ be isomorphic and J and K be cogredient.*

THEOREM 11. *Any automorphism G in $\mathfrak{S}(\mathfrak{A}, J)$ has the form $a^G = g^{-1} a g$ where g is a J-orthogonal element.*

THEOREM 12. *Any restricted derivation in $\mathfrak{S}(\mathfrak{A}, J)$ is inner.*

If $p = 2$ we cannot have an isomorphism between $\bar{\mathfrak{S}}(\mathfrak{A}, J)$ and $\bar{\mathfrak{S}}(\mathfrak{B}, K)$ unless either $\bar{\mathfrak{S}}(\mathfrak{A}, J)_\Omega = \bar{\mathfrak{S}}(\mathfrak{B}, K) = \bar{\mathfrak{S}}(\Omega_n, J_1)$ or $\bar{\mathfrak{S}}(\mathfrak{A}, J)_\Omega = \bar{\mathfrak{S}}(\mathfrak{B}, K)_\Omega = \bar{\mathfrak{S}}(\Omega_n, J_2)$. In the former case an isomorphism between the two algebras becomes an automorphism in $\bar{\mathfrak{S}}(\Omega_n, J_1)$ and may have the form $a \to g^{-1} a g + \mathrm{tr}' a = b$. But then $\mathrm{tr}' a = \mathrm{tr}' b$ since $\nu \equiv 0 \pmod 2$ and hence by combining the isomorphism with the automorphism $b \to b + \mathrm{tr}' b$ we obtain the isomorphism $a \to g^{-1} a g$. In a similar fashion we may treat the other case. We obtain the following results:

THEOREM 13. *If $p = 2$, $\bar{\mathfrak{S}}(\mathfrak{A}, J)$ and $\bar{\mathfrak{S}}(\mathfrak{B}, K)$ are isomorphic if and only if $\mathfrak{A}$ and $\mathfrak{B}$ are isomorphic and J and K are cogredient.*

THEOREM 14. *If $p = 2$ the following are the automorphisms of $\bar{\mathfrak{S}}(\mathfrak{A}, J)$*

$a \to g^{-1} a g$, g J-orthogonal if $\mathfrak{S}_\Omega = \mathfrak{S}(\Omega_{2\nu}, J_1)$ and $\nu \not\equiv 0 \pmod 2$ or if $\mathfrak{S}_\Omega = \mathfrak{S}(\Omega_n, J_2)$ and $n \not\equiv 0 \pmod 2$

$a \to g^{-1} a g$, $a \to g^{-1} a g + \mathrm{tr}' a$ if $\mathfrak{S}_\Omega = \mathfrak{S}(\Omega_{2\nu}, J_1)$ and $\nu \equiv 0 \pmod 2$.
$a \to g^{-1} a g$, $a \to g^{-1} a g + \mathrm{tr}\, a$ if $\mathfrak{S}_\Omega = \mathfrak{S}(\Omega_n, J_2)$ and $n \equiv 0 \pmod 2$.

THEOREM 15. *The restricted derivations of* $\bar{\mathfrak{S}}(\mathfrak{A}, J)$ *are inner if* $p = 2$ *and* $\mathfrak{S}(\mathfrak{A}, J)_\Omega = \mathfrak{S}(\Omega_n, J_1)$. *For the other case* $\bar{\mathfrak{S}}(\mathfrak{A}, J)$ *any restricted derivation is induced by a derivation of* $\mathfrak{A}$.

If $\mathfrak{S}_\Omega = \mathfrak{S}(\Omega_n, J_1)$ and $\nu \not\equiv 0 \pmod 2$ we have $\bar{\mathfrak{S}}(\mathfrak{A}, J) = \bar{\mathfrak{S}}(\mathfrak{A}, J)' \oplus (1)$. Any automorphism (derivation) in $\bar{\mathfrak{S}}'$ is induced by an automorphism (derivation) of $\bar{\mathfrak{S}}$. Hence these have the form $a \rightarrow g^{-1}ag$ ($a \rightarrow [a, d]$, d in $\mathfrak{A}$). If $\mathfrak{S}_\Omega = \mathfrak{S}(\Omega_n, J_2)$ and $n \not\equiv 0 \pmod 2$, $\bar{\mathfrak{S}}(\mathfrak{A}, J) = \bar{\mathfrak{S}}(\mathfrak{A}, J)' \oplus (1)$. The automorphisms have the form $a \rightarrow g^{-1}ag$ and the derivations $a \rightarrow [a, d]$, d in $\bar{\mathfrak{S}}'$.

We remark finally that Theorem 9 for $p \neq 2$ or $p = 2$ and $\mathfrak{S}_\Omega = \mathfrak{S}(\Omega_n, J_1)$, Theorems 10, 11 and the remarks of the last paragraph hold when these algebras are regarded as ordinary Lie algebras rather than restricted Lie algebras. For these algebras have no centers.

2.7. Completeness. We wish to determine the restricted Lie algebras $\mathfrak{L}$ such that $\mathfrak{L}_\Omega$ is one of the simple algebras $\mathfrak{S}(\Omega_n, J_i)$, $i = 1, 2, 3$ for $p \neq 2$, or $\bar{\mathfrak{S}}(\Omega_n, J_1)'$ for $p = 2$, $\nu \not\equiv 0 \pmod 2$, $\bar{\mathfrak{S}}(\Omega_n, J_2)'$, $p = 2$, $n \not\equiv 0 \pmod 2$. If $a_1, a_2, \cdots, a_n$ is a basis for $\mathfrak{L}$ over Φ we may suppose that the elements $\Sigma\, a_j\phi_j$ of $\mathfrak{L}$ are subsets of the corresponding $\mathfrak{S}$ or $\bar{\mathfrak{S}}'$ such that the elements of $\mathfrak{S}$ or $\bar{\mathfrak{S}}'$ are representable uniquely in the form $\Sigma\, a_j\omega_j$. The elements of the matrices a_j and hence of $\Sigma\, a_j\phi_j$ determine a finite extension Γ of Φ. Thus the e_a, h basis of $\mathfrak{S}$ or $\bar{\mathfrak{S}}'$ is expressible in terms of the a_j with coefficients in Γ and so we may replace Ω by Γ: $\mathfrak{L}_\Gamma = \mathfrak{S}(\Gamma_n, J_i)$ or $\bar{\mathfrak{S}}(\Gamma_n, J_i)'$. Let $[a_j, a_k] = \Sigma\, a_p\alpha_{pjk}$, $a_j{}^p = \Sigma\, a_q\beta_{qj}$, $\alpha, \beta \in \Phi$ be the multiplication table of the a's.

We suppose first that $\Gamma = \Phi(x_1, x_2, \cdots, x_r, y)$ is a direct product of $\Phi(x_1, \cdots, x_r)$, where $x_i{}^p = \xi_i \in \Phi$ and $\Phi(y)$ is a normal separable field over Φ. Let $\mathfrak{g} = (i, s, t, \cdots)$ be the Galois group of Γ over $\Phi(x_1, \cdots, x_r)$ and d a derivation in Γ such that the d-constants are the elements of $\Phi(y)$.[16] Thus the elements of Φ are the only ones which satisfy $\alpha^d = 0$, $\alpha^s = \alpha$ for all s in $\mathfrak{g}$. If $c = (\gamma_{ij})$ is any element of Γ_n the correspondence $c \rightarrow c^d = (\gamma_{ij}{}^d)$ is a derivation in Γ_n over Φ and $c \rightarrow c^s = (\gamma_{ij}{}^s)$ is an automorphism in Γ_n over Φ. Since the matrices s_i which define the involution in Γ_n have elements in Φ these correspondences induce, respectively, a restricted derivation and an automorphism in $\mathfrak{L}_\Gamma = \mathfrak{S}(\Gamma_n, J_i)$ or $\bar{\mathfrak{S}}(\Gamma_n, J_i)'$ over Φ. Since the constants of multiplication, $\alpha, \beta \in \Phi$ the correspondence $\Sigma\, a_j\gamma_j \rightarrow \Sigma\, a_j{}^d\gamma_j = aD$ is a restricted derivation in $\mathfrak{L}_\Gamma$ over Γ and hence is induced by a derivation D in the associative algebra Γ_n over Γ. Similarly $a = \Sigma\, a_j\gamma_j \rightarrow \Sigma\, a_j{}^s\gamma_j = a^S$ is induced by an automorphism S in Γ_n over Γ. Let $D_1 = d - D$, $S_1 = sS^{-1}$.

These are respectively derivations and automorphisms in Γ_n over Φ. Hence the elements b such that $bD_1 = 0$, $bS_1 = b$ for all S_1 form an algebra $\mathfrak{A}$ over Φ. Since $(\Sigma a_j\gamma_j)D_1 = \Sigma a_j\gamma_j^d$, $(\Sigma a_j\gamma_j)S_1 = \Sigma a_j\gamma_j^s$, $\mathfrak{L}_\Gamma \wedge \mathfrak{A} = \mathfrak{L}$. Since the enveloping algebra of $\mathfrak{L}_\Gamma$ is Γ_n we may obtain a basis for Γ_n of the form $a_1, a_2, \cdots, a_{n^2}$ where a_i is a product of the a_j's, $j = 1, \cdots, m$. Each $a_i \in \mathfrak{A}$. Hence if $(\Sigma a_i\gamma_i)D_1 = \Sigma a_i\gamma_i^d = 0$ and $(\Sigma a_i\gamma_i)S_1 = \Sigma a_i\gamma_i$, $\gamma_i = \phi_i \in \Phi$. $\mathfrak{A}$ therefore consists of the set $\Sigma a_i\phi_i$ and $\mathfrak{A}_\Gamma = \Gamma_n$. Thus $\mathfrak{A}$ is normal simple. Since the $a_i = a_{k_1} \cdots a_{k_t}$, $a_i^{J_i} = (-1)^t a_{k_t} \cdots a_{k_1} \in \mathfrak{A}$ and J_i induces an involution in $\mathfrak{A}$. Since the a_j form a basis for the J_i-skew elements of Γ_n or for $\bar{\mathfrak{S}}(\Gamma_n, J_i)'$ they also form a basis for those sets in $\mathfrak{A}$. Thus $\mathfrak{L} = \mathfrak{S}(\mathfrak{A}, J_i)$ or $\bar{\mathfrak{S}}(\mathfrak{A}, J_i)'$.

Now let Γ be any finite extension and Δ the maximal separable subfield of Γ. There is a chain of fields $\Delta_0 = \Delta, \Delta_1, \cdots, \Delta_u = \Gamma$ where

$$\Delta_i = \Delta_{i-1}(x_1, \cdots, x_{r_i}) > \Delta_{i-1}, \qquad x_j^p = \xi_j \in \Delta_{i-1}.$$

Now

$$(\mathfrak{L}_{\Delta_{u-1}})_{\Delta_u} = \mathfrak{S}(\Delta_{u,n}, J_i) \quad \text{or} \quad \bar{\mathfrak{S}}(\Delta_{u,n}, J_i)'.$$

Hence by what has been proved $\mathfrak{L}_{\Delta_{u-1}} = \mathfrak{S}(\mathfrak{A}_u, J)$ or $\bar{\mathfrak{S}}(\mathfrak{A}_u, J)'$. Let $\bar{\Gamma}$ be a separable extension of Δ_{u-1}. It is known that if $\bar{\Delta}$ is the maximal separable over Φ subfield of $\bar{\Gamma}$ then there is a chain of subfields of the above type such that $\bar{\Gamma} = \bar{\Delta}_{u-1}$.[17] If $\mathbf{E}$ is a splitting field for $\mathfrak{A}_u$ over Δ_u, $\mathfrak{L}_\mathbf{E} = \mathfrak{S}(\mathbf{E}_n, J)$ or $\bar{\mathfrak{S}}(\mathbf{E}_n, J)'$ and the involution J has the form $a \to s^{-1}a's$ where $s' = \pm s$. If $p \neq 2$ and $s' = -s$ or $p = 2$ and $s' = s$ is alternate we may replace s by the cogredient matrix $\begin{pmatrix} 0 & 1_v \\ -1_v & 0 \end{pmatrix}$ or $\begin{pmatrix} 0 & 1_v \\ 1_v & 0 \end{pmatrix}$. Thus $\mathfrak{L}_\mathbf{E} = \mathfrak{S}(\mathbf{E}_n, J_1)$ or $\bar{\mathfrak{S}}(\mathbf{E}_n, J_1)'$. If $p \neq 2$ and $s' = s$ we adjoin certain square roots to $\mathbf{E}$ and obtain s cogredient to s_2 or s_3 of (9). Hence by first choosing $\mathbf{E}$ to be separable we may obtain a separable $\bar{\Gamma} = \bar{\Delta}_{u-1}$ over Δ_{u-1} such that $\mathfrak{L}_{\bar{\Gamma}} = \mathfrak{S}(\bar{\Gamma}_n, J_i)$ or $\mathfrak{L}_{\bar{\Gamma}} = \bar{\mathfrak{S}}(\bar{\Gamma}_n, J_1)'$. If $p = 2$ and $s' = s$ there is, by a lemma we shall prove below, a $\bar{\Gamma}$ of the form $\bar{\Delta}_{u-1}$ over $\Phi(b_1, b_2 \cdots b_t)$, b_i^p in Φ such that $\mathfrak{L}_{\bar{\Gamma}} = \mathfrak{S}(\Gamma_n, J_2)$. By continuing this process we obtain an extension Γ^* of Φ which is either separable or is a separable extension of a $\Phi(b_1, \cdots, b_v)$, $b_i^p \in \Phi$ such that $\mathfrak{L}_{\Gamma^*} = \mathfrak{S}(\Gamma^*_n, J_i)$ or $\bar{\mathfrak{S}}(\Gamma^*_n, J_i)'$. If Γ^* is separable we replace it by a separable normal field. In the other case Γ^* is a direct product of $\Phi(b_k)$ and a separable field over Φ and the latter may be replaced by a separable normal field. In either case we obtain from $\mathfrak{L}_{\Gamma^*} = \mathfrak{S}$ or $\bar{\mathfrak{S}}'$ that $\mathfrak{L} = \mathfrak{S}(\mathfrak{A}, J)$ or $\bar{\mathfrak{S}}(\mathfrak{A}, J)'$ where $\mathfrak{A}$ is normal simple over Φ.

LEMMA 14. *If $\mathfrak{L}$ is a restricted Lie algebra over Φ of characteristic 2 such that there exists an extension P of Φ with the property $\mathfrak{L}_{\mathrm{P}} = \mathfrak{S}(\mathrm{P}_n, J)'$ where n is odd and J is the involution $a \rightarrow s^{-1}a's$, $s' = s$, not alternate, then there is an extension Σ of the form $\Sigma = \Phi(b_1, \cdots, b_t, \mathrm{P})$ where $b_j^2 = \beta_j \in \Phi$ such that $\mathfrak{L}_\Sigma = \mathfrak{S}(\Sigma_n, J_2)'$.*

Without loss of generality we may suppose that

$$s = \{\alpha_1, \alpha_2, \cdots, \alpha_n\}.$$

Then $\mathfrak{S}(\mathrm{P}_n, J)$ has the basis $h_i = e_{ii}$ and $e_{\lambda_i+\lambda_j} = e_{ij} + e_{ji}\alpha_i\alpha_j^{-1}$ with $i < j$ and the multiplication table

$$[e_a, h] = e_a\alpha \text{ if } h = \Sigma h_i\lambda_i$$

$$[e_\beta, e_a] = \begin{cases} 0 \text{ if } \alpha + \beta \text{ is not a root } \lambda_i + \lambda_j \\ e_{\beta+a}N_{a\beta} \neq 0 \text{ if } \alpha + \beta \text{ is a root} \end{cases}$$

$$h_i^2 = h_i, \qquad e_{\lambda_i+\lambda_j}{}^2 = (h_i + h_j)\alpha_i\alpha_j^{-1}.$$

Since $[[e_\beta, e_a], e_a] = [e_\beta, e_a{}^2] = e_\beta\alpha_i\alpha_j^{-1}$ if $\alpha = \lambda_i + \lambda_j$ and $\alpha + \beta$ is a root $(\neq 0)$ we have $N_{a\beta}N_{a+\beta,a} = \alpha_i\alpha_j^{-1}$. The matrix of $\Sigma h_i\lambda_i + \Sigma e_a\mu_a$ in the adjoint representation has the form

$$\begin{pmatrix} 0 & X \\ 0 & M \end{pmatrix} \tag{16}$$

relative to the basis h_i and e_a where the 0-block in the upper corner has n rows and columns. M has α's down the diagonal and $\mu_aN_{a\beta}$ in the intersection of the "βth" row and $\alpha + \beta$-column if $\alpha + \beta$ is a root, otherwise the elements of M are 0. Thus if $\alpha + \beta$ is a root we also have $(\alpha + \beta) + \alpha$ as a root and hence $\mu_aN_{a+\beta,a}$ is in the $(\alpha + \beta, \beta)$-place. We note also that there are $2(n-2)$ roots β such that $\alpha + \beta$ is a root for a fixed α. It follows that the characteristic polynomial of (16) is

$$\lambda^n(\lambda^m + \lambda^{m-2}\phi_2 + \cdots), \qquad (m = n(n-1)/2)$$

where

$$\phi_2(\lambda_i, \mu_a) = \Sigma (\lambda_1 + \lambda_2)(\lambda_1 + \lambda_3) + (n-2) \Sigma \mu^2{}_{\lambda_i+\lambda_j}\alpha_i\alpha_j^{-1}$$
$$= \Sigma (\lambda_1 + \lambda_2)(\lambda_1 + \lambda_3) + \Sigma \mu_{ij}{}^2\alpha_i\alpha_j^{-1}.$$

We understand by the first summation the sum of all products of pairs of distinct form $\lambda_i + \lambda_j$, $\lambda_k + \lambda_l$ and we have set $\mu_{\lambda_i+\lambda_j} = \mu_{ij}$. Since $\mathfrak{S}(\mathrm{P}_n, J) = \mathfrak{S}(\mathrm{P}_n, J)' \oplus (1)$ the characteristic polynomial of $\Sigma h_i\lambda_i + \Sigma e_a\mu_a$ in $\mathfrak{S}'$ is $\lambda^{n-1}(\lambda^m + \lambda^{m-2}\phi_2 + \cdots)$. Thus we obtain the characteristic polynomial in the adjoint representation of $\mathfrak{S}'$ by setting $\Sigma \lambda_i = 0$ or $\lambda_n = \lambda_1 + \cdots + \lambda_{n-1}$. Then ϕ_2 becomes

$$\phi'_2 = \sum_{i=1}^{n-1} \lambda_i^2 + \sum_{i<j=1}^{n-1} \lambda_i \lambda_j + \Sigma \mu_{ij}^2 \alpha_i \alpha_j^{-1}$$

For

$$\Sigma(\lambda_1 + \lambda_2)(\lambda_1 + \lambda_3) = \tfrac{1}{2}(n-2)(n-1) \sum_1^n \lambda_i^2 + n(n-2) \sum_{i<j=1}^{n} \lambda_i \lambda_j$$

and this becomes $\sum_1^{n-1} \lambda_i^2 + \sum_{i<j=1}^{n-1} \lambda_i \lambda_j$ when we set $\lambda_n = \lambda_1 + \cdots + \lambda_{n-1}$. We may apply Albert's theory of quadratic forms mod 2.[22] The matrix of the form in the λ's is

$$A = \begin{pmatrix} 1 & 0 & \cdot & \cdot & 0 \\ 1 & 1 & \cdot & \cdot & \cdot \\ \cdot & \cdot & \cdot & \cdot & \cdot \\ \cdot & \cdot & \cdot & 1 & 0 \\ 1 & \cdot & \cdot & 1 & 1 \end{pmatrix}.$$

Since $A + A' = 1 + B$ where B is the matrix all of whose elements are 1 we have $(1+B)^2 = 1 + B^2$ and since $(n-1)$ is even $B^2 = 0$. Thus $A + A'$ is non-singular. Hence the form in the λ's may be reduced to $\Sigma \lambda'^2_i \gamma'_i + \lambda'_1 \lambda'_{q+1} + \lambda'_2 \lambda'_{q+2} + \cdots \lambda'_q \lambda'_{2q}$ where $q = (n-1)/2$ and a reduction of ϕ'_2 is obtained by adding $\Sigma \mu_{ij}^2 \alpha_i \alpha_j^{-1}$.

Now we have assumed that $\mathfrak{S}(\mathrm{P}_n, J)' = \mathfrak{L}_{\mathrm{P}}$ where $\mathfrak{L}$ is a Lie algebra over Φ. This means that $\mathfrak{S}'$ has a basis a_j with a multiplication table in Φ and if we use this basis to determine the characteristic polynomial of the general element $\Sigma a_j \nu_j$ we obtain for the second coefficient a quadratic form $\psi_2(\nu)$ with coefficients in Φ equivalent to ϕ'_2. Since $n-1$ is an invariant ψ'_2 may be reduced to $\sum_1^{n-1} \nu'^2_i \gamma_i + \nu'_1 \nu'_{q+1} + \cdots + \nu'_q \nu'_{2q} + \sum_n^m \nu'^2_j \beta_j$, $\gamma \in \Phi$. It follows that $\Sigma \mu_{ij}^2 \alpha_i \alpha_j^{-1}$ and $\sum_n^m \nu'^2_j \beta_j$ are equivalent. If we adjoin $\sqrt{\beta_j}$ to Φ we obtain $\Sigma \mu_{ij}^2 \alpha_i \alpha_j^{-1}$, equivalent to $\Sigma \nu'^2_j$. Hence each $\alpha_i \alpha_j^{-1}$ is a square in $\Sigma = \Phi(\mathrm{P}, b_1, \cdots, b_t)$, $b_k^2 = \beta_k$, $t = n - m$ and the matrix $s\alpha_1^{-1}$ is cogredient to 1. Hence $\mathfrak{L}_\Sigma = \mathfrak{S}(\Sigma_n, J_2)'$. We have therefore proved the lemma and with it the

THEOREM 16. *If $\mathfrak{L}$ is a restricted Lie algebra over Φ such that $\mathfrak{L}_\Omega$ is one of the simple algebras $\mathfrak{S}(\Omega_n, J_i)$, $i = 1, 2, 3$, for $p \neq 2$ or $\bar{\mathfrak{S}}(\Omega_n, J_1)'$ for $p = 2$ and $\nu \not\equiv 0 \pmod 2$ or $\mathfrak{S}(\Omega_n, J_2)'$ for $p = 2$ and $n \not\equiv 0 \pmod 2$ then $\mathfrak{L}$ is isomorphic to an $\mathfrak{S}(\mathfrak{A}, J)$ to an $\bar{\mathfrak{S}}(\mathfrak{A}, J)'$ or to an $\mathfrak{S}(\mathfrak{A}, J)'$ respectively.*

[22] *Loc. cit.*[18], p. 400.

3

III. APPLICATIONS.

3. 1. Isomorphisms between algebras of different types. Suppose that $p \neq 2$ and that there is an isomorphism between $\mathfrak{S}(\Omega_m, J_i)$ and Ω'_{nl} $n \not\equiv 0 \pmod{p}$. This isomorphism maps the maximum commutative subalgebra $\mathfrak{H}$ of $\mathfrak{S}$ into a maximal commutative subalgebra $\bar{\mathfrak{H}}$ of Ω'_{nl}. By modifying the isomorphism we may suppose that $\bar{\mathfrak{H}}$ is contained in the subalgebra with basis $e_{ii} - e_{nn}$, $i = 1, \cdots, n-1$. In order to be maximal $\bar{\mathfrak{H}}$ must coincide with this algebra. Thus in the case $\mathfrak{S}(\Omega_m, J_1)$, $m = 2\mu$, we therefore have $n - 1 = \mu$ and hence the orders $n^2 - 1 = \mu(2\mu + 1)$ so that $\mu = 1$. For $\mathfrak{S}(\Omega_m, J_2)$, $m = 2\mu + 1$, we obtain the one possibility $\mu = 1$ and for $\mathfrak{S}(\Omega_m, J_3)$, $m = 2\mu$, the only possibility is $\mu = 3$. In these particular cases it is readily seen that isomorphisms exist between $\mathfrak{S}(\Omega_2, J_1)$ and Ω'_{2l} between $\mathfrak{S}(\Omega_3, J_2)$ and Ω'_{2l} and between $\mathfrak{S}(\Omega_6, J_3)$ and Ω'_{4l}.

Consider next $\Omega'_{nl}/(1)$ with $n \equiv 0 \pmod{p}$. Any restricted derivation D in Ω_{nl} is inner and induces a restricted derivation in $\Omega'_{nl}/(1)$. In order that D be 0 in the difference algebra it is necessary that $[a, b]D = 1\rho \in \Omega$ for all a and b in Ω_{nl}. It follows that if $u \in \Omega'_{nl}$ then $u^p D = 0$. Using $u = e_{ii} - e_{nn}$ we obtain that $xD = [x, h_0]$, h_0 in $\mathfrak{H}$. Then $[e_a, h_0] = e_a \alpha_0 = \rho_a$ implies that $h_0 = 1\sigma$, $D = 0$. Hence the algebra $\Omega'_{nl}/(1)$ has outer derivations and can not be isomorphic to any $\mathfrak{S}(\Omega_m, J_i)$. A similar discussion for the case $p = 2$ yields no isomorphic pairs.

If we use the completeness theorems we see that for $p \neq 2$ any restricted Lie algebra $\mathfrak{S}(\mathfrak{A}, J)$ such that $\mathfrak{S}(\mathfrak{A}, J)_\Omega = \mathfrak{S}(\Omega_6, J_3)$ is isomorphic to a $\mathfrak{B}'_l$, $\mathfrak{B}$ normal simple of order 16 or to an $\mathfrak{S}(\mathfrak{B}, K)'$ where $\mathfrak{B}$ is an involutorial algebra of order 16 over the center $\mathrm{P} = \Phi(q)$ and K is an involution of the second kind. Conversely any Lie algebra of one of the last two types is isomorphic to an $\mathfrak{S}(\mathfrak{A}, J)$. Similarly any $\mathfrak{S}(\mathfrak{A}, J)$ with $\mathfrak{S}(\mathfrak{A}, J)_\Omega = \mathfrak{S}(\Omega_3, J_2)$ or $\mathfrak{S}(\Omega_5, J_2)$ is isomorphic to an $\mathfrak{S}(\mathfrak{B}, K)$ with $\mathfrak{S}(\mathfrak{B}, K)_\Omega = \mathfrak{S}(\Omega_2, J_1)$ or $\mathfrak{S}(\Omega_4, J_1)$. Any restricted Lie algebra of order 3 in the classes enumerated here is isomorphic to a $\mathfrak{B}'_l$. Our determination of the automorphism groups gives the well-known isomorphism theorems between certain projective orthogonal and full projective groups.[23] Aside from the cases we have noted here, no other isomorphisms exist between the Lie algebras enumerated here. For any isomorphism may be extended to an isomorphism between algebras $\mathfrak{S}(\Omega_n, J_i)$ or between these and Ω'_{ml}'s. Whether or not this holds for the groups of automorphisms is an open question.

---

[23] Cf. van der Waerden, *Gruppen von linearen Transformationen*, Berlin, 1935, pp. 18-28.

3.2. Extension to non-normal algebras. If $\mathfrak{L}$ is an (ordinary) simple Lie algebra over Φ the enveloping algebra of the linear transformations $x \to [x, a] = xA$ is a matrix algebra Σ_m where Σ is a field of finite order over Φ.[24] $\mathfrak{L}$ may be regarded as an algebra over Σ and when this is done $(\mathfrak{L} \text{ over } \Sigma)_\Omega$, Ω the algebraic closure of Σ, is simple. Σ is the only field of operators containing Φ for which this holds. It is known that any automorphisms S of $\mathfrak{L}$ over Φ induces an automorphism s in Σ over Φ such that $(x\xi)^S = x^S\xi^s$ if $\xi \in \Sigma$. Now consider a derivation D in $\mathfrak{L}$ over Φ. If the A_i are defined as above: $[x, a_i] = xA_i$ then

$$x(\Sigma A_1A_2 \cdots A_r)D = (xD) \Sigma A_1A_2 \cdots A_r + x(\Sigma A'_1A_2 \cdots A_r + \Sigma A_1A'_2 \cdots A_r + \cdots + \Sigma A_1A_2 \cdots A'_r)$$

where A'_i is the transformation $x \to [x, a_iD]$. Hence if $\Sigma A_1A_2 \cdots A_r = \Sigma B_1B_2 \cdots B_s$ then

$$\Sigma A'_1A_2 \cdots A_r + \cdots + \Sigma A_1A_2 \cdots A'_r = \Sigma B'_1B_2 \cdots B_s + \cdots + \Sigma B_1B_2 \cdots B'_s.$$

Thus the correspondence

$$\Sigma A_1A_2 \cdots A_1 \to \Sigma A'_1A_2 \cdots A_r + \cdots + \Sigma A_1A_2 \cdots A'_r$$

is a derivation in the associated algebra Σ_m and induces a derivation d in Σ over Φ such that $(x\xi)D = (xD)\xi + x(\xi^d)$. It follows that the derivation algebra $\mathfrak{D}$ of $\mathfrak{L}$ over Φ contains as an ideal $\mathfrak{J}$ the set of derivations of $\mathfrak{L}$ over Σ and that $\mathfrak{D}/\mathfrak{J}$ is isomorphic to a subalgebra of the algebra of derivations of the field Σ over Φ.

For example let $\Sigma = \Phi(b_1, b_2, \cdots, b_m)$ where $b_i^p = \beta_i$ is in Φ and $\mathfrak{L} = \Sigma'_{nl}$ with $n \not\equiv 0 \pmod{p}$. If d is any derivation in Σ over Φ the correspondence $(\xi_{ij}) \to (\xi_{ij}{}^d)$ is a derivation in $\mathfrak{L}$ over Φ which induces the derivation d in Σ. Hence $\mathfrak{D}/\mathfrak{J}$ is isomorphic to the complete algebra of derivations of Σ over Φ. It is known that the latter is a simple Lie algebra.[25] Since we have shown above that $\mathfrak{J}$ is the algebra of inner derivations of Σ'_{nl}, the outer derivation algebra of $\mathfrak{L}$ is not solvable and this contradicts a conjecture recently made by Zassenhaus.[2]

UNIVERSITY OF NORTH CAROLINA
ON LEAVE AT
THE JOHNS HOPKINS UNIVERSITY.

[24] See the author's "A note on non-associative algebras," *Duke Mathematical Journal*, vol. 3 (1937), pp. 544-548.

[25] See [6], p. 218.

CLASSES OF RESTRICTED LIE ALGEBRAS OF CHARACTERISTIC p, II

By N. Jacobson

1. The class of algebras considered in this paper is obtained as follows: Let Φ be a field of characteristic p and let $\mathfrak{A} = \Phi(x_1, \cdots, x_m)$ be the commutative associative algebra with the basis $x_1^{\alpha_1} \cdots x_m^{\alpha_m}$, $0 \leq \alpha_i < p$, where $x_i^0 = 1$ and $x_i^p = \xi_i$ is in Φ. Let $\mathfrak{L} = \mathfrak{D}(\mathfrak{A})$ be the restricted Lie algebra of derivations of $\mathfrak{A}$, i.e., the set of transformations d of $\mathfrak{A}$ that satisfy

$$(x + y)d = xd + yd, \qquad (x\alpha)d = (xd)\alpha, \qquad (xy)d = x(yd) + (xd)y$$

for x, y in $\mathfrak{A}$ and α in Φ. The fundamental operations in $\mathfrak{L}$ are addition, scalar multiplication, commutation, $[d, e] = de - ed$, and p-exponentiation, d^p. (We shall show in §9 that our results are valid also when we drop the operation $d \to d^p$ and consider $\mathfrak{L}$ as a Lie algebra in the ordinary sense.) The case in which $\mathfrak{A}$ is a field has been considered by the author in a previous paper [3] and the algebra $\mathfrak{L}$ obtained when $m = 1$ and $\xi = 1$ is equivalent to one discovered by Witt and studied by Zassenhaus [8] and by Ho-Jui Chang [2]. We shall show that for any m and ξ_i, $\mathfrak{L}$ is normal simple unless $m = 1$, $p = 2$, and we obtain the derivation algebra of $\mathfrak{L}$. The automorphisms of $\mathfrak{L}$ and conditions that two algebras $\mathfrak{L}_1$ and $\mathfrak{L}_2$ be isomorphic are given for $p \geq 5$.

Since the x's generate $\mathfrak{A}$, any derivation d is determined by its effect on the x_i. Moreover, we may choose elements $y_1, \cdots, y_m$ arbitrarily in $\mathfrak{A}$ and obtain a derivation d such that $x_i d = y_i$, see [3; 217]. Thus, we have a 1-1 correspondence between the elements of $\mathfrak{L}$ and vectors $(y_1, \cdots, y_m)$, where y_i ranges over $\mathfrak{A}$. If $d \to (y_1, \cdots, y_m)$ and $c \to (z_1, \cdots, z_m)$, then $d + c \to (y_1 + z_1, \cdots, y_m + z_m)$ and $d\alpha \to (y_1\alpha, \cdots, y_m\alpha)$, α in Φ. Hence, the correspondence is linear and so the dimensionality of $\mathfrak{L}$ over Φ is mp^m. We note also that $[d, c] \to (w_i)$, where

$$w_i = \sum_k \left(\frac{\partial y_i}{\partial x_k} z_k - \frac{\partial z_i}{\partial x_k} y_k\right).$$

An explicit formula for the vector corresponding to d^p would be rather difficult to write and so we shall be content to note that the component y_{pi} of this vector is obtained by the recursion formula

$$y_{pi} = \sum_{k_1, \cdots, k_{p-1}} \left(\frac{\partial}{\partial x_{k_{p-1}}} \cdots \left(\frac{\partial}{\partial x_{k_2}} \left(\frac{\partial y_i}{\partial x_{k_1}} y_{k_1}\right) y_{k_2}\right) \cdots\right) y_{k_{p-1}}.$$

Received July 15, 1942; presented to the American Mathematical Society, April 3, 1942. Part I appeared in [5]; we follow the notation and terminology of that paper.

When $m = 1$, this reduces to

$$y_p = (\cdots((y'y)'y)'y\cdots)'y, \quad \overleftrightarrow{p-1}$$

where $y' = y(x)'$ is the ordinary derivative.

2. Suppose now that all the $\xi_i = 1$. Then the basis $x_1^{\alpha_1} \cdots x_m^{\alpha_m}$ constitutes a group under multiplication, which is the direct product of m cyclic groups of order p, and $\mathfrak{A}$ is the group algebra of this group.

In place of the generators x_i of $\mathfrak{A}$ we may use 1 and $z_i = x_i - 1$. Then we obtain the basis $1, z_1^{\alpha_1}, \cdots, z_m^{\alpha_m}, 0 \leq \alpha_i \leq p-1, (\alpha_1, \cdots, \alpha_m) \neq (0, \cdots, 0)$. Since $z_i^p = 0$, the elements $z_1^{\alpha_1} \cdots z_m^{\alpha_m}$ are in the radical $\mathfrak{R}$ of $\mathfrak{A}$ and $\mathfrak{A} = (1) + \mathfrak{R}$. If $\sum \alpha_i = k$, then the base element $z_1^{\alpha_1}, \cdots, z_m^{\alpha_m}$ belongs to $\mathfrak{R}^k$. Since $\mathfrak{R}^k$ is generated by products of k elements of any basis for $\mathfrak{R}$, it follows that the monomials $z_1^{\alpha_1} \cdots z_m^{\alpha_m}$ of degree $\geq k$ form a basis for $\mathfrak{R}^k$. Hence, the dimensionality of $(\mathfrak{R}^k - \mathfrak{R}^{k+1})$ is the number r_k of monomials of degree k. In particular, $r_1 = m$ and the cosets of $z_1, \cdots, z_m$ form a basis for $\mathfrak{R} - \mathfrak{R}^2$.

In general, any set of elements $u_1, \cdots, u_m$ whose cosets form a basis for $\mathfrak{R} - \mathfrak{R}^2$ generates $\mathfrak{R}$ by multiplication and linear combination. For, any element b_1 of $\mathfrak{R} \equiv \sum u_i\beta_i \ (\mathfrak{R}^2)$. Hence, any b_2 in $\mathfrak{R}^2 \equiv \sum u_iu_j\beta_{ij} \ (\mathfrak{R}^3)$, etc. Hence, any element of $\mathfrak{R}$ is a linear combination of the elements $u_1^{\alpha_1} \cdots u_m^{\alpha_m}$, where $0 \leq \alpha_i \leq p-1$, $(\alpha_1, \cdots, \alpha_m) \neq (0, \cdots, 0)$, and since the number of these elements is $p^m - 1$, they form a basis for $\mathfrak{R}$.

It follows readily that any set of elements $u_0, \cdots, u_m$ whose cosets form a basis for $\mathfrak{A} - \mathfrak{R}^2$ generates $\mathfrak{A}$. Since any basis of $\mathfrak{A}$ contains $m+1$ elements such as the u_i, it follows that any basis contains $m+1$ elements that generate $\mathfrak{A}$.

3. We return to the original basis $x_1^{\alpha_1} \cdots x_m^{\alpha_m}$ with $x_i^p = 1$. For any linear form $\alpha = \sum \alpha_i\lambda_i$, $\alpha_i = 0, 1, \cdots, p-1$, we define the derivation $e_\alpha^{(i)}$ in $\mathfrak{A}$ by the equation

$$x_r e_\alpha^{(i)} = \delta_{ir}(x_1^{\alpha_1} \cdots x_m^{\alpha_m})x_i .$$

The mp^m derivations obtained in this way form a basis for $\mathfrak{L} = \mathfrak{D}(\mathfrak{A})$, and the following is the multiplication table:

(1) $[e_\alpha^{(i)}, e_\beta^{(j)}] = e_{\alpha+\beta}^{(i)}\alpha_j - e_{\alpha+\beta}^{(j)}\beta_i$,

(2) $(e_\alpha^{(i)})^p = e_0^{(i)}$ if $\alpha_i = 0$,
$(e_\alpha^{(i)})^p = 0$ if $\alpha_i \neq 0$.

Set $e_0^{(i)} = h_i$, $\sum h_i\lambda_i = h$. Then $(x_1^{\alpha_1} \cdots x_m^{\alpha_m})h = (x_1^{\alpha_1} \cdots x_m^{\alpha_m})\alpha$. Special cases of (1) and (2) are

(3) $[e_\alpha^{(i)}, h] = e_\alpha^{(i)}\alpha$,

(4) $[e_\alpha^{(i)}, e_{-\alpha}^{(j)}] = h_i\alpha_j + h_j\alpha_i$,

(5) $h_i^p = h_i$.

Suppose that Φ is infinite. Then we may choose the λ_i in Φ so that the p^m linear forms $\sum \alpha_i\lambda_i$ are all different. Then the minimum equation of the linear transformation h is $\prod_\alpha (h - \alpha) = 0$. Since α ranges over the linear forms $\sum \alpha_i\lambda_i$, this equation has the form

$$h^{p^m} + h^{p^{m-1}}\beta_1 + \cdots + h\beta_m = 0, \tag{6}$$

where β_i is a homogeneous polynomial of degree $(p^m - p^{m-i})$ in the λ's. Thus, the subalgebra $\mathfrak{H}$ of $\mathfrak{L}$ generated by h has dimensionality m and the basis h, $h^p, \cdots, h^{p^{m-1}}$. Since $h = \sum h_i\lambda_i$, $h^p = \sum h_i\lambda_i^p, \cdots$, a second basis for this algebra is $h_1, h_2, \cdots, h_m$.

Let $\mathfrak{M}_\alpha$ be the subspace of $\mathfrak{L}$ of elements of the form $\sum e_\alpha^{(i)}\rho_i$, ρ_i in Φ. Then $\mathfrak{M}_0 = \mathfrak{H}$ and $\mathfrak{M}_\alpha$ is an invariant subspace relative to the adjoint transformations $x \to [x, h]$, x in $\mathfrak{L}$, i.e., $[\mathfrak{M}_\alpha, \mathfrak{H}] \leq \mathfrak{M}_\alpha$. Actually we have $[\mathfrak{M}_\alpha, \mathfrak{H}] = \mathfrak{M}_\alpha$ if $\alpha \neq 0$ since $[e_\alpha, h] = e_\alpha\alpha$ for e_α in $\mathfrak{M}_\alpha$. Since $\mathfrak{L} = \mathfrak{M}_0 \oplus \mathfrak{M}_\alpha \oplus \mathfrak{M}_\beta \oplus \cdots$, any x has a unique representation in the form $h_0 + e_\alpha + e_\beta + \cdots$, h_0 in $\mathfrak{H}$, e_α in $\mathfrak{M}_\alpha$, $\alpha \neq 0$. It follows that $\mathfrak{M}_\alpha$ may be characterized as the complete set of elements satisfying the equation $[e_\alpha, h] = e_\alpha\alpha$. This (or equations (1) and (2)) implies that

$$[\mathfrak{M}_\alpha, \mathfrak{M}_\beta] \leq \mathfrak{M}_{\alpha+\beta}, \qquad \mathfrak{M}_\alpha^p \leq \mathfrak{H}.$$

If $\alpha \neq \beta$, let r be an index such that $\alpha_r \neq \beta_r$. Then $[e_\alpha^{(r)}, e_\beta^{(r)}] = e_{\alpha+\beta}^{(r)}(\alpha_r - \beta_r)$ and so $e_{\alpha+\beta}^{(r)}$ is in $[\mathfrak{M}_\alpha, \mathfrak{M}_\beta]$. Either $\alpha_r \neq 0$ or $\beta_r \neq 0$. Assume the former. Then $[e_\alpha^{(i)}, e_\beta^{(r)}] = e_{\alpha+\beta}^{(i)}\alpha_r - e_{\alpha+\beta}^{(r)}\beta_i$ and hence $e_{\alpha+\beta}^{(i)}$ is in $[\mathfrak{M}_\alpha, \mathfrak{M}_\beta]$. Thus,

$$[\mathfrak{M}_\alpha, \mathfrak{M}_\beta] = \mathfrak{M}_{\alpha+\beta} \qquad \text{if } \alpha \neq \beta. \tag{7}$$

The forms α will be called the *roots* of h. We may now prove

THEOREM 1. *Unless $p = 2$, $m = 1$, $\mathfrak{L} = \mathfrak{D}(\mathfrak{A})$ is a simple Lie algebra.*

In this theorem we drop the operation $d \to d^p$ in $\mathfrak{L}$ and regard $\mathfrak{L}$ as an ordinary Lie algebra. The theorem is therefore stronger than the statement that $\mathfrak{L}$ is a simple restricted Lie algebra. Let $\mathfrak{B} \neq 0$ be an ideal in $\mathfrak{L}$ and $b = \sum e_\alpha \neq 0$ an element in $\mathfrak{B}$. Suppose that $e_{\alpha^*} \neq 0$ in this expression. Since $\mathfrak{B}$ is an ideal, it contains

$$\overleftarrow{\quad r \quad}\!\!\!\!\!\!\!\!\!\!\!\!\overrightarrow{\quad\quad}$$
$$[b, h], h] \cdots, h] = \sum e_\alpha\alpha^r.$$

Since the α are distinct, we may choose a suitable linear combination of these elements to obtain $e_{\alpha^*} = \sum e_{\alpha^*}^{(i)}\rho_i \neq 0$ as an element in $\mathfrak{B}$. Now $[e_{\alpha^*}, e_{-\alpha^*}^{(j)}] = k\alpha_j^* + h_j\rho$, where $k = \sum h_i\rho_i$ and $\rho = \sum \alpha_i^*\rho_i$. If $\alpha^* \neq 0$, there exists a j such that $k\alpha_j^* + h_j\rho \neq 0$. For this is clear if $\rho = 0$ and $\rho \neq 0$, $k\alpha_j^* + h_j\rho = 0$, implies that $h_j = -k\alpha_j^*\rho^{-1}$ and hence that $m = 1$. Then $[e_{\alpha^*}, e_{-\alpha^*}] \neq 0$ unless $p = 2$. Thus, we may suppose that $\mathfrak{B}$ contains $h_0 = \sum h_i\lambda_i^0 \neq 0$. There is a root α such that $\alpha_0 = \sum \alpha_i\lambda_i^0 \neq 0$. Then $[\mathfrak{M}_\alpha, h_0] = \mathfrak{M}_\alpha \leq \mathfrak{B}$. If γ is a root $\neq 2\alpha$, $\alpha + \beta = \gamma$, where $\beta = \gamma - \alpha \neq \alpha$. Hence, by (7), $\mathfrak{M}_\gamma = [\mathfrak{M}_\alpha, \mathfrak{M}_\beta]$ is in $\mathfrak{B}$.

Now, unless $p = 2, 3$ and $m = 1$, there is a root $\delta \neq 0, \alpha, -\alpha$. Then, if $p \neq 2$, $\alpha + \delta \neq \alpha - \delta$ and so $\mathfrak{M}_{2\alpha} = [\mathfrak{M}_{\alpha+\delta}, \mathfrak{M}_{\alpha-\delta}]$ is in $\mathfrak{B}$ also. Thus, we have proved that $\mathfrak{B}$ contains all $\mathfrak{M}_\alpha$, $\alpha \neq 0$, except in the cases $p = 2, 3$ and $m = 1$. It follows then that $\mathfrak{B}$ contains $h_i\alpha_j + h_j\alpha_i$ for all $\alpha_i, \alpha_j = 0, \cdots, p-1$, and hence that $\mathfrak{B}$ contains $\mathfrak{H}$ and $\mathfrak{B} = \mathfrak{L}$. If $p = 3$, $m = 1$, the simplicity may be proved directly, and if $p = 2$, $m = 1$, $\mathfrak{L}$ is solvable.

If $\mathfrak{A} = \Phi(x_1, \cdots, x_m)$ with Φ arbitrary and $x_i^p = \xi_i$, let Ω be the algebraic closure of Φ and consider $\mathfrak{A}_\Omega$. In this extension we replace x_i by $x_i' = x_i + 1$ if $\xi_i = 0$ and by $x_i' = x_i\xi_i^{-1/p}$ if $\xi_i \neq 0$. Then the x_i' generate $\mathfrak{A}_\Omega$ and $(x_i')^p = 1$ and so $\mathfrak{A}_\Omega$ is of the type considered above. We recall that $\mathfrak{L}_\Omega$ is the derivation algebra of $\mathfrak{A}_\Omega$. Hence, $\mathfrak{L}_\Omega$ is simple and we have

THEOREM 2. *If $\mathfrak{A} = \Phi(x_1, \cdots, x_m)$, $x_i^p = \xi_i$ and either $p > 2$ or $m > 1$, then the derivation algebra $\mathfrak{L} = \mathfrak{D}(\mathfrak{A})$ is a normal simple Lie algebra.*

4. We suppose again that $\xi_i = 1$ and that Φ is infinite. Let d be any derivation having distinct characteristic roots all contained in Φ. Then there is a basis $u_0, \cdots, u_{p^m-1}$ of $\mathfrak{A}$ such that $u_id = u_i\mu_i$, where $\mu_i \neq \mu_j$ if $i \neq j$. Since $1d = 0$, we may suppose that $\mu_0 = 0$ and $u_0 = 1\gamma$. Let $u_0, u_1, \cdots, u_m$ be elements of this basis such that the cosets $u_0 + \mathfrak{N}^2, \cdots, u_m + \mathfrak{N}^2$ form a basis for $\mathfrak{A} - \mathfrak{N}^2$. Then the u_i are generators of $\mathfrak{A}$ and u_0, $u_1^{\alpha_1} \cdots u_m^{\alpha_m}$, $0 \leq \alpha_i \leq p-1$, $(\alpha_1, \cdots, \alpha_m) \neq (0, \cdots, 0)$ is a basis for $\mathfrak{A}$. Since d is a derivation and $u_id = u_i\mu_i$, $i = 1, \cdots, m$, $(u_1^{\alpha_1} \cdots u_m^{\alpha_m})d = (u_1^{\alpha_1} \cdots u_m^{\alpha_m})(\sum \alpha_i\mu_i)$. Hence, the characteristic roots of d are the linear forms $\sum \alpha_i\mu_i$ in the μ_i and the characteristic polynomial of d is the p-polynomial

$$\prod (\lambda - \sum \alpha_i\mu_i) = \lambda^{p^m} + \lambda^{p^{m-1}}\delta_1 + \cdots + \lambda\delta_m. \tag{8}$$

Consider now the characteristic polynomial $f(\lambda, \tau)$ of the "general" transformation $\sum e_\alpha^{(i)}\tau_\alpha^{(i)}$, where the $\tau_\alpha^{(i)}$ are indeterminates. We have seen that the $\tau_\alpha^{(i)}$ may be specialized in Φ to give a derivation, namely h, with distinct characteristic roots. Hence, the discriminant $\delta(t)$ of $f(\lambda, \tau)$ is an element $\neq 0$ in the extension field $\Phi(\tau_\alpha^{(i)})$. It follows that in a suitable extension P of $\Phi(\tau_\alpha^{(i)})$, the polynomial $f(\lambda, \tau)$ has distinct roots. Since $\sum e_\alpha^{(i)}\tau_\alpha$ may be extended to a derivation in $\mathfrak{A}_P$, we see that $f(\lambda, \tau)$, its characteristic polynomial, is a p-polynomial. By specializing the $\tau_\alpha^{(i)}$ in Φ we obtain that the characteristic polynomial of any derivation in $\mathfrak{A}$ is a p-polynomial. By using the extension $\mathfrak{A}_\Omega$, Ω, the algebraic closure, we may prove this result also for any $\mathfrak{A}$ with $x_i^p = \xi_i$ arbitrary.

THEOREM 3. *If $\mathfrak{A} = \Phi(x_i)$, $x_i^p = \xi_i$, the characteristic polynomial of any derivation in $\mathfrak{A}$ is a p-polynomial.*

Since the dimensionality of the subalgebra $(d, d^p, \cdots)$ generated by d is the degree of the minimum p-polynomial satisfied by d, this theorem together with the Hamilton-Cayley theorem shows that the maximum dimensionality of a "cyclic" algebra $(d, d^p, \cdots)$ is m.

If the characteristic roots of d are distinct (in an extension field), then the characteristic polynomial $f(\lambda)$ of d is its minimum polynomial and hence its minimum p-polynomial. In this case $\delta_m \neq 0$ in (8), i.e., 0 is not a multiple root. Conversely, if $\delta_m \neq 0$, then $f'(\lambda) = \delta_m$ and $(f(\lambda), f'(\lambda)) = 1$ and so $f(\lambda)$ has distinct characteristic roots. We call a derivation d *regular* if $\delta_m \neq 0$. Since there exist regular derivations, e.g., h determined above, in $\mathfrak{L}_\Omega$, Ω the algebraic closure of Φ, the usual specialization argument shows that if Φ is infinite, there exist regular derivations in $\mathfrak{L}$. Evidently the zeros of a regular derivation form a one-dimensional space over Φ. Since $1d = 0$ for any d, we have the following theorem, due to Baer [1] for $\mathfrak{A}$ a field:

THEOREM 4. *If Φ is an infinite field, there exist derivations d in $\mathfrak{A}$ over Φ whose d-constants (zeros) consist of the multiples 1α, α in Φ.*

5. We consider now the automorphisms of $\mathfrak{L}$ regarded as a restricted Lie algebra. It is readily verified that if G is any automorphism of $\mathfrak{A}$, then the correspondence $d \to G^{-1}dG$ is an automorphism in $\mathfrak{L}$. We wish to show that if $p \geq 5$, any automorphism S of $\mathfrak{L}$ has this form. We assume first that $\xi_i = 1$ and Φ is algebraically closed. Let h be the derivation defined in §3. Since $(h^p)^S = (h^S)^p, \cdots$, the minimum p-polynomial satisfied by h^S is (6). Since the roots of $f(\lambda) = \lambda^{p^m} + \lambda^{p^{m-1}}\beta_1 + \cdots + \lambda\beta_m$ are distinct, h^S is regular and its roots are in Φ. Hence, there is a basis $1, u_1^{\alpha_1} \cdots u_m^{\alpha_m}$ such that $(u_1^{\alpha_1} \cdots u_m^{\alpha_m})h^S = (u_1^{\alpha_1} \cdots u_m^{\alpha_m})(\sum \alpha_i\mu_i)$, where $\mu_1, \cdots, \mu_m$ are linear forms in the λ's that form a basis for the set of roots $\rho = \sum \rho_i\lambda_i$.

LEMMA. *If $p \geq 5$, each $u_i^p \neq 0$.*

We define the derivation $f_\alpha^{(i)}$ by $1f_\alpha^{(i)} = 0$ and

$$\begin{aligned} u_r f_\alpha^{(i)} &= \delta_{ir}(u_1^{\alpha_1} \cdots u_m^{\alpha_m})u_i \qquad \text{if } \alpha_i \neq p-1 \\ &= \delta_{ir}u_1^{\alpha_1} \cdots u_{i-1}^{\alpha_{i-1}}u_{i+1}^{\alpha_{i+1}} \cdots u_m^{\alpha_m} \qquad \text{if } \alpha_i = p-1. \end{aligned}$$

Denote the space with the basis $f_\alpha^{(1)}, \cdots, f_\alpha^{(m)}$ by $\mathfrak{N}_\alpha$ and set $\mathfrak{N}_0 = \mathfrak{K}$. Evidently $\mathfrak{K} = \mathfrak{H}^S$. If we set $k_i = f_0^{(i)}$ and $k = \sum k_i\lambda_i$, then we have $[f_\alpha, k] = f_\alpha\alpha$ for any f_α in $\mathfrak{N}_\alpha$. Since the elements h_i^S form a basis for $\mathfrak{H}^S$, $k = \sum h_i^S\lambda_i'$, where λ_i' is a linear form in the λ's. Since the λ''s form a basis for the linear forms, each $\alpha = \sum \alpha_i\lambda_i = \sum \alpha_i'\lambda_i'$. Hence, $\mathfrak{N}_\alpha = \mathfrak{M}_{\alpha'}^S$, $\alpha' = \sum \alpha_i'\lambda_i'$. Now suppose that $u_1^p = 0$. Then we may verify that $[f_{(p-2)\lambda_1}^{(i)}, f_{(p-3)\lambda_1}^{(j)}] = 0$ for all i, j and so there exist two distinct spaces $\mathfrak{M}_{\alpha'}$, $\mathfrak{M}_{\beta'}$ such that $[\mathfrak{M}_{\alpha'}^S, \mathfrak{M}_{\beta'}^S] = 0$. It follows that $[\mathfrak{M}_{\alpha'}, \mathfrak{M}_{\beta'}] = 0$ and this is impossible.

Since $u_i^p = \sigma_i \neq 0$, we may replace these elements by $v_i = u_i\sigma_i^{-1/p}$ and obtain $v_i^p = 1$ and $v_ih^S = v_i\mu_i$, $(v_1^{\alpha_1} \cdots v_m^{\alpha_m})h^S = (v_1^{\alpha_1} \cdots v_m^{\alpha_m})(\sum \alpha_i\mu_i)$. We may choose α_{ji} such that $\sum \alpha_{ji}\mu_i = \lambda_j$. Then, if $w_j = v_1^{\alpha_{j1}} \cdots v_m^{\alpha_{jm}}$, we have $w_jh^S = w_j\lambda_j$ and the elements $w_1^{\alpha_1} \cdots w_m^{\alpha_m}$ form a basis for $\mathfrak{A}$ also. Since $w_i^p = 1$, the mapping $\sum x_1^{\alpha_1} \cdots x_m^{\alpha_m}\gamma_{\alpha_1 \cdots \alpha_m} \to \sum w_1^{\alpha_1} \cdots w_m^{\alpha_m}\gamma_{\alpha_1 \cdots \alpha_m}$ is an automorphism G in $\mathfrak{A}$. Since $xG^{-1}h^SG = x_i\lambda_i$, $G^{-1}h^SG = h$. The mapping $d \to G^{-1}d^SG^{-1} \equiv d^T$

is therefore an automorphism in $\mathfrak{A}$ that leaves the elements of $\mathfrak{H}$ invariant. We consider automorphisms of this type in the following

THEOREM 5. *If either* $p \neq 2$ *or* $p = 3$, $m > 1$, *then any automorphism* T of $\mathfrak{L}$ *that leaves the elements of* $\mathfrak{H}$ *invariant is the identity.*

Our assumption implies that $\mathfrak{M}_\alpha^T = \mathfrak{M}_\alpha$ and so T induces a non-singular linear transformation in $\mathfrak{M}_\alpha$. We prove first that T is a multiple of the identity in each $\mathfrak{M}_\alpha$. If $m = 1$, this is trivial and so we assume that $m > 1$. Set

$$e_{\lambda_1}^{(i)T} = \sum_k e_{\lambda_1}^{(k)} \rho_{ki}, \qquad e_{\lambda_2}^{T} = \sum e_{\lambda_2}^{(k)} \sigma_{ki},$$

$$e_{\lambda_1+\lambda_2}^{(i)T} = \sum e_{\lambda_1+\lambda_2}^{(k)} \tau_{ki}.$$

Then the matrices (ρ), (σ) and (τ) are non-singular. From the relations

$$\begin{aligned} &[e_{\lambda_1}^{(2)}, e_{\lambda_2}^{(1)}] = e_{\lambda_1+\lambda_2}^{(2)} - e_{\lambda_1+\lambda_2}^{(1)}, \\ &[e_{\lambda_1}^{(i)}, e_{\lambda_2}^{(1)}] = e_{\lambda_1+\lambda_2}^{(i)} \qquad \text{if } i \neq 2, \\ &[e_{\lambda_1}^{(2)}, e_{\lambda_2}^{(j)}] = -e_{\lambda_1+\lambda_2}^{(j)} \qquad \text{if } j \neq 1, \\ &[e_{\lambda_1}^{(i)}, e_{\lambda_2}^{(j)}] = 0 \qquad \text{in all other cases,} \end{aligned} \tag{9}$$

we obtain

$$[e_{\lambda_1}^{(i)}, e_{\lambda_2}^{(j)}]^T = e_{\lambda_1+\lambda_2}^{(1)}(\rho_{1i}\sigma_{1j} - \rho_{2i}\sigma_{1j}) + e_{\lambda_1+\lambda_2}^{(2)}(\rho_{2i}\sigma_{1j} - \rho_{2i}\sigma_{2j}) + \sum_{2<k} e_{\lambda_1+\lambda_2}^{(k)}(\rho_{ki}\sigma_{1j} - \rho_{2i}\sigma_{kj}).$$

Hence,

$$\begin{aligned} &\tau_{11} = (\rho_{11} - \rho_{21})\sigma_{11}, && \tau_{21} = \rho_{21}(\sigma_{11} - \sigma_{21}), && \tau_{k1} = \rho_{k1}\sigma_{11} - \rho_{21}\sigma_{k1}, \\ &\tau_{12} = (\rho_{22} - \rho_{12})\sigma_{12}, && \tau_{22} = \rho_{22}(\sigma_{22} - \sigma_{12}), && \tau_{k2} = \rho_{22}\sigma_{k2} - \rho_{k2}\sigma_{12}, \\ &\tau_{1l} = (\rho_{1l} - \rho_{2l})\sigma_{11}, && \tau_{2l} = \rho_{2l}(\sigma_{11} - \sigma_{21}), && \tau_{kl} = \rho_{kl}\sigma_{11} - \rho_{2l}\sigma_{k1}, \\ &\quad\ = (\rho_{22} - \rho_{12})\sigma_{1l}, && \quad\ = \rho_{22}(\sigma_{2l} - \sigma_{1l}), && \quad\ = \rho_{22}\sigma_{kl} - \rho_{k2}\sigma_{1l}, \end{aligned} \tag{10}$$

where $k, l > 2$. Since

$$[e_{\lambda_1}^{(i)}, e_{\lambda_2}^{(j)}]^T = 0$$

if $i = 1, j \neq 1$; $i \neq 2, j = 2$ and for $i, j > 2$ and

$$[e_{\lambda_1}^{(2)}, e_{\lambda_2}^{(1)}]^T + [e_{\lambda_1}^{(1)}, e_{\lambda_2}^{(1)}]^T + [e_{\lambda_1}^{(2)}, e_{\lambda_2}^{(2)}]^T = 0,$$

we have

$$(\rho_{1i} - \rho_{2i})\sigma_{1j} = \rho_{2i}(\sigma_{1j} - \sigma_{2j}) = \rho_{ki}\sigma_{1j} - \rho_{2i}\sigma_{kj} = 0 \tag{11}$$

for $i = 1, j \neq 1$; $i \neq 2, j = 1$ and for $i, j > 2$ and

$$\begin{aligned} \tau_{11} - \tau_{12} + (\rho_{12} - \rho_{22})\sigma_{11} &= \tau_{21} - \tau_{22} + \rho_{22}(\sigma_{11} - \sigma_{21}) \\ &= \tau_{k1} - \tau_{k2} + \rho_{k2}\sigma_{11} - \rho_{22}\sigma_{k1} = 0. \end{aligned} \tag{12}$$

Now suppose that $m > 1$. We assert that $\rho_{2i} = 0$ for $i \neq 2$. For, otherwise, $\sigma_{2j} = \sigma_{1j}$ for all $j > 1$ and $\sigma_{kj} = \gamma_k \sigma_{1j}$, where $\gamma_k = \rho_{2i}^{-1} \rho_{ki}$ for $j > 1$ and $k > 2$. If $m > 2$, this is impossible since (σ) is non-singular. If we use λ_j, $j = 3, 4, \cdots$, in place of λ_2, we obtain in a similar manner that $\rho_{ji} = 0$ if $i \neq j$ and $j \neq 1$. Similarly, $\sigma_{ki} = 0$ if $i \neq k$ and $k \neq 2$. Since

$$[e_{\lambda_1}^{(1)}, e_{\lambda_1+\lambda_2}^{(1)}] = 0, \qquad [e_{\lambda_1}^{(1)}, e_{\lambda_1+\lambda_2}^{(2)}] = -e_{2\lambda_1+\lambda_2}^{(2)},$$

$$[e_{\lambda_1}^{(2)}, e_{\lambda_1+\lambda_2}^{(1)}] = e_{2\lambda_1+\lambda_2}^{(2)} - e_{2\lambda_1+\lambda_2}^{(1)}, \qquad [e_{\lambda_1}^{(2)}, e_{\lambda_1+\lambda_2}^{(2)}] = -e_{2\lambda_1+\lambda_2}^{(2)},$$

$$[e_{\lambda_1}^{(1)}, e_{\lambda_1+\lambda_2}^{(i)}] = [e_{\lambda_1}^{(2)}, e_{\lambda_1+\lambda_2}^{(i)}] = -[e_{\lambda_1}^{(i)}, e_{\lambda_1+\lambda_2}^{(1)}] = -e_{2\lambda_1+\lambda_2}^{(i)}$$

for $i > 2$ and

$$[e_{\lambda_1}^{(i)}, e_{\lambda_1+\lambda_2}^{(j)}] = 0$$

for all other i, j, we obtain

$$[e_{\lambda_1}^{(1)}, e_{\lambda_1+\lambda_2}^{(2)}]^T = [e_{\lambda_1}^{(2)}, e_{\lambda_1+\lambda_2}^{(2)}]^T$$

and this implies

$$\tau_{12}(\rho_{21} - \rho_{22}) = 0, \qquad (\rho_{11} + \rho_{21} - \rho_{12} - \rho_{22})\tau_{22} = 0. \tag{13}$$

If we interchange the rôles of λ_1 and λ_2, we obtain the corresponding equations

$$\tau_{21}(\sigma_{12} - \sigma_{11}) = 0, \qquad (\sigma_{22} + \sigma_{12} - \sigma_{21} - \sigma_{11})\tau_{11} = 0. \tag{14}$$

Now, if $m = 2$ and $\rho_{21} \neq 0$, by (11) we obtain $\sigma_{12} = \sigma_{22}$. Hence, $\tau_{22} = 0$ and since (τ) is non-singular, $\tau_{12}\tau_{21} = \sigma_{12}\rho_{21}(\sigma_{11} - \sigma_{21})(\rho_{22} - \rho_{12}) \neq 0$. Thus, $\sigma_{12} \neq 0$ and so by (11), $\rho_{11} = \rho_{21}$ and $\tau_{11} = 0$. From (13) and (14) we obtain $\rho_{21} = \rho_{22}$ and $\sigma_{12} = \sigma_{11}$ and by substituting in (12):

$$(\rho_{12} - \rho_{22})(\sigma_{11} + \sigma_{12}) = 0 = (\rho_{21} + \rho_{22})(\sigma_{11} - \sigma_{21}).$$

Hence $\rho_{21} = -\rho_{22}$ and $\sigma_{11} = -\sigma_{12}$ and since $p \neq 2$, $\rho_{21} = \rho_{22} = 0$ and $\sigma_{11} = \sigma_{12} = 0$, which is impossible since (ρ) and (σ) are non-singular. Similarly, the assumption $\sigma_{12} \neq 0$ is impossible.

Thus, in all cases we have proved that $\rho_{j1} = 0 = \sigma_{i2}$ if $j \neq 1$ and $i \neq 2$. This implies that $\tau_{12} = \tau_{21} = 0$ and $\tau_{11} = \rho_{11}\sigma_{11}$, $\tau_{22} = \rho_{22}\sigma_{22}$. Since (ρ) and (σ) are non-singular, $\tau_{11} \neq 0$ and $\tau_{22} \neq 0$ and so by (13) and (14)

$$\rho_{11} - \rho_{12} - \rho_{22} = 0 = \sigma_{22} - \sigma_{21} - \sigma_{11}.$$

On the other hand, by (12),

$$\rho_{11} + \rho_{12} - \rho_{22} = 0 = \sigma_{22} + \sigma_{21} - \sigma_{11}.$$

Hence, $\rho_{12} = \sigma_{21} = 0$, $\rho_{11} = \rho_{22}$, $\sigma_{11} = \sigma_{22}$. If we use the other indices, we obtain in a similar fashion

$$\rho_{ij} = 0 = \sigma_{ij} \quad \text{if } i \neq j, \qquad \rho_{ii} = \rho, \qquad \sigma_{ii} = \sigma.$$

Next we prove that the transformation induced by T in $\mathfrak{M}_{\lambda_i}$ is $1\rho_{\lambda_i}$ $(\rho_{\lambda_1} = \rho,$

$\rho_{\lambda_2} = \sigma$). It follows that T induces the transformation $1\rho_\alpha$ in $\mathfrak{M}_\alpha$ and $\rho_{\alpha+\beta} = \rho_\alpha\rho_\beta$ if $\alpha \neq \beta$ and $\rho_0 = 1$.

If $m = 1$ and $p > 3$,

$$\rho_{3\lambda_1} = \rho_{2\lambda_1}\rho_{\lambda_1}, \quad \cdots, \quad \rho_{k\lambda_1} = \rho_{2\lambda_1}\rho_{\lambda_1}^{k-2}.$$

Since $\rho_{p\lambda_1} = 1$,

$$\rho_{2\lambda_1}\rho_{\lambda_1}^{p-2} = 1, \quad \rho_{2\lambda_1} = \rho_{\lambda_1}^{2-p}, \quad \rho_{k\lambda_1} = \rho_{\lambda_1}^{k-p}.$$

If we use $1 = \rho_{3\lambda_1}\rho_{(p-3)\lambda_1}$, we obtain $\rho_{\lambda_1}^{3-p}\rho_{\lambda_1}^{-3} = 1 = \rho_{\lambda_1}^{-p}$. Hence, $\rho_{\lambda_1} = 1$ and hence $\rho_{k\lambda_1} = 1$. If $m > 1$ and α is any root, we choose a β not a multiple of α. Then

$$\overbrace{[\cdots[\mathfrak{M}_\beta\mathfrak{M}_\alpha]\cdots\mathfrak{M}_\alpha]}^{\longleftarrow p \longrightarrow} = \mathfrak{M}_\beta$$

and hence $\rho_\beta\rho_\alpha^p = \rho_\beta$ and so $\rho_\alpha = 1$. Hence, in all cases $\rho_\alpha = 1$ and the theorem is proved.

Since we have seen that if S is any automorphism of $\mathfrak{L}$, then there exists an automorphism G of $\mathfrak{A}$ such that $d^T = Gd^SG^{-1}$ is an automorphism that leaves all the elements of $\mathfrak{H}$ invariant, we have

THEOREM 6. *If Φ is algebraically closed and $p \geq 5$, any automorphism of $\mathfrak{L}$ has the form $d \to G^{-1}dG$, where G is an automorphism of $\mathfrak{A}$.*

6. We again assume that Φ is algebraically closed and that h is the element determined above. Suppose that D is a restricted derivation in $\mathfrak{L}$. Then, by (6), we obtain

$$\overbrace{[\cdots[h'h]\cdots h]}^{\leftarrow p^m-1 \rightarrow} + \overbrace{[[h'h]\cdots h]}^{\leftarrow p^{m-1}-1 \rightarrow}\beta_1 + \cdots + h'\beta_m = 0,$$

where $h' = hD$. Hence, if we denote the inner derivation $d \to [d, h]$ by H,

$$h'(H^{p^m-1} + H^{p^{m-1}-1}\beta_1 + \cdots + \beta_m) = 0.$$

Since $\lambda^{p^m} + \lambda^{p^{m-1}}\beta_1 + \cdots + \lambda\beta_m = \prod(\lambda - \alpha)$, this may be written as

$$h'\prod_{\alpha\neq 0}(H - \alpha) = 0.$$

Since $\mathfrak{M}_\alpha$ is the set of elements e_α of $\mathfrak{L}$ such that $[e_\alpha, h] \equiv e_\alpha H = e_\alpha\alpha$, it follows that h' belongs to $\sum_{\alpha\neq 0}\mathfrak{M}_\alpha$, say $h' = \sum_{\alpha\neq 0} e_\alpha$. Hence, if we denote the inner derivation $d \to [d, b]$, where $b = -\sum e_\alpha\alpha^{-1}$, by B, then $hD = hB$ and so the restricted derivation $E = D - B$ maps $\mathfrak{H}$ into 0. It follows that $\mathfrak{M}_\alpha^E \leq \mathfrak{M}_\alpha$.

We set $e_{\lambda_1}^{(1)E} = \sum e_{\lambda_1}^{(i)}\mu_i$. Since $e_{\lambda_1}^{(1)p} = 0$ and

$$((e_{\lambda_1}^{(1)})^p)^E = \overbrace{[\cdots[e_{\lambda_1}^{(1)E}, e_{\lambda_1}^{(1)}]\cdots e_{\lambda_1}^{(1)}]}^{\leftarrow p - 1 \rightarrow},$$

we have

$$\sum[\cdots[e_{\lambda_1}^{(1)}, e_{\lambda_1}^{(1)}], e_{\lambda_1}^{(1)}], \cdots, e_{\lambda_1}^{(1)}]\mu_i = 0.$$

Since $[e_{k\lambda_1}^{(i)}, e_{\lambda_1}^{(1)}] = e_{(k+1)\lambda_1}^{(i)}k$ if $i \neq 1$, this reduces to $\sum h_i\mu_i$ $((p-1)! = -1)$. Hence, $\mu_i = 0$ if $i \neq 1$ and $e_{\lambda_1}^{(1)E} = e_{\lambda_1}^{(1)}\mu_1$. Similarly, $e_{\lambda_i}^{(i)E} = e_{\lambda_i}^{(i)}\mu_i$. If we subtract the inner derivation $x \to [x, k]$, where $k = \sum h_i\mu_i$, from E, we obtain the derivation F such that $h^F = 0$ and $e_{\lambda_i}^{(i)F} = 0$. Let $e_{-\lambda_1}^{(j)F} = \sum e_{-\lambda_1}^{(k)}\mu_{kj}$. Then, from the relations $[e_{\lambda_1}^{(1)}, e_{-\lambda_1}^{(1)}] = 2h_1$, $[e_{\lambda_1}^{(1)}, e_{-\lambda_1}^{(j)}] = h_j$ if $j \neq 1$, we obtain $[e_{\lambda_1}^{(1)}, \sum e_{-\lambda_1}^{(k)}\mu_{kj}] = 0$ and so $2h_1\mu_{1j} + \sum_{k \neq 1} h_k\mu_{kj} = 0$. Thus, if $p \neq 2$, $\mu_{1j} = \mu_{kj} = 0$ and so $e_{-\lambda_1}^{(j)F} = 0$. If $p = 2$, $\mu_{kj} = 0$ if $k \neq 1$ and so $e_{-\lambda_1}^{(j)F} = e_{-\lambda_1}^{(1)}\mu_{1j}$. Since $(e_{-\lambda_1}^{(j)2})^F = [e_{-\lambda_1}^{(j)F}, e_{-\lambda_1}^{(j)}] = 0$, we have $[e_{-\lambda_1}^{(1)}, e_{-\lambda_1}^{(j)}]\mu_{1j} = h_j\mu_{1j} = 0$ for $j \neq 1$. Hence $\mu_{1j} = 0$ and since in this case $-\lambda_1 = \lambda_1$, we have $e_{\lambda_1}^{(j)F} = 0$ and $e_{\lambda_i}^{(j)F} = 0$ for all i, j. Now, if $p \neq 2$, we may reverse the rôles of λ_1 and $-\lambda_1$ and obtain $e_{\lambda_1}^{(j)F} = 0$ and similarly $e_{\lambda_i}^{(j)F} = 0$. It follows readily that if $m > 1$, $F = 0$. If $m = 1$, $e_{\alpha\lambda_1}^F = e_{\alpha\lambda_1}\nu_\alpha$, where $\nu_1 = 0$. Since $[e_{\alpha\lambda_1}, e_{\beta\lambda_1}] = e_{(\alpha+\beta)\lambda_1}(\alpha - \beta)$, $\nu_{\alpha+\beta} = \nu_\alpha + \nu_\beta$ if $\alpha \neq \beta$ and so $\nu_2 = \nu_3 = \nu_4 = \cdots$. Since $[e_{(p-1)\lambda_1}, e_{\lambda_1}] = -h(p-2)$, we obtain $\nu_{p-1} = 0$ if $p \neq 2$. Hence, $\nu_2 = \nu_3 = \cdots = 0$. Thus, $F = 0$ and $D = B + K$ is inner.

THEOREM 7. *If Φ is algebraically closed, any restricted derivation of $\mathfrak{L}$ is inner.*

7. We shall now extend the results of the last two sections to arbitrary fields Φ of characteristic p. For this purpose we require the following lemma in the algebraically closed case.

LEMMA. *If R is a linear transformation in $\mathfrak{A}$ that commutes with all the derivations of $\mathfrak{A}$ then $R = 1\rho$, ρ in Φ.*

Since the derivation h satisfies $(x_1^{\alpha_1} \cdots x_m^{\alpha_m})h = (x_1^{\alpha_1} \cdots x_m^{\alpha_m})\alpha$ and the α's are distinct, R satisfies $(x_1^{\alpha_1} \cdots x_m^{\alpha_m})R = x_1^{\alpha_1} \cdots x_m^{\alpha_m}\rho_{\alpha_1\cdots\alpha_m}$, $\rho_{\alpha_1\cdots\alpha_m}$ in Φ. If $(\alpha_1, \cdots, \alpha_m) \neq (0, \cdots, 0)$, the element $x_1^{\alpha_1} \cdots x_m^{\alpha_m}$ may be used as a generator of $\mathfrak{A}$. Hence, there is a derivation e such that $(x_1^{\alpha_1} \cdots x_m^{\alpha_m})e = x_1^{\beta_1} \cdots x_m^{\beta_m}$. Since $Re = eR$, $\rho_{\alpha_1\cdots\alpha_m} = \rho_{\beta_1\cdots\beta_m} = \rho$.

Now let $\mathfrak{A} = \Phi(y_1, \cdots, y_m)$, $y_i^p = \eta_i$ and $\mathfrak{B} = \Phi(z_1, \cdots, z_m)$, $z_i^p = \zeta_i$, Φ arbitrary of characteristic $p \geq 5$. Suppose that the restricted Lie algebras of derivations of these algebras $\mathfrak{L}$ and $\mathfrak{L}^*$ respectively are isomorphic under the correspondence $d \to d^*$. If Ω is the algebraic closure of Φ, $\mathfrak{A}_\Omega \equiv \mathfrak{B}_\Omega$ and both of the algebras $\mathfrak{A}$ and $\mathfrak{B}$ may be regarded as subsets of the same algebra $\Omega(x_1, \cdots, x_m)$, $x_i^p = 1$. Moreover, any basis for $\mathfrak{A}$ over Φ ($\mathfrak{B}$ over Φ) is a basis for $\Omega(x_i)$. If $d_1, \cdots, d_n$ ($n = mp^m$) is a basis for the derivation algebra $\mathfrak{L}$ of $\mathfrak{A}$, it is also a basis for the derivation algebra $\mathfrak{L}_\Omega$ of $\Omega(x_i)$, see [3; 213]. This, of course, holds also for $\mathfrak{L}^*$. Thus, the correspondence $\sum d_i\omega_i \to \sum d_i^*\omega_i$ is an automorphism in the derivation algebra of $\Omega(x_i)$. Hence, there exists an automorphism G in $\Omega(x_i)$ such that $d^* = G^{-1}dG$ for all d in $\mathfrak{L}$. If $y_1, \cdots, y_{p^m}$ is a basis for $\mathfrak{A}$ over Φ and $y_id = \sum y_j\mu_{ji}$ and if $z_1, \cdots, z_{p^m}$ is a basis for $\mathfrak{B}$ over Φ and $z_id^* = \sum z_j\mu_{ji}^*$, then we have the relation $(\mu^*) = (\sigma)^{-1}(\mu)(\sigma)$, where (σ) is a matrix with elements in Ω. Since the matrices (μ) and (μ^*) have elements in Φ, there exists a matrix (τ) with elements in Φ such that $(\mu^*) = (\tau)^{-1}(\mu)(\tau)$, see

[7]. We replace the basis $y_1, \cdots, y_{p^m}$ by $y_i^* = \sum y_j\tau_{ji}$ and obtain $y_i^*d = \sum y_j^*\mu_{ji}^*$. If K denotes the linear transformation $\sum y_i^*\omega_i \to \sum z_i\omega_i$, then K maps $\mathfrak{A}$ into $\mathfrak{B}$ and $d^* = K^{-1}dK$. Hence, $K = G\rho$. Since G is an automorphism, $1G = 1$ and hence $1K = \rho$. Since this element is in $\mathfrak{B}$, ρ belongs to Φ. Hence, G maps $\mathfrak{A}$ into $\mathfrak{B}$ and is therefore an isomorphism between $\mathfrak{A}$ and $\mathfrak{B}$.

THEOREM 8. *If $\mathfrak{A} = \Phi(y_i)$, $y_i^p = \eta_i$ and $\mathfrak{B} = \Phi(z_i)$, $z_i^p = \mathfrak{Z}_i$, $p \geq 5$, have isomorphic derivation algebras, then $\mathfrak{A}$ and $\mathfrak{B}$ are isomorphic.*

THEOREM 9. *Let $\mathfrak{A} = \Phi(x_i)$, $x_i^p = \xi_i$, $p \geq 5$, and let $\mathfrak{L}$ be the restricted Lie algebra of derivations of $\mathfrak{A}$. Then any automorphism of $\mathfrak{L}$ over Φ has the form $d \to G^{-1}dG$, where G is an automorphism of $\mathfrak{A}$ over Φ.*

If G_1 and G_2 are two automorphisms in $\mathfrak{A}$ over Φ such that $G_1^{-1}dG_1 = G_2^{-1}dG_2$ for all d in $\mathfrak{L}$, then $G = G_1^{-1}G_2$ commutes with all d. The extension of the linear transformation G to $\mathfrak{A}_\Omega$ commutes with all the derivations in $\mathfrak{A}_\Omega$. Hence, $G = 1\sigma$ and $G_2 = G_1\sigma$. Since $1G_2 = 1G_1 = 1$, $\sigma = 1$ and $G_2 = G_1$. This shows that the correspondence between the automorphism G in $\mathfrak{A}$ and the automorphism $d \to G^{-1}dG$ is an isomorphism between these groups of automorphisms.

THEOREM 10. *The group of automorphisms of $\mathfrak{L}$ over Φ is isomorphic to the group of automorphisms of $\mathfrak{A}$ over Φ.*

8. Since any element x of $\mathfrak{A}$ satisfies an equation of the form $x^p = \xi$ in Φ, the only idempotent element in $\mathfrak{A}$ or in any difference algebra $\mathfrak{A} - \mathfrak{N}$, $\mathfrak{N}$ an ideal, is the identity. Hence, if $\mathfrak{N}$ is the radical, $\mathfrak{A} - \mathfrak{N}$ is simple and is therefore a field. Evidently $\mathfrak{A} - \mathfrak{N}$ has the form $\Phi(y_1, \cdots, y_r)$, where $y_i^p = \eta_i$. It is readily seen that the only automorphism of a field of this structure is the identity. Now if G is an automorphism in $\mathfrak{A}$, G induces an automorphism G' in $\mathfrak{N}$ and the correspondence between G and G' is a homomorphism between the group $\mathfrak{U}$ of automorphisms of $\mathfrak{A}$ over Φ and a subgroup $\mathfrak{V}'$ of the group of automorphisms of $\mathfrak{N}$ over Φ. Evidently $\mathfrak{V}'$ consists of the automorphisms of $\mathfrak{N}$ that can be extended to automorphisms of $\mathfrak{A}$. The subgroup $\mathfrak{U}_0$ mapped into 1 by our homomorphism consists of the automorphisms of $\mathfrak{A}$ that leave the elements of $\mathfrak{N}$ invariant and $\mathfrak{V}' \cong \mathfrak{U}/\mathfrak{U}_0$.

We suppose now that the generators $x_1, \cdots, x_n$ of $\mathfrak{A}$ have been arranged so that $\xi_1, \cdots, \xi_r$ are p-independent but that $\xi_{r+1}, \cdots, \xi_m$ are p-dependent on the ξ_i, $i < r$. (A set of elements $\xi_1, \cdots, \xi_r$ is p-independent if no $\xi_i \in \Phi^p(\xi_1, \cdots, \xi_{i-1}, \xi_{i+1}, \cdots, \xi_r)$, where Φ^p is the subfield of p-th powers in Φ. An element ξ is p-dependent on $\xi_1, \cdots, \xi_r$ if $\xi \in \Phi^p(\xi_1, \cdots, \xi_r)$.) Then $\mathfrak{C} = \Phi(x_1, \cdots, x_r)$ is a subfield of $\mathfrak{A}$. For, otherwise, there is a nilpotent element $\sum x_1^{\alpha_1} \cdots x_r^{\alpha_r}\tau_{\alpha_1 \cdots \alpha_r} \neq 0$, $\alpha_i < p$. Then $(\sum x_1^{\alpha_1} \cdots x_r^{\alpha_r}\tau_{\alpha_1 \cdots \alpha_r})^p = \sum \xi_1^{\alpha_1} \cdots \xi_r^{\alpha_r}\tau_{\alpha_1 \cdots \alpha_r}^p = 0$. This equation yields an equation $g(\xi) = 0$ for one of the ξ's, say ξ_r, where $f(\lambda)$ is a polynomial $\neq 0$ of degree $< p$ and with coefficients in the field $\Gamma = \Phi^p(\xi_1, \cdots, \xi_r)$. Since $\xi_r^p \in \Gamma$, it follows that ξ_r is in Γ, contrary to the p-independence of $\xi_1, \cdots, \xi_r$. Suppose that $\xi_k = \sum \xi_1^{\alpha_1} \cdots \xi_r^{\alpha_r}\tau_{k,\alpha_1 \cdots \alpha_r}^p$, $0 \leq \alpha_i < p$ and $k > r$.

Then, if we replace the generator x_k by $x_k' = x_k - \sum x_1^{\alpha_1} \cdots x_r^{\alpha_r} \tau_{\alpha_1 \cdots \alpha_r}$, we obtain $(x_k')^p = 0$. We return to the original notation and suppose then that $x_1, \cdots, x_r$ generate a field $\mathfrak{C}$ and $x_k^p = 0$ if $k > r$. Hence, $\mathfrak{A} = \mathfrak{C} + \mathfrak{N}$. Since $\mathfrak{N}\mathfrak{C} = \mathfrak{N}$ and $\mathfrak{C}$ includes the identity, we may regard $\mathfrak{N}$ as an algebra over $\mathfrak{C}$. In this sense $\mathfrak{N}$ is generated by the elements $x_{r+1}, \cdots, x_m$. Since $p^m - p^r = (\mathfrak{N} : \Phi) = (\mathfrak{N} : \mathfrak{C})(\mathfrak{C} : \Phi) = (\mathfrak{N} : \mathfrak{C})p^r$, $(\mathfrak{N} : \mathfrak{C}) = p^{m-r} - 1$. Hence, the elements $x_{r+1}^{\alpha_{r+1}} \cdots x_m^{\alpha_m}$, $0 \leq \alpha_i < p$, $(\alpha_{r+1}, \cdots, \alpha_m) \neq (0, \cdots, 0)$ form a basis for $\mathfrak{N}$ over $\mathfrak{C}$. Thus, if we regard $\mathfrak{A}$ as an algebra over $\mathfrak{C}$, we see that it has the same structure as the algebra considered in §2. We note that if (c_{kl}), $k, l = r + 1, \cdots, m$, is an arbitrary non-singular matrix of $(m - r)$-rows and columns, then the mapping $c \to c$, $x_k \to \sum x_l c_{lk}$ determines an automorphism of $\mathfrak{A}$ over $\mathfrak{C}$.

Now let $G_0 \in \mathfrak{U}_0$. Then, $r^{G_0} = r$ for any r in $\mathfrak{N}$ and $c^{G_0} = c + r$. Hence, if H_0 is also in $\mathfrak{U}_0$, $r^{H_0} = r$, $c^{H_0} = c + r'$, r' in $\mathfrak{N}$, and $H_0 G_0 = G_0 H_0$. Then, $\mathfrak{U}_0$ is a commutative group.

We consider next the structure of $\mathfrak{B}$, the group of automorphisms of $\mathfrak{N}$ over Φ. Any element G of $\mathfrak{B}$ induces an automorphism in $\mathfrak{N}^s$ and hence also an automorphism in $\mathfrak{N} - \mathfrak{N}^s$. Let $\mathfrak{B}_s$, $s = 2, \cdots, p$, denote the subgroup of $\mathfrak{B}$ of elements whose induced automorphism in $\mathfrak{N} - \mathfrak{N}^s$ is the identity. Then, if G_s is in $\mathfrak{B}_s$, $r^{G_s} \equiv r\ (\mathfrak{N}^s)$. It follows that if $r_t \in \mathfrak{N}^t$, $r_t^{G_s} \equiv r_t\ (\mathfrak{N}^{s+t-1})$. Since G_s^{-1} is in $\mathfrak{B}_s$, we have $r^{G_s} = r + r_s$ and $r^{G_s^{-1}} = r + r_s'$, where r_s, r_s' are in $\mathfrak{N}_s$. Since $r = r^{G_s G_s^{-1}} = r + r_s + r_s' + r_{2s-1}''$, where r_{2s-1}'' is in $\mathfrak{N}_{2s-1}$, $r_s + r_s' \in \mathfrak{N}_{2s-1}$. Suppose that H_t is in $\mathfrak{B}_t$ and $r^{H_t} = r + u_t$, $r^{H_t^{-1}} = r + u_t'$. Then a simple computation shows that $r^{G_s^{-1} H_t^{-1} G_s H_t} \equiv r\ (\mathfrak{N}_{2l-1})$, where $l = \min(s, t)$. Thus, $G_s^{-1} H_t^{-1} G_s H_t$ is in $\mathfrak{B}_{2l-1}$. Since $\mathfrak{B}_p$ is the identity, this shows that $\mathfrak{B}_2$ is a solvable group. This discussion of $\mathfrak{B}$ is applicable to any nilpotent associative algebra.

We consider finally the structure of $\mathfrak{B}'$. If $G' \in \mathfrak{B}'$, there exists a G in $\mathfrak{U}$ such that $r^G = r^{G'}$ for all r in $\mathfrak{N}$. Since $c^G = c + r'$, r' in $\mathfrak{N}$, for any c, $(rc)^{G'} \equiv r^{G'} c$ $(\mathfrak{N}^2)$ and so the automorphism induced in $\mathfrak{N} - \mathfrak{N}^2$ by G' is a linear transformation of this space over $\mathfrak{C}$. We have also noted above that any linear transformation in $(\mathfrak{N} - \mathfrak{N}^2)$ over $\mathfrak{C}$ can be induced by an automorphism in $\mathfrak{N}$. Hence, the induced group $\mathfrak{U}^*$ in $\mathfrak{N} - \mathfrak{N}^2$ is isomorphic to the complete group of linear transformations in the space $(\mathfrak{N} - \mathfrak{N}^2)$ over $\mathfrak{C}$ and hence to the group $L(\mathfrak{C}, m - r)$ of non-singular $(m - r)$-rowed matrices with elements in $\mathfrak{C}$. Now the correspondence between $\mathfrak{B}'$ and $\mathfrak{U}^*$ is a homomorphism such that the elements mapped into the identity are those in $\mathfrak{B}_2' \equiv \mathfrak{B}' \wedge \mathfrak{B}_2$. Hence, $\mathfrak{B}'/\mathfrak{B}_2' \cong \mathfrak{U}^*$. Since $\mathfrak{B}_2'$ is solvable, this together with the earlier result that $\mathfrak{U}/\mathfrak{U}_0 \cong \mathfrak{B}'$ implies

THEOREM 11. *The group $\mathfrak{U}$ of automorphisms of $\mathfrak{A}$ over Φ contains a solvable invariant subgroup $\mathfrak{U}_1$ such that $\mathfrak{U}/\mathfrak{U}_1$ is isomorphic to $L(\mathfrak{C}, m - r)$.*

9. The question of the restricted derivations of $\mathfrak{L}$ is settled by

THEOREM 12. *The only restricted derivations in $\mathfrak{L}$ over Φ are the inner derivations.*

If $\mathfrak{D}$ is the algebra of restricted derivations of $\mathfrak{L}$ and $\mathfrak{J}$ is the ideal of inner derivations, then $\mathfrak{D}_\Omega$ and $\mathfrak{J}_\Omega$ are respectively the algebra of restricted derivations and of inner derivations of $\mathfrak{L}_\Omega$. Hence, $\mathfrak{D}_\Omega = \mathfrak{J}_\Omega$, by Theorem 7, and so $\mathfrak{D} = \mathfrak{J}$.

We remark finally that the theorems of §§7-9 hold also when the algebra $\mathfrak{L}$ is regarded as an ordinary Lie algebra. This follows from the fact that the algebras $\mathfrak{L}$ have no centers. Hence, any isomorphism between them regarded as ordinary Lie algebras is an isomorphism between them regarded as restricted Lie algebras [6]. Any derivation of $\mathfrak{L}$ is restricted.

10. Suppose that $\mathfrak{L}$ is a restricted Lie algebra over Φ such that $\mathfrak{L}_P$ is the derivation algebra of the algebra $P(y_1, \cdots, y_m)$, $y_i^p = 1$, $p \geq 5$, and P an extension of the field Φ. Then, it seems likely that $\mathfrak{L}$ itself is the derivation algebra of an algebra $\Phi(x_1, \cdots, x_m)$, with $x_i^p = \xi_i$. We shall prove that this is the case when (1) P is separable and (2) P is a purely inseparable field of the form $\Phi(\eta_1, \cdots, \eta_s)$, where $\eta_i^p = \eta_{i0} \in \Phi$.

If P is separable, we may extend P further to a separable, normal field over Φ. Hence, we may suppose at the outset that P is separable and normal. The proof of the theorem in this case is identical with the proof of the corresponding theorem for Lie algebras of type G given in an earlier paper [4]. We shall therefore omit the proof here. If $P = \Phi(\eta_1, \cdots, \eta_s)$, $\eta_i^p = \eta_{i0}$ in Φ, then we shall see that the argument parallels that of the normal case. Here the derivations play the rôle of the automorphisms of the former case.

LEMMA. *Let $\mathfrak{S}$ be a vector space of r dimensions over* P *and D a differential transformation in $\mathfrak{S}$: $(x + y)D = xD + yD$, $(x\rho)D = (xD)\rho + x\rho^d$, where $\rho \to \rho^d$ is a derivation in* P *over Φ. Suppose that the only d-constants are the elements of Φ and that the minimum polynomial of d is $\lambda^{p^s} + \lambda^{p^{s-1}}\beta_1 + \cdots + \lambda\beta_s$. Then, if $D^{p^s} + D^{p^{s-1}}\beta_1 + \cdots + D\beta_s = 0$, the dimensionality over Φ of the space $\mathfrak{S}_0$ of D-constants is r and $\mathfrak{S} = \mathfrak{S}_0$*P.

(We are assuming that $(P : \Phi) = p^s$. The elements β_i are in Φ. Cf. [3; 218].)

The mappings $\sum_0^{p^s-1} D^i\rho_i$ are linear transformations of the rp^s-dimensional vector space $\mathfrak{S}$ over Φ. The mappings $\sum_0 d^i\rho_i$ are linear transformations of P over Φ and we have shown previously that this set comprises all the linear transformations of the p^s-dimensional space P over Φ. Since d and D satisfy the same equation and $\rho d = d\rho + \rho^d$, $\rho D = D\rho + \rho^d$, the correspondence $\sum d^i\rho_i \to \sum D^i\rho_i$ is a representation such that $1 \to 1$. Since the former algebra is isomorphic to the matrix algebra P_{p^s}, the representation is completely reducible into r irreducible components which are similar to the original representation in P. Thus, we may decompose $\mathfrak{S}$ as $\mathfrak{S}_1 \oplus \cdots \oplus \mathfrak{S}_r$, where $\mathfrak{S}_i E \leq \mathfrak{S}_i$ for any $E = \sum_0 D^i\rho_i$. If $x = x_1 + \cdots + x_r$, x_i in $\mathfrak{S}_i$ is a D-constant, each x_i is. Moreover, because of the similarity $d \to D$ and the fact that the d-constants of P form a 1-dimensional space, the D-constants in $\mathfrak{S}_i$ are the Φ-multiples of a fixed vector $z_i \neq 0$. Thus, $\mathfrak{S}_0$ is r-dimensional over Φ. Since $z_iE = z_i\rho_0$, the images of z_i under the transformations E is the set $z_i\text{P}$. Since $\mathfrak{S}_i$ is irreducible, $z_i\text{P} = \mathfrak{S}_i$. Hence, $\mathfrak{S}_0\text{P} = z_1\text{P} \oplus \cdots \oplus z_r\text{P} = \mathfrak{S}$.

Let $y_1, \cdots, y_r$ $(r = p^m)$ be a basis for $\text{P}(y_i)$, $y_i^p = 1$, over P and let d^* be a derivation in P over Φ such that the only d^*-constants are the elements of Φ. The mapping $y = \sum y_i\rho_i \to \sum y_i\rho_i^d = yD$ is a derivation in $\text{P}(x_i)$ over Φ and $(y\rho)D = (yD)\rho + y\rho^d$. It follows that, if e is a derivation of $\text{P}(x_i)$ over P, then $[e, D]$ is also a derivation of $\text{P}(x_i)$ over P. Hence, if the elements $d_1, \cdots, d_n$, $n = mp^m$, form a basis for $\mathfrak{L}$ and hence for the derivation algebra $\mathfrak{L}_\text{P}$ of $\text{P}(x_i)$ over P, the mapping $\sum d_j\rho_j \to \sum [d_j, D]\rho_j$ is a restricted derivation in $\mathfrak{L}_\text{P}$ over P and so there exists an element d such that $\sum [d_j, D]\rho_j = [\sum d_j\rho_j, d]$. Set $D_1 = D - d$. Then D_1 commutes with all the d_j and hence with every element $\sum d_j\varphi_j$, $\varphi_j \in \Phi$, of $\mathfrak{L}$. Evidently D_1 is also a derivation of $\text{P}(x_i)$ over Φ and $(y\rho)D_1 = (yD_1)\rho + y\rho^d$. It follows that $E = D_1^{p^s} + D_1^{p^{s-1}}\beta_1 + \cdots + D_1\beta_s$ is a derivation of $\text{P}(y_i)$ over P. Since E commutes with all the d_j, $E = 0$ and so $D_1^{p^s} + D_1^{p^{s-1}}\beta_1 + \cdots + D_1\beta_s = 0$.

We may now apply the lemma. We then obtain that the set of D_1-constants is an r-dimensional space over Φ and $\mathfrak{A}\text{P} = \text{P}(y_i)$. Since D_1 is a derivation, it is clear that $\mathfrak{A}$ is closed under multiplication. Thus $\mathfrak{A}$ is an algebra over Φ whose extension $\mathfrak{A}_\text{P} = \text{P}(y_i)$. Now, if $x_0 = 1, x_1, \cdots, x_{r-1}$ is a basis for $\mathfrak{A}$ over Φ and hence for $\text{P}(y_i)$ over P we may suppose that the cosets $x_0 + \mathfrak{R}^2, \cdots, x_m + \mathfrak{R}^2$, $\mathfrak{R}$ the radical of $\text{P}(y_i)$, forms a basis for $\text{P}(y_i) - \mathfrak{R}^2$. Hence, as we have seen, the elements $x_0, \cdots, x_m$ are generators for $\text{P}(y_i)$, i.e., $x_0, x_1^{\alpha_1} \cdots x_m^{\alpha_m}$, $0 \leq \alpha_i < p$, $(\alpha_1, \cdots, \alpha_m) \neq (0, \cdots, 0)$ form a basis for $\text{P}(y_i)$ over P. Thus, these elements also form a basis for $\mathfrak{A}$ over Φ and $x_i^p = \xi_i$ are in Φ. We have, therefore, shown that $\mathfrak{A} = \Phi(x_i)$, $x_i^p = \xi_i$.

Since D_1 commutes with all the elements $\sum d_j\rho_j$ of $\mathfrak{L}$, the latter map $\mathfrak{A}$ into itself and hence induce derivations in $\mathfrak{A}$ over Φ. Since the derivation algebra of $\mathfrak{A}$ over Φ is n-dimensional, it coincides with the set $\mathfrak{L}$. Hence, we have

THEOREM 13. *If $\mathfrak{L}$ is a restricted Lie algebra over Φ and $\mathfrak{L}_\text{P}$ is the derivation algebra of $\text{P}(y_i)$, $y_i^p = 1$, $p \geq 5$, and either* P *is separable or* $\text{P} = \Phi(\eta_1, \cdots, \eta_s)$, $\eta_i^p = \eta_{i0}$ *in Φ, then $\mathfrak{L}$ is the derivation algebra of an algebra* $\mathfrak{A} = \Phi(x_i)$, $x_i^p = \xi_i$.

11. We now consider some consequences of our results. The first of these is a corollary to Theorem 3.

Let Γ be a field of characteristic p and $\mathfrak{A} = \Gamma(x)$ a simple transcendental extension. If we set $\Phi = \Gamma(\xi)$, where $\xi = x^p$, $\mathfrak{A}$ has the structure considered above, over Φ. Any derivation d is determined by $xd = y$ and, as we have seen, the element y may be chosen arbitrarily in $\mathfrak{A}$. By Theorem 3 and the Hamilton-Cayley theorem, there exists an element η in Φ such that $d^p = d\eta$. On the other hand, by direct computation we obtain $xd^p = y_p$, where

$$y_1 = y, \qquad y_2 = y'y, \qquad \cdots, \qquad y_k = y'_{k-1}y, \qquad \cdots$$

and where the prime denotes ordinary differentiation of the rational function

$y(x)$. It follows that $d^p = dy'_{p-1}$ and unless $d = 0$ $(y = 0)$, $y'_{p-1} = \eta$ is in Φ. Evidently, if y is a polynomial in x, y'_{p-1} is also a polynomial. This result may be stated in the following more elementary form:

THEOREM 14. *Let $y(x)$ be a polynomial with rational integral coefficients and define $y_1(x) = y(x)$, $y_k(x) = y_{k-1}(x)'y_1(x)$. Then, for any prime p, $y_{p-1}(x)' \equiv \eta(x^p)$, a polynomial in x^p,* mod p.

A similar statement may be made for rational functions with integral coefficients. We may also obtain a generalization of this theorem to m indeterminates by applying Theorem 3 to $\mathfrak{A} = \Gamma(x_1, \cdots, x_m)$, $\Phi = \Gamma(x_1^p, \cdots, x_m^p)$.

We obtain next a matrix theorem that is equivalent to the lemma of the preceding section. Let D be a differential transformation in $\mathfrak{S}$ over P with associated derivation d: $(x + y)D = xD + yD$, $(x\rho)D = (xD)\rho + x\rho^d$. If $u_1, \cdots, u_r$ is a basis for $\mathfrak{S}$ and $u_iD = \sum u_j\mu_{ji}$, then (μ) is the *matrix of D relative to the basis $u_1, \cdots, u_r$*. If $v_1, \cdots, v_r$ is a second basis for $\mathfrak{S}$, $v_i = \sum u_j\beta_{ji}$, and $v_iD = \sum v_j\nu_{ji}$, then $(\nu) = (\beta)^{-1}(\mu)(\beta) + (\beta)^{-1}(\beta^d)$ $((\beta^d) \equiv (\beta_{ij}^d))$. From $u_iD = \sum u_j\mu_{ji}$ we obtain $u_iD^2 = \sum u_j\mu_{ji}^{[2]}$, where $(\mu^{[2]}) = (\mu^d) + \mu^2$ and in general $u_iD^k = \sum u_j\mu_{ji}^{[k]}$, $(\mu^{[k]}) = (\mu^{[k-1]d}) + (\mu)(\mu^{[k-1]})$. The transformation D^p is a differential transformation with d^p as its induced derivation, i.e., $(x + y)D^p = xD^p + yD^p$, $(x\rho)D^p = (xD)\rho + x\rho^{d^p}$. The matrix of D^p relative to the basis $u_1, \cdots, u_r$ is $(\mu^{[p]})$. Since the induced derivation d satisfies the equation $d^{p^s} + d^{p^{s-1}}\beta_1 +, \cdots + d\beta_s$, the transformation $D^{p^s} + D^{p^{s-1}}\beta_1 + \cdots + D\beta_s$ is linear. Its matrix relative to the u's is $(\mu^{[p^s]}) + (\mu^{[p^{s-1}]})\beta_1 + \cdots + (\mu)\beta_s$. Thus, the condition that $D^{p^s} + D^{p^{s-1}}\beta_1 + \cdots + D\beta_s = 0$ is that the matrix (μ) satisfy

$$(\mu^{[p^s]}) + (\mu^{[p^{s-1}]})\beta_1 + \cdots + (\mu)\beta_s = 0. \tag{15}$$

If this condition is satisfied, by the lemma of the preceding section, there exists a basis $v_1, \cdots, v_r$ for $\mathfrak{S}$ over P such that $v_iD = 0$. Hence, the matrix $(\mu) = (\beta)^{-1}(\beta^d)$, where (β) is the matrix expressing the v's in terms of the u's. The converse is clear. For, if $(\mu) = (\beta)^{-1}(\beta^d)$, there exists a basis v_i such that $v_iD = 0$ and $v_i(D^{p^s} + D^{p^{s-1}}\beta_1 + \cdots + D\beta_s) = 0$. Since $D^{p^s} + \cdots + D\beta_s$ is a linear transformation, it follows that $D^{p^s} + \cdots + D\beta_s = 0$ and hence (15) holds.

THEOREM 15. *Let $\mathrm{P} = \Phi(\eta_i)$, $\eta_i^p = \eta_{i0}$ in Φ be a field of dimensionality p^s over Φ of characteristic p and let d be a derivation in P having Φ as its set of constants. Suppose that $d^{p^s} + d^{p^{s-1}}\beta_1 + \cdots + d\beta_s = 0$, β_i in Φ, is the minimum equation of d. Then if (μ) is an $r \times r$-matrix with elements in P, (μ) is a logarithmic derivative $((\mu) = (\beta)^{-1}(\beta^d))$ if and only if* (15) *holds. Here $(\mu^{[k]})$ is defined by $(\mu^{[1]}) = (\mu)$, $(\mu^{[k]}) = (\mu^{[k-1]d}) + (\mu)(\mu^{[k-1]})$.*

The case of 1×1-matrices, i.e., elements in P was treated in a previous paper [3; 224]. It has been shown also in this case

$$\mu^{[p^j]} = \sum_{k=0}^{j} \mu^l, \qquad l = d^{p^k-1}p^{j-k},$$

that $v(\mu) \equiv \mu^{[p^s]} + \mu^{[p^{s-1}]}\beta_1 + \cdots + \mu\beta_s$ is an element of Φ and that $v(\mu)$ is an additive function of μ.

BIBLIOGRAPHY

1. R. BAER, *Algebraische Theorie der differentiierbaren Funktionenkörper* I, Sitzungsberichte Heidelberger Akad., 1927, pp. 15–32.
2. HO-JUI CHANG, *Ueber Wittsche Lie-Ringe*, Abhandlungen aus dem Mathematischen Seminar der Hansischen Universität, vol. 14(1941), pp. 151–184.
3. N. JACOBSON, *Abstract derivation and Lie algebras*, Transactions of the American Mathematical Society, vol. 42(1937), pp. 206–224.
4. N. JACOBSON, *Cayley numbers and simple Lie algebras of type G*, this Journal, vol. 5(1939), p. 782.
5. N. JACOBSON, *Classes of restricted Lie algebras of characteristic p*. I, American Journal of Mathematics, vol. 63(1941), pp. 481–515.
6. N. JACOBSON, *Restricted Lie algebras of characteristic p*, Transactions of the American Mathematical Society, vol. 50(1941), p. 23.
7. B. L. VAN DER WAERDEN, *Gruppen von linearen Transformationen*, Ergebnisse der Mathematik, vol. 4(1935), p. 70.
8. H. ZASSENHAUS, *Ueber Lie'sche Ringe mit Primzahlcharakteristik*, Abhandlungen aus dem Mathematischen Seminar der Hansischen Universität, vol. 13(1940), pp. 1–100.

PRE-FLIGHT SCHOOL, CHAPEL HILL, N. C.

Reprinted from
Duke Mathematical Journal
March 1943.

A Personal History and Commentary

1943-1946

Hopkins had a small but distinguished faculty in mathematics when I joined it in 1943. Professor F. D. Murnaghan was chairman, Zariski was a full professor, and Aurel Wintner was an associate professor. E. R. Van Kampen, who held a title equivalent to assistant professor, had died shortly before I came to Hopkins. We had become close friends during my visit to Hopkins in 1940–1941. There were serious internal tensions in the department, and the department was not in the good graces of the president, Isaiah Bowman, since it was not popular with the undergraduates, who constituted the constituency he wished to cultivate. When I received an offer of an associate professorship from Yale in 1947 at a salary of $6000 ($1000 more than my Hopkins salary) the president tried to persuade me to stay by offering me a full professorship at a salary matching Yale's offer. In my interview with him, I learned why he was anxious for me to stay at Hopkins: I had the reputation of being an excellent undergraduate teacher! Nevertheless, Hopkins provided an excellent atmosphere for research, and my four years there were among the most productive of my career.

A major conference, The Third Conference on Algebra, was held in Eckhart Hall at the University of Chicago, July 15–19, 1946. The list of speakers for the conference in the order of their presentations were Marshall Stone, Irving Segal, Samuel Eilenberg, Oscar Zariski, Otto Schilling, Nathan Jacobson, Irving Kaplansky, Adrian Albert, Garrett Birkhoff, R. H. Bruck, Reinhold Baer, Richard Brauer, and Saunders MacLane. In addition to these, I. S. Cohen, E. Hewitt, G. P. Hochschild, G. Kalisch, A. L. Putnam, S. Sherman, P. Whitman, and D. Zelinsky served as discussion leaders. The title of my talk was Structure Theory of Rings without Finiteness Conditions.

An important event that also occurred in 1946 was the Princeton Bicentennial Celebration. This consisted of a series of conferences in various fields, emphasizing the open problems in these fields. The conference on mathematics was an exceptionally fine one since it drew a number of the top mathematicians from all over the world. In his foreword to the pamphlet on the conference, *Problems of Mathematics*, Professor Solomon Lefschetz set the tone in these words:

> Owing to the spiritual and intellectual ravage caused by the war years, it seemed exceedingly desirable to have as many participants from abroad as possible. As the list of members shows, considerable success was obtained in this. Our conference became, as it were, the first international gathering of mathematicians in a long and terrible decade. The manifold contacts and friendships renewed on this occasion will, we all hope, in the words of the Bicentennial announcement, "contribute to the advancement of the unity of all nations and to building a free and peaceful world."

Among the foreign participants were G. Ancochea, K. Borsuk, H. Cramer, P. A. M. Dirac, V. Hlavaty, W. V. D. Hodge, H. Hopf, L. K. Hua, L. J. Gärding, M. H. A. Newman, M. Riesz, and J. H. C. Whitehead.

The dates for the conference were December 17 and 18, 1946, and the mathematical program was presented in nine sections: Algebra, Algebraic Geometry, Differential Geometry, Mathematical Logic, Topology, New Fields, Mathematical Probability, Analysis, and Analysis in the Large. The Algebra Section had E. Artin as Chairman,

G. P. Hochschild as reporter, and G. Birkhoff, R. Brauer, and N. Jacobson as Discussion Leaders. Undoubtedly, one of the high points of the conference was Brauer's announcement of his recent results on induced characters: (1) Every character χ of a finite group G is an integral linear combination of characters induced from linear characters of subgroups, (2) If q is the exponent of G, then every representation of G can be written in the field of qth roots of unity. The first of these results has applications to zeta functions of number fields.

The conference concluded with a dinner at which Weyl was a principal speaker. In his speech, as he had on several earlier occasions, he sounded a pessimistic note on the future of mathematics:

> At a conference in Bern in 1931 I said: "Before one can generalize, formalize, or axiomatize, there must be a mathematical substance. I am afraid that the mathematical substance in the formalization of which we have exercised our powers in the last two decades shows signs of exhaustion. Thus I foresee that the coming generation will have a hard lot in mathematics." That challenge, I am afraid, has only been partially met in the intervening fifteen years. There were plenty of encouraging signs in this conference. But the deeper one drives the spade the harder the digging gets; maybe it has become too hard for us unless we are given some outside help, be it even by such devilish devices as high speed computing machines.

Although *Problems of Mathematics* lists Wedderburn as a participant, I do not believe that he attended any of the scientific sessions or the closing dinner. He was on leave from Princeton during the academic year 1944–1945, and he had retired shortly before the conference began "for reasons of ill health." He died in October 1948 in his home on Mercer Street in Princeton.

My research at Hopkins during the years 1943–1946 was mainly in the areas of Galois theory and the general structure theory of rings. I also wrote my first paper on Jordan rings during this period. The papers dealing with Galois theory are [25], [27], [28], [29], and [36]. The classical Galois theory establishes a (1-1) correspondence between the set of subfields of a given field P over which P is finite. dimensional separable and normal and the set of finite groups of automorphisms of P. For a given finite group of automorphisms G, the corresponding subfield Φ_G of P is the set of G-fixed elements, and for a subfield Φ such that P/Φ is finite dimensional separable and normal the corresponding group is the Galois group $\mathrm{Gal}\,\mathrm{P}/\Phi$. In [25]. I developed a Galois theory based on the concept of a self-representation of a field P defined to be a homomorphism E of P into a matrix ring P_m.[5] The corresponding module concept is that of a P–P bimodule $\mathscr{R}$ ("double module" in the terminology of [25]) in which the right dimensionality $[\mathscr{R}:\mathrm{P}]_r = m$. Associated with a self-representation P we have the composite $(\mathrm{P}^E\mathrm{P}, E, R)$, where $\mathrm{P}^E\mathrm{P}$ is the P span of the set of representing matrices α^E, and R is the map $\alpha \to \alpha 1$ where 1 is the m-rowed unit matrix. In general we define a (P, P) composite as a triple (K, S, T), where K is a ring and S and T are homomorphisms of P into K such that $\alpha^S\beta^T = \beta^T\alpha^S$, $\alpha, \beta \in \mathrm{P}$, $1^S = 1^T$, and $K = \mathrm{P}^S\mathrm{P}^T$. These imply that K is commutative and $1^S = 1^T$ is the unit. As in the composite $(\mathrm{P}^E\mathrm{P}, E, R)$, it is assumed throughout [25] that $[K:\mathrm{P}^T] < \infty$. The composite (K, S, T) is called a cover of the composite (K', S', T') ($(K, S, T) \geq (K', S', T')$) if there exists a homomorphism

[5] This constituted the major part of my Princeton bicentennial talk.

$k \to k'$ of K into K' such that $(\alpha^S)' = \alpha^{S'}$, $(\alpha^T)' = \alpha^{T'}$. If $k \to k'$ is an isomorphism then (K, S, T) and (K', S', T') are called equivalent.

Any self-representation E determines a subfield Φ_E of P as the set of elements α such that $\alpha^E = \alpha 1$ (1 the unit matrix). If E and F are self-representations, then we obtain a self-representation $E \times F$, whose degree is the product of the degrees of E and F, by substituting for the (i, j) entry of α^E the matrix representing this element in F. The self-representation E is called closed if $(\mathrm{P}^E\mathrm{P}, E, R) \geq (\mathrm{P}^{E \times E}\mathrm{P}, E \times E, R)$. For such a self-representation, P is finite dimensional over Φ_E. Moreover, any subfield of finite codimension in P can be obtained in this way and $\Phi_E = \Phi_{E'}$ for the self-representation E and E' if and only if (P^E, E, R) and $(\mathrm{P}^{E'}, E', R)$ are equivalent. In this way we obtain a (1-1) correspondence between the set of subfields of P of finite codimension in P and the set of equivalence classes of composites $(\mathrm{P}^E\mathrm{P}, E, R)$ of closed self-representations E.

This can also be formulated wholly in terms of composites. An important concept for the proof is that of the relations space of a self-representation (or a composite). For a self-representation E let E_{ij} denote the map $\alpha \to (i, j)$ entry of α^E. Then E_{ij} is an endomorphism of the additive group of P, and the relations space is $\Sigma\, \mathrm{P}E_{ij}$, the P span of the E_{ij}. The self-representation E is closed if and only if its relations space is a subring of the ring of endomorphisms of E. The theory also provides conditions that P/Φ_E is separable or normal. In the separable case one associates with Φ_E a hypergroup (definition in [25], p. 25) of simple composites where simplicity means that $\mathrm{P}^E\mathrm{P}$ is a field. This is similar to an earlier result due to Kaloujnine (reference in [25]).

Papers [27], [28], [29] constitute a continuation of [25]. The first of these develops the properties of the hypergroup associated with a subfield Φ such that P/Φ is finite dimensional separable. Among other results, it is shown that the hypergroup associated with P/Φ is isomorphic to the hypergroup of double cosets of a subgroup of a finite group. In [11] I had developed a Galois theory for purely inseparable extensions P of a field Φ of characteristic p: $\mathrm{P} = \Phi(c_1, \ldots, c_m)\, c_i^p = \gamma_i \in \Phi$, and I had established a (1-1) correspondence between the set of intermediate fields Σ and the set of restricted subalgebras $\mathfrak{E}$ of the derivation algebra $\mathfrak{D}(\mathrm{P})$ that are P algebras in the sense that $\mathrm{P}_R\mathfrak{E} \subset \mathfrak{E}$. If Σ is given, then $\mathfrak{E}$ is the set of derivations of P/Σ, and if $\mathfrak{E}$ is given, then Σ is the set of $\mathfrak{E}$ constants ($D\alpha = 0$, $D \in \mathfrak{E}$). Using the theory of self-representations, this result was improved in [28] by showing that if $\mathfrak{C}$ is any restricted P—Lie algebra of derivations of a field P of characteristic p such that $[\mathfrak{E}:\mathrm{P}_R]_r < \infty$ then P is finite dimensional purely inseparable of exponent one over the subfield Φ of $\mathfrak{C}$ constants and $\mathfrak{C} = \mathfrak{D}(\mathrm{P}/\Phi)$. In [29] I was interested in analyzing the structure of a central simple algebra $\mathfrak{A}$ of degree n over a field Φ ($[\mathfrak{A}:\Phi] = n^2$) relative to a subfield P that is separable and n dimensional over Φ. As the example of $\mathfrak{A}$ over an algebraically closed field shows, no such P may exist. On the other hand, such P's do exist in abundance if $\mathfrak{A}$ is a division algebra. I was interested in studying $\mathfrak{A}$ with respect to P intrinsically, that is, without recourse to extension of the base field.[6] Then algebra $\mathfrak{A}$ is a P–P bimodule, where the actions of P on $\mathfrak{A}$ are the multiplications. I showed that the self-representations of P obtained by regarding $\mathfrak{A}$ as bimodule in this way are the regular representations of P over Φ. Hence if $(x_1, \ldots, x_n)$ is a right base for $\mathfrak{A}$ over P, then for any $\alpha \in \mathrm{P}$, $\alpha x_i = \Sigma x_j(\alpha E_{ji})$, where $\alpha \to \alpha^E = (\alpha E_{ij})$ is a regular representation of P over Φ. We have $x_i x_j = \Sigma x_\mu \sigma_{\mu ij}$, $\sigma_{\mu ij} \in \mathrm{P}$, and the product of any two elements in $\mathfrak{A}$ is determined by the compositions in P, the self-representation E, and the $\sigma_{kij} \in \mathrm{P}$.

[6] For comparison with the field extension method, see e.g. [98]

These σ's satisfy certain conditions that come from the associativity conditions: $(x_i, x_j)x_k = x_i(x_j, x_k)$, $\rho(x_i x_j) = (\rho x_i)x_j$. A set $\sigma = \{\sigma_{ijk}\}$ satisfying these conditions is called a factor set for P and E. Conversely, given a P–P bimodule $\mathfrak{A}$ having E as a corresponding self-representation and a factor set σ, we can use these data to define a product in $\mathfrak{A}$ so that $\mathfrak{A}$ becomes an associative algebra containing P as a subalgebra with the given actions of P as multiplications. Conditions that $\mathfrak{A}$ is simple and that $\mathfrak{A}$ is split ($\simeq M_n(\Phi)$) can be derived. Also I obtained a matrix analog of Hilbert's Satz 90 for the self-representation E.

The purpose of the paper [36] was to obtain a Galois theory for division rings that combined Emmy Noether's Galois theory of groups of inner automorphisms and the outer automorphism Galois theory of [20]. The results of this paper were obtained independently about the same time by H. Cartan ("Théorie de Galois pour les corps non-commutatifs," *Annales Scientifiques de l'École Normale Supérieure* **65**, 1948, pp. 60–77). If G is a group of automorphisms of a division ring P, then we have the normal subgroup H of G consisting of the inner automorphisms contained in G, and G is called closed if H is the set of inner automorphisms defined by the non-zero elements of a division subalgebra A over the center Γ. The division algebra A is uniquely determined by G (or H) and is called the division subalgebra *associated* with the closed group G. The *reduced order* of G is defined to be $[A:\Gamma](G:H)$, where $(G:H)$ is the index of H in G. The main theorem concerns closed groups of automorphisms of finite reduced order: Let G be such a group in the division ring P. To each closed subgroup K of G we associate the division subring Ψ of P of elements invariant (fixed) under every $\sigma \in K$, and to each division subring Ψ of P containing $\Phi = \mathrm{P}(G)$ we associate the closed subgroup $K = G(\Psi)$ of automorphisms having the elements of Ψ as invariants. Then these two correspondences are inverses, so each is (1-1) between the set of closed subgroups of G and the set of division subrings of P containing Φ.

The proof of this result depends on a generalization of the Dedekind independence property of automorphisms and on the Jacobson–Bourbaki theorem, which is proved in this paper. The latter result establishes a (1-1) correspondence between the set of division subrings Φ of a division ring P such that $[\mathrm{P}:\Phi]_l < \infty$ and the set of rings of endomorphisms $\mathfrak{A}$ of the additive group of P such that (1) $\mathfrak{A} \supset \mathrm{P}_r$ (the set of right multiplications) and (2) $[\mathfrak{A}:\mathrm{P}_r]_r < \infty$. Given Φ, we take $\mathfrak{A}$ to be the ring of linear transformations of P regarded as a left vector space over Φ, and given $\mathfrak{A}$, we take Φ to be the set of $\gamma \in \mathrm{P}$ such that $[\gamma_l, A] = 0$ for all $A \in \mathfrak{A}$. The proof of this theorem made use of the Chevalley–Jacobson density theorem that I shall consider in a moment.

In the field case, the ring $\mathfrak{A}$ is the relations space of a closed self-representation as defined in [25]. Several papers related to [25] were published by Hochschild, by Dieudonné and by Nakayama. Hochschild's paper, "Double vector spaces over division rings" (*American Journal of Mathematics* **71**, 1949, pp. 443–460) considered D–E bimodules for division rings D and E. The finiteness conditions imposed in this paper were removed in Dieudonné's paper, "Linearly compact spaces and double vector spaces over fields" (*American Journal of Mathematics* **73**, 1951, pp. 13–24). Both Hochschild's and Dieudonné's papers make use of the forms of the Jacobson–Bourbaki correspondence. Nakayama's papers dealt with bimodules for arbitrary rings. These papers are "Non-normal Galois theory for non-commutative and non-semisimple rings" (*Canadian Journal of Mathematics* **3**, 1951, pp. 208–218) and "Note on double-modules over arbitrary rings" (*American Journal of Mathematics* **74**, 1952, pp. 645–655).

The series of papers [31]–[35] develop a structure theory of rings without finiteness conditions, such as Wedderburn's condition of finiteness of dimensionality over a field, or the chain conditions on one-sided ideals that were the keystone of Artin's extension of Wedderburn's theory. The first paper of the series deals with simple associative and non-associative algebras. The key result on which most of the others depend is the Chevalley–Jacobson density theorem: Any irreducible ring of endomorphisms $\mathfrak{A}$ of a commutative group $\mathscr{R}$ (written additively) is a dense ring of linear transformations on $\mathscr{R}$ regarded as a vector space over the centralizer $\mathfrak{D}$ of $\mathfrak{A}$ in the ring of endomorphisms of $\mathscr{R}$ (which is a division ring by Schur's lemma). This means that given any pair of ordered finite subsets of $\mathscr{R}$, $(x_1,\ldots,x_r)$ and $(y_1,\ldots,y_r)$ with the x_i linearly independent, there exists an $l \in \mathfrak{A}$ such that $x_i l = y_i$, $1 \leq i \leq r$.[7] The main result of the first part of [31] is the following structure theorem. An associative ring is simple and contains a minimal right ideal if and only if it is isomorphic to a dense ring of linear transformations $\mathfrak{A}$ of a vector space $\mathscr{R}$ over a division ring $\mathfrak{D}$ such that every $l \in \mathfrak{A}$ is of finite dimensional range ($\mathscr{R}l$ finite dimensional). The Wedderburn–Artin structure theorem on simple rings is an easy consequence. The main result is supplemented by an isomorphism theorem on dense rings of linear transformations with finite dimensional ranges: If $a_1 \to a_2$ is an isomorphism between two such rings, then there exists an isomorphism $x_1 \to x_2$ between the additive groups of the underlying vector spaces $\mathscr{R}_1$ and $\mathscr{R}_2$ and an isomorphism $\alpha_1 \to \alpha_2$ between the division rings $\mathfrak{D}_1$ and $\mathfrak{D}_2$ of $\mathscr{R}_1$ and $\mathscr{R}_2$ such that $x_1\alpha_1 \to x_2\alpha_2$ and $x_1a_1 \to x_2a_2$. This specializes to give the form of the automorphisms of a dense ring of finite ranged linear transformations. An example is given of a differential polynomial ring that is simple but contains no minimal right ideals. The isomorphism theorem that holds when the rings contain minimal right ideals fails in this example.

The second part of the paper deals with arbitrary non-associative rings. The results generalize some of those appearing in [10] and they overlap with some contained in Nakayama's paper cited above. If $\mathfrak{A}$ is a non-associative ring, the ring of endomorphisms of the additive group of $\mathfrak{A}$ generated by the left and right multiplications a_l, a_r, $a \in \mathfrak{A}$ is called the multiplication ring $\mathfrak{M}$ of $\mathfrak{A}$. The centralizer of $\mathfrak{M}$ in the ring of endomorphism of $\mathfrak{A}$ is called the multiplication centralizer $\mathfrak{C}$ of $\mathfrak{A}$. If $\mathfrak{A}$ is simple ($\mathfrak{A}^2 \neq 0$), then $\mathfrak{C}$ is a field and $\mathfrak{M}$ is a dense ring of linear transformations in $\mathfrak{A}$ over $\mathfrak{C}$, and $\mathfrak{A}$ can be regarded as an algebra over $\mathfrak{C}$. The center $\mathfrak{C}_0$ of $\mathfrak{A}$ is defined to be the set of elements $c \in \mathfrak{A}$ that commute with every $x \in \mathfrak{A}$ and associate with every x and y in the sense that $(xy)c = x(yc)$, $(xc)y = x(cy)$, $(cx)y = c(xy)$. This is a commutative and associative subring of $\mathfrak{A}$. If $\mathfrak{A}$ is simple, then either $\mathfrak{C}_0 = 0$ or $\mathfrak{A}$ has a unit and $\mathfrak{S}$ is the set of right (or left) multiplications by the elements of $\mathfrak{S}_0$.

The last part of the paper is mainly concerned with isomorphism theorems on rings of linear transformations. Since these are rather weak compared to results obtained later (see Section 11 of *Structure of Rings*), I shall forego discussing them here.

Paper [32] is the one in which the radical $\mathscr{R}$ of a ring $\mathfrak{A}$, which has become known as the Jacobson radical, was introduced. The definition given in this paper is based on an element property called quasi-regularity that appeared in a characterization of the radical of a finite dimensional algebra given by S. Perlis, with acknowledgment of a suggestion by M. Hall ("A characterization of the radical of an algebra," *Bulletin of*

[7] Chevalley's result was published in a paper by Nakayama, "Über einfache distributive Systeme unendliche Ränge," *Proceedings Imperial Academy Tokyo* **20**, 1944, pp. 61–66.

the American Mathematical Society **48**, 1942, pp. 128–132). A second procedure for defining $\mathscr{R}$ is the one given in *Structure of Rings*, namely, $\mathscr{R}$ is the intersection of the kernels of the irreducible representations of $\mathfrak{A}$. An element z is called right quasi-regular if there exists a $z' \in \mathfrak{A}$ such that $z + z' + zz' = 0$ and a right ideal $\mathfrak{J}$ is quasi-regular if every $z \in \mathfrak{J}$ is right quasi-regular. According to the definition of [32], $\mathscr{R} = \Sigma\mathfrak{J}$, where summation is taken over all quasi-regular right ideals. Alternatively, $\mathscr{R} = \bigcap \operatorname{ann}\mathfrak{M}$, where $\mathfrak{M}$ is an irreducible (right) $\mathfrak{A}$ module, $\operatorname{ann}\mathfrak{M} = \{z \in \mathfrak{A} \mid \mathfrak{M}z = 0\}$ and the intersection is taken over all irreducible $\mathfrak{M}$.

A ring $\mathfrak{A}$ is called primitive if it has a faithful irreducible representation, that is, there exists an irreducible module $\mathfrak{M}$ for $\mathfrak{A}$ such that $\operatorname{ann}\mathfrak{M} = 0$. By the density theorem, this is the case if and only if $\mathfrak{A}$ is isomorphic to a dense ring of linear transformations in a vector space over a division ring. A ring is called semi-primitive (semi-simple in the terminology of [32]) if $\mathscr{R} = 0$. This is the case if and only if $\mathfrak{A}$ is a subdirect product (subdirect sum in the terminology of *Structure of Rings*) of primitive rings. For any $\mathfrak{A}$, $\mathfrak{A}/\mathscr{R}$ is semi-primitive.

If $\mathfrak{A}$ is right (left) Artinian ($\mathfrak{A}$ satisfies the minimum condition for right (left) ideals) then $\mathscr{R}$ is nilpotent. It is easy to derive the Wedderburn–Artin structure theorem for semi-simple Artinian rings from the results we have indicated.

I also attempted to generalize a classical result of Krull's on commutative Noetherian rings: $\mathfrak{R}^{\omega} \equiv \bigcap_1^{\infty} \mathfrak{R}^i = 0$ for the radical $\mathfrak{R}$ of such a ring. I showed that if $\mathfrak{A}$ is right Noetherian, then $\mathfrak{R}^{\alpha} = 0$ for some transfinite α. The proof made use of a lemma (Theorem 10) that if $\mathfrak{N}$ is a right ideal $\neq 0$ contained in $\mathfrak{R}$ and $\mathfrak{N}$ is finitely generated, then $\mathfrak{N}\mathfrak{R} \neq \mathfrak{N}$. An example was given by Herstein in "A Counterexample in Noetherian Rings," *Proceedings of the National Academy of Sciences* **54**, 1965, pp. 1036–1037 of a right Noetherian ring in which $\mathfrak{R}^{\omega} \neq 0$. On the other hand, in *Structure of Rings*, p. 200, the following question is raised: Do the maximum conditions on both left and right ideals imply $\mathfrak{R}^{\omega} = 0$? This is still an open question at the present writing. The lemma stated above is often called "Nakayama's lemma." (See M. Nagata, *Local Rings*, Interscience Publishers, 1962, p. 213.)

A somewhat improved exposition of these results is given in *Structure of Rings*.

An algebra $\mathfrak{A}$ over a field Φ is called algebraic if every $a \in \mathfrak{A}$ is algebraic in the sense that there exists a polynomial $f(\lambda) \neq 0$ such that $f(a) = 0$. The Jacobson radical of an algebraic algebra is a nil algebra ([32], p. 308). A question raised by Kurosch in 1941 was: Is every algebraic algebra $\mathfrak{A}$ locally finite in the sense that the subalgebras generated by finite subsets of $\mathfrak{U}$ are finite dimensional over the base field? In [33] I considered this problem for algebraic algebras that are of bounded degree in the sense that the degrees of the minimum polynomials of the elements are bounded. By using the structure theory of [31] and [32], I showed that Kurosch's problem has an affirmative answer for such an algebra that contains no nil ideals, and I reduced the problem to nil algebras of bounded degree. This was solved by J. Levitzki in "On a Problem of Kurosch," *Bulletin of the American Mathematical Society* **52**, 1946, pp. 1033–1035, which implied an affirmative answer to Kurosch's problem for algebraic algebras of bounded degree.

Another application of the structure theory given in [33] is to the proof of the theorem that if $\mathfrak{A}$ is a ring in which for every $a \in \mathfrak{A}$ there exists an integer $n(a) > 1$ such that $a^{n(a)} = a$, then $\mathfrak{A}$ is commutative. In 1937 Marshall Stone showed that a Boolean ring $\mathfrak{B}$, defined by the condition $a^2 = a$ for all a is necessarily commutative. This led to attempts to generalize, and it was conjectured that if $a^n = a$ for all a and a fixed $n > 1$ in a ring $\mathfrak{A}$, then $\mathfrak{A}$ is commutative. The case $a^p = a$, $pa = 0$ for a prime p could be settled elementarily, but to go beyond this appeared difficult. The result I proved showed this and more. Many generalizations of the $a^{n(a)} = a$ commutativity theorem have been proved; the earlier ones appear in Chapter X of *Structure of Rings*.

Let $\mathscr{R}$ and $\mathscr{R}'$ be a right and left vector space, respectively, over a division ring $\mathfrak{D}$. Suppose $\mathscr{R}$ and $\mathscr{R}'$ are paired by a bilinear function $(x', x) \in \mathfrak{D}$, that is, (x', x) for $x \in \mathfrak{R}$, $x' \in \mathfrak{R}$ is additive in both arguments and $(\alpha x', x) = \alpha(x', x)$, $(x', x\alpha) = (x', x)\alpha$, $\alpha \in \mathfrak{D}$. We say that $\mathscr{R}$ and $\mathscr{R}'$ are dual relative to $(\ ,\ )$ if $(z', \mathscr{R}) = 0 \Rightarrow z' = 0$ and $(\mathscr{R}', z) = 0 \Rightarrow z = 0$. Suppose this is the case. Then we say that a linear transformation A in $\mathscr{R}$ has an adjoint A' in $\mathscr{R}'$ if A' is a map $\mathscr{R}' \to \mathscr{R}'$ such that $(x', xA) = (x'A', x)$ for all $x \in \mathscr{R}$, $x' \in \mathscr{R}'$. Then A' is linear, and it can be shown that A has an adjoint if and only if A is a continuous map relative to a suitable topology of $\mathscr{R}$. The set of linear transformations of $\mathscr{R}$ that have adjoints in $\mathscr{R}'$ form a subring $\mathfrak{A}$ of the ring of linear transformations. A simple verification shows that for any $x' \in \mathscr{R}'$ and $x \in \mathscr{R}$ the map $y \to x(x', y)$ is a linear transformation with adjoint $y' \to (y', x)x'$. Hence, if $x'_1, \ldots, x'_r \in \mathscr{R}'$ and $x_1, \ldots, x_r \in \mathscr{R}$, then the map $y \to \sum x_i(x'_i, y)$ is contained in $\mathfrak{A}$. This map evidently has finite dimensional range, and it can be shown that the set $\mathfrak{F}$ of these maps is the subring of $\mathfrak{A}$ of linear transformations contained in $\mathfrak{A}$ having finite dimensional ranges. The ring $\mathfrak{F}$ and, á fortiori, $\mathfrak{A}$ is a dense ring of linear transformations. Hence $\mathfrak{A}$ is primitive and $\mathfrak{F}$ is simple with minimal right ideals that are also minimal right ideals of $\mathfrak{A}$. Conversely, any primitive ring containing minimal right ideals can be obtained in this way.

The study of simple rings containing minimal right ideals by means of dual vector spaces is due to Dieudonné in his paper "Sur le socle d'un anneu et les anneaux simple infinis" (*Bulletin Société Mathématique de France* **71**, 1943, pp. 1–30). The extension of this to primitive rings containing minimal right ideals was given in [35]. The theory is developed more fully in Chapter IV of *Structure of Rings*.

In [34], I introduced a topology for the set $S = S(\mathfrak{A})$ of primitive ideals for an arbitrary ring $\mathfrak{A}$. Here $\mathfrak{B}$ is called a primitive ideal in $\mathfrak{A}$ if $\mathfrak{A}/\mathfrak{B}$ is a primitive ring. The topology I defined was by a closure operator: If $A \subset S$, then $\bar{A}$ is defined as the set of primitive ideals $\mathfrak{C}$ that contain the intersection of all the primitive ideals in A. This method of defining a topology was first used by Marshall Stone (reference in [34]) for the set of ideals in a Boolean ring. Moreover, as the reader may be aware, it is the method used in defining the prime spectrum, spec $\mathfrak{A}$, of a commutative ring, which is a central concept in modern commutative ideal theory. The set S with the topology is called the structure space of the ring $\mathfrak{A}$. Since the primitive ideals of a commutative ring $\mathfrak{A}$ are just the maximal ideals of $\mathfrak{A}$, the structure space for $\mathfrak{A}$ commutative is the space max spec $\mathfrak{A}$, the subspace of spec $\mathfrak{A}$ consisting of the maximal ideals of $\mathfrak{A}$. Structure spaces have been used very effectively by Kaplansky, especially in his work on Kurosch's problem (see the references in [85]).

One can reduce the study of structure spaces to the case in which $\mathfrak{A}$ is semi-primitive since $S(\mathfrak{A})$ for any $\mathfrak{A}$ is homeomorphic to $S(\bar{\mathfrak{A}})$, where $\bar{\mathfrak{A}} = \mathfrak{A}/\text{rad}\,\mathfrak{A}$, rad $\mathfrak{A}$, the Jacobson radical of $\mathfrak{A}$. The topological space $S(\mathfrak{A})$ is compact if $\mathfrak{A}$ has a unit, and if $\mathfrak{A}$ is semi-primitive with unit then $S(\mathfrak{A})$ is disconnected if and only if $\mathfrak{A}$ is a direct sum of two non-zero ideals. Further results on structure spaces are given in [34] and in Chapter 9 of *Structure of Rings*.

In the paper "Semi-automorphisms of division algebras" (*Annals of Mathematics* **48**, 1947, pp. 147–154), G. Ancochea defined a semi-isomorphism S of an associative ring $\mathfrak{A}$ onto an associative ring $\mathfrak{B}$ as a bijective additive map of $\mathfrak{A}$ onto $\mathfrak{B}$ such that

$(ab + ba)S = (aS)(bS) + (bS)(aS)$. Any isomorphism or anti-isomorphism is a semi-isomorphism, and Ancochea proved the converse if $\mathfrak{A}$ and $\mathfrak{B}$ are simple of finite dimensionality over their centers of characteristic $\neq 2$.[8]

The concept of a semi-isomorphism for rings that are uniquely divisible by 2 in the sense that for any a, $2x = a$ has a unique solution $\frac{1}{2}a$ can be formulated in terms of Jordan rings. For such a ring we have an associated Jordan ring $\mathfrak{A}_j$ in which the given associative multiplication ab is replaced by Jordan multiplication $a \cdot b = \frac{1}{2}(ab + ba)$. Then S is a semi-isomorphism of $\mathfrak{A}$ onto $\mathfrak{B}$ if and only if S is an isomorphism of $\mathfrak{A}_j$ onto $\mathfrak{B}_j$, and Ancochea's theorem is that if $\mathfrak{A}$ and $\mathfrak{B}$ are simple of finite dimensionality over their centers of characteristic $\neq 2$ (in which case unique divisibility by 2 holds), then any isomorphism of $\mathfrak{A}_j$ onto $\mathfrak{B}_j$ is either an isomorphism or an anti-isomorphism of $\mathfrak{A}$ onto $\mathfrak{B}$.

Now, suppose $\mathfrak{A}$ has an involution J. Then the set $\mathfrak{H}(\mathfrak{A}, J)$ of J-symmetric elements $(h^J = h)$ is a subalgebra of $\mathfrak{A}_j$. In [37], I considered the isomorphisms among the Jordan rings $\mathfrak{A}_j$ and $\mathfrak{H}(\mathfrak{A}, J)$, where the $\mathfrak{A}$ are simple and finite dimensional over their centers of characteristic $\neq 2$.

The condition $(ab + ba)S = (aS)(bS) + (bS)(aS)$ is equivalent to $(a^2S) = (aS)^2$ if the additive group of $\mathfrak{A}$ has no 2-torsion. On the other hand, it is stronger if $\mathfrak{A}$ has 2-torsion, and if the characteristic is 2 $(2\mathfrak{A} = 0)$, then, as in [23], we can define the restricted Lie algebra $\mathfrak{A}_l$, and if $\mathfrak{A}$ has an involution we have the restricted Lie algebra $\mathfrak{S}(\mathfrak{A}, J) = \mathfrak{H}(\mathfrak{A}, J)$. Isomorphisms for these Lie algebras for $\mathfrak{A}$ finite dimensional normal simple were determined in [23].

The paper [37] was devoted to the study of these problems. The results I obtained were improved in the later papers [40], [44], and [48] (see also *Structure and Representations of Jordan Algebras*, p. 141f).

For any non-associative ring $\mathfrak{R}$ with product $a \cdot b$, one defines the center $\mathfrak{C}(\mathfrak{R})$ as the set of $c \in \mathfrak{R}$ such that $c \cdot a = a \cdot c$, $(a \cdot b)c = a \cdot (b \cdot c)$, $(a \cdot c) \cdot b = a \cdot (c \cdot b)$, $(c \cdot a) \cdot b = c \cdot (a \cdot b)$ for all a, $b \in \mathfrak{R}$. This is an associative and commutative subring of $\mathfrak{A}$ (see [31] or Nakayama's paper referred to in [31]). If $\mathfrak{A}$ is an associative ring, $\mathfrak{A}$ defines a Jordan ring $\mathfrak{A}_j$ and a Lie ring $\mathfrak{A}_l$, where $\mathfrak{A}_j$ $(\mathfrak{A}_l)$ is obtained by replacing the given associative product by $\{ab\} = ab + ba$ $([ab] = ab - ba)$. In [38], I considered the centers $\mathfrak{C} = \mathfrak{C}(\mathfrak{A})$, $\mathfrak{C}_j = \mathfrak{C}(\mathfrak{A}_j)$ and $\mathfrak{C}_l = \mathfrak{C}(\mathfrak{A}_l)$ for any associative ring $\mathfrak{A}$. I showed that $\mathfrak{C}_j = \{c \in \mathfrak{A} \mid [[ca]b] = 0,\ a, b \in \mathfrak{A}\}$ and $\mathfrak{C}_l = \{c \in \mathfrak{A} \mid 2[ca] = 0 = [[ca]b],\ a, b \in \mathfrak{A}\}$. Hence, $\mathfrak{C} \subseteq \mathfrak{C}_l \subseteq \mathfrak{C}_j$. I also considered the subring $\mathfrak{B} = \mathfrak{B}(\mathfrak{A}) = \{c \in \mathfrak{A} \mid [c[ab]] = 0;\ a, b \in \mathfrak{A}\}$. This was introduced by Kaplansky in a paper cited in [38]. Relations among the rings $\mathfrak{B}(\mathfrak{A})$, $\mathfrak{C}(\mathfrak{A})$, $\mathfrak{C}_j$, and $\mathfrak{C}_l$ are given under mild conditions on $\mathfrak{A}$, and examples of inequalities among these are given. I also considered the center of a simple subalgebra $\mathfrak{U}$ of $\mathfrak{A}_j$, $\mathfrak{A} = \Phi_n$, where Φ is a field of characteristic 0.

The two papers [26] and [30] do not fit into any of the major fields of my research and so are listed under "Miscellaneous". The paper [26] gives a new proof and extension of a theorem of Schur on commuting complex matrices. Schur showed that the maximum number $N(n)$ of linearly independent commuting $n \times n$ complex matrices is $[n^2/4] + 1$, and for even n there is one similarity class of linear spans of such matrices, and for odd n there are two similarity classes. In [26], I extended Schur's results to arbitrary base fields with the exception of imperfect fields of characteristic two. This exception was removed by W. H. Gustafson in "On maximal commutative

[8] Actually Ancochea stated this only for semi-automorphisms, that is, in the case where $\mathfrak{B} = \mathfrak{A}$.

algebras of linear transformations," *Journal of Algebra* **42**, 1976, pp. 557–563. In the paper "Commutative subalgebras of semisimple Lie algebras" (in Russian, translated as Number 40, 1951, pp. 215–227, in the American Mathematical Society translation series), A. I. Malcev extended Schur's results to all complex semisimple Lie algebras. It is natural to ask if Malcev's results can be extended to split semi-simple Lie algebras (definition in *Lie Algebras*, p. 108) over arbitrary fields. As far as I know, this has not been done.

In a paper appearing in the *Bulletin of the American Mathematical Society* (loc. cit in [30], R. E. Johnson considered the solvability of equations of the form $\chi\alpha = \gamma\chi + \beta$ in a division algebra that is separable algebraic over its center. He distinguished two a cases: (1) α and γ are not conjugate (by an inner automorphism), and (2) α and γ are conjugate. The second case can be reduced to an equation of the form $xd - dx = b$. I considered this type of equation for an arbitrary associative algebra assuming d is algebraic over the base field Φ. Conditions for the existence of a solution and the question of uniqueness were considered in [30]. These generalize Johnson's results.

AN EXTENSION OF GALOIS THEORY TO NON-NORMAL AND NON-SEPARABLE FIELDS.*[1]

By N. Jacobson.

As is well known, modern Galois theory is a theory of the automorphisms of an arbitrary field P. Its principal result is the establishment of a $(1-1)$ correspondence between the finite groups $\mathfrak{g}$ of automorphisms in P and the subfields $\Phi_{\mathfrak{g}}$ over which P is finite, separable and normal, such that 1) $\mathfrak{g}_1 \geqq \mathfrak{g}_2$ if and only if $\Phi_{\mathfrak{g}_1} \leqq \Phi_{\mathfrak{g}_2}$ and 2) $\mathfrak{g}_2$ is invariant in $\mathfrak{g}_1$ if and only if $\Phi_{\mathfrak{g}_2}$ is normal over $\Phi_{\mathfrak{g}_1}$. A second type of Galois theory has been given by the author in a previous paper.[2] It associates the subfields Φ of a field P of characteristic p over which P has the form $\Phi(x_1, \cdots, x_m)$ where $x_i^p = \xi_i$ in Φ with certain algebras of derivations in P. In seeking an extension of these theories we have been led to a study of the *self-representations* of P, i. e. the representations of P by matrices with elements in this field.[3] One is led to seek an extension along these lines by the following two remarks. First, if S is an automorphism, then $\alpha \rightarrow \alpha^S$ is a self-representation by 1-dimensional matrices and second, if D is a derivation, then

$$\alpha \rightarrow \begin{pmatrix} \alpha & \alpha D \\ 0 & \alpha \end{pmatrix}$$

is a representation by 2-rowed matrices.

Now by our definition of a self-representation any isomorphism of P into a subfield is a self-representation. It is therefore desirable to restrict the

* Received July 12, 1943.

[1] Presented to the Society, September 11, 1943.

[2] "Abstract derivation and Lie algebras," *Transactions of the American Mathematical Society*, vol. 43 (1937), p. 220.

[3] A Galois theory of separable extensions which makes use of the concept of self-representation has been announced recently by L. Kaloujnine, "Sur la théorie de Galois des corps non galoisiens séparables," *Comptes Rendus de l'Académie des Sciences*, vol. 214 (1942), pp. 597-599. (Abstracted in *Mathematical Reviews*, vol. 4 (1943), p. 130). It appears that Kaloujnine's results are special cases of those obtained here. See **12.** It should be mentioned also that Galois theory for separable extensions based on the classical theory for separable normal extensions had been developed previously. See Krasner, M. "Sur la théorie de la ramification des ideaux des corps non-galoisiens de nombres algebriques," *Thèse*, Paris, 1938. The present theory (and apparently that of Kaloujnine) is independent of the classical theory.

class of self-representations in such a way that any self-representation in this class having rank one (one-rowed) is an automorphism of P. This has been done by defining *non-singular* self-representations. There is a natural way to define the product of two self-representations: One substitutes for the elements of the matrices of one representation the matrices representing these elements in the second representation. This, of course, is a generalization of the product of isomorphisms.

With any self-representation—or more exactly, with any class of similar self-representations—we may associate a double P-module $\mathfrak{R}$ having the property that its right dimensionality $(\mathfrak{R}:P_r) = m < \infty$. If, in addition, $(\mathfrak{R}:P_l) = m$, then $\mathfrak{R}$ is said to be non-singular. Modules of this type correspond to non-singular self-representations. One may also define a product for double modules corresponding to the product of self-representations.

Now with each self-representation we may associate a composite (ring) of the field P with itself. This consists of the ring of transformations in P generated by the scalar multiplications $x \to \alpha x$ and $x \to x\alpha$. These composites may also be defined abstractly: A system (K, S, T) is a *composite* of P with itself if K is a ring and S and T are isomorphisms between P and subfields P^S and P^T of K such that 1) $K = P^S P^T$, 2) K is commutative, 3) $1^S = 1^T$, 4) $(K:P^T) < \infty$. If (K', S', T') is a second composite and the mapping $\Sigma\alpha^S\beta^T \to \Sigma\alpha'^S\beta^{T'}$ is a homomorphism, then (K, S, T) is a *cover* of (K', S', T') $((K, S, T) \geq (K', S', T'))$. If this mapping is an isomorphism, the two composites are equivalent. Throughout our discussion equivalent composites are identified. One may define non-singular composites and the product $(K, S, T) \times (L, U, V)$ of any two composites. A composite (K, S, T) is called a *Galois composite* if $(K, S, T) \geq (K, S, T) \times (K, S, T)$, (K, S, T) $\geq$ the identity composite (P^U, U, U) and $(K, S, T) = (K, T, S)$. Our main result is the establishment of a $(1-1)$ correspondence between the Galois composites Γ of P and the subfields Φ_Γ of P over which P is finite, such that $\Gamma_1 \geq \Gamma_2$ if and only if $\Phi_{\Gamma_1} \leq \Phi_{\Gamma_2}$.

Any composite may be decomposed in a certain sense into indecomposable composites (K_i, S_i, T_i). Conditions that (K, S, T) be a Galois composite may be given in terms of the components (K_i, S_i, T_i). If each K_i is a field, (K, S, T) is called *semi-simple*. In this case the condition that (K, S, T) be a Galois composite is that the components (K_i, S_i, T_i) form a hypergroup relative to the product $(K_i, S_i, T_i)(K_j, S_j, T_j)$ defined to be the set of components of the composite $(K_i, S_i, T_i) \times (K_j, S_j, T_j)$. Our fundamental correspondence between composites and subfields induces a $(1-1)$ correspondence between semi-simple Galois composites and subfields Φ over which

P is finite and separable. We investigate also the fields Φ over which P is normal. Combining our results for normal and separable fields we obtain a (1 — 1) correspondence between the semi-simple composites whose hypergroups are groups and the subfields Φ over which P is finite, separable and normal. Finally, it is easy to obtain the connection between these composites and finite groups of automorphisms and this gives the classical theorem.

Our results could have been formulated in terms of self-representations but this seems to be unnatural except in the case where P has a primitive element over Φ. In a later paper we hope to investigate in greater detail the theory for subfields Φ over which P is purely inseparable.

1. Self-representations of fields. If P is an arbitrary field, we define a *self-representation* of P as an isomorphism between P and a field of m-rowed matrices with elements in P such that $1 \to 1$ the identity matrix. The integer m is the *rank* of the representation. If $E\colon \alpha \to \alpha^E$ in P_m is a self-representation, we define E_{ij} as the transformation sending α into the element in the i-th row and j-th column of α^E.[4] Thus $\alpha^E = (\alpha E_{ij})$ and the E_{ij} satisfy the following conditions:

$$(1) \qquad (\alpha+\beta)E_{ij} = \alpha E_{ij} + \beta E_{ij}, \qquad (\alpha\beta)E_{ij} = \Sigma_\lambda(\alpha E_{i\lambda})(\beta E_{\lambda j}), \qquad 1E_{ij} = \delta_{ij}$$

Conversely any m^2 transformations E_{ij} in P that satisfy (1) determine a self-representation $\alpha \to (\alpha E_{ij})$ of P.

We shall call γ a *fixed element* under E if γ^E is the diagonal matrix $\{\gamma, \cdots, \gamma\}$. Hence γ is fixed if $\gamma E_{ii} = \gamma$ and $\gamma E_{ij} = 0$ for $i \neq j$. Evidently the set of fixed elements is a subfield of P.

We shall recall now some of the important concepts from general representation theory. Two self-representations E and F are *similar* if there exists a fixed non-singular matrix $S = (\sigma_{ij})$ such that $\alpha^F = S^{-1}\alpha^E S$ for all α. Similar representations evidently have the same fixed elements. A representation is *reducible* if it is similar to a representation F in which $F_{ij} = 0$ for $i > r > 0$ and $j \leq r$. In this case the correspondences $\alpha \to (\alpha F_{kl})$ $k, l = 1, \cdots, r$ and $\alpha \to (\alpha F_{pq})$, $p, q = r+1, \cdots, m$ are self-representations. The first of these is a *subrepresentation* and the second a *difference representation* of E. If besides $F_{ij} = 0$ for $i > r$ and $j \leq r$ we have $F_{ij} = 0$ for $i \leq r$ and $j > r$, then E is *decomposable* into the *components* $\alpha \to (\alpha F_{kl})$, $\alpha \to (\alpha F_{pq})$. In a similar manner decomposability into more than two components may be de-

[4] P_m denotes the ring of $m \times m$ matrices with elements in P.

fined. We shall write $E = E_1 + E_2 + \cdots + E_s$ when E is decomposable into the components E_i. If $E = E_1 + \cdots + E_s$ where the E_i are irreducible, E is said to be *completely reducible.* If an element γ of P is a fixed element relative to E, it is fixed under any subrepresentation and under any difference representation of E.

2. Double P-modules. We shall now give a module formulation of the representation problem. We define a *double* P-*module* [5] as a commutative group $\mathfrak{R}$ which is both a right P-module and a left P-module such that

1. $$1x = x = x1.$$
2. $$(\alpha x)\beta = \alpha(x\beta).$$

Thus if we denote the endomorphism $x \to \alpha x$ by α_l and the endomorphism $x \to x\alpha$ by α_r then $\alpha_l\beta_r = \beta_r\alpha_l$ for all α, β in P. We denote the field of endomorphisms $\alpha_l(\alpha_r)$ by $\mathrm{P}_l(\mathrm{P}_r)$. For our purposes it suffices to consider only double P-modules that satisfy the following condition

3. $$(\mathfrak{R} : \mathrm{P}_r) = m < \infty.$$

Hence from now on we use the term " double P-module " for " double P-module satisfying condition 3."

Let $x_1, \cdots, x_m$ be a basis for $\mathfrak{R}$ over P_r. Then $\alpha x_i = \Sigma x_j \alpha_{ji}$ and it is readily verified that the correspondence $\alpha \to (\alpha_{ij})$ is a self-representation E of P. The rank of E is m. Conversely if E is any self-representation, then we let $\mathfrak{R}$ be a right P-module such that $(\mathfrak{R} : \mathrm{P}_r) = m$. If $x_1, \cdots, x_m$ is a basis for $\mathfrak{R}$, we write $x = \Sigma x_i \xi_i$ and define $\alpha x = \Sigma x_j \eta_j$ where $\eta_j = \Sigma(\alpha E_{ij})\xi_i$. Then it is readily verified that $\mathfrak{R}$ is a double P-module and $\alpha x_i = \Sigma x_j(\alpha E_{ji})$. Thus any self-representation is obtained from some double P-module in the manner indicated. If $y_1, \cdots, y_m$, $y_i = \Sigma x_j \sigma_{ji}$ is a second basis over P_r of the double P-module $\mathfrak{R}$, then $\alpha y_i = \Sigma y_j(\alpha F_{ji})$ and $(\alpha F_{ij}) = \alpha^F = S^{-1}\alpha^E S$ where $S = (\sigma_{ij})$. Hence the different right bases for $\mathfrak{R}$ correspond to the different self-representations similar to E.

It is clear that an element γ of P is a fixed element under E if and only if $\gamma_l = \gamma_r$.

If $\mathfrak{S}$ is a submodule of $\mathfrak{R}$, αy and $y\alpha \in \mathfrak{S}$ for any y in $\mathfrak{S}$ and any α in P.

[5] Cf. E. Noether, " Hyperkomplexe Grössen und Darstellungstheorie," *Mathematische Zeitschrift*, vol. 30 (1929), p. 669; v. d. Waerden, *Moderne Algebra*, vol. 2, p. 131; or Jacobson, *The Theory of Rings*, New York, 1943, p. 95 (referred to hereafter as *R*).

We may choose a basis $y_1, \cdots, y_m$ of $\mathfrak{R}$ over P_r such that $y_1, \cdots, y_r$ is a basis for $\mathfrak{S}$ over P_r. Then $\alpha y_k = \Sigma y_l \alpha_{lk}$, $k, l = 1, \cdots, r$. Hence E is reducible and the representation F determined by $\mathfrak{S}$ is a subrepresentation of E. Also we have $\alpha y_p \equiv \Sigma y_q \alpha_{qp} \pmod{\mathfrak{S}}$ for $p, q = r+1, \cdots, m$. Hence the representation G determined by the difference P-module $\mathfrak{R} - \mathfrak{S}$ is a difference representation of E. In a similar manner we see that the condition that E be decomposable is that $\mathfrak{R}$ be a direct sum of submodules $\neq 0$ and the condition that E be completely reducible is that $\mathfrak{R}$ be a completely reducible double module.

3. Composites. If $\mathfrak{R}$ is a double P-module, the set of transformations that are finite sums of products $\alpha_l \beta_r$ is a ring $\mathrm{P}_l \mathrm{P}_r = \mathrm{P}_r \mathrm{P}_l$. The elements of $\mathrm{P}_l \mathrm{P}_r$ belong to $\mathfrak{L}$ the algebra of linear transformations of $\mathfrak{R}$ over P_r. Now we recall that the scalar multiplications in the algebra $\mathfrak{L}$ are the mappings $A \rightarrow A\alpha_r = \alpha_r A$ and that the dimensionality of $\mathfrak{L}$ over P_r is m^2. Since $\mathrm{P}_l \mathrm{P}_r \geqq \mathrm{P}_r$, $\mathrm{P}_l \mathrm{P}_r$ is a subalgebra of $\mathfrak{L}$. Hence $(\mathrm{P}_l \mathrm{P}_r : \mathrm{P}_r) = q \leqq m^2$.

We now define a *composite* of a field with itself as a system (K, S, T) consisting of a ring K and two isomorphisms S and T between P and subfields P^S and P^T of K such that the following conditions hold:

1. $K = \mathrm{P}^S \mathrm{P}^T$
2. $\alpha^S \beta^T = \beta^T \alpha^S$ for all α, β
3. $1^S = 1^T$
4. $(K : \mathrm{P}^T) < \infty$.

Evidently by 1. and 2., K is a commutative ring and by 1. and 3. $1^S = 1^T$ is an identity element 1 in K. We have seen that any double P-module $\mathfrak{R}$ determines a composite $(\mathrm{P}_l \mathrm{P}_r, L, R)$ where L denotes the isomorphism $\alpha \rightarrow \alpha_l$ and R denotes the isomorphism $\alpha \rightarrow \alpha_r$.

In all of our work we shall identify composites (K, S, T) and (K', S', T') that are *equivalent* in the sense that there exists an isomorphism $k \rightarrow k'$ between K and K' such that $(\alpha^S)' = \alpha^{S'}$, $(\alpha^T)' = \alpha^{T'}$. The composite (K, S, T) will be called a *cover* of (K', S', T') $((K, S, T) \geqq (K', S', T'))$ if there exists a homomorphism $k \rightarrow k'$ between K and K' such that $(\alpha^S)' = \alpha^{S'}$, $(\alpha^T)' = \alpha^{T'}$. Since $K = \mathrm{P}^S \mathrm{P}^T$, it is clear that if such a homomorphism exists, it is unique. In fact, it may be characterized by the fact that it maps $\Sigma \alpha^S \beta^T$ into $\Sigma \alpha^{S'} \beta^{T'}$. This, of course, implies that the mapping $\Sigma \alpha^S \beta^T \rightarrow \Sigma \alpha^{S'} \beta^{T'}$ is single-valued. Thus if $\Sigma \alpha^S \beta^T = 0$, then $\Sigma \alpha^{S'} \beta^{T'} = 0$. Conversely if the latter condition holds, the mapping $\Sigma \alpha^S \beta^T \rightarrow \Sigma \alpha^{S'} \beta^{T'}$ is a homomorphism and hence (K, S, T) $\geqq (K', S', T')$. The condition that (K, S, T) and (K', S', T') be equivalent is that $\Sigma\alpha^S\beta^T = 0$ holds in K if and only if $\Sigma\alpha^{S'}\beta^{T'} = 0$ holds in K'. Thus $(K, S, T) = (K', S', T')$ if and only if $(K, S, T) \geq (K', S', T')$ and $(K', S', T') \geq (K, S, T)$. If $(K, S, T) \geq (K', S', T')$ and $(K', S', T') \geq (K'', S'', T'')$, then $(K, S, T) \geq (K'', S'', T'')$. By the fundamental theorem of homomorphisms of rings any composite covered by (K, S, T) is equivalent to a composite $(K - B, \bar{S}, \bar{T})$ where B is an ideal in K and $\bar{S}$ and $\bar{T}$ are obtained from S and T by applying the natural homomorphism that maps the element k of K into its coset $\bar{k} = k + B$.

Any composite is equivalent to a composite determined by a double P-module. For we may take $\mathfrak{R}$ to be K and define $\alpha x = \alpha^S x = x\alpha^S$ and $x\alpha = \alpha^T x = x\alpha^T$ for any x in $\mathfrak{R}$ and any α in P. Then $(\mathfrak{R} : \mathrm{P}_r) = (K : \mathrm{P}^T)$ is finite. It is readily seen that the composite $(\mathfrak{R}, L, R)$ of $\mathfrak{R}$ is equivalent to (K, S, T).

If $\mathfrak{S}$ is a submodule or a difference module of a double P-module $\mathfrak{R}$, then the correspondence between $\Sigma\alpha_l\beta_r$ in $\mathfrak{R}$ and $\Sigma\alpha_l\beta_r$ in $\mathfrak{S}$ is clearly a homomorphism. Thus the composite $(\mathrm{P}_l\mathrm{P}_r, L, R; \mathfrak{R})$ associated with $\mathfrak{R}$ is a cover of the composite $(\mathrm{P}_l\mathrm{P}_r, L, R; \mathfrak{S})$. If $\mathfrak{R} = \mathfrak{R}_1 \oplus \mathfrak{R}_2$ then if $\Sigma\alpha_l\beta_r = 0$ in both $\mathfrak{R}_1$ and $\mathfrak{R}_2$, $\Sigma\alpha_l\beta_r = 0$ in $\mathfrak{R}$. It follows that $(\mathrm{P}_l\mathrm{P}_r, L, R; \mathfrak{R})$ is a *least common cover* of the composites $(\mathrm{P}_l\mathrm{P}_r, L, R; \mathfrak{R}_i)$, $i = 1, 2$, in the sense that any cover of the latter two composites is a cover of $(\mathrm{P}_l\mathrm{P}_r, L, R; \mathfrak{R})$. We may also define the least common cover $(K, S, T) + (L, U, V)$ for any two composites (K, S, T) and (L, U, V) abstractly: For let $K \oplus L$ be the direct sum of the rings K and L. The elements of $K \oplus L$ are uniquely representable in the form, $k + l$, $k \epsilon K$ and $l \epsilon L$ and $(k + l)(k' + l') = kk' + ll'$. We set $\alpha^X = \alpha^S + \alpha^U$ and $\alpha^Y = \alpha^T + \alpha^V$. Then the set of elements $\alpha^X(\alpha^Y)$ is a field $\mathrm{P}^X(\mathrm{P}^Y)$ isomorphic to P. If $k_1, \cdots, k_q$ is a basis for K over P^T and $l_1, \cdots, l_r$ is a basis for L over P^V, the elements $k_1, \cdots, k_q; l_1, \cdots, l_r$ form a basis for $K \oplus L$ over P^Y. Thus $((K + L) : \mathrm{P}^Y)$ is finite. Hence if $M = \mathrm{P}^X\mathrm{P}^Y$, we may conclude also that $(M : \mathrm{P}^Y)$ is finite. Thus (M, X, Y) is a composite of P with itself. Since $\Sigma\alpha^X\beta^Y = \Sigma(\alpha^S + \alpha^U)(\beta^T + \beta^V) = \Sigma(\alpha^S\beta^T + \alpha^U\beta^V) = 0$ if and only if $\Sigma\alpha^S\beta^T = 0$ and $\Sigma\alpha^U\beta^V = 0$, (M, X, Y) is a least common cover of (K, S, T) and (L, U, V). The least common cover is uniquely determined in the sense of equivalence. In a similar fashion we may define a least common cover for any finite number of composites.

We have seen that the fixed elements of a self-representation E are the elements γ such that $\gamma_l = \gamma_r$ in the double P-module associated with E. Evidently these elements are determined by the composite $(\mathrm{P}_l\mathrm{P}_r, L, R)$. Accordingly we define an element γ of P to be a *fixed element* of a composite $\Gamma = (K, S, T)$ if $\gamma^S = \gamma^T$. These elements form a subfield Φ_Γ of P.

4. The relations space of a composite. Let $\Gamma = (K, S, T)$ be a composite of P with itself and suppose that $(K : P^T) = q$. Since $K = P^S P^T$, there exists a basis for K over P^T consisting of elements $\alpha_1^S, \cdots, \alpha_q^S$ in P^S. For any α in P we may write

$$\alpha^S = \alpha_1^S \mu_1(\alpha)^T + \cdots + \alpha_q^S \mu_q(\alpha)^T. \tag{2}$$

Since this expression is unique, the mapping $M_i : \alpha \to \mu_i(\alpha)$ is single-valued. Since $(\alpha + \beta)^S = \alpha^S + \beta^S$, $\mu_i(\alpha + \beta)^T = \mu_i(\alpha)^T + \mu_i(\beta)^T$ and hence

$$(\alpha + \beta) M_i = \alpha M_i + \beta M_i. \tag{3}$$

If γ is a fixed element under Γ, $(\alpha\gamma)^S = \alpha^S \gamma^S = \alpha^S \gamma^T$. Hence

$$(\alpha\gamma) M_i = (\alpha M_i)\gamma. \tag{4}$$

Now suppose that

$$\alpha_i^S \alpha_j^S = \Sigma \alpha_k^S \epsilon_{kij}^T. \tag{5}$$

Then

$$\alpha^S \beta^S = (\Sigma \alpha_i^S \mu_i(\alpha)^T)(\Sigma \alpha_j^S \mu_j(\beta)^T) = \Sigma \alpha_k^S \epsilon_{kij}^T \mu_i(\alpha)^T \mu_j(\beta)^T.$$

On the other hand, $\alpha^S \beta^S = (\alpha\beta)^S = \Sigma \alpha_i^S \mu_i(\alpha\beta)^T$ so that

$$(\alpha\beta) M_i = \Sigma_{\lambda,\mu} \epsilon_{i\lambda\mu} (\alpha M_\lambda)(\beta M_\mu). \tag{6}$$

These equations may be written in a more useful form as follows: If β is any element of P, we let $\bar{\beta}$ denote the multiplication $\alpha \to \alpha\beta = \beta\alpha$. With this notation equation (4) becomes

$$\bar{\gamma} M_i = M_i \bar{\gamma} \tag{4'}$$

and (6) becomes

$$\bar{\beta} M_i = \Sigma M_\lambda \bar{\zeta}_{\lambda i} \tag{6'}$$

where $\zeta_{\lambda i} = \Sigma_\mu \epsilon_{i\lambda\mu} (\beta M_\mu)$. Equation (3) states that M_i is an endomorphism of the additive group of P. Of course, the multiplications $\bar{\beta}$ are also endomorphisms. Hence this is true for any transformation of the form $\Sigma M_i \bar{\rho}_i$, ρ_i in P.

We suppose now that $\beta_1, \cdots, \beta_q$ is a second set of elements such that

$\beta_1{}^S, \cdots, \beta_q{}^S$ is a basis for K over P^T. Then $\beta_i{}^S = \Sigma \alpha_j{}^S \rho_{ji}{}^T$ where $(\rho_{ij}) = R$ is a non-singular matrix. If $R^{-1} = (\sigma_{ij})$ and

$$\alpha^S = \beta_1{}^S \nu_1(\alpha)^T + \cdots + \beta_q{}^S \nu_q(\alpha)^T$$

then $\mu_i(\alpha)^T = \Sigma \rho_{ij}{}^T \nu_j(\alpha)^T$ and $\nu_i(\alpha)^T = \Sigma \sigma_{ij}{}^T \mu_j(\alpha)^T$. Thus

$$M_i = \Sigma N_j \bar{\rho}_{ij} \quad \text{and} \quad N_i = \Sigma M_j \bar{\sigma}_{ij}. \tag{7}$$

Hence the totality $\mathfrak{A}$ of endomorphisms (of the additive group of P) of the form $\Sigma M_i \bar{\rho}_i$ is independent of the choice of the basis. It is clear also that any two equivalent composites determine the same sets $\mathfrak{A}$. The set $\mathfrak{A}$ is evidently closed under multiplication on the right by the multiplications $\bar{\rho}$. By (6′) $\mathfrak{A}$ is also closed under multiplication by $\bar{\rho}$ on the left. We assert now that the M_i are right linearly independent relative to the $\bar{\rho}$. For by (2), $\alpha_i M_j = \delta_{ij}$. Hence $\alpha_i \Sigma M_j \bar{\rho}_j = \rho_i$ and so if $\Sigma M_j \bar{\rho}_j = 0$, $\rho_i = 0$ and $\bar{\rho}_i = 0$. We shall call $\mathfrak{A}$ the *relations space* of the composite (K, S, T) and we shall denote the set of multiplications $\bar{\rho}$ by $\bar{\mathrm{P}}$. We have therefore proved the following

THEOREM 1. *If $\mathfrak{A}$ is the relations space of the composite (K, S, T) and $\bar{\mathrm{P}}$ is the set of multiplications, then $\mathfrak{A}\bar{\mathrm{P}} \leq \mathfrak{A}$ and $\bar{\mathrm{P}}\mathfrak{A} \leq \mathfrak{A}$. The right dimensionality* $(\mathfrak{A} : \bar{\mathrm{P}})_r = (K : \mathrm{P}^T)$.[6]

5. Conditions for covering composites. If $\mathfrak{R}$ is a two-sided P-module we have defined the composite of $\mathfrak{R}$ as $(\mathrm{P}_l\mathrm{P}_r, L, R)$ where L is the correspondence $\alpha \to \alpha_l$ and R is the correspondence $\alpha \to \alpha_r$. Now let $x_1, \cdots, x_m$ be a basis for $\mathfrak{R}$ over P_r and set $\alpha x_i = \Sigma x_j(\alpha E_{ij})$. As we have seen the correspondence $\alpha \to \alpha^E = (\alpha E_{ij})$ is a self-representation. By the well-known isomorphism between linear transformations and matrices, we may substitute for α_l the matrix α^E and for α_r the matrix $\alpha^D = \{\alpha, \cdots, \alpha\}$ and obtain in this way a new composite equivalent to $(\mathrm{P}_l\mathrm{P}_r, L, R)$. We denote this composite as $(\mathrm{P}^E\mathrm{P}^D, E, D)$ where E is the mapping $\alpha \to \alpha^E$ and D is the mapping $\alpha \to \alpha^D$.

Consider now the transformations E_{ij}. By (1) E_{ij} is an endomorphism of the additive group of P and the set of E_{ij} satisfy the following conditions:

---

[6] Using (5) and (6′) one can show that the self-representation determined by the basis M_i of the double P-module $\mathfrak{A}$ is E' the transposed representation of the self-representation E determined by the basis $\alpha_1 S, \cdots, \alpha_q S$ of K. See a forthcoming paper of the author's entitled "Construction of central simple associative algebras."

$$\bar{\gamma}E_{ij} = E_{ij}\bar{\gamma} \tag{8}$$

if $\gamma \in \Phi$ the field of fixed elements and

$$\bar{\beta}E_{ij} = \Sigma E_{i\lambda}(\overline{\beta E_{\lambda j}}). \tag{9}$$

As in the preceding section we determine the transformations M_i by choosing elements $\alpha_1, \cdots, \alpha_q$ such that each $\alpha_l = \alpha_{1l}\mu_1(\alpha)_r + \cdots \alpha_{ql}\mu_q(\alpha)_r$ uniquely. Then $\alpha^E = \alpha_1{}^E\mu_1(\alpha)^D + \cdots + \alpha_q{}^E\mu_q(\alpha)^D$ and

$$\alpha E_{ij} = \mu_1(\alpha)(\alpha_1 E_{ij}) + \cdots + \mu_q(\alpha)(\alpha_q E_{ij}).$$

Thus

$$E_{ij} = M_1(\overline{\alpha_1 E_{ij}}) + M_2(\overline{\alpha_2 E_{ij}}) + \cdots + M_q(\overline{\alpha_q E_{ij}}) \tag{10}$$

is in the relations space $\mathfrak{A}$ of our composite. It follows that the set of endomorphisms of the form $\Sigma E_{ij}\bar{\rho}_{ij}$, where the $\bar{\rho}$ are arbitrary multiplications, is a subspace $\mathfrak{B}$ of $\mathfrak{A}$ regarded relative to $\bar{\mathrm{P}}$ on the right. We assert that $\mathfrak{B} = \mathfrak{A}$. For otherwise there exist $p < q$ endomorphisms $N_1, \cdots, N_p$ in $\mathfrak{A}$ such that

$$E_{ij} = N_1\bar{\sigma}_{ij1} + \cdots + N_p\bar{\sigma}_{ijp},$$

or $\alpha E_{ij} = (\alpha N_1)\sigma_{ij1} + \cdots + (\alpha N_p)\sigma_{ijp}$. Thus each matrix α^E is a linear combination of the p matrices $\sigma_k = (\sigma_{ijk})$, $k = 1, \cdots, p$, with coefficients $\alpha N_k{}^D$ and this contradicts the fact that $\alpha_1{}^E, \cdots, \alpha_q{}^E$ are linearly independent over P^D. This proves

THEOREM 2. *If E_{ij} are the E's determined by a self-representation E, then the set of endomorphisms $\Sigma E_{ij}\bar{\rho}_{ij}$ is the complete relations space $\mathfrak{A}$ of the composite associated with E.*

Suppose now that $\Gamma = (K, S, T)$ is a cover of the composite $\Gamma' = (K', S', T')$. We determine the M_i for K by equation (2). Then

$$\alpha^{S'} = \alpha_1{}^{S'}\mu_1(\alpha)^{T'} + \cdots + \alpha_q{}^{S'}\mu_q(\alpha)^{T'}$$

and every element of K' is a linear combination of the $\alpha_i{}^{S'}$ with coefficients in $\mathrm{P}^{T'}$. Now let $\beta_1, \cdots, \beta_p$, $p \leq q$ be elements of P such that the $\beta_k{}^{S'}$ form a basis over $\mathrm{P}^{T'}$ of K'. Then we have $\alpha_i{}^{S'} = \Sigma\beta_k{}^{S'}\rho_{ki}{}^{T'}$. Hence

$$\alpha^{S'} = \beta_1{}^{S'}\nu_1(\alpha)^{T'} + \cdots + \beta_p{}^{S'}\nu_p(\alpha)^{T'}$$

where $\nu_k(\alpha)^{T'} = \Sigma\mu_i(\alpha)^{T'}\rho_{ki}{}^{T'}$. The mappings N_k: $\alpha \to \nu_k(\alpha)$ form a basis for the relations space $\mathfrak{A}(\Gamma')$ of Γ' and we have shown that $N_k = \Sigma M_i \bar{\rho}_{ki}$. Thus $\mathfrak{A}(\Gamma') \leq \mathfrak{A}(\Gamma)$.

Now suppose conversely that $\mathfrak{A}(\Gamma') \leq \mathfrak{A}(\Gamma)$. Let $\mathfrak{R}$ and $\mathfrak{S}$ be double P-modules that have composites equivalent to (K, S, T) and (K', S', T') respectively. Let E be a self-representation determined by $\mathfrak{R}$ and F one determined by $\mathfrak{S}$. Since $\mathfrak{A}(\Gamma') \leq \mathfrak{A}(\Gamma)$ it follows from Theorem 2 that there exist elements $\rho_{ij,kl}$ such that $F_{kl} = \Sigma E_{ij} \bar{\rho}_{ji,kl}$ for all k and l. These equations may be written in the form

$$\alpha F_{kl} = \mathrm{tr}(\alpha^E \rho_{kl}) \tag{11}$$

where ρ_{kl} is the matrix $(\rho_{ij,kl})$. Now suppose that we have a relation of the form $\Sigma \alpha^E \beta^D = 0$ where $\beta^D = \{\beta, \cdots, \beta\}$. Then it follows readily from (11) that $\Sigma \alpha^F \beta^D = 0$. Thus (K, S, T) is a cover for (K', S', T'). This completes the proof of the following important

THEOREM 3. *A necessary and sufficient condition that the composite* $\Gamma \geq \Gamma'$ *is that the relations space* $\mathfrak{A}(\Gamma) \geq \mathfrak{A}(\Gamma')$.

This implies

THEOREM 4. *A necessary and sufficient condition that* $\Gamma = \Gamma'$ *is that* $\mathfrak{A}(\Gamma) = \mathfrak{A}(\Gamma')$.

6. Non-singular composites. If (K, S, T) is a composite such that $(K : \mathrm{P}^S) < \infty$, it defines a new composite (K, T, S). It may happen that both $(K : \mathrm{P}^S)$ and $(K : \mathrm{P}^T)$ are finite but that these dimensionalities are not equal. For example let $K = \mathrm{P} = \Phi(t)$ the field of rational functions in one indeterminate over Φ. Let T be the identity and S the automorphism defined by $t^S = t^2$. Then the self-representation determined by K is the isomorphism $\alpha(t) \to \alpha(t^2)$ that maps P into a proper subfield. We may avoid this type of situation by restricting our attention to composites (K, S, T) that are *non-singular* in the sense that $(K : \mathrm{P}^S) = (K : \mathrm{P}^T)$. If (K, S, T) is non-singular, the composite (K, T, S) will be called the *inverse* of (K, S, T). If (K, S, T) is equivalent to (K, T, S) we shall call this composite *symmetric.*

Corresponding to the concept of non-singular composite, we define a double P-module $\mathfrak{R}$ to be *non-singular* if $(\mathfrak{R} : \mathrm{P}_l) = (\mathfrak{R} : \mathrm{P}_r)$. Suppose that $\mathfrak{R}$ has this property. We prove first the following

LEMMA. *There exists a set of elements* $y_1, \cdots, y_m$ *that constitute both a left and a right basis for* $\mathfrak{R}$ *over* P.

The element y_1 may be taken to be any element $\neq 0$ in $\mathfrak{R}$. Then $y_1\alpha \neq 0$ and $\alpha y_1 \neq 0$ for all $\alpha \neq 0$. Now suppose that $y_1, \cdots, y_r$ are elements of $\mathfrak{R}$ which are left linearly independent and right linearly independent over P. Then if $r < m = (\mathfrak{R} : \mathrm{P}_l) = (\mathfrak{R} : \mathrm{P}_r)$, there is an element y not of the form $\Sigma\alpha_i y_i$. If y does not have the form $\Sigma y_i \alpha'_i$ either, y may be taken to be the element y_{m+1}. Hence suppose that $y = \Sigma y_i \alpha'_i$. Similarly we choose an element z not of the form $\Sigma y_i \beta_i$ and we may suppose that $z = \Sigma\beta'_i y_i$. Now form $w = y + z$. Then if $w = \Sigma y_i \alpha_i$, $z = w - y = \Sigma y_i(\alpha_i - \alpha'_i)$ contrary to assumption. Similarly, w cannot be represented in the form $\Sigma\beta_i y_i$. Hence we may take $y_{m+1} = w$. This process leads to a basis of the required type.

The same method may be used to show that if $\mathfrak{S}$ is a non-singular submodule of $\mathfrak{R}$, there exists a two-sided basis for $\mathfrak{R}$ that includes a two-sided basis for $\mathfrak{S}$.

Now if $\mathfrak{R}$ is a non-singular double P-module, we shall define the inverse $\mathfrak{R}^{-1}$ of $\mathfrak{R}$ to be the module whose elements may be put in $(1-1)$ correspondence with the elements of $\mathfrak{R}$ in such a way that if $x \to x'$ in $\mathfrak{R}^{-1}$,

$$
\begin{aligned}
(x + y)' &= x' + y' \\
(\alpha x)' &= x'\alpha \\
(x\alpha)' &= \alpha x'.
\end{aligned} \tag{12}
$$

Thus $\mathfrak{R}^{-1}$ is obtained from $\mathfrak{R}$ by interchanging the roles of P_l and P_r. If $x_1, \cdots, x_m$ is a left (right) basis for $\mathfrak{R}$, then $x'_1, \cdots, x'_m$ is a right (left) basis for $\mathfrak{R}^{-1}$. Let $y_1, \cdots, y_m$ be a two-sided basis for $\mathfrak{R}$ and $\alpha y_i = \Sigma y_j(\alpha E_{ji})$, $y_i\alpha = \Sigma(\alpha E^*_{ji})y_j$. Then we recall that the correspondence $\alpha \to \alpha^E = (\alpha E_{ij})$ is a self-representation determined by $\mathfrak{R}$. In $\mathfrak{R}^{-1}$ we have $\alpha y'_i = \Sigma y'_j(\alpha E^*_{ji})$ and $y'_i\alpha = \Sigma(\alpha E_{ji})y'_j$. Thus $\alpha \to \alpha^{E^*} = (\alpha E^*_{ij})$ is the self-representation associated with $\mathfrak{R}^{-1}$. We note that

$$
\begin{aligned}
\alpha y_i &= \Sigma y_j(\alpha E_{ji}) = \Sigma(\alpha E_{ji}E^*_{kj})y_k \\
y_i\alpha &= \Sigma(\alpha E^*_{ji})y_j = \Sigma y_k(\alpha E^*_{ji}E_{kj}).
\end{aligned}
$$

Hence $\alpha\delta_{ik} = \Sigma\alpha E_{ji}E^*_{kj}$ and $\alpha\delta_{ik} = \Sigma\alpha E^*_{ji}E_{kj}$ or

$$
\Sigma E_{ji}E^*_{kj} = \delta_{ik}, \qquad \Sigma E^*_{ji}E_{kj} = \delta_{ik}. \tag{13}
$$

These equations may be written in a simpler form by introducing the ring $\mathfrak{E}_m$ of m-rowed matrices with elements in the ring $\mathfrak{E}$ of endomorphisms of the additive group of P. For we set $(E) = (E_{ij})$ and call this matrix in $\mathfrak{E}_m$ the matrix of the *self-representation*. If $(E^*) = (E^*_{ij})$ then (13) may be replaced by the single equation

(14) $$(E)'(E^*)' = 1 = (E^*)'(E)'$$

where the prime denotes the transposed matrix. The representation $\alpha \to (\alpha E^*_{ij})$ will be called the *inverse* E^{-1} *of the representation* E.

If (K, S, T) is a non-singular composite, we have seen that we may take K to be a double P-module by setting $\alpha x = \alpha^S x = x\alpha^S$ and $x\alpha = \alpha^T x = x\alpha^T$. The dimensionality $(\mathfrak{R}: \mathrm{P}_l) = (K: \mathrm{P}^S) = (K: \mathrm{P}^T) = (\mathfrak{R}: \mathrm{P}_r)$. Hence $\mathfrak{R}$ is non-singular and its inverse is $\mathfrak{R}^{-1} = (K, T, S)$ with P_l here $= \mathrm{P}^T$ and $\mathrm{P}_r = \mathrm{P}^S$. If we choose two-sided bases in these modules we obtain a pair of inverse self-representations of P.

If (K, S, T) and (L, U, V) are non-singular composites, their least common cover (M, X, Y) is also non-singular. For, let $\mathfrak{R}$ be a double P-module such that $\mathfrak{R} = \mathfrak{R}_1 \oplus \mathfrak{R}_2$ where the $\mathfrak{R}_i$ are non-singular submodules and $\mathfrak{R}_1$ has the composite (K, S, T) and $\mathfrak{R}_2$ has the composite (L, U, V). Then $(\mathfrak{R}: \mathrm{P}_l) = (\mathfrak{R}_1: \mathrm{P}_l) + (\mathfrak{R}_2: \mathrm{P}_l) = (\mathfrak{R}_1: \mathrm{P}_r) + (\mathfrak{R}_2: \mathrm{P}_r) = (\mathfrak{R}: \mathrm{P}_r)$ and so $\mathfrak{R}$ is non-singular. Thus (M, X, Y) is non-singular. Its inverse (M, Y, X) is evidently the least common cover of (K, T, S) and (L, V, U). It follows from this that if (K, S, T) is an arbitrary non-singular composite, then $(K, S, T) + (K, T, S)$ is a symmetric composite.

7. The product of self-representations. If E and F are two self-representations of rank m and r respectively, they determine a self-representation of rank mr obtained by substituting for the elements αE_{ij} of α^E the matrices $(\alpha E_{ij})^F$ that represent these elements under F. We shall call this representation the *product* $E \times F$ *of the representations* E and F. The product $E \times F$ is an associative one. Moreover if 1 denotes the identity self-representation $\alpha \to (\alpha)$ of one row, then $E \times 1 = E = 1 \times E$ for all E. Hence the set of self-representations forms a semi-group with an identity relative to our composition. If (E) is the matrix of E and (F) is the matrix of F, then the matrix (P) of $P = E \times F$ is the (right) direct product $(E) \times (F)$, i. e. $P_{(i-1)r+k,\, (j-1)r+l} = E_{ij}F_{kl}$.

We shall define next a product of double P-modules that corresponds to the product of self-representations. Let $\mathfrak{R}$ and $\mathfrak{S}$ be two double P-modules. We wish to construct a double P-module $\mathfrak{P}$ and a function xy of x in $\mathfrak{R}$ and of y in $\mathfrak{S}$ such that the following conditions obtain:

1. $\alpha(xy) = (\alpha x)y$
2. $(xy)\alpha = x(y\alpha)$
3. $(x\alpha)y = x(\alpha y)$
4. $x(y + y') = xy + xy'$

5. $(x+x')y = xy + x'y$
6. Any element of $\mathfrak{P}$ has the form Σxy for suitable x in $\mathfrak{R}$ and y in $\mathfrak{S}$.
7. $(\mathfrak{P}: \mathrm{P}_r) = (\mathfrak{R}: \mathrm{P}_r)(\mathfrak{S}: \mathrm{P}_r)$.

Let $x_1, \cdots, x_m$ be a right basis for $\mathfrak{R}$ and $\alpha x_i = \Sigma x_j(\alpha E_{ji})$ and let $y_1, \cdots, y_r$ be a right basis for $\mathfrak{S}$ and $\alpha y_k = \Sigma y_l(\alpha F_{lk})$. Then if $x = \Sigma x_i \xi_i$ and $y = \Sigma y_k \eta_k$ conditions 1.-5. imply that

$$xy = \Sigma x_j y_l (\xi_j F_{lk}) \eta_k. \tag{15}$$

By 6. every element of $\mathfrak{P}$ is a linear combination of the elements $x_i y_k$. Hence by 7. these elements form a right basis for $\mathfrak{P}$ over P. The procedure for defining $\mathfrak{P}$ and xy is now clear. We let $\mathfrak{P}$ be a right module of dimensionality mr over P and define xy by (15) where $x_i y_k = z_{(i-1)r+k}$ form a basis for $\mathfrak{P}$. Then 2., 3., 4., 5. and 6. are valid. Next we define α_l to be the linear transformation in $\mathfrak{P}$ over P_r that sends $x_i y_k$ into $\Sigma x_j y_l (\alpha E_{ji} F_{lk})$. We set $\alpha z = \alpha_l z$ and we observe that $\alpha(xy) = \Sigma x_\lambda y_\sigma (\alpha E_{\lambda j} F_{\sigma l})(\xi_j F_{lk})\eta_k$. We may now verify that 1. holds and hence $\mathfrak{P}$ is a double P-module and xy is a function satisfying our conditions.

If $x'_1, \cdots, x'_m$ and $y'_1, \cdots, y'_r$, respectively, form a second pair of right bases for $\mathfrak{R}$ and for $\mathfrak{S}$, $x = \Sigma x'_i \xi'_i$ and $y = \Sigma y'_k \eta'_k$. Hence $xy = \Sigma x'_j y'_l (\alpha F'_{lk}) \eta'_k$. Thus all the xy are linear combinations of the elements $x'_i y'_k$ and so these elements form a second right basis for $\mathfrak{P}$. Now suppose that we construct a second space $\mathfrak{P}'$ and a second function xy which we shall denote as $[xy]$ by using the bases x'_i and y'_k in place of the x_i and the y_k. Then it is readily verified that the mapping $\Sigma x'_i y'_k \xi_{ik} \to \Sigma [x'_i y'_k] \xi_{ik}$ is an isomorphism between the double P-modules $\mathfrak{P}$ and $\mathfrak{P}'$. In this sense the module $\mathfrak{P}$ does not depend on the choice of the bases x_i and y_k. We shall therefore denote this module as $\mathfrak{R} \times \mathfrak{S}$ and shall call it the *product of the modules* $\mathfrak{R}$ and $\mathfrak{S}$.

By definition

$$\alpha(x_i y_k) = \Sigma x_j y_l (\alpha E_{ji} F_{lk}).$$

Hence the representation determined by the basis z_s, $z_{(i-1)r+k} = x_i y_k$ is the product $E \times F$. The independence of $\mathfrak{P} = \mathfrak{R} \times \mathfrak{S}$ of the bases of $\mathfrak{R}$ and $\mathfrak{S}$ may be stated as

THEOREM 5. *If E' and F' are self-representations similar respectively to E and F, then $E' \times F'$ is similar to $E \times F$.*

If $x'_1, \cdots, x'_p$ are right linearly independent in $\mathfrak{R}$ and $y'_1, \cdots, y'_q$ are right linearly independent in $\mathfrak{S}$ then the elements $x'_j y'_l$ are right linearly independent in $\mathfrak{P}$. For we may supplement the x' and the y' to obtain bases $x'_1, \cdots, x'_m$; $y'_1, \cdots, y'_r$. Then we have seen that the elements $x'_i y'_k$ are linearly independent. Now suppose that the elements $x'_1, \cdots, x'_p$ form a right basis for a submodule $\mathfrak{R}'$ of $\mathfrak{R}$ and the elements $y'_1, \cdots, y'_q$ form a right basis for a submodule $\mathfrak{S}'$ of $\mathfrak{S}$. Then the elements $x'_j y'_l$ form a basis for a submodule $\mathfrak{P}'$ of $\mathfrak{P}$ and $\mathfrak{P}'$ is essentially the module $\mathfrak{R}' \times \mathfrak{S}'$. This proves

THEOREM 6. *If E' is a subrepresentation of E and F' is a subrepresentation of F, then $E' \times F'$ is a subrepresentation of $E \times F$.*

In a similar manner we may prove

THEOREM 7. *If E is decomposable into the components E_i and F is decomposable into the components F_j, then $E \times F$ is decomposable into the components $E_i \times F_j$.*

We suppose now that $\mathfrak{R}$ and $\mathfrak{S}$ are non-singular double P-modules and that the x_i and y_k are two-sided bases. Let the elements x'_i and y'_k form the corresponding two-sided bases for $\mathfrak{R}^{-1}$ and $\mathfrak{S}^{-1}$. If $\alpha x_i = \Sigma x_j(\alpha E_{ji})$, $\alpha x'_i = \Sigma x'_j(\alpha E^*_{ji})$ where the self-representation E^* thus determined is the inverse of E. Similarly $\alpha y_k = \Sigma y_l(\alpha F_{lk})$ and $\alpha y'_k = \Sigma y'_l(\alpha F^*_{lk})$ where F^* is the inverse of F. Now $x_i \alpha = \Sigma(\alpha E^*_{ji})x_j$ and $y_k \alpha = \Sigma(\alpha F^*_{lk})y_l$ and so $(x_i y_k)\alpha = \Sigma(\alpha F^*_{lk}E^*_{ji})x_j y_l$. Hence every element of $\mathfrak{R} \times \mathfrak{S}$ is a left linear combination of the elements $x_i y_k$. These form a basis. For if $\Sigma \mu_{ik} x_i y_k = 0$, $\Sigma x_j y_l \mu_{ik} E_{ji} F_{lk} = 0$ and hence $\Sigma \mu_{ik} E_{ji} F_{lk} = 0$ for all j and l. It follows that $\mu_{i'k'} = \Sigma_{l,j} \mu_{ik} E_{ji} F_{lk} F^*_{k'l} E^*_{i'j} = 0$. We have therefore proved that $\mathfrak{R} \times \mathfrak{S}$ is non-singular and that the elements $x_i y_k$ form a two-sided basis for this module. The elements $(x_i y_k)'$ form a two-sided basis for the inverse module and we have $\alpha(x_i y_k)' = \Sigma(x_j y_l)'(\alpha F^*_{lk} E^*_{ji})$. Our argument shows also that the elements $y'_k x'_i$ form a two-sided basis for $\mathfrak{S}^{-1} \times \mathfrak{R}^{-1}$ and we may verify that $\alpha y'_k x'_i = \Sigma y'_l x'_j(\alpha F^*_{lk} E^*_{ji})$. Hence the modules $(\mathfrak{R} \times \mathfrak{S})^{-1}$ and $\mathfrak{S}^{-1} \times \mathfrak{R}^{-1}$ are isomorphic.

8. The product of composites. It is easy to see from Theorems 2 and 4 that the composite determined by $\mathfrak{R} \times \mathfrak{S}$ depends only on the composite of $\mathfrak{R}$ and the composite of $\mathfrak{S}$. For the composite of $\mathfrak{R} \times \mathfrak{S}$ has as its relations space the smallest subspace over $\bar{\text{P}}$ (on the right) containing all of the products MN where M is in the relations space of the composite of $\mathfrak{R}$ and N is in the relations space of the composite of $\mathfrak{S}$. Thus the composite of $\mathfrak{R} \times \mathfrak{S}$ may be regarded as the product of the composite of $\mathfrak{R}$ by the composite of $\mathfrak{S}$. It is of interest to give a direct definition of the product of composites. We shall obtain such a definition by introducing a type of three-fold composite of P with itself. Aside from its use in the present connection, the concept of a three-fold composite will play an important role in the proof in **9** of one of the main results of the present theory.

We define first a *three-fold composite* of P with itself as a system (Δ, A, B, C) consisting of a ring Δ and the isomorphisms A, B, C of P into subfields P^A, P^B, P^C of Δ such that

1. $\Delta = P^A P^B P^C$
2. Δ is commutative
3. $1^A = 1^B = 1^C$.

Evidently by 1. and 3., $1^A = 1^B = 1^C$ is the identity 1 of Δ.

Covers of three-fold composites are defined in the obvious way: (Δ, A, B, C) is a cover of (Δ', A', B', C') if there is a homomorphism $a \to a'$ between Δ and Δ' such that $(\alpha^A)' = \alpha^{A'}$, $(\alpha^B)' = \alpha^{B'}$, $(\alpha^C)' = \alpha^{C'}$. Equivalence is defined in a similar manner.

Now suppose that (K, S, T) and (L, U, V) are arbitrary (two-fold) composites. We wish to construct a three-fold composite (Δ, A, B, C) having the following properties:

4. $(P^A P^B, A, B)$ is equivalent to (K, S, T) and $(P^B P^C, B, C)$ is equivalent to (L, U, V).
5. $(\Delta : P^C) = (K : P^T)(L : P^V)$.

We shall show first that any two three-fold composites having these properties are equivalent. For this purpose let $\alpha_1, \cdots, \alpha_q$ be a set of elements of P such that $\alpha_1{}^S, \cdots, \alpha_q{}^S$ form a basis for K over P^T and let $\beta_1, \cdots, \beta_{q'}$ be elements such that $\beta_1{}^U, \cdots, \beta_{q'}{}^U$ form a basis for L over P^V. As before, we determine the endomorphisms $\alpha \to \mu_i(\alpha) \equiv \alpha M_i$ and $\alpha \to \nu_j(\alpha) \equiv \alpha N_j$ by writing

$$\alpha^S = \alpha_1{}^S \mu_1(\alpha)^T + \cdots + \alpha_q{}^S \mu_q(\alpha)^T$$
$$\alpha^U = \beta_1{}^U \nu_1(\alpha)^V + \cdots + \beta_{q'}{}^T \nu_{q'}(\alpha)^V.$$

Suppose also that

$$\alpha_i{}^S \alpha_{i'}{}^S = \Sigma \alpha_\lambda{}^S \epsilon_{\lambda i i'}{}^T, \qquad \beta_j{}^U \beta_{j'}{}^U = \Sigma \beta_\nu{}^U \eta_{\nu j j'}{}^V.$$

In Δ we have $\alpha^A = \Sigma \alpha_i{}^A \mu_i(\alpha)^B$, $\alpha^B = \Sigma \beta_j{}^B \nu_j(\alpha)^C$ and $\alpha_i{}^A \alpha_{i'}{}^A = \Sigma \alpha_\lambda{}^A \epsilon_{\lambda i i'}{}^B$, $\beta_j{}^B \beta_{j'}{}^B = \Sigma \beta_\nu{}^B \eta_{\nu j j'}{}^C$. Hence

$$(16) \quad \alpha_i{}^A \alpha_{i'}{}^A = \Sigma \alpha_\lambda{}^A \epsilon_{\lambda i i'}{}^B = \Sigma \alpha_\lambda{}^A \beta_j{}^B \nu_j(\epsilon_{\lambda i i'})^C, \quad \beta_j{}^B \beta_{j'}{}^B = \Sigma \beta_\nu{}^B \eta_{\nu j j'}{}^C.$$

Let $\bar{\Delta}$ denote the totality of elements of the form $\Sigma \alpha_i{}^A \beta_j{}^B \zeta_{ij}{}^C$. By (16), $\bar{\Delta}$ is a subring of Δ. Since

$$(17) \qquad \alpha^A = \Sigma \alpha_i{}^A \mu_i(\alpha)^B = \Sigma \alpha_i{}^A \beta_j{}^B \nu_j(\mu_i(\alpha))^C,$$

$P^A \leq \bar{\Delta}$. Hence 1 and $P^C \leqq \bar{\Delta}$. Moreover, $\beta_j{}^B = (\alpha_i{}^A)^{-1}(\alpha_i{}^A \beta_j{}^B) \in \bar{\Delta}$ and so each $\alpha^B = \Sigma \beta_j{}^B \nu_j(\alpha)^C$ is in $\bar{\Delta}$. Thus $\bar{\Delta} = \Delta$ and every element of Δ is a linear combination with coefficients in P^C of the qq' elements $\alpha_i{}^A \beta_j{}^B$. Hence by 5. the elements $\alpha_i{}^A \beta_j{}^B$ form a basis for Δ over P^C.

Now suppose that (Δ', A', B', C') is a second three-fold composite having the properties 4. and 5. Then the mapping $\Sigma \alpha_i{}^A \beta_j{}^B \zeta_{ij}{}^C \rightarrow \Sigma \alpha_i{}^{A'} \beta_j{}^{B'} \zeta_{ij}{}^{C'}$ is clearly $(1-1)$ and by (16), it is an isomorphism. It is evident from the above considerations that this isomorphism maps α^A into $\alpha^{A'}$, α^B into $\alpha^{B'}$ and α^C into $\alpha^{C'}$.

We now construct the composite (Δ, A, B, C). Suppose that the elements $\alpha_i{}^S$ and $\beta_j{}^U$ have been chosen in the way indicated above. We shall suppose in addition that $\alpha_1{}^S = 1^S$ and $\beta_1{}^U = 1^U$. Now let P^C be a field isomorphic to P under the isomorphism $\alpha \rightarrow \alpha^C$ and let Δ be the algebra over P^C with the basis $\alpha_i{}^A \beta_j{}^B$ and the multiplication table

$$(18) \qquad (\alpha_i{}^A \beta_j{}^B)(\alpha_{i'}{}^A \beta_{j'}{}^B) = \Sigma \alpha_\lambda{}^A \beta_\mu{}^B \eta_{\mu k \nu}{}^C \eta_{\nu j j'}{}^C \nu_k(\epsilon_{\lambda i i'})^C.$$

We shall show that Δ is associative and that Δ determines a composite satisfying our conditions. We note first that $\alpha_1{}^A \beta_1{}^B$ acts as the identity 1 of Δ. The elements $\beta_j{}^B \equiv \alpha_1{}^A \beta_j{}^B$ satisfy the multiplication table $\beta_j{}^B \beta_{j'}{}^B = \Sigma \beta_\nu{}^B \eta_{\nu j j'}{}^C$ and these elements are linearly independent over P^C. It follows that the set of elements $\alpha^B = \Sigma \beta_j{}^B \nu_j(\alpha)^C$ is a subfield P^B of Δ isomorphic under the correspondence $\alpha^B \rightarrow \alpha$ to P. The totality of elements $\Sigma \beta_j{}^B \rho_j{}^C$ where the ρ_j are arbitrary defines a composite $(P^B P^C, B, C)$ equivalent to (L, U, V). We note next that if we set $\alpha_i{}^A \equiv \alpha_i{}^A \beta_1{}^B$ the product of this element with $\beta_j{}^B$ (in either order) is the element $\alpha_i{}^A \beta_j{}^B$ of Δ. The elements $\alpha_i{}^A$ satisfy the relations

$$\alpha_i{}^A \alpha_{i'}{}^A = \Sigma \alpha_\lambda{}^A \beta_\mu{}^B \nu_\mu(\epsilon_{\lambda i i'})^C = \Sigma \alpha_\lambda{}^A \epsilon_{\lambda i i'}{}^B$$

and these elements are linearly independent over P^B. If we make use of equations for the ν_k analogous to (6), we may prove that $(\alpha_i{}^A \rho^B)(\alpha_{i'}{}^A \sigma^B) = \Sigma \alpha_\lambda{}^A \epsilon_{\lambda i i'}{}^B \rho^B \sigma^B$. It follows that the totality of elements of the form $\Sigma \alpha_i{}^A \rho_i{}^B$ is a ring isomorphic to K. The set of elements $\alpha^A = \Sigma \alpha_i{}^A \mu_i(\alpha)^B$ is a field P^A isomorphic under the correspondence $\alpha^A \rightarrow \alpha$ to P and the set of elements $\Sigma \alpha_i{}^A \rho_i{}^B$ determines a composite $(P^A P^B, A, B)$ equivalent to (K, S, T). Now the elements of Δ may be written in one and only one way in the form $\Sigma \alpha_i{}^A l_i$ where $l_i \in P^B P^C$. We may prove that $(\alpha_i{}^A l)(\alpha_{i'}{}^A m) = \Sigma \alpha_\lambda{}^A (\epsilon_{\lambda i i'}{}^B lm)$ and this implies that Δ is associative. Since Δ is generated by P^A, P^B and

$\mathrm{P}^C \equiv 1\mathrm{P}^C$, Δ is commutative. Conditions 4. and 5. are evidently satisfied. Hence (Δ, A, B, C) is the required three-fold composite.

We now define the *product of the composites* $(K, S, T) \times (L, U, V)$ as the composite $(\mathrm{P}^A\mathrm{P}^C, A, C)$ determined from the three-fold composite (Δ, A, B, C) that we have constructed. This product is uniquely determined. It can be seen that if (K, S, T) is a cover of (K', S', T') and (L, U, V) is a cover of (L', U', V'), then the three-fold composite (Δ, A, B, C) is a cover of the three-fold composite (Δ', A', B', C') constructed from (K', S', T') and (L', U', V'). It follows that $(K, S, T) \times (L, U, V)$ is a cover of $(K', S', T') \times (L', U', V')$. If (K, S, T) and (L, U, V) are non-singular, the three-fold composite (Δ, C, B, A) has the same relation to (L, V, U) and (K, T, S) as (Δ, A, B, C) has to (K, S, T) and (L, U, V). Hence $(L, V, U) \times (K, T, S) = (\mathrm{P}^A\mathrm{P}^C, C, A)$ and the latter is the inverse of $(\mathrm{P}^A\mathrm{P}^C, A, C)$. It is not difficult to give a direct proof of the associative law for multiplication of composites. For this purpose it is necessary to construct of four-fold composite (E, A, B, C, D) such that $(\mathrm{P}^A\mathrm{P}^B\mathrm{P}^C, A, B, C)$ is equivalent to (Δ, A, B, C) and $(\mathrm{P}^B\mathrm{P}^C\mathrm{P}^D, B, C, D)$ is equivalent to the composite $(\Delta_1, A_1, B_1, C_1)$ determined by (L, U, V) and (M, X, Y). We shall not carry through this proof but instead we shall obtain the associative law indirectly by determining the relations space of the product of the composites (K, S, T) and (L, U, V).

Let $\delta_1{}^A, \cdots, \delta_r{}^A$ be a basis for $\mathrm{P}^A\mathrm{P}^C$ over P^C and suppose that

$$\alpha^A = \delta_1{}^A \pi_1(\alpha)^C + \cdots + \delta_r{}^A \pi_r(\alpha)^C. \tag{19}$$

Then the endomorphisms $\alpha \to \pi_k(\alpha) \equiv \alpha P_k$ form a basis over $\overline{\mathrm{P}}$ of the relations space of the composite $(\mathrm{P}^A\mathrm{P}^C, A, C)$. Since $\delta_k{}^A = \Sigma \alpha_i{}^A \mu_i(\delta_k)^B$, $\alpha^A = \Sigma \alpha_i{}^A \mu_i(\delta_k)^B \pi_k(\alpha)^C$ and since $\mu_i(\delta_k)^B = \Sigma \beta_j{}^B \nu_j(\mu_i(\delta_k))^C$,

$$\alpha^A = \Sigma \alpha_i{}^A \beta_j{}^B \nu_j(\mu_i(\delta_k))^C \pi_k(\alpha)^C$$

Hence by (17) $\nu_j(\mu_i(\alpha)) = \Sigma \nu_j(\mu_i(\delta_k)) \pi_k(\alpha)$, or

$$M_i N_j = \Sigma P_k \overline{\nu_j(\mu_i(\delta_k))}$$

is in the relations space of $(\mathrm{P}^A\mathrm{P}^C, A, C)$.

Now let $R_1, \cdots, R_s$ be a basis for the space over $\overline{\mathrm{P}}$ (on the right) generated by the product $M_i N_j$ and write $\alpha R_k = \rho_k(\alpha)$. Then by expressing the elements $M_i N_j$ in terms of the R_k we may replace the relations $\alpha^A = \Sigma \alpha_i{}^A \beta_j{}^B \nu_j(\mu_i(\alpha))^C$ by

$$\alpha^A = \sum_1^s \zeta_k \rho_k(\alpha)^C \tag{20}$$

where the $\zeta_k \in \Delta$. We wish to show that the $\zeta_k \in \mathrm{P}^A\mathrm{P}^C$. For this purpose we require the

2

LEMMA. *If $R_1, \cdots, R_s$ are endomorphisms in P that are linearly independent over $\bar{\mathrm{P}}$, then there exist elements λ_k, $k = 1, \cdots, s$ such that the matrix $(\lambda_k R_l)$ is non-singular.*

Suppose that we have already determined r elements λ_k such that the r vectors $(\lambda_k R_1, \cdots, \lambda_k R_s)$ are linearly independent. Then we assert that if $r < s$, there exists an element λ_{r+1} such that the $(r+1)$ vectors $(\lambda_k R_l)$ are linearly independent. For otherwise for each λ we have

$$\lambda R_l = \lambda_1 R_l \sigma_1(\lambda) + \cdots + \lambda_r R_l \sigma_r(\lambda), \qquad (l = 1, \cdots, s). \tag{21}$$

If $\Sigma \lambda_k R_l \sigma_k = 0$ for all l, by the linear independence of the vectors $(\lambda_k R_l)$ for $k = 1, \cdots, r$, each $\sigma_k = 0$. It follows that the σ in (21) are uniquely determined and so the correspondence $\lambda \rightarrow \sigma_k(\lambda) \equiv \lambda S_k$ is single valued. It is clear that S_k is an endomorphism of P and the relation (21) states that each R_l is a linear combination with coefficients in $\bar{\mathrm{P}}$ of the r endomorphisms S_k. This contradicts the linear independence of the R's and proves the lemma.

Now choose the λ's as in the lemma and write $\lambda_l{}^A = \Sigma \zeta_k \rho_k(\lambda_l)^C$. Since $(\rho_k(\lambda_l))$ is non-singular, there exist $\tau_{lk'}$ such that $\Sigma \rho_k(\lambda_l)^C \tau_{lk'}{}^C = \delta_{kk'}$. Then $\zeta_{k'} = \Sigma \lambda_l{}^A \tau_{lk'}{}^C$ is in $\mathrm{P}^A \mathrm{P}^C$. We now write $\zeta_k = \Sigma \delta_l{}^A \beta_{lk}{}^C$ and substitute in (20). This yields $\alpha^A = \Sigma \delta_l{}^A \beta_{lk}{}^C \rho_k(\alpha)^C$ and so by (19) $\pi_l(\alpha) = \Sigma \beta_{lk} \rho_k(\alpha)$. Thus $P_l = \Sigma R_k \beta_{lk}$ is in the space generated by the products $M_i N_j$. This completes the proof of

THEOREM 8. *The relations space of a product of two composites is the smallest space over $\bar{\mathrm{P}}$ containing all products. MN where M is in the relations space of the first composite and N is in the relations space of the second composite.*

This along with Theorem 4 proves that multiplication of composites is associative. Now it is readily seen that the relations space of the least common cover $(K, S, T) + (L, U, V)$ is the join of their respective spaces. It follows from Theorem 8 that the distributive laws hold: $[(K, S, T) + (L, U, V)] \times (M, X, Y) = (K, S, T) \times (M, X, Y) + (L, U, V) \times (M, X, Y)$ and $(M, X, Y) \times [(K, S, T) + (L, U, V)] = (M, X, Y) \times (K, S, T) + (M, X, Y) \times (L, U, V)$. We note also that any two composites (P^U, U, U) are identical and (P^U, U, U) satisfies the equation $(K, S, T) \times (\mathrm{P}^U, U, U) = (K, S, T) = (\mathrm{P}^U, U, U) \times (K, S, T)$. For this reason we shall call (P^U, U, U) the *identity* composite.

If we refer to the preceding section we see that the composite of a product of two representations or modules is the product of the corresponding composites.

9. Closed composites. We shall call a composite $\Gamma = (K, S, T)$ *closed* (under multiplication) if $(K, S, T) \geq (K, S, T) \times (K, S, T)$. By Theorem 3 and the considerations of the preceding section, (K, S, T) is closed if and only if its relations space $\mathfrak{A}(\Gamma)$ is a ring. A composite (K, S, T) will be called a *Galois composite* if 1) (K, S, T) is closed, 2) (K, S, T) is a cover of the identity composite (P^U, U, U) and 3) $(K, S, T) = (K, T, S)$. It will be a consequence of the results of **9** and **10** that any closed composite is a Galois composite. Hence the conditions 2) and 3) can be deleted in the definition of a Galois composite. We have preferred, however, to state them as separate conditions since they correspond to the existence of the identity and of inverses in a group.

We suppose now that $\Gamma = (K, S, T)$ is a closed composite and we let Φ_Γ denote the set of fixed elements relative to Γ. Let $\alpha_1, \cdots, \alpha_q$ be elements of P such that $\alpha_1^S, \cdots, \alpha_q^S$ form a basis for K over P^T. We assert that these elements form a basis for P over Φ_Γ. To prove this we consider the three-fold composite (Δ, A, B, C) that satisfies conditions 4. and 5. for (K, S, T) and $(L, U, V) = (K, S, T)$. In (K, S, T) we have $\alpha^S = \alpha_1^S \mu_1(\alpha)^T + \cdots + \alpha_q^S \mu_q(\alpha)^T$. Hence in Δ we have the relation $\alpha^A = \Sigma \alpha_i^A \mu_i(\alpha)^B$. Since (K, S, T) is closed, $(\mathrm{P}^A \mathrm{P}^C, A, C)$ is covered by (K, S, T). Hence $\alpha^A = \Sigma \alpha_i^A \mu_i(\alpha)^C$. Thus $\Sigma \alpha_i^A (\mu_i(\alpha)^B - \mu_i(\alpha)^C) = 0$. We have seen that the elements α_i^A are independent over $\mathrm{P}^B \mathrm{P}^C$. Hence this relation implies that $\mu_i(\alpha)^B = \mu_i(\alpha)^C$ and so each $\mu_i(\alpha) \in \Phi_\Gamma$. Thus each α in P is a linear combination with coefficients in Φ_Γ of the elements $\alpha_1, \cdots, \alpha_q$. Since the elements α_i^S are linearly independent over P^T, the elements α_i are linearly independent over Φ_Γ.

THEOREM 9. *Let $\Gamma = (K, S, T)$ be a closed composite, $(K : \mathrm{P}^T) = q$ and let Φ_Γ denote the field of fixed elements relative to Γ. Then $(\mathrm{P} : \Phi_\Gamma) = q$ and if $\alpha_1, \cdots, \alpha_q$ are q elements of P, they form a basis for P over Φ_Γ if and only if the elements $\alpha_1^S, \cdots, \alpha_q^S$ form a basis for K over P^T.*

The necessity of the condition is trivial. For if $\alpha_1, \cdots, \alpha_q$ from a basis for P over Φ_Γ, each $\alpha^S = \alpha_1^S \gamma_1^T + \cdots + \alpha_q^S \gamma_q^T$ where the $\gamma_i^T \in \Phi_\Gamma^T = \Phi_\Gamma^S$. Hence the α_i^S are linearly independent over P^T.

We consider now the relations ring $\mathfrak{A}(\Gamma)$. We have seen that its dimensionality over $\bar{\mathrm{P}}$ is q. Hence the dimensionality of $\mathfrak{A}(\Gamma)$ over $\bar{\Phi}_\Gamma$ is q^2. On the other hand we have seen that the transformations belonging to $\mathfrak{A}(\Gamma)$ commute with the elements of $\bar{\Phi}_\Gamma$, i. e., they are linear transformations in the space P over Φ_Γ. Since $(\mathrm{P} : \Phi_\Gamma) = q$, it follows that $\mathfrak{A}(\Gamma)$ is the complete set of linear transformations of P over Φ_Γ.

THEOREM 10. *If* $\Gamma = (K, S, T)$ *is a closed composite and* Φ_Γ *is its field of fixed elements, the relations space* $\mathfrak{A}(\Gamma)$ *is the complete ring of linear transformations of* P *over* Φ_Γ.

Suppose now that $\Gamma' = (K', S', T')$ is any composite of P which leaves the elements of Φ_Γ fixed. Then if N is any element of the relations space $\mathfrak{A}(\Gamma')$, N is a linear transformation in P over Φ_Γ. Hence by Theorem 10, $N \epsilon \mathfrak{A}(\Gamma)$, and so $\mathfrak{A}(\Gamma') \leq \mathfrak{A}(\Gamma)$. This implies the following

THEOREM 11. *Let* Γ *be a closed composite and* Φ_Γ *its field of fixed elements. Then if* Γ' *is any composite such that* $\Phi_{\Gamma'} \geq \Phi_\Gamma$, $\Gamma' \leq \Gamma$.

This, of course, implies that there is only one (in the sense of equivalence) closed composite having $\Phi = \Phi_\Gamma$ as its field of fixed elements.

It may be remarked that if $\Delta = (L, U, V)$ is any composite such that $(P : \Phi_\Delta) = r < \infty$, then (L, U, V) is non-singular. For if we set $\Phi_\Delta = \Phi$, then $(P^U : \Phi^U) = r = (P^V : \Phi^V)$. Hence if $(L : P^V) = q'$, $(L : \Phi^V) = rq'$ and since $\Phi^V = \Phi^U$, $(L : \Phi^U) = rq'$. It follows that $(L : P^U) = q'$. A similar argument may be used to show that if $\mathfrak{S}$ is any double P-module with composite equivalent to Δ, then $\mathfrak{S}$ is non-singular. This may be stated as follows:

THEOREM 12. *If* F *is a self-representation of* P *such that the dimensionality of* P *over the field of fixed elements is finite, then* F *is similar to a non-singular self-representation.*

We again suppose that $\Delta = (L, U, V)$ is a composite such that $(P : \Phi_\Delta) = r < \infty$. If Δ is not closed, we form the least common cover $\Delta^{(2)} = \Delta + \Delta \times \Delta$. Then the relations space $\mathfrak{A}(\Delta^{(2)}) > \mathfrak{A}(\Delta)$. It is readily seen that the field of fixed elements $\Phi_{\Delta^{(2)}} = \Phi_\Delta$ so that $(P : \Phi_{\Delta^{(2)}}) = r$ also. If $\Delta^{(2)}$ is not closed we set $\Delta^{(3)} = \Delta + (\Delta \times \Delta) + (\Delta \times \Delta \times \Delta)$ and note that $\mathfrak{A}(\Delta^{(3)}) > \mathfrak{A}(\Delta^{(2)})$ and $\Phi_{\Delta^{(3)}} = \Phi_{\Delta^{(2)}}$. Since $\mathfrak{A}(\Delta)$, $\mathfrak{A}(\Delta^{(2)})$, $\mathfrak{A}(\Delta^{(3)})$ all have dimensionalities over $\bar{\Phi}_\Delta \leq r^2$, the dimensionality over $\bar{\Phi}_\Delta$ of the complete ring of linear transformations of P over Φ_Δ, this process leads after a finite number of steps to a closed composite $\Gamma = \Delta + (\Delta \times \Delta) + \cdots + (\Delta \times \cdots \times \Delta)$. By Theorem 10 we obtain

THEOREM 13. *Suppose that* $\alpha \rightarrow (\alpha F_{kl})$ *is a self-representation of* P *such that* $(P : \Phi) < \infty$ *for* Φ *the field of fixed elements. Then if* L *is a linear transformation of* P *over* Φ, L *is expressible as a polynomial in the transformations* F_{kl} *with coefficients in* $\bar{P}$ *the field of multiplications in* P.

10. The fundamental Galois correspondence. We suppose now that P is an arbitrary field and that Φ is a subfield of P such that $(P:\Phi) = q < \infty$. We take two copies P^S and P^T of P and form the direct product K of these fields relative to Φ.[7] Thus K is the set of sums $\Sigma\alpha^S\beta^T$, α^S in P^S and β^T in P^T where $\gamma^S = \gamma^T$ if $\gamma \epsilon \Phi$ and addition and multiplication are defined in the obvious way. The dimensionality $(K:\Phi^S) = q^2$ and $(K:P^S) = q = (K:P^T)$. We recall also that $P^S \wedge P^T = \Phi^S = \Phi^T$. It is evident that $\Gamma = (K, S, T)$, where S is the mapping $\alpha \rightarrow \alpha^S$ and T is the mapping $\alpha \rightarrow \alpha^T$, is a composite of P with itself. Since the mapping $\alpha^S \rightarrow \alpha^T$, $\alpha^T \rightarrow \alpha^S$ is an automorphism in K, (K, S, T) is symmetric. Since $P^S \wedge P^T = \Phi^S = \Phi^T$, the field Φ_Γ of fixed elements is Φ. Now let (L, U, V) be any composite of P leaving the elements of Φ fixed. Suppose that $\Sigma\alpha^S\beta^T = 0$ in K. Then if $\alpha_1, \cdots, \alpha_q$ form a basis for P over Φ, the elements $\alpha_i{}^S\alpha_j{}^T$ form a basis for K over $\Phi^S = \Phi^T$ and the elements $\alpha_i{}^U\alpha_j{}^V$ are generators of L over $\Phi^U = \Phi^V$. We replace α^S by $\Sigma\gamma_i{}^S\alpha_i{}^S$, γ_i in Φ and β^T by $\Sigma\delta_j{}^T\alpha_j{}^T$, δ_j in Φ and substitute in the relation $\Sigma\alpha^S\beta^T = 0$. Then the coefficients of the products $\alpha_i{}^S\alpha_j{}^T$ in the resulting expression are all 0. Hence if we substitute $\alpha^U = \Sigma\gamma_i{}^U\alpha_i{}^U$ and $\beta^V = \Sigma\delta_j{}^V\alpha_j{}^V$ in $\alpha^U\beta^V$, we see that $\Sigma\alpha^U\beta^V = 0$. We have therefore proved that (K, S, T) is a cover for (L, U, V). This shows, in particular, that $(K, S, T) \geqq (K, S, T) \times (K, S, T)$ and $(K, S, T) \geqq (P^U, U, U)$.

THEOREM 14. *If* P *is a field and* Φ *a subfield such that* $(P:\Phi) = q < \infty$, *then the direct product of* P *over* Φ *with itself determines a Galois composite whose field of fixed elements is* Φ.

Now suppose that Γ is any closed composite and let Φ_Γ be the subfield of P of fixed elements under Γ. We have seen that $(P:\Phi_\Gamma) < \infty$ and hence we may form the direct product Γ' of P over Φ_Γ by itself. By Theorem 11, $\Gamma' \leqq \Gamma$ and by what we have just proved $\Gamma \leqq \Gamma'$. Hence Γ is equivalent to Γ' and so we have proved

THEOREM 15. *Any closed composite is a Galois composite.*

As a consequence of Theorems 9, 11, and 14 we have the following fundamental

THEOREM 16. *Let* P *be a fixed field and* Γ *a Galois composite of* P. *If* Φ_Γ *denotes the field of fixed elements under* Γ, *then* P *is finite over* Φ_Γ *and the correspondence* $\Gamma \rightarrow \Phi_\Gamma$ *is* $(1-1)$ *between the Galois composites and the subfields* Φ *over which* P *is finite.* $\Gamma_1 \geqq \Gamma_2$ *if and only if* $\Phi_{\Gamma_1} \leqq \Phi_{\Gamma_2}$.

---

[7] For the definition and properties of the direct product see, for example, *R*, p. 88.

11. Decomposition of composites. Let $\Gamma = (K, S, T)$ be an arbitrary composite. Since K is a commutative algebra with a finite basis over P^T, we may decompose K as a direct sum

$$K = K_1 \oplus \cdots \oplus K_s \tag{21}$$

of indecomposable algebras K_i. The K_i are uniquely determined. Moreover if B is any ideal in K, $B = B_1 \oplus \cdots \oplus B_s$ where $B_i = B \wedge K_i$. Hence if N is the radical of K, $N = N_1 \oplus \cdots \oplus N_s$ where $N_i = N \wedge K_i$ is the radical of K_i. We recall now that any indecomposable commutative algebra is completely primary in the sense that the difference algebra with respect to its radical is a field.[8] Any ideal of such an algebra is contained in the radical. It follows that the homomorphic image of such an algebra is also completely primary and hence is also indecomposable. These results apply in particular to the algebras K_i.

Corresponding to the decomposition (21) we write

$$1 = e_1 + \cdots + e_s$$

where

$$e_i^2 = e_i, \quad e_i e_j = 0 \text{ if } i \neq j.$$

Then $K_i = Ke_i$. The mapping $k \to ke_i$ is a homomorphism E_i of K into K_i. We set $S_i = SE_i$, $T_i = TE_i$. These are isomorphisms between P and subfields P^{S_i} and P^{T_i} of K_i. Since $K = \mathrm{P}^S \mathrm{P}^T$, $K_i = \mathrm{P}^{S_i}\mathrm{P}^{T_i}$. Evidently $(K : \mathrm{P}^T) = \Sigma(K_i : \mathrm{P}^T)$ and since $k_i \alpha^T = k_i \alpha^{T_i}$ if $k_i \in K_i$, $(K_i : \mathrm{P}^T) = (K_i : \mathrm{P}^{T_i})$. Hence $(K_i : \mathrm{P}^{T_i})$ is finite. Thus (K_i, S_i, T_i) is a composite of P. It is evident that (K, S, T) is the least common cover of these composites. We shall now show that the composites (K_i, S_i, T_i) are *disjoint* in the sense that there exists no composite (L, U, V) which is covered by both (K_i, S_i, T_i) and the least common cover of all the (K_j, S_j, T_j) for $j \neq i$. For let (L, U, V) be such a composite. Suppose that $e_i = \Sigma \alpha^S \beta^T$. Then $e_i = e_i^2 = \Sigma \alpha^{S_i} \beta^{T_i}$ and $0 = e_i e_j = \Sigma \alpha^{S_j} \beta^{T_j}$ for all $j \neq i$. In L we have $1^U = \Sigma \alpha^U \beta^V$ and $0 = \Sigma \alpha^U \beta^V$ and this contradiction proves our assertion.

If B is an ideal in K and $k \to \bar{k}$ denotes the natural homomorphism between K and $\bar{K} = K - B$, it is clear that $(\bar{K}, \bar{S}, \bar{T})$ where $\alpha^{\bar{S}} = \overline{\alpha^S}$, $\beta^{\bar{T}} = \overline{\beta^T}$ is a composite of P covered by (K, S, T) and, as we have remarked, any composite covered by (K, S, T) can be obtained in this way. If C is an ideal containing B, it is readily seen that the composite $(K - B, \bar{S}, \bar{T})$ is a cover of $(K - C, \bar{S}, \bar{T})$. Now since any ideal B_i of K_i is contained in the radical

[8] Cf. Theorem 3, Chapter 4 of *R*.

N_i, this shows that any composite covered by (K_i, S_i, T_i) is a cover of the composite $(K_i - N_i, \bar{S}_i, \bar{T}_i)$. Thus it is impossible to find disjoint composites which are covered by (K_i, S_i, T_i).

From the decomposition $K = K_1 \oplus \cdots \oplus K_s$ we obtain $K - B = \bar{K} = \bar{K}_1 + \cdots + \bar{K}_s$. Also $B = B_1 \oplus \cdots \oplus B_s$ where $B_i = B \wedge K_i$. Hence if $\bar{k}_1 + \bar{k}_2 + \cdots + \bar{k}_s = 0$, $k_1 + k_2 + \cdots + k_s = b_1 + b_2 + \cdots + b_s$ where the $b_i \in B_i$. Then $k_i = b_i$ is in B_i and $\bar{k}_i = 0$. Thus $\bar{K} = \bar{K}_1 \oplus \cdots \oplus \bar{K}_s$. Each $\bar{K}_i$ is indecomposable since it is the homomorphic image of the indecomposable algebra K_i. We suppose that $\bar{K}_i \neq 0$ if $i \leq r$ and that $\bar{K}_i = 0$ if $i > r$. We now denote the mapping $k \to \bar{k}$ by H, the mapping $\bar{k} \to \bar{k}_i$ by $\bar{E}_i$ and we set $\bar{S}_i = SH\bar{E}_i$, $\bar{T}_i = TH\bar{E}_i$. Then $(\bar{K}_i, \bar{S}_i, \bar{T}_i)$ is a composite of P with itself. Moreover, these composites are obtained from $(\bar{K}, \bar{S}, \bar{T})$ in the same way that the (K_i, S_i, T_i) are obtained from (K, S, T). We assert now that $(\bar{K}_i, \bar{S}_i, \bar{T}_i)$ is covered by (K_i, S_i, T_i). For suppose that $\Sigma\alpha^{S_i}\beta^{T_i} = 0$. Then if $k = \Sigma\alpha^S\beta^T$, $k_i = 0$. Now in the decomposition $\bar{k} = \bar{k}_1 + \cdots + \bar{k}_s$, $\bar{k}_i$ is the coset of k_i. Hence $\bar{k}_i = 0$. Thus $\Sigma\alpha^{\bar{S}_i}\beta^{\bar{T}_i} = (\Sigma\alpha^S\beta^T)^{HE_i} = \bar{k}_i = 0$ and this proves our assertion. In particular we see that if $(\bar{K}, \bar{S}, \bar{T})$ is indecomposable, i. e. if $r = 1$, then $(\bar{K}, \bar{S}, \bar{T})$ is covered by one of the composites (K_i, S_i, T_i).

Now suppose that (K, S, T) is the least common cover of the composites (K'_j, S'_j, T'_j), $j = 1, \cdots, s'$ which are disjoint and *indecomposable* in the sense that they are not least common covers of disjoint composites. Then each algebra K'_j is indecomposable and since (K, S, T) is a cover of (K'_j, S'_j, T'_j) one of the (K_i, S_i, T_i) is a cover of (K'_j, S'_j, T'_j). Since the (K_i, S_i, T_i) are disjoint, only one of these, say $(K_{\bar{j}}, S_{\bar{j}}, T_{\bar{j}})$, is a cover for (K'_j, S'_j, T'_j). Since no two composites covered by (K_j, S_j, T_j) are disjoint, it follows that if $j \neq k$, then $\bar{j} \neq \bar{k}$. If we re-arrange the (K_j, S_j, T_j), we may suppose that $\bar{j} = j$. We wish to show that (K_j, S_j, T_j) and (K'_j, S'_j, T'_j) are equivalent and so we assume that this is not the case. Then there is a relation $\Sigma\lambda^{S'_j}\mu^{T'_j} = 0$ such that $\Sigma\lambda^{S_j}\mu^{T_j} \neq 0$. If $e_j = \Sigma\alpha^S\beta^T$, $e_j = \Sigma\alpha^{S_j}\beta^{T_j}$ and $\Sigma\lambda^{S_j}\alpha^{S_j}\mu^{T_j}\beta^{T_j} \neq 0$. Since $e_je_k = 0$ if $j \neq k$, $\Sigma\alpha^{S_k}\beta^{S_k} = 0$ and hence $\Sigma\lambda^{S_k}\alpha^{S_k}\mu^{T_k}\beta^{T_k} = 0$. Thus we may suppose at the start that $\Sigma\lambda^{S_k}\mu^{T_k} = 0$ for $k \neq j$. It follows that $\Sigma\lambda^{S'_k}\mu^{T'_k} = 0$ and hence that $\Sigma\lambda^S\mu^T = 0$ and this contradicts the fact that $\Sigma\lambda^{S_j}\mu^{T_j} \neq 0$. Now it is clear that $s' = s$. For otherwise there exists a (K_i, S_i, T_i) which is covered by the least common cover of the (K_j, S_j, T_j) for $j \neq i$. We have therefore proved the following:

THEOREM 17. *Any composite (K, S, T) is expressible in one and only one way as the least common cover of disjoint indecomposable composites.*

We shall call the *indecomposable* composites (K_i, S_i, T_i) the *indecomposable components* of (K, S, T). The argument given above proves

THEOREM 18. *If (L, U, V) is covered by (K, S, T) then each indecomposable component of (L, U, V) is covered by an indecomposable component of (K, S, T).*

Suppose now that (K, S, T) is a Galois composite. Then we have seen that any composite covered by (K, S, T) is non-singular. This applies in particular to the indecomposable components (K_i, S_i, T_i). Consider now the inverses (K_i, T_i, S_i). It is clear that these are indecomposable and disjoint and that their least common cover is (K, T, S). Since $(K, S, T) = (K, T, S)$ it follows that the set of inverses (K_i, T_i, S_i) coincides with the set (K_i, S_i, T_i). We observe next that since $(K, S, T) \times (K, S, T) \leq (K, S, T)$ and since $(K, S, T) \times (K, S, T)$ is a cover of every product $(K_i, S_i, T_i) \times (K_j, S_j, T_j)$, the indecomposable components of these products are covered by suitable ones of the (K_k, S_k, T_k). We note finally that since (K, S, T) is a cover of (P^U, U, U) one of the (K_i, S_i, T_i) is a cover of this composite. Conversely suppose that (K, S, T) is any composite whose indecomposable components (K_i, S_i, T_i) satisfy

1. Each (K_i, S_i, T_i) is non-singular,
2. One of the (K_i, S_i, T_i), say (K_1, S_1, T_1), is a cover of the identity composite (P^U, U, U),
3. For each (K_i, S_i, T_i) the inverse (K_i, T_i, S_i) is an indecomposable component,
4. The indecomposable components of $(K_i, S_i, T_i) \times (K_j, S_j, T_j)$ are covered by the (K_k, S_k, T_k).

Then since $(K, S, T) \times (K, S, T)$ is the least common cover of the products $(K_i, S_i, T_i) \times (K_j, S_j, T_j)$, (K, S, T) is a cover of $(K, S, T) \times (K, S, T)$. It is evident also that $(K, S, T) \geq (P^U, U, U)$ and that (K, S, T) is symmetric.

THEOREM 19. *Conditions 1.-4. on the indecomposable components of a composite are necessary and sufficient that the composite be a Galois composite.*

12. Separable fields. We shall call a composite (K, S, T) *simple* if K is a field. In this case the only composite covered by (K, S, T) is (K, S, T) itself. (K, S, T) is *semi-simple* if its indecomposable components are simple. If the components of (K, S, T) are (K_i, S_i, T_i), $i = 1, \cdots, s$, it follows that any composite covered by (K, S, T) is the least common cover of certain of the (K_i, S_i, T_i).

We shall prove first the following

THEOREM 20. *Let* P *be a field,* Φ *a subfield such that* $(\mathrm{P}:\Phi) = q < \infty$ *and let* $\Gamma = (K, S, T)$ *be the Galois composite whose field of fixed elements* $\Phi_\Gamma = \Phi$. *Then a necessary and sufficient condition that* P *be separable over* Φ *is that* Γ *be semi-simple.*

Sufficiency. Let α be any element of P and set $\Sigma = \Phi(\alpha^p)$ where p is the characteristic of Φ. Consider the Galois composite $\Delta = (L, U, V)$ of P having $\Sigma = \Phi_\Delta$ as its field of fixed elements. Since Δ is covered by Γ, it is semi-simple. Now $(\alpha^U)^p = (\alpha^p)^U = (\alpha^p)^V = (\alpha^V)^p$. Hence $(\alpha^U - \alpha^V)^p = 0$ and so $\alpha^U = \alpha^V$. Thus $\alpha \in \Sigma = \Phi(\alpha^p)$ and hence α is separable.

Necessity. Suppose that P is separable over Φ. Then there exists a primitive element α in P such that $\mathrm{P} = \Phi(\alpha)$. Let (L, U, V) be an indecomposable composite and let $\phi^V(\lambda)$ be the minimum polynomial of α^U over the field P^V. Since L is an indecomposable algebra $\phi^V(\lambda) = \psi^V(\lambda)^e$ where $\psi^V(\lambda)$ is irreducible. For if $\phi^V(\lambda) = \psi_1{}^V(\lambda)\psi_2{}^V(\lambda)$ where $(\psi_1{}^V(\lambda), \psi_2{}^V(\lambda)) = 1$, $\psi_1{}^V(\alpha^U)$ is a zero-divisor $\neq 0$ in L and since any zero-divisor of a completely primary algebra is contained in the radical $\psi_1{}^V(\alpha^U)^f = 0$. This implies that $\psi_1{}^V(\lambda)^f$ is divisible by $\psi_1{}^V(\lambda)\psi_2{}^V(\lambda)$ contrary to $(\psi_1{}^V(\lambda), \psi_2{}^V(\lambda)) = 1$. Now let $\mu(\lambda)$ be the minimum polynomial of α over Φ. Then $\mu^V(\lambda) = \phi^V(\lambda)\phi_1{}^V(\lambda) = \psi^V(\lambda)^e\psi_1{}^V(\lambda)$. It follows from this that $(\mu(\lambda), \mu'(\lambda)) \neq 1$ if $e > 1$. Thus $e = 1$ and α^U satisfies an irreducible polynomial over P^V. Hence $L = \mathrm{P}^V(\alpha^U)$ is a field. We have therefore proved that any indecomposable composite of P is irreducible and hence any composite is semi-simple.

We now recall the definition of a *hypergroup* (with an identity) H as a system in which a product of pairs equal to a subset of H is defined. If A and B are subsets of H, we define AB to be the set of elements contained in all products ab, a in A and b in B. It is assumed that 1) the product is associative: the set $(ab)c = a(bc)$, 2) there exists an identity 1 in H such that $a1 = 1a = a$ for all a in H and 3) for each a there is an inverse b such that ab and ba contain 1.

Suppose now that H is a set of non-singular simple composites of P having the following properties: 1) If Γ_1 and Γ_2 belong to H, then $\Gamma_1 \times \Gamma_2$ is the least common cover of certain Γ_i in H. 2) H contains the identity composite. 3) H contains the inverse Γ^{-1} of any Γ in H. Then if we define $\Gamma_1\Gamma_2$ to be the set of simple composites contained in H and covered by $\Gamma_1 \times \Gamma_2$, it is easy to see that H is a hypergroup. We shall therefore call a set of non-singular simple composites with the properties 1), 2) and 3) a *hypergroup of simple composites.*

Now let H be a finite hypergroup of simple composites and let $\Gamma = (K, S, T)$ be the least common cover of all the Γ_i in H. If (K_i, S_i, T_i) are the indecomposable components of Γ, it is evident that each Γ_i is covered by one of the (K_i, S_i, T_i) no two Γ_i are covered by the same (K_i, S_i, T_i) and no two (K_i, S_i, T_i) cover the same Γ_i. It follows that the Γ_i coincide with the (K_i, S_i, T_i). Hence by Theorem 19, (K, S, T) is a Galois composite and since its indecomposable components are simple, (K, S, T) is semi-simple. Let Φ_H be the field of fixed elements relative to the Γ_i. Then $\Phi_H = \Phi_\Gamma$ and hence P is finite and separable over Φ_H. Conversely suppose that P is finite and separable over Φ and let Γ be the Galois composite such that $\Phi_\Gamma = \Phi$. Then we have seen that Γ is semi-simple and it follows from Theorem 19 that the indecomposable components Γ_i of Γ form a finite hypergroup H of simple composites. Evidently $\Phi_H = \Phi$. We have therefore proved

THEOREM 21. *Let* P *be a fixed field and* H *a finite hypergroup of simple composites of* P. *Then if* Φ_H *denotes the field of fixed elements under all the composites in* H, P *is finite and separable over* Φ_H. *The correspondence* $H \rightarrow \Phi_H$ *is* $(1 - 1)$ *between the finite hypergroups of simple composites of* P *and the subfields* Φ_H *over which* P *is finite and separable.* $H_1 \geq H_2$ *if and only if* $\Phi_{H_1} \leq \Phi_{H_2}$.[9]

13. Normal fields. We shall call a composite (L, U, V) *one-dimensional* if $(L : \mathrm{P}^V) = 1$. Evidently this implies that (L, U, V) is simple. With this definition we have

THEOREM 22. *A necessary and sufficient condition that a field* P *finite over a subfield* Φ *be normal over* Φ *is that every simple composite* (L, U, V), *whose field of fixed elements contains* Φ, *is one-dimensional.*

Suppose that P is normal over Φ and that (L, U, V) is a simple composite whose field of fixed elements contains Φ. Let α be an element of P and let $\mu(\lambda)$ be its minimum polynomial over Φ. Since P is normal, $\mu(\lambda) = (\lambda - \alpha_1) \cdots (\lambda - \alpha_m)$, $\alpha_1 = \alpha$, in $\mathrm{P}[\lambda]$. Hence $\mu^V(\lambda) = \Pi(\lambda - \alpha_i{}^V)$. Since $\mu^V(\alpha^U) = 0$, $\Pi(\alpha^U - \alpha_i{}^V) = 0$ and since L is a field $\alpha^U = \alpha_i{}^V$ for a suitable i. Thus $\mathrm{P}^U \leq \mathrm{P}^V$ and $(L : \mathrm{P}^V) = 1$.

Next suppose that for every simple (L, U, V) whose field of fixed elements contains Φ, $(L : \mathrm{P}^V) = 1$. Consider the Galois composite $\Gamma = (K, S, T)$ such

[9] Cf. Kaloujnine, *lcc. cit.*[3].

that $\Phi_\Gamma = \Phi$. Let $K = K_1 \oplus \cdots \oplus K_s$ where the K_i are indecomposable. If N is the radical of K, $N = N_1 \oplus \cdots \oplus N_s$, N_i the radical of K_i. We define the S_i, T_i as in **11** and consider the simple composite $(K_i - N_i, \bar{S}_i, \bar{T}_i)$ where $\bar{S}_i$, $\bar{T}_i$ are obtained by first applying S_i, T_i and then the natural homomorphism between K_i and $K_i - N_i$. By our assumption for any α in P, $\alpha^{\bar{S}_i} = \beta_i^{\bar{T}_i}$ for a suitable β_i. Hence $(\alpha^{S_i} - \beta_i^{T_i})^{n_i} = 0$ for some n_i. We assume that n_i is minimal and we may assume also that the notation has been chosen so that $\beta_1 = \cdots = \beta_{r_1} \equiv \delta_1 \neq \beta_{r_1+1} = \cdots = \beta_{r_1+r_2} = \delta_2 \neq \cdots$ and that $m_1 = n_1 \geq n_2 \geq \cdots \geq n_{r_1}$; $m_2 = n_{r_1+1} \geq n_{r_1+2} \geq \cdots \geq n_{r_1+r_2}; \cdots$. Consider the polynomial $\mu^T(\lambda) = (\lambda - \delta_1{}^T)^{m_1} \cdots (\lambda - \delta_t{}^T)^{m_t}$. We wish to prove that $\mu^T(\lambda)$ is the minimum polynomial over P^T satisfied by α^S. We note first that $(\alpha^{S_i} - \delta_1{}^{T_i})^{m_1} \cdots (\alpha^{S_i} - \delta_t{}^{T_i})^{m_t}$ contains the factor $(\alpha^{S_i} - \beta_i{}^{T_i})^{n_i} = 0$. Since (K, S, T) is the least common cover of the (K_i, S_i, T_i) this implies that $\mu^T(\alpha^S) = 0$. Now let $\nu^T(\lambda)$ be the minimum polynomial of α^S over P^T. We recall that if $1, \alpha, \cdots, \alpha^{r-1}$ are linearly independent over Φ, then $1, \alpha^S, \cdots, (\alpha^S)^{r-1}$ are linearly independent over P^T. This implies that $\nu^T(\lambda)$ has coefficients in $\Phi^T = \Phi^S$. Hence $\nu(\lambda)$ is the minimum polynomial of α over Φ. Since $\nu^T(\alpha^S) = 0$, $\nu^{T_i}(\alpha^{S_i}) = 0$. This implies that $\nu^{T_i}(\lambda)$ is divisible by $(\lambda - \beta_i{}^{T_i})^{n_i}$ and hence that $\nu^T(\lambda)$ is divisible by $(\lambda - \beta_i{}^T)^{n_i}$. It follows that $\nu^T(\lambda)$ is divisible by $\mu^T(\lambda)$ and since $\nu^T(\lambda)$ is minimal, $\mu^T(\lambda) = \nu^T(\lambda)$. From the factorization $\mu^T(\lambda) = \Pi(\lambda - \delta_j{}^T)^{m_j}$ we obtain $\mu(\lambda) = \Pi(\lambda - \delta_j)^{m_j}$. Thus the minimum polynomial of α over Φ factors into linear factors in $\mathrm{P}[\lambda]$ and so P is normal over Φ.

Let (L, U, V) be any non-singular one dimensional composite of P. Then $L = \mathrm{P}^V$. Since $(L : \mathrm{P}^U) = 1$ also, $L = \mathrm{P}^U$. If we apply the isomorphism V^{-1} to L, we see that (L, U, V) is equivalent to the composite $(P, A, 1)$ where $A = UV^{-1}$ is an automorphism in P and 1 is the identity automorphism. Conversely if A is any automorphism, $(\mathrm{P}, A, 1)$ is a non-singular composite. It is clear that $(\mathrm{P}, A, 1) = (P, 1, A^{-1})$. Hence the inverse of $(\mathrm{P}, A, 1)$ is $(\mathrm{P}, A^{-1}, 1) = (\mathrm{P}, 1, A)$. It follows directly from the definition of the product that $(\mathrm{P}, A, 1) \times (\mathrm{P}, B, 1) = (\mathrm{P}, A, 1) \times (\mathrm{P}, 1, B^{-1}) = (\mathrm{P}, A, B^{-1}) = (\mathrm{P}, AB, 1)$. Thus the totality of non-singular one-dimensional composites of P is a group under multiplication isomorphic to the group of automorphisms of P.

Suppose now that H is a finite hypergroup of one dimensional composites. Then it is clear that H is a group and by Theorems 21 and 22, P is finite, separable and normal over Φ_H. Conversely if P is finite, separable and normal over Φ, then the indecomposable components of the Galois composite Γ such that $\Phi_\Gamma = \Phi$ form a group H under multiplication. If $(\mathrm{P}, A, 1)$ is one of the simple components of Γ then A is an automorphism of P that leaves the elements of Φ fixed. On the other hand if A is such an automorphism of P, $(P, A, 1)$ is a composite leaving the elements of Φ fixed. Hence $(P, A, 1)$ is covered by Γ and therefore $(P, A, 1)$ is one of the indecomposable components of Γ. This completes the proof of the classical correspondence:

THEOREM 23. *If* P *is an arbitrary field, there is a* $(1-1)$ *correspondence between the finite groups* H *of automorphisms in* P *and the subfields* Φ_H *over which* P *is finite, separable and normal, namely,* Φ_H *is the set of fixed elements under the automorphisms of* H *and* H *is the complete set of automorphisms leaving the elements of* Φ *fixed.*

We recall that $(K : P^T) = \Sigma(K_i : P^{T_i})$. If P is separable and normal over Φ_Γ, each $(K_i : P^{T_i}) = 1$ and hence $(P : \Phi_\Gamma) = (K : P^T)$ is the number of components (K_i, S_i, T_i). It follows that $(P : \Phi_\Gamma)$ is the order of the Galois group of P over Φ_Γ.

14. Simple extensions. Suppose that $P = \Phi(\alpha)$ a simple extension of Φ and let $\mu(\lambda)$ be the minimum polynomial of α over Φ. Let $\Gamma = (K, S, T)$ be the Galois composite whose field of fixed elements $\Phi_\Gamma = \Phi$. Then the minimum polynomial of α^S over P^T is $\mu^T(\lambda)$ and the degree of $\mu^T(\lambda)$ is the dimensionality q of K over P^T. If $\Delta = (L, U, V)$ is any composite such that $\Phi_\Delta \geqq \Phi$, Γ is a cover of Δ and hence L is generated by α^U over P^V. Thus if $\nu^V(\lambda)$ is the minimum polynomial of α^U over P^V and r is the degree of $\nu^V(\lambda)$ then the elements $1, \alpha^U, \cdots, (\alpha^U)^{r-1}$ form a basis for L over P^V. Since $\mu^V(\alpha^U) = 0$, $\nu^V(\lambda)$ is a factor of $\mu^V(\lambda)$ and hence the polynomial $\nu(\lambda)$ in $P[\lambda]$ is a factor of $\mu(\lambda)$. Thus we have associated with the composite Δ the factor $\nu(\lambda)$ of $\mu(\lambda)$ in $P[\lambda]$. The composite $\Delta_1 \geqq \Delta_2$ if and only if the associated factor $\nu_1(\lambda)$ is divisible by $\nu_2(\lambda)$. We shall show next that every factor $\nu(\lambda)$ of $\mu(\lambda)$ in $P[\lambda]$ arises in this way from some composite. For if $\mu(\lambda) = \nu(\lambda)\nu_1(\lambda)$ and B is the ideal in K generated by $\nu^T(\alpha^S)$ then $(K - B, \bar{S}, \bar{T})$ is the required composite. The correspondence between composites Δ such that $\Phi_\Delta \geqq \Phi$ and the factors $\nu(\lambda)$ of $\mu(\lambda)$ in $P[\lambda]$ is therefore $(1-1)$. It is readily seen that the indecomposable components of Γ correspond to the prime power factors $\pi_i(\lambda)^{e_i}$ of $\mu(\lambda) = \pi_1(\lambda)^{e_1} \cdots \pi_s(\lambda)^{e_s}$ in $P[\lambda]$.

Let $\nu(\lambda) = \lambda^m - \beta_1\lambda^{m-1} - \cdots - \beta_m$ be a factor of $\mu(\lambda)$ and let (L, U, V) be the corresponding composite. Then if we turn L into a double P-module by defining $\alpha x = \alpha^U x = x\alpha^U$ and $x\alpha = \alpha^V x = x\alpha^V$, it may be verified that the matrix of α^U relative to the basis $1, \alpha^U, \cdots, (\alpha^U)^{m-1}$ is

$$\alpha^F = \begin{pmatrix} 0 & \cdot & \cdot & \cdot & \cdot & \beta_m \\ 1 & & & & & \cdot \\ & \cdot & & & & \cdot \\ & & \cdot & & & \cdot \\ & & & \cdot & & \cdot \\ & & & & 1 & \beta_1 \end{pmatrix}. \tag{22}$$

If $\pi(\lambda) = \lambda^n - \delta_1\lambda^{n-1} - \cdot \cdot \cdot - \delta_n$ is a second factor of $\mu(\lambda)$, it determines in the same way a composite and a self-representation in which α is represented by the matrix

$$\alpha^G = \begin{pmatrix} 0 & \cdot & \cdot & \cdot & \cdot & \delta_n \\ 1 & & & & & \cdot \\ & \cdot & & & & \cdot \\ & & \cdot & & & \cdot \\ & & & \cdot & & \cdot \\ & & & & 1 & \delta_1 \end{pmatrix}.$$

The elements β_i are polynomials in α with coefficients in Φ. If we replace the β_i in α^F by the corresponding polynomials in α^G, we obtain the matrix $\alpha^{F \times G}$ representing α in the product representation. If $\rho(\lambda)$ is the minimum polynomial of $\alpha^{F \times G}$, $\rho(\lambda)$ is a factor of $\mu(\lambda)$ and the composite associated with $\rho(\lambda)$ is the product of the composites associated with $\nu(\lambda)$ and the composite associated with $\pi(\lambda)$.

If we apply the theory of a single linear transformation we see that any self-representation $\alpha \to \alpha^E$ is decomposable into "cyclic" self-representations E_i where α^{E_i} has the form (22).

THE JOHNS HOPKINS UNIVERSITY.

Reprinted from
American Journal of Mathematics
January 1944.

SCHUR'S THEOREMS ON COMMUTATIVE MATRICES

N. JACOBSON

In 1905 I. Schur[1] proved that the maximum number $N(n)$ of linearly independent commutative matrices of n rows and columns is given by the formula $N(n)=[n^2/4]+1=\nu^2+1$ if $n=2\nu$ and $=\nu(\nu-1)+1$ if $n=2\nu-1$. Schur also determined the sets of linearly independent commutative matrices containing $N(n)$ elements. In this note we give a simpler derivation of Schur's results and an extension of these results from algebraically closed fields to arbitrary fields.

If $A_1, \cdots, A_{N(n)}$ is a set of linearly independent commutative matrices, the set $\mathfrak{A}$ of matrices $\sum A_i\phi_i$ where ϕ_i is arbitrary in the underlying field Φ is a commutative subalgebra containing the identity of the matrix algebra Φ_n. Hence $N(n)$ is the maximal dimensionality of commutative subalgebras of Φ_n. It is easy to see that $N(n)\geqq[n^2/4]+1$. For consider the set $\mathfrak{Z}_n$ of matrices

$$\begin{pmatrix} 0 & A \\ 0 & 0 \end{pmatrix} \tag{1}$$

where if $n=2\nu$, A is arbitrary in Φ_ν and if $n=2\nu-1$, A is an arbitrary matrix of ν rows and $\nu-1$ columns. Thus dim $\mathfrak{Z}_n=[n^2/4]$. It may be verified that $\mathfrak{Z}_n$ is a zero algebra. Hence the algebra $\mathfrak{B}_n$ obtained by adjoining 1 to $\mathfrak{Z}_n$ is a commutative algebra of dimensionality $[n^2/4]+1$. We remark also that if $n=2\nu-1$ we may replace $\mathfrak{Z}_n$ by the algebra $\overline{\mathfrak{Z}}_n$ of matrices of the form (1) in which A is an arbitrary matrix of $\nu-1$ rows and ν columns. We denote by $\overline{\mathfrak{B}}_n$ the extension of $\overline{\mathfrak{Z}}_n$ obtained by adjoining 1.

To prove that $N(n)\leqq[n^2/4]+1$ it suffices to assume that Φ is algebraically closed. For if $A_1, \cdots, A_{N(n)}$ are linearly independent and commutative in Φ_n, then they have these properties in Σ_n for any extension field Σ of the field Φ. Thus $N(n, \Phi)\leqq N(n, \Sigma)$. We shall therefore assume that Φ is algebraically closed. Let $\mathfrak{A}$ be a commutative subalgebra of Φ_n containing the identity and let N be the dimensionality of $\mathfrak{A}$ over Φ. We suppose first that $\mathfrak{A}$ is an indecomposable algebra of matrices. Then it is known that by replacing $\mathfrak{A}$ by a similar set we may suppose that the matrices of $\mathfrak{A}$ have the form

Received by the editors January 14, 1944.

[1] *Zur Theorie vertauschbaren Matrizen*, J. Reine Angew. Math. vol. 130 (1905) pp. 66–76.

$$(2)\qquad \begin{pmatrix} \alpha & & & * \\ & \alpha & & \\ & & \ddots & \\ 0 & & & \alpha \end{pmatrix}.$$

Thus $\mathfrak{A}=(1)+\mathfrak{N}$ where $\mathfrak{N}$ is a nilpotent algebra of matrices in proper triangular form, that is, of the form (2) in which $\alpha=0$. Evidently $\dim \mathfrak{N}=N-1$.

Let the k_1th column $(k_1>1)$ be the first column for which there exists a matrix U_{1k_1} in $\mathfrak{N}$ with element in the $(1, k_1)$ position not equal to 0. We may suppose that the element in the $(1, k_1)$ position of U_{1k_1} is 1. We normalize U_{1k_1} further by using the following lemma.

LEMMA 1. *Let $U\in\Phi_n$ and let V be the matrix obtained from U by adding the kth column multiplied by θ to the lth column $(k\neq l)$ and then subtracting the lth row multiplied by θ from the kth row. Then U and V are similar.*

We have $V=S^{-1}US$ where $S=1+e_{kl}\theta$, e_{kl} the matrix with 1 in the (k, l) position and 0's elsewhere.

We may apply this lemma to U_{1k_1} and replace it by a matrix whose first row is $e_{k_1}=(0, \cdots, 1, 0, \cdots, 0)$ where the 1 is in the k_1th column. The operations required for this purpose are additions of multiples of the k_1th column to later columns and additions to the k_1th row of later rows. These operations replace $\mathfrak{N}$ by a properly triangular set of matrices $\mathfrak{N}'$ similar to $\mathfrak{N}$ such that all the elements in the $(1, j)$ position with $j<k_1$ in $\mathfrak{N}'$ are 0 and such that $\mathfrak{N}'$ contains a matrix V_{1k_1} (similar to U_{1k_1}) whose first row is e_{k_1}. Now let $\mathfrak{P}'$ be the subspace of $\mathfrak{N}'$ of matrices in which the elements in the $(1, k_1)$ position are 0 and suppose that the k_2th column $(k_2>k_1)$ is the first column for which there is a matrix U_{1k_2} in $\mathfrak{P}'$ with element in the $(1, k_2)$ place not equal to 0. Evidently any matrix in $\mathfrak{N}'$ has the form $V_{1k_1}\beta_1+P'$, P' in $\mathfrak{P}'$. We now apply to U_{1k_2} the process used before for U_{1k_1} and replace it by a matrix V_{1k_2} similar to it and having e_{k_2} for first row. The set $\mathfrak{N}'$ will be transformed into a set $\mathfrak{N}''$ of properly triangular matrices and V_{1k_1} changed into a new matrix which we shall again denote as V_{1k_1} with first row e_{k_1}. Any matrix in $\mathfrak{N}''$ has the form $A=V_{1k_1}\beta_1+P''$, P'' in $\mathfrak{P}''$, the transform of the set $\mathfrak{P}'$. It is clear that the elements in the $(1, j)$ position, $j<k_2$, for any matrix in $\mathfrak{P}''$ are 0. Hence $A=V_{1k_1}\beta_1+V_{1k_2}\beta_2+S''$ where S'' is in the subspace $\mathfrak{S}''$ of $\mathfrak{N}''$ of matrices having 0 in the $(1, j)$ position with $j\leq k_2$. This process may be continued and proves the following lemma.

LEMMA 2. *The set $\mathfrak{N}$ is similar to a set $\mathfrak{N}^{(r)}$ of properly triangular* *matrices that contain matrices* $V_{1k_1}, \cdots, V_{1k_r}$ *such that the first row of* V_{1k_i} is e_{k_i}, $1<k_1<k_2<\cdots<k_r$, *and such that any matrix in* $\mathfrak{N}^{(r)}$ *has the form* $\sum V_{1k_i}\beta_i+Z$, *where* Z *has first row* 0.

Now let $\mathfrak{N}_2$ be the subset of $\mathfrak{N}^{(r)}$ of matrices Z having first row 0. Evidently $\mathfrak{N}^{(r)}=\{V_{1k_1}, \cdots, V_{1k_r}\}+\mathfrak{N}_2$ and the V_{1k_i} are linearly independent. Hence dim $\mathfrak{N}^{(r)}=N-1=r+\dim \mathfrak{N}_2$. Now we note that if $Z\in\mathfrak{N}_2$, the first row of $V_{1k_i}Z$ is the k_ith row of Z and the first row of ZV_{1k_i} is 0. Hence the k_ith row of every matrix Z in $\mathfrak{N}_2$ is 0.

We now repeat the argument for $\mathfrak{N}_2$. Then $\mathfrak{N}_2$ may be replaced by a set $\mathfrak{N}_2^{(s)}$ similar to $\mathfrak{N}_2$ such that (1) $\mathfrak{N}_2^{(s)}$ is properly triangular, (2) $\mathfrak{N}_2^{(s)}$ contains matrices $V_{2l_1}, \cdots, V_{2l_s}$ having first row 0 and second row $e_{l_1}, \cdots, e_{l_s}$, respectively, such that any matrix in $\mathfrak{N}_2^{(s)}$ has the form $\sum V_{2l_i}\beta_i+Z$ where Z is a matrix with first two rows 0. Let $\mathfrak{N}_3$ denote the set of matrices Z. We assert that if $s=l_i$ or $s=k_j$ then the sth row of $\mathfrak{N}_3$ is 0. This is clear if $s=l_i$. Hence suppose that $s=k_j\neq$any l_i. Then the matrices of $\mathfrak{N}_2$ all have k_jth row 0 and the operations performed in passing from $\mathfrak{N}_2$ to $\mathfrak{N}_2^{(s)}$ do not affect this row. Hence the k_jth row of every matrix in $\mathfrak{N}_2^{(s)}$ is 0. Evidently $N-1=r+s+\dim \mathfrak{N}_3$.

We now write $k_i=k_{1i}$, $l_i=k_{2i}$, $r=r_1$, $s=r_2$. Then if we continue this process we see that $N-1$ is equal to the number of matrices in the following set

$$
\begin{array}{ll}
 & e_{1k_{11}}, \cdots, e_{1k_1r_1} \\
(3) & e_{2k_{21}}, \cdots, e_{2k_2r_2} \\
 & \cdot\ \cdot\ \cdot\ \cdot\ \cdot\ \cdot\ \cdot
\end{array}
$$

where $1<k_{11}<\cdots<k_{1r_1}$, $2<k_{21}<k_{22}<\cdots<k_{2r_2}, \cdots$, and $r_i=0$ if $i=k_{jl}$ with $j<i$. Let $s_1, s_2, \cdots, s_m$ be the complete set of integers k_{ij} arranged in increasing order. Then it is clear that $N-1 \leqq N(s_1, s_2, \cdots, s_m)$, the number of matrices in the set

$$
\begin{array}{ll}
 & e_{1s_1}, e_{2s_1}, \cdots, e_{s_1-1,s_1} \\
(4) & e_{1s_2}, e_{2s_2}, \cdots, e_{s_1-1,s_2}, e_{s_1+1,s_2}, \cdots, e_{s_2-1,s_2} \\
 & \cdot\ \cdot\ \cdot\ \cdot\ \cdot\ \cdot\ \cdot\ \cdot\ \cdot\ \cdot\ \cdot\ \cdot\ \cdot\ \cdot\ \cdot\ \cdot\ \cdot\ \cdot\ \cdot\ \cdot .
\end{array}
$$

Evidently

$$
\begin{aligned}
N(s_1, s_2, \cdots, s_m) &= (s_1-1)+(s_2-2)+\cdots+(s_m-m) \\
&= \sum s_i - m(m+1)/2.
\end{aligned} \tag{5}
$$

Hence we have

$$N - 1 \leqq N(s_1, \cdots, s_m) \leqq N(n - m + 1, \cdots, n) = m(n - m). \tag{6}$$

Now $m(n-m)$ attains its maximum value for $m = [n/2]$. If $n = 2\nu$ this maximum is ν^2 and if $n = 2\nu - 1$, it is $\nu(\nu-1)$. Thus the maximum value is $[n^2/4]$. This proves for indecomposable algebras $\mathfrak{A}$ the following theorem.

THEOREM 1. *If $\mathfrak{A}$ is a commutative subalgebra of* Φ_n, dim $\mathfrak{A} \leqq [n^2/4] + 1$.

If $\mathfrak{A}$ is decomposable we suppose that the matrices of $\mathfrak{A}$ have the form

$$\begin{pmatrix} A & 0 \\ 0 & B \end{pmatrix}$$

where $A \in \Phi_{n_1}$ and $B \in \Phi_{n_2}$, $n_i \geqq 1$, $n_1 + n_2 = n$. We may assume that the theorem holds for the Φ_{n_i}.

Case 1. $n = 2\nu - 1$, $n_1 = 2\nu_1 - 1$, $n_2 = 2\nu_2$. Here $\nu = \nu_1 + \nu_2$ and $N \leqq \nu_1(\nu_1 - 1) + 1 + \nu_2^2 + 1 \leqq \nu(\nu - 1) + 1$. Equality holds between the last two terms only when $n = 3$.

Case 2. $n = 2\nu$, $n_1 = 2\nu_1 - 1$, $n_2 = 2\nu_2 - 1$. Here $\nu = \nu_1 + \nu_2 - 1$ and $N \leqq \nu_1(\nu_1 - 1) + 1 + \nu_2(\nu_2 - 1) + 1 \leqq \nu^2 + 1$. Equality holds only if $n = 2$.

Case 3. $n = 2\nu$, $n_1 = 2\nu_1$, $n_2 = 2\nu_2$. Here $\nu = \nu_1 + \nu_2$ and $N = \nu_1^2 + 1 + \nu_2^2 + 1 < \nu^2 + 1$. Thus the theorem is proved.

We have also proved the following theorem.

THEOREM 2. *The maximum number $N(n)$ of linearly independent commutative matrices of n rows and columns is given by the formula* $N(n) = [n^2/4] + 1$.

We shall investigate next the form of commutative subalgebras $\mathfrak{A}$ of Φ_n of the maximum dimensionality $N(n)$. Suppose first that $\mathfrak{A}$ has the structure $\mathfrak{A} = (1) + \mathfrak{N}$ where $\mathfrak{N}$ is a nilpotent algebra. Then it is known that by replacing $\mathfrak{A}$ by a similar set we may suppose that the matrices of $\mathfrak{N}$ are properly triangular. We may apply the above considerations to $\mathfrak{N}$. By (3), (4), (5) and (6) we see that if $n = 2\nu$ we must have $k_{11} = k_{21} = \cdots = k_{\nu 1} = \nu + 1, \cdots, k_{1\nu} = k_{2\nu} = \cdots = k_{\nu\nu} = n$ as the set of k's in (3). If $n = 2\nu - 1$ the set of k's is either $k_{11} = \cdots = k_{\nu 1} = \nu + 1, \cdots, k_{1\,\nu-1} = \cdots = k_{\nu\,\nu-1} = n$ or $k_{11} = \cdots = k_{\nu-1\,1} = \nu, \cdots, k_{1\nu} = \cdots = k_{\nu-1\,\nu} = n$. Suppose first that n is even. Let $\mathfrak{N}^{(r)}$ $(r = \nu)$ and $\mathfrak{N}_2$ be determined as before. It is clear that $\mathfrak{N}^{(r)}$ is similar to $\mathfrak{N}$ by a matrix in Φ_n and we need not assume here that Φ is algebraically closed. The matrices of $\mathfrak{N}_2$ have the form

$$B = \left(\begin{array}{c|c} 0 \cdots 0 & \overbrace{0 \cdots 0}^{\nu} \\ R & A \\ \hline 0 & 0 \end{array}\right) \Big\}\nu\,. \tag{7}$$

Since $k_{21}=\nu+1$ it is clear that the second row of R is 0. Moreover the operations used to pass from $\mathfrak{N}_2$ to $\mathfrak{N}_3$ affect only the last ν rows and last ν columns of $\mathfrak{N}_2$. Hence the third row of R is the same as the third row of the corresponding matrix in $\mathfrak{N}_3$. Since $k_{31}=\nu+1$ the third row of R is 0. Similarly the other rows of R are 0, and $R=0$ in (7). Now $\dim \mathfrak{N}_2=\nu^2-\nu$. Hence $\mathfrak{N}_2$ consists of all matrices of the form (7) in which $R=0$ and A is arbitrary. Let

$$V_{1j} = \left(\begin{array}{c|c} 0 & 0 \cdots 0 \;\; 1 \;\; 0 \cdots 0 \\ & V_j \\ \hline & T_j \end{array}\right), \qquad j = \nu+1, \cdots, n,$$

where the 1 is in the jth column and T_j is a properly triangular matrix. Since $V_{1j}B=BV_{1j}$ the following holds in Φ_ν:

$$\begin{pmatrix} 0 \cdots 0 \\ A \end{pmatrix} T_j = 0.$$

Since A is arbitrary, $T_j=0$. Thus $\mathfrak{N}^{(r)}$ is the set $\mathfrak{Z}_n$ and $\mathfrak{A}$ is similar to the algebra $\mathfrak{B}_n$ defined before. If n is odd a similar argument shows that $\mathfrak{A}$ is similar either to $\mathfrak{B}_n$ or to $\overline{\mathfrak{B}}_n$.

We suppose now that $\mathfrak{A}$ is arbitrary. Evidently $\mathfrak{A}$ contains the identity matrix. Since $n>3$ by the proof of Theorem 1, $\mathfrak{A}$ is indecomposable. Moreover if Ω is the algebraic closure of Φ then $\mathfrak{A}_\Omega$ is an indecomposable algebra containing the identity. It follows that $\mathfrak{A}_\Omega$ is similar to a set of matrices of the form (1). Hence $\mathfrak{A}_\Omega=(1)+\mathfrak{Z}$ where $\mathfrak{Z}$ is nilpotent and so $\mathfrak{A}_\Omega$ is similar to either $\mathfrak{B}_n(\Omega)$ or $\overline{\mathfrak{B}}_n(\Omega)$. Thus $\mathfrak{Z}$ is a zero algebra. Now let $\mathfrak{N}$ be the radical of the algebra $\mathfrak{A}$ and consider the semi-simple algebra $\overline{\mathfrak{A}}=\mathfrak{A}-\mathfrak{N}$. The extension $\overline{\mathfrak{A}}_\Omega$ is a homomorphic image of $\mathfrak{A}_\Omega$. Hence $\overline{\mathfrak{A}}_\Omega=(1)+\overline{\mathfrak{Z}}$ where $\overline{\mathfrak{Z}}$ is a zero algebra. The structure of $\overline{\mathfrak{A}}$ is given by the following lemma.

LEMMA 3. *If $\overline{\mathfrak{A}}$ is a semi-simple commutative algebra such that $\overline{\mathfrak{A}}_\Omega=(1)+\overline{\mathfrak{Z}}$ where $\overline{\mathfrak{Z}}$ is a zero algebra, then either $\overline{\mathfrak{A}}=(1)$ or Φ is an imperfect field of characteristic 2 and $\overline{\mathfrak{A}}=\Phi(x)$ where $x^2=\xi$, a non-square in Φ.*

Since $\mathfrak{A}$ is semi-simple, $\overline{\mathfrak{A}}$ is a direct sum of fields, but since $\overline{\mathfrak{A}}_\Omega$ has only one idempotent element, $\overline{\mathfrak{A}}$ is a field. Let $\overline{\mathfrak{A}}>(1)$. Then $\overline{\mathfrak{A}}$ has no separable subfields, for if Σ were such a subfield Σ_Ω is a direct sum of fields and $\overline{\mathfrak{A}}_\Omega$ would contain more than one idempotent element. Thus Φ has characteristic $p\neq 0$ and $\overline{\mathfrak{A}}$ contains an element x such that $x^p=\xi$ is in Φ where ξ is not a pth power in Φ. Now there exists an element η in Ω such that $\eta^p=\xi$ and hence the element $z=x-\eta$ in $\overline{\mathfrak{A}}_\Omega$ is nilpotent of index p. Since $\overline{\mathfrak{Z}}$ is a zero algebra, $p=2$. It follows readily that in this case $\mathfrak{A}=\Phi(x)$, $x^2=\xi$.

This lemma shows that unless Φ is an imperfect field of characteristic 2 any commutative subalgebra $\mathfrak{A}$ of $\Phi_n(n>3)$ of maximum dimensionality has a difference algebra with respect to its radical $\mathfrak{R}$ of dimensionality 1. Since $\mathfrak{A}$ contains the identity, $\mathfrak{A}=(1)+\mathfrak{R}$. As we have seen, this implies that $\mathfrak{A}$ is similar to either $\mathfrak{B}_n$ or to $\overline{\mathfrak{B}}_n$.

THEOREM 3. *Suppose that Φ is not an imperfect field of characteristic* 2 *and let $n>3$. Then if $\mathfrak{A}$ is a subalgebra of Φ_n of maximum dimensionality $N(n)$, $\mathfrak{A}$ is similar to $\mathfrak{B}_n$ if $n=2\nu$ and $\mathfrak{A}$ is similar to either $\mathfrak{B}_n$ or $\overline{\mathfrak{B}}_n$ if $n=2\nu-1$.*[2]

As a consequence we have the following theorem.

THEOREM 4. *Let Φ, n and $\mathfrak{A}$ be as in Theorem* 3. *Then $\mathfrak{A}=(1)+\mathfrak{R}$ where $\mathfrak{R}$ is a zero algebra.*

We remark finally that if n is odd the sets $\mathfrak{B}_n$ and $\overline{\mathfrak{B}}_n$ are not similar. This may be seen by considering the sets $\mathfrak{Z}_n$ and $\overline{\mathfrak{Z}}_n$. Let $\mathfrak{S}(\overline{\mathfrak{S}})$ be the space determined by the columns of the matrices of $\mathfrak{Z}_n(\overline{\mathfrak{Z}}_n)$. Then $\dim \mathfrak{S}=\nu$ and $\dim \overline{\mathfrak{S}}=\nu-1$. On the other hand if $\mathfrak{Z}_n$ were similar to $\overline{\mathfrak{Z}}_n$ we would have $\dim \mathfrak{S}=\dim \overline{\mathfrak{S}}$. It follows that $\mathfrak{Z}_n$ and $\overline{\mathfrak{Z}}_n$ are not similar and hence $\mathfrak{B}_n$ and $\overline{\mathfrak{B}}_n$ are not similar. Thus in this case there are for $n=2\nu-1>3$ two distinct classes in the sense of similarity of commutative subalgebras of dimensionality $N(n)$.

JOHNS HOPKINS UNIVERSITY

---

[2] If $n=2, 3$, $\mathfrak{A}$ may be decomposable. The determination of these algebras is readily obtained.

Reprinted from
Bulletin of the American Mathematical Society
June 1944.

RELATIONS BETWEEN THE COMPOSITES OF A FIELD AND THOSE OF A SUBFIELD.*

By N. JACOBSON.

The present note is an addendum to a recent paper appearing in this *Journal* on a Galois theory for arbitrary fields.[1] We recall that the fundamental concept of the general theory is that of a composite of a field P with itself defined to be a system $\Gamma = (K, S, T)$ consisting of a ring K and two isomorphisms S and T of P into subfields P^S and P^T of K such that 1) K is commutative, 2) $K = P^S P^T$, 3) $1^S = 1^T$, and 4) $(K : P^T)$ is finite. Now let Σ be a subfield of finite index (i. e., $(P : \Sigma)$ finite) in P; then it is readily seen that Γ determines a composite $\Gamma(\Sigma) = (\Sigma^S\Sigma^T, S, T)$ of Σ with itself. In this paper we shall investigate relations between Γ and $\Gamma(\Sigma)$. In the special case where $P \geq \Sigma \geq \Phi$ and P is finite and separable over Φ, the correspondence between Γ and $\Gamma(\Sigma)$ induces a homomorphism between the hypergroup $H_{P|\Phi}$ of P over Φ [2] and the hypergroup $H_{\Sigma|\Phi}$ of Σ over Φ. We show also that the hypergroup $H_{\Sigma|\Phi}$ is isomorphic to the hypergroup $H_{P|\Phi} /\!\!/ H_{P|\Sigma}$ of double cosets of $H_{P|\Sigma}$ in $H_{P|\Phi}$. This implies that the hypergroup of a separable field is isomorphic to the hypergroup of double cosets of a group and hence is completely regular in the sense of Dresher and Ore.[3] In the last section of this paper we give an independent proof of this fact by deriving certain properties of inverses of self-representation that may be of intrinsic interest.

1. Composites and self-representations induced in a subfield. Let P be an arbitrary field and let $\Gamma = (K, S, T)$ be a composite of P with itself. Suppose that Σ is a subfield of finite index in P. Then if $(K : P^T) = m$ and $(P : \Sigma) = q$, $(K : \Sigma^T) = mq$ and so $(\Sigma^S\Sigma^T : \Sigma^T) \leq mq$. It follows that

* Received April 27, 1944.

[1] "An extension of Galois theory to non-normal and non-separable fields," vol. 66 (1944), pp. 1-29, referred to as E.

[2] In a slightly different form this hypergroup was first defined by Kaloujnine in "Sur la théorie de Galois des corps nongaloisiens séparables," *Comptes Rendus de l'Académie des Sciences*, vol. 214 (1942), pp. 597-599. Cf. also E. pp. 24-26.

[3] "Theory of multigroups," this *Journal*, vol. 60 (1938), pp. 705-733.

636

$(\Sigma^S\Sigma^T, S, T)$ is a composite of Σ with itself. We shall denote this composite as $\Gamma(\Sigma)$ and shall call it the *contraction* of Γ to the subfield Σ.

Suppose next that E is a self-representation of P. Then if $\gamma \epsilon \Sigma$, the correspondence $\gamma \rightarrow \gamma^E$ is a representation of Σ by matrices with elements in P. Let R be a regular representation of P over Σ. Then the elements αR_{pq} of the representing matrices $\alpha^R = (\alpha R_{pq})$ are in Σ. Now we may replace the elements γE_{ij} of $\gamma^E = (\gamma E_{ij})$ by the matrices $\gamma E_{ij}{}^R$ and obtain in this way a self-representation $G = E \times R$ of Σ. Let $\mathfrak{R}$ be the double P-module and $x_1, \cdots, x_n$ the right basis of $\mathfrak{R}$ that gives rise to E. If $\rho_1, \cdots, \rho_q$ is a basis for P over Σ that gives rise to the regular representation R, then the vectors $x_1\rho_1, \cdots, x_1\rho_q$; $x_2\rho_1, \cdots, x_2\rho_q; \cdots; \cdots, x_n\rho_q$ form a right Σ-basis for $\mathfrak{R}$. Then $\mathfrak{R}$ may be regarded as a double Σ-module $\mathfrak{R}(\Sigma)$ and it is readily seen that the self-representation obtained from the basis $x_i\rho_j$ in the order given is $G = E \times R$. If $\Gamma = (\mathrm{P}^E\mathrm{P}^D, E, D)$ is the composite of E, we know that $\Gamma = (\mathrm{P}_l\mathrm{P}_r, L, R)$.[4] Since the composite of $\mathfrak{R}(\Sigma)$ is $(\Sigma_l\Sigma_r, L, R)$, it is clear that the composite of the self-representation G of Σ is the contraction of the composite of E.

Now let Γ' be a given composite of Σ with itself and let G be a self-representation of Σ having Γ' as its composite. Again let R be a regular representation of P over Σ. For the element αR_{pq} of the matrix α^R we now substitute $\alpha R_{pq}{}^G$ and we obtain in this way a new self-representation E of P such that the elements αE_{ij} of the matrices α^E all lie in Σ. For γ in Σ we have

$$\gamma^E = \left.\begin{pmatrix} \gamma^G & & & \\ & \cdot & & \\ & & \cdot & \\ & & & \cdot \\ & & & & \gamma^G \end{pmatrix}\right\} q \quad = \quad 1_q \times \gamma^G.$$

Hence in the self-representation $E \times R$ of Σ we have

$$\gamma^{E\times R} = \begin{pmatrix} \gamma^G \times 1_q & & & \\ & \cdot & & \\ & & \cdot & \\ & & & \cdot \\ & & & & \gamma^G \times 1_q \end{pmatrix}$$

and this matrix is similar to $1_{q^2} \times \gamma^G$. Thus the self-representation $E \times R$ of Σ is similar to a multiple of the given self-representation G. Hence the composite Γ' of G is the contraction $\Gamma(\Sigma)$ of the composite Γ of E.

---

[4] As in E equivalent composites are identified. The symbol $=$ is used for equivalence and we write $\Gamma_1 \geq \Gamma_2$ for "Γ_1 is a cover of Γ_2."

We recall that a composite $\Gamma = (K, S, T)$ of P is simple if $K = \mathrm{P}^S\mathrm{P}^T$ is a field. Then if Σ is a subfield of finite index in P, K is finite over Σ^T and hence $\Sigma^S\Sigma^T$ is a field. Thus the contraction $\Gamma(\Sigma)$ is simple. Conversely let Γ' be a given simple composite of Σ with itself and let Γ be a composite of P with itself such that $\Gamma(\Sigma) = \Gamma'$. Suppose that $\mathfrak{R}$ is a double P-module having the composite Γ and let $\mathfrak{S}$ be an irreducible submodule of $\mathfrak{R}$. Then if $\bar{\Gamma}$ is the composite of $\mathfrak{S}$, $\bar{\Gamma}(\Sigma) \leq \Gamma(\Sigma) = \Gamma'$. Since Γ' is simple, $\bar{\Gamma}(\Sigma) = \Gamma'$. We summarize these results in the following

THEOREM 1. *If Σ is a subfield of finite index in P and Γ' is a composite of Σ with itself, then there exists a composite Γ of P with itself whose contraction $\Gamma(\Sigma) = \Gamma'$. Moreover, if Γ' is simple, Γ may be taken to be simple.*

2. Conditions for equivalence. We recall that the composite of a self-representation E is determined by the set of endomorphisms $\Sigma E_{ij}\bar{\rho}_{ij}$ where $\bar{\rho}$ denotes the multiplication $\xi \to \xi\rho$ in P. Since $\bar{\rho}E_{ij} = \Sigma E_{ia}\bar{\rho}_{aj}$, it follows that a necessary and sufficient condition that $\alpha \to \alpha^E$ and $\alpha \to \alpha^F = (\alpha F_{kl})$ have equivalent composites is that each F_{kl} be expressible in the form $\Sigma E_{ij}\bar{\rho}_{ij,kl}$ and each E_{ij} have the form $\Sigma F_{kl}\bar{\tau}_{kl,ij}$. We recall also that if R is the regular self-representation of P over Σ, its composite Δ is closed and the set $\Sigma R_{pq}\bar{\rho}_{pq}$ is the complete ring of linear transformations of P over Σ. If $\gamma \in \Sigma$, $\gamma R_{pq} = \delta_{pq}\gamma$ and hence the totality $\Sigma R_{pq}\bar{\sigma}_{pq}$, σ in Σ, reduces to the set of multiplications $\bar{\sigma}$ if these endomorphisms are restricted to act in Σ. We may now prove the following

THEOREM 2. *Let Γ_1 and Γ_2 be two composites of P, Σ a subfield of finite index in P and Δ the closed composite of P over Σ. Then if*

$$\Gamma_1(\Sigma) = \Gamma_2(\Sigma), \qquad \Delta \times \Gamma_1 \times \Delta = \Delta \times \Gamma_2 \times \Delta.$$

Let E_1 and E_2 be self-representations of P with composites Γ_1 and Γ_2 respectively and let R be a regular representation of P over Σ. Then the composite of R is Δ. Suppose now that $\Gamma_1(\Sigma) = \Gamma_2(\Sigma)$. Then the induced representations $G_1 = E_1 \times R$ and $G_2 = E_2 \times R$ of Σ have the same com-composite. Hence if $E_{ij}{}^{(1)}, E_{kl}{}^{(2)}, R_{pq}$ denote the endomorphisms associated with E_1, E_2 and R respectively, there exist elements μ, ν in Σ such that

$$\gamma E_{ij}{}^{(1)} R_{pq} = \gamma(\Sigma E_{kl}{}^{(2)} R_{p'q'}\bar{\mu}_{klp'q',ijpq})$$
$$\gamma E_{kl}{}^{(2)} R_{pq} = \gamma(\Sigma E_{ij}{}^{(1)} R_{p'q'}\bar{\nu}_{ijp'q',klpq})$$

for all γ in Σ. Since for any α in P, αR_{rs} is in Σ, it follows that these equations are valid when γ is replaced by αR_{rs}. The resulting equations show that $R \times E_1 \times R$ and $R \times E_2 \times R$ have the same composite. Thus

$$\Delta \times \Gamma_1 \times \Delta = \Delta \times \Gamma_2 \times \Delta.$$

3. Combinatorial properties of composites. It follows directly from the definitions that if Γ_1 and Γ_2 are composites of P and Γ_1 is a cover of $\Gamma_2 (\Gamma_1 \geq \Gamma_2)$ then the contraction $\Gamma_1(\Sigma) \geq \Gamma_2(\Sigma)$. If $\Gamma_1 + \Gamma_2$ denotes the least common cover of Γ_1 and Γ_2 then

$$(\Gamma_1 + \Gamma_2)(\Sigma) = \Gamma_1(\Sigma) + \Gamma_2(\Sigma).$$

Concerning multiplication of composites we have the following

THEOREM 3. *Let Γ_1, Γ_2, Σ and Δ be as in Theorem 2. Then*

$$(\Gamma_1 \times \Delta \times \Gamma_2)(\Sigma) = \Gamma_1(\Sigma) \times \Gamma_2(\Sigma).$$

Let E_1, E_2 and R be determined as before, and let $G_1 = E_1 \times R$, $G_2 = E_2 \times R$ be the induced representations of Σ. Then it is immediate that $G_1 \times G_2 = (E_1 \times R \times E_2) \times R$ is the self-representation of Σ induced by $E_1 \times R \times E_2$. It follows that

$$\Gamma_1(\Sigma) \times \Gamma_2(\Sigma) = (\Gamma_1 \times \Delta \times \Gamma_2)(\Sigma).$$

COROLLARY 1. *For any composites Γ_1, Γ_2 we have*

$$(\Gamma_1 \times \Gamma_2)(\Sigma) \leq \Gamma_1(\Sigma) \times \Gamma_2(\Sigma).$$

Since Δ contains the identity composite, $\Gamma_1 \times \Gamma_2 \leq \Gamma_1 \times \Delta \times \Gamma_2$. The corollary then follows from Theorem 3.

We also have the following partial converse of Theorem 2.

COROLLARY 2. *If Γ_1 and Γ_2 are simple composites such that*

$$\Delta \times \Gamma_1 \times \Delta = \Delta \times \Gamma_2 \times \Delta$$

then

$$\Gamma_1(\Sigma) = \Gamma_2(\Sigma).$$

For

$$(\Delta \times \Gamma_1 \times \Delta)(\Sigma) \leq \Delta(\Sigma) \times \Gamma_1(\Sigma) \times \Delta(\Sigma) = \Gamma_1(\Sigma).$$

Suppose now that $\Gamma = (K, S, T)$ is a simple non-singular composite of P. Then $(K : \mathrm{P}^S) = (K : \mathrm{P}^T)$ and so $(K : \Sigma^S) = (K : \Sigma^T)$. Since $\Sigma^S \Sigma^T$ is a subfield we have

$$(K:\Sigma^S) = (K:\Sigma^S\Sigma^T)(\Sigma^S\Sigma^T:\Sigma^S)$$
$$(K:\Sigma^T) = (K:\Sigma^S\Sigma^T)(\Sigma^S\Sigma^T:\Sigma^T).$$

Hence $(\Sigma^S\Sigma^T:\Sigma^S) = (\Sigma^S\Sigma^T:\Sigma^T)$. This proves that $\Gamma(\Sigma)$ is non-singular. It is evident from the definition that $\Gamma(\Sigma) = \Gamma^{-1}(\Sigma)$.

THEOREM 4. *If Γ is a non-singular simple composite, then the contraction $\Gamma(\Sigma)$ is non-singular and simple and $\Gamma(\Sigma)^{-1} = \Gamma^{-1}(\Sigma)$.*

4. The hypergroup of a separable field. We suppose now that P is finite and separable over a subfield Φ and we let $H_{P|\Phi}$ denote the hypergroup of simple composites of P over Φ (i. e., leaving the elements of Φ fixed). We recall that $H_{P|\Phi}$ is finite and the least common cover of all the Γ_i in $H_{P|\Phi}$ is the closed composite Γ of P over Φ. Any composite of P over Φ is the least common cover $\Sigma_k\Gamma_{i_k}$ of certain of the Γ_i in $H_{P|\Phi}$ and by the distributive law we have $(\Sigma\Gamma_{i_k}) \times (\Sigma\Gamma_{j_l}) = \Sigma\Gamma_{i_k} \times \Gamma_{j_l}$. If $\Gamma_1 \times \Gamma_2 = \Sigma\Gamma_{i_k}$, we define the set $\Gamma_{i_1}, \Gamma_{i_2}, \cdots$ to be the product $\Gamma_1\Gamma_2$ of the simple composites Γ_1 and Γ_2. This operation is the hypergroup operation in $H_{P|\Phi}$.

Let Σ be a field between P and Φ and let $H_{P|\Sigma}$ be the subhypergroup of $H_{P|\Phi}$ of simple composites of P over Σ. We recall that any subhypergroup of $H_{P|\Phi}$ is an $H_{P|\Sigma}$ and that this correspondence is $(1-1)$. If Δ is the closed composite corresponding to Σ, $\Delta = \Sigma\Delta_i$, Δ_i in $H_{P|\Sigma}$.

Now if $\Gamma_1 \in H_{P|\Phi}$, the contraction $\Gamma'_1 = \Gamma_1(\Sigma)$ is in the hypergroup $H_{\Sigma|\Phi}$ of Σ over Φ. By Theorem 1 any Γ'_1 in $H_{\Sigma|\Phi}$ may be obtained in this way and by Corollary 1 $(\Gamma_1 \times \Gamma_2)' \leq \Gamma'_1 \times \Gamma'_2$. Thus the correspondence $\Gamma_1 \rightarrow \Gamma'_1$ is a homomorphism between the hypergroups $H_{P|\Phi}$ and $H_{\Sigma|\Phi}$. Evidently the kernel of this homomorphism is $H_{P|\Sigma}$. Now let Γ_1 and Γ_2 be two elements of $H_{P|\Phi}$, such that $\Gamma'_1 = \Gamma'_2$. Then by Theorem 2,

$$\Delta \times \Gamma_1 \times \Delta = \Delta \times \Gamma_2 \times \Delta.$$

Since $\Delta \times \Gamma_1 \times \Delta = \Sigma(\Delta_i \times \Gamma_1 \times \Delta_j)$ for Δ_i, Δ_j in $H_{P|\Sigma}$, $\Delta \times \Gamma_1 \times \Delta$ is the least common cover of all the elements in the double coset $H_{P|\Sigma}\Gamma_1 H_{P|\Sigma}$. Hence

$$\Delta \times \Gamma_1 \times \Delta = \Delta \times \Gamma_2 \times \Delta$$

implies that $H_{P|\Sigma}\Gamma_1 H_{P|\Sigma} = H_{P|\Sigma}\Gamma_2 H_{P|\Sigma}$. Next let $\Gamma_2 \in H_{P|\Sigma}\Gamma_1 H_{P|\Sigma}$. Then $\Gamma_2 \leq \Delta \times \Gamma_1 \times \Delta$ and $\Gamma'_2 \leq \Gamma'_1$. Since Γ'_1 is simple, $\Gamma'_2 = \Gamma'_1$ and so $H_{P|\Sigma}\Gamma_1 H_{P|\Sigma} = H_{P|\Sigma}\Gamma_2 H_{P|\Sigma}$. This shows that two double cosets are either identical or their intersection is vacuous. The double cosets of $H_{P|\Sigma}$ give a division of $H_{P|\Phi}$ into mutually exclusive sets and those cosets define a factor hypergroup $H_{P|\Phi} /\!/ H_{P|\Sigma}$. It is clear also that the double cosets of

$H_{\Sigma|\Phi}$ are in $(1-1)$ correspondence with the elements of $H_{P|\Sigma}$, namely, each double coset is the inverse image of an element of $H_{\Sigma|\Phi}$ relative to the homomorphism between $H_{P|\Phi}$ and $H_{\Sigma|\Phi}$. We consider now the product $(H_{P|\Sigma}\Gamma_1 H_{P|\Sigma})(H_{P|\Sigma}\Gamma_2 H_{P|\Sigma})$. The least common cover of the elements of this set is the composite $\Delta\times\Gamma_1\times\Delta\times\Gamma_2\times\Delta$. By Theorem 3,

$$(\Delta\times\Gamma_1\times\Delta\times\Gamma_2\times\Delta)'=(\Delta\times\Gamma_1)'\times(\Delta\times\Gamma_2)'$$

and since $(\Delta\times\Gamma_1)'=\Gamma'_1$ and $(\Delta\times\Gamma_2)'=\Gamma'_2$ by the simplicity of Γ_1 and Γ_2, $(\Delta\times\Gamma_1\times\Delta\times\Gamma_2\times\Delta)'=\Gamma'_1\times\Gamma'_2$. It follows that the double cosets contained in the product $(H_{P|\Sigma}\Gamma_1 H_{P|\Sigma})(H_{P|\Sigma}\Gamma_2 H_{P|\Sigma})$ are the double cosets corresponding to the elements of $\Gamma'_1\Gamma'_2$. Hence we have proved that $(H_{P|\Phi}/\!/H_{P|\Sigma})$ is isomorphic to $H_{\Sigma|\Phi}$.

THEOREM 5. *Let* P *be finite and separable over* Φ *and let* Σ *be a field between* P *and* Φ. *Then the correspondence* $\Gamma_i\to\Gamma'_i=\Gamma_i(\Sigma)$ *is a homomorphism between the hypergroups* $H_{P|\Phi}$ *and* $H_{\Sigma|\Phi}$. *The kernel of this homomorphism is* $H_{P|\Sigma}$ *and the double cosets of* $H_{P|\Sigma}$ *form a hypergroup* $H_{P|\Phi}/\!/H_{P|\Sigma}$ *isomorphic to* $H_{\Sigma|\Phi}$.

As has been shown by Dresher and Ore, $H_{P|\Phi}/\!/H_{P|\Sigma}$ is a group if and only if $H_{P|\Sigma}$ is strongly normal in $H_{P|\Phi}$ in the sense that for any Γ_1, $\Gamma_1^{-1}H_{P|\Sigma}\Gamma_1=H_{P|\Sigma}$. We recall also that $H_{\Sigma|\Phi}$ is a group if and only if Σ is normal over Φ. Hence we have

THEOREM 6. *Let* $P\geq\Sigma\geq\Phi$ *where* P *is separable and finite over* Φ. *Then* Σ *is normal over* Φ *if and only if* $H_{P|\Sigma}$ *is strongly normal in* $H_{P|\Phi}$. *When the condition is satisfied* $H_{P|\Phi}/\!/H_{P|\Sigma}$ *is isomorphic to the Galois group* $H_{\Sigma|\Phi}$ *of* Σ *over* Φ.

Theorem 5 enables us to obtain very precise information on the nature of the hypergroup $H_{\Sigma|\Phi}$ of a finite separable extension Σ of Φ. For we may extend Σ to the finite separable and normal extension P over Φ. Then $H_{P|\Phi}$ is a group isomorphic to the Galois group of P over Σ and $H_{P|\Sigma}$ is the subgroup of $H_{P|\Phi}$ corresponding to the Galois group of P over Σ. Thus the hypergroup $H_{\Sigma|\Phi}$ is a hypergroup of double cosets of a finite group. Such hypergroups are known to have many important special properties. They are, for example, completely regular in the sense that for any Δ_1 in $H_{\Sigma|\Phi}$ the only solutions of either of the equations $\Delta_1\bar{\Delta}=\{1,\cdots\}$ or $\bar{\Delta}\Delta_1=\{1,\cdots\}$ is $\bar{\Delta}=\Delta_1^{-1}$. In the remainder of this paper we shall give a direct proof of this property based on some general theorems on inverses of self-representation.

5. Properties of the inverse of a self-representation. Let E be a non-singular self-representation of P of rank n and let E^* be its inverse. Then if $\alpha^E = (\alpha E_{ij})$ and $\alpha^{E^*} = (\alpha E^*_{ij})$ we have the defining relations

$$\sum_k E_{ki} E^*_{jk} = \delta_{ij}, \qquad \sum E^*_{ki} E_{jk} = \delta_{ij}$$

where δ_{ij} is the 0 endomorphism or the identity according as $i \neq j$ or $i = j$. Suppose that $\mathfrak{R}$ is the double P-module and $x_1, \cdots, x_n$ a right P-basis giving rise to E, so that $\alpha x_i = \Sigma x_j (\alpha E_{ji})$. Then $x_1, \cdots, x_n$ is also a left P-basis. Similarly in the inverse double P-module $\mathfrak{R}^{-1}$ we have a basis $x^*_1, \cdots, x^*_n$ such that $\alpha x^*_i = \Sigma x^*_j (\alpha E^*_{ji})$. Finally we let $\mathfrak{R}'$ be the double P-module corresponding to the transposed representation E' and let $x'_1, \cdots, x'_n$ be a right P-basis such that $\alpha x'_i = \Sigma x'_j (\alpha E_{ij})$. Consider the product module $\mathfrak{P} = \mathfrak{R}' \times \mathfrak{R}^{-1}$. A right basis for this module is $x'_i x^*_j$, $i, j = 1, \cdots, n$. Thus the element $u = \Sigma x'_i x^*_i \neq 0$ in $\mathfrak{P}$. Moreover the following relations hold:

$$\alpha u = \alpha \Sigma x'_i x^*_i = \Sigma x'_j \alpha E_{ij} x^*_i = \Sigma x'_j x^*_k \alpha E_{ij} E^*_{ki} = (\Sigma x'_i x^*_i) \alpha = u\alpha.$$

Evidently the vector $u_\rho = \Sigma (x'_i \rho) x^*_i$ also satisfies the relation $\alpha u_\rho = u_\rho \alpha$. Suppose now that (K, T, S) is the composite of E^* and let $(K : P^S) = m$. Then there exist m elements $\rho_1, \cdots, \rho_m$ in P such that the matrices $\rho_1^{E^*}, \cdots, \rho_m^{E^*}$ are linearly independent over P. We assert that the vectors

$$u_{\rho_l} = \Sigma (x'_i \rho_l) x^*_i = \Sigma x'_i x^*_j (\rho_l E^*_{ji})$$

are linearly independent. For if $\Sigma u_{\rho_l} \zeta_l = 0$, $\Sigma (\rho_l E^*_{ji}) \zeta_l = 0$ for all i, j and so $\Sigma \rho_l^{E^*} \zeta_l = 0$. Hence each $\zeta_l = 0$. If we choose a right basis of P to be $u_{\rho_1}, \cdots, u_{\rho_m}$ and $n^2 - m$ other vectors, the associated self-representation is

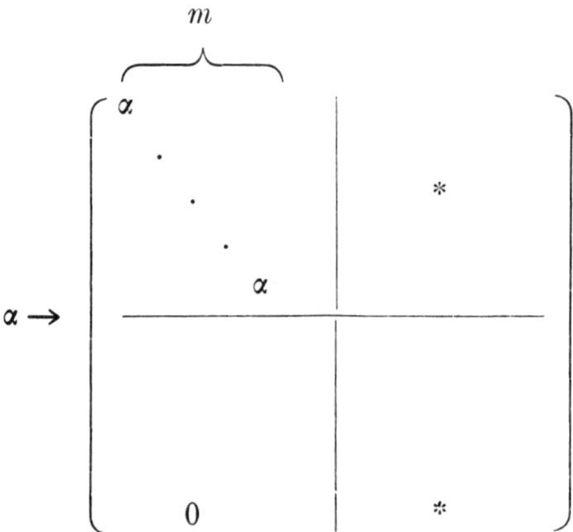

and this representation is similar to $E' \times E^*$.

THEOREM 6. *Let E be a non-singular self-representation of a field* P, *E^* its inverse and E' its transposed. Then if $(K : \mathrm{P}^S) = m$ for the composite (K, T, S) of E^*, the identity self-representation occurs as an irreducible component of $E' \times E^*$ with a multiplicity $r \geq m$.*

We suppose now that P is finite and separable over Φ and let E be an irreducible self-representation of P over Φ (leaving the elements of Φ fixed). We know that the composite $\Gamma_1 = (K_1, S_1, T_1)$ associated with E is simple, and it follows from this that E is similar to the self-representation obtained by regarding K_1 as a double P-module. Hence rank $E = (K_1 : P^{T_1}) \equiv m$ and since Γ_1 is non-singular, $(K_1 : \mathrm{P}^{T_1}) = (K_1 : \mathrm{P}^{S_1})$.

We assume now that E is non-singular and let R be the regular representation of P over Φ. Consider the self-representation $R \times E$. Since the elements $\alpha R_{pq} \epsilon \Phi$, $\alpha^{R \times E} = \alpha^R \times 1_m$ so that $\alpha^{R \times E}$ is similar to the direct sum of R with itself taken m times. We recall that R is completely reducible. that any irreducible self-representation of P over Φ is similar to one of the components of R and that no two components of R are similar. It follows from this that the identity has the multiplicity m in $R \times E$. On the other hand, by Theorem 6, $(E^*)' \times E$ contains the identity with multiplicity $r \geq m$. Since $(E^*)'$ may be taken to be one of the irreducible components of R, we see that the multiplicity of the identity self-representation in $(E^*)' \times E$ is m and if G is any irreducible self-representation of P over Φ such that $G \times E$ contains the identity, then G is similar to $(E^*)'$ and hence to E^*.

Now let E be an arbitrary irreducible self-representation of P over Φ and let F be any irreducible self-representation of P over Φ with composite Γ_1^{-1}. Then by replacing E and F by similar self-representations, we see that $F \times E$ contains the identity as an m-fold component and if G is irreducible with composite $\neq \Gamma_1^{-1}$, $G \times E$ does not contain the identity as a component. By symmetry we see also that $E \times F$ contains the identity as an m-fold component and if G does not have the composite Γ_1^{-1}, $E \times G$ does not contain the identity. This proves the following theorem.

THEOREM 7. *Let* P *be finite and separable over Φ and let E be an irreducible self-representation of rank m of* P *over Φ. Then if F is an irreducible self-representation of* P *over Φ whose composite is the inverse of the composite of E, $E \times F$ and $F \times E$ contain the identity as an m-fold component. If G is an irreducible self-representation of* P *over Φ with composite different from the inverse of the composite of E, then $E \times G$ and $G \times E$ do not contain the identity as a component.*

THEOREM 8. *Let* P *be finite and separable over* Φ *and let* $\Gamma_1 \in H_{P|\Phi}$. *Then* $1 \in \Gamma_1\Gamma_1^{-1}$ *and* $1 \in \Gamma_1^{-1}\Gamma_1$ *and if* Γ_2 *is any element of* $H_{P|\Phi} \neq \Gamma_1^{-1}$, *neither* $\Gamma_1\Gamma_2$ *nor* $\Gamma_2\Gamma_1$ *contain the identity.*

This result amounts to the statement that $H_{P|\Phi}$ is a completely regular hypergroup.

THE JOHNS HOPKINS UNIVERSITY.

Reprinted from
American Journal of Mathematics
October 1944.

GALOIS THEORY OF PURELY INSEPARABLE FIELDS OF EXPONENT ONE.*

By N. Jacobson.

Let P be a purely inseparable extension of finite dimensionality of a field Φ. Then $P = \Phi(x_1, \cdots, x_m)$, where $x_i^p = \xi_i$ is in Φ and p is the characteristic of Φ. Without loss of generality we may assume that $(P : \Phi) = p^m$. Let $\mathfrak{D}(\Phi)$ denote the set of derivations of P over Φ.[1] Then $\mathfrak{D}$ is a vector space of dimensionality m over $\bar{P}$ the set of multiplications $\bar{\beta} : \xi \to \xi\beta$ in P and $\mathfrak{D}$ is closed under commutation and under the operation of taking p-th powers. We call any set $\mathfrak{E}$ of derivations in P having these closure properties a *restricted* P-*Lie* ring of derivations. We shall call $\mathfrak{E}$ *finite* if $(\mathfrak{E} : \bar{P})$ is finite. In a previous paper[2] we set up a $(1-1)$ correspondence between the fields Σ between P and Φ and the restricted P-Lie rings of derivations contained in $\mathfrak{D}$. In this note we obtain the following results: 1) the only elements of P that are constants relative to all the derivations in $\mathfrak{D}$ are the elements of Φ; and 2) if $\mathfrak{D}$ is any finite restricted P-Lie ring of derivations in P and Φ is the subfield of $\mathfrak{D}$-constants, then P is finite and purely inseparable of exponent 1 over Φ and $\mathfrak{D}$ is the complete set of derivations of P over Φ. Thus we have an order anti-isomorphism between the subfields Φ over which P is finite and purely inseparable of exponent 1 and the finite restricted P-Lie rings $\mathfrak{D}$ of derivations in P. This, of course, contains our earlier result. The improvement obtained here, analogous to that of Artin and Baer in the ordinary Galois theory, consists in showing that $\mathfrak{D}$ and Φ serve equally well as starting points of the Galois theory. We remark that the determination of the structure of finite restricted P-Lie rings of derivations is a consequence of our results. Our proofs are based on theorems on self-representations of fields recently obtained by the author.[3]

---

* Received April 27, 1944.

[1] For the definition of a derivation and the properties quoted in this paragraph, see the author's "Abstract derivation and Lie algebras," *Transactions of the American Mathematical Society*, vol. 42 (1937), pp. 206-224. This paper is referred to as D.

[2] D. p. 220.

[3] See "An extension of Galois theory to non-normal and non-separable fields," this *Journal*, vol. 66 (1944), pp. 1-29. We refer to this paper as E.

1. Let $P = \Phi(x_1, \cdots, x_m)$, $x_i^p = \xi_i$ in Φ and let $(P : \Phi) = p^m$ where p is the characteristic of Φ, and let $\mathfrak{D}(\Phi)$ be the restricted P-Lie ring of derivations of P over Φ. If $D \in \mathfrak{D}$ the correspondence

$$\alpha \rightarrow \begin{pmatrix} \alpha & \alpha D \\ 0 & \alpha \end{pmatrix} \tag{1}$$

is a self-representation of P whose field of fixed elements contains Φ. We recall that there exists a derivation D such that the field of D-constants is Φ.[4] Let D have this property. Then the subfield of fixed elements under (1) is precisely Φ. Then by Theorem 13 of E.[3] any linear transformation in P over Φ is a polynomial in D with coefficients in $\bar{P}$. Consider the sequence $1, D, D^2, \cdots$ and let D^r be the first of these transformations that is right linearly dependent over $\bar{P}$ on $1, D, \cdots, D^{r-1}$. Then any polynomial in D and hence any linear transformation in P over Φ may be written in one and only one way in the form $\bar{\beta}_0 + D\bar{\beta}_1 + \cdots + D^{r-1}\bar{\beta}_{r-1}$. Thus if $\mathfrak{L}$ denotes the complete set of linear transformations of P over Φ, $(\mathfrak{L} : \bar{P}) = r$. Since $(P : \Phi) = p^m$ this gives $(\mathfrak{L} : \bar{\Phi}) = p^m r$. On the other hand, as is well known, $(P : \Phi) = p^m$ implies that $(\mathfrak{L} : \bar{\Phi}) = p^{2m}$. Hence $r = p^m$. Now the linear transformations

$$E = D\bar{\beta}_1 + D^p\bar{\beta}_p + D^{p^2}\bar{\beta}_{p^2} + \cdots + D^{p^{m-1}}\bar{\beta}_{p^{m-1}} \tag{2}$$

are derivations and since their totality is an m-dimensional space over $\bar{P}$, this totality coincides with $\mathfrak{D}$. We have therefore proved

LEMMA 1. *If D is a derivation in P over Φ such that the only D-constants are the elements in Φ then any derivation E in P over Φ is a p-polynomial* (2) *in D with coefficients* (*on the right*) *in* P.[5]

We shall require also the following

LEMMA 2. *If $\mathfrak{E}$ is a restricted P-Lie subring of $\mathfrak{D}$, there exists a derivation E in $\mathfrak{E}$ such that any derivation in $\mathfrak{D}$ is a p-polynomial over $\bar{P}$ in E.*[6]

As a consequence we have

LEMMA 3. *If $\mathfrak{E}$ is a restricted P-Lie subring of $\mathfrak{D}$ whose field of constants is Φ, then $\mathfrak{E} = \mathfrak{D}$.*

---

[4] This result is due to Baer, "Algebraische Theorie der differentierbaren Funktionenkorper. I," *Sitzungsberichte Heidelberger Akad.*, 1927, pp. 15-32. Cf. also the author's paper "Classes of restricted Lie algebras of characteristic p, II," *Duke Mathematical Journal*, vol. 10 (1943), p. 111.

[5] This is proved in D. p. 218 by using the theory of linear differential equations.

[6] D. p. 219.

For the field of $\mathfrak{E}$-constants is evidently the same as the field of E-constants. Thus E is a derivation of P whose field of constants is Φ and by Lemma 1 any derivation in P over Φ is a p-polynomial in E.

2. We suppose now that P is any field of characteristic $p \neq 0$ and let $\mathfrak{D}$ be any finite restricted P-Lie ring of derivations in P. Let $D_1, \cdots, D_m$ be a (right) $\bar{\text{P}}$-basis of $\mathfrak{D}$ and let $\mathfrak{A}$ be the smallest ordinary ring of endomorphisms in P containing $\mathfrak{D}$ and $\bar{\text{P}}$. Then we have

LEMMA 4. *$\mathfrak{A}$ is finite dimensional over* $\bar{\text{P}}$.

Proof. Let $\mathfrak{A}'$ denote the totality of endomorphisms of the form $\Sigma D_1^{k_1} D_2^{k_2} \cdots D_m^{k_m} \bar{\beta}_{k_1 \ldots k_m}$ where $0 \leqq k_i < p$ and we set $D_i^0 = 1$. Evidently $\mathfrak{A} \geq \mathfrak{A}' \geq \mathfrak{D}, \text{P}$. Since D_i is a derivation for any β, $\bar{\beta} D_i = D_i \bar{\beta} + \overline{\beta D_i}$ and since $\mathfrak{D}$ is a restricted P-Lie ring, $D_i D_j = D_j D_i + \Sigma D_k \bar{\gamma}_{kij}$ and $D_i^p = \Sigma D_j \bar{\mu}_{ji}$. It is readily seen from these relations that $\mathfrak{A}'$ is a ring. Hence $\mathfrak{A}' = \mathfrak{A}$ and since any element in this ring is a linear combination of the p^m elements $D_1^{k_1} D_2^{k_2} \cdots D_m^{k_m}$, $(\mathfrak{A} : \bar{\text{P}}) \leq p^m$.

We may now prove the following

THEOREM. *Let $\mathfrak{D}$ be a finite restricted* P*-Lie ring of derivations in* P *and let Φ be the subfield of $\mathfrak{D}$-constants. Then* P *is finite and purely inseparable of exponent* 1 *over Φ and $\mathfrak{D}$ is the complete set of derivations of* P *over Φ.*

As above we let $D_1, \cdots, D_m$ be a $\bar{\text{P}}$-basis of $\mathfrak{D}$ and we form the self-representation E:

$$\alpha \to \begin{pmatrix} \alpha & \alpha D_1 & & & & & \\ 0 & \alpha & & & & & \\ & & \alpha & \alpha D_2 & & & \\ & & 0 & \alpha & & & \\ & & & & \ddots & & \\ & & & & & \alpha & \alpha D_m \\ & & & & & 0 & \alpha \end{pmatrix}. \tag{3}$$

The field of fixed elements relative to this representation is Φ. By substituting for the elements β of the matrices in (3) the matrices β^E we obtain the self-representation $E \times E$. We write the resulting matrices as $\alpha^{E \times E} = (\alpha E^{(2)}{}_{kl})$ and note that the endomorphisms $E^{(2)}{}_{kl}$ include $1, D_i, D_i D_j, i, j = 1, \cdots, m$. Similarly we form the product $E \times E \times E$ and obtain $\alpha^{E \times E \times E} = (\alpha E^{(3)}{}_{pq})$

where the $E^{(3)}{}_{pq}$ include $1, D_i, D_iD_j, D_iD_jD_k$. Continuing in this way we obtain finally a self-representation $F = E \times \cdots \times E$ whose endomorphisms include all the elements $D_1^{k_1} \cdots D_m^{k_m}$, $0 \leq k_i < p$. Since these elements generate the ring $\mathfrak{A}$ it follows that the composite associated with F is closed and hence P is finite over the field of fixed elements under F.[7] Since this field is the same as the field Φ of fixed elements under E, $(\mathrm{P} : \Phi)$ is finite. Now let x be any element of p and let $D \in \mathfrak{D}$. Then $x^pD = 0$. Hence $x^p = \xi$ is in Φ and P has exponent 1 over Φ. This proves the first part of the theorem. The second part of the theorem is an immediate consequence of Lemma 3.

We remark that by the above theorem and a previous result[8] we obtain the

Corollary. *Let $\mathfrak{D}$ be a finite restricted P-Lie ring of derivations in P and let Φ be the subfield of $\mathfrak{D}$-constants. Then unless $\mathrm{P} = \Phi(x)$, $x^2 = \xi$ in Φ, $\mathfrak{D}$ is a simple Lie algebra over Φ.*

Now by Lemma 1, if $\mathrm{P} = \Phi(x_1, \cdots, x_m)$, $x_i^p = \xi_i$ in Φ and $\mathfrak{D}$ is the derivation ring of P over Φ, then Φ is the field of $\mathfrak{D}$-constants. This completes the proof of the $(1-1)$ correspondence between the subfields Φ of P over which P is finite and purely inseparable of exponent 1 and the finite restricted P-Lie rings $\mathfrak{D}(\Phi)$ in P. Evidently if $\mathrm{P} \geq \Sigma \geq \Phi$ then $\mathfrak{D}(\Sigma) \leq \mathfrak{D}(\Phi)$ and conversely.

The Johns Hopkins University.

Reprinted from
American Journal of Mathematics
October 1944.

---

[7] E. p. 19.

[8] D. p. 218.

ANNALS OF MATHEMATICS
Vol. 45, No. 4, October, 1944

CONSTRUCTION OF CENTRAL SIMPLE ASSOCIATIVE ALGEBRAS

BY N. JACOBSON

(Received December 20, 1943)

It is well known that any central division algebra $\mathfrak{A}$ over Φ contains a subfield P of dimensionality n if n^2 is the dimensionality of $\mathfrak{A}$ over Φ. Moreover the field P may be chosen to be separable over Φ and there exist algebras of the form $\mathfrak{B} = \mathfrak{A} \times \Phi_r$ that contain normal and separable subfields of dimensionality nr over Φ. The algebra $\mathfrak{B}$ is a normal crossed product and the theory of these algebras has proved to be of great importance in the study of central division algebras. Nevertheless the passage from $\mathfrak{A}$ to $\mathfrak{B}$ is somewhat artificial. A more direct procedure would be to study $\mathfrak{A}$ relative to any one of its maximal subfields P or, at any rate, relative to a maximal separable subfield. This is true especially since it is not known whether or not $\mathfrak{A}$ possesses a normal separable subfield of dimensionality n. The beginning of the study of an algebra relative to an arbitrary subfield was made by Professor Wedderburn in a beautiful paper appearing in 1921.[1] Here Wedderburn showed that $\mathfrak{A}$ can be regarded as a vector space over P and that one obtains in this way a representation of P by matrices with elements in P.

In a recent paper[2] the present author has considered the general theory of self-representations of a field. As we show here, this theory can be applied to investigate the structure of a central simple algebra of dimensionality n^2 relative to any separable subfield of dimensionality n. We shall show that the self-representations of $\mathfrak{A}$ defined by P are the regular self-representations and that $\mathfrak{A}$ may be written as a type of crossed product (P, E, σ) where E denotes a regular self-representation and σ is a factor set. We obtain associativity conditions, conditions that $\mathfrak{A}$ be simple and that $\mathfrak{A}$ be the complete matrix algebra Φ_n. As an application of the theorem that the automorphisms of Φ_n are all inner we obtain necessary and sufficient conditions that an element of P_n have the form $\alpha^E\alpha^{-1}$, α in P and $\alpha \rightarrow \alpha^E$ a regular self-representation.

1. Self-representations of fields

In this section we collect some of the results of the theory of self-representations that we require in the sequel.[3] We recall first the definition of a double P-module $\mathfrak{A}$, P a field, to be a left and a right P-module satisfying the following conditions:

1. $$1x = x = x1$$
2. $$(\alpha x)\beta = \alpha(x\beta)$$
3. $$(\mathfrak{A}: \mathrm{P}_R) = n < \infty.$$

[1] *On division algebras*, Trans. Am. Math. Soc. *22* (1921), 129–135.

[2] *An extension of Galois theory to non-normal and non-separable fields*, Am. Jour. of Math. vol. 66 (1944), pp. 1–29 referred to hereafter as E.

[3] Cf. E for the results stated in this section.

Here P_R denotes the set of right transformations $x \to x\alpha$. If we choose a right basis $x_1, \cdots, x_n$ for $\mathfrak{A}$, we may write $\alpha x_i = \sum x_\lambda(\alpha E_{\lambda i})$. Then the correspondence $\alpha \to \alpha^E = (\alpha E_{\lambda i})$ is a self-representation of P, i.e. a representation of P by matrices with elements in P. Conversely any self-representation may be obtained in this way. A change of basis from $x_1, \cdots, x_n$ to $y_1, \cdots, y_n$ where $y_i = \sum x_\lambda \mu_{\lambda i}$ replaces the self-representation E by the self-representation F where $\alpha^F = M^{-1}\alpha^E M$, $M = (\mu_{ij})$. We call F and E similar.

If $\alpha \to \alpha^E$ is a self-representation, then $\alpha \to \alpha^{E'} \equiv (\alpha^E)'$ the transposed of α^E is also a self-representation. Let Φ be the subfield of P of elements γ that are fixed in the sense that γ^E is the diagonal matrix $\{\gamma, \cdots, \gamma\}$ and suppose that P is finite and separable over Φ. Then $\mathrm{P} = \Phi(\theta)$ and the self-representations E and E' are completely determined by the matrix θ^E. Since θ^E is similar to $\theta^{E'}$ the self-representations E and E' are similar.

We recall also the definition of a composite (K, S, T) of P with itself as a system consisting of a commutative ring K and two isomorphisms S and T between P and subfields P^S and P^T of K such that

$$1)\ \ 1^S = 1^T, \qquad 2)\ \ K = \mathrm{P}^S\mathrm{P}^T, \qquad 3)\ \ (K\colon \mathrm{P}^T) \text{ is finite.}$$

If $\mathfrak{A}$ is the module associated with the self-representation E, then $\mathfrak{A}$ determines a composite $(\mathrm{P}_L\mathrm{P}_R, L, R)$ where L is the correspondence between α and the left multiplication $x \to \alpha x$ and R is the correspondence between α and $x \to x\alpha$. On the other hand if (K, S, T) is a composite, K is a double P-module module relative to the multiplications $\alpha x \equiv \alpha^S x = x\alpha^S$ and $x\alpha \equiv \alpha^T x = x\alpha^T$.

There is also a second way in which a composite (K, S, T) determines a double P-module. We choose a basis $\alpha_1^S, \cdots, \alpha_n^S$ for K over P^T. Then for any α in P we may write

$$\alpha^S = \alpha_1^S\mu_1(\alpha)^T + \cdots + \alpha_q^S\mu_q(\alpha)^T \tag{1}$$

and the mappings $M_i : \alpha \to \mu_i(\alpha)$ are single valued. These transformations are endomorphisms in the additive group of P and it is readily seen that the totality of endomorphisms $\sum M_i\bar{\rho}_i$ where $\bar{\rho}$ is the multiplication $\alpha \to \alpha\rho$ in P is an invariant of (K, S, T), i.e. the sets of this type determined by any pair of bases $\alpha_1^S, \cdots, \alpha_n^S$ and $\beta_1^S, \cdots, \beta_n^S$ are identical. We call the set of endomorphisms $\sum M_i\bar{\rho}_i$ the relations space $\mathfrak{S}$ of the composite.

Suppose that the multiplication table of the α_i^S is

$$\alpha_i^S\alpha_j^S = \sum \alpha_k^S\epsilon_{ij}^T\,. \tag{2}$$

Then a simple computation shows that

$$\bar{\rho}M_i = \sum M_j\bar{\epsilon}_{ij\lambda}(\bar{\rho}M_\lambda). \tag{3}$$

Thus $\mathfrak{S}$ is also invariant under left multiplication by the elements $\bar{\rho}$. It can be shown that $\mathfrak{S}$ has dimensionality n over P (on the right). Hence if we set $\rho M \equiv \bar{\rho}M$ and $M\rho \equiv M\bar{\rho}$ for M in $\mathfrak{S}$, we may consider $\mathfrak{S}$ as a double P-module. The matrix corresponding to ρ in the self-representation determined by the basis

$M_1, \cdots, M_n$ of $\mathfrak{S}$ is because of (3), $\rho^G = (\rho G_{ij})$ where $\rho G_{ji} = \sum \epsilon_{ij\lambda}(\rho M_\lambda)$. On the other hand, if we regard K as a double P-module as before and we use the right basis $\overset{S}{\alpha_1}, \cdots, \overset{S}{\alpha_n}$ for P we have for any ρ in P

$$\rho\overset{S}{\alpha_j} \equiv \overset{S}{\rho}\overset{S}{\alpha_j} = \sum \overset{S}{\alpha_\lambda}\mu_\lambda(\rho)^T\overset{S}{\alpha_j}$$
$$= \sum \overset{S}{\alpha_i}\overset{T}{\epsilon_{i\lambda j}}(\rho M_\lambda)^T.$$

Thus the self-representation determined by this basis is $\rho^E = (\rho E_{ij})$ where $\rho E_{ij} = \sum \epsilon_{i\lambda j}(\rho M_\lambda)$. Since P is commutative $\epsilon_{i\lambda j} = \epsilon_{ij\lambda}$. Hence $G_{ji} = E_{ij}$.

THEOREM 1. *The self-representations determined by the relations space of a composite* (K, S, T) *are the transposed representations of the representations determined by* K.

2. A special generation of Φ_n

A composite of particular interest is obtained by forming the direct product K over Φ of P by itself. Here Φ is a subfield of P such that $(\mathrm{P}: \Phi) = n < \infty$ and $K = \mathrm{P}^S \times \mathrm{P}^T$ where $\gamma^S = \gamma^T$ if $\gamma \in \Phi$ and $(K: \mathrm{P}^T) = (K: \mathrm{P}^S) = (\mathrm{P}: \Phi)$. If $\alpha_1, \cdots, \alpha_n$ is a basis for P over Φ then $\overset{S}{\alpha_1}, \cdots, \overset{S}{\alpha_n}$ is a basis for K over P^T. Thus if $\alpha^S = \sum \overset{S}{\alpha_i}\mu_i(\alpha)^T$ then the elements $\mu_i(\alpha)$ are in Φ. The representations determined by K are called *regular representations* and those determined by the bases $\overset{S}{\alpha_1}, \cdots, \overset{S}{\alpha_n}$ in which $\overset{S}{\alpha_1} = 1^S$ are the *special* regular representations of P. The matrices of a special regular representation belong to Φ_n.

A characteristic property of the direct product is that its relations space $\mathfrak{S}$ is a ring.[4] Moreover it can be shown that $\mathfrak{S}$ is the complete set of endomorphisms in P commutative with the endomorphisms $\bar{\gamma}$, γ in Φ. It follows that $\mathfrak{S} \cong \Phi_n$.

Now let $\alpha_1, \cdots, \alpha_n$ be a basis of P such that $\alpha_1 = 1$. Then for any γ in Φ, $\gamma^S = 1^S\gamma^T = \overset{S}{\alpha_1}\gamma^T$ and hence $\gamma M_1 = \gamma$ and $\gamma M_j = 0$ if $j \neq 1$. Since $\alpha M_i \in \Phi$ for any α it follows that,

$$M_iM_1 = M_i \qquad M_iM_j = 0 \text{ if } j \neq 1. \tag{4}$$

As before for any ρ in P we have $\bar{\rho}M_i = \sum M_j(\overline{\rho E_{ij}})$ where $\rho \rightarrow \rho^E$ is the special regular representation determined by the basis $\overset{S}{\alpha_1}, \cdots, \overset{S}{\alpha_n}$ of K. By means of the isomorphism $\mathfrak{S} \cong \Phi_n$ this proves

THEOREM 2. *Let* P *be an arbitrary field over* Φ *such that* $(\mathrm{P}: \Phi) = n$ *and let* $\rho \rightarrow \rho^E = (\rho E_{ij})$ *be a special regular representation of* P *over* Φ. *Then* P *may be regarded as a subfield of* Φ_n *and there exist elements* $y_1, \cdots, y_n$ *in* Φ_n *such that every element in* Φ_n *may be written in one and only one way in the form* $\sum y_i\rho_i$, ρ_i *in* P *where the* y_i *satisfy*

$$\rho y_i = \sum y_j(\rho E_{ij}), \qquad y_iy_j = \delta_{1j}y_i. \tag{5}$$

3. Maximal subfields of a central simple algebra

We suppose now that $\mathfrak{A}$ is any central simple algebra of dimensionality n^2 over Φ and that $\mathfrak{A}$ contains a subfield P of dimensionality n. The latter as-

[4] E, pp. 19–21.

sumption is always satisfied when $\mathfrak{A}$ is a division algebra. We turn $\mathfrak{A}$ into a double P-module by defining αx and $x\alpha$ to be the ordinary products of α in P by x in $\mathfrak{A}$. Since $(\mathrm{P}:\Phi) = n$, $(\mathfrak{A}:\mathrm{P}_R) = n = (\mathfrak{A}:\mathrm{P}_L)$. If a is any element in $\mathfrak{A}$ we denote the left multiplication $x \to ax$ by a_l and the right multiplication $x \to xa$ by a_r. It is known that if $a_1, \cdots, a_s$ are elements of $\mathfrak{A}$ that are independent over Φ then $\sum a_{il}b_{ir} = 0$ only if all the $b_i = 0$. In particular if $\alpha_1, \cdots, \alpha_n$ form a basis for P over Φ then $\sum \alpha_{il}\beta_{ir} = 0$ only if all the $\beta_i = 0$. This implies that the composite $(\mathrm{P}_L\mathrm{P}_R, L, R)$ is equivalent to $(\mathrm{P}^S \times \mathrm{P}^T, S, T)$[5].

Suppose now that P is separable over Φ. Then the regular self-representation of P over Φ is completely reducible into dissimilar irreducible components. Since the self-representation determined by $\mathfrak{A}$ has a composite equivalent to the direct product, it is completely reducible into irreducible components that include all of the components of the regular representation. A comparison of the ranks of these two self-representations shows that they are similar.

THEOREM 3. *Let $\mathfrak{A}$ be a central simple algebra of dimensionality n^2 over Φ and let* P *be a separable subfield of dimensionality n. Then the self-representations of* P *obtained by regarding $\mathfrak{A}$ as a double* P*-module are similar to the regular self-representations of* P *over Φ.*

Now let $x_1, \cdots, x_n$ be a right basis of $\mathfrak{A}$ over P. Then for every α in P we have $\alpha x_i = \sum x_j(\alpha E_{ji})$, where the self-representation $\alpha \to \alpha^E$ is regular. Moreover, if

$$x_i x_j = \sum x_\mu \sigma_{\mu i j} \tag{6}$$

then the product of any two elements of $\mathfrak{A}$ is completely determined by the multiplication of elements in P, by the self-representation E and by the *factor set* $\sigma = \{\sigma_{ijk}\}$. Since $(x_ix_j)x_k = x_i(x_jx_k)$

$$\sum_{\lambda,\mu} \sigma_{\rho\lambda\mu}(\sigma_{\lambda ij}E_{\mu k}) = \sum_{\lambda} \sigma_{\rho i\lambda}\,\sigma_{\lambda jk} \tag{7}$$

and since $\rho(x_ix_j) = (\rho x_i)x_j$ for all ρ in P

$$\sum_{\lambda,\mu} \sigma_{\nu\lambda\mu}(\rho E_{\lambda i}E_{\mu j}) = \sum (\rho E_{\nu\lambda})\sigma_{\lambda ij}. \tag{8}$$

4. Crossed products

We suppose now that P is any finite separable extension of Φ of dimensionality n and let $\mathfrak{A}$ be a double P-module whose representations are regular. Let $x_1, \cdots, x_n$ be a right P-basis for $\mathfrak{A}$ and for $x = \sum x_i\rho_i$, $x' = \sum x_j\rho_j'$ define

$$xx' = \sum x_\mu\sigma_{\mu i\lambda}(\rho_iE_{\lambda j})\rho_j' \tag{9}$$

where the elements σ_{ijk} are arbitrary elements in P that satisfy (7) and (8). We wish to show that $\mathfrak{A}$ is an associative algebra over Φ. We note first that by (9)

[5] Equivalence of composites is defined as for field composites. Cf. E, p. 5.

$$(10) \qquad (xx')\rho = x(x'\rho).$$

Moreover $(\sum x_i\rho_i)(\sum x_j\rho_j') = \sum x_\mu\sigma_{\mu i\lambda}(\rho_i E_{\lambda j})\rho_j' = \sum (x_i x_\lambda)\rho_i E_{\lambda j}\rho_j'$. Since $\rho \to \rho^E$ is a representation this implies that

$$(11) \qquad (x\rho)x' = x(\rho x').$$

We note also that $\rho(x_i x_j) = (\rho x_i)x_j$ and using these relations we may prove that

$$(x_i\rho)((x_j\sigma)x_k\tau) = \sum_{\lambda,\mu,\nu} x_i(x_\mu x_\nu)(\rho E_{\mu j} E_{\nu\lambda}\sigma E_{\lambda k}\tau)$$

and

$$((x_i\rho)(x_j\sigma))x_k\tau = \sum_{\lambda,\mu,\nu} (x_i x_\mu)x_\nu(\rho E_{\mu j} E_{\nu\lambda}\sigma E_{\lambda k}\tau).$$

Since by (7) $x_i(x_\mu x_\nu) = (x_i x_\mu)x_\nu$ this proves that $\mathfrak{A}$ is associative. If $\gamma \in \Phi$, $\gamma x = x\gamma$ and hence

$$(xx')\gamma = x(x'\gamma) = (x\gamma)x'$$

so that $\mathfrak{A}$ is an algebra over Φ. Evidently $(\mathfrak{A}:\Phi) = (\mathfrak{A}:\mathrm{P})(\mathrm{P}:\Phi) = n^2$.

THEOREM 4. *Let* P *be a separable extension of dimensionality n over* Φ, $\alpha \to \alpha^E$ *a regular self-representation of* P *over* Φ *and* $\{\sigma_{ijk}\}$ *a set of elements in* P *satisfying* (7) *and* (8). *Then the totality of elements* $\sum x_i\rho_i$ *of the n-dimensional space over* P *is an algebra if multiplication is defined by* (9).

We shall call $\mathfrak{A}$ the *crossed product* of P with regular self-representation E and factor set $\sigma = \{\sigma_{ijk}\}$ and we write $\mathfrak{A} = (\mathrm{P}, E, \sigma)$. The conditions (7) and (8) will be called the *associativity conditions* on the factor set σ. By the preceding section we have

THEOREM 5. *Let* $\mathfrak{A}$ *be a central simple algebra of dimensionality* n^2 *over* Φ *and let* P *be a separable subfield of* $\mathfrak{A}$ *of dimensionality* n. *Then* $\mathfrak{A}$ *is a crossed product* (P, E, σ).

The associativity conditions (8) take on a simpler form if E is a special regular representation. For in this case ρE_{ij} are in Φ and hence $\rho E_{ij}E_{kl} = \delta_{kl}(\rho E_{ij})$. The equations (8) reduce to

$$(12) \qquad \sum \sigma_{\nu\lambda j}(\rho E_{\lambda i}) = \sum (\rho E_{\nu\lambda})\sigma_{\lambda ij}.$$

If we set $\sigma_j = (\sigma_{\nu\lambda j})$ this is equivalent to

$$(13) \qquad \sigma_j\rho^E = \rho^E\sigma_j.$$

LEMMA. *If* A *is a matrix in* P_n *that commutes with all the matrices* ρ^E, A *is a linear combination of the matrices* ρ^E *with coefficients in* P.

A matrix A in P_n commutative with all the ρ^E corresponds to an endomorphism $\bar{A}$ in the direct product $K = \mathrm{P}^S \times \mathrm{P}^T$ that commutes with all the multiplications $x \to x\rho^S$ and all of the multiplications $x \to x\rho^T$ and hence with all of the multiplications $x \to xa$ for arbitrary a in K. Since K is commutative, $\bar{A}$ is a multiplication. It follows that A is a linear combination of the matrices ρ^E using coefficients in P.

Thus the elements σ_j that satisfy (13) are linear combinations of matrices ρ^E. Each σ_j will therefore depend on n parameters in P and hence the factor set σ depends on n^2 parameters.

We suppose again the E is an arbitrary regular representation. Then we may verify that

$$\rho(xx') = (\rho x)x'. \tag{14}$$

In particular if $y_1, \cdots, y_n$ is a second right P-basis for $\mathfrak{A}$ then $\rho(y_iy_j) = (\rho y_i)y_j$. Hence if $y_iy_j = \sum y_\lambda\tau_{\lambda ij}$ and $\rho y_i = \sum y_j(\rho F_{ji})$ then the τ's and the F_{ij} satisfy equations analogous to (7) and (8). If $y_i = \sum x_\lambda\mu_{\lambda i}$ we know that $\alpha^F = M^{-1}\alpha^E M$, $M = (\mu_{ij})$. We may verify also that

$$\tau_{kij} = \sum \bar{\mu}_{k\rho}\sigma_{\rho\lambda\mu}(\mu_{\lambda i}E_{\mu\nu})\mu_{\nu j} \tag{15}$$

where $(\bar{\mu}_{ij}) = M^{-1}$.

THEOREM 6. *If M is any non-singular matrix in P_n, then the crossed products* (P, E, σ) *and* (P, F, τ) *where $\alpha^F = M^{-1}\alpha^E M$ and τ satisfies* (15) *are isomorphic.*

5. Conditions for simplicity

We consider now the conditions that $\mathfrak{A} = (P, E, \sigma)$ be central and simple. For this purpose we investigate the structure of the double P-module, $\mathfrak{A}$. By definition $\mathfrak{A}$ is isomorphic to the double P-module $K = P^S \times P^T$ obtained from the direct product (K, S, T). Hence we consider K. The submodules of K are the (two-sided) ideals of the ring K. Since P is separable over Φ, $K = K_1 \oplus K_2 \oplus \cdots \oplus K_s$ where the K_i are fields. The self-representations determined by the K_i are irreducible and dissimilar and any irreducible self-representation of P over Φ is similar to one of these. In particular one of the K_i, say K_1, is associated with the identity self-representation $\alpha \rightarrow \alpha$. Any submodule of K is a direct sum of certain of the K_i.

If $\alpha \rightarrow \alpha^{E_1}$ and $\alpha \rightarrow \alpha^{E_2}$ are self-representations of P the product $E_1 \times E_2$ is the self-representation obtained by replacing the elements αE_{ij1} of α^{E_1} by the matrices $\alpha E_{ij1}^{E_2}$ that are associated with these elements in E_2.[6] Let $\mathfrak{A}_1$ be the double P-module associated with E_1 and $x_1, \cdots, x_m$ a basis such that $\alpha x_i = \sum x_\lambda(\alpha E_{\lambda i1})$ and similarly let $\mathfrak{A}_2$ be a double P-module with basis $y_1, \cdots, y_r$ such that $\alpha y_j = \sum y_\mu(\alpha E_{\mu j2})$. Then it is possible to define a double P-module $\mathfrak{A}_1 \times \mathfrak{A}_2$ having a right basis $x_i \times y_j$ such that $\alpha(x_i \times y_j) = \sum x_\lambda \times y_\mu(\alpha E_{\lambda i1}E_{\mu j2})$.[7] This product module is independent of the choice of the bases in $\mathfrak{A}_i$. It follows from this that if F_1 is a self-representation similar to E_1 and F_2 is similar to E_2 then $F_1 \times F_2$ is similar to $E_1 \times E_2$.

Now let E_i be an irreducible self-representation associated with K_i and E_j one associated with K_j and form $E_i \times E_j$. This representation is completely

[6] Cf. E. p. 12 or L. KALUJNINE, *Sur la theorie de Galois de Corps non-galoisiennes separables*, Comptes Rendus de l'Acad. des Sciences, *214* (1942), 597–599.

[7] E, p. 12.

reducible into irreducible representations determined by, say, $K_{i_1}, \cdots, K_{i_r}$. If we replace E_i and E_j by F_i and F_j two other self-representations associated with K_i and K_j respectively, we may associate with $F_i \times F_j$ the same set $K_{i_1}, \cdots, K_{i_r}$. Hence we may define the set $(K_{i_1}, \cdots, K_{i_r})$ to be the product $K_i \times K_j$ of K_i and K_j. It is known that relative to this product the set of K_i forms a hypergroup $\mathfrak{H}$: the product is associative, K_i acts as an identity and for each K_i there exists a $K_{i'}$ such that $K_i \times K_{i'} = (K_1, \cdots)$ and an element $K_{i''}$ such that $K_{i''} \times K_i = (K_1, \cdots)$.[8]

We now apply these results to $\mathfrak{A} = (\mathrm{P}, E, \sigma)$. Then we see that $\mathfrak{A} = \mathfrak{A}_1 \oplus \mathfrak{A}_2 \oplus \cdots \oplus \mathfrak{A}_s$ where $\mathfrak{A}_i$ is the double P-module corresponding to K_i. Any submodule of $\mathfrak{A}$ is a direct sum of certain of the $\mathfrak{A}_i$. We now define the *product* $\mathfrak{A}_i\mathfrak{A}_j$ to be the set of sums $\sum a_i a_j$ where $a_i \in \mathfrak{A}_i$ and $a_j \in \mathfrak{A}_j$. By (10) and (14) $\mathfrak{A}_i\mathfrak{A}_j$ is a submodule of $\mathfrak{A}$. Thus $\mathfrak{A}_i\mathfrak{A}_j$ is a direct sum of certain of the $\mathfrak{A}_i$. Let $x_1, \cdots, x_m$ be a right basis for $\mathfrak{A}_i$ and $y_1, \cdots, y_r$ a right basis for $\mathfrak{A}_j$. Any element of $\mathfrak{A}_i\mathfrak{A}_j$ is a sum of elements of the form $(\sum x_i\xi_i)(\sum y_j\eta_j)$ and this may be written as $\sum x_i y_j \zeta_{ij}$ if we use (11). Thus the right module generated by the elements $x_i y_j$ is the whole module $\mathfrak{A}_i\mathfrak{A}_j$. We consider now the module $\mathfrak{A}_i \times \mathfrak{A}_j$. It is readily verified that the mapping $\sum (x_i \times y_j)\zeta_{ij} \rightarrow \sum x_i y_j \zeta_{ij}$ is a homomorphism between the P-modules $\mathfrak{A}_i \times \mathfrak{A}_j$ and $\mathfrak{A}_i\mathfrak{A}_j$. Hence the irreducible components of $\mathfrak{A}_i\mathfrak{A}_j$ are contained among the irreducible components of $\mathfrak{A}_i \times \mathfrak{A}_j$. It follows from our definition of $K_i \times K_j$ that if $K_i \times K_j = (K_{i_1}, \cdots, K_{i_r})$, then $\mathfrak{A}_i\mathfrak{A}_j = \mathfrak{A}_{i_1} \oplus \cdots \oplus \mathfrak{A}_{i_t}$, $t \leqq r$ if the K's are suitably ordered. Since $K_1 \times K_j = K_j = K_j \times K_1$, $\mathfrak{A}_1\mathfrak{A}_j = 0$ or $\mathfrak{A}_j$ and $\mathfrak{A}_j\mathfrak{A}_1 = 0$ or $\mathfrak{A}_j$.

We assume now that $\mathfrak{A}_j\mathfrak{A}_1 = \mathfrak{A}_j = \mathfrak{A}_1\mathfrak{A}_j$ for all $\mathfrak{A}_j$. In particular $\mathfrak{A}_1^2 = \mathfrak{A}_1$. Hence if u is an element $\neq 0$ in $\mathfrak{A}_1$, $u^2 = u\delta$ where $\delta \neq 0$. Then $e = u\delta^{-1}$ is idempotent. Since $\mathfrak{A}_1\mathfrak{A} = \mathfrak{A} = \mathfrak{A}\mathfrak{A}_1$, $e\mathfrak{A} = \mathfrak{A} = \mathfrak{A}e$. Hence e is the identity 1 of $\mathfrak{A}$. Now $\mathfrak{A}$ contains the subfield $\mathfrak{A}_1 = 1\mathrm{P}$ isomorphic to P. Since $\alpha x = \alpha(1x) = (\alpha 1)x = (1\alpha)x$ and $x\alpha = (x1)\alpha = x(1\alpha)$ the module operations in $\mathfrak{A}$ are the left and the right multiplications by the elements of $1\mathrm{P}$. If $\mathfrak{B}$ is a two-sided ideal, $\mathfrak{B}$ is a submodule of $\mathfrak{A}$ and hence $\mathfrak{B}$ is a sum of certain of the $\mathfrak{A}_i$. This proves

THEOREM 7. *If $\mathfrak{A}$ has an identity there are only a finite number of two-sided ideals in $\mathfrak{A}$. Any two-sided ideal is a sum of certain of the submodules $\mathfrak{A}_i$.*

If we choose a basis $u_1, \cdots, u_n$ of $\mathfrak{A}$ composed of bases for the $\mathfrak{A}_i$, a condition that $\mathfrak{A}$ have an identity may be obtained on the elements μ_{kij} such that $u_i u_j = \sum u_k \mu_{kij}$. This is that the matrices (μ_{ki1}) and (μ_{k1j}) be non-singular. For this implies that $\mathfrak{A}_i\mathfrak{A}_1 \neq 0$ and $\mathfrak{A}_1\mathfrak{A}_i \neq 0$.

We note next that the only elements of $\mathfrak{A}$ that commute with all the elements of $1\mathrm{P}$ are the elements in this field. For if u is such an element then $\alpha u = u\alpha$ for all α. Hence $u \in \mathfrak{A}_1 = 1\mathrm{P}$. Since the only elements of P that are fixed under any regular self-representation are the elements of Φ[9], the only elements

[8] E, p. 25 or KALUJNINE loc. cit. in 6.
[9] E, p. 21.

of $1\mathrm{P}$ that commute with all the elements x_i of a P-basis of $\mathfrak{A}$ are the elements in 1Φ. It follows that 1Φ is the center of $\mathfrak{A}$.

THEOREM 8. *If $\mathfrak{A}$ has an identity, $\mathfrak{A}$ is a central algebra.*

We assume next that for each module $\mathfrak{A}_i$ there exists an $\mathfrak{A}_j$ such $\mathfrak{A}_j\mathfrak{A}_i$ contains $\mathfrak{A}_1$. Let $\mathfrak{B}$ be a two-sided ideal $\neq 0$ in $\mathfrak{A}$. Then $\mathfrak{B}$ contains one of the $\mathfrak{A}_i$ and hence $\mathfrak{B}$ contains $\mathfrak{A}_j\mathfrak{A}_i$. It follows that $1 \in \mathfrak{B}$ and $\mathfrak{B} = \mathfrak{A}$.

THEOREM 9. *The algebra $\mathfrak{A}$ is simple if* 1) $\mathfrak{A}_1\mathfrak{A}_i = \mathfrak{A}_i = \mathfrak{A}_i\mathfrak{A}_1$ *for all $\mathfrak{A}_i$ and* 2) *for each $\mathfrak{A}_i$ there exists an $\mathfrak{A}_j$ such that $\mathfrak{A}_j\mathfrak{A}_i$ contains $\mathfrak{A}_1$.*

Conversely suppose that $\mathfrak{A}$ is simple. Then $\mathfrak{A}$ contains an identity. Since $\alpha 1 = 1\alpha$, $1 \in \mathfrak{A}_1$. Hence $\mathfrak{A}_1\mathfrak{A}_i = \mathfrak{A}_i = \mathfrak{A}_i\mathfrak{A}_1$.

We consider now the set $\mathfrak{M} = \mathfrak{A}_L\mathrm{P}_R = \mathrm{P}_R\mathfrak{A}_L$ of endomorphisms of the form $a_{1l}\rho_{1r} + \cdots + a_{sl}\rho_{sr}$ where a_l is a left multiplication $x \to ax$, a in $\mathfrak{A}$ and ρ_r is a right multiplication by an element of P. We recall that if $a_1, \cdots, a_s$ are linearly independent over Φ then $a_{1l}\rho_{1r} + \cdots + a_{sl}\rho_{sr} \neq 0$ unless all the ρ_i are 0. Thus $(\mathfrak{M}: \mathrm{P}_R) = (\mathfrak{A}: \Phi) = n^2$. On the other hand the elements of $\mathfrak{M}$ are endomorphisms in $\mathfrak{A}$ that commute with all the elements of P_R. Since the dimensionality of $\mathfrak{A}$ over P_R is n, it follows that $\mathfrak{M}$ is the complete set of endomorphisms in $\mathfrak{A}$ commutative with the elements of P_R. This implies that $\mathfrak{M}$ is an irreducible set of endomorphisms.

Now consider any one of the $\mathfrak{A}_i$. Form $\mathfrak{A}\mathfrak{A}_i$. This is a submodule of $\mathfrak{A}$ and a left ideal. Hence $\mathfrak{A}\mathfrak{A}_i$ is a subgroup $\neq 0$ of $\mathfrak{A}$ which is invariant under all the endomorphisms in $\mathfrak{M}$. Since $\mathfrak{M}$ is irreducible $\mathfrak{A}\mathfrak{A}_i = \mathfrak{A}$. It follows that there exists an $\mathfrak{A}_j$ such that $\mathfrak{A}_j\mathfrak{A}_i$ contains $\mathfrak{A}_1$. Similarly there exists an $\mathfrak{A}_k$ such that $\mathfrak{A}_i\mathfrak{A}_k$ contains $\mathfrak{A}_1$. This proves

THEOREM 10. *If $\mathfrak{A}$ is a simple algebra, the irreducible modules $\mathfrak{A}_i$ form a hypergroup H under multiplication.*

We have also noted above the relation between H and the hypergroup $\mathfrak{H}$ of the K_i, namely, there is a (1–1) correspondence $K_i \to \mathfrak{A}_i$ between $\mathfrak{H}$ and H such that if $K_i \times K_j = (K_{i_1}, \cdots, K_{i_r})$ then $\mathfrak{A}_i\mathfrak{A}_j = (\mathfrak{A}_{i_1}, \cdots, \mathfrak{A}_{i_t})$ where $t \leqq r$.

We consider now the special case where P is normal. Here the irreducible self-representations of P are of rank 1 and hence they are identical with the automorphism $\alpha \to \alpha^S$ of the Galois group $\mathfrak{G}$ of P over Φ. The modules $\mathfrak{A}_i$ are one dimensional over P and are in (1–1) correspondence with the elements S. We may denote the correspondent of S by $\mathfrak{A}_S$. Then the condition that $\mathfrak{A}$ be simple is that $\mathfrak{A}_S\mathfrak{A}_T = \mathfrak{A}_{ST}$. We choose any element $u_S \neq 0$ in $\mathfrak{A}_S$. Then the condition that $\mathfrak{A}$ be simple is that $u_Su_T = u_{ST}\mu_{S,T}$ where $\mu_{S,T} \neq 0$. These conditions for normal crossed products are well-known, as are also the associativity conditions on the $\mu_{S,T}$ that one obtains from (7).

6. Conditions that $\mathfrak{A} \cong \Phi_n$

We have seen that a change of basis from $x_1, \cdots, x_n$ to $y_1, \cdots, y_n$ in a crossed product $\mathfrak{A}$ replaces the self-representation E associated with the x's by a similar self-representation $\alpha \to \alpha^F = M^{-1}\alpha^E M$ and replaces the x-factor set σ by a factor set τ defined by (15). If M is a matrix commutative with all the α^E then $F = E$. In this case we shall say that the factor sets σ and τ are *associates* ($\sigma \sim \tau$).

In order to determine whether or not a central simple $\mathfrak{A}$ is the complete matrix algebra, it suffices to consider $\mathfrak{A}$ in the form (P, E', σ) where E' is the transposed of a special regular representation. Now we have seen in Theorem 2 that if 1 denotes the factor set $\{\lambda_{ijk}\}$ where $\lambda_{ijk} = \delta_{k1}\delta_{ij}$ then (P, E', 1) $\cong \Phi_n$ or (P, E', 1) $\sim$ 1. The usual proof making use of the fact that an isomorphism between simple subalgebras of a central simple algebra $\mathfrak{A}$ may be extended to an automorphism in $\mathfrak{A}$ shows that conversely if (P, E', σ) $\sim$ 1 then $\sigma \sim 1$.

THEOREM 11. *A necessary and sufficient condition that a central simple algebra* $\mathfrak{A} = (\mathrm{P}, E', \sigma) \sim 1$ *is that* $\sigma \sim 1$.

Suppose now that $\mathfrak{A} = (\mathrm{P}, E', 1)$ and that $y_1, \cdots, y_n$ is a P-basis such that $\rho y_i = \sum y_j(\rho E_{ij})$ and $y_i y_j = \delta_{1j} y_i$. Let $z_1, \cdots, z_n$ be a second right P-basis such that $\rho z_i = \sum z_j(\rho E_{ij})$ and $z_i z_j = \delta_{1j} z_i$. Then $z_i = \sum y_j \mu_{ji}$ where the matrix M satisfies the following two conditions:

$$\alpha^{E'} M = M\alpha^{E'} \quad \text{for all } \alpha, \tag{16}$$

$$\delta_{j1}\,\delta_{ik} = \sum_{\rho,\nu} \bar{\mu}_{k\rho}(\mu_{\rho i}\, E_{\nu 1})\mu_{\nu j}. \tag{17}$$

For (17) is obtained from (15) by setting $\tau_{kij} = \lambda_{kij}$, $\sigma_{\rho\lambda\mu} = \lambda_{\rho\lambda\mu}$ and by replacing $E_{\mu\nu}$ by $E_{\nu\mu}$. From (17) we obtain

$$(\mu_{\rho i} E_{\nu 1})\mu_{\nu j} = \delta_{j1}\mu_{\rho i} \tag{18}$$

and conversely (18) implies (17). Now let M be any non-singular matrix in P_n satisfying (16) and (18). Then the elements $z_i = \sum y_j \mu_{ji}$ will satisfy the same multiplication table as the y_i. Thus (16) and (18) are the conditions that the mapping $\sum y_i \rho_i \to \sum z_i \rho_i$ be an automorphism A in (P, E', 1). It is clear that A leaves the elements of P invariant. Hence A is an inner automorphism of the form $x \to \beta x \beta^{-1}$. In particular $z_i = \beta y_i \beta^{-1} = \sum y_j(\beta E_{ij})\beta^{-1} = \sum y_j \mu_{ji}$ and so $M = \beta^{E'}\beta^{-1}$. Conversely if M has this form then (16) and (18) hold.

THEOREM 12. *Let* P *be an extension of* Φ *such that* $(\mathrm{P}:\Phi) = n$ *and let* E *be a special regular representation of* P *over* Φ. *Then the conditions* (16) *and* (18) *are necessary and sufficient that the non-singular matrix* M *in* P_n *have the form* $\beta^{E'}\beta^{-1}$.

We recall that (16) holds if and only if M is a linear combination of matrices $\alpha^{E'}$ using coefficients in P. It may be remarked also that separability is not required in Theorem 11 since P was arbitrary in Theorem 2.

THE JOHNS HOPKINS UNIVERSITY.

Reprinted from *Bulletin of the American Mathematical Society*, December 1944.

THE EQUATION $x' \equiv xd - dx = b$

N. JACOBSON

Let $\mathfrak{A}$ be an associative algebra with a possibly infinite basis over a field Φ. Then if d is a fixed element in $\mathfrak{A}$, it is well known that the mapping $x \rightarrow x' \equiv [x, d] = xd - dx$ is a derivation[1] in $\mathfrak{A}$; that is,

$$(x + y)' = x' + y', \qquad (x\alpha)' = x'\alpha, \qquad (xy)' = x'y + xy'$$

for all x, y in $\mathfrak{A}$ and all α in Φ. The constants relative to such a derivation are the elements of $\mathfrak{A}$ that commute with d. We shall call an element b a *d-integral* if $b = a'$ for some element a in $\mathfrak{A}$, that is, if the equation $x' = xd - dx = b$ has a solution in $\mathfrak{A}$. Clearly if a is a solution of this equation then the totality of solutions is the set $\{a + c\}$ where c ranges over the set of d-constants. In a recent paper appearing in this Bulletin, R. E. Johnson obtained a necessary and sufficient condition that an element b be a d-integral under the assumption that $\mathfrak{A}$ is a separable algebraic division ring.[2] In this note we allow $\mathfrak{A}$ to be an arbitrary algebra but we make the assumption that d is an algebraic element in the sense that it satisfies a polynomial equation with coefficients in Φ. We obtain a necessary condition, which is equivalent to Johnson's condition when $\mathfrak{A}$ is a division ring, that b be a d-integral. If the minimum polynomial $\mu(\lambda)$ of d is relatively prime to its derivative $\mu'(\lambda)$, then it is easy to see that the condition is also sufficient and one may give an explicit formula for a solution of the equation $x' = b$. If we assume that $\mathfrak{A}$ is a simple algebra satisfying the descending chain condition for left ideals then we can show that our condition is also sufficient when $\mu(\lambda)$ is a product of distinct irreducible factors in $\Phi[\lambda]$ and in certain other cases. Here, however, we do not display a solution but merely prove its existence. Our results include, of course, Johnson's result for algebraic division rings, since the minimum polynomial of an element in such a ring is irreducible. No assumption about separability is required.

In order to obtain a condition for the solvability of the equation $x' = b$ we consider the matrices

$$u = \begin{pmatrix} d & 0 \\ 0 & d \end{pmatrix}, \qquad v = \begin{pmatrix} d & b \\ 0 & d \end{pmatrix} \tag{1}$$

Received by the editors May 19, 1944.

[1] Cf. the author's paper *Abstract derivation and Lie algebras*, Trans. Amer. Math. Soc. vol. 42 (1937) pp. 206–224.

[2] *On the equation* $\chi\alpha = \gamma\chi + \beta$ *over an algebraic division ring*, Bull. Amer. Math. Soc. vol. 50 (1944) pp. 202–208.

in the matrix algebra $\mathfrak{A}_2$ of two-rowed matrices with elements in $\mathfrak{A}$. If x is any element in $\mathfrak{A}$

$$\begin{pmatrix} 1 & -x \\ 0 & 1 \end{pmatrix} = \begin{pmatrix} 1 & x \\ 0 & 1 \end{pmatrix}^{-1}$$

and the equation $x' = xd - dx = b$ is equivalent to the matrix equation

$$\begin{pmatrix} 1 & x \\ 0 & 1 \end{pmatrix}\begin{pmatrix} d & 0 \\ 0 & d \end{pmatrix}\begin{pmatrix} 1 & x \\ 0 & 1 \end{pmatrix}^{-1} = \begin{pmatrix} d & b \\ 0 & d \end{pmatrix}. \tag{2}$$

Thus if b is a d-integral the matrices (1) are similar in $\mathfrak{A}_2$. We suppose now that d is an algebraic element and let

$$\phi(\lambda) = \lambda^m + \alpha_1\lambda^{m-1} + \cdots \tag{3}$$

be the minimum polynomial of d over Φ. Then it is clear that $\phi(u) = 0$. Hence a necessary condition that b be a d-integral is that $\phi(v) = 0$. Now

$$v^r = \begin{pmatrix} d^r & B_r \\ 0 & d^r \end{pmatrix}$$

where $B_r = \sum_{k=0}^{r-1} d^k b d^{r-k-1}$. Hence the condition that $\phi(v) = 0$ is that

$$B_m + \alpha_1 B_{m-1} + \cdots + \alpha_{m-1}B_1 = 0. \tag{4}$$

As we have shown elsewhere[3]

$$B_r = C_{r,1}d^{r-1}b + C_{r,2}d^{r-2}b' + \cdots + b^{(r-1)}$$

where $b^{(k)} = (b^{(k-1)})'$. Hence if we define

$$\phi_h(\lambda) = C_{m,h}\lambda^{m-h} + C_{m-1,h}\alpha_1\lambda^{m-h-1} + \cdots \equiv \phi^{(h)}(\lambda)/h!$$

we may write (4) in the more useful form

$$\phi_1(d)b + \phi_2(d)b' + \cdots + \phi_m(d)b^{(m-1)} = 0. \tag{5}$$

We suppose now that $\phi_1(\lambda) \equiv \phi'(\lambda)$ is relatively prime to $\phi(\lambda)$.[4] Then $\phi_1(d)$ is a regular element in $\mathfrak{A}$. Hence if b is an element such that (5) holds,

$$b = -\phi_1(d)^{-1}\phi_2(d)b' - \cdots - \phi_1(d)^{-1}\phi_m(d)b^{(m-1)} = x'$$

where

$$x = -\phi_1(d)^{-1}\phi_2(d)b - \cdots - \phi_1(d)^{-1}\phi_m(d)b^{(m-2)}. \tag{6}$$

---

[3] Loc. cit. footnote 1, p. 209.

[4] This condition will be satisfied if d generates a separable algebraic field over Φ.

This proves the following theorem.

THEOREM 1. *Let $\mathfrak{A}$ be an arbitrary algebra and let d be an algebraic element of $\mathfrak{A}$ having a minimum polynomial $\phi(\lambda)$ relatively prime to its derivative. Then* (5) *is a necessary and sufficient condition in order that the element b be a d-integral. When the condition holds, $b=x'$ where x is given by* (6).

We suppose now that $\mathfrak{A}$ is a simple algebra with an identity satisfying the descending chain condition for left (right) ideals. Then $\mathfrak{A}=\mathfrak{D}_h$, a matrix algebra of h rows over the (not necessarily finite) division algebra $\mathfrak{D}$ and conversely any algebra of this form satisfies our condition. As before let d be an algebraic element of $\mathfrak{A}$ and let $\phi(\lambda)$ be its minimum polynomial. Let b be an element of $\mathfrak{A}$ such that (5) holds. Then (4) holds and hence the minimum polynomial of v as well as of u is $\phi(\lambda)$. Since d, $b\in\mathfrak{A}=\mathfrak{D}_h$ the matrices u and $v\in\mathfrak{D}_{2h}$ and these may be regarded as the matrices of linear transformations in a $2h$-dimensional vector space $\mathfrak{R}$ over $\mathfrak{D}$. Let T be the linear transformation corresponding to v. Then according to the form of v we have an h-dimensional subspace $\mathfrak{S}$ of $\mathfrak{R}$ invariant under T such that the matrix of T in $\mathfrak{S}$ is d and the matrix of T in the difference space $\mathfrak{R}-\mathfrak{S}$ is also d.

We suppose now that $\phi(\lambda)$ is a product of irreducible factors in $\Phi[\lambda]$. In this case the linear transformation is completely reducible.[5] Hence there exists a subspace $\mathfrak{S}'$ invariant under T such that $\mathfrak{R}=\mathfrak{S}+\mathfrak{S}'$, $\mathfrak{S}\cap\mathfrak{S}'=0$ and such that the matrix of T in $\mathfrak{S}'$ is also d. Let $x_1, \cdots, x_h, x_{h+1}, \cdots, x_{2h}$ be the original basis of $\mathfrak{R}$ relative to which T has the matrix v so that $x_1, \cdots, x_h$ is a basis for $\mathfrak{S}$. Corresponding to the decomposition $\mathfrak{R}=\mathfrak{S}+\mathfrak{S}'$ we have the basis $x_1, \cdots, x_h, x'_{h+1}, \cdots, x'_{2h}$. The matrix relating this basis to the original one has the form

$$\begin{pmatrix} 1 & p \\ 0 & q \end{pmatrix},$$

where p, $q\in\mathfrak{D}_h$ and the matrix of T relative to the basis $x_1, \cdots, x_h, x'_{h+1}, \cdots, x'_{2h}$ is u. Hence we have the equation

$$\begin{pmatrix} d & 0 \\ 0 & d \end{pmatrix} = \begin{pmatrix} 1 & p \\ 0 & q \end{pmatrix}^{-1} \begin{pmatrix} d & b \\ 0 & d \end{pmatrix} \begin{pmatrix} 1 & p \\ 0 & q \end{pmatrix}.$$

This implies that $dq=qd$ and that $bq=pd-dp$. Since the matrix

[5] See the author's paper *Pseudo-linear transformations*, Ann. of Math. vol. 38 (1937) p. 498.

$$\begin{pmatrix} 1 & p \\ 0 & q \end{pmatrix}$$

is regular, q is regular and hence we have the relation $b = xd - dx$ where $x = pq^{-1}$.

THEOREM 2. *Let $\mathfrak{A} = \mathfrak{D}_h$ where $\mathfrak{D}$ is a division algebra over Φ and let d be an algebraic element of $\mathfrak{A}$. Then if the minimum polynomial $\phi(\lambda)$ of d is a product of distinct irreducible factors in $\Phi[\lambda]$, the condition* (5) *is necessary and sufficient in order that the element b of $\mathfrak{A}$ be a d-integral.*

We next let $\mathfrak{A} = \Phi_h$, the matrix algebra of h rows over Φ. Let d be a non-derogatory matrix in Φ_h. Thus d has only one invariant factor $\phi(\lambda) \neq 1$ and $\phi(\lambda)$ is the minimum polynomial of d. Let b be a matrix such that (5) holds and consider the matrix v as before. The minimum polynomial of v is $\phi(\lambda)$. If T is the linear transformation in the $2h$-dimensional space over Φ associated with the matrix v then $\mathfrak{R}$ contains an invariant subspace $\mathfrak{S}$ whose matrix is d. Since d is non-derogatory, $\mathfrak{S}$ is a cyclic subspace and its order is the minimum polynomial of T in $\mathfrak{R}$. Now it is known that this implies that $\mathfrak{R} = \mathfrak{S} + \mathfrak{S}'$, $\mathfrak{S} \cap \mathfrak{S}' = 0$ where $\mathfrak{S}'$ is also invariant relative to T.[6] A repetition of the argument used to prove Theorem 2 will now yield the following theorem.

THEOREM 3. *Let d be a non-derogatory matrix in the matrix algebra Φ_h and let $\phi(\lambda)$ be its minimum polynomial. Then the condition* (5) *is necessary and sufficient that the matrix b be a d-integral.*

We give finally an example in which the condition (5) is not sufficient to insure that an element be a d-integral. Let

$$d = \begin{pmatrix} 0 & 1 & 0 \\ 0 & 0 & 0 \\ 0 & 0 & 0 \end{pmatrix}, \qquad b = \begin{pmatrix} 0 & 0 & 0 \\ 0 & 0 & 0 \\ 0 & 0 & 1 \end{pmatrix}.$$

Then the minimum polynomial of d is $\phi(\lambda) = \lambda^2$. Since $bd = db = 0$, b satisfies (5). On the other hand, the invariant factors of the matrices u and v here are respectively $\lambda^2, \lambda^2, \lambda, \lambda$ and $\lambda^2, \lambda^2, \lambda^2$. It follows that these matrices are not similar and hence b is not a d-integral.

THE JOHNS HOPKINS UNIVERSITY

[6] See van der Waerden's *Moderne Algebra*, vol. 2, pp. 129–130. The proof given there of this theorem for ordinary finite groups is also valid for vector spaces relative to a single linear transformation.

STRUCTURE THEORY OF SIMPLE RINGS WITHOUT FINITENESS ASSUMPTIONS

BY

N. JACOBSON

It is the purpose of this paper to lay the foundations of a general theory of simple rings, both associative and non-associative. In part I we obtain the structure of simple associative rings that either contain minimal right ideals or contain maximal right ideals. In either case we obtain a realization of our ring $\mathfrak{A}$ as a certain type of ring of linear transformations in a vector space over a division ring. If $\mathfrak{A}$ has a minimal right ideal this realization is essentially unique and we can prove a converse theorem that rings of linear transformations having certain properties are simple and contain minimal right ideals. Thus the theory here is essentially as complete as that of the classical case of rings that satisfy the descending chain condition for right ideals. Our structure theory hinges on a general theorem (Theorem 6) on the structure of irreducible rings of endomorphisms of commutative groups. This theorem may be regarded as an extension of Burnside's theorem on irreducible algebras of matrices. The classical theory of simple rings with descending chain condition is a simple consequence of our results and we believe that the present treatment is more transparent than the methods of proof previously given(1).

In studying an arbitrary non-associative ring $\mathfrak{A}$ one is led to consider the associative ring $\mathfrak{M}$ generated by the left and the right multiplications $x\rightarrow ax$ and $x\rightarrow xa$ acting in $\mathfrak{A}$. If $\mathfrak{A}$ is simple, $\mathfrak{M}$ is an irreducible ring of endomorphisms. In part II we describe the structure of $\mathfrak{M}$ in terms of the multiplication centralizer $\mathfrak{C}$ of $\mathfrak{A}$ defined to be the totality of endomorphisms γ in $\mathfrak{A}$ such that $(xy)\gamma=(x\gamma)y=x(y\gamma)$. We define the concepts of center, central algebra and extension of the underlying field of an algebra.

In Part III we investigate a special type of simple associative algebra that may be regarded as a generalization of the concept of the complete algebra of linear transformations in a vector space over a field, or equivalently, of the concept of the complete matrix algebra. There are many points of contact between the discussion here (and in other parts of the paper) and recent work in the theory of rings of transformations in Banach spaces and in vector spaces over the fields of real and complex numbers(2).

Presented to the Society, August 12, 1944; received by the editors June 24, 1944.

(1) Cf. Jacobson [1, chap. 4] and Artin and Whaples [1]. Numbers in brackets refer to the Bibliography at the end of the paper.

(2) Cf. Arnold [1] and the references given in this paper.

I. Structure of simple rings

1. Dense rings of linear transformations. Let $\mathfrak{R}$ be an arbitrary vector space over a division ring $\mathfrak{D}$. We recall the definition: $\mathfrak{R}$ is a commutative group written additively and there is singled out a division ring $\mathfrak{D}$ of endomorphisms $\alpha, \beta, \cdots$ acting in $\mathfrak{R}$ that includes the identity transformation. We make no assumption as to the dimensionality of $\mathfrak{R}$ over $\mathfrak{D}$. Let $\mathfrak{L}$ be the ring of linear transformations (l.t.) in $\mathfrak{R}$ over $\mathfrak{D}$. Thus $\mathfrak{L}$ is the totality of endomorphisms in $\mathfrak{R}$ that commute with every $\alpha \in \mathfrak{D}$.

Let $\mathfrak{A}$ be a set of l.t. Then we shall call $\mathfrak{A}$ *k-fold transitive* if for any two ordered sets of vectors $x_1, \cdots, x_k; y_1, \cdots, y_k$ such that the x's are linearly independent, there exists an $A \in \mathfrak{A}$ such that $x_iA = y_i$, $i=1, \cdots, k$. It is obvious that if $\mathfrak{A}$ is k-fold transitive, then $\mathfrak{A}$ is l-fold transitive for every $l<k$ and it is easy to see that if $\mathfrak{A}$ is n-fold transitive and the dimensionality $(\mathfrak{R}:\mathfrak{D})=n$, then $\mathfrak{A}=\mathfrak{L}$. If $\mathfrak{A}$ is 1-fold transitive, $\mathfrak{A}$ is an irreducible set of endomorphisms in the sense that the only subgroups of $\mathfrak{R}$ invariant under every $A \in \mathfrak{A}$ are $\mathfrak{R}$ and 0. Conversely if $\mathfrak{A} \neq 0$ is a subring of $\mathfrak{L}$ and is an irreducible set of endomorphisms, then $\mathfrak{A}$ is 1-fold transitive. For if $x \neq 0$ is in $\mathfrak{R}$ the totality $x\mathfrak{A}$ of vectors xA, $A \in \mathfrak{A}$, is a subgroup invariant under $\mathfrak{A}$. If $x\mathfrak{A}=0$ the space $\{x\}$ of multiples $x\alpha$ is invariant and not equal to 0. It follows that $\{x\}=\mathfrak{R}$ and this contradicts $\mathfrak{A} \neq 0$. Hence $x\mathfrak{A}=\mathfrak{R}$ and any y in $\mathfrak{R}$ has the form xA for a suitable A in $\mathfrak{A}$.

If $\mathfrak{A}$ is a set of l.t. that is k-fold transitive for each $k=1, 2, 3, \cdots$, then we shall say that $\mathfrak{A}$ is *dense* (in $\mathfrak{L}$). If $\mathfrak{R}$ is finite-dimensional and $\mathfrak{A}$ is dense, $\mathfrak{A}=\mathfrak{L}$. We also have the following extension of this result: $\mathfrak{A}$ is dense if and only if for any finite dimensional subspace $\mathfrak{S}$ of $\mathfrak{R}$ and any l.t. B in $\mathfrak{S}$ there exists an $A \in \mathfrak{A}$ that coincides with B in $\mathfrak{S}$. This follows directly from the definition.

By using the well-ordering theorem we can prove the existence of a complement $\mathfrak{S}'$ of any subspace $\mathfrak{S}$, that is, a subspace $\mathfrak{S}'$ such that $\mathfrak{R}=\mathfrak{S}+\mathfrak{S}'$, $\mathfrak{S} \wedge \mathfrak{S}'=0$(3). Now if $\mathfrak{S}$ is finite-dimensional and B is any l.t. in $\mathfrak{S}$ we can extend B to an l.t. A in $\mathfrak{R}$ by defining $xA=xB$ for $x \in \mathfrak{S}$ and $x'B=0$ for $x' \in \mathfrak{S}'$. This proves that $\mathfrak{L}$ itself is dense. Moreover, it should be noted that A is *finite valued* in the sense that the image space $\mathfrak{R}A$ is finite-dimensional. Hence we have also shown that the totality $\mathfrak{F}$ of finite valued l.t. is a dense set.

To construct a third example of a dense set of l.t. we assume that $\mathfrak{R}$ has a denumerable basis $u_1, u_2, \cdots$. Let $\mathfrak{U}_n$ be the totality of l.t. that map $\mathfrak{R}_n=\{u_1, \cdots, u_n\}$ into itself and map $\mathfrak{R}_n'=\{u_{n+1}, u_{n+2}, \cdots\}$ into 0. Let $\mathfrak{U}$ be the join of all the $\mathfrak{U}_n$. Then if $x_1, \cdots, x_k$ and $y_1, \cdots, y_k$ are given and the x's are linearly independent, the x's and the y's are contained in a suitable $\mathfrak{R}_N$. Hence there is an A in $\mathfrak{U}_N$ such that $x_iA=y_i$. This proves that $\mathfrak{U}$ is dense.

---

(3) See, for example, Jacobson [1, p. 18].

If F_1 and $F_2 \in \mathfrak{F}$ and $A \in \mathfrak{L}$, $\mathfrak{R}F_1 + \mathfrak{R}F_2$, $\mathfrak{R}AF_1$ and $\mathfrak{R}F_1A$ are all finite dimensional. Thus $F_1 + F_2$, F_1A, $AF_1 \in \mathfrak{F}$ and so $\mathfrak{F}$ is a two-sided ideal in the ring $\mathfrak{L}$. It is evident that if $\mathfrak{R}$ is infinite-dimensional, $\mathfrak{F} < \mathfrak{L}$ since $1 \notin \mathfrak{F}$. We shall now show that any dense subring of $\mathfrak{F}$ is simple. For this purpose we prove first the following lemma.

Lemma 1. *Let $\mathfrak{A}$ be an arbitrary dense ring of l.t. and let $\mathfrak{B}$ be a two-sided ideal in $\mathfrak{A}$ that contains a finite valued l.t. $B \neq 0$. Then if $\mathfrak{S}$ is any finite-dimensional subspace of $\mathfrak{R}$, there exists a projection $E \in \mathfrak{B}$ of $\mathfrak{R}$ on $\mathfrak{S}$.*

Let $y_1, \cdots, y_n$ be a basis for $\mathfrak{R}B$ and let x_1 be a vector such that $x_1B = y_1$. We determine an l.t. $A_1 \in \mathfrak{A}$ such that $y_1A_1 = x_1$ and $y_iA_1 = 0$ for $i \neq 1$. Then $BA_1 = E_1 \in \mathfrak{B}$, $\mathfrak{R}E_1 = \{x_1\}$ and $x_1E_1 = x_1$. Hence E_1 is a projection of $\mathfrak{R}$ on $\{x_1\}$. Now let x be any vector not equal to 0 and let C and D be determined in $\mathfrak{A}$ so that $xC = x_1$ and $x_1D = x$. Then it can be verified that $E = CE_1D$ is a projection of $\mathfrak{R}$ on $\{x\}$. Since $E \in \mathfrak{B}$ this proves the lemma for one-dimensional subspaces. We now assume it true for $k-1$-dimensional subspaces and let $\mathfrak{S}$ be k-dimensional. Let $\mathfrak{S}_{k-1}$ be a $k-1$-dimensional subspace of $\mathfrak{S}$ and let $E_{k-1} \in \mathfrak{B}$ be a projection of $\mathfrak{R}$ on $\mathfrak{S}_{k-1}$. We may decompose $\mathfrak{S}$ as $\mathfrak{S}_{k-1} \oplus \mathfrak{S}'$ where $\mathfrak{S}'$ is the one-dimensional subspace such that $\mathfrak{S}'E_{k-1} = 0$. Let $E' \in \mathfrak{B}$ be a projection of $\mathfrak{R}$ on $\mathfrak{S}'$ and set $E = E_{k-1} + E' - E_{k-1}E'$. Then if $x_{k-1} \in \mathfrak{S}_{k-1}$, $x_{k-1}E = x_{k-1}$, if $x' \in \mathfrak{S}'$, $x'E = x'$ and if x is arbitrary, then $xE \in \mathfrak{S}$. Since $E \in \mathfrak{B}$, E is the required projection.

We may now prove the following theorem.

Theorem 1. *Any dense ring $\mathfrak{A}$ of finite valued linear transformations in a vector space over a division ring is simple.*

Let $\mathfrak{B}$ be a two-sided ideal not equal to 0 in $\mathfrak{A}$ and let A be an arbitrary element of $\mathfrak{A}$. We determine a projection E in $\mathfrak{B}$ of $\mathfrak{R}$ on $\mathfrak{R}A$. Then $A = AE \in \mathfrak{B}$ and so $B \in \mathfrak{A}$.

This proves in particular that $\mathfrak{F}$ and $\mathfrak{U}$ are simple rings. Concerning $\mathfrak{L}$ we have the following theorem.

Theorem 2. *Any two-sided ideal $\mathfrak{B} \neq 0$ of $\mathfrak{L}$ contains $\mathfrak{F}$.*

We note first that $\mathfrak{B} \wedge \mathfrak{F} \neq 0$. For otherwise $\mathfrak{B}\mathfrak{F} = \mathfrak{F}\mathfrak{B} = 0$. Then if B is any l.t. in $\mathfrak{B}$ and u is a fixed vector not equal to 0, any vector x has the form uF for some F in $\mathfrak{F}$. Then $xB = uFB = 0$. Hence $B = 0$ so that $\mathfrak{B} = 0$ contrary to hypothesis. Thus we have proved that $\mathfrak{B}$ contains finite valued l.t. not equal to 0. The proof of Theorem 1 then shows that $\mathfrak{B}$ contains $\mathfrak{F}$.

2. **One-sided ideals in dense rings of linear transformations.** If $\mathfrak{S}$ is a subspace of $\mathfrak{R}$ we shall call a linear transformation an *annihilator* of $\mathfrak{S}$ if $\mathfrak{S}A = 0$ and we shall call A a *retraction* on $\mathfrak{S}$ if $\mathfrak{R}A \leqq \mathfrak{S}$. Now if $\mathfrak{A}$ is a subring of $\mathfrak{L}$, it is clear that the totality of annihilators of $\mathfrak{S}$ contained in $\mathfrak{A}$ is a right ideal $\mathfrak{J}_r(\mathfrak{S})$. The totality of retractions on $\mathfrak{S}$ contained in $\mathfrak{A}$ is a left ideal

$\mathfrak{J}_l(\mathfrak{S})$. We note that $\mathfrak{J}_l(\mathfrak{S})\mathfrak{J}_r(\mathfrak{S})=0$ and if $\mathfrak{S}_1\leqq\mathfrak{S}_2$ then $\mathfrak{J}_l(\mathfrak{S}_1)\leqq\mathfrak{J}_l(\mathfrak{S}_2)$ and $\mathfrak{J}_r(\mathfrak{S}_1)\geqq\mathfrak{J}_r(\mathfrak{S}_2)$.

We suppose now that $\mathfrak{A}$ is dense and that $\mathfrak{S}$ is finite-dimensional. Then if $\mathfrak{S}\neq\mathfrak{R}$, $\mathfrak{J}_r(\mathfrak{S})\neq 0$. For if $x_1, \cdots, x_m$ is a basis for $\mathfrak{S}$ and x_{m+1} is a vector not in $\mathfrak{S}$, $x_1, \cdots, x_{m+1}$ are linearly independent. Hence there is an $A\in\mathfrak{A}$ such that $x_1A=\cdots=x_mA=0$ but $x_{m+1}A\neq 0$. Thus $A\neq 0$ and $A\in\mathfrak{J}_r(\mathfrak{S})$. Our argument shows also that if $\mathfrak{S}_1$ and $\mathfrak{S}_2$ are finite-dimensional and $\mathfrak{S}_2>\mathfrak{S}_1$, then $\mathfrak{J}_r(\mathfrak{S}_1)>\mathfrak{J}_r(\mathfrak{S}_2)$. We note next that if $(\mathfrak{S}_2:\mathfrak{D})=(\mathfrak{S}_1:\mathfrak{D})+1$, then $\mathfrak{J}_r(\mathfrak{S}_1)$ is *prime over* $\mathfrak{J}_r(\mathfrak{S}_2)$ in the sense that there is no right ideal $\mathfrak{J}$ in $\mathfrak{A}$ such that $\mathfrak{J}_r(\mathfrak{S}_1)>\mathfrak{J}>\mathfrak{J}_r(\mathfrak{S}_2)$. For suppose that $\mathfrak{J}_r(\mathfrak{S}_1)\geqq\mathfrak{J}>\mathfrak{J}_r(\mathfrak{S}_2)$ and let $B\in\mathfrak{J}$, $B\notin\mathfrak{J}_r(\mathfrak{S}_2)$. Then if $x\in\mathfrak{S}_2$, $x\notin\mathfrak{S}_1$, $xB\neq 0$. Hence if C is any l.t. in $\mathfrak{J}_r(\mathfrak{S}_1)$, there exists a U in $\mathfrak{A}$ such that $(xB)U=xC$. Thus $C-BU$ annihilates x and since $C-BU\in\mathfrak{J}_r(\mathfrak{S}_1)$, $C-BU\in\mathfrak{J}_r(\mathfrak{S}_2)$. Since $BU\in\mathfrak{J}$ this proves that $C\in\mathfrak{J}$ and so $\mathfrak{J}=\mathfrak{J}_r(\mathfrak{S}_1)$.

Now $\mathfrak{J}_r(0)=\mathfrak{A}$. Hence we have shown in particular that $\mathfrak{J}_r(\{x\})$ is a maximal right ideal in $\mathfrak{A}$. If $\mathfrak{S}$ is m-dimensional, there is a chain of subspaces $0=\mathfrak{S}_0<\mathfrak{S}_1<\cdots<\mathfrak{S}_m=\mathfrak{S}$ such that $(\mathfrak{S}_i:\mathfrak{D})=(\mathfrak{S}_{i-1}:\mathfrak{D})+1$. Correspondingly we have a chain of right ideals $\mathfrak{A}=\mathfrak{J}_0>\mathfrak{J}_1>\cdots>\mathfrak{J}_m$ where $\mathfrak{J}_j=\mathfrak{J}_r(\mathfrak{S}_j)$ is prime over $\mathfrak{J}_{j+1}$. If $\mathfrak{R}$ itself is finite-dimensional, the sequence of right ideals leading to $\mathfrak{J}_r(\mathfrak{R})$ is a composition series and it is well known that this implies that $\mathfrak{A}(=\mathfrak{L})$ satisfies the chain conditions for right ideals. On the other hand if $\mathfrak{R}$ is infinite-dimensional, we can find an infinite properly ascending sequence of finite-dimensional subspaces $\mathfrak{S}_1<\mathfrak{S}_2<\cdots$ in $\mathfrak{R}$. Then $\mathfrak{J}_r(\mathfrak{S}_1)>\mathfrak{J}_r(\mathfrak{S}_2)>\cdots$. This proves the following theorem.

THEOREM 3. *Let $\mathfrak{A}$ be a dense ring of linear transformations in $\mathfrak{R}$ over $\mathfrak{D}$. Then a necessary and sufficient condition that $\mathfrak{A}$ satisfy the descending chain condition for right ideals is that $\mathfrak{R}$ be finite-dimensional over $\mathfrak{D}$.*

We have noted also that $\mathfrak{J}_r\{x\}$ is maximal if $x\neq 0$. It is evident that the intersection of all the ideals $\mathfrak{J}_r\{x\}$ is the 0-ideal. Hence we have the following theorem.

THEOREM 4. *If $\mathfrak{A}$ is a dense ring of linear transformations, $\mathfrak{A}$ contains maximal right ideals and the intersection of all the maximal right ideals in $\mathfrak{A}$ is the 0-ideal.*

We assume next that $\mathfrak{A}$ is dense and that $\mathfrak{A}$ contains finite valued transformations not equal to 0. Then if $\mathfrak{S}$ is any finite-dimensional subspace of $\mathfrak{R}$, by Lemma 1, $\mathfrak{A}$ contains a projection E of $\mathfrak{R}$ on $\mathfrak{S}$. If $\mathfrak{S}'$ is the totality of vectors y' such that $y'E=0$, we have $\mathfrak{R}=\mathfrak{S}+\mathfrak{S}'$, $\mathfrak{S}\wedge\mathfrak{S}'=0$. We consider now $\mathfrak{J}_r(\mathfrak{S}')$. By definition $E\in\mathfrak{J}_r(\mathfrak{S}')$. If $x=y+y'$ where $y\in\mathfrak{S}$, $y'\in\mathfrak{S}'$ and $B\in\mathfrak{J}_r(\mathfrak{S}')$ then $xB=yB=xEB$. Hence $B=EB$ and so $\mathfrak{J}_r(\mathfrak{S}')=E\mathfrak{A}$. Suppose now that $\mathfrak{S}=\{u\}$ is one-dimensional and let $B\neq 0$ and C be arbitrary elements of $\mathfrak{J}_r(\mathfrak{S}')$. Then $uB\neq 0$ and there is an l.t. U in $\mathfrak{A}$ such that

$(uB)U = uC$. It follows that $C = BU$ and this implies that $\mathfrak{J}_r(\mathfrak{S}')$ is a minimal right ideal.

THEOREM 5. *Let $\mathfrak{A}$ be a dense ring of linear transformations containing finite valued transformations not equal to* 0. *Then $\mathfrak{A}$ contains minimal right ideals.*

The theory of the left ideals $\mathfrak{J}_l(\mathfrak{S})$ is considerably less satisfactory than the foregoing. We note only the following result: If $\mathfrak{A}$ is dense and contains finite valued l.t. not equal to 0 and if $\mathfrak{S}$ is a finite-dimensional subspace of $\mathfrak{R}$, $\mathfrak{A}$ contains a projection E of $\mathfrak{R}$ on $\mathfrak{S}$. Hence $\mathfrak{J}_l(\mathfrak{S}) = \mathfrak{A}E \neq 0$. The join $\mathfrak{R}\mathfrak{J}_l(\mathfrak{S})$ of the spaces $\mathfrak{R}B$ where $B \in \mathfrak{J}_l(\mathfrak{S})$ is the whole space $\mathfrak{S}$. Hence if $\mathfrak{S}_1 < \mathfrak{S}_2$ then $\mathfrak{J}_l(\mathfrak{S}_1) < \mathfrak{J}_l(\mathfrak{S}_2)$.

3. **Irreducible rings of endomorphisms.** We shall now shift our emphasis somewhat and begin with an arbitrary commutative group $\mathfrak{R}$ and with an irreducible ring $\mathfrak{A}$ of endomorphisms acting in $\mathfrak{R}$. Let $\mathfrak{D}$ be the centralizer of $\mathfrak{A}$ in the complete ring of endomorphisms in $\mathfrak{R}$. Thus $\mathfrak{D}$ consists of the endomorphisms α that commute with every A in $\mathfrak{A}$. We recall the well known lemma.

LEMMA 2 (SCHUR). *$\mathfrak{D}$ is a division ring*(4).

We now regard $\mathfrak{R}$ as a vector space over $\mathfrak{D}$ and we shall prove the following fundamental theorem.

THEOREM 6. *Let $\mathfrak{A} \neq 0$ be an irreducible ring of endomorphisms acting in a commutative group $\mathfrak{R}$. Then if $\mathfrak{D}$ is the division ring of endomorphisms that commute with the elements of $\mathfrak{A}$, $\mathfrak{A}$ is dense in $\mathfrak{R}$ over $\mathfrak{D}$.*

Proof. Let $x_1, \cdots, x_k$ be any k $\mathfrak{D}$-independent vectors and let $y_1, \cdots, y_k$ be arbitrary. We have to show that there is an A in $\mathfrak{A}$ such that $x_iA = y_i$, $i = 1, \cdots, k$. The case $k = 1$ has been proved in §1. We assume now that $k = 2$.

$k = 2$. We prove first that there is an A_2 in $\mathfrak{A}$ such that $x_1A_2 = 0$, $x_2A_2 \neq 0$. If this is not the case, then any equation of the form $x_1X = x_1Y$ for $X, Y \in \mathfrak{A}$ entails $x_2X = x_2Y$. Hence the mapping $y = x_1X \rightarrow x_2X = y'$ is single-valued and since every y is representable in the form x_1X, this mapping is defined in the whole of $\mathfrak{R}$. It is clear that

$$(y_1 + y_2)' = y_1' + y_2', \qquad (yB)' = y'B,$$

for all $B \in \mathfrak{A}$. Hence the mapping $y \rightarrow y'$ is in $\mathfrak{D}$ and there is a γ in $\mathfrak{D}$ such that $x_2X = (x_1X)\gamma$ for all X. Thus $(x_2 - x_1\gamma)X = 0$ and this implies that $x_2 = x_1\gamma$, contrary to the linear independence of x_1 and x_2. This contradiction shows that there is an A_2 in $\mathfrak{A}$ such that $x_1A_2 = 0$, $x_2A_2 \neq 0$. Similarly we have an A_1 such that $x_1A_1 \neq 0$, $x_2A_1 = 0$. We now choose B_1 and B_2 so that $x_1A_1B_1 = y_1$ and $x_2A_2B_2 = y_2$. Then $A = A_1B_1 + A_2B_2$ has the required properties.

(4) Jacobson [1, p. 57].

k arbitrary. We assume the theorem for $k-1$ and as in the case $k=2$ it suffices to prove that there is an A_k in $\mathfrak{A}$ such that $x_1A_k = \cdots = x_{k-1}A_k = 0$, $x_kA_k \neq 0$. By the induction assumption there is a B in $\mathfrak{A}$ such that $x_1B = \cdots = x_{k-2}B = 0$ but $x_kB \neq 0$. Let $\mathfrak{B}$ be the totality of mappings in $\mathfrak{A}$ that have this property. Assume first that $x_{k-1}B$ and x_kB are $\mathfrak{D}$-independent. Then by the case $k=2$ there is a C such that $x_{k-1}BC=0$, $x_kBC\neq 0$ and $A_k=BC$ is of the required type. Next let $x_{k-1}B$ and x_kB be linearly dependent and we may assume that $x_{k-1}B=(x_kB)\beta$ where $\beta\neq 0$ is in $\mathfrak{D}$. We now choose a D in $\mathfrak{A}$ such that $x_1D = \cdots = x_{k-2}D=0$ but $(x_{k-1}-x_k\beta)D\neq 0$. If $x_kD=0$ and $x_{k-1}D\neq 0$ there is a U in $\mathfrak{A}$ such that $x_{k-1}DU=x_{k-1}B$. Then for $A_k=B-DU$ we have $x_1A_k = \cdots = x_{k-1}A_k=0$ and $x_kA_k=x_kB\neq 0$. If $x_kD\neq 0$, $D\in\mathfrak{B}$. If either $x_{k-1}D=0$ or $x_{k-1}D$ and x_kD are $\mathfrak{D}$-independent, then, as before, we may modify D to obtain an A_k of the required type. Hence we suppose that $x_{k-1}D=(x_kD)\delta$ where $\delta\neq 0$. Evidently $\delta\neq\beta$. Now there exists a U in $\mathfrak{A}$ such that $x_{k-1}DU=x_{k-1}B$. Then $x_kDU=(x_{k-1}D)\delta^{-1}U=(x_{k-1}DU)\delta^{-1}$ $=(x_{k-1}B)\delta^{-1}=(x_kB)\beta\delta^{-1}$, that is, $x_k(B\beta\delta^{-1}-DU)=0$. Since $\beta\delta^{-1}\neq 1$, if we set $A_k=B-DU$ we have $x_kA_k\neq 0$, and so A_k has the required properties.

As a first application of Theorem 6 we shall prove that if $\mathfrak{A}$ is a 2-fold transitive ring of linear transformations in a vector space $\mathfrak{R}$ over $\mathfrak{D}$, then $\mathfrak{A}$ is dense. This will follow from the following theorem.

THEOREM 7. *If $\mathfrak{A}$ is a 2-fold transitive ring of linear transformations in a vector space $\mathfrak{R}$ over $\mathfrak{D}$ then $\mathfrak{D}$ is the centralizer of $\mathfrak{A}$ in the complete ring of endomorphisms of $\mathfrak{R}$.*

If $(\mathfrak{R}:\mathfrak{D})=1$ and $x\neq 0$ is in $\mathfrak{R}$, $\mathfrak{A}$ consists of the mappings $y=x\xi\rightarrow(x\alpha)\xi=y'$. $\mathfrak{A}$ is then a division ring anti-isomorphic to $\mathfrak{D}$ and the theorem is readily verified in this case(5). Now let $(\mathfrak{R}:\mathfrak{D})\geqq 2$ and let B be an endomorphism in $\mathfrak{R}$ that commutes with every A in $\mathfrak{A}$. Then we assert that for every x in $\mathfrak{R}$, x and xB are $\mathfrak{D}$-dependent. For otherwise there is an A in $\mathfrak{A}$ such that $xA=0$ but $xBA\neq 0$ and this contradicts $AB=BA$. Hence if $x\neq 0$, $xB=x\beta_x$, β_x in $\mathfrak{D}$. Now let x and y be linearly independent. Then $xB=x\beta_x$ and $yB=y\beta_y$. There is an A in $\mathfrak{A}$ such that $xA=y$. Since $AB=BA$ this implies that $\beta_x=\beta_y=\beta$. Since $x\alpha$ and y are linearly independent we have $(x\alpha)B=(x\alpha)\beta$ and this shows that $B=\beta\in\mathfrak{D}$.

As an immediate consequence of Theorems 6 and 7 we have the following theorem.

THEOREM 8. *If $\mathfrak{A}$ is a 2-fold transitive ring of linear transformations in $\mathfrak{R}$ over $\mathfrak{D}$, $\mathfrak{A}$ is a dense ring.*

4. **Structure of simple rings that contain minimal right ideals.** Let $\mathfrak{A}$ be an abstract simple ring that contains a minimal right ideal $\mathfrak{J}$. We assume

(5) Jacobson [1, p. 22].

also that $\mathfrak{A}$ is not a zero ring. Hence $\mathfrak{A}^2=\mathfrak{A}\neq 0$. It follows that $\mathfrak{J}\mathfrak{A}\neq 0$. For otherwise $\mathfrak{B}=\mathfrak{A}\mathfrak{J}+\mathfrak{J}$ is a nilpotent two-sided ideal and $\mathfrak{B}=\mathfrak{A}$ contrary to $\mathfrak{A}^2\neq 0$. It is known that $\mathfrak{J}=e\mathfrak{A}$ where e is an idempotent element(6).

Each element a of $\mathfrak{A}$ defines an endomorphism $\bar{a}$ in $\mathfrak{J}$ by the equation $x\bar{a}=xa$, $x\in\mathfrak{J}$. The set $\overline{\mathfrak{A}}$ of these endomorphisms is a subring not equal to 0 of the ring of endomorphisms of $\mathfrak{J}$ and the correspondence $a\rightarrow\bar{a}$ is a homomorphism. Since $\mathfrak{A}$ is simple, $a\rightarrow\bar{a}$ is an isomorphism. Since $\mathfrak{J}$ is minimal, $\overline{\mathfrak{A}}$ is an irreducible ring of endomorphisms. Hence if $\mathfrak{D}$ denotes the division ring consisting of the endomorphisms $\alpha, \beta, \cdots$ that commute with all the $\bar{a}$, $\overline{\mathfrak{A}}$ is a dense ring of linear transformations in $\mathfrak{J}$ regarded as a vector space over $\mathfrak{D}$.

We shall show next that $\overline{\mathfrak{A}}$ consists of finite valued l.t. We prove first that $(\mathfrak{J}\bar{e}:\mathfrak{D})=1$. For otherwise we should have two $\mathfrak{D}$-independent elements x, y in $\mathfrak{J}$ such that $x\bar{e}=x$, $y\bar{e}=y$. Then there is a $\bar{d}\in\overline{\mathfrak{A}}$ such that $x\bar{d}=0$ and $y\bar{d}\neq 0$ and if $\bar{b}=\bar{e}\bar{d}$, $x\bar{b}=0$, $y\bar{b}\neq 0$. Since $\bar{b}\in\bar{e}\overline{\mathfrak{A}}=\overline{\mathfrak{J}}$, this shows that $\overline{\mathfrak{J}}\wedge\overline{\mathfrak{J}}_r\{x\}\neq 0$, where $\overline{\mathfrak{J}}_r\{x\}$ denotes as usual the annihilator of the subspace $\{x\}$. Since $\overline{\mathfrak{J}}$ is minimal, $\overline{\mathfrak{J}}\wedge\overline{\mathfrak{J}}_r\{x\}=\overline{\mathfrak{J}}$ and this contradicts the equation $x\bar{e}=x$. We have therefore proved our assertion that $(\mathfrak{J}\bar{e}:\mathfrak{D})=1$. Thus $\bar{e}$ is finite valued. Now the subset of finite valued transformations belonging to $\overline{\mathfrak{A}}$ is a two-sided ideal. Since we have shown that this subset is not equal to 0 it follows that $\overline{\mathfrak{A}}$ contains only finite valued l.t. This together with Theorems 1 and 5 proves the following structure theorem.

THEOREM 9. *If $\mathfrak{A}$ is a simple ring (not a zero ring) that contains a minimal right ideal, then $\mathfrak{A}$ is isomorphic to a dense ring of finite valued linear transformations in a vector space over a division ring. Conversely, any dense ring of finite valued linear transformations in a vector space over a division ring is simple and contains minimal right ideals.*

As a corollary of this theorem and of Theorem 3 we have the fundamental Wedderburn-Artin structure theorem.

THEOREM 10. *If $\mathfrak{A}$ is a simple ring (not a zero ring) that satisfies the descending chain condition for right ideals, then $\mathfrak{A}$ is isomorphic to the complete ring of linear transformations in a suitable finite-dimensional vector space over a division ring, and conversely.*

Another consequence of the main theorem is a result noted previously by Artin and Whaples, namely, if $\mathfrak{A}$ is simple and contains a minimal right ideal, then $\mathfrak{A}$ satisfies the descending chain condition for r'ght ideals if and only if $\mathfrak{A}$ has an identity(7). This is clear since the identity 1 of $\mathfrak{A}$ is finite valued if and only if the space is finite dimensional. Other simple consequences of the

(6) Jacobson [1, p. 64].
(7) Artin and Whaples [1, p. 91].

main theorem and of the discussion of §2 are that if $\mathfrak{A}$ is simple and contains a minimal right ideal, then $\mathfrak{A}$ contains maximal right ideals and the intersection of all the maximal right ideals of $\mathfrak{A}$ is the 0-ideal.

We consider now the question of the uniqueness of the representation of $\mathfrak{A}$ as a dense ring of linear transformations. We suppose that $\mathfrak{R}$ is an arbitrary commutative group, that $\bar{\mathfrak{A}}$ is an irreducible subring not equal to 0 of the ring of endomorphisms of $\mathfrak{R}$ and that $\bar{\mathfrak{A}}$ is a homomorphic image of $\mathfrak{A}$. Let $a \rightarrow \bar{a}$ be the correspondence between $\mathfrak{A}$ and $\bar{\mathfrak{A}}$. Since $\mathfrak{A}$ is simple and $\bar{\mathfrak{A}} \neq 0$ our correspondence is in reality an isomorphism. Let $\mathfrak{J}$ be a minimal right ideal in $\mathfrak{A}$. Then there is an x in $\mathfrak{R}$ such that $x\bar{\mathfrak{J}} \neq 0$. Since the set $x\bar{\mathfrak{J}}$ is a subgroup invariant under $\mathfrak{A}$, $x\bar{\mathfrak{J}} = \mathfrak{R}$ and since $\mathfrak{J}$ is minimal, $x\bar{b} = 0$ for b in $\mathfrak{J}$ only if $b = 0$. Thus the correspondence $b \rightarrow x\bar{b} \equiv b'$ is an isomorphism between $\mathfrak{J}$ and $\mathfrak{R}$. Since $\overline{ba} = \bar{b}\bar{a}$, $(ba)' = b'\bar{a}$ and so $b \rightarrow b'$ is an operator isomorphism. It follows from this that if $\mathfrak{D}$ and $\bar{\mathfrak{D}}$, respectively, are the centralizers of the multiplications $b \rightarrow ba$ in $\mathfrak{J}$ and of the ring $\bar{\mathfrak{A}}$ in $\mathfrak{R}$, then there is an isomorphism $\alpha \rightarrow \bar{\alpha}$ between $\mathfrak{D}$ and $\bar{\mathfrak{D}}$ such that $(b\alpha)' = b'\bar{\alpha}$ for all b. By Theorem 6, we know that $\bar{\mathfrak{A}}$ is dense in $\mathfrak{R}$ over $\bar{\mathfrak{D}}$. Moreover, by our isomorphisms, and the fact that the multiplications in $\mathfrak{J}$ are all finite valued, we see that the l.t. of $\bar{\mathfrak{A}}$ are all finite valued.

Now suppose that $\mathfrak{R}_1$ and $\mathfrak{R}_2$ are vector spaces over $\mathfrak{D}_1$ and $\mathfrak{D}_2$, respectively, and that $\mathfrak{A}_1$ and $\mathfrak{A}_2$ are isomorphic dense rings of finite valued transformations in these spaces. Let $a_1 \rightarrow a_2$ be a definite isomorphism between these rings. We may regard $\mathfrak{A}_1$ and $\mathfrak{A}_2$ as isomorphic images of the same simple ring $\mathfrak{A}$ that contains a minimal right ideal $\mathfrak{J}$. Let the isomorphism between $\mathfrak{A}$ and $\mathfrak{A}_i$, $i = 1, 2$, be $a \rightarrow a_i$. By Theorem 6, $\mathfrak{D}_i$ is the centralizer of $\mathfrak{A}_i$. Hence by the above argument there is an isomorphism $b \rightarrow b_i$ between $\mathfrak{J}$ and $\mathfrak{R}_i$ and an isomorphism $\alpha \rightarrow \alpha_i$ between $\mathfrak{D}$, the centralizer of the multiplications in $\mathfrak{J}$, and $\mathfrak{D}_i$ such that $b\alpha \rightarrow b_i\alpha_i$ and $ba \rightarrow b_ia_i$. It follows that the correspondences $b_1 \rightarrow b_2$, $b_i \in \mathfrak{R}_i$, $\alpha_1 \rightarrow \alpha_2$, $\alpha_i \in \mathfrak{D}_i$, are isomorphisms and that $b_1\alpha_1 \rightarrow b_2\alpha_2$ and $b_1a_1 \rightarrow b_2a_2$. This proves the following important uniqueness theorem.

THEOREM 11. *Let $\mathfrak{R}_1$ and $\mathfrak{R}_2$ be vector spaces over $\mathfrak{D}_1$ and $\mathfrak{D}_2$, respectively, and let $\mathfrak{A}_1$ and $\mathfrak{A}_2$ be isomorphic dense rings of finite valued linear transformations acting in these spaces. Then if $a_1 \rightarrow a_2$ is a given isomorphism between $\mathfrak{A}_1$ and $\mathfrak{A}_2$, there exists an isomorphism $x_1 \rightarrow x_2$ between $\mathfrak{R}_1$ and $\mathfrak{R}_2$ and an isomorphism $\alpha_1 \rightarrow \alpha_2$ between $\mathfrak{D}_1$ and $\mathfrak{D}_2$ such that $x_1\alpha_1 \rightarrow x_2\alpha_2$ and $x_1a_1 \rightarrow x_2a_2$.*

Theorem 11 shows that two dense rings of finite valued l.t. are not isomorphic if the vector spaces in which they act are not isomorphic. If the spaces are isomorphic, we may without loss of generality identify them. Then our theorem assumes the following form:

THEOREM 12. *Let $\mathfrak{R}$ be a vector space over a division ring $\mathfrak{D}$ and let $\mathfrak{A}_1$ and $\mathfrak{A}_2$ be two isomorphic dense rings of finite valued linear transformations in $\mathfrak{R}$ over $\mathfrak{D}$.*

Then if $a_1 \to a_2$ is a given isomorphism between $\mathfrak{A}_1$ and $\mathfrak{A}_2$, there exists a semi-linear transformation $\mathfrak{S}$ in $\mathfrak{R}$ over $\mathfrak{D}$ such that $a_2 = S^{-1}a_1S$ for all a_1(8).

If $\mathfrak{A}$ is a simple, nonzero ring that contains a minimal left ideal, it is known that $\mathfrak{A}$ contains a minimal right ideal(9). Our discussion is therefore applicable to these rings also.

5. **Simple rings that contain maximal right ideals.** We shall suppose now that $\mathfrak{A}$ is a simple ring that contains a maximal right ideal $\mathfrak{J}$. By the well ordering theorem we may prove that any ring with an identity has this property. As before, we consider the right multiplications $x \to xa$ for x in $\mathfrak{A}$ and a in $\mathfrak{A}$. These mappings are endomorphisms that induce endomorphisms in $\mathfrak{J}$ and also in the difference group $\mathfrak{R} = \mathfrak{A} - \mathfrak{J}$. We denote the endomorphism induced by $x \to xa$ in $\mathfrak{R}$ by $\bar{a}$ and we denote the set of $\bar{a}$ by $\overline{\mathfrak{A}}$. Since $\mathfrak{J}$ is maximal, $\overline{\mathfrak{A}}$ is an irreducible set of endomorphisms. If $\overline{\mathfrak{A}} = 0$, $\mathfrak{A}^2 \leqq \mathfrak{J}$, and this implies that $\mathfrak{A}$ is a zero-ring. We exclude this case from further consideration. Then $\overline{\mathfrak{A}}$ is a ring not equal to 0 and the correspondence $a \to \bar{a}$ is an isomorphism between $\mathfrak{A}$ and $\overline{\mathfrak{A}}$.

Now let $\mathfrak{D}$ be the centralizer of $\overline{\mathfrak{A}}$ in the ring of endomorphisms of $\mathfrak{R}$. Then by Theorem 6, $\overline{\mathfrak{A}}$ is a dense ring of linear transformations in $\mathfrak{R}$ over $\mathfrak{D}$. Since $\mathfrak{A}$ is simple, we have the following two possibilities: either $\overline{\mathfrak{A}}$ consists exclusively of finite valued l.t. or every l.t. not equal 0 in $\overline{\mathfrak{A}}$ is infinite valued. We have seen that the former case obtains if and only if $\mathfrak{A}$ has a minimal right ideal. This proves the following theorem.

THEOREM 13. *Let $\mathfrak{A}$ be a simple ring (not a zero ring) that contains a maximal right ideal. Then $\mathfrak{A}$ is isomorphic to a dense ring $\overline{\mathfrak{A}}$ of linear transformations in a suitable vector space over a division ring. If $\mathfrak{A}$ contains a minimal right ideal, $\overline{\mathfrak{A}}$ consists exclusively of finite valued linear transformations, otherwise every linear transformation not equal to 0 in $\overline{\mathfrak{A}}$ is infinite valued.*

If $\mathfrak{A}$ has an identity, the division ring $\mathfrak{D}$ obtained from the maximal right ideal $\mathfrak{J}$ may be obtained in the following manner. We have $\mathfrak{R} = [1]\mathfrak{A}$ where $[1]$ denotes the coset $1 + \mathfrak{J}$. Let $\beta \in \mathfrak{D}$ and suppose that $[1]\beta = [b] = b + \mathfrak{J}$. Then if $[x]$ is any coset, we have $[x] = [1]x$ and $[x]\beta = ([1]x)\beta = ([1]\beta)x = [b]x = [bx]$. Since $[0]\beta = [0]$, the elements $bx \in \mathfrak{J}$ for every x in $\mathfrak{J}$. Now we shall define the *normalizer*(10) $\mathfrak{B}$ of the right ideal $\mathfrak{J}$ to be the totality of elements b such that $b\mathfrak{J} \leqq \mathfrak{J}$. $\mathfrak{B}$ is a subring of $\mathfrak{A}$ containing $\mathfrak{J}$ and is in fact the largest subring of $\mathfrak{A}$ in which $\mathfrak{J}$ is a two-sided ideal. Our computation above shows that the elements of $\mathfrak{D}$ are of the form $[x] \to [bx]$ where $b \in \mathfrak{B}$. Conversely, it is readily seen that if b is any element of $\mathfrak{B}$ the mappings of

(8) A transformation S in a vector space is called semi-linear if it is an endomorphism and $(x\alpha)S = (xS)\bar{\alpha}$ where $\alpha \to \bar{\alpha}$ is an automorphism in $\mathfrak{D}$.

(9) Artin and Whaples [1, p. 92].

(10) This is the "Eigenring" of an ideal first defined by Ore [1]. Cf. also Fitting [1].

this form are single valued and belong to $\mathfrak{D}$. Two elements b_1 and b_2 give rise to the same mapping $\beta \in \mathfrak{D}$ if and only if $b_1 - b_2 \in \mathfrak{J}$. It follows from this that $\mathfrak{D}$ is anti-isomorphic to the difference ring $\mathfrak{B} - \mathfrak{J}$.

We shall give now an example of a simple ring that contains no minimal right ideals.

Example. Let Φ be a differential field of characteristic 0 such that the field Φ_0 of constants is a proper subfield of Φ. Let $\mathfrak{A}$ be the domain of differential polynomials $D^m a_1 + D^{m-1} a_1 + \cdots + a_m$ where the $a_i \in \Phi$. We have the commutation rule $aD = Da + a'$ where a' is the derivative of a. Then it is known that every right ideal and every left ideal in $\mathfrak{A}$ is principal. Suppose now that $\mathfrak{J}^*$ is a two-sided ideal not equal to 0. Then it is easily shown that $\mathfrak{J}^* = \mathfrak{A}A^* = A^*\mathfrak{A}$ where $A^* = D^m a_0 + D^{m-1} a_1 + \cdots + a_m$, $a_0 \neq 0$(11). Thus if B is any element of A, BA^* is divisible on the left by A^*, say $BA^* = A^*\bar{B}$. In particular, for $B = b \in \Phi$, considerations of degree show that $\bar{b} \in \Phi$. A comparison of leading coefficients shows that $b = \bar{b}$ and so A^* commutes with all the elements of Φ. Now $bD^k = D^k b + C_{k,1} D^{k-1} b' + \cdots + b^{(k)}$. Hence $0 = bA^* - A^* b = C_{m,1} D^{m-1} b' a_0 + \cdots$. If $m > 0$ this implies that $b' = 0$ for all b in Φ contrary to assumption. Thus $A^* = a \in \Phi$ and $\mathfrak{J}^* = \mathfrak{A}$. This proves the simplicity of $\mathfrak{A}$. If $\mathfrak{J} = A\mathfrak{A}$ is any right ideal not equal to 0 and B is an element of positive degree in $\mathfrak{A}$, $\mathfrak{J} > AB\mathfrak{A} > 0$. This shows that $\mathfrak{A}$ contains no minimal right ideals. It is not difficult to construct maximal ideals $\mathfrak{J}_1$ and $\mathfrak{J}_2$ in $\mathfrak{A}$ such that if $\mathfrak{B}_1$ and $\mathfrak{B}_2$ are the corresponding normalizers, then $\mathfrak{B}_1 - \mathfrak{J}_1$ and $\mathfrak{B}_2 - \mathfrak{J}_2$ are not isomorphic(12). Thus no uniqueness theorem of the form of Theorem 11 is valid in this case.

II. Non-associative simple rings

6. **Multiplication ring of a non-associative simple ring.** If $\mathfrak{A}$ is an arbitrary non-associative ring(13) we define the *right multiplication* a_r as the mapping $x \to xa$, x in $\mathfrak{A}$, and the left multiplication a_l as $x \to ax$. These are endomorphisms in the additive group of $\mathfrak{A}$ and they generate a subring $\mathfrak{M}$ of the complete ring of endomorphisms of $\mathfrak{A}$. We call $\mathfrak{M}$ the *multiplication ring of* $\mathfrak{A}$. A subgroup $\mathfrak{B}$ of $\mathfrak{A}$ is a two-sided ideal if $\mathfrak{B}$ is invariant under all the right and left multiplications, or, what is equivalent, if $\mathfrak{B}$ is invariant under the endomorphisms belonging to $\mathfrak{M}$. Thus $\mathfrak{A}$ is simple if and only if $\mathfrak{M}$ is an irreducible ring of endomorphisms.

We shall call the centralizer $\mathfrak{C}$ of $\mathfrak{M}$ in the ring of endomorphisms of $\mathfrak{A}$ the *multiplication centralizer* of $\mathfrak{A}$. The elements of $\mathfrak{C}$ are the endomorphisms γ that satisfy

$$(xy)\gamma = (x\gamma)y = x(y\gamma). \tag{1}$$

(11) Jacobson [1, p. 37].

(12) This is indicated in Jacobson [3, p. 506].

(13) By a non-associative ring we, of course, mean a ring which is not necessarily associative.

Hence, if $\gamma, \delta \in \mathfrak{C}$

$$(xy)\gamma\delta = ((x\gamma)y)\delta = (x\gamma)(y\delta)$$
$$= (x(y\gamma))\delta = (x\delta)(y\gamma)$$

and this proves that $(xy)\gamma\delta = (xy)\delta\gamma$. Thus if $\mathfrak{A}^2$ is the subring consisting of the finite sums of products ab, and $\mathfrak{A}^2 = \mathfrak{A}$, then $\gamma\delta = \delta\gamma$. This shows that in this case $\mathfrak{C}$ is a commutative ring.

Now let $\mathfrak{A}$ be a non-associative simple ring that is not a zero ring. Then $\mathfrak{M}$ is an irreducible ring of endomorphisms. Hence $\mathfrak{C}$ is a division ring. Since $\mathfrak{A}^2$ is a two-sided ideal not equal to 0, $\mathfrak{A}^2 = \mathfrak{A}$. Hence $\mathfrak{C}$ is a field. By Theorem 6, $\mathfrak{A}$ is a dense ring of linear transformations in $\mathfrak{A}$ over $\mathfrak{C}$. This proves the following theorem.

THEOREM 14. *Let $\mathfrak{A}$ be a non-associative ring that is simple and not a zero ring. Then the multiplication centralizer $\mathfrak{C}$ of $\mathfrak{A}$ is a field and the multiplication ring $\mathfrak{M}$ is a dense ring of linear transformations in $\mathfrak{A}$ over $\mathfrak{C}$*(14).

By equation (1), $\mathfrak{A}$ may be regarded as an algebra over $\mathfrak{C}$. Obviously $\mathfrak{A}$ is a simple algebra.

A similar discussion applies to non-associative algebras. If Φ is the underlying field, then $\Phi \leqq \mathfrak{C}$. Now by (1), $(a\gamma)_r = a_r\gamma$ and $(a\gamma)_l = a_l\gamma$. Hence $\mathfrak{M}\mathfrak{C} = \mathfrak{C}\mathfrak{M} \leqq \mathfrak{M}$ and in particular $\mathfrak{M}$ is closed under multiplication by the elements of Φ. It follows that if x is any element of $\mathfrak{A}$, the totality $x\mathfrak{M}$ of elements xM, M in $\mathfrak{M}$, is a subspace over Φ as well as an ideal in $\mathfrak{A}$. If $\mathfrak{A}$ is a simple algebra that is not a zero algebra, we can show that if $x\mathfrak{M} = 0$ then $x = 0$. Hence $x\mathfrak{M} = \mathfrak{A}$ for every $x \neq 0$ and this proves that $\mathfrak{A}$ is a simple ring. The converse is obvious. Theorem 14 may now be stated for algebras also.

We suppose now that $(\mathfrak{A}:\Phi) = n < \infty$. In this case $\mathfrak{M}$ and $\mathfrak{C}$ are subalgebras of the complete algebra of linear transformations in $\mathfrak{A}$ regarded as a space over Φ. Hence $(\mathfrak{M}:\Phi)$ and $(\mathfrak{C}:\Phi) \leqq n^2$. If $(\mathfrak{C}:\Phi) = r$ we know that $n = mr$ and $(\mathfrak{A}:\mathfrak{C}) = m$. By Theorem 14, $\mathfrak{M}$ is the complete set of linear transformations in $\mathfrak{A}$ over $\mathfrak{C}$. Thus $(\mathfrak{M}:\mathfrak{C}) = m^2$ and $(\mathfrak{M}:\Phi) = m^2r$. This proves the following known(15) theorem.

THEOREM 15. *Let A be a simple, nonzero, non-associative algebra with a finite basis over a field Φ and let $(\mathfrak{A}:\Phi) = n$. Then if $\mathfrak{C}$ is the multiplication centralizer, $\mathfrak{C}$ is a finite extension field over Φ and $(\mathfrak{C}:\Phi)$ is a divisor of n. The multiplication algebra $\mathfrak{M}$ is the complete set of linear transformations of $\mathfrak{A}$ over $\mathfrak{C}$.*

It is clear from this theorem that in the finite basis case, $\mathfrak{C}$ is a subset of $\mathfrak{M}$. Hence $\mathfrak{C}$ is the center of $\mathfrak{M}$.

7. **The center of a non-associative simple ring.** We shall define the *center* $\mathfrak{C}_0$ of a non-associative ring $\mathfrak{A}$ to be the totality of elements c that commute

(14) This theorem for associative algebras is due to Artin and Whaples [1, p. 98].

(15) Jacobson [2].

with every element of $\mathfrak{A}$ and that *associate* with every pair of elements x, y in $\mathfrak{A}$ in the sense that

$$(2) \qquad (xy)c = x(yc), \qquad (xc)y = x(cy), \qquad (cx)y = c(xy).$$

We note that the elements of $\mathfrak{C}_0$ may also be characterized by the equation $xc=cx$ and

$$(3) \qquad (xy)c = x(yc) = (xc)y.$$

For by (2) we have $(xy)c=x(yc)$ and $(xc)y=x(cy)=x(yc)$ since $yc=cy$. Hence (3) holds. Conversely, if (3) holds and c commutes with every x, then $(xy)c=c(xy)=x(yc)=x(cy)=(xc)y=(cx)y$. Hence (2) is satisfied and in fact all the products in (2) are equal. Thus the conditions that $c\in\mathfrak{C}_0$ are that $c_r=c_l$ and that this endomorphism belongs to the multiplication centralizer $\mathfrak{C}$. If $c_1, c_2\in\mathfrak{C}_0$ it is clear that $c_1\pm c_2\in\mathfrak{C}_0$ and $(c_1\pm c_2)_r=c_{1r}\pm c_{2r}$. Also $(xc_1)c_2=x(c_1c_2)$ so that $(c_1c_2)_r=c_{1r}c_{2r}$ and similarly $(c_1c_2)_l=(c_2c_1)_l=c_{1l}c_{2l}=c_{1r}c_{2r}=(c_1c_2)_r$. Evidently $c_{1r}c_{2r}\in\mathfrak{C}$. Hence we have proved that $c_1c_2\in\mathfrak{C}_0$. Now let γ be an arbitrary element of $\mathfrak{C}$. Then if $c\in\mathfrak{C}_0$, $x(c\gamma)=(x\gamma)c=(xc)\gamma=(cx)\gamma=(c\gamma)x$ and so $(c\gamma)_r=(c\gamma)_l=c_r\gamma$. This proves that $c\gamma\in\mathfrak{C}_0$. Thus $\mathfrak{C}_0$ is a commutative and associative subring of $\mathfrak{A}$ and $\mathfrak{C}_0$ is invariant under the endomorphisms $\gamma\in\mathfrak{C}$. We have also shown that the correspondence $c\rightarrow c_r$ is a homomorphism that commutes with all γ in $\mathfrak{C}$ between $\mathfrak{C}_0$ and a subring $\mathfrak{C}_{0r}$ of $\mathfrak{C}$.

We suppose now that $\mathfrak{A}$ is simple and not a zero ring. Then we note first that the correspondence $c\rightarrow c_r$ is an isomorphism. For if $c_r=0$ then $\{c\}$, the additive group generated by c, is a two-sided ideal in $\mathfrak{A}$. If $\mathfrak{A}=\{c\}$, $\mathfrak{A}$ is a zero ring. Hence $\{c\}=0$ and $c=0$. We remark next that $\mathfrak{C}_{0r}$ is an ideal in $\mathfrak{C}$ since $c_r\gamma=(c\gamma)_r$. Since, by Theorem 14, $\mathfrak{C}$ is a field, either $\mathfrak{C}_{0r}=0$ or $\mathfrak{C}_{0r}=\mathfrak{C}$. In the former case $\mathfrak{C}_0=0$. In the latter case $\mathfrak{C}_0$ is a field and if 1 is the identity of $\mathfrak{C}_0$, then $1_r=1_l$ is the identity transformation. Thus 1 is the identity of $\mathfrak{A}$ also. We have therefore proved the following theorem.

THEOREM 16. *If $\mathfrak{A}$ is a simple, nonzero, non-associative ring, then a necessary and sufficient condition that $\mathfrak{A}$ have an identity is that its center $\mathfrak{C}_0\neq 0$. If the condition is satisfied, $\mathfrak{C}_0$ is a field and the multiplication centralizer consists of the multiplications by the elements of $\mathfrak{C}_0$.*

These results are valid for algebras also since, as we have seen, a nonzero algebra $\mathfrak{A}$ is simple if and only if $\mathfrak{A}$ is a simple ring.

8. **Extension of the field of a vector space.** One of the most powerful methods of studying a non-associative algebra $\mathfrak{A}$ of a finite dimensionality is to consider the extension algebras $\mathfrak{A}_\Sigma$, Σ a field containing the underlying field Φ. It is often possible to determine the structure of a certain extension $\mathfrak{A}_\Sigma$ and then to use this information to obtain the structure of $\mathfrak{A}$. This method has been particularly successful in the study of simple Lie algebras. In the

next section we shall lay the foundations for a theory of extensions of simple algebras of infinite dimensionality. As a necessary preliminary we consider here the definition of an extension $\mathfrak{R}_\Sigma$ of a vector space $\mathfrak{R}$ over Φ by means of the extension field Σ of Φ.

Let $\mathfrak{R}$ be a vector space over a field Φ and let Σ be an abstract field that contains a subfield isomorphic to Φ. Now we wish to construct a vector space $\mathfrak{R}_\Sigma$ over a field (isomorphic to) Σ with the following properties:

1. $\mathfrak{R}_\Sigma$ regarded as a space over Φ contains a subspace (isomorphic to) $\mathfrak{R}$.
2. Any elements $x_1, \cdots, x_n$ of $\mathfrak{R}$ that are Φ-independent are also Σ-independent.
3. The smallest subspace over Σ of $\mathfrak{R}_\Sigma$ that contains $\mathfrak{R}$ is $\mathfrak{R}_\Sigma$ itself.

The definition of an extension $\mathfrak{R}_\Sigma$ by means of a basis is well known and in a recent paper Artin and Whaples(16) have given a definition without using bases. We follow their method here. We consider first the totality $\mathfrak{U}$ of forms $x_1\sigma_1+x_2\sigma_2+\cdots+x_m\sigma_m$ of arbitrary length m in which the $x_i\in\mathfrak{R}$ and the $\sigma_i\in\Sigma$. The order of the terms in $\sum x_i\sigma_i$ is immaterial and we define the sum of two forms $\sum_1^m x_i\sigma_i+\sum_{m+1}^n x_j\sigma_j$ as $\sum_1^n x_k\sigma_k$. Relative to this definition $\mathfrak{U}$ is a commutative semi-group. Now we call $\sum x_i\sigma_i$ a *null-form* if there exists a substitution $x_i=\sum y_k\alpha_{ki}$, α_{ki} in Φ, so that the resulting form $\sum y_k(\alpha_{ki}\sigma_i)$ has 0 coefficients, that is, $\sum\alpha_{ki}\sigma_i=0$. It can be shown that $\sum x_i\sigma_i$ is a null form if and only if there exists a substitution $\sigma_i=\sum\beta_{ik}\tau_k$, β_{ik} in Φ, so that the coefficients $\sum x_i\beta_{ik}$ of the σ_k in the form $\sum x_i\beta_{ik}\sigma_k$ are all 0. If the x's are linearly independent in $\mathfrak{R}$ and not all the σ's are 0 *or* if the σ's are linearly independent in Σ (over Φ) and not all the x's are 0 then $\sum x_i\sigma_i$ is not a null-form.

It is readily seen that the totality $\mathfrak{N}$ of null-forms is a subsemi-group of $\mathfrak{U}$. Now we define $-x$ for $x=\sum x_i\sigma_i$ to be the form $\sum x_i(-\sigma_i)$ and we define $x\equiv y(\text{mod }\mathfrak{N})$ if $x+(-y)\in\mathfrak{N}$. It can be shown that this notion of congruence is an equivalence relation and that congruences can be added. Hence if we let $\mathfrak{R}_\Sigma$ denote the set of congruence classes, $\mathfrak{R}_\Sigma$ is also a commutative semi-group. Moreover, it can be seen that $\mathfrak{R}_\Sigma$ is a group with $\mathfrak{N}$ as its zero and the class of $-x$ as the negative of the class of x.

We shall denote the elements of $\mathfrak{R}_\Sigma$ by their representatives $\sum x_i\sigma_i$ in $\mathfrak{U}$. When this is done we have the following criterion: $\sum x_i\sigma_i=\sum y_j\tau_j$ if and only if when the x's and the y's are expressed in terms of the same set of linearly independent vectors $u_1, \cdots, u_r$, the resulting coefficient of each u_k on the left-hand side is the same as that on the right-hand side. In $\mathfrak{R}_\Sigma$ we have $\sum_1^m x_i\sigma_i+\sum_{m+1}^n x_j\sigma_j=\sum_1^n x_k\sigma_k$.

It is now easy to verify that for each $\sigma\in\Sigma$, the mapping $x=\sum x_i\sigma_i\to\sum x_i(\sigma_i\sigma)$ is single valued and an endomorphism in $\mathfrak{R}_\Sigma$. The totality of these endomorphisms is a subfield containing 1 of the ring of endomorphisms of $\mathfrak{R}_\Sigma$. This subfield is isomorphic to Σ. We shall identify it with Σ and we shall write $\sum x_i(\sigma_i\sigma)$ as $(\sum x_i\sigma_i)\sigma$. Then $\mathfrak{R}_\Sigma$ is a vector space over Σ.

(16) Artin and Whaples [1, pp. 95–97].

Consider for the moment $\mathfrak{R}_\Sigma$ as a space over the subfield Φ of Σ. Thus the endomorphisms of Φ are the mappings $x=\sum x_i\sigma_i \rightarrow \sum x_i(\sigma_i\alpha)$, α in Φ. Let $\mathfrak{R}_1$ denote the totality of vectors $x1$, $x \in R$, $1 \in \Sigma$. Since $(x1)\alpha = x(\alpha) = (x\alpha)(1)$, $\mathfrak{R}_1$ is a subspace of $\mathfrak{R}_\Sigma$ over Φ. This equation and the criterion for equality show that the correspondences $x \rightarrow x1$ and $\alpha \rightarrow \alpha$ define an isomorphism between the spaces $\mathfrak{R}$ and $\mathfrak{R}_1$. We shall now identify $\mathfrak{R}_1$ with $\mathfrak{R}$ and write $x1 = x$.

Condition 1 for extension spaces has therefore been verified. The test for equality shows that condition 2 holds. Since any $x=\sum x_i\sigma_i$ may be written in the form $\sum (x_i)\sigma_i$, x_i in $\mathfrak{R}$, condition 3 is also valid and so $\mathfrak{R}_\Sigma$ has the required properties.

We suppose now that $\mathfrak{R}^{(1)}$ and $\mathfrak{R}^{(2)}$ are spaces over $\Phi^{(1)}$ and $\Phi^{(2)}$, respectively, and that we have a homomorphism $x^{(1)} \rightarrow x^{(2)}$ of $\mathfrak{R}^{(1)}$ on a subspace $\mathfrak{S}^{(2)}$ of $\mathfrak{R}^{(2)}$ and an isomorphism $\alpha^{(1)} \rightarrow \alpha^{(2)}$ between $\Phi^{(1)}$ and $\Phi^{(2)}$ such that $x^{(1)}\alpha^{(1)} \rightarrow x^{(2)}\alpha^{(2)}$. Furthermore let $\mathfrak{R}^{(1)}_{\Sigma^{(1)}}$ and $\mathfrak{R}^{(2)}_{\Sigma^{(2)}}$ be two spaces over fields $\Sigma^{(1)}$ and $\Sigma^{(2)}$, respectively, satisfying the conditions 1, 2, 3. We assume finally that $\Sigma^{(1)}$ and $\Sigma^{(2)}$ are isomorphic under a correspondence $\sigma^{(1)} \rightarrow \sigma^{(2)}$ that is an extension of the correspondence $\alpha^{(1)} \rightarrow \alpha^{(2)}$.

Consider now the correspondence $\sum x_i^{(1)}\sigma_i^{(1)} \rightarrow \sum x_i^{(2)}\sigma_i^{(2)}$. We assert that it is single valued. For if $\sum x_i^{(1)}\sigma_i^{(1)} = \sum y_j^{(1)}\tau_j^{(1)}$, we may choose vectors $u_1^{(1)}, \cdots, u_r^{(1)}$ in $\mathfrak{R}^{(1)}$ which are linearly independent over $\Phi^{(1)}$ and write $x_i^{(1)} = \sum u_p^{(1)}\alpha_{pi}^{(1)}$, $y_j^{(1)} = \sum u_q^{(1)}\beta_{qj}^{(1)}$. Then we have $\sum \alpha_{pi}^{(1)}\sigma_i^{(1)} = \sum \beta_{pj}^{(1)}\tau_j^{(1)}$. Hence $\sum \alpha_{pi}^{(2)}\sigma_i^{(2)} = \sum \beta_{pj}^{(2)}\tau_j^{(2)}$, $x_i^{(2)} = \sum u_p^{(2)}\alpha_{pi}^{(2)}$ and $y_j^{(2)} = \sum u_q^{(2)}\beta_{qj}^{(2)}$, and these equations imply that $\sum x_i^{(2)}\sigma_i^{(2)} = \sum y_j^{(2)}\sigma_j^{(2)}$. This proves the assertion. It is clear that (1) the correspondence $\sum x_i^{(1)}\sigma_i^{(1)} \rightarrow \sum x_i^{(2)}\sigma_i^{(2)}$ is a homomorphism, (2) $(\sum x_i^{(1)}\sigma_i^{(1)})\sigma^{(1)} \rightarrow (\sum x_i^{(2)}\sigma_i^{(2)})\sigma^{(2)}$ and (3) the correspondence is an extension of the original correspondence between $\mathfrak{R}^{(1)}$ and $\mathfrak{S}^{(2)}$. It is also clear that there is only one extension of the original correspondence having properties (1) and (2). If the original correspondence is an isomorphism between $\mathfrak{R}^{(1)}$ and $\mathfrak{S}^{(2)}$ the above argument shows that the extension is an isomorphism. If $\mathfrak{S}^{(2)} = \mathfrak{R}^{(2)}$ then the extension maps $\mathfrak{R}^{(1)}_{\Sigma^{(1)}}$ on the whole of $\mathfrak{R}^{(2)}_{\Sigma^{(2)}}$. Hence as a special case of these results we see that any two extension spaces of the same space by the same field are isomorphic.

We shall suppose now that $\mathfrak{R}^{(1)} = \mathfrak{R}^{(2)} = \mathfrak{R}$, $\Phi^{(1)} = \Phi^{(2)} = \Phi$, $\Sigma^{(1)} = \Sigma^{(2)} = \Sigma$ and let the automorphism in Σ be the identity. Then our results may be stated in the following form: If A is a linear transformation in $\mathfrak{R}$ over Φ, A has a unique extension, which we may also denote as A, to a linear transformation in $\mathfrak{R}_\Sigma$ over Σ. If A is (1-1) between $\mathfrak{R}$ and a subspace (the whole space) then its extension is (1-1) between $\mathfrak{R}_\Sigma$ and a subspace over Σ (the whole space).

These results are all well known in the finite-dimensional case. We recall also that if $(\mathfrak{R}:\Phi) = n$ then $(\mathfrak{R}_\Sigma:\Sigma) = n$, as is clear from 1, 2 and 3.

If $\mathfrak{A}$ $(=\mathfrak{R})$ is an algebra over Φ, the extension space $\mathfrak{A}_\Sigma$ can be made into an algebra by defining $(\sum x_i\sigma_i)(\sum y_j\tau_j) = \sum (x_iy_j)(\sigma_i\tau_j)$. It is easy to see that the product thus defined is single valued and distributive, and that the left and the right multiplications commute with the endomorphisms $\sigma \in \Sigma$. If $\mathfrak{A}$ is associative (a Lie algebra), then $\mathfrak{A}_\Sigma$ is associative (a Lie algebra).

9. **Central simple non-associative algebras.** We consider now the relation between the multiplication algebra $\mathfrak{M}$ of an arbitrary algebra $\mathfrak{A}$ over a field Φ and the multiplication algebra of the extension $\mathfrak{A}_\Sigma$. As in the case of vector spaces, $\mathfrak{A}$ is a subset of $\mathfrak{A}_\Sigma$. If $a \in \mathfrak{A}$, it is evident that the right (left) multiplication a_r (a_l) in $\mathfrak{A}_\Sigma$ is the extension of the linear transformation a_r (a_l) in $\mathfrak{A}$. If $\sum a_i\sigma_i$, $a_i \in \mathfrak{A}$, $\sigma_i \in \Sigma$, is any element of $\mathfrak{A}_\Sigma$, then $(\sum a_i\sigma_i)_r = \sum (a_i)_r\sigma_i$, $(\sum a_i\sigma_i)_l = \sum (a_i)_l\sigma_i$. It follows that the multiplication algebra of $\mathfrak{A}_\Sigma$ is the totality of linear combinations with coefficients in Σ of the extensions of the multiplications of $\mathfrak{A}$. Thus $\mathfrak{M}(\mathfrak{A}_\Sigma) = \mathfrak{M}\Sigma$ where the $\mathfrak{M}$ on the right-hand side denotes the set of extensions of the multiplications of $\mathfrak{A}$.

We shall call a non-associative algebra $\mathfrak{A}$ over Φ *central* if its multiplication centralizer $\mathfrak{C} = \Phi$. If $\mathfrak{A}$ is an arbitrary, nonzero, simple algebra (or ring), we have seen that we may regard $\mathfrak{A}$ as an algebra over its multiplication centralizer. When this is done, $\mathfrak{A}$ becomes a central simple algebra. We may now prove the main theorem on extensions.

THEOREM 17. *Any extension $\mathfrak{A}_\Sigma$ of a central simple, nonzero, non-associative algebra is central simple.*

It suffices to show that $\mathfrak{M}(\mathfrak{A}_\Sigma)$ is a dense set of linear transformations in $\mathfrak{A}_\Sigma$ over Σ. For then $\mathfrak{M}(\mathfrak{A}_\Sigma)$ is irreducible and so $\mathfrak{A}_\Sigma$ is simple. Moreover by Theorem 7, the centralizer of $\mathfrak{M}(\mathfrak{A}_\Sigma)$ is Σ. Hence the multiplication centralizer $\mathfrak{C}(\mathfrak{A}_\Sigma) = \Sigma$. That $\mathfrak{M}(\mathfrak{A}_\Sigma)$ is dense is an immediate consequence of the above remarks and of the following lemma.

LEMMA 3. *Let $\mathfrak{A}$ be a dense set of linear transformations in $\mathfrak{R}$ over Φ and let $\mathfrak{A}\Sigma$ be the totality of Σ-linear combinations of the extensions of the linear transformations of $\mathfrak{A}$ to linear transformations in $\mathfrak{R}_\Sigma$. Then $\mathfrak{A}\Sigma$ is dense in $\mathfrak{R}_\Sigma$ over Σ.*

We have to show that if any finite-dimensional subspace $\mathfrak{S}$ of $\mathfrak{R}_\Sigma$ and any linear transformation B in $\mathfrak{S}$ are given, then there is a l.t. $A \in \mathfrak{A}\Sigma$ which coincides with B in $\mathfrak{S}$. Let $y_1, y_2, \cdots, y_m$ be a basis for $\mathfrak{S}$ and let $u_1, \cdots, u_n$ be vectors of $\mathfrak{R}$ in terms of which all the y's can be expressed. Then $\mathfrak{S}$ is a subspace of the extension $\mathfrak{S}_{0\Sigma}$, $\mathfrak{S}_0 = \{u_1, \cdots, u_n\}$. Now there exist l.t. E_{ij}, $i, j = 1, \cdots, n$, in $\mathfrak{A}$ such that $u_rE_{ij} = 0$ if $r \neq i$ and $u_iE_{ij} = u_j$. It follows that there is a l.t. A in $\mathfrak{A}\Sigma$ that induces any l.t. in $\mathfrak{S}_{0\Sigma}$. Hence there is an A that induces the linear transformation B in the subspace $\mathfrak{S}$ of $\mathfrak{S}_0$.

III. SPECIAL ALGEBRAS OF LINEAR TRANSFORMATIONS

10. **Linear transformations in vector spaces over a field.** In this part we shall be concerned mainly with dense algebras of finite valued linear transformations in a vector space $\mathfrak{R}$ over a field Φ. As in §1 we let $\mathfrak{L}$ denote the

complete set of l.t. and let $\mathfrak{F}$ be the subset of finite valued l.t. Both $\mathfrak{L}$ and $\mathfrak{F}$ are algebras over Φ, the set of multiplications by the elements of the underlying field Φ. We have seen that if $\mathfrak{A}$ is a dense subalgebra of $\mathfrak{F}$, then $\mathfrak{A}$ is simple and $\mathfrak{A}$ contains minimal right ideals. We prove next that $\mathfrak{A}$ is central by proving the following lemma.

LEMMA 4. *If $\mathfrak{A}$ is a dense ring of linear transformations in a vector space $\mathfrak{R}$ over a field Φ, then the transformations of the multiplication centralizer $\mathfrak{C}$ of $\mathfrak{A}$ are multiplications by the elements of Φ.*

In order to avoid confusion we write the effect of $\gamma\in\mathfrak{C}$ on $X\in\mathfrak{A}$ as X^γ. Then we have $(XY)^\gamma=X^\gamma Y=X(Y^\gamma)$. We choose a vector $u\neq 0$ in $\mathfrak{R}$ and suppose that X is a l.t. in $\mathfrak{A}$ such that $uX=0$. Then $uX^\gamma=0$ also. For otherwise there is a $Y\in\mathfrak{A}$ such that $(uX^\gamma)Y\neq 0$ contrary to $uX^\gamma Y=(uX)Y^\gamma=0$. It follows from this that the mapping $v=uX\rightarrow uX^\gamma=v'$ is single valued and defined in the whole of $\mathfrak{R}$. It is clear that $v\rightarrow v'$ is an endomorphism and since $(uX)A\rightarrow u(XA)^\gamma=(uX^\gamma)A$, $v\rightarrow v'$ is in the centralizer of $\mathfrak{A}$. Hence by Theorem 7 there is a $\gamma\in\Phi$ such that $uX^\gamma=u(X\gamma)$ for all $X\in\mathfrak{A}$. Replacing X by AX yields the equation $u(AX)^\gamma=u(AX\gamma)$ so that $(uA)X^\gamma=(uA)(X\gamma)$. As A ranges over $\mathfrak{A}$, uA ranges over $\mathfrak{R}$. Hence $X^\gamma=X\gamma$.

If $\mathfrak{A}$ is an algebra, $\mathfrak{C}\geqq\Phi$. Hence if $\mathfrak{A}$ is a dense algebra, $\mathfrak{C}=\Phi$ and $\mathfrak{A}$ is central.

We shall call an algebra $\mathfrak{A}$ over Φ *algebraic* if every element of $\mathfrak{A}$ satisfies some non-trivial polynomial equation, or what is the same thing, if every element generates an algebra of finite dimensionality. We shall call $\mathfrak{A}$ *locally finite* if every finite subset of $\mathfrak{A}$ generates a finite-dimensional subalgebra. It is an open question whether or not simple algebras that are algebraic are necessarily locally finite. We shall show next that any dense algebra of finite valued l.t. in $\mathfrak{R}$ over Φ is locally finite. For this purpose we require the following lemma.

LEMMA 5. *If $A_1, A_2, \cdots, A_n$ are finite valued linear transformations in $\mathfrak{R}$, there exists a decomposition $\mathfrak{R}=\mathfrak{S}\oplus\mathfrak{N}$ where $\mathfrak{S}$ is a finite-dimensional subspace invariant under all the A_i and $\mathfrak{N}$ is annihilated by all the A_i.*

We shall show first that if $\mathfrak{R}_1$ is any subspace of $\mathfrak{R}$, $\mathfrak{R}_1=\mathfrak{Z}_1\oplus\mathfrak{M}_1$ where $\mathfrak{Z}_1A_1=0$ and $\mathfrak{M}_1$ is finite-dimensional. Now $\mathfrak{R}_1A_1$ is finite-dimensional so that $\mathfrak{R}_1A_1=\{y_1, \cdots, y_r\}$ where the y_i are linearly independent. We may choose vectors $x_i\in\mathfrak{R}_1$ such that $x_iA_1=y_i$ and we set $\mathfrak{M}_1=\{x_1, \cdots, x_r\}$. Then if x is any vector in $\mathfrak{R}_1$, $xA_1=\sum y_i\alpha_i=(\sum x_i\alpha_i)A_1$. Hence if $z=x-\sum x_i\alpha_i$, $zA_1=0$ and $x=z+\sum x_i\alpha_i\in\mathfrak{Z}_1+\mathfrak{M}_1$ where $\mathfrak{Z}_1$ is the subspace of $\mathfrak{R}_1$ of vectors annihilated by A_1. If $(\sum x_i\alpha_i)A_1=0$, $\sum y_i\alpha_i=0$ and all the $\alpha_i=0$. This shows that $\mathfrak{Z}_1\wedge\mathfrak{M}_1=0$ and so $\mathfrak{R}_1=\mathfrak{Z}_1\oplus\mathfrak{M}_1$. We now begin with $\mathfrak{R}_1=\mathfrak{R}$ and write $\mathfrak{R}=\mathfrak{Z}_1\oplus\mathfrak{M}_1$, where $\mathfrak{Z}_1A_1=0$ and $\mathfrak{M}_1$ is finite-dimensional. We then repeat the argument with $\mathfrak{Z}_1$ and A_2 and obtain $\mathfrak{Z}_1=\mathfrak{Z}_2\oplus\mathfrak{M}_2$ where $\mathfrak{Z}_2A_1=\mathfrak{Z}_2A_2=0$ and

$\mathfrak{M}_2$ is finite-dimensional. Then $\mathfrak{R}=\mathfrak{Z}_2\oplus\mathfrak{M}_1\oplus\mathfrak{M}_2$. Continuing in this way we obtain finally that $\mathfrak{R}=\mathfrak{Z}\oplus\mathfrak{M}$ where $\mathfrak{Z}A_i=0$, $i=1, \cdots, n$, and $\mathfrak{M}$ is finite-dimensional. Since $\mathfrak{Z}A_i=0$, we have $\mathfrak{R}A_i=\mathfrak{M}A_i$. Hence $\mathfrak{S}=\mathfrak{M}+\mathfrak{M}A_1+\cdots+\mathfrak{M}A_n$ is a finite-dimensional subspace invariant under all the A_i. We have $\mathfrak{S}=\mathfrak{Z}'\oplus\mathfrak{M}$ where $\mathfrak{Z}'=\mathfrak{S}\wedge\mathfrak{Z}$ and we may decompose $\mathfrak{Z}$ as $\mathfrak{Z}=\mathfrak{Z}'\oplus\mathfrak{N}$. Then $\mathfrak{R}=\mathfrak{S}\oplus\mathfrak{N}$ where $\mathfrak{N}A_i=0$ as required.

Now by Lemma 5 we see that the linear transformations $A_1, \cdots, A_n$ generate an algebra which is isomorphic to an algebra of linear transormations in a finite-dimensional space $\mathfrak{S}$. Hence the algebra generated by the A_i is finite-dimensional over Φ. We summarize our results in the following theorem:

THEOREM 18. *Let $\mathfrak{R}$ be a vector space over a field Φ and let $\mathfrak{A}$ be a dense algebra of finite valued linear transformations in $\mathfrak{R}$ over Φ. Then $\mathfrak{A}$ is central simple and locally finite.*

11. Isomorphisms and automorphisms of algebras of linear transformations. Let $\mathfrak{A}_1$ and $\mathfrak{A}_2$ be two dense algebras in $\mathfrak{R}$ over Φ which are isomorphic under a correspondence $A_1\rightarrow A_2$. By an isomorphism here we mean an algebra isomorphism, that is, $A_1\alpha\rightarrow A_2\alpha$ for all α in Φ. By Theorem 12 there is a semi-linear transformation S in $\mathfrak{R}$ over Φ such that $A_2=S^{-1}A_1S$. If $\alpha\rightarrow\bar{\alpha}$ is the automorphism in Φ associated with $\mathfrak{S}$ we have $\bar{\alpha}=S^{-1}\alpha S$. Since $S^{-1}(A_1\alpha)S=A_2\alpha$ it follows that $\bar{\alpha}=\alpha$ and hence S is a linear transformation. This theorem may be stated in the following form:

THEOREM 19. *If $\mathfrak{A}_1$ and $\mathfrak{A}_2$ are isomorphic dense algebras of linear transformations in $\mathfrak{R}$ over Φ any isomorphism between $\mathfrak{A}_1$ and $\mathfrak{A}_2$ can be extended to an inner automorphism in $\mathfrak{L}$, the complete algebra of linear transformations.*

This, of course, gives the following corollary.

COROLLARY. *If $\mathfrak{A}$ is a dense subalgebra of $\mathfrak{L}$, any automorphism in $\mathfrak{A}$ can be extended to an inner automorphism in $\mathfrak{L}$.*

We may now prove the following theorem.

THEOREM 20. *Any automorphism in $\mathfrak{L}$ is inner.*

We remark first that if $X\rightarrow X^G$ is an automorphism in $\mathfrak{L}$, then $\mathfrak{F}^G=\mathfrak{F}$. This is clear since $\mathfrak{F}$, and hence $\mathfrak{F}^G$, is contained in every two-sided ideal not equal to 0 in $\mathfrak{L}$. Thus G induces an automorphism in $\mathfrak{F}$. Hence there is an S in $\mathfrak{L}$ such that $F^G=S^{-1}FS$ for all F in $\mathfrak{F}$. Now consider the automorphism $X\rightarrow X^H=SX^GS^{-1}$. We have $F^H=F$ for all F in $\mathfrak{F}$. If x is any vector there is an F in $\mathfrak{F}$ such that $xF=x$. Hence $xX=xFX=x(FX)^H=xF^HX^H=xX^H$ and so $X^H=X$. This proves that H is the identity and G is the inner automorphism $X\rightarrow S^{-1}XS$.

Bibliography

B. H. Arnold

1. *Rings of operators on vector spaces*, Ann. of Math. vol. 45 (1944) pp. 24–49.

E. Artin and G. Whaples

1. *The theory of simple rings*, Amer. J. Math. vol. 65 (1943) pp. 87–107.

H. Fitting

1. *Primarkomponentenzerlegung in nichtkommutativen Ringe*, Math. Ann. vol. 111 (1935) pp. 19–41.

N. Jacobson

1. *The theory of rings*, Mathematical Surveys, vol. 2, New York, 1943.
2. *A note on non-associative algebras*, Duke Math. J. vol. 3 (1937) pp. 544–548.
3. *Pseudo-linear transformations*, Ann. of Math. vol. 38 (1937) pp. 484–507.

E. Noether

1. *Nichtkommutativ Algebra*, Math. Zeit. vol. 37 (1933) pp. 514–541.

O. Ore

1. *Formale Theorie lineare Differentialgleichungen*. II. Journal für Mathematik vol. 168 (1932) pp. 233–252.

The Johns Hopkins University,
Baltimore, Md.

Reprinted from
Transactions of the American Mathematical Society
March 1945.

THE RADICAL AND SEMI-SIMPLICITY FOR ARBITRARY RINGS.* [1]

By N. JACOBSON.

The radical of an algebra with a finite basis, or, more generally, of a ring $\mathfrak{A}$ that satisfies the descending chain condition is defined to be the join of the nil right (left) ideals of $\mathfrak{A}$. The importance of the radical for the structure theory of these rings is due to the facts that 1) the radical $\mathfrak{N}$ is a two-sided ideal whose difference ring $\mathfrak{A}-\mathfrak{N}$ is semi-simple in the sense that its radical is 0, and 2) the structure of semi-simple rings satisfying the descending chain condition can be subjected to a thorough analysis that leads in many important cases to a complete classification. Several investigations of nil ideals in arbitrary rings have been made recently but none of these has led to a structure theory for general semi-simple rings.[2] This is one of a number of indications that in order to develop a satisfactory structure theory for arbitrary rings it is necessary to abandon the concept of a nil ideal in defining the radical.

Other possibilities for defining a radical are afforded by two important characterizations of the radical $\mathfrak{N}$ of an algebra $\mathfrak{A}$ with a finite basis. One of these, due to Perlis, makes use of the notion of quasi-regularity.[3] An element z of $\mathfrak{A}$ is right quasi-regular if there exists a z in $\mathfrak{A}$ such that $z+z'+zz'=0$. Perlis has shown that $z\in\mathfrak{N}$ if and only if $u+z$ is right quasi-regular for all right quasi-regular u. A second characterization of $\mathfrak{N}$ for algebras with an identity is that $\mathfrak{N}$ is the intersection of the maximal right (left) ideals of $\mathfrak{A}$.[4] A start in the investigation of the first characterization as a possibility for defining a radical for an arbitrary ring $\mathfrak{A}$ was made by Baer, who showed that the totality $\mathfrak{R}$ of elements z that generate right ideals containing only right quasi-regular elements is a right ideal.[5]

The point of departure of the present investigation is the observation that $\mathfrak{R}$ is a two-sided ideal and that $\mathfrak{R}$ coincides with the two-sided ideal

* Received September 6, 1944; Revised January 5, 1945.

[1] Presented to the Society October 28, 1944.

[2] Baer [1] and Levitzki [1] and [2].

[3] Perlis [1].

[4] Jacobson [1], p. 66. Cf. also G. Birkhoff [1].

[5] Baer [1], p. 562. This definition of the radical has been independently proposed by Hille and Zorn who proved that $\mathfrak{R}$ is a two-sided ideal and that if $\mathfrak{A}$ has an identity, $\mathfrak{R}$ is the intersection of the maximal right (left) ideals of $\mathfrak{A}$. These results were announced by Professor Hille in his Colloquium Lectures in August 1944.

defined in a similar manner using left quasi-regularity and left ideals. We define the radical of $\mathfrak{A}$ to be the ideal $\mathfrak{R}$. In the first part of the paper we establish a number of "radical-like" properties of $\mathfrak{R}$ and we investigate $\mathfrak{R}$ in several important special cases. For rings that satisfy the descending chain condition $\mathfrak{R} = \mathfrak{N}$ where $\mathfrak{N}$ is defined as above. If $\mathfrak{A}$ is a commutative normed ring $\mathfrak{R}$ coincides with the radical as defined by Gelfand to be the totality of generalized nilpotent elements.[6] A similar characterization holds for non-commutative normed rings.

In the latter half of the paper we investigate the relation between $\mathfrak{R}$ and the intersection of the maximal right ideals of $\mathfrak{A}$. We show that if $\mathfrak{A} \neq \mathfrak{R}$ then $\mathfrak{R} \geq \Pi\mathfrak{J}$ for the maximal right ideals $\mathfrak{J}$ of $\mathfrak{A}$ and $\Pi\mathfrak{J} \geq \mathfrak{A}\mathfrak{R}$. If $\mathfrak{A}$ has an identity, $\mathfrak{R} = \Pi\mathfrak{J}$. For arbitrary rings that contain maximal right ideals we show that $\mathfrak{R}$ is the intersection of certain two-sided ideals $\mathfrak{B}$ whose difference rings are of a special type called primitive.

This result implies that any semi-simple ring is isomorphic to a subring of the complete direct sum of primitive rings. To a certain extent this reduces the study of semi-simple rings to that of primitive rings. A tool for studying the latter is a representation theorem for these rings that states that a primitive ring is isomorphic to a dense ring of linear transformations in a vector space over a division ring. These theorems are analogues of the fundamental Wedderburn-Artin structure theorems for semi-simple rings that satisfy the descending chain condition. The Wedderburn-Artin theorems can be deduced quite simply from our results. Moreover, our results contain as special cases Stone's theorem on Boolean rings and other known results on the representability of rings as subrings of direct sums of fields.[7] We can also obtain from our theory a theory of algebraic algebras that is fairly conclusive for algebras with elements of bounded degree. These results will be published in a subsequent paper.

I am indebted to R. Baer for a number of important suggestions that led to simplifications of some of the proofs and to extensions of several of the theorems from rings with an identity to arbitrary rings.

1. Definition of the radical. Let $\mathfrak{A}$ be a ring with an identity and let z be an element of $\mathfrak{A}$ such that $1 + z$ has a right inverse u. We write $u = 1 + z'$. Then $(1 + z)(1 + z') = 1$ implies

$$z + z' + zz' = 0. \tag{1}$$

Conversely if z is an element for which there exists an element z' satisfying (1) then $1 + z$ has the right inverse $1 + z'$. This leads to

---

[7] Stone [1], McCoy and Montgomery [1], McCoy [1] and [2].

[6] Gelfand [1], p. 10.

9

Definition 1. An element z of an arbitrary ring $\mathfrak{A}$ is *right quasi-regular* if there exists an element z' in $\mathfrak{A}$ such that $z + z' + zz' = 0$. The element z satisfying this equation is called a *right quasi-inverse* of z.

We have noted that if $\mathfrak{A}$ has an identity then z is right quasi-regular with right inverse z' if, and only if, $1 + z$ has the right inverse $1 + z'$.

If z is any element of $\mathfrak{A}$ the totality $\{x + zx\}$ of elements $x + zx$ where x ranges over $\mathfrak{A}$ is a right ideal. If z is right quasi-regular with right quasi-inverse z' then $-z = z' + zz' \in \{x + zx\}$. Hence $zx \in \{x + zx\}$ and $x \in \{x + zx\}$. Then $\{x + zx\} = \mathfrak{A}$. On the other hand if $\{x + zx\} = \mathfrak{A}$ then $-z = z' + zz'$ for a suitable z' and z is right quasi-regular. Hence we have the alternative

Definition 1'. An element z of a ring $\mathfrak{A}$ is *right quasi-regular* if the totality of elements $\{x + zx\} = \mathfrak{A}$.[8]

A right ideal $\mathfrak{J}$ will be called *quasi-regular* if all the elements of $\mathfrak{J}$ are right quasi-regular. Let $\mathfrak{J}_1$ and $\mathfrak{J}_2$ be quasi-regular right ideals and let $z_i \in \mathfrak{J}_i$, $i = 1, 2$. There exists an element z'_1 such that $z_1 + z'_1 + z_1 z'_1 = 0$. Since $z_2 + z_2 z'_1 \in \mathfrak{J}_2$ this element has a right quasi-inverse w' such that

$$(z_2 + z_2 z'_1) + w' + (z_2 + z_2 z'_1) w' = 0.$$

Hence

$$\begin{aligned}(z_1 + z_2) &+ (z'_1 + w' + z'_1 w') + (z_1 + z_2)(z'_1 + w' + z'_1 w') \\ &= [z_1 + z'_1 + z_1 z'_1] + [(z_2 + z_2 z'_1) + w' + (z_2 + z_2 z') w'] \\ &\quad + [(z_1 + z'_1 + z_1 z'_1) w'] = 0.\end{aligned}$$

Thus $z'_1 + w' + z'_1 w'$ is a right quasi-inverse of $z_1 + z_2$. Hence $\mathfrak{J}_1 + \mathfrak{J}_2$ is quasi-regular.

Now let $\mathfrak{R}$ be the join of all the quasi-regular right ideals of $\mathfrak{A}$. Since the right ideal generated by an element z is the totality of elements $zi + za$ where i is an integer and $a \in \mathfrak{A}$, it is clear that $\mathfrak{R}$ is the totality of elements z such that $zi + za$ is right quasi-regular for all integral i and all a in $\mathfrak{A}$. The above result shows that $\mathfrak{R}$ is a right ideal. We wish to show that $\mathfrak{R}$ is a two-sided ideal. Let $z \in \mathfrak{R}$ and $b \in \mathfrak{A}$. Then $zb \in \mathfrak{R}$ and there exists an element w' such that $zb + w' + (zb) w' = 0$. Then

$$bz + (-bz - bw'z) + bz(-bz - bw'z) = -b(w' + zb + zbw')z = 0$$

and so $-(bz + bw'z)$ is a right quasi-inverse for bz. Similarly if i is an integer and $a \in \mathfrak{A}$, $(bz)i + (bz)a = b(zi + za)$ is right quasi-regular. Hence $bz \in \mathfrak{R}$ and we have proved

[8] This definition of quasi-regularity is due to Baer.

THEOREM 1. *If $\mathfrak{A}$ is an arbitrary ring the join $\mathfrak{R}$ of all the quasi-regular right ideals of $\mathfrak{A}$ is a (right) quasi-regular two-sided ideal.*

Definition 2. The *radical* of a ring is the join of all the quasi-regular right ideals of the ring.

We have seen that $\mathfrak{R}$ is the set of elements z such that $zi + za$ is right quasi-regular for all integers i and all a in $\mathfrak{A}$. If $\mathfrak{A}$ is a ring with an identity, the connection between quasi-regularity and regularity shows that $\mathfrak{R}$ is the totality of elements z such that $1 + za$ has a right inverse for every a in $\mathfrak{A}$.

A similar discussion holds for left ideals. We define *left quasi-regularity, left quasi-inverse, quasi-regular left ideal* and left radical $\mathfrak{R}'$ in a manner analogous to the above. As for ordinary inverses we say that an element is *quasi-regular* if it is both right and left quasi-regular.

LEMMA 1. *If z is quasi-regular any right (left) quasi-inverse is a left (right) quasi-inverse, is uniquely determined and commutes with z.*

Let z' be a right quasi-inverse and z'' a left quasi-inverse. Then

$$z + z' + zz' = 0 \quad \text{and} \quad z + z'' + z''z = 0.$$

Hence

$$\begin{aligned} z'' &= z'' + (z + z' + zz') + z''(z + z' + zz') \\ &= z' + (z + z'' + z''z) + (z + z'' + z''z)z' = z'. \end{aligned}$$

This proves the first two statements. Since

$$z + z' + zz' = 0 = z + z' + z'z, \qquad zz' = z'z.$$

We call z' *the quasi-inverse* of z.

Now let $z \in \mathfrak{R}$. Then z has a right quasi-inverse $z' = -z - zz'$. Since $\mathfrak{R}$ is an ideal $z' \in \mathfrak{R}$. Hence z' has a right quasi-inverse. Since z is a left quasi-inverse of z' it follows by Lemma 1 that z' is quasi-regular and z is its quasi-inverse. Hence z is quasi-regular. Since $\mathfrak{R}$ is a left ideal $\mathfrak{R} \leq$ the left radical $\mathfrak{R}'$. By symmetry $\mathfrak{R}' \leq \mathfrak{R}$. This proves

THEOREM 2. *The radical of a ring $\mathfrak{A}$ is the join of all the quasi-regular left ideals of $\mathfrak{A}$.*

By the Lemma the elements of $\mathfrak{R}$ are quasi-regular.

2. Elementary properties of the radical. If $\mathfrak{A}$ is an arbitrary ring we know that we can imbed $\mathfrak{A}$ in a ring with an identity $\mathfrak{A}^*$ such that $\mathfrak{A}^* = \mathfrak{A} + (1)$, $\mathfrak{A} \wedge (1) = 0$ where (1), the ring generated by 1, is isomorphic to the ring of integers.[9] At first we assume only that $\mathfrak{A}^* = \mathfrak{A} + (1)$. Let $\mathfrak{R}(\mathfrak{A}^*)$ be the radical of $\mathfrak{A}^*$ and let $z \in \mathfrak{R}(\mathfrak{A}^*) \wedge \mathfrak{A}$. Then z has a quasi-inverse z' in $\mathfrak{A}^*$. Since $z' = -z - zz'$, $z' \in \mathfrak{A}$ and z is quasi-regular in $\mathfrak{A}$. Hence $(\mathfrak{R}(\mathfrak{A}^*) \wedge \mathfrak{A}) \leq \mathfrak{R}(\mathfrak{A})$. Since $\mathfrak{A}^* = \mathfrak{A} + (1)$ any right ideal of $\mathfrak{A}$ is a right ideal of $\mathfrak{A}^*$. Hence $\mathfrak{R}(\mathfrak{A}) \leq \mathfrak{R}(\mathfrak{A}^*)$. Thus $\mathfrak{R}(\mathfrak{A}) = \mathfrak{R}(\mathfrak{A}^*) \wedge \mathfrak{A}$.

Suppose now that $\mathfrak{A} \wedge (1) = 0$ and that (1) is isomorphic to the ring of integers. Then if $z^* \in \mathfrak{R}(\mathfrak{A}^*)$ the coset $\bar{z}^*$ of z^* in the difference ring $\mathfrak{A}^* - \mathfrak{A}$ is in the radical of this ring. Since the radical of the ring of integers is 0, $\bar{z}^* = 0$ and $z^* \in \mathfrak{A}$. Hence $\mathfrak{A} \geq \mathfrak{R}(\mathfrak{A}^*)$. From the equation $\mathfrak{R}(\mathfrak{A}) = \mathfrak{R}(\mathfrak{A}^*) \wedge \mathfrak{A}$ we obtain $\mathfrak{R}(\mathfrak{A}) = \mathfrak{R}(\mathfrak{A}^*)$.

THEOREM 3. *Let $\mathfrak{A}$ be an arbitrary ring and let $\mathfrak{A}^*$ be a ring with an identity containing $\mathfrak{A}$ such that $\mathfrak{A}^* = \mathfrak{A} + (1)$. Then the radical $\mathfrak{R}(\mathfrak{A}) = \mathfrak{R}(\mathfrak{A}^*) \wedge \mathfrak{A}$. If in addition $\mathfrak{A} \wedge (1) = 0$ and (1) is isomorphic to the ring of integers then $\mathfrak{R}(\mathfrak{A}) = \mathfrak{R}(\mathfrak{A}^*)$.*

We shall call $\mathfrak{A}$ a *radical ring* if $\mathfrak{A} = \mathfrak{R}$. If $\mathfrak{R} = 0$, $\mathfrak{A}$ is *semi-simple*. For many problems the following theorem effects a reduction to the consideration of these two extreme types of rings.

THEOREM 4. *If $\mathfrak{R}$ is the radical of $\mathfrak{A}$, $\bar{\mathfrak{A}} = \mathfrak{A} - \mathfrak{R}$ is semi-simple.*

Let $\bar{z}$ be an element of the radical of $\bar{\mathfrak{A}}$ and let z be an element in the coset $\bar{z}$. Then there exists an element z' such that $z + z' + zz' = u \in \mathfrak{R}$. Also there exists a u' such that $u + u' + uu' = 0$. Hence

$$\begin{aligned} 0 &= (z + z' + zz') + u' + (z + z' + zz')u' \\ &= z + (z' + u' + z'u') + z(z' + u' + z'u'). \end{aligned}$$

Thus z is right quasi-regular. Since the totality of elements z in the cosets $\bar{z}$ of $\mathfrak{R}(\bar{\mathfrak{A}})$ is an ideal, this totality is a quasi-regular ideal. Hence $z \in \mathfrak{R}$ and $\bar{z} = 0$.

If z is a nilpotent element of index n, $z' = \sum_{1}^{n-1} (-1)^i z^i$ is a quasi-inverse of z. Hence we have the following

THEOREM 5. *The radical of a ring contains every nil right (left) ideal of the ring.*

As a consequence of Theorems 4 and 5 we may prove the

COROLLARY. *If z is an element such that $\mathfrak{A} z \mathfrak{A} \leq \mathfrak{R}$ then $z \in \mathfrak{R}$.*

[9] Albert [1], p. 22.

For if $\mathfrak{J}$ is the right ideal generated by z, $\mathfrak{J}^3 \leq \mathfrak{A}z\mathfrak{A} \leq \mathfrak{R}$. Hence $\bar{\mathfrak{J}} = (\mathfrak{J} + \mathfrak{R}) - \mathfrak{R}$ is nilpotent in the semi-simple ring $\bar{\mathfrak{A}} = \mathfrak{A} - \mathfrak{R}$. Hence $\bar{\mathfrak{J}} = 0$, $\mathfrak{J} \leq \mathfrak{R}$ and $z \in \mathfrak{R}$.

This corollary implies that $z \in \mathfrak{R}$ if and only if za (az) is right (left) quasi-regular for all a in $\mathfrak{A}$.

The elements of the radical need not be nilpotent. This can be seen in the example: $\mathfrak{A}$ the ring of p-adic integers. By applying the definition directly or by using Corollary 2 to Theorem 18 we can show that $\mathfrak{R} = p\mathfrak{A}$. However, no element of $\mathfrak{A}$ is nilpotent. Some information on the nature of the elements of $\mathfrak{R}$ can be obtained from the following

THEOREM 6. *If $\mathfrak{N}$ is a subring of $\mathfrak{R}$ and $z \in \mathfrak{N}$, then for any positive integer h either $z^{h-1}\mathfrak{N} > z^h\mathfrak{N}$ or $z^h = 0$.*

Evidently $z^{h-1}\mathfrak{N} \geq z^h\mathfrak{N}$. Suppose that $z^{h-1}\mathfrak{N} = z^h\mathfrak{N}$. Then $z^h = z^h y$ for some y in $\mathfrak{N}$. Let y' be a quasi-inverse of $-y$. Then

$$0 = z^h - z^h y + z^h y' - z^h y y' = z^h + z^h(-y + y' - yy') = z^h.$$

This theorem implies that the radical contains no idempotent element $\neq 0$. It is known that if $\mathfrak{A}$ is a ring with an identity whose lattice of right ideals is completely reducible, then every right ideal $\mathfrak{J}$ in $\mathfrak{A}$ has the form $e\mathfrak{A}$ where e is an idempotent element in $\mathfrak{J}$.[10] Hence we have the following

THEOREM 7. *If $\mathfrak{A}$ is a ring with an identity whose lattice of right (left) ideals is completely reducible, then $\mathfrak{A}$ is semi-simple.*

We shall show next that any ring which is regular[11] in the sense of von Neumann is semi-simple. We recall the definition: $\mathfrak{A}$ is regular if every element a of $\mathfrak{A}$ has a *relative inverse* u such that $aua = a$. Suppose that $a \in \mathfrak{R}$. Then $-ua$ has a quasi-inverse v such that $-ua + v - uav = 0$. Hence

$$0 = -aua + av - auav = -a + av - av = -a.$$

THEOREM 8. *Any regular ring is semi-simple.*

3. The radical of a ring satisfying the descending chain condition. We suppose now that $\mathfrak{A}$ is a ring for which the descending chain condition for right (left) ideals holds. Let $\mathfrak{N}$ be a two-sided ideal contained in $\mathfrak{R}$ and suppose that $\mathfrak{N}^2 = \mathfrak{N}$. Then if $\mathfrak{N} \neq 0$ there exists a minimum right ideal $\mathfrak{J}$ of $\mathfrak{A}$ with the properties 1) $\mathfrak{J} \leq \mathfrak{N}$, 2) $\mathfrak{J}\mathfrak{N} \neq 0$.

---

[10] See, for example, Jacobson [1], p. 65.

[11] Von Neumann [1].

Let b be an element of $\mathfrak{J}$ such that $b\mathfrak{N} \neq 0$. Then $(b\mathfrak{N})\mathfrak{N} = b\mathfrak{N} \neq 0$ and since $b\mathfrak{N} \leq \mathfrak{J}$ we have $b\mathfrak{N} = \mathfrak{J}$ by the minimality of $\mathfrak{J}$. Since $b \in \mathfrak{J}$ there is an element y in $\mathfrak{N}$ such that $by = b$. As in the proof of Theorem 6 this leads to $b = 0$ contrary to $b\mathfrak{N} \neq 0$. Thus $\mathfrak{N} = 0$. Now the positive integral powers $\mathfrak{R}^k$ of $\mathfrak{R}$ are two-sided ideals and $\mathfrak{R} \geq \mathfrak{R}^2 \geq \cdots$. Hence there is an integer ρ such that $\mathfrak{R}^\rho = \mathfrak{R}^{\rho+1}$. Then for $\mathfrak{N} = \mathfrak{R}^\rho$ we have $\mathfrak{N}^2 = \mathfrak{N}$. Hence $\mathfrak{N} = \mathfrak{R}^\rho = 0$. This proves

THEOREM 9. *If $\mathfrak{A}$ is a ring that satisfies the descending chain condition for right (left) ideals, then the radical of $\mathfrak{A}$ is nilpotent.*

Since any nil ideal is contained in the radical this proves that any nil ideal in a ring that satisfies the descending chain condition for right (left) ideals is nilpotent. It is clear also that $\mathfrak{R}$ coincides with the usual radical defined as the join of all nilpotent ideals.

4. Finitely generated ideals contained in $\mathfrak{R}$.

THEOREM 10. *If $\mathfrak{N}$ is a right ideal with a finite basis contained in the radical $\mathfrak{R}$ then either $\mathfrak{N}\mathfrak{R} < \mathfrak{N}$ or $\mathfrak{N} = 0$.*

Since $\mathfrak{N}$ is a right ideal either $\mathfrak{N}\mathfrak{R} < \mathfrak{N}$ or $\mathfrak{N}\mathfrak{R} = \mathfrak{N}$. Assume that $\mathfrak{N}\mathfrak{R} = \mathfrak{N}$. Let $y_1, \cdots, y_n$ be a basis for $\mathfrak{N}$. Then every element of $\mathfrak{N}$ has the form $\Sigma y_i a_i + \Sigma y_i j_i$ where the $a_i \in \mathfrak{A}$ and the j_i are integers. Since $\mathfrak{N}\mathfrak{R} = \mathfrak{N}$ every element of $\mathfrak{N}$ also has the form $\Sigma y_i z_i$, z_i in $\mathfrak{R}$. In particular $y_1 = \Sigma y_i z_i$. Let z'_1 be the quasi-inverse of $-z_1$. Then we have

$$y_1 = y_1 + y_1(-z_1 + z'_1 - z_1 z'_1)$$
$$= (y_1 - y_1 z_1) + (y_1 - y_1 z_1) z'_1 = \sum_2^n y_j(z_j + z_j z'_1).$$

Hence y_1 can be eliminated from the basis. Similarly every y_i can be eliminated and so $\mathfrak{N} = 0$.

If $\mathfrak{A}$ is an arbitrary ring we define the transfinite powers $\mathfrak{A}^\alpha$ by the conditions 1) $\mathfrak{A}^1 = \mathfrak{A}$, 2) $\mathfrak{A}^{\alpha+1} = \mathfrak{A}^\alpha\mathfrak{A}$, 3) if α is the limit ordinal $\mathfrak{A}^\alpha$ is the join of all $\mathfrak{A}^\beta$ with $\beta < \alpha$. There exists a least ordinal ρ such that $\mathfrak{A}^\rho = \mathfrak{A}^{\rho+1}$. We shall call ρ the *index* of $\mathfrak{A}$ and we shall say that $\mathfrak{A}$ is *transfinite nilpotent* if $\mathfrak{A}^\rho = 0$. Now suppose that $\mathfrak{A}$ is a ring that satisfies the ascending chain condition for right ideals. We recall that this condition is equivalent to the requirement that every ideal has a finite basis. Let $\mathfrak{N} = \mathfrak{R}^\rho$ where ρ is the index of the radical $\mathfrak{R}$ of $\mathfrak{A}$. Then $\mathfrak{N}\mathfrak{R} = \mathfrak{N}$ and so, by Theorem 10, $\mathfrak{N} = \mathfrak{R}^\rho = 0$.

THEOREM 11. *The radical of a ring that satisfies the ascending chain condition for right ideals is transfinite nilpotent.*

5. The radical of a matrix ring. If $\mathfrak{A}$ is an arbitrary ring we denote as usual the ring of $n \times n$ matrices with elements in $\mathfrak{A}$ by $\mathfrak{A}_n$. If $\mathfrak{B}$ is a subring (ideal) in $\mathfrak{A}$ then $\mathfrak{B}_n$ is a subring (ideal) in $\mathfrak{A}_n$. Let $\mathfrak{R}$ be the radical of $\mathfrak{A}$. Then we wish to show that $\mathfrak{R}_n$ is the radical $\mathfrak{R}(\mathfrak{A}_n)$ of $\mathfrak{A}_n$.

LEMMA 2. *Any matrix $z = (z_{ij})$ of $\mathfrak{A}_n$ in which z_{11} is right quasi-regular and the $z_{ij} = 0$ for $i > 1$ is right quasi-regular in $\mathfrak{A}_n$.*

Since z_{11} is right quasi-regular we know that the ideal $\{x + z_{11}x\} = \mathfrak{A}$. Hence there exist elements z'_{1i} such that $z'_{1i} + z_{11}z'_{1i} = -z_{1i}$. Then if we set $z'_{ij} = 0$ for $i > 1$ we may verify directly that $z' = (z'_{ij})$ is a right quasi-inverse of z.

Consider now the totality $\mathfrak{J}_1$ of matrices with first row consisting of elements in $\mathfrak{R}$ and other rows 0. Then $\mathfrak{J}_1$ is a right ideal. Hence by the lemma $\mathfrak{J}_1 \leq \mathfrak{R}(\mathfrak{A}_n)$. Similarly the totality $\mathfrak{J}_j$ of matrices with j-th row consisting of elements of $\mathfrak{R}$ and other rows 0 is contained in $\mathfrak{R}(\mathfrak{A}_n)$. Since

$$\mathfrak{R}_n = \mathfrak{J}_1 + \cdots + \mathfrak{J}_n, \qquad \mathfrak{R}_n \leq \mathfrak{R}(\mathfrak{A}_n).$$

Conversely let $Y = (y_{ij}) \in \mathfrak{R}(\mathfrak{A}_n)$. If a is any element of $\mathfrak{A}$ we denote the matrix that has a in the (i, j) position and 0's elsewhere by A_{ij}. Let a and b be arbitrary and form the matrix $D = \sum_k A_{kp}YB_{qk}$. Then D is the diagonal matrix $\{ay_{pq}b, \cdots, ay_{pq}b\}$ and $D \in \mathfrak{R}(\mathfrak{A}_n)$. If $D' = (d'_{ij})$ is a right quasi-inverse of D it is easy to see that $d' = d'_{11}$ is a right quasi-inverse of $d = ay_{pq}b$. Evidently this implies that for arbitrary c in $\mathfrak{A}$, and arbitrary integral i, $dc + di$ is right quasi-regular. Hence $d \in \mathfrak{R}$. Since a and b are arbitrary the Corollary to Theorem 5 shows that $y_{pq} \in \mathfrak{R}$ and $Y \in \mathfrak{R}_n$.

THEOREM 12. *If $\mathfrak{A}$ is an arbitrary ring and $\mathfrak{A}_n$ is the ring of $n \times n$ matrices with elements in $\mathfrak{A}$, then the radical $\mathfrak{R}(\mathfrak{R}_n) = \mathfrak{R}_n$, $\mathfrak{R}$ the radical of $\mathfrak{A}$.*

6. The radical of an algebra. Let $\mathfrak{A}$ be an algebra of possibly infinite order over a field Φ. By an ideal in the algebra $\mathfrak{A}$ we mean, of course, an ideal of the ring $\mathfrak{A}$ that is invariant under the scalar multiplications $x \to x\alpha$, α in Φ. If $\mathfrak{A}$ has an identity, $x\alpha = x(1\alpha) = (1\alpha)x$ and so any ideal of the ring $\mathfrak{A}$ is an ideal of the algebra $\mathfrak{A}$.

The above discussion is valid without change for an arbitrary algebra $\mathfrak{A}$. Thus the radical $\mathfrak{R}$ of $\mathfrak{A}$ can be defined to be the join of all the quasi-regular right ideals of $\mathfrak{A}$. $\mathfrak{R}$ is also the maximum quasi-regular left ideal of $\mathfrak{A}$. All of our theorems hold for algebras.

An element a of an algebra $\mathfrak{A}$ is *algebraic* if it satisfies a non-trivial algebraic equation with coefficients in the underlying field Φ. An equivalent condition is that a generates a subalgebra A with a finite basis. As in the special case of a field, an element which is not algebraic will be called *transcendental*. If every element of $\mathfrak{A}$ is algebraic, then $\mathfrak{A}$ is algebraic. If a is an algebraic element and A is the subalgebra generated by a then there exists a positive integer h such that $a^{h-1}A = a^hA$. Hence if a is in the radical, by Theorem 6, a is nilpotent. This proves

THEOREM 13. *If $\mathfrak{A}$ is an algebra over a field, the elements of the radical $\mathfrak{R}$ of $\mathfrak{A}$ are either nilpotent or transcendental over Φ.*

If $\mathfrak{A}$ is algebraic every z in $\mathfrak{R}$ is nilpotent. Hence, since $\mathfrak{R}$ contains every nil ideal, we have the following

THEOREM 14. *The radical of an algebraic algebra $\mathfrak{A}$ is the join of all nil right (left) ideals of $\mathfrak{A}$.*

If $\mathfrak{A}$ is commutative, an element z generates a nil ideal if and only if it is nilpotent. Hence we have the

COROLLARY. *If $\mathfrak{A}$ is a commutative algebraic algebra, the radical of $\mathfrak{A}$ is the totality of its nilpotent elements.*

7. The radical of a normed ring. We suppose now that $\mathfrak{A}$ is a normed ring, i. e., $\mathfrak{A}$ is an algebra over the field Φ of complex numbers and for each a in $\mathfrak{A}$ there is defined a non-negative real norm $\| a \|$ such that

1. $\| a \| > 0$ if $a \neq 0$ $\quad \| 0 \| = 0$
2. $\| a + b \| \leq \| a \| + \| b \|$
3. $\| a\alpha \| = \| a \| \cdot | \alpha |$ if $\alpha \in \Phi$
4. $\| ab \| \leq \| a \| \cdot \| b \|$
5. $\mathfrak{A}$ has an identity and $\| 1 \| = 1$
6. $\mathfrak{A}$ is complete relative to the metric $D(a, b) = \| a - b \|$

Following Gelfand we call an element of $\mathfrak{A}$ a *generalized nilpotent element* if $\lim \| z^n \|^{1/n} = 0$.[12] For commutative normed rings Gelfand has defined the radical to be the totality of generalized nilpotent elements. We shall show that this set coincides with the radical as defined here and we shall obtain a similar characterization of the radical for non-commutative normed rings. We prove first the following

THEOREM 15. *The radical of a normed ring $\mathfrak{A}$ is the totality of elements z such that $(za)^n \to 0 ((az)^n \to 0)$ for every a in $\mathfrak{A}$.*

[12] Gelfand [1], p. 10.

Since $\mathfrak{A}$ has an identity $\mathfrak{R}$ is the totality of elements z such that $1 + za$ has an inverse for every a in $\mathfrak{A}$. Now suppose that z is an element such that $(za)^n \to 0$ for every a. Then for any α in Φ $\| (z\alpha)^n \| < 1$ for n sufficiently large. Thus $\| z^n \| < \beta^n$ where $\beta = 1/|\alpha|$. We choose α so that $|\alpha| > 1$. The series $1 - z + z^2 - \cdots$ is ultimately dominated by the convergent series $1 + \beta + \beta^2 + \cdots$. Hence $1 - z + z^2 - \cdots$ exists in $\mathfrak{A}$ and this element is the inverse of $1 + z$. Similarly we can show that if $z' = za$, then $1 + z'$ has an inverse. Hence $z \in \mathfrak{R}$. Conversely suppose that $z \in \mathfrak{R}$. Then $1 + z\alpha$ has an inverse $(1 + z\alpha)^{-1}$ for every α in Φ. Using the fact that $(1 + z\alpha)^{-1}$ is an analytic function of α we may prove, exactly as Gelfand has done in the commutative case, that $(z\alpha)^n \to 0$.[13] In particular $z^n \to 0$ and since $\mathfrak{R}$ is an ideal $(za)^n \to 0$ for every a.

We shall call an ideal a *generalized nil ideal* if all of its elements are generalized nilpotent elements. If $z \in \mathfrak{R}$, $(z\alpha)^n \to 0$. Then for n sufficiently large $\| (z\alpha)^n \| < 1$ and $\| z^n \| < \beta^n$ for $\beta = 1/|\alpha|$. Hence $0 \leq \| z^n \|^{1/n} < \beta$. Since β is arbitrary $\lim \| z^n \|^{1/n} = 0$. Thus z is a generalized nilpotent element and $\mathfrak{R}$ is a generalized nil ideal. Next let $\mathfrak{J}$ be an arbitrary generalized nil right ideal. Then if $y \in \mathfrak{J}$, $\| y^n \|^{1/n} \to 0$. Hence $y^n \to 0$. Since $\mathfrak{J}$ is a right ideal, $y' = ya \in \mathfrak{J}$ and $(y')^n \to 0$. By Theorem 15, $y \in \mathfrak{R}$ and so $\mathfrak{J} \leq \mathfrak{R}$. This proves

THEOREM 16. *The radical of a normed ring is a generalized nil ideal that contains every generalized nil right (left) ideal of the ring.*

Let $\mathfrak{A}$ be commutative. Then if z and $a \in \mathfrak{A}$,

$$\| (za)^n \| = \| z^n a^n \| \leq \| z^n \| \cdot \| a^n \| \leq \| z^n \| \cdot \| a \|^n.$$

Hence $\| (za)^n \|^{1/n} \leq \| z^n \|^{1/n} \cdot \| a \|$ and if z is a generalized nilpotent element then za is a generalized nilpotent element. Thus any generalized nilpotent element generates a generalized nil ideal and $\mathfrak{R}$ is the totality of generalized nilpotent elements.

8. Quotient ideals. We return to the consideration of an arbitrary ring $\mathfrak{A}$. The results that we obtain are also valid for algebras but we shall not state them explicitly for these.

Let $\mathfrak{J}$ be a right ideal in $\mathfrak{A}$. Then if $a \in \mathfrak{A}$ the right multiplication $x \to xa$ determined by a induces an endomorphism $\bar{a}$ in the difference group $\mathfrak{M} = \mathfrak{A} - \mathfrak{J}$. The mapping $\bar{a}$ sends the coset $x + \mathfrak{J}$ into $xa + \mathfrak{J}$. The totality of elements $\bar{a}$ is a subring $\bar{\mathfrak{A}}$ of the ring of endomorphisms of $\mathfrak{M}$ and the

[13] Gelfand [1], p. 10.

correspondence $a \to \bar{a}$ is a homomorphism between $\mathfrak{A}$ and $\bar{\mathfrak{A}}$. The kernel of this homomorphism is a two-sided ideal $\mathfrak{J}:\mathfrak{A}$ which we shall call the *quotient* of $\mathfrak{J}$ relative to $\mathfrak{A}$. Evidently $\mathfrak{A}(\mathfrak{J}:\mathfrak{A}) \leq \mathfrak{J}$ and if $\mathfrak{A}$ has an identity, $(\mathfrak{J}:\mathfrak{A})$ is the largest two-sided ideal of $\mathfrak{A}$ contained in $\mathfrak{J}$. By the fundamental theorem on homomorphisms $\bar{\mathfrak{A}} \cong \mathfrak{A} - (\mathfrak{J}:\mathfrak{A})$.

Let $\mathfrak{J}$ be maximal, i. e., $\mathfrak{A} > \mathfrak{J}$ and there is no right ideal $\mathfrak{J}'$ such that $\mathfrak{A} > \mathfrak{J}' > \mathfrak{J}$. Then $\mathfrak{M} = \mathfrak{A} - \mathfrak{J}$ is irreducible relative to $\bar{\mathfrak{A}}$. As usual we call a ring of endomorphisms $\bar{\mathfrak{A}}$ irreducible if the group $\mathfrak{M}$ in which $\bar{\mathfrak{A}}$ acts is irreducible. Let $\bar{\mathfrak{A}} \neq 0$ have this property. Then the totality $\mathfrak{Z}$ of elements z in $\mathfrak{M}$ such that $z\bar{\mathfrak{A}} = 0$ is a subgroup of $\mathfrak{M}$ invariant under $\mathfrak{A}$. Hence either $\mathfrak{Z} = 0$ or $\mathfrak{Z} = \mathfrak{M}$. Since $\bar{\mathfrak{A}} \neq 0$, $\mathfrak{Z} \neq \mathfrak{M}$ and so $\mathfrak{Z} = 0$. It follows that if x is any element $\neq 0$ of $\mathfrak{M}$ then $x\bar{\mathfrak{A}} \neq 0$. Since $x\bar{\mathfrak{A}}$ is a subgroup invariant under $\bar{\mathfrak{A}}$, $x\bar{\mathfrak{A}} = \mathfrak{M}$. We use this to prove

THEOREM 17. *Any irreducible ring of endomorphisms is semi-simple.*

Let $\bar{z}$ be an element of the radical of $\mathfrak{A}$ and let $x \neq 0$ be arbitrary in $\mathfrak{M}$. If $x\bar{z} \neq 0$, $(x\bar{z})\bar{\mathfrak{A}} = \mathfrak{M}$. Hence there is an $\bar{a}$ in $\bar{\mathfrak{A}}$ such that $x\bar{z}\bar{a} = x$. The element $-\bar{z}\bar{a}$ has a quasi-inverse $\bar{z}'$. Hence

$$x = x - (x\bar{z}\bar{a} - x\bar{z}' + x\bar{z}\bar{a}\bar{z}') = x - x\bar{z}\bar{a} + (x - x\bar{z}\bar{a})\bar{z}' = 0.$$

This contradiction shows that $x\bar{z} = 0$ for all x. Thus $\bar{z} = 0$ and $\bar{\mathfrak{A}}$ is semi-simple.

This theorem has the following important consequence.

COROLLARY. *If $\mathfrak{J}$ is a maximal right ideal $(\mathfrak{J}:\mathfrak{A})$ contains the radical $\mathfrak{R}$ of $\mathfrak{A}$.*

If $(\mathfrak{J}:\mathfrak{A}) = \mathfrak{A}$ there is nothing to prove. Hence suppose that $(\mathfrak{J}:\mathfrak{A}) \neq \mathfrak{A}$. Then $\bar{\mathfrak{A}} \cong \mathfrak{A} - (\mathfrak{J}:\mathfrak{A})$ is an irreducible ring of endomorphisms $\neq 0$. If $z \in \mathfrak{R}$ the coset $\bar{z} = z + (\mathfrak{J}:\mathfrak{A})$ is in the radical of $\mathfrak{A} - (\mathfrak{J}:\mathfrak{A})$. Since $\bar{\mathfrak{A}}$ is semi-simple, $\bar{z} = 0$. Hence $z \in (\mathfrak{J}:\mathfrak{A})$ and $\mathfrak{R} \leq (\mathfrak{J}:\mathfrak{A})$.

9. The radical as intersection of maximal right ideals. Let $\mathfrak{A}$ be any ring that is not a radical ring. Then $\mathfrak{A}$ contains an element a which is not right quasi-regular. Hence the right ideal $\{x + ax\}$ does not contain a. Moreover, if $\mathfrak{J}$ is a right ideal that contains a and contains the ideal $\{x + ax\}$ then $\mathfrak{J} = \mathfrak{A}$. By using Zorn's maximum principle we may prove

LEMMA 3. *If a is an element of $\mathfrak{A}$ that is not right quasi-regular the right ideal $\{x + ax\}$ can be imbedded in a maximal right ideal.*

This shows, of course, that any ring that is not a radical ring contains maximal right ideals. We assume now that $\mathfrak{A}$ is any ring that contains maximal right ideals. Consider the intersection $\Pi\mathfrak{J}$ of the maximal right ideals $\mathfrak{J}$ of $\mathfrak{A}$. Let $y \in \Pi\mathfrak{J}$. Then y is right quasi-regular. For otherwise $\{x + yx\}$ can be imbedded in the maximal right ideal $\mathfrak{J}$. Then $y \in \mathfrak{J}$ and $\mathfrak{J} = \mathfrak{A}$ contrary to the maximality of $\mathfrak{J}$. Thus every element of $\Pi\mathfrak{J}$ is quasi-regular and since $\Pi\mathfrak{J}$ is a right ideal, $\Pi\mathfrak{J} \leqq \mathfrak{R}$. On the other hand, by the corollary to Theorem 17, $\mathfrak{R} \leq (\mathfrak{J}:\mathfrak{A})$. Hence $\mathfrak{A}\mathfrak{R} \leq \mathfrak{A}(\mathfrak{J}:\mathfrak{A}) \leq \mathfrak{J}$ and $\mathfrak{A}\mathfrak{R} \leq \Pi\mathfrak{J}$. This proves

THEOREM 18. *If $\mathfrak{A}$ is a ring that contains maximal right ideals and $\Pi\mathfrak{J}$ is the intersection of these maximal right ideals, then $\Pi\mathfrak{J} \leq \mathfrak{R}$ and $\mathfrak{A}\mathfrak{R} \leq \Pi\mathfrak{J}$.*

COROLLARY 1. *If $\mathfrak{A}$ is not a radical ring and $\Pi\mathfrak{J}$ is the intersection of the maximal right ideals of $\mathfrak{A}$ then $\Pi\mathfrak{J} \leq \mathfrak{R}$ and $\mathfrak{A}\mathfrak{R} \leq \Pi\mathfrak{J}$.*

COROLLARY 2. *If $\mathfrak{A}$ is a ring with an identity the radical of $\mathfrak{A}$ is the intersection of the maximal right ideals of $\mathfrak{A}$.*

For $\mathfrak{A}\mathfrak{R} = \mathfrak{R}$. Hence $\mathfrak{R} \leq \Pi\mathfrak{J}$ as well as $\Pi\mathfrak{J} \leq \mathfrak{R}$.

The following results also are consequences of Theorem 18:

If $\mathfrak{A}$ is a ring that contains maximal right ideals then $\Pi\mathfrak{J}$ is a two-sided ideal.

For $\mathfrak{A}(\Pi\mathfrak{J}) \leq \mathfrak{A}\mathfrak{R} \leq \Pi\mathfrak{J}$. Hence $\Pi\mathfrak{J}$ is a left ideal as well as a right ideal.

The radical of a normed ring is a closed ideal.

It is known that any maximal ideal $\mathfrak{J}$ is closed.[14] Hence $\mathfrak{R} = \Pi\mathfrak{J}$ is closed.

If $\mathfrak{A}$ is a semi-simple ring, $\Pi\mathfrak{J} = 0$. Suppose, in addition, that $\mathfrak{A}$ satisfies the descending chain condition for right ideals. Then we can find a finite number of maximal right ideals $\mathfrak{J}_j$, $j = 1, \cdots, n$ such that $\Pi\mathfrak{J}_j = 0$. We may suppose that the set $\{\mathfrak{J}_j\}$ is minimal in the sense that $\prod_{k \neq i} \mathfrak{J}_k = \mathfrak{J}'_i \neq 0$ for every i. Then $\mathfrak{J}_i \wedge \mathfrak{J}'_i = 0$ and $\mathfrak{J}'_i \nleqq \mathfrak{J}_i$. Since $\mathfrak{J}_i$ is maximal it follows that $\mathfrak{A} = \mathfrak{J}_i + \mathfrak{J}'_i$. Hence $\mathfrak{J}'_i$ is isomorphic to the difference $\mathfrak{A}$-group $\mathfrak{A} - \mathfrak{J}_i$ and $\mathfrak{J}'_i$ is minimal. Using a simple lattice-theoretic argument we can conclude that $\mathfrak{A} = \mathfrak{J}'_1 \oplus \cdots \oplus \mathfrak{J}'_n$.[15] This proves the well-known

THEOREM 19. *If $\mathfrak{A}$ is a semi-simple ring that satisfies the descending chain condition for right ideals, $\mathfrak{A}$ is a direct sum of a finite number of minimal right ideals.*

---

[14] Gelfand [1], p. 8.

[15] Jacobson [1], p. 35.

Suppose again that $\mathfrak{A}$ is any ring that contains maximal right ideals. Then, by the corollary to Theorem 17, $\mathfrak{R} \leqq \Pi(\mathfrak{J}:\mathfrak{A})$ for all maximal $\mathfrak{J}$. Conversely let $y \in \Pi(\mathfrak{J}:\mathfrak{A})$. Then $\mathfrak{A}y \leq \Pi\mathfrak{J} \leq \mathfrak{R}$. Hence $\mathfrak{A}y\mathfrak{A} \leq \mathfrak{R}$ and this implies that $y \in \mathfrak{R}$. We have therefore proved

THEOREM 20. *Let $\mathfrak{A}$ be an arbitrary ring that contains maximal right ideals. Then the radical of $\mathfrak{A}$ is the intersection $\Pi(\mathfrak{J}:\mathfrak{A})$ where $\mathfrak{J}$ ranges over the maximal right ideals of $\mathfrak{A}$.*

COROLLARY. *If $\mathfrak{A}$ is not a radical ring, $\mathfrak{R} = \Pi(\mathfrak{J}:\mathfrak{A})$ where $\mathfrak{J}$ ranges over the maximal right ideals.*

The results of this section hold also for left ideals. An interesting consequence of the second corollary to Theorem 18 is that if $\mathfrak{A}$ is a ring with an identity then the intersection of the maximal right ideals of $\mathfrak{A}$ coincides with the intersection of the maximal left ideals of $\mathfrak{A}$.

10. Primitive rings.

Definition 3. A ring $\mathfrak{A}$ is *primitive* if $\mathfrak{A}$ contains a maximal right ideal $\mathfrak{J}$ whose quotient $(\mathfrak{J}:\mathfrak{A}) = 0$.

Let $\mathfrak{A}$ be of this type and as before let $\bar{\mathfrak{A}}$ denote the ring of endomorphisms $x + \mathfrak{J} \to xa + \mathfrak{J}$ in $\mathfrak{M} = \mathfrak{A} - \mathfrak{J}$. Then $\bar{\mathfrak{A}}$ is irreducible and $\bar{\mathfrak{A}} \cong \mathfrak{A} - (\mathfrak{J}:\mathfrak{A}) = \mathfrak{A}$. Thus any primitive ring is isomorphic to an irreducible ring of endomorphisms.

Conversely suppose that $\bar{\mathfrak{A}} \neq 0$ is an irreducible ring of endomorphisms acting in $\mathfrak{M}$. Let x be an element $\neq 0$ in $\mathfrak{M}$ and let $\mathfrak{J}_x$ denote the totality of elements b of $\bar{\mathfrak{A}}$ such that $xb = 0$. Evidently $\mathfrak{J}_x$ is a right ideal in $\bar{\mathfrak{A}}$. We have seen that $x\bar{\mathfrak{A}} = \mathfrak{M}$. Hence $\mathfrak{J}_x < \bar{\mathfrak{A}}$. Let $\bar{a}$ be an element of $\bar{\mathfrak{A}}$ not in $\mathfrak{J}_x$ and let $\bar{c}$ be arbitrary. Then $x\bar{a} \neq 0$. Hence $(x\bar{a})\bar{\mathfrak{A}} = \mathfrak{M}$ and so there exists an element $\bar{u}$ such that $x\bar{a}\bar{u} = x\bar{c}$. Thus $\bar{c} - \bar{a}\bar{u} \in \mathfrak{J}_x$ and $\bar{\mathfrak{A}} = \mathfrak{J}_x + \bar{a}\bar{\mathfrak{A}}$ for any $\bar{a}$ not in $\mathfrak{J}_x$. This proves that $\mathfrak{J}_x$ is maximal. If $\bar{c} \neq 0 \in \bar{\mathfrak{A}}$ there is a y in $\mathfrak{M}$ such that $y\bar{c} \neq 0$. There exists an $\bar{a}$ such that $x\bar{a} = y$. It follows that $x\bar{a}\bar{c} \neq 0$. Thus for any $\bar{c} \neq 0$ there exists an $\bar{a}$ such that $\bar{a}\bar{c} \notin \mathfrak{J}_x$. On the other hand, if $\bar{c} \in (\mathfrak{J}_x:\bar{\mathfrak{A}})$, $\bar{a}\bar{c} \in \mathfrak{J}_x$ for all $\bar{a}$. Hence $\bar{c} = 0$ and $(\mathfrak{J}_x:\bar{\mathfrak{A}}) = 0$.

THEOREM 21. *A necessary and sufficient condition that $\mathfrak{A}$ be primitive is that $\mathfrak{A}$ be isomorphic to an irreducible ring of endomorphisms.*

If $\bar{\mathfrak{B}}$ is a two-sided ideal $\neq 0$ in the irreducible ring $\bar{\mathfrak{A}}$, the argument that we have used before for $\bar{\mathfrak{A}}$ shows that $x\bar{\mathfrak{B}} = \mathfrak{M}$ for any $x \neq 0$. Thus $\bar{\mathfrak{B}}$ is irreducible. Hence we have the following

THEOREM 22. *Any two-sided ideal in a primitive ring is primitive.*

For a deeper study of primitive rings we require a theorem on irreducible rings of endomorphisms recently proved by the author. We recall first that if $\bar{\mathfrak{A}}$ is irreducible in $\mathfrak{M}$ then the totality of endomorphisms of $\mathfrak{M}$ that commute with all the $\bar{a} \in \bar{\mathfrak{A}}$ is a division subring $\mathfrak{D}$ of the ring of endomorphisms of $\mathfrak{M}$.[16] Since $\mathfrak{D}$ contains the identity mapping, we may regard $\mathfrak{M}$ as a vector space over $\mathfrak{D}$.

We call a set $\bar{\mathfrak{A}}$ of linear transformations in a vector space $\mathfrak{M}$ over $\mathfrak{D}$ a *dense set* if for any finite set $x_1, \cdots, x_n$ of $\mathfrak{D}$-independent vectors and any finite set $y_1, \cdots, y_n$ of arbitrary vectors there exists a linear transformation $\bar{a} \in \bar{\mathfrak{A}}$ such that $x_i\bar{a} = y_i$. An equivalent condition is that if $\mathfrak{S}$ is any finite dimensional subspace of $\mathfrak{M}$ and A is any linear transformation in $\mathfrak{S}$ over $\mathfrak{D}$, then A can be extended to a linear transformation $\bar{a} \in \bar{\mathfrak{A}}$ of $\mathfrak{M}$ over $\mathfrak{D}$.[17] Hence if $\mathfrak{M}$ is finite dimensional $\bar{\mathfrak{A}} = \mathfrak{L}$ the complete ring of linear transformations in $\mathfrak{M}$ over $\mathfrak{D}$. We have proved elsewhere the following

THEOREM 23. *If $\bar{\mathfrak{A}} \neq 0$ is an irreducible ring of endomorphisms in the commutative group $\mathfrak{M}$ and $\mathfrak{D}$ is the division ring of endomorphisms commutative with the $\bar{a} \in \bar{\mathfrak{A}}$, then $\bar{\mathfrak{A}}$ is a dense ring of linear transformations in the vector space $\mathfrak{M}$ over $\mathfrak{D}$.*

The converse is trivial. Thus the concepts of primitive ring, irreducible ring of endomorphisms and dense ring of linear transformations may be used interchangeably.

If $\bar{\mathfrak{A}}$ is a dense ring of linear transformations in $\mathfrak{M}$ over $\mathfrak{D}$ and $x_1, x_2, x_3, \cdots$ are $\mathfrak{D}$-independent, then

$$\mathfrak{J}_{x_1} > (\mathfrak{J}_{x_1} \wedge \mathfrak{J}_{x_2}) > (\mathfrak{J}_{x_1} \wedge \mathfrak{J}_{x_2} \wedge \mathfrak{J}_{x_3}) > \cdots$$

where $\mathfrak{J}_x$ denotes the right ideal of annihilators of x. Hence if $\mathfrak{M}$ is infinite dimensional, then $\bar{\mathfrak{A}}$ does not satisfy the descending chain condition for right ideals. Thus if $\bar{\mathfrak{A}}$ satisfies the descending chain condition $\mathfrak{M}$ is finite dimensional over $\mathfrak{D}$ and therefore $\bar{\mathfrak{A}} = \mathfrak{L}$ the complete ring of linear transformations in $\mathfrak{M}$ over $\mathfrak{D}$. This proves

THEOREM 24. *Any primitive ring that satisfies the descending chain condition for right ideals is isomorphic to the complete ring of linear transformations of a suitable finite dimensional vector space over a division ring.*

---

[16] This is Schur's lemma. See Jacobson [1], p. 57.

[17] For the definitions and results on dense rings of linear transformations quoted here, see Jacobson [2].

Since the complete ring of linear transformations in a finite dimensional vector space is simple we have the

COROLLARY. *Any primitive ring that satisfies the descending chain condition for right ideals is simple.*

11. Structure of semi-simple rings. We have seen that if $\mathfrak{A}$ is not a radical ring, the radical $\mathfrak{R} = \Pi(\mathfrak{J} : \mathfrak{A})$ for the maximal right ideals $\mathfrak{J}$. The difference rings $\mathfrak{A} - (\mathfrak{J} : \mathfrak{A})$ are isomorphic to irreducible rings of endomorphisms. Hence they are primitive. In particular we have the following

THEOREM 25. *If $\mathfrak{A}$ is a semi-simple ring $\mathfrak{A}$ contains two-sided ideals $\mathfrak{B}$ such that* 1) *each $\mathfrak{A} - \mathfrak{B}$ is primitive and* 2) $\Pi\mathfrak{B} = 0$.

The converse holds also. For if $\mathfrak{B}$ is a two-sided ideal in a ring and $\mathfrak{A} - \mathfrak{B}$ is semi-simple then $\mathfrak{B} \geq \mathfrak{R}$. Since any primitive ring is semi-simple each ideal $\mathfrak{B}$ such that $\mathfrak{A} - \mathfrak{B}$ is primitive contains $\mathfrak{R}$. Hence if $\Pi\mathfrak{B} = 0$ $\mathfrak{R} = 0$.

Theorem 25 reduces, to a certain extent, the study of semi-simple rings to that of primitive rings and hence to that of dense rings of linear transformations in a vector space over a division ring. This will be clearer in the next section where we derive a certain alternate form of this theorem. Before doing this we obtain some consequences of the theorem in the present form. We note first

THEOREM 26. *Any two-sided ideal in a semi-simple ring is semi-simple.*

Let $\mathfrak{A}_1$ be a two-sided ideal in the semi-simple ring $\mathfrak{A}$ and let $\{\mathfrak{B}\}$ be a set of two-sided ideals in $\mathfrak{A}$ such that 1) each $\mathfrak{A} - \mathfrak{B}$ is primitive and 2) $\Pi\mathfrak{B} = 0$. Let $\mathfrak{B}_1 = \mathfrak{A}_1 \wedge \mathfrak{B}$. Then $\mathfrak{A}_1 - (\mathfrak{A}_1 \wedge \mathfrak{B}) \cong (\mathfrak{A}_1 + \mathfrak{B}) - \mathfrak{B}$ a two-sided ideal in the primitive ring $\mathfrak{A} - \mathfrak{B}$. It follows that $\mathfrak{A}_1 - \mathfrak{B}_1$ is primitive. Evidently $\Pi\mathfrak{B} = 0$. Hence $\mathfrak{A}_1$ is semi-simple.

Next let $\mathfrak{A}$ be a semi-simple ring that satisfies the descending chain condition for right ideals. Let $\{\mathfrak{B}\}$ be a set of two-sided ideals satisfying the conditions of Theorem 25. We may replace the set $\{\mathfrak{B}\}$ by a finite set $\mathfrak{B}_i$, $i = 1, \cdots, s$, and we may suppose that the set is minimal in the sense that $\mathfrak{B}'_i = \prod_{j \neq i} \mathfrak{B}_j \neq 0$. Since $\mathfrak{A} - \mathfrak{B}$ is primitive and satisfies the descending chain condition for right ideals, $\mathfrak{A} - \mathfrak{B}$ is simple and so $\mathfrak{B}$ is a maximal two-sided ideal in $\mathfrak{A}$. It follows that $\mathfrak{B}'_i + \mathfrak{B}_i = \mathfrak{A}$. Hence $\mathfrak{B}'_i \cong \mathfrak{A} - \mathfrak{B}_i$ is simple. As for one-sided ideals we can prove that $\mathfrak{A} = \mathfrak{B}'_1 \oplus \cdots \oplus \mathfrak{B}'_s$. This proves the Wedderburn-Artin structure theorem:

THEOREM 27. *Any semi-simple ring that satisfies the descending chain condition for right ideals is isomorphic to a direct sum of a finite number of simple rings.*

By the preceding section the structure of the component ring $\mathfrak{B}'_i$ is that of a ring of linear transformations in a finite dimensional vector space over a division ring. This is the second fundamental Wedderburn-Artin structure theorem.

12. Infinite direct sums. Let S be an arbitrary set. We call the elements of S points and we suppose that to each point P there is associated a ring $\mathfrak{E}_P$. Now let $\mathfrak{S} = (S, \mathfrak{E}_P)$ be the totality of functions $f(P)$ with domain S and with $f(P)$ in $\mathfrak{E}_P$. If $f(P)$ and $g(P) \in \mathfrak{S}$ we define $f + g$ by $(f + g)(P) = f(P) + g(P)$ and we define $fg(P) = f(P)g(P)$. Then it is readily verified that $\mathfrak{S}$ is a ring. We shall call it the *complete direct sum* of the rings $\mathfrak{E}_P$. Let $\mathfrak{S}_0$ be the subset of functions $f(P)$ such that $f(P) = 0$ for all but a finite number of the points P. Then $\mathfrak{S}_0$ is a subring of $\mathfrak{S}$. We call it the (discrete) *direct sum* of the rings $\mathfrak{E}_P$. Let $\overline{\mathfrak{E}}_Q$ be the subset of $\mathfrak{S}_0$ of elements $f(P)$ such that $f(P) = 0$ for all $P \neq Q$. Then 1) $\overline{\mathfrak{E}}_Q$ is a two-sided ideal in $\mathfrak{S}_0$, 2) $\mathfrak{S}_0 = \Sigma\overline{\mathfrak{E}}_Q$ where $\Sigma\overline{\mathfrak{E}}_Q$ denotes the totality of finite sums of the elements of the ideals $\overline{\mathfrak{E}}_Q$ and 3) $\overline{\mathfrak{E}}_Q \wedge \sum_{P \neq Q} \overline{\mathfrak{E}}_P = 0$. Conversely these conditions are sufficient that a ring be isomorphic to a direct sum of rings isomorphic to the ideals $\overline{\mathfrak{E}}_Q$.

If $\mathfrak{A}$ is a subring of the complete direct sum $\mathfrak{S}$ we define the *component* $\mathfrak{A}_P$ *of* $\mathfrak{A}$ *at* P to be the totality of elements $f(P)$ of $\mathfrak{E}_P$ for f in $\mathfrak{A}$. It is clear that $\mathfrak{A}_P$ is a subring of $\mathfrak{E}_P$ and the correspondence $f \to f(P)$ is a homomorphism between $\mathfrak{A}$ and $\mathfrak{A}_P$. The kernel of this homomorphism is the two-sided ideal $\mathfrak{B}_P$ of elements g such that $g(P) = 0$. Hence $\Pi\mathfrak{B}_P = 0$. Moreover, $\mathfrak{A}_P \cong \mathfrak{A} - \mathfrak{B}_P$.

We suppose, conversely, that S is a set of two-sided ideals $P = \mathfrak{B}$ such that $\Pi\mathfrak{B} = 0$. Let $\mathfrak{A}_P = \mathfrak{A} - \mathfrak{B}$ and form the complete direct sum $\mathfrak{S} = (S, \mathfrak{A}_P)$. Then each a in $\mathfrak{A}$ determines an element $\bar{a}$ in $\mathfrak{S}$ such that $\bar{a}(P)$ is the coset $a + \mathfrak{B}$. The set $\overline{\mathfrak{A}}$ of the $\bar{a}$ is a subring of $\mathfrak{S}$ and the correspondence $a \to \bar{a}$ is a homomorphism. It is clear that component $\overline{\mathfrak{A}}_P = \mathfrak{A}_P$ and that $\bar{a}(P) = 0$ if and only if $a \in \mathfrak{B}$. Thus $\bar{a} = 0$ if and only if $a \in \Pi\mathfrak{B}$ and since $\Pi\mathfrak{B} = 0$ we see that the correspondence $a \to \bar{a}$ is, in fact, an isomorphism between $\mathfrak{A}$ and $\overline{\mathfrak{A}}$. Using these remarks we may formulate Theorem 25 and its converse as follows:

THEOREM 28. *A necessary and sufficient condition that a ring $\mathfrak{A}$ be semi-*

simple is that $\mathfrak{A}$ *be isomorphic to a subring* $\overline{\mathfrak{A}}$ *of a complete direct sum* $(S, \overline{\mathfrak{A}}_P)$ *where the components* $\overline{\mathfrak{A}}_P$ *of* $\overline{\mathfrak{A}}$ *are primitive.*

This theorem serves as a substitute in the general theory for the Wedderburn-Artin Theorem 27. A number of special cases of the theorem are well-known. We cite, for example, Stone's theorem on the representability of a Boolean ring as a subring of a direct sum of the fields $\mathfrak{J}_2$ of residues mod 2 and Montgomery and McCoy's generalization of this theorem to commutative rings in which every element a satisfies the equations $pa = 0$, $a^p = a$ for a fixed prime p. These results follow from Theorem 28 and the easily proved fact that any primitive commutative ring is a field.

13. Primitive rings that contain minimal ideals. If $\mathfrak{A}$ is an arbitrary ring it is well-known that any minimal right ideal $\mathfrak{J}$ in $\mathfrak{A}$ either has the form $e\mathfrak{A}$ where e is an idempotent element or $\mathfrak{J}^2 = 0$.[18] If $\mathfrak{J}$ is a minimal right ideal and a is arbitrary $a\mathfrak{J}$ is either 0 or a minimal right ideal. It follows that the join $\Sigma\mathfrak{J}$ of all the minimal right ideals is a two-sided ideal.[19]

If $\overline{\mathfrak{A}}$ is an irreducible ring of endomorphisms in the commutative group $\mathfrak{M}$ and $\overline{\mathfrak{B}}$ is a two-sided ideal $\neq 0$ in $\overline{\mathfrak{A}}$ we have seen that $x\overline{\mathfrak{B}} = \mathfrak{M}$ for every $x \neq 0$ in $\mathfrak{M}$. $\overline{\mathfrak{B}}$ is primitive also. If $\overline{\mathfrak{B}}_1$ and $\overline{\mathfrak{B}}_2$ are two-sided ideals $\neq 0$ then the equation $x\overline{\mathfrak{B}}_1 = \mathfrak{M}$ implies that $\overline{\mathfrak{B}}_1\overline{\mathfrak{B}}_2 \neq 0$. We state this as

LEMMA 4. *If* $\mathfrak{B}_1, \mathfrak{B}_2$ *are two-sided ideals* $\neq 0$ *in the primitive ring* $\mathfrak{A}$, $\mathfrak{B}_1\mathfrak{B}_2 \neq 0$.

This, of course, implies that $\mathfrak{B}_1 \wedge \mathfrak{B}_2 \neq 0$.

Suppose now that $\mathfrak{A}$ is a primitive ring that contains minimal right ideals. Since $\mathfrak{A}$ is semi-simple any minimal right ideal $\mathfrak{J}$ of $\mathfrak{A}$ has the form $e\mathfrak{A}$, $e^2 = e \neq 0$. Let $\mathfrak{F}$ be the two-sided ideal $\Sigma\mathfrak{J}$ for all minimal right ideals $\mathfrak{J}$. We assert that $\mathfrak{F}$ is a minimal two-sided ideal. For let $\mathfrak{F}_1$ be a two-sided ideal such that $\mathfrak{F}_1 < \mathfrak{F}$. There exists a minimal right ideal $\mathfrak{J} \nleqq \mathfrak{F}_1$. Since $\mathfrak{J} \wedge \mathfrak{F}_1$ is a right ideal and $\mathfrak{J}$ is minimal it follows that $\mathfrak{J} \wedge \mathfrak{F}_1 = 0$. Hence also $\mathfrak{J}\mathfrak{F}_1 = 0$. Since $\mathfrak{J}$ is not nilpotent $\mathfrak{A}\mathfrak{J} \neq 0$. Evidently $\mathfrak{A}\mathfrak{J}$ is a two-sided ideal and since $(\mathfrak{A}\mathfrak{J})\mathfrak{F}_1 = 0$, $\mathfrak{F}_1 = 0$ by the lemma. The minimality of $\mathfrak{F}$ has therefore been established. It follows from Lemma 4 that $\mathfrak{F}$ is contained in every two-sided ideal $\mathfrak{B} \neq 0$ of $\mathfrak{A}$. We wish to show next that $\mathfrak{F}$ is a simple ring. Let $\mathfrak{F}_1 < \mathfrak{F}$ now denote a two-sided ideal of $\mathfrak{F}$. Then $\mathfrak{F}\mathfrak{F}_1\mathfrak{F} \leq \mathfrak{F}_1 < \mathfrak{F}$ and $\mathfrak{F}\mathfrak{F}_1\mathfrak{F}$ is a two-sided ideal of $\mathfrak{A}$. Hence $\mathfrak{F}\mathfrak{F}_1\mathfrak{F} = 0$ and since $\mathfrak{F}$ is primitive $\mathfrak{F}_1 = 0$.

[18] See, for example, Jacobson [1], p. 64.

[19] This ideal has been called the anti-radical by Baer. See Baer [1], p. 544.

THEOREM 29. *Let $\mathfrak{A}$ be a primitive ring that contains minimal right ideals and let $\mathfrak{F} = \Sigma\mathfrak{J}$ the join of all the minimal right ideals of $\mathfrak{A}$. Then $\mathfrak{F}$ is a two-sided ideal that is contained in every two-sided ideal $\neq 0$ of $\mathfrak{A}$ and $\mathfrak{F}$ is a simple ring.*

If $\mathfrak{B}$ is a two-sided ideal $\neq 0$ of $\mathfrak{A}$, $\mathfrak{B}$ contains every minimal right ideal $\mathfrak{J}$ of $\mathfrak{A}$. It is easy to see by the above reasoning that $\mathfrak{J}$ is minimal in $\mathfrak{B}$. Hence $\mathfrak{B}$ satisfies the same conditions as $\mathfrak{A}$.

Again let $\bar{\mathfrak{A}}$ be an irreducible ring of endomorphisms in a commutative group $\mathfrak{M}$ and let $\mathfrak{D}$ be the division ring of endomorphisms in $\mathfrak{M}$ commutative with the elements of $\bar{\mathfrak{A}}$. We call a linear transformation A in a vector space $\mathfrak{M}$ over $\mathfrak{D}$ *finite valued* if the image space $\mathfrak{M}A$ is finite dimensional. We have shown elsewhere that if $\bar{\mathfrak{A}}$ contains finite valued linear transformations $\neq 0$ then $\bar{\mathfrak{A}}$ contains minimal right ideals.[20] Conversely suppose that $\bar{\mathfrak{A}}$ contains a minimal right ideal $\mathfrak{J}$. Then $\mathfrak{J} = \bar{e}\bar{\mathfrak{A}}$, $\bar{e}^2 = \bar{e} \neq 0$. We wish to show that $\mathfrak{M}\bar{e}$ is one dimensional. For, if not, there exist two $\mathfrak{D}$-independent elements x, y of $\mathfrak{M}$ such that $x\bar{e} = x$, $y\bar{e} = y$. By the density of $\bar{\mathfrak{A}}$ there exists a $\bar{b}$ in $\bar{\mathfrak{A}}$ such that $x\bar{b} = 0$, $y\bar{b} \neq 0$. Then $\bar{c} = \bar{e}\bar{b}$ satisfies $x\bar{c} = 0$, $y\bar{c} \neq 0$ and $\bar{c} \in \mathfrak{J}$. Hence if $\mathfrak{J}_x$ denotes the annihilator of x, $\mathfrak{J}_x \wedge \mathfrak{J} \neq 0$. Since $\mathfrak{J}$ is minimal $\mathfrak{J} = \mathfrak{J} \wedge \mathfrak{J}_x$. But this contradicts the fact that $\bar{e} \in \mathfrak{J}$ and $x\bar{e} \neq 0$.

We have shown also that if $\bar{\mathfrak{A}}$ contains finite valued transformations $\neq 0$ the totality $\bar{\mathfrak{F}}$ of these transformations is a two-sided ideal contained in every two-sided ideal $\neq 0$ of $\mathfrak{A}$.[21] Thus $\bar{\mathfrak{F}} = \Sigma\mathfrak{J}$ for all minimal right ideals $\mathfrak{J}$. This completes the proof of the following

THEOREM 30. *Let $\bar{\mathfrak{A}} \neq 0$ be an irreducible ring of endomorphisms in $\mathfrak{M}$ and let $\mathfrak{D}$ be the division ring of endomorphisms commutative with the elements of $\bar{\mathfrak{A}}$. Then $\bar{\mathfrak{A}}$ contains minimal right ideals if and only if it contains finite valued linear transformations $\neq 0$ of $\mathfrak{M}$ over $\mathfrak{D}$. If the condition is satisfied the minimal two-sided ideal $\bar{\mathfrak{F}}$ coincides with the totality of finite valued transformations in $\bar{\mathfrak{A}}$.*

We recall that a homomorphism $a \rightarrow \bar{a}$ between an abstract ring $\mathfrak{A}$ and a ring of endomorphisms $\bar{\mathfrak{A}}$ in a commutative group $\mathfrak{M}$ is called a *representation* of $\mathfrak{A}$. The representation is *irreducible* if $\bar{\mathfrak{A}}$ is irreducible. Two representations are *equivalent* if the groups in which the endomorphisms act are $\mathfrak{A}$-isomorphic. We have seen that the fundamental fact about a primitive ring is that it possesses a $(1 - 1)$ irreducible representation.

[20] Jacobson [2], p. 232.

[21] Jacobson [2], p. 230.

10

Suppose now that $\mathfrak{A}$ is a primitive ring that contains a minimal right ideal $\mathfrak{J}$ and let $a \rightarrow \bar{a}$ be any $(1-1)$ irreducible representation of $\mathfrak{A}$. Let $\bar{\mathfrak{A}}$ be the corresponding ring of endomorphisms and $\mathfrak{M}$ the group in which they act. The image $\bar{\mathfrak{J}}$ of $\mathfrak{J}$ is $\neq 0$. Hence there exists an x in $\mathfrak{M}$ such that $x\bar{\mathfrak{J}} \neq 0$. Because of the irreducibility of $\mathfrak{M}$, $x\bar{\mathfrak{J}} = \mathfrak{M}$. It can be verified directly that the correspondence $u \rightarrow xu$, u in $\mathfrak{J}$, is an $\mathfrak{A}$-isomorphism between $\mathfrak{J}$ and $\mathfrak{M}$. Thus the representation in $\mathfrak{M}$ is equivalent to the representation by the right multiplications in $\mathfrak{J}$. This proves

THEOREM 31. *If $\mathfrak{A}$ is a primitive ring that contains minimal right ideals, any two $(1-1)$ irreducible representations of $\mathfrak{A}$ are equivalent.*

Now let $\bar{\mathfrak{A}}_1$ and $\bar{\mathfrak{A}}_2$ be dense rings of linear transformations in $\mathfrak{M}_1$ over $\mathfrak{D}_1$ and $\mathfrak{M}_2$ over $\mathfrak{D}_2$ respectively. Suppose that the $\bar{\mathfrak{A}}_i$ are isomorphic under a correspondence $\bar{a}_1 \rightarrow \bar{a}_2$ and that the $\bar{\mathfrak{A}}_i$ contain minimal right ideals. Then if $\mathfrak{A}$ is the abstract primitive ring isomorphic to the $\bar{\mathfrak{A}}_i$ we have two irreducible representations $a \rightarrow \bar{a}_1$, $a \rightarrow \bar{a}_2$ of $\mathfrak{A}$. Since these representations are equivalent we have a $(1-1)$ correspondence $S : x_1 \rightarrow x_2 = x_1 S$ between $\mathfrak{M}_1$ and $\mathfrak{M}_2$ such that

$$(x_1 + y_1)S = x_1 S + y_1 S, \qquad (x_1 \bar{a}_1)S = (x_1 S)\bar{a}_2.$$

The latter equation may be written in the form $\bar{a}_2 = S^{-1}\bar{a}_1 S$. Now it is easy to see that $\mathfrak{D}_i$ is the complete set of endomorphisms in $\mathfrak{M}_i$ commutative with $\bar{\mathfrak{A}}_i$.[22] Hence the correspondence $\alpha_1 \rightarrow S^{-1}\alpha_1 S = \alpha_2$ for α_1 in $\mathfrak{D}_1$ is an isomorphism between $\mathfrak{D}_1$ and $\mathfrak{D}_2$. This shows that the two vector spaces are isomorphic. If we identify these spaces, setting $\mathfrak{M}_i = \mathfrak{M}$, $\mathfrak{D}_i = \mathfrak{D}$, then S becomes a semi-linear transformation in $\mathfrak{M}$ over $\mathfrak{D}$.

THEOREM 32. *Two dense rings of linear transformations that contain minimal right ideals can not be isomorphic unless the spaces in which they act are isomorphic. If $\bar{\mathfrak{A}}_1$ and $\bar{\mathfrak{A}}_2$ are two dense rings in $\mathfrak{M}$ over $\mathfrak{D}$ and the $\bar{\mathfrak{A}}_i$ contain minimal right ideals and are isomorphic under a correspondence $\bar{a}_1 \rightarrow \bar{a}_2$ then there exists a $(1-1)$ semi-linear transformation S in $\mathfrak{M}$ over $\mathfrak{D}$ such that $\bar{a}_2 = S^{-1}\bar{a}_1 S$ for all $\bar{a}_1$.*

This result shows that for primitive rings that contain minimal right ideals we have precisely the same uniqueness of representation as a dense ring of linear transformations as we have in the classical case of simple rings that satisfy the descending chain condition.

[22] Jacobson [2], p. 233.

14. Atomic semi-simple rings. Let $\mathfrak{A}$ be any ring that contains minimal right ideals and let $\mathfrak{F}$ be the join $\Sigma\mathfrak{J}$ of all the minimal right ideals $\mathfrak{J}$ of $\mathfrak{A}$. The right multiplications $u \rightarrow ua$ in $\mathfrak{J}$ form an irreducible ring and therefore a semi-simple ring. It follows that if $\mathfrak{R}$ is the radical of $\mathfrak{A}$ then $\mathfrak{J}\mathfrak{R} = 0$. Hence $\mathfrak{F}\mathfrak{R} = 0$.

We shall call a ring $\mathfrak{A}$ *atomic* if $\mathfrak{A} = \mathfrak{F}$. If $\mathfrak{A}$ is atomic $\mathfrak{A}$ is semi-simple if and only if $\mathfrak{A}$ contains no nilpotent ideals. Let $\mathfrak{A}$ be of this type and let $\mathfrak{J}$ be a minimal right ideal in $\mathfrak{A}$. Then we have the following

LEMMA 5. *If $\mathfrak{A}$ is atomic and semi-simple and $\mathfrak{J}$ is a minimal right ideal in $\mathfrak{A}$ then $\mathfrak{A}\mathfrak{J} \geq \mathfrak{J}$ and $\mathfrak{A}\mathfrak{J}$ is a simple ring.*

If $\mathfrak{B}$ is any two-sided ideal of $\mathfrak{A}$ either $\mathfrak{B} \geq \mathfrak{J}$ or $\mathfrak{J} \wedge \mathfrak{B} = 0 = \mathfrak{J}\mathfrak{B}$. Now if $\mathfrak{J}(\mathfrak{A}\mathfrak{J}) = 0$, $\mathfrak{J}^3 = 0$ contrary to the semi-simplicity. Hence $\mathfrak{A}\mathfrak{J} \geq \mathfrak{J}$. If $\mathfrak{B}$ is any two-sided ideal of $\mathfrak{A}$ properly contained in $\mathfrak{A}\mathfrak{J}$, $\mathfrak{B} \not\geq \mathfrak{J}$. Hence $\mathfrak{J}\mathfrak{B} = 0$, $(\mathfrak{A}\mathfrak{J})\mathfrak{B} = 0$ and $\mathfrak{B}^2 = 0$. Hence $\mathfrak{B} = 0$. Thus $\mathfrak{A}\mathfrak{J}$ is a minimal two-sided ideal of $\mathfrak{A}$. It follows readily that $\mathfrak{A}\mathfrak{J}$ is a simple ring.

Let $\{\mathfrak{A}_\alpha\}$ be the set of distinct two-sided ideals $\mathfrak{A}\mathfrak{J}$, $\mathfrak{J}$ minimal. Each $\mathfrak{A}_\alpha$ is simple. Hence if $\alpha \neq \beta$, $\mathfrak{A}_\alpha\mathfrak{A}_\beta = 0$. It follows that $\mathfrak{A}_\alpha \wedge \sum_{\beta\neq\alpha} \mathfrak{A}_\beta = 0$. By the lemma $\mathfrak{A} = \Sigma\mathfrak{A}_\alpha$. Hence we have proved

THEOREM 33. *If $\mathfrak{A}$ is atomic and semi-simple $\mathfrak{A}$ is isomorphic to a direct sum of simple rings.*

Each component $\mathfrak{A}\mathfrak{J}$ is a join of minimal right ideals $a\mathfrak{J}$ of $\mathfrak{J}$. It is easy to see that these ideals are also minimal in $\mathfrak{A}\mathfrak{J}$. Hence by Theorem 29 each $\mathfrak{A}\mathfrak{J}$ is isomorphic to a dense ring of finite-valued linear transformations.

THE JOHNS HOPKINS UNIVERSITY.

BIBLIOGRAPHY

A. A. Albert:
[1] *Modern Higher Algebra*, Chicago, 1937.

R. Baer:
[1] "Radical ideals," *American Journal of Mathematics*, vol. LXV (1943), pp. 537-568.

G. Birkhoff:
[1] "The radical of a group with operators," *Bulletin of the American Mathematical Society*, vol. 49 (1943), pp. 751-753.
[2] "Subdirect unions in universal algebra," *Bulletin of the American Mathematical Society*, vol. 50 (1944), pp. 764-769.

I. Gelfand:
[1] "Normierte Ringe," *Math. Sbornik*, vol. 9 (51, 1941), pp. 1-23.

N. Jacobson:
[1] "The Theory of Rings," *Mathematical Surveys*, vol. 2 (New York, 1943).
[2] "Structure theory of simple rings without finiteness assumptions," *Transactions of the American Mathematical Society*, vol. 57 (1945), pp. 228-245.

J. Levitzki:
[1] "On the radical of a general ring," *Bulletin of the American Mathematical Society*, vol. 49 (1943), pp. 461-466.
[2] "Semi-nilpotent ideals," *Duke Mathematical Journal*, vol. 10 (1943), pp. 553-556.

N. H. McCoy:
[1] "Subrings of direct sums," *American Journal of Mathematics*, vol. LX (1938), pp. 374-382.
[2] "Subrings of infinite direct sums," *Duke Mathematical Journal*, vol. 4 (1938), pp. 486-494.

N. H. McCoy and D. Montgomery:
[1] "A representation of generalized Boolean rings," *Duke Mathematical Journal*, vol. 3 (1937), pp. 455-459.

J. von Neumann:
[1] "On regular rings," *Proceedings of the National Academy of Sciences*, vol. 22 (1936), pp. 707-713.

S. Perlis:
[1] "A characterization of the radical of an algebra," *Bulletin of the American Mathematical Society*, vol. 48 (1942), pp. 128-132.

M. H. Stone:
[1] "The theory of representations for Boolean algebras," *Transactions of the American Mathematical Society*, vol. 40 (1936), pp. 37-111.

Reprinted from
American Journal of Mathematics
April 1945.

Annals of Mathematics
Vol. 46, No. 4, October, 1945

STRUCTURE THEORY FOR ALGEBRAIC ALGEBRAS OF BOUNDED DEGREE[1]

By N. Jacobson

(Received January 23, 1945)
(Revised March 13, 1945)

In a recent paper[2] we defined the radical for arbitrary rings and algebras and we developed a theory of semi-simple rings without chain conditions. In the present paper we specialize the theory to the case of algebras that are *algebraic* in the sense that every element satisfies a non-trivial equation with coefficients in the underlying field. For such an algebra we have shown that the radical is a nil ideal that contains all nil right (left) ideals. The present theory is primarily concerned with semi-simple algebras. The results are relatively complete for algebras in which the degrees of the elements have a finite maximum. Both in this case and in the more general case in which the degrees of the nilpotent elements are bounded the discussion can be reduced to that of division algebras. The theory of division algebras with elements of bounded degree can be reduced to the classical theory of division algebras with finite bases over the underlying field.

There is one other case in which we obtain fairly conclusive results, namely, that in which the underlying field Φ is finite. Here we prove that any algebraic division algebra over Φ is commutative. This leads to results on the commutativity of certain rings. One of these is the following generalization of the well-known theorem that any Boolean ring is commutative: Any ring in which every element satisfies an equation of the form $a^{n(a)} = a$, $n(a) > 1$, is commutative.

A second application of our theory is to the following question recently proposed by Kurosch:[3] Does every algebraic algebra with a finite number of generators have a finite basis? This problem is analogous to a well-known problem of Burnside's in the theory of groups. An affirmative answer to this question would have important consequences in the theory of algebraic algebras. It is easy to see that it would imply that if $\mathfrak{A}$ is algebraic then the matrix algebra $\mathfrak{A}_n$ is algebraic and the extension algebra $\mathfrak{A}_\Sigma$ is algebraic over Σ for any extension Σ of the underlying field Φ of $\mathfrak{A}$. All of these problems seem to be difficult.

Kurosch has proved that if $\mathfrak{A}$ is finitely generated and is algebraic with elements of degree $\leqq 3$ then $\mathfrak{A}$ has a finite basis. His proof is elementary. In the present paper we apply our structure theory to prove that any finitely generated semi-simple algebra with elements of bounded degree has a finite basis. Our results also enable us to reduce Kurosch's problem for algebras of bounded degree to this problem for the special case of nil algebras. The latter problem may

---

[1] Presented to the Society Oct. 28, 1944.
[2] Jacobson [3].
[3] Kurosch [1].

be reduced to the same problem for certain algebras that can be defined quite concretely. Moreover, we can show that an affirmative answer to Burnside's question on groups implies an affirmative answer to Kurosch's question for nil algebras over a field of characteristic $p \neq 0$.

1. General structure theory of algebras

In this section and the next we summarize the results of the structure theory of algebras that will be required in the present paper. The results stated here have been derived in the author's papers [2] and [3].

Let $\mathfrak{A}$ be an algebra of not necessarily finite dimensionality over a field Φ. We introduce in $\mathfrak{A}$ a new operation called *quasi-addition* defined as $a \oplus b = a + b + ab$. It is easy to verify that this operation is associative and that 0 acts as an identity relative to $\oplus$. An element z of $\mathfrak{A}$ is called *right quasi-regular* if it has a right inverse z' relative to $\oplus$, i.e. $z \oplus z' = 0$. If $\mathfrak{A}$ has an identity this amounts to saying that $1 + z$ has the right inverse $1 + z'$. A right ideal $\mathfrak{J}$ of $\mathfrak{A}$ is *quasi-regular* if all of its elements are right quasi-regular. It can be shown that the join $\mathfrak{N}$ of all the quasi-regular right ideals of $\mathfrak{A}$ is a two-sided ideal. We call $\mathfrak{N}$ the *radical* of $\mathfrak{A}$. We know also that $\mathfrak{N}$ is the join of all the quasi-regular left ideals if left quasi-regularity and quasi-regular left ideals are defined in the obvious way.

An algebra is called a *radical algebra* if $\mathfrak{A} = \mathfrak{N}$. If $\mathfrak{A}$ has this property, $\mathfrak{A}$ is a group relative to quasi-addition. If $\mathfrak{A}$ is not a radical algebra $\mathfrak{A}$ contains maximal right (left) ideals.

An algebra $\mathfrak{A}$ is *semi-simple* if its radical $\mathfrak{N} = 0$. If $\mathfrak{N}$ is the radical of any algebra $\mathfrak{A}$, then the difference algebra $\bar{\mathfrak{A}} = \mathfrak{A} - \mathfrak{N}$ is semi-simple. The radical $\mathfrak{N}$ itself is a radical algebra and so to a certain extent we can reduce problems on algebras to the consideration of the two extreme cases: radical algebras and semi-simple algebras.

Any semi-simple algebra contains maximal right ideals. Moreover, it is known that the intersection $\Pi\mathfrak{J}$ of all the maximal right ideals of $\mathfrak{A}$ is the 0-ideal. A similar result holds for left ideals.

If $\mathfrak{J}$ is a right ideal of $\mathfrak{A}$ we define the *quotient* $(\mathfrak{J}:\mathfrak{A})$ to be the totality of elements b such that $xb \in \mathfrak{J}$ for all x in $\mathfrak{A}$. $(\mathfrak{J}:\mathfrak{A})$ is a two-sided ideal and if $\mathfrak{A}$ has an identity then $(\mathfrak{J}:\mathfrak{A})$ is the largest two-sided ideal of $\mathfrak{A}$ contained in $\mathfrak{J}$.

Since any ideal in an algebra is a Φ-subspace it is clear that the difference group $\mathfrak{M} = \mathfrak{A} - \mathfrak{J}$ is a space over Φ. If $a \in \mathfrak{A}$ we define $\bar{a}$ to be the mapping $[x] = x + \mathfrak{J} \rightarrow [x]\bar{a} = xa + \mathfrak{J}$. Since $\mathfrak{J}$ is a right ideal $\bar{a}$ is single-valued. The mapping $\bar{a}$ is a linear transformation in $\mathfrak{M}$ over Φ and the totality $\bar{\mathfrak{A}}$ of these mappings is a subalgebra of the complete algebra of linear transformations in $\mathfrak{M}$. The correspondence $a \rightarrow \bar{a}$ is a homomorphism between $\mathfrak{A}$ and $\bar{\mathfrak{A}}$. The kernel of this homomorphism is $(\mathfrak{J}:\mathfrak{A})$. Hence $\bar{\mathfrak{A}} \cong \mathfrak{A} - (\mathfrak{J}:\mathfrak{A})$.

An algebra $\mathfrak{A}$ is called *primitive* if $\mathfrak{A}$ contains a maximal right ideal $\mathfrak{J}$ whose quotient $(\mathfrak{J}:\mathfrak{A}) = 0$. If $\mathfrak{A}$ is primitive $\mathfrak{A} \cong \bar{\mathfrak{A}}$ an irreducible algebra of linear transformations. The converse holds also: any irreducible algebra of linear transformations is primitive. This implies that if $\mathfrak{A}$ is an arbitrary algebra and $\mathfrak{J}$ is a maximal right ideal then $\mathfrak{A} - (\mathfrak{J}:\mathfrak{A})$ is primitive.

If $\mathfrak{A}$ is semi-simple it is known that the intersection $\Pi(\mathfrak{J}:\mathfrak{A})$ where $\mathfrak{J}$ ranges over the maximal ideals is the 0-ideal. Each $\mathfrak{A} - (\mathfrak{J}:\mathfrak{A})$ is primitive. This result yields the necessity part of the following:

THEOREM 1. *A necessary and sufficient condition that an algebra $\mathfrak{A}$ be semi-simple is that $\mathfrak{A}$ contain a set of two-sided ideals $\mathfrak{B}$ such that* 1) *each $\mathfrak{A} - \mathfrak{B}$ is primitive and* 2) *the intersection $\Pi\,\mathfrak{B} = 0$.*

This theorem can also be formulated as a theorem on direct sums. We recall the definition of this concept. Let S be an arbitrary set whose elements P, Q, $\cdots$ we call points, and suppose that with each $P \in S$ there is associated an algebra $\mathfrak{E}_P$ over Φ. Let $(S, \mathfrak{E}_P)$ be the totality of functions $f(P)$ where $P \in S$ and $f(P) \in \mathfrak{E}_P$. We define $f + g$ by $(f + g)\,(P) = f(P) + g(P)$, $f\alpha$ for α in Φ by $(f\alpha)(P) = f(P)\alpha$ and fg by $(fg)(P) = f(P)g(P)$. It is easy to verify that $(S, \mathfrak{E}_P)$ is an algebra. We call this algebra the *complete direct sum* of the algebras $\mathfrak{E}_P$.

If $\mathfrak{A}$ is a subalgebra of $(S, \mathfrak{E}_P)$ we define the *component* $\mathfrak{A}_P$ of $\mathfrak{A}$ at P to be the totality of elements $f(P)$ such that $f \in \mathfrak{A}$. Evidently $\mathfrak{A}_P$ is a subalgebra of $\mathfrak{E}_P$. Clearly $\mathfrak{A}$ is also a subalgebra of the complete direct sum $(S, \mathfrak{A}_P)$. Hence in studying a single subalgebra $\mathfrak{A}$ of a complete direct sum we can always suppose that the components $\mathfrak{A}_P$ of $\mathfrak{A}$ are the algebras used to construct the direct sum. The correspondence $f \rightarrow f(P)$ is a homomorphism between $\mathfrak{A}$ and $\mathfrak{A}_P$. We may now state Theorem 1 in the following alternate form.

THEOREM 1′. *A necessary and sufficient condition that $\mathfrak{A}$ be semi-simple is that $\mathfrak{A}$ be isomorphic to a subalgebra $\bar{\mathfrak{A}}$ of a complete direct sum $(S, \bar{\mathfrak{A}}_P)$ in which the components $\bar{\mathfrak{A}}_P$ of $\bar{\mathfrak{A}}$ are primitive.*

In this form the theorem appears to be an extension of Wedderburn's first structure theorem on semi-simple algebras with a finite basis. To a certain extent this result reduces the study of semi-simple algebras to that of primitive algebras.

2. Structure of primitive algebras

We have noted that $\mathfrak{A}$ is primitive if and only if $\mathfrak{A}$ is isomorphic to an irreducible algebra $\bar{\mathfrak{A}}$ of linear transformations in a vector space $\mathfrak{M}$ over Φ. The assumption that $\mathfrak{M}$ is an irreducible vector space relative to $\bar{\mathfrak{A}}$ means that there are no proper subspaces of $\mathfrak{M}$ invariant under all the $\bar{a} \in \bar{\mathfrak{A}}$. Thus there are no proper subgroups of $\mathfrak{M}$ invariant under all the scalar multiplications and under all the elements of $\bar{\mathfrak{A}}$. Hence if $\bar{\mathfrak{A}}^*$ denotes the algebra that is obtained by adjoining the scalar multiples of the identity mapping to $\bar{\mathfrak{A}}$, then there are no proper subgroups of $\mathfrak{M}$ invariant under $\bar{\mathfrak{A}}^*$.

If $\mathfrak{A}$ has an identity we may suppose that $\bar{\mathfrak{A}}$ contains the identity mapping and therefore that $\bar{\mathfrak{A}} = \bar{\mathfrak{A}}^*$. If $\mathfrak{A}$ does not have an identity, $\bar{\mathfrak{A}} < \bar{\mathfrak{A}}^*$. It is easy to see that here $\bar{\mathfrak{A}}^*$ is isomorphic to the algebra $\mathfrak{A}^* = \mathfrak{A} + (1)$ obtained by adjoining an identity to $\mathfrak{A}$. Since $\mathfrak{A}^* \cong \bar{\mathfrak{A}}^*$ and the latter is an irreducible algebra of linear transformations, $\mathfrak{A}^*$ is primitive. We remark also that $\mathfrak{A}$ is a two-sided ideal in $\mathfrak{A}^*$ and that $\mathfrak{A}^* - \mathfrak{A} \cong \Phi$.

Let $\mathfrak{D}$ be the totality of endomorphisms in $\mathfrak{M}$ that commute with the elements of $\bar{\mathfrak{A}}^*$. By Schur's lemma $\mathfrak{D}$ is a division ring. Since $\mathfrak{D} \geqq \Phi \geqq 1$ and Φ is in the center of $\mathfrak{D}$, $\mathfrak{D}$ is a subalgebra of the complete algebra of linear transformations of $\mathfrak{M}$ over Φ. The space $\mathfrak{M}$ may be regarded as a vector space over $\mathfrak{D}$. When this is done it is known that $\bar{\mathfrak{A}}^*$ is a *dense* set of linear transformations in the following sense. If $u_1, \cdots, u_k$ are any k $\mathfrak{D}$-independent vectors and $v_1, \cdots, v_k$ are arbitrary vectors in $\mathfrak{M}$, then there exists a linear transformation $\bar{a}^*$ in $\bar{\mathfrak{A}}^*$ such that $u_i\bar{a}^* = v_i$, $i = 1, \cdots, k$. We therefore have the following structure theorem for primitive algebras:

THEOREM 2. *If $\mathfrak{A}$ is a primitive algebra with an identity, $\mathfrak{A}$ is isomorphic to a dense algebra of linear transformations in a vector space $\mathfrak{M}$ over a division algebra $\mathfrak{D}$. If $\mathfrak{A}$ is primitive and does not have an identity, then the algebra $\mathfrak{A}^* = \mathfrak{A} + (1)$ is primitive and has the described structure.*

If $\mathfrak{M}$ is finite dimensional over $\mathfrak{D}$, the density of $\bar{\mathfrak{A}}^*$ implies that $\bar{\mathfrak{A}}^* = \mathfrak{L}$ the complete algebra of linear transformations in $\mathfrak{M}$ over $\mathfrak{D}$. It is known that $\mathfrak{L}$ is isomorphic to the matrix algebra $\mathfrak{D}'_h$ where $\mathfrak{D}'$ is the algebra anti-isomorphic to $\mathfrak{D}$ and h is the dimensionality of $\mathfrak{M}$ over $\mathfrak{D}$. Since $\mathfrak{D}'_h$ is simple and $\bar{\mathfrak{A}}$ is two-sided ideal in $\bar{\mathfrak{A}}^*$, $\mathfrak{A} = \mathfrak{L}$. This situation always obtains for algebras with a finite basis and Theorem 2 therefore specializes to Wedderburn's second structure theorem.

If $\mathfrak{M}$ is infinite dimensional over $\mathfrak{D}$, let $u_1, u_2, \cdots$ be an infinite sequence of linearly independent elements of $\mathfrak{M}$ and let $\mathfrak{M}_k$ denote the $\mathfrak{D}$-subspace generated by $u_1, \cdots, u_k$. Let $\bar{\mathfrak{A}}_k^*$ be the subalgebra of $\bar{\mathfrak{A}}^*$ that leaves the subspace $\mathfrak{M}_k$ invariant. Then the density of $\bar{\mathfrak{A}}^*$ implies that the elements of $\bar{\mathfrak{A}}_k^*$ induce every linear transformation in $\mathfrak{M}_k$ over $\mathfrak{D}$. It follows that $\bar{\mathfrak{A}}_k^*$ is homomorphic to $\mathfrak{L}_k$ the complete algebra of linear transformations in $\mathfrak{M}_k$ over $\mathfrak{D}$. Hence $\mathfrak{A}_k^*$ is homomorphic to $\mathfrak{D}'_k$. Thus for each $k = 1, 2, 3, \cdots$ $\bar{\mathfrak{A}}^*$ contains a subalgebra homomorphic to the matrix algebra $\mathfrak{D}'_k$. This proves the following

THEOREM 3. *If $\mathfrak{A}$ is a primitive algebra with an identity there exists a division algebra $\mathfrak{D}'$ such that either $\mathfrak{A} \cong \mathfrak{D}'_h$ for some h or for each $k = 1, 2, \cdots$ $\mathfrak{A}$ contains a subalgebra homomorphic to $\mathfrak{D}'_k$. If $\mathfrak{A}$ is primitive without an identity there exists a division algebra $\mathfrak{D}'$ such that for each k, $\mathfrak{A}^* = \mathfrak{A} + (1)$ contains a subalgebra homomorphic to $\mathfrak{D}'_k$.*

3. Algebraic algebras

We suppose now that $\mathfrak{A}$ is an algebraic algebra. Then it is known that the radical $\mathfrak{N}$ of $\mathfrak{A}$ is a nil ideal that contains every nil right (left) ideal of $\mathfrak{A}$. $\mathfrak{A}$ is semi-simple if and only if it contains no nil right (left) ideals.

If $\mathfrak{A}$ has an identity we define the *minimum polynomial* $\mu(\lambda)$ of the element a to be the polynomial of least degree of the form $\lambda^m + \lambda^{m-1}\alpha_1 + \cdots + \alpha_m$ having a for a root. If $\mathfrak{A}$ does not have an identity $\mu(\lambda)$ is the polynomial of least degree of the form $\lambda^m + \lambda^{m-1}\alpha_1 + \cdots + \lambda\alpha_{m-1}$ having a for a root. In either case if $\nu(\lambda)$ is a polynomial such that $\nu(a) = 0$ then $\mu(\lambda)$ is a factor of $\nu(\lambda)$. The degree of $\mu(\lambda)$ is called the *degree of the element a*. Thus if $\mathfrak{A}$ has an identity, the degree of a is the dimensionality of the algebra A^* generated by a and 1 and if $\mathfrak{A}$ does not have an identity, the degree of a exceeds by 1 the dimensionality of the algebra A generated by a.

Evidently if $\mathfrak{A}$ is algebraic, any subalgebra $\mathfrak{B}$ of $\mathfrak{A}$ and any homomorphic image $\bar{\mathfrak{A}}$ of $\mathfrak{A}$ is algebraic. Consider now the algebra $\mathfrak{A}^* = \mathfrak{A} + (1)$ obtained by adjoining an identity to $\mathfrak{A}$. If $a^* \epsilon \mathfrak{A}^*$, $a^* = a + 1\alpha$ where $a \epsilon \mathfrak{A}$. Hence if $\mu(\lambda)$ is a polynomial having a for a root, then $\mu(\lambda - \alpha)$ has a^* for a root. Thus $\mathfrak{A}^*$ is algebraic also.

These remarks show that if $\mathfrak{A}$ is a semi-simple algebraic algebra then the primitive components $\bar{\mathfrak{A}}_P$ of Theorem 1′ are also algebraic. If $\mathfrak{A}$ is primitive and algebraic the matrix algebras $\mathfrak{D}_h'$ and $\mathfrak{D}_k'$ of Theorem 3 are homomorphic images of subalgebras of $\mathfrak{A}$ or of $\mathfrak{A}^*$. Hence these matrix algebras are algebraic. Consequently $\mathfrak{D}'$ is an algebraic algebra.

We shall call an algebra $\mathfrak{A}$ of *bounded degree* if the degrees of its elements have a maximum value. We denote this bound as $B(\mathfrak{A})$. It is clear that if $\mathfrak{B}$ is a subalgebra and $\bar{\mathfrak{A}}$ is a homomorphic image, then $\mathfrak{B}$ and $\bar{\mathfrak{A}}$ are bounded and $B(\mathfrak{B}) \leqq B(\mathfrak{A})$ and $B(\bar{\mathfrak{A}}) \leqq B(\mathfrak{A})$. Also the algebra $\mathfrak{A}^*$ is bounded and $B(\mathfrak{A}^*) = B(\mathfrak{A})$.

Though the most extensive part of our discussion will be concerned with algebraic algebras of bounded degree, a substantial part will be valid for a wider class of algebraic algebras, namely, those algebras in which the index of nilpotency (i.e. degree) of the nilpotent elements are bounded. Such an algebra will be called an algebra of *bounded index* and we shall denote the maximum index of nilpotency as $N(\mathfrak{A})$. As in the case of algebras of bounded degree it is evident that if $\mathfrak{A}$ is of bounded index then so is any subalgebra $\mathfrak{B}$ and if $\mathfrak{A}$ does not have an identity then the algebra $\mathfrak{A}^* = \mathfrak{A} + (1)$ is of bounded index. Here, too, $N(\mathfrak{B}) \leqq N(\mathfrak{A})$ and $N(\mathfrak{A}^*) = N(\mathfrak{A})$. We now wish to prove the following

LEMMA 1. *If $\mathfrak{A}$ is algebraic of bounded index then any homomorphic image $\bar{\mathfrak{A}}$ of $\mathfrak{A}$ is algebraic of bounded index. The index $N(\bar{\mathfrak{A}}) \leqq N(\mathfrak{A})$.*

To prove this lemma we require the concept of the *index* of an arbitrary element a of an algebraic algebra. We define this to be the exponent r of the highest power of λ that divides the minimum polynomial $\mu(\lambda)$ of a. Let $\mu(\lambda) = \nu(\lambda)\lambda^r$ and let $z = \phi(a)$ where $\phi(\lambda) = \nu(\lambda)\lambda$. Then $z^r = 0$ and since $\phi(\lambda)^{r-1}$ is not divisible by $\mu(\lambda)$, $z^{r-1} \neq 0$. Thus if $\mathfrak{A}$ contains any element of index r, it contains a nilpotent element of index r. Hence if $\mathfrak{A}$ is of bounded index, the indices of all the elements of $\mathfrak{A}$ are $\leqq N(\mathfrak{A})$. It follows that the indices of all the elements of $\bar{\mathfrak{A}}$ are $\leqq N(\mathfrak{A})$ and this proves the lemma.

If z is a nilpotent element of index 1, $z = 0$ and conversely. This shows that $\mathfrak{A}$ has no nilpotent elements $\neq 0$ if and only if it is of bounded index with bound $N(\mathfrak{A}) = 1$. Hence we have the

COROLLARY. If $\mathfrak{A}$ *is an algebraic algebra without nilpotent elements* $\neq 0$, *then any homomorphic image* $\bar{\mathfrak{A}}$ *of* $\mathfrak{A}$ *is algebraic and has no nilpotent elements* $\neq 0$.

4. Semi-simple algebraic algebras of bounded index

THEOREM 4. *If $\mathfrak{A}$ is a primitive algebraic algebra of bounded index, $\mathfrak{A} \cong \mathfrak{D}'_h$ where $\mathfrak{D}'$ is an algebraic division algebra and $h \leqq N(\mathfrak{A})$.*

By Theorem 2 $\mathfrak{A}$ has an identity. For otherwise $\mathfrak{A}^* = \mathfrak{A} + (1)$ contains for each positive integer k a subalgebra homomorphic to $\mathfrak{D}'_k$, $\mathfrak{D}'$ a division algebra. By Lemma 1, $N(\mathfrak{D}'_k) \leqq N(\mathfrak{A})$. Hence also for the matrix algebra Φ_k, $N(\Phi_k) \leqq N(\mathfrak{A})$. On the other hand $N(\Phi_k) = k$ and this contradicts the fact that k is arbitrary. The same argument shows that $\mathfrak{A}$ does not contain a subalgebra homomorphic to $\mathfrak{D}'_k$ with $k > N(\mathfrak{A})$. The only other alternative by Theorem 2 is that $\mathfrak{A} \cong \mathfrak{D}'_h$ with $h \leqq N(\mathfrak{A})$.

This theorem implies that any primitive algebraic algebra of bounded index contains an identity and is simple. Conversely if $\mathfrak{A}$ is a simple algebraic algebra and $\mathfrak{A}$ is not a nil algebra, then $\mathfrak{A}$ contains a maximal right ideal $\mathfrak{J}$. Then $(\mathfrak{J}:\mathfrak{A}) = 0$. For otherwise $(\mathfrak{J}:\mathfrak{A}) = \mathfrak{A}$ and $\mathfrak{A}^2 \leqq \mathfrak{J} < \mathfrak{A}$ so that $\mathfrak{A}^2 = 0$. Thus $\mathfrak{A}$ is primitive. The structure of arbitrary simple nil algebras or even of simple nil algebras of bounded index has yet to be determined. Also it is not known whether or not the converse of Theorem 4 holds, namely, if $\mathfrak{A} = \mathfrak{D}'_h$ where $\mathfrak{D}'$ is an algebraic division algebra, then is $\mathfrak{A}$ necessarily algebraic of bounded index? A special case of Theorem 4 is the

COROLLARY. *Any primitive algebraic algebra without nilpotent elements $\neq 0$ is a division algebra.*

Theorem 4 applies also to algebras of bounded degree. In this case $\mathfrak{D}'$ has bounded degree. This proves

THEOREM 5. *If $\mathfrak{A}$ is a primitive algebraic algebra of bounded degree, $\mathfrak{A} \cong \mathfrak{D}'_h$ where $\mathfrak{D}'$ is an algebraic division algebra of bounded degree with $B(\mathfrak{D}') \leqq B(\mathfrak{A})$ and $h \leqq N(\mathfrak{A})$.*

By Theorems 1′ and 4 we have the following

THEOREM 6. *If $\mathfrak{A}$ is a semi-simple algebraic algebra of bounded index, $\mathfrak{A}$ is isomorphic to a subalgebra $\bar{\mathfrak{A}}$ of a complete direct sum $(S, \bar{\mathfrak{A}}_P)$ where each component $\bar{\mathfrak{A}}_P$ of $\mathfrak{A}$ is a matrix algebra $\mathfrak{D}'_h$, $\mathfrak{D}'$ a division algebra and $h \leqq N(\mathfrak{A})$.*

COROLLARY. *Any algebraic algebra without nilpotent elements $\neq 0$ is a subalgebra of a complete direct sum of division algebras.*

It is easy to see that the converses of these results do not hold.

5. Algebraic division algebras

Let $\mathfrak{D}'$ be an algebraic division algebra of bounded degree. If a is an element $\neq 0$ in $\mathfrak{D}'$ the algebra A generated by a contains 1. Hence degree a = dimensionality of A.

We suppose first that $\mathfrak{D}'$ is central in the sense that the center of $\mathfrak{D}'$ is Φ. As usual if $\mathfrak{B}$ is a subalgebra of $\mathfrak{D}'$ we shall call the subalgebra $\bar{\mathfrak{B}}$ of elements $\bar{b}$ that commute with every $b \in \mathfrak{B}$ the *centralizer of* $\mathfrak{B}$. We wish to prove the following

LEMMA 2. *$\mathfrak{D}'$ contains a finite and separable subfield $\mathfrak{F}$ whose centralizer $\bar{\mathfrak{F}} = \mathfrak{F}$.*

PROOF. If the characteristic of Φ is 0 we take $\mathfrak{F} = \Phi(u)$ where u is an element of $\mathfrak{D}'$ of maximum degree. Then $\bar{\mathfrak{F}} = \mathfrak{F}$. For otherwise there is a $v \in \bar{\mathfrak{F}}, \notin \mathfrak{F}$

and $\mathfrak{F}(v)$ is a field properly containing $\mathfrak{F}$. Since $\mathfrak{F}(v)$ is separable over Φ, $\mathfrak{F}(v) = \Phi(w)$ and the degree of w exceeds that of u contrary to assumption. Now let the characteristic of Φ be $p \neq 0$. We prove first that if $\mathfrak{D}' > \Phi$ then $\mathfrak{D}'$ contains a separable field $\Phi(u_1) > \Phi$. For otherwise any subfield properly containing Φ is purely inseparable and hence $\mathfrak{D}'$ contains a subfield $\Phi(v) > \Phi$ where $v^p = \gamma \in \Phi$. If u is any element of $\mathfrak{D}'$ we define $u' = [u, v] = uv - vu$ and if $k > 1$ we define $u^{(k)} = (u^{(k-1)})'$. Since $v \notin \Phi$ there exists an element w such that $w' \neq 0$. Since $v^p = \gamma$, $w^{(p)} = [\cdots \overbrace{[w, v] \cdots v]}^{p} = [w, v^p] = 0$.[4] Hence we may determine the integer $k > 1$ such that $w^{(k)} = 0$ but $w^{(k-1)} \neq 0$. If we set $y = w^{(k-2)}(w^{(k-1)})^{-1}$ we find that $y' = (w^{(k-2)})'(w^{(k-1)})^{-1} + w^{(k-2)}((w^{(k-1)})^{-1})' = w^{(k-1)}(w^{(k-1)})^{-1} + 0 = 1$. This implies that if $z = yv$ then $z' = v$. Thus $zv - vz = v$ and $v^{-1}zv = z + 1$. This shows that the field $\Phi(z)$ has a non-trivial automorphism and therefore it can not be purely inseparable over Φ. We have therefore proved that $\mathfrak{D}'$ contains a separable subfield $\Phi(u_1) > \Phi$. We now consider the centralizer $\bar{\mathfrak{B}}_1$ of $\mathfrak{B}_1 = \Phi(u_1)$. It is known that $\bar{\bar{\mathfrak{B}}}_1 = \mathfrak{B}_1$.[5] Hence the center of $\bar{\mathfrak{B}}_1$ is $\mathfrak{B}_1$. If $\bar{\mathfrak{B}}_1 > \mathfrak{B}_1$ we may repeat our argument to obtain a field $\mathfrak{B}_2 > \mathfrak{B}_1$ which is finite and separable over $\mathfrak{B}_1$. Hence $\mathfrak{B}_2 = \Phi(u_2)$ and the degree of u_2 exceeds that of u_1. Since the degrees of the elements of $\mathfrak{D}'$ are bounded, this process leads in a finite number of steps to a field $\mathfrak{F} = \Phi(u_m)$ finite and separable over Φ such that $\bar{\mathfrak{F}} = \mathfrak{F}$.

Now it has been proved by Artin and Whaples[6] that if $\mathfrak{F}$ is a finite dimensional subalgebra of a central division algebra $\mathfrak{D}'$ over Φ, then the right (left) dimensionality $(\mathfrak{D}' : \bar{\mathfrak{F}}) = (\mathfrak{F} : \Phi)$. Hence if $\mathfrak{D}'$ is algebraic of bounded degree and $\mathfrak{F}$ is chosen as in the lemma, then $(\mathfrak{D}' : \mathfrak{F}) = (\mathfrak{F} : \Phi)$. Then $(\mathfrak{D}' : \Phi) = (\mathfrak{D}' : \mathfrak{F})(\mathfrak{F} : \Phi) = (\mathfrak{F} : \Phi)^2$. This proves

THEOREM 7. *Let $\mathfrak{D}'$ be an algebraic division algebra of bounded degree over Φ. Then if $\mathfrak{D}'$ is central, $\mathfrak{D}'$ is finite dimensional over Φ.*

If $\mathfrak{D}'$ is any division algebra, its center $\mathfrak{C}$ is a field and $\mathfrak{D}'$ may be regarded as a central algebra over $\mathfrak{C}$. Evidently if $\mathfrak{D}$ is algebraic of bounded degree over Φ it has these properties over $\mathfrak{C}$ and $\mathfrak{C}$ has these properties over Φ. Now it is easy to see that a field $\mathfrak{C}$ which is algebraic of bounded degree is either finite dimensional or is inseparable. Hence we have the

COROLLARY. *If $\mathfrak{D}'$ is an algebraic division algebra of bounded degree over a perfect field Φ, then $\mathfrak{D}'$ has a finite basis over Φ.*

We now drop the restriction that $\mathfrak{D}'$ has bounded degree but we assume that Φ is a finite field. Then we have the following theorem:

THEOREM 8. *Any algebraic division algebra $\mathfrak{D}'$ over a finite field is commutative.*

Let $\mathfrak{C}$ be the center of $\mathfrak{D}'$. If $\mathfrak{D}' > \mathfrak{C}$ let x be an element of $\mathfrak{D}'$ not contained in $\mathfrak{C}$. Then $\mathfrak{C}(x)$ is a field properly containing $\mathfrak{C}$. Hence also $\Phi(x) > \Phi$ and since Φ is finite and x is algebraic over Φ, $\Phi(x)$ is cyclic over Φ with a generating

[4] Cf. Jacobson [1] p. 102.

[5] See for example Jacobson [1] p. 104.

[6] Artin and Whaples [1] p. 104.

automorphism of the form $a \to a^{p^e}$. It follows that $\mathfrak{C}(x)$ is cyclic over $\mathfrak{C}$ with a generating automorphism U of the form $a \to a^{p^f}$. Now it is known that U can be extended to an inner automorphism in $\mathfrak{D}'$.[7] Thus there exists an element y in $\mathfrak{D}'$ such that $y^{-1}xy = x^{p^f}$. It follows that the subalgebra $\mathfrak{B} = \Phi(x, y)$ generated by x and y has a finite basis consisting of certain of the products $y^i x^j$. Thus $\mathfrak{B}$ has a finite number of elements. Since $\mathfrak{B}$ is not commutative this contradicts a known theorem due to Wedderburn. Hence $\mathfrak{D}' = \mathfrak{C}$ is commutative.

6. Commutativity of certain algebras

The preceding theorem and the corollary to Theorem 6 imply

THEOREM 9. *Any algebraic algebra without nilpotent elements $\neq 0$ over a finite field is commutative.*

Let Ω be the algebraic closure of the finite field Φ and consider the representation of $\mathfrak{A}$ as a subalgebra $\bar{\mathfrak{A}}$ of $(S, \bar{\mathfrak{A}}_P)$ where the $\bar{\mathfrak{A}}_P$ are algebraic fields over Φ. The $\bar{\mathfrak{A}}_P$ may be regarded as subalgebras of Ω. Hence $\bar{\mathfrak{A}}$ may be regarded as a subalgebra of the complete direct sum $\mathfrak{S} = (S, \Omega)$. Here $\mathfrak{S}$ is the totality of functions that have S for domain and Ω for range. If S is infinite $\mathfrak{S}$ is not an algebraic algebra. However, the totality $\mathfrak{S}_0$ of algebraic elements of $\mathfrak{S}$ is a subalgebra and $\bar{\mathfrak{A}}$ is a subalgebra of $\mathfrak{S}_0$. The converse is clear. Hence we have the following

THEOREM 10. *A necessary and sufficient condition that an algebraic algebra $\mathfrak{A}$ over a finite field Φ have no nilpotent elements $\neq 0$ is that $\mathfrak{A}$ be isomorphic to a subalgebra $\bar{\mathfrak{A}}$ of the algebra $\mathfrak{S}_0$ of algebraic elements in the complete direct sum $\mathfrak{S} = (S, \Omega)$ where Ω is the algebraic closure of Φ.*

We have seen that $\mathfrak{A}$ contains no nilpotent elements $\neq 0$ if and only if every element of $\mathfrak{A}$ is a root of a polynomial of the form $\mu(\lambda) = \lambda^m + \cdots + \lambda\alpha_{m-1}$ in which $\alpha_{m-1} \neq 0$. It is easy to see that such a polynomial is a factor of some polynomial of the form $\lambda^n - \lambda$. Hence each a satisfies an equation of the form $a^n = a$, $n > 1$. Theorem 9 may therefore be formulated in the alternate form: *If $\mathfrak{A}$ is an algebra over a finite field and every element of $\mathfrak{A}$ satisfies an equation of the form $a^n = a$, $n > 1$, then $\mathfrak{A}$ is commutative.* In this form the theorem holds also for rings.

THEOREM 11. *Let $\mathfrak{R}$ be a ring in which every element satisfies an equation of the form $a^{n(a)} = a$, $n(a)$ an integer > 1. Then every element of $\mathfrak{R}$ has finite additive order and $\mathfrak{R}$ is commutative.*[8]

PROOF. If $a^n = a$, $a^{2n-1} = a, \cdots, a^{l(n-1)+1} = a$. Hence if b is a second element of $\mathfrak{R}$ and $b^m = b$, then $a^r = a$ and $b^r = b$ for $r = (m-1)(n-1) + 1$. We apply this to $b = ka$ where k is a positive integer. Then $(ka) = (ka)^r = k^r a^r = k^r a$. Hence $(k^r - k)a = 0$ and so the order of a in the additive group of $\mathfrak{R}$ is finite. If $h(a)$ is the order of a, $h(a)$ is a factor of $k^r - k$. Here k is arbitrary

---

[7] See for example Jacobson [1] p. 46.

[8] In an earlier version of this paper the present theorem was stated for rings in which every element satisfies an equation $a^n = a$, n fixed. I am indebted to Dr. I. Kaplansky for pointing out to me that the present more general theorem can be proved by our method.

but r depends on k. Since $p^s - p$ is divisible by p to the first power only, $h(a)$ is a product of distinct primes. Now let $\mathfrak{R}_i$ be the subgroup of $\mathfrak{R}$ consisting of the elements of order p_i, p_i the ith prime. Since every element of $\mathfrak{R}$ has an order of the form $p_1p_2 \cdots p_s$ where the p_i are distinct, $\mathfrak{R}$ is a direct sum of the $\mathfrak{R}_i$. Each $\mathfrak{R}_i$ is a two-sided ideal and satisfies the same conditions as $\mathfrak{R}$. Since $p_ia = 0$ for every a in $\mathfrak{R}_i$, $\mathfrak{R}_i$ may be regarded as an algebra over $\mathfrak{J}_i$ the field of residues mod p_i. By the alternate form of Theorem 9 $\mathfrak{R}_i$ is commutative. Since $\mathfrak{R}$ is the direct sum in the ordinary sense of the rings $\mathfrak{R}_i$, $\mathfrak{R}$ is commutative.

7. Semi-simple algebraic algebras of bounded degree

We suppose that $\mathfrak{A}$ is any semi-simple algebraic algebra of bounded degree and we consider any representation of $\mathfrak{A}$ as a subalgebra $\bar{\mathfrak{A}}$ of the complete direct sum $(S, \bar{\mathfrak{A}}_P)$ of primitive algebras $\bar{\mathfrak{A}}_P$. We have seen that the components $\bar{\mathfrak{A}}_P$ have the form $\mathfrak{D}'_h$, $\mathfrak{D}'$ a division algebra and $h \leqq B(A)$. By Theorem 7 $\mathfrak{D}'$ has a finite basis over its center $\mathfrak{C}$. Finally $\mathfrak{C}$ is a field which is algebraic and of bounded degree over the underlying field Φ.

We consider the homomorphism $a \rightarrow \bar{a}(P)$ between $\mathfrak{A}$ and $\bar{\mathfrak{A}}_P$. Let $\mathfrak{B}_P$ be the kernel of this homomorphism. Then $\mathfrak{A} - \mathfrak{B}_P \cong \bar{\mathfrak{A}}_P = \mathfrak{D}'_h$. Since the correspondence $a \rightarrow \bar{a}$ is an isomorphism $\Pi\mathfrak{B}_P = 0$. Now let $\mathfrak{B}_1, \cdots, \mathfrak{B}_k$ be k distinct ideals in the set $\{\mathfrak{B}_P\}$. We wish to prove the following

LEMMA 3. *$\mathfrak{A}_k = \mathfrak{A} - (\mathfrak{B}_1 \wedge \cdots \wedge \mathfrak{B}_k)$ is a direct sum of k primitive algebras which are isomorphic to the algebras $\mathfrak{A} - \mathfrak{B}_i$, $i = 1, \cdots, k$.*

This is true for $k = 1$ and we assume that it has already been established for $k - 1$. Thus $\mathfrak{A}_{k-1}$ is a direct sum of $k - 1$ primitive algebras $\mathfrak{P}_j \cong \mathfrak{A} - \mathfrak{B}_j$, $j = 1, \cdots, k - 1$. It follows that any two-sided ideal in $\mathfrak{A}_{k-1}$ has the form $\mathfrak{P}_{i_1} + \cdots + \mathfrak{P}_{i_r}$. Hence there are $k - 1$ maximal ideals in $\mathfrak{A}_{k-1}$ and it is clear that these ideals are the ideals $\mathfrak{B}_j - (\mathfrak{B}_1 \wedge \cdots \wedge \mathfrak{B}_{k-1})$. This implies that $\mathfrak{B}_k \ngeqq (\mathfrak{B}_1 \wedge \cdots \wedge \mathfrak{B}_{k-1})$. For if $\mathfrak{B}_k \geqq (\mathfrak{B}_1 \wedge \cdots \wedge \mathfrak{B}_{k-1})$, $\mathfrak{B}_k - (\mathfrak{B}_1 \wedge \cdots \wedge \mathfrak{B}_{k-1})$ is a maximal ideal in $\mathfrak{A}_{k-1}$, and so $\mathfrak{B}_k - (\mathfrak{B}_1 \wedge \cdots \wedge \mathfrak{B}_{k-1}) = \mathfrak{B}_j - (\mathfrak{B}_1 \wedge \cdots \wedge \mathfrak{B}_{k-1})$ for some $j < k$. Then $\mathfrak{B}_k = \mathfrak{B}_j$ contrary to the assumption that the $\mathfrak{B}$'s are distinct. Now since $\mathfrak{B}_k \ngeqq (\mathfrak{B}_1 \wedge \cdots \wedge \mathfrak{B}_{k-1})$, $\mathfrak{B}_k + (\mathfrak{B}_1 \wedge \cdots \wedge \mathfrak{B}_{k-1}) > \mathfrak{B}_k$. Consequently $(\mathfrak{B}_k + (\mathfrak{B}_1 \wedge \cdots \wedge \mathfrak{B}_{k-1})) - \mathfrak{B}_k$ is an ideal $\neq 0$ in $\mathfrak{A} - \mathfrak{B}_k$. Hence because of the simplicity of $\mathfrak{A} - \mathfrak{B}_k$, $\mathfrak{A} = \mathfrak{B}_k + (\mathfrak{B}_1 \wedge \cdots \wedge \mathfrak{B}_{k-1})$ and this implies that $\mathfrak{A}_k = \mathfrak{A} - (\mathfrak{B}_1 \wedge \cdots \wedge \mathfrak{B}_k) = (\mathfrak{B}_k - (\mathfrak{B}_1 \wedge \cdots \wedge \mathfrak{B}_k)) + ((\mathfrak{B}_1 \wedge \cdots \wedge \mathfrak{B}_{k-1}) - (\mathfrak{B}_1 \wedge \cdots \wedge \mathfrak{B}_k))$. The first component of this direct sum is isomorphic to $\mathfrak{A}_k - ((\mathfrak{B}_1 \wedge \cdots \wedge \mathfrak{B}_{k-1}) - (\mathfrak{B}_1 \wedge \cdots \wedge \mathfrak{B}_k)) \cong \mathfrak{A}_{k-1}$. The second component $(\mathfrak{B}_1 \wedge \cdots \wedge \mathfrak{B}_{k-1}) - (\mathfrak{B}_1 \wedge \cdots \wedge \mathfrak{B}_k) \cong \mathfrak{A}_k - (\mathfrak{B}_k - (\mathfrak{B}_1 \wedge \cdots \wedge \mathfrak{B}_k)) \cong \mathfrak{A} - \mathfrak{B}_k$. This completes the proof.

We now write $\mathfrak{A}_k = \mathfrak{P}_1 \oplus \cdots \oplus \mathfrak{P}_k$ where $\mathfrak{P}_i \cong \mathfrak{A} - \mathfrak{B}_i$. Let e_i denote the identity of $\mathfrak{P}_i$. We assume now that Φ is infinite. Then we can choose k distinct elements ρ_i in Φ and form the element $w = \Sigma e_i\rho_i$. Then w is of degree k over Φ. On the other hand $\mathfrak{A}_k$ is a homomorphic image of $\mathfrak{A}$. Hence $\mathfrak{A}_k$ is algebraic and its elements have degree $\leqq B = B(\mathfrak{A})$. Hence $k \leqq B$. Thus there exist at most B distinct ideals $\mathfrak{B}_\alpha$ and we denote these as $\mathfrak{B}_1, \cdots, \mathfrak{B}_g$, $g \leqq B$. Since $\Pi\mathfrak{B}_\alpha = 0$, $\Pi\mathfrak{B}_i = 0$. Then $\mathfrak{A} = \mathfrak{A}_g$ and so we have proved the following

THEOREM 12. *Let Φ be an infinite field and let $\mathfrak{A}$ be a semi-simple algebraic algebra of bounded degree over Φ. Then $\mathfrak{A}$ is a direct sum of a finite number of primitive algebras.*

We have already noted that each $\mathfrak{P}_j$ has the form $\mathfrak{D}'_h$, $\mathfrak{D}'$ a division algebra. Using this fact and the corollary to Theorem 7 we obtain the following

COROLLARY. *Any semi-simple algebraic algebra of bounded degree over an infinite perfect field Φ has a finite basis over Φ.*

We consider next the structure of semi-simple algebraic algebras of bounded degree over a finite field.

THEOREM 13. *A necessary and sufficient condition that an algebra $\mathfrak{A}$ over a finite field Φ be semi-simple and algebraic of bounded degree is that $\mathfrak{A}$ be isomorphic to a subalgebra $\bar{\mathfrak{A}}$ of the complete direct sum $(S, \bar{\mathfrak{A}}_P)$ where the components $\bar{\mathfrak{A}}_P$ of $\bar{\mathfrak{A}}$ are primitive subalgebras of a fixed algebra $\mathfrak{E}$ that has a finite basis over Φ.*

REMARK. The representation of $\mathfrak{A}$ as a subalgebra $\bar{\mathfrak{A}}$ of $(S, \bar{\mathfrak{A}}_P)$ evidently leads to a representation of $\mathfrak{A}$ as a subalgebra of $(S, \mathfrak{E})$, the algebra of all functions having the set S for domain and the algebra $\mathfrak{E}$ for range.

To prove the necessity of the condition we consider the representation of $\mathfrak{A}$ as a subalgebra of a complete direct sum $(S, \bar{\mathfrak{A}}_P)$, $\bar{\mathfrak{A}}_P$ primitive. We know that $\bar{\mathfrak{A}}_P = \mathfrak{D}'_h$, $h \leqq B = B(\mathfrak{A})$ and $\mathfrak{D}'$ a division algebra with a finite basis over its center $\mathfrak{C}$. Since Φ is perfect, $\mathfrak{C}$ has a finite basis over Φ. Hence $\mathfrak{D}'$ has a finite basis over Φ. It follows by Wedderburn's theorem on finite division rings that $\mathfrak{D}' = \mathfrak{C}$. Thus $\mathfrak{D}'$ is a field over Φ. Since $\mathfrak{D}'$ is separable its dimensionality $\leqq B$. Hence $\mathfrak{D}'$ is isomorphic to a subalgebra of the field Δ of dimensionality $B!$ over Φ. Then $\bar{\mathfrak{A}}_P$ is isomorphic to a subalgebra of $\mathfrak{E} = \Delta_B$.

Conversely let $\mathfrak{A}$ be a subalgebra of $(S, \mathfrak{A}_P)$ in which the components $\mathfrak{A}_P$ of $\mathfrak{A}$ are primitive and subalgebras of a fixed algebra $\mathfrak{E}$ that has a finite basis over Φ. By Theorem 1′ $\mathfrak{A}$ is semi-simple. Since $\mathfrak{E}$ has a finite basis over Φ it has a finite number of elements. Hence there exists a single polynomial $\phi(\lambda)$ in $\Phi[\lambda]$ such that $\phi(e) = 0$ for all e in $\mathfrak{E}$. It is clear that this implies that $\phi(x) = 0$ for all x in $(S, \mathfrak{E})$. Since $\mathfrak{A}$ may be regarded as a subalgebra of $(S, \mathfrak{E})$, $\phi(a) = 0$ for all a in $\mathfrak{A}$. This proves that $\mathfrak{A}$ is algebraic of bounded degree.

We have also proved for semi-simple algebras the following

THEOREM 14. *If $\mathfrak{A}$ is an algebraic algebra of bounded degree over a finite field Φ, there exists a polynomial $\phi(\lambda)$ with coefficients in Φ such that $\phi(a) = 0$ for all a in $\mathfrak{A}$.*

By what we have proved we know that there exists a polynomial $\psi(\lambda)$ such that $\psi(a) = z \in \mathfrak{N}$ for all a. Since $\mathfrak{N}$ is a nil algebra of bounded degree there is an integer h such that $z^h = 0$ for all z in $\mathfrak{N}$. Hence $[\psi(a)]^h = 0$ for all a.

8. Kurosch's problem

We shall call an algebra $\mathfrak{A}$ over Φ *locally finite* if every finite set of elements in $\mathfrak{A}$ generates an algebra with a finite basis over Φ. Evidently if $\mathfrak{A}$ is locally finite it is algebraic. Is the converse true? This question, recently raised by Kurosch, is analogous to Burnside's problem in the theory of periodic groups. In this section we consider Kurosch's problem for algebraic algebras of bounded degree. In the next section we point out a connection between Burnside's problem and Kurosch's problem. We prove first the following theorem which enables us to reduce Kurosch's problem for algebraic algebras (of bounded degree) to the same problem for the two extreme types of these algebras, namely, semi-simple algebras and nil algebras.

THEOREM 15. *Let $\mathfrak{A}$ be an algebra that contains a two-sided ideal $\mathfrak{B}$ that is locally finite and whose difference algebra $\bar{\mathfrak{A}} = \mathfrak{A} - \mathfrak{B}$ is locally finite. Then $\mathfrak{A}$ is locally finite.*

Let $x_1, \cdots, x_r$ be r arbitrary elements in $\mathfrak{A}$. We consider the cosets $\bar{x}_1, \cdots, \bar{x}_r$ of these elements in the difference algebra $\bar{\mathfrak{A}} = \mathfrak{A} - \mathfrak{B}$. Since $\bar{\mathfrak{A}}$ is locally finite we can supplement the cosets $\bar{x}_1, \cdots, \bar{x}_r$ by certain products $\bar{x}_{r+1}, \cdots, \bar{x}_m$ of these cosets in such a way that the products $\bar{x}_i\bar{x}_j$ are all linear combinations of the $\bar{x}_i$. We choose representatives $x_{r+1}, \cdots, x_n$ of the cosets $\bar{x}_{r+1}, \cdots, \bar{x}_n$. Then $x_ix_j = \sum x_k\gamma_{kij} + z_{ij}$, $i, j = 1, \cdots, m$, where $z_{ij} \in \mathfrak{B}$. Let X be the algebra generated by all the x's. Evidently X contains the elements z_{ij} and hence also the elements x_kz_{ij}, $z_{ij}x_k$, $x_kz_{ij}x_l$. These elements generate a subalgebra M of X. Now x_sz_{ij}, $z_{ij}x_s \in M$ and $x_s(x_kz_{ij}) = (x_sx_k)z_{ij} = \sum x_pz_{ij}\gamma_{psk} + z_{sk}z_{ij} \in M$. In a similar manner we see that $x_s(z_{ij}x_k)$, $x_s(x_kz_{ij}x_l)$, $(x_kz_{ij})x_s$, $(z_{ij}x_k)x_s$ and $(x_kz_{ij}x_l)x_s \in M$. It follows that M is a two-sided ideal in X. Since the $z_{ij} \in \mathfrak{B}$, M is a subalgebra of $\mathfrak{B}$. Hence M has a finite basis over Φ. Since $x_ix_j = \sum x_k\gamma_{kij} \pmod{M}$ it is clear that a subset of the cosets $x_i + M$ forms a basis for $\bar{X} = X - M$. Thus $\bar{X}$ and consequently X has a finite basis. Hence the algebra generated by the x's has a finite basis.

We prove next that Kurosch's question has an affirmative answer for semi-simple algebraic algebras of bounded degree.

THEOREM 16. *If $\mathfrak{A}$ is a semi-simple algebraic algebra of bounded degree, $\mathfrak{A}$ is locally finite.*

We suppose first that the underlying field Φ is infinite. Then $\mathfrak{A} = \mathfrak{P}_1 \oplus \cdots \oplus \mathfrak{P}_g$ where $\mathfrak{P}_i$ is primitive. Hence it suffices to prove that any primitive algebra is locally finite. We have seen that if $\mathfrak{A}$ is primitive then $\mathfrak{A} = \mathfrak{D}'_n$ where $\mathfrak{D}'$ is a division algebra that has a finite basis over its center $\mathfrak{C}$. Now $\mathfrak{C}$ is also the center of $\mathfrak{A}$ and $\mathfrak{A}$ has a finite basis over $\mathfrak{C}$. Hence if $x_1, \cdots, x_r$ are arbitrary in $\mathfrak{A}$ we can supplement these elements by elements $x_{r+1}, \cdots, x_m$ to obtain a set of x's such that $x_ix_j = \sum x_kc_{kij}$, c_{kij} in $\mathfrak{C}$. Since $\mathfrak{C}$ is commutative it is locally finite over Φ. Hence we can find elements $c_1, \cdots, c_l$ that include 1 and the c_{kij} such that $c_pc_q = \sum c_s\gamma_{spq}$, γ in Φ. Then it is clear that the elements x_ic_p generate an algebra $\mathfrak{B}$ with a finite basis over Φ. Since $\mathfrak{B}$ contains the x's, the x's generate an algebra with a finite basis.

Now let Φ be finite. Then $\mathfrak{A}$ may be taken to be a subalgebra of a complete direct sum $\mathfrak{S} = (S, \mathfrak{E})$ where $\mathfrak{E}$ is an algebra with a finite basis over Φ. Our theorem will be proved by showing that $\mathfrak{S}$ is locally finite. For this purpose let $x_1, \cdots, x_r$ be any r elements of $\mathfrak{S}$ and let $\mathfrak{B}$ be the algebra generated by the

x's. We introduce an equivalence relation in the set S by defining $P \sim Q$ if $x_i(P) = x_i(Q)$ for $i = 1, \cdots, r$. Since $\mathfrak{C}$ is finite we obtain in this way a finite number of equivalence classes $C_1, \cdots, C_q$ in S. Let $x_{ij} = x_i(P)$ for P in the class C_j. We form the direct sum $\bar{\mathfrak{S}} = (T, \mathfrak{C})$ where T is the set $1, 2, \cdots q$. Consider the elements $\bar{x}_i$ of $\bar{\mathfrak{S}}$ such that $\bar{x}_i(j) = x_{ij}$. Then the $\bar{x}_i$ generate a subalgebra $\bar{\mathfrak{B}}$ of $\bar{\mathfrak{S}}$ which is readily seen to be homomorphic under the correspondence $\bar{x}_i \rightarrow x_i$ with $\mathfrak{B}$. Since $\bar{\mathfrak{S}}$ has a finite basis over Φ, $\bar{\mathfrak{B}}$ and $\mathfrak{B}$ have finite bases and the proof is complete.

9. Nil algebras

Theorems 15 and 16 reduce Kurosch's problem for algebraic algebras of bounded degree to the special case of nil algebras of bounded degree. The latter problem may be re-stated as follows: Let $\mathfrak{N}$ be an algebra with a finite number of generators and having the property that $z^n = 0$ for all z and a fixed integer n. Then does $\mathfrak{N}$ necessarily have a finite basis? This is analogous to the following special case of Burnside's problem: Let $\mathfrak{G}$ be a finitely generated group in which every element satisfies the equation $g^n = 1$ for a fixed n. Then is $\mathfrak{G}$ necessarily finite? We shall now point out a connection between these problems.

We suppose first that $\mathfrak{N}$ is any radical algebra. We have seen that $\mathfrak{N}$ is a group relative to the operation $\oplus$ where $z_1 \oplus z_2 = z_1 + z_2 + z_1 z_2$. If we denote $\overbrace{z \oplus z \oplus \cdots \oplus z}^{n}$ by $[n]\, z$ we can verify that $[n]\, z = \sum_1^n \binom{n}{i} z^i$.

Now let Φ have characteristic $p \neq 0$. Then if z is nilpotent of index n and $p^e \geqq n$, $[p^e]\, z = 0$. Thus if N is a nil algebra the elements of the group $\mathfrak{N}$ are all of finite order.

Suppose now that $\mathfrak{N}$ is a nil algebra over a field of characteristic $p \neq 0$ and suppose that Burnside's question has an affirmative answer. Then if $z_1, \cdots, z_r$ are arbitrary elements of $\mathfrak{N}$ we can supplement these by elements $z_{r+1}, \cdots, z_m$ in such a way that $z_i \oplus z_j$ is one of the z_k for every i, j. Since $z_i z_j = (z_i \oplus z_j) - z_i - z_j$ it follows that the totality of linear combinations of the elements $z_1, \cdots, z_m$ is a subalgebra of $\mathfrak{N}$. Hence $\mathfrak{N}$ is locally finite. In a similar manner we see that if every finitely generated group in which every element satisfies $g^{p^e} = 1$ is finite, then every algebra $\mathfrak{N}$ over a field of characteristic p in which every element satisfies $z^n = 0$ with $n \leqq p^e$ is locally finite.

We wish to show next that Kurosch's problem for nil algebras of bounded degree needs to be decided only for a certain class of algebras that can be defined quite concretely. For this purpose we introduce the free algebra $\mathfrak{F}_{(r)}$ generated by the elements $u_1, \cdots, u_r$. Then if $\mathfrak{N}$ is an algebra satisfying our conditions the correspondence $\sum u_{i_1} \cdots u_{i_k} \rho_{i_1 \cdots i_k} \rightarrow \sum z_{i_1} \cdots z_{i_k} \rho_{i_1 \cdots i_k}$ is a homomorphism between $\mathfrak{F}_{(r)}$ and $\mathfrak{N}$. The kernel of this homomorphism contains the two-sided ideal $\mathfrak{F}_{(r)}^{(n)}$ generated by the nth powers of the elements $\mathfrak{F}_{(r)}$. Hence $\mathfrak{N}$ is a homomorphic image of $\mathfrak{N}(r, n) = \mathfrak{F}_{(r)} - \mathfrak{F}_{(r)}^{(n)}$ and $\mathfrak{N}$ will have finite dimensionality if $\mathfrak{N}(r, n)$ has. Since $\mathfrak{N}(r, n)$ satisfies the same conditions as $\mathfrak{N}$, our problem is reduced to that of determining whether or not the dimensionality $d(r, n)$ of $\mathfrak{N}(r, n)$ is finite. We may also ask for the value of $d(r, n)$.

We shall discuss here only the simplest case $n = 2$. Then $u^2 = 0$ for every u in $\mathfrak{N}(r, 2)$ and this implies that $uv = -vu$ for any pair of elements u, v. It follows readily that $\mathfrak{N}(r, 2)$ is the algebra with the basis $u_1^{\epsilon_1} \cdots u_r^{\epsilon_r}$ where $\epsilon_1 = 0, 1$ and $(\epsilon_1, \cdots, \epsilon_r) \neq (0, \cdots, 0)$. The multiplication table of the basis may be deduced from the relations $u_i^2 = 0$ $u_i u_j = -u_j u_i$ if $i \neq j$. Thus $\mathfrak{N}(r, 2)$ is the so-called algebra of dual quantities and $d(r, 2) = 2^r - 1$. Since Kurosch has shown that any algebraic algebra with elements of degree $\leqq 3$ is locally finite we know that $d(r, 3)$ is finite. It would be interesting to determine its value.

THE JOHNS HOPKINS UNIVERSITY.

BIBLIOGRAPHY

E. ARTIN AND G. WHAPLES

[1] *The theory of simple rings*, Amer. Jour. of Math., vol. 65 (1943), pp. 87–107.

N. JACOBSON

[1] *The theory of rings*, Math. Surveys II, New York 1943.

[2] *Structure theory of simple rings without finiteness assumptions*, Trans. Amer. Math. Soc., vol. 57 (1945), pp. 228–245.

[3] *The radical and semi-simplicity for arbitrary rings*, Amer. Jour. of Math., vol. 67 (1945), pp. 300–320.

A. KUROSCH

[1] *Ringtheoretische Probleme die mit dem Burnsideschen Problem uber periodische Gruppen in Zusamenhang stehen*, Bull. Acad. Sci. URSS., Math. vol. 5 (1941), pp. 233–240. (Russian with a summary in German.)

Reprinted from the Proceedings of the NATIONAL ACADEMY OF SCIENCES,
Vol. 31, No. 10, pp. 333–338. October, 1945

A TOPOLOGY FOR THE SET OF PRIMITIVE IDEALS IN AN ARBITRARY RING

BY N. JACOBSON

DEPARTMENT OF MATHEMATICS, JOHNS HOPKINS UNIVERSITY

Communicated July 11, 1945

1. In a recent paper we have called a ring $\mathfrak{A}$ primitive if $\mathfrak{A}$ contains a maximal right ideal $\mathfrak{J}$ whose quotient $\mathfrak{J}:\mathfrak{A} = 0$.[1] In general if $\mathfrak{J}$ is any right ideal $\mathfrak{J}:\mathfrak{A}$ is the totality of elements b such that $xb \in \mathfrak{J}$ for all x in $\mathfrak{A}$. $\mathfrak{J}:\mathfrak{A}$ is a two-sided ideal and if $\mathfrak{A}$ has an identity, $\mathfrak{J}:\mathfrak{A}$ is the largest two-sided ideal of $\mathfrak{A}$ contained in $\mathfrak{J}$. The primitive rings appear to play the same rôle in the general structure theory of rings that is played by simple rings in the classical theory of rings that satisfy the descending chain condition for one-sided ideals. Corresponding to the Wedderburn-Artin structure theorem on simple rings satisfying the descending chain condition we have the theorem that if $\mathfrak{A}$ is a primitive ring $\neq 0$, $\mathfrak{A}$ is isomorphic to a dense ring of linear transformations in a suitable vector space over a division ring.

We shall call a two-sided ideal $\mathfrak{B}$ in $\mathfrak{A}$ a *primitive ideal* if $\mathfrak{B} \neq \mathfrak{A}$ and $\mathfrak{A} - \mathfrak{B}$ is a primitive ring. It is known that if $\mathfrak{A}$ is not a radical ring then $\mathfrak{A}$ contains primitive ideals. Moreover, in this case the intersection $\Pi\mathfrak{B}$ of all the primitive ideals in $\mathfrak{A}$ coincides with the radical $\mathfrak{R}$ of $\mathfrak{A}$. In particular if $\mathfrak{A}$ is semi-simple, $\Pi\mathfrak{B} = 0$. If $\mathfrak{A}$ is a ring with an identity, $\mathfrak{A}$ is not a radical ring. Hence if $\mathfrak{C}$ is any two-sided ideal $\neq \mathfrak{A}$ in $\mathfrak{A}$, $\mathfrak{A} - \mathfrak{C}$ contains a primitive ideal $\mathfrak{B} - \mathfrak{C}$. Since $\mathfrak{A} - \mathfrak{B} \cong (\mathfrak{A} - \mathfrak{C}) - (\mathfrak{B} - \mathfrak{C})$, $\mathfrak{B}$ is primitive in $\mathfrak{A}$. Thus if $\mathfrak{A}$ is a ring with an identity any two-sided ideal $\mathfrak{C} \neq \mathfrak{A}$ can be imbedded in a primitive ideal.

Any commutative primitive ring is a field. Consequently any primitive ideal in a commutative ring is maximal. On the other hand, in a non-commutative ring there may exist primitive ideals that are not maximal. For example, the ring $\mathfrak{L}$ of all linear transformations in an infinite dimensional vector space $\mathfrak{R}$ over a division ring is primitive. Hence (0) is a primitive ideal. However, $\mathfrak{L}$ contains as a proper two-sided ideal the set $\mathfrak{F}$ of finite valued linear transformations in $\mathfrak{R}$.

A simple ring is either primitive or a radical ring. In particular any simple ring with an identity is primitive. These remarks imply that a maximal two-sided ideal $\mathfrak{B}$ such that $\mathfrak{A} - \mathfrak{B}$ is not a radical ring is primitive. If $\mathfrak{A}$ is a ring with an identity, any maximal two-sided ideal in $\mathfrak{A}$ is primitive.

In this note we shall define a topology for the set of primitive ideals of any ring. The space determined in this way appears to be an important invariant of the ring. We hope to discuss its rôle in the general structure theory in greater detail at a later date.

2. Let S be the set of primitive ideals in the ring $\mathfrak{A}$. S is vacuous if and only if $\mathfrak{A}$ is a radical ring. If A is a non-vacuous subset of S we let $\mathfrak{D}_A$ denote the intersection of all the primitive ideals $\mathfrak{B} \in A$. We now define the closure $\bar{A}$ of A to be the totality of primitive ideals $\mathfrak{C}$ such that $\mathfrak{C} \geqq \mathfrak{D}_A$. It is clear that

1. $\bar{A} \geqq A$.
2. $\bar{\bar{A}} = \bar{A}$.

We wish to prove next that

3. $\overline{A \vee B} = \bar{A} \vee \bar{B}$.[2]

Proof. Let $\mathfrak{B} \in \bar{A} \vee \bar{B}$, say $\mathfrak{B} \in \bar{A}$. Then $\mathfrak{B} \geqq \mathfrak{D}_A$. Hence $\mathfrak{B} \geqq \mathfrak{D}_A \wedge \mathfrak{D}_B = \mathfrak{D}_C$ where $C = A \vee B$. Thus $\mathfrak{B} \in \overline{A \vee B}$. Suppose next that $\mathfrak{B} \,\bar{\in}\, \bar{A} \vee \bar{B}$. Then $\mathfrak{B} \not\geqq \mathfrak{D}_A$ and $\mathfrak{B} \not\geqq \mathfrak{D}_B$. We consider now the primitive ring $\mathfrak{A} - \mathfrak{B}$. The two-sided ideals $(\mathfrak{D}_A + \mathfrak{B}) - \mathfrak{B}$ and $(\mathfrak{D}_B + \mathfrak{B}) - \mathfrak{B}$ are $\neq 0$ in $\mathfrak{A} - \mathfrak{B}$. Hence the product $[(\mathfrak{D}_A + \mathfrak{B}) - \mathfrak{B}][(\mathfrak{D}_B + \mathfrak{B}) - \mathfrak{B}] \neq 0$.[3] It follows that $\mathfrak{D}_A\mathfrak{D}_B \not\leqq \mathfrak{B}$. Hence also $\mathfrak{D}_A \wedge \mathfrak{D}_B \not\leqq \mathfrak{B}$ and so $\mathfrak{B} \,\bar{\in}\, \overline{A \vee B}$.

For the vacuous set ω we define

4. $\bar{\omega} = \omega$.

The properties 1–4 show that *S is a topological space relative to the closure operation $A \to \bar{A}$.* In the special case of a Boolean ring this topology is due to Stone.[4] It has also been introduced by Gelfand and Silov in commutative normed rings.[5] We shall call the topological space S the *structure space* of the ring $\mathfrak{A}$.

If $\mathfrak{B}$ is a primitive ideal in $\mathfrak{A}$, $\mathfrak{B}$ is a point in S. The closure of the set $\{\mathfrak{B}\}$ is the totality of primitive ideals $\mathfrak{C}$ such that $\mathfrak{C} \geqq \mathfrak{B}$. Hence if $\overline{\{\mathfrak{B}_1\}} = \overline{\{\mathfrak{B}_2\}}$ then $\mathfrak{B}_1 = \mathfrak{B}_2$. This shows that S is a T_0-space. In general S is not a T_1-space. For we have seen that there exist primitive rings $\mathfrak{A}$ that are not simple. If $\mathfrak{A}$ is a ring with an identity of this type and $\mathfrak{B}$ is a proper two-sided ideal in $\mathfrak{A}$, $\mathfrak{B}$ can be imbedded in a primitive ideal $\mathfrak{C}$. Then $\overline{\{(0)\}}$ contains $\mathfrak{C} \neq 0$. The subspace M of S of primitive ideals that are maximal is clearly a T_1-space. If $\mathfrak{A}$ is commutative $S = M$ is a T_1-space. However, even in this case S need not be a T_2- (or Hausdorff) space. An example of a normed ring of this type has been given by Gelfand and Silov.[5] A simpler one is the following:

Example. Let $\mathfrak{A} = J$ the ring of integers. The primitive ideals are the prime ideals (p). Since the intersection of an infinite number of prime ideals is the 0-ideal, $\bar{A} = S$ for any infinite set A. If A is finite, $\bar{A} = A$. Hence the open sets $\neq \omega$, $\neq S$ are the complements of finite sets. Any two open sets $\neq \omega$ have a non-vacuous intersection and so the Hausdorff separation property does not hold.

3. Let $\mathfrak{A}_1$ be an arbitrary two-sided ideal in $\mathfrak{A}$ and let S_1 denote the closed set in S consisting of the primitive ideals $\mathfrak{B}$ of $\mathfrak{A}$ that contain $\mathfrak{A}_1$. If $\mathfrak{B} \in S_1$ then $\mathfrak{B} - \mathfrak{A}_1$ is a primitive ideal in $\mathfrak{A} - \mathfrak{A}_1$ and any primitive ideal in $\mathfrak{A} - \mathfrak{A}_1$ is obtained in this way. The correspondence $\mathfrak{B} \to \mathfrak{B} - \mathfrak{A}_1$ is $(1 - 1)$ between the subspace S_1 of S and the structure space T of the ring $\mathfrak{A} - \mathfrak{A}_1$. Since this correspondence preserves intersection it is a homeomorphism between S_1 and T.

Suppose in particular that $\mathfrak{A}_1 = \mathfrak{R}$ the radical of $\mathfrak{A}$. Then every primitive ideal of $\mathfrak{A}$ contains $\mathfrak{R}$. Hence $S_1 = S$, and we see that *the structure space of* $\mathfrak{A}$ *is homeomorphic to the structure space of the semi-simple ring* $\mathfrak{A} - \mathfrak{R}$.

We return to the general case in which $\mathfrak{A}_1$ is arbitrary and we now consider the open set S_1' of primitive ideals that do not contain $\mathfrak{A}_1$. Let $\mathfrak{C} \in S_1'$. Then $\mathfrak{C} \wedge \mathfrak{A}_1$ is a two-sided ideal $\neq \mathfrak{A}_1$ in $\mathfrak{A}_1$ and $\mathfrak{A}_1 - (\mathfrak{C} \wedge \mathfrak{A}_1) \cong (\mathfrak{C} + \mathfrak{A}_1) - \mathfrak{C}$. The latter ring is a two-sided ideal in the primitive ring $\mathfrak{A} - \mathfrak{C}$. Hence it is primitive.[6] Thus $\mathfrak{C} \wedge \mathfrak{A}_1$ is in the structure space U of the ring $\mathfrak{A}_1$. Let A be a subset of S_1' and let $\mathfrak{C} \in \bar{A} \wedge S_1'$ so that $\mathfrak{C}$ is in the closure of A in the subspace S_1'. If $\mathfrak{D}_A$ is the intersection of the primitive ideals in A then $\mathfrak{C} \geqq \mathfrak{D}_A$ but $\mathfrak{C} \not\geqq \mathfrak{A}_1$. Let B be the subset of U of ideals $\mathfrak{B} \wedge \mathfrak{A}_1$ where $\mathfrak{B} \in A$. The intersection of all of these ideals is the ideal $\mathfrak{D}_A \wedge \mathfrak{A}_1$. Since $(\mathfrak{C} \wedge \mathfrak{A}_1) \geqq (\mathfrak{D}_A \wedge \mathfrak{A}_1)$, $\mathfrak{C} \wedge \mathfrak{A}_1$ is in $\bar{B}$. Hence the mapping that we have defined between S_1' and the subspace of U is a continuous one.

4. Let $\{F_\alpha\}$ be a set of closed sets in the space S. As before let $\mathfrak{D}_{F_\alpha}$ denote the two-sided ideal of elements common to all the $\mathfrak{B} \in F_\alpha$. Suppose that the intersection $\Pi F_\alpha \neq \omega$ and let $\mathfrak{B}$ be a point in this intersection. Then $\mathfrak{B} \geqq \mathfrak{D}_{F_\alpha}$ for all α. Hence $\mathfrak{B}$ contains the two-sided ideal $\Sigma\mathfrak{D}_{F_\alpha}$ generated by the $\mathfrak{D}_{F_\alpha}$. The converse follows by retracing the steps of this argument. We therefore have the

LEMMA 1. *If* $\{F_\alpha\}$ *is a set of closed sets in* S, $\Pi F_\alpha \neq \omega$ *if and only if* $\Sigma\mathfrak{D}_{F_\alpha}$ *can be imbedded in a primitive ideal.*

If $\mathfrak{A}$ is a ring with an identity any two-sided ideal $\neq \mathfrak{A}$ can be imbedded in a primitive ideal. Hence we have the

Corollary. If $\mathfrak{A}$ *is a ring with an identity and* $\{F_\alpha\}$ *is a set of closed sets in* S *then* $\Pi F_\alpha \neq \omega$ *if and only if* $\Sigma\mathfrak{D}_{F_\alpha} \neq \mathfrak{A}$.

If $\mathfrak{A}$ has an identity and $\Sigma\mathfrak{D}_{F_\alpha} = \mathfrak{A}$, $1 = d_1 + \ldots + d_r$ where $d_i \in \mathfrak{D}_{F_i}$. Hence also $\Sigma\mathfrak{D}_{F_i} = \mathfrak{A}$. The corollary therefore shows that if $\Pi F_\alpha = \omega$

then there is a finite set of closed sets F_i such that $\Pi F_i = \omega$. We therefore have the following

THEOREM 1. *The structure space of a ring with an identity is bicompact.*

Suppose now that $\mathfrak{A}$ is a semi-simple ring. Let $\mathfrak{A}$ be decomposable as a direct sum $\mathfrak{A}_1 \oplus \mathfrak{A}_2$ of the two-sided ideals $\mathfrak{A}_i \neq 0$. Let S_i be the closed subset of S consisting of the primitive ideals $\mathfrak{B}$ containing the ideal $\mathfrak{A}_i$. Since $\mathfrak{A} - \mathfrak{A}_1 \cong \mathfrak{A}_2$ a semi-simple ring, $S_1 \neq \omega$. Similarly $S_2 \neq \omega$. Also the semi-simplicity of $\mathfrak{A} - \mathfrak{A}_i$ implies that $\mathfrak{A}_i$ is the intersection of all the $\mathfrak{B}$ in S_i. Since $\mathfrak{A}_1 + \mathfrak{A}_2 = \mathfrak{A}$ the lemma implies that $S_1 \wedge S_2 = \omega$. Now let $\mathfrak{B}$ be any primitive ideal in $\mathfrak{A}$. We assert that either $\mathfrak{B} \geqq \mathfrak{A}_1$ or $\mathfrak{B} \geqq \mathfrak{A}_2$. For otherwise $\mathfrak{A}_i + \mathfrak{B} > \mathfrak{B}$. The ideals $(\mathfrak{A}_i + \mathfrak{B}) - \mathfrak{B}$ are $\neq 0$ in the primitive ring $\mathfrak{A} - \mathfrak{B}$. Since $[(\mathfrak{A}_1 + \mathfrak{B}) - \mathfrak{B}][(\mathfrak{A}_2 + \mathfrak{B}) - \mathfrak{B}] = 0$ this is impossible. We have therefore proved our assertion. Evidently it is equivalent to the relation $S_1 \vee S_2 = S$. Thus S is disconnected into the two components S_1 and S_2.

Conversely suppose that $S = S_1 \vee S_2$ where the S_i are closed sets $\neq \omega$ such that $S_1 \wedge S_2 = \omega$. We assume also that $\mathfrak{A}$ has an identity. Let $\mathfrak{A}_i = \mathfrak{D}_{S_i}$. Then $\mathfrak{A}_1 + \mathfrak{A}_2 = \mathfrak{A}$. Since $\mathfrak{A}$ is semi-simple $\mathfrak{A}_1 \wedge \mathfrak{A}_2 = \mathfrak{D}_S = 0$. If $\mathfrak{A}_i = 0$, $S = \bar{S}_i = S_i$ contrary to $S_1 \neq \omega$ and $S_2 \neq \omega$. A part of our result is the following

THEOREM 2. *If $\mathfrak{A}$ is a semi-simple ring with an identity, $\mathfrak{A} = \mathfrak{A}_1 \oplus \mathfrak{A}_2$ where the $\mathfrak{A}_i$ are two-sided ideals $\neq 0$ if and only if S is not connected.*

5. If S' is a completely regular bicompact space and $\mathfrak{S}$ is the ring of real-valued (complex valued) continuous functions on S' then it has been shown by Gelfand and Silov that the structure space S of $\mathfrak{S}$ is homeomorphic to S'.[7] If S' is a bicompact totally disconnected space and $\mathfrak{S}$ is the ring of continuous functions on S' having values in the field of residues mod 2, then by a result of Stone's the structure space of $\mathfrak{S}$ is homeomorphic to S'.[8] We conclude this note by giving another example of this type based on an arbitrary totally disconnected bicompact space S' and an arbitrary division ring $\mathfrak{K}'$.

We consider any decomposition of S' into a finite number of components (non-overlapping open and closed sets) S'_i and we choose corresponding elements $k_i \in \mathfrak{K}'$. We define a function $f(x)$ by setting $f(x_i) = k_i$ for x_i in S'_i. A function of this type will be called a *finite decomposition function.* The totality $\mathfrak{S}$ of these functions is a ring under the ordinary operations of addition and multiplication. $\mathfrak{S}$ contains the subring $\mathfrak{K}$ of constant functions, isomorphic to $\mathfrak{K}'$ and $\mathfrak{S}$ is commutative if and only if $\mathfrak{K}'$ is commutative. The constant 1 acts as an identity in $\mathfrak{S}$. Since S' is totally disconnected, for any two points $a \neq b$ in S' there is an $f(x) \in \mathfrak{S}$ such that $f(a) \neq f(b)$. Let $\mathfrak{A}$ be any subring of $\mathfrak{S}$ having this property and containing $\mathfrak{K}$. We shall sketch a proof of the fact that the space M of maximal two-sided ideals of $\mathfrak{A}$ is homeomorphic to S'.

Lemma 2. *If F' is a closed subset of S' and a is a point $\bar{\epsilon}$ F' then there exists a function $\varphi(x) \in \mathfrak{A}$ such that $\varphi(y) = 0$ for all $y \in F'$ but $(\varphi a) \neq 0$.*

Our assumptions imply that for each $y \in F'$ there is an $f_y \in \mathfrak{A}$ such that $f_y(y) = 0$ but $f_y(a) \neq 0$. The set $Z'(f_y)$ of zeros of f_y is open and the totality of these sets covers F'. Let $Z'(f_{y_1}), \ldots, Z'(f_{y_m})$ be a finite subset of these sets covering F'. Then $\varphi(x) = f_{y_1}(x) \ldots f_{y_m}(x)$ has the required properties.

If $a \in S'$ we let $\mathfrak{B}_a$ denote the totality of functions $g(x) \in \mathfrak{A}$ such that $g(a) = 0$. It is easy to see that $\mathfrak{B}_a$ is a maximal two-sided ideal in $\mathfrak{A}$ and that $\mathfrak{A} - \mathfrak{B}_a \cong \mathfrak{K}$. Our conditions imply that if $a \neq b$ then $\mathfrak{B}_a \neq \mathfrak{B}_b$.

Lemma 3. *If $\mathfrak{B}$ is a two-sided ideal $\neq \mathfrak{A}$, then there exists a point a such that $g(a) = 0$ for all $g \in \mathfrak{B}$.*

Let $Z'(f)$ be the set of zeros of f. $Z'(f)$ is closed. If the lemma is false, the intersection $\Pi Z'(f)$ for all $f \in \mathfrak{B}$ is vacuous. Hence there is a finite, number of functions $f_1, \ldots, f_r$ in $\mathfrak{B}$ such that $\Pi Z'(f_i) = \omega$. Hence if a is any point of S' at least one of the functions f_i does not vanish at a. Let $k_{1i}, \ldots, k_{n_i i}$ be the non-zero values taken on by f_i and form the function $\psi_i(x) = (f_i(x) - k_{1i})(f_i(x) - k_{2i}) \ldots (f_i(x) - k_{n_i i})$ and $\psi(x) = \psi_1(x)\psi_2(x)\ldots \psi_r(x)$. Then $\psi(x) \equiv 0$. The form of $\psi(x)$ shows that $0 = \psi(x) = k + g(x)$ where $k \neq 0$ and $g(x) \in \mathfrak{B}$. Hence $\mathfrak{B}$ contains a constant function $\neq 0$ and $\mathfrak{B} = \mathfrak{A}$ contrary to asumption.

This lemma shows that $\mathfrak{B} \leqq \mathfrak{B}_a$ for some a. If $\mathfrak{B}$ is maximal $\mathfrak{B} = \mathfrak{B}_a$. Thus the correspondence $a \rightarrow \mathfrak{B}_a$ is $(1 - 1)$ between S' and the space M of maximal two-sided ideals in $\mathfrak{A}$.

Let F' be a closed subset of S' and let F be the corresponding set in M. The intersection $\mathfrak{D}_F = \Pi\mathfrak{B}_a$ for all $a \in F'$ is the set of functions f such that $f(a) = 0$ for all $a \in F$. Let $\mathfrak{B}_b$ be a maximal two-sided ideal containing $\mathfrak{D}_F$. If $b \bar{\epsilon} F'$ there is a function ϕ such that $\phi(a) = 0$ for all a in F' but $\phi(b) \neq 0$. Then $\phi \in \mathfrak{D}_F$ but $\bar{\epsilon}\, \mathfrak{B}_b$ contrary to $\mathfrak{B}_b \geqq \mathfrak{D}_F$. Hence $b \in F'$ and $\mathfrak{B}_b \in F$. Thus F is closed. Conversely let F be any closed set in M and let F' be the corresponding set in S'. It is easy to see that F' is the set of points y such that $f(y) = 0$ for all f in $\mathfrak{D}_F$. Thus F' is the intersection of closed sets and is therefore closed. Hence the correspondence $a \rightarrow \mathfrak{B}_a$ is a homeomorphism.

Theorem 3. *Let S' be a totally disconnected bicompact space, $\mathfrak{K}'$ a division ring and $\mathfrak{A}$ a ring of $\mathfrak{K}'$-valued decomposition functions such that (1) $\mathfrak{A}$ contains the constants and (2) if $a \neq b$ in S', then $\mathfrak{A}$ contains a function f such that $f(a) \neq f(b)$. Then the space M of maximal two-sided ideals of $\mathfrak{A}$ is homeomorphic to S'.*

If $\mathfrak{K}'$ is commutative, $M = S$. It is an open question whether or not this is true in general. However, it is clear that $\bar{M} = S$ since the intersection of the set of maximal two-sided ideals is (0).

Theorem 3 shows that two rings $\mathfrak{A}_1$ and $\mathfrak{A}_2$ of the type considered are not isomorphic unless the spaces S_1' and S_2' are homeomorphic. Since the division ring $\mathfrak{K}_i'$ is isomorphic to $\mathfrak{A}_i - \mathfrak{B}_i$ for any maximal two-sided ideal

$\mathfrak{B}_i$ in $\mathfrak{A}_i$, the isomorphism of $\mathfrak{A}_1$ and $\mathfrak{A}_2$ implies that of $\mathfrak{R}_1'$ and $\mathfrak{R}_2'$. If $\mathfrak{A}_i = \mathfrak{S}_i$ the complete ring of finite decomposition functions then the converse holds. Hence necessary and sufficient conditions that $\mathfrak{S}_1$ and $\mathfrak{S}_2$ be isomorphic are that S_1' and S_2' be homeomorphic and that $\mathfrak{R}_1'$ and $\mathfrak{R}_2'$ be isomorphic.

[1] "The Radical and Semi-simplicity for Arbitrary Rings," *Amer. Jour. Math.*, **67**, 300–320 (1945). The results that we state without proof in this section can be found in the above-mentioned paper.

[2] We use the notations $\vee$ and $\wedge$, respectively, for the logical sum and logical product of a finite number of sets.

[3] Loc. cit. in reference 1, p. 316.

[4] "Applications of the Theory of Boolean Rings to General Topology," *Trans. Amer. Math. Soc.*, **41**, 375–481 (1937).

[5] "Über verschiedene Methoden der Einführing der Topologie in die Menge der maximalen ideale eines normierten Ringes," *Math. Sbornik*, **51**, 25–39 (1941).

[6] Loc. cit. in reference 1, p. 313.

[7] Loc. cit. in reference 5, pp. 27–30.

[8] Loc. cit. in reference 4, p. 380. It is assumed here that the field of residues mod 2 is endowed with a T_1-topology. This means that each point is both open and closed.

ANNALS OF MATHEMATICS
Vol. 48, No. 1, January, 1947

ON THE THEORY OF PRIMITIVE RINGS[1]

BY N. JACOBSON

(Received April 22, 1946)

In a recent paper[2] we have called a ring $\mathfrak{A}$ *primitive* if $\mathfrak{A}$ contains a maximal right ideal $\mathfrak{J}$ such that the quotient $(\mathfrak{J}:\mathfrak{A}) = (0)$. The quotient $(\mathfrak{J}:\mathfrak{A})$ is defined to be the largest two-sided ideal of $\mathfrak{A}$ that satisfies the condition $\mathfrak{A}(\mathfrak{J}:\mathfrak{A}) \subseteq \mathfrak{J}$. Thus if $\mathfrak{A}$ has an identity, $(\mathfrak{J}:\mathfrak{A})$ is the largest two-sided ideal of $\mathfrak{A}$ contained in $\mathfrak{J}$. Primitive rings appear to play a rôle in the general structure theory of rings analogous to that played by simple rings in the Wedderburn-Artin theory. Primitive rings are also fundamental in general representation theory since it is known that a necessary and sufficient condition that a ring $\mathfrak{A}$ be primitive is that $\mathfrak{A}$ be isomorphic to an irreducible ring $\bar{\mathfrak{A}}$ of endomorphisms in a commutative group $\mathfrak{R}$. If $\bar{\mathfrak{A}}$ is irreducible in $\mathfrak{R}$ and $\mathfrak{D}$ is the division ring of endomorphisms commutative with the elements of $\bar{\mathfrak{A}}$, then $\bar{\mathfrak{A}}$ is a dense ring of linear transformations in $\mathfrak{R}$ regarded as a vector space over $\mathfrak{D}$.[3]

We shall call $\mathfrak{A}$ a *left primitive* ring if $\mathfrak{A}$ contains a maximal left ideal $\mathfrak{J}'$ such that the quotient $(\mathfrak{J}':\mathfrak{A})_l = (0)$. Here $(\mathfrak{J}':\mathfrak{A})_l$ is the largest two-sided ideal in $\mathfrak{A}$ having the property $(\mathfrak{J}':\mathfrak{A})_l\, \mathfrak{A} \subseteq \mathfrak{J}'$. Evidently, a ring is left primitive if and only if it is anti-isomorphic to an irreducible ring of endomorphisms. It is an open question whether or not a primitive ring is necessarily left primitive. This is indeed the case if $\mathfrak{A}$ contains minimal ideals. In many other respects the theory of primitive rings that contain minimal ideals constitutes the most satisfactory part of the theory of primitive rings. An appropriate tool for studying rings of this type is a generalization of the duality theory previously used by Dieudonné in studying simple rings that possess minimal ideals.[4]

In this note we develop this generalization of Dieudonné's results. We also investigate certain natural topologies that can be defined in any primitive ring.

1. Primitive rings and matrix rings

Any primitive ring $\mathfrak{A}$ is isomorphic to a dense ring $\bar{\mathfrak{A}}$ of linear transformations (l.t.) in a vector space $\mathfrak{R}$ over a division ring $\mathfrak{D}$. The density of $\bar{\mathfrak{A}}$ means that if $x_1, x_2, \cdots, x_m$ are m (arbitrary) linearly independent vectors in $\mathfrak{R}$, and $y_1, y_2, \cdots, y_m$ are arbitrary then there exists an l.t. A in $\bar{\mathfrak{A}}$ such that $x_iA = y_i$, $i = 1, 2, \cdots m$. Let $\mathfrak{S}$ be a finite dimensional subspace of $\mathfrak{R}$ and let $\bar{\mathfrak{B}}$ be the subring of $\bar{\mathfrak{A}}$ of the l.t. that sends $\mathfrak{S}$ into itself. Then the density of $\bar{\mathfrak{A}}$ implies that $\bar{\mathfrak{B}}$ induces every l.t. in $\mathfrak{S}$. Hence if $\bar{\mathfrak{U}}$ is the subring of l.t. that annihilate $\mathfrak{S}$, then $\bar{\mathfrak{B}}/\bar{\mathfrak{U}}$ is isomorphic to the complete ring of l.t. in $\mathfrak{S}$. Hence $\bar{\mathfrak{B}}/\bar{\mathfrak{U}} \cong$

---

[1] Presented to the Society December 26, 1946.
[2] [4] of the Bibliography.
[3] [3], p. 232.
[4] [2].

$\mathfrak{D}'_m$ where $\mathfrak{D}'$ is anti-isomorphic to $\mathfrak{D}$ and $m = \dim \mathfrak{S}$. The ring $\bar{\mathfrak{A}}$ satisfies the descending chain condition for right ideals if and only if $\mathfrak{R}$ is finite dimensional.[5] This leads to the following

THEOREM 1. *If $\mathfrak{A}$ is a primitive ring that satisfies the descending chain condition for right ideals, then $\mathfrak{A}$ is isomorphic to a matrix ring $\mathfrak{D}'_n$, $\mathfrak{D}'$ a division ring. If $\mathfrak{A}$ is primitive but does not satisfy the chain condition, then for each $m = 1, 2, \cdots$, $\mathfrak{A}$ contains a subring homomorphic to a matrix ring $\mathfrak{D}'_m$.*[6]

AN APPLICATION. It is clear that if $\mathfrak{D}$ is of characteristic $p \neq 0$ then $pa = 0$ for all a in $\mathfrak{A}$. If $\mathfrak{D}$ is of characteristic 0, then $ma \neq 0$ if $a \neq 0$ in $\mathfrak{A}$ and m is an integer $\neq 0$. Suppose now that $\mathfrak{A}$ is a primitive ring for which there exists an equation

$$a^r + m_1 a^{r-1} + \cdots + m_{r-1} a = 0 \tag{1}$$

for all a and fixed integers m_i. Then this same equation holds for all the elements of the rings $\mathfrak{D}'_n$, $\mathfrak{D}'_m$ of Theorem 1. We may write the relation (1) in the form

$$a^r + m_1 a^{r-1} + \cdots + m_{r-k} a^k = 0 \tag{1'}$$

where m_{r-k} is not a multiple of the characteristic of $\mathfrak{D}$ (and of $\mathfrak{A}$). Now suppose that every element of $\mathfrak{D}'_m$ satisfies (1). Then we assert that $m \leqq k$. For we know that $\mathfrak{D}'_m$ contains an element z such that $z^m = 0$, $z^{m-1} \neq 0$. On the other hand, if $m > k$, multiplication of (1′) for z by z^{m-k-1} gives

$$m_{r-k} z^{m-1} = -z^{m+r-k-1} - m_1 z^{m+r-k-2} - \cdots = 0.$$

This gives the contradiction $z^{m-1} = 0$. Thus $m \leqq k$. It is also easy to see that $\mathfrak{D}'$ must be a finite field.[7] It now follows from Theorem 1 that any primitive ring in which a relation of the form (1′) holds is a matrix ring $\mathfrak{D}'_n$, $n \leqq k$, $\mathfrak{D}'$ a finite field.

2. One-sided ideals in dense rings of l. t.

Let $\bar{\mathfrak{A}}$ be a dense ring of l.t. in $\mathfrak{R}$ and let $\bar{\mathfrak{J}}'$ be a left ideal in $\bar{\mathfrak{A}}$. If x is any vector in $\mathfrak{R}$ the totality $x\bar{\mathfrak{J}}'$ of vectors xB, $B \in \bar{\mathfrak{J}}'$ is a subgroup of $\mathfrak{R}$. Moreover, if $\alpha \in \mathfrak{D}$ then there is an l. t. $A \in \bar{\mathfrak{A}}$ such that $xA = x\alpha$. Then $xAB = (x\alpha)B = (xB)\alpha$. Since $AB \in \bar{\mathfrak{J}}'$, $(xB)\alpha \in x\bar{\mathfrak{J}}'$. Hence $x\bar{\mathfrak{J}}'$ is a subspace of $\mathfrak{R}$.

Now let $\mathfrak{R}\bar{\mathfrak{J}}'$ denote the totality of vectors of the form $x_1B_1 + x_2B_2 + \cdots + x_kB_k$, x_i in $\mathfrak{R}$ and B_i in $\bar{\mathfrak{J}}'$. Then $\mathfrak{R}\bar{\mathfrak{J}}'$ is the join of all the spaces $x\bar{\mathfrak{J}}'$, x in $\mathfrak{R}$. We assert that if x is any vector $\neq 0$ then $x\bar{\mathfrak{J}}' = \mathfrak{R}\bar{\mathfrak{J}}'$. For if $x \neq 0$ and the x_i are arbitrary we can find $A_i \in \bar{\mathfrak{A}}$ such that $xA_i = x_i$. Then $x_1B_1 + \cdots + x_kB_k = x(\sum A_iB_i) = xB'$ where $B' = \sum A_iB_i \in \bar{\mathfrak{J}}'$. This proves our assertion.

If $x \neq 0$ is in $\mathfrak{R}\bar{\mathfrak{J}}'$, $x\bar{\mathfrak{J}}' = \mathfrak{R}\bar{\mathfrak{J}}'$. Hence $\bar{\mathfrak{J}}'$ is an irreducible ring of endo-

[5] [4] p. 313.

[6] Cf. [5] p. 698.

[7] The center $\mathfrak{C}$ of $\mathfrak{D}'$ is finite since an equation of finite degree can not have an infinite number of distinct roots in a field. The result $\mathfrak{C} = \mathfrak{D}'$ then follows from [5] p. 701.

morphisms in $\mathfrak{S} = \mathfrak{R}\mathfrak{J}'$. We wish to show that $\mathfrak{J}'$ is dense in $\mathfrak{S}$. It suffices to assume that $\mathfrak{J} \neq 0$ and to prove that if $x_1, x_2, \cdots, x_m$ are any linearly independent vectors then there exists a $B \in \mathfrak{J}'$ such that $x_1B \neq 0$ but $x_2B = \cdots = x_mB = 0$. Now there is an $A \in \mathfrak{A}$ such that $x_1A = 0$, $x_iA = 0, i > 1$. Since $x_1A \neq 0$ and $\mathfrak{J}' \neq 0$ there is a $C \in \mathfrak{J}'$ such that $(x_1A)C \neq 0$. Then $B = AC \in \mathfrak{J}'$ and has the required properties.

THEOREM 2. *Let $\mathfrak{J}'$ be a left ideal in the dense ring of* l.t. $\mathfrak{A}$. *Then $\mathfrak{J}'$ induces a dense ring of l.t. in the subspace $\mathfrak{S} = \mathfrak{R}\mathfrak{J}'$.*

Let Z belong to the kernel of the homomorphism between $\mathfrak{J}'$ and the induced mappings in $\mathfrak{S}$. Then $\mathfrak{S}Z = 0$ and $\mathfrak{R}\mathfrak{J}'Z = 0$. Hence $\mathfrak{J}'Z = 0$. Thus $\mathfrak{Z}$ the totality of Z is the ideal of right annihilators of $\mathfrak{J}'$. Since $\mathfrak{J}'/\mathfrak{Z}$ is isomorphic to an irreducible ring of endomorphisms, $\mathfrak{Z}$ is the radical of $\mathfrak{J}'$. Hence we have the

COROLLARY. *Let $\mathfrak{J}'$ be a left ideal in the primitive ring $\mathfrak{A}$. Then the radical $\mathfrak{Z}$ of $\mathfrak{J}'$ is the right annihilator of $\mathfrak{J}'$ and $\mathfrak{J}'/\mathfrak{Z}$ is primitive.*

We consider next an arbitrary right ideal $\mathfrak{J}$ in the dense ring $\mathfrak{A}$. Let $\mathfrak{N}$ be the subspace of elements $z \in \mathfrak{R}$ such that $zB = 0$ for all $B \in \mathfrak{J}$. The elements of $\mathfrak{J}$ induce l.t. in the factor space $\mathfrak{S} = \mathfrak{R}/\mathfrak{N}$. If $\bar{x}$ is a coset of $\mathfrak{S}$ that is $\neq 0$, $\bar{x}\mathfrak{J} \neq 0$ and $x\mathfrak{J} \neq 0$ is invariant under $\mathfrak{A}$. Hence $x\mathfrak{J} = \mathfrak{R}$ and $\bar{x}\mathfrak{J} = \mathfrak{S}$. Thus $\mathfrak{J}$ is irreducible in $\mathfrak{S}$. We shall now show that $\mathfrak{J}$ induces a dense ring of l.t. in $\mathfrak{S}$. It suffices to show that $\mathfrak{J}$ is two-fold transitive in $\mathfrak{S}$.[8] Let $\bar{x}_1, \bar{x}_2$ be linearly independent in $\mathfrak{S}$. Then the only element in the space $[x_1, x_2]$ generated by x_1, x_2 that is annihilated by all $B \in \mathfrak{J}$ is the vector 0. We shall show that there is a $B \in \mathfrak{J}$ such that $x_1B = x_1$, $x_2B = x_2$. Consider the elements $C \in \mathfrak{J}$ such that $[x_1, x_2]C \neq 0$. The set of these C is not vacuous. If C is of this type and x_1C, x_2C are linearly independent we can find an $A \in \mathfrak{A}$ such that $x_1CA = \bar{x}_1$ and $x_2CA = x_2$. Then CA is the required l.t. B. Suppose that x_1C and x_2C are linearly dependent, say $(x_1C)\gamma_1 + (x_2C)\gamma_2 = 0$. Hence $(x_1\gamma_1 + x_2\gamma_2)C = 0$ where $(\gamma, \gamma_2) \neq (0, 0)$. The vectors $y \in [x_1, x_2]$ such that $yC = 0$ are multiples of $x_1\gamma_1 + x_2\gamma_2$. If $(x_1\gamma_1 + x_2\gamma_2)D = 0$ for all D in $\mathfrak{J}$ such that $[x_1, x_2]$ $D \neq 0$ then $(x_1\gamma_1 + x_2\gamma_2)B = 0$ for all $B \in \mathfrak{J}$. This contradicts our assumption. Hence there is a D such that $(x_1\gamma_1 + x_2\gamma_2)D \neq 0$. If x_1D, x_2D are linearly independent we argue as before. Hence we suppose that there is an $x_1\delta_1 + x_2\delta_2 \neq 0$ such that $(x_1\delta_1 + x_2\delta_2)D = 0$. Since $[x_1, x_2]$ $C \neq 0$, $(x_1\delta_1 + x_2\delta_2)C \neq 0$. Thus the two vectors $y_1 = x_1\gamma_1 + x_2\gamma_2$ and $y_2 = x_1\delta_1 + x_2\delta_2$ form a basis for $[x_1, x_2]$ and

$$y_1C = 0, \; y_2C \neq 0; \qquad y_1D \neq 0, \; y_2D = 0.$$

We can find an A_1 such that $(y_1D)A_1 = y_1$ and an A_2 such that $(y_2C)A_2 = y_2$. Then $B = DA_1 + CA_2$ has the required property.

Now if B is chosen so that $x_1B = x_1$, $x_2B = x_2$ and u_1 and u_2 are arbitrary, we can find $A \in \mathfrak{A}$ such that $x_1A = u_1$, $x_2A = u_2$. Hence $x_1BA = u_1$, $x_2BA = u_2$ and $BA \in \mathfrak{J}$. Also $\bar{x}_1BA = \bar{u}_1$, $\bar{x}_2BA = \bar{u}_2$ and $\mathfrak{J}$ is dense in $\mathfrak{R}/\mathfrak{N}$.

---

[8] [3] p. 233.

THEOREM 3. *Let $\bar{\mathfrak{A}}$ be a dense ring of l.t. in $\mathfrak{R}$ and let $\bar{\mathfrak{J}}$ be a right ideal in $\bar{\mathfrak{A}}$. Then if $\mathfrak{N}$ is the totality of vectors annihilated by all $B \in \bar{\mathfrak{J}}$, $\bar{\mathfrak{J}}$ induces a dense ring of l.t. in the factor space $\mathfrak{R}/\mathfrak{N}$.*

The kernel $\bar{\mathfrak{Z}}$ of the homomorphism between $\bar{\mathfrak{J}}$ and the induced l.t. in $\mathfrak{R}/\mathfrak{N}$ is the totality of left annihilators of $\bar{\mathfrak{J}}$ and $\bar{\mathfrak{Z}}$ is the radical of $\bar{\mathfrak{J}}$.

COROLLARY. *Let $\mathfrak{A}$ be a primitive ring and let $\mathfrak{J}$ be a right ideal in $\mathfrak{A}$. Then the radical of the ring $\mathfrak{J}$ is the totality $\mathfrak{Z}$ of left annihilators of $\mathfrak{J}$ and $\mathfrak{J}/\mathfrak{Z}$ is primitive.*

Now let $\bar{\mathfrak{B}}$ be a two-sided ideal. $\mathfrak{R}\bar{\mathfrak{B}}$ is invariant relative to $\bar{\mathfrak{A}}$. Hence either $\mathfrak{R}\bar{\mathfrak{B}} = 0$ or $\mathfrak{R}\bar{\mathfrak{B}} = \mathfrak{R}$. In the former case $\bar{\mathfrak{B}} = 0$ and in the latter, by Theorem 2, $\bar{\mathfrak{B}}$ is dense in $\mathfrak{R}$.

3. Finite topology in rings of linear transformations

In the ring $\bar{\mathfrak{L}}$ of all linear transformations of a vector space $\mathfrak{R}$ over $\mathfrak{D}$ we may introduce, as usual, the finite topology. We recall the definition: If $\Sigma = \{A\}$ is a set of l.t. the closure Σ^* is specified to be the totality of l.t. B that have the property that if $x_1, x_2, \cdots, x_m$ is any finite set of vectors, then there exists an $A \in \Sigma$ such that $x_iA = x_iB$, $i = 1, 2, \cdots, m$. This topology can also be obtained by selecting as neighborhoods of the 0 of $\bar{\mathfrak{L}}$ the sets $\mathfrak{K} = \mathfrak{K}_{x_1} \frown \mathfrak{K}_{x_2} \frown \cdots \frown \mathfrak{K}_{x_m}$ where $\mathfrak{K}_x$ denotes the set of l.t. C such that $xC = 0$. The neighborhoods of any element B are taken to be the cosets $B + \mathfrak{K}$. Now it is clear that a subring $\bar{\mathfrak{A}}$ of $\bar{\mathfrak{L}}$ is dense in the sense defined above if and only if it is topologically dense in the finite topology.

The topology in $\bar{\mathfrak{L}}$ induces a topology in $\bar{\mathfrak{A}}$. The neighborhoods of B are the cosets $B + \bar{\mathfrak{J}}$ where $\bar{\mathfrak{J}} = \bar{\mathfrak{J}}_{x_1} \frown \cdots \frown \bar{\mathfrak{J}}_{x_m}$ and $\bar{\mathfrak{J}}_x = \bar{\mathfrak{K}}_x \frown \bar{\mathfrak{A}}$. If $x = 0$ it is clear that $\bar{\mathfrak{J}}_x = \bar{\mathfrak{A}}$ and if $x \neq 0$ it is known that $\bar{\mathfrak{J}}_x$ is a maximal ideal. Evidently $\bigcap \bar{\mathfrak{J}}_x = 0$. We note also that if A is any element of $\bar{\mathfrak{A}}$ the totality $(\bar{\mathfrak{J}}_x : \bar{\mathfrak{A}})$ of l.t. Z in $\bar{\mathfrak{A}}$ such that $AZ \leqq \bar{\mathfrak{J}}_x$ coincides with the right ideal $\bar{\mathfrak{J}}_{xA}$. Hence either $(\bar{\mathfrak{J}}_x : \bar{\mathfrak{A}}) = \bar{\mathfrak{A}}$ or $(\bar{\mathfrak{J}}_x : \bar{\mathfrak{A}})$ is a maximal right ideal. These remarks may be used to prove that $\bar{\mathfrak{A}}$ is a topological ring relative to the finite topology. We consider now a generalization of these considerations.

4. Right ideal topologies in an arbitrary ring

Let $\mathfrak{A}$ be an arbitrary ring. If $\mathfrak{J}$ is a right ideal and $a \in \mathfrak{A}$ we define $(\mathfrak{J} : a)$ to be the totality of elements z such that $az \in \mathfrak{J}$. Thus $(\mathfrak{J} : a)$ is the annihilator of the coset $a + \mathfrak{J}$ in the difference $\bar{\mathfrak{A}}$-group $\mathfrak{A}/\mathfrak{J}$. Evidently $(\mathfrak{J} : a)$ is a right ideal containing the two-sided ideal $(\mathfrak{J} : \mathfrak{A})$.

We suppose now that Γ is any set of right ideals having the following properties:

1. $\mathfrak{A} \in \Gamma$.
2. If $\mathfrak{J} \neq \mathfrak{A} \in \Gamma$ and a is arbitrary in $\mathfrak{A}$, $(\mathfrak{J} : a) \in \Gamma$.
3. $\bigcap_\Gamma \mathfrak{J} = (0)$.

We use Γ to introduce a *Γ-topology* in $\mathfrak{A}$ by specifying as neighborhoods of b the cosets $b + (\mathfrak{J}_1 \frown \cdots \frown \mathfrak{J}_m)$ where $\mathfrak{J}_j \in \Gamma$. We have the following

THEOREM 4. *$\mathfrak{A}$ is a topological ring relative to the Γ-topology and $\mathfrak{A}$ is a totally disconnected Hausdorff space.*

Since $\bigcap_\Gamma \mathfrak{J} = (0)$, $\mathfrak{A}$ is a T_1-space. Since the $\mathfrak{J}$ are subgroups of $\mathfrak{A}$, $\mathfrak{A}$ is a topological group. The neighborhoods $\mathfrak{J}_1 \frown \mathfrak{J}_2 \frown \cdots \frown \mathfrak{J}_m$ are subgroups. Hence they are closed as well as open and $\mathfrak{A}$ is totally disconnected Hausdorff space. In order to show that $\mathfrak{A}$ is a topological ring it remains to show that xy is a continuous function. Now if $\mathfrak{J} \epsilon \Gamma$

$$xy - ab = (x - a)y + a(y - b)$$

$\epsilon \mathfrak{J}$ provided that $x - a \epsilon \mathfrak{J}$ and $y - b \epsilon (\mathfrak{J}:a)$. Continuity at the point ab follows.

SPECIAL CASES. The Γ's that we consider here will consist exclusively of maximal right ideals and of $\mathfrak{A}$ itself. We assume, moreover, that $\mathfrak{A}$ is semi-simple.[9]

CASE 1. $\Gamma = \Omega$ the complete set of maximal right ideals together with the ideal $\mathfrak{A}$. We have seen that if $\mathfrak{J}$ is maximal, then either $(\mathfrak{J}:a) = \mathfrak{A}$ or $(\mathfrak{J}:a)$ is maximal. Since $\mathfrak{A}$ is semi-simple $\bigcap_\Omega \mathfrak{J} = (0)$.[10] Hence Ω may be used to define an *Ω-topology.*

CASE 2. Let $\mathfrak{A}$ be primitive. Then $\mathfrak{A}$ is semi-simple and the above topology may be used. A second topology that is conceivably stronger can be obtained by using a particular $(1 - 1)$ representation of $\mathfrak{A}$ as a dense ring of linear transformations in a vector space $\mathfrak{R}$. Let $\bar{\mathfrak{J}}_x$ be defined, as before, as the annihilator of the vector $x \epsilon \mathfrak{R}$ and let $\mathfrak{J}_x$ be the corresponding right ideal in $\mathfrak{A}$. If Φ denotes the totality of $\mathfrak{J}_x$, Φ satisfies our conditions. The *Φ-topology* thus defined corresponds to the finite topology in $\bar{\mathfrak{A}}$.

CASE 3. Let Δ denote the set consisting of $\mathfrak{A}$ and of all the maximal right ideals $\mathfrak{J}$ such that $(\mathfrak{J}:\mathfrak{A}) = 0$ in the primitive ring $\mathfrak{A}$. As has been shown, $\Delta \supseteq \Phi$ where Φ is determined by any $(1 - 1)$ representation of $\mathfrak{A}$ as a dense ring of l.t. Hence $\bigcap_\Delta \mathfrak{J} = (0)$. Now let $\mathfrak{J} \neq \mathfrak{A}$ be in Δ. Consider the difference $\mathfrak{A}$-group $\mathfrak{A}/\mathfrak{J}$. Since $\mathfrak{J}$ is maximal, $\mathfrak{A}/\mathfrak{J}$ is irreducible. The elements b of $\mathfrak{A}$ induce endomorphisms $x + \mathfrak{J} \rightarrow xb + \mathfrak{J}$ in $\mathfrak{A}/\mathfrak{J}$. The totality of these mappings is an irreducible ring $\bar{\mathfrak{A}}$ of endomorphisms, a homomorphic image of $\mathfrak{A}$. The kernel of the homomorphisms between $\mathfrak{A}$ and $\bar{\mathfrak{A}}$ is $(\mathfrak{J}:\mathfrak{A})$. Hence $\mathfrak{A} \cong \bar{\mathfrak{A}}$. Now if a is any element of $\mathfrak{A}$, $(\mathfrak{J}:a)$ corresponds to the annihilator $\bar{\mathfrak{J}}_{\bar{a}}$ of $\bar{a} = a + \mathfrak{J}$ in $\mathfrak{A}/\mathfrak{J}$. Hence $(\mathfrak{J}:a) = \mathfrak{A}$ or $(\mathfrak{J}:a)$ is maximal. Moreover it is known that if $\bar{\mathfrak{J}}_{\bar{a}}$ is a maximal right ideal then $(\bar{\mathfrak{J}}_{\bar{a}}:\bar{\mathfrak{A}}) = 0$.[11] Hence if $(\mathfrak{J}:a) \neq \mathfrak{A}$, this ideal is in Δ. This proves that Δ satisfies our conditions.

Evidently if $\mathfrak{A}$ is a semi-simple ring such that the Γ-topology, Γ a set of maximal ideals plus $\mathfrak{A}$, is discrete then (0) is an intersection of a finite number of maximal ideals. This implies that the descending chain condition holds and conversely.

---

[9] We assume the definition of semi-simplicity given in [4].

[10] [4] p. 311.

[11] [4] p. 312.

It need hardly be added that the above discussion can be duplicated for left ideals.

5. Rings containing minimal ideals

The following lemmas are well known.

LEMMA 1. *If* $\mathfrak{J} = e\mathfrak{A}$, $e^2 = e \neq 0$ *then the ring of endomorphisms of the* $\mathfrak{A}$*-group* $\mathfrak{J}$ *is isomorphic to* $e\mathfrak{A}e$.

LEMMA 2. *Any minimal non-nilpotent right ideal in a ring* $\mathfrak{A}$ *has the form* $e\mathfrak{A}$ *where* $e^2 = e \neq 0$.

We may now prove

THEOREM 5. *If* $\mathfrak{A}$ *is a ring that contains a minimal right ideal but no nilpotent ideals, then* $\mathfrak{A}$ *contains a minimal left ideal.*[12]

Let $\mathfrak{J}$ be a minimal right ideal. Then $\mathfrak{J} = e\mathfrak{A}$, $e^2 = e \neq 0$ and the ring of endomorphisms of $\mathfrak{J}$ is isomorphic to $e\mathfrak{A}e$. Since $\mathfrak{J}$ is minimal $e\mathfrak{A}e$ is a division ring. Set $\mathfrak{J}' = \mathfrak{A}e$. This is a left ideal $\neq 0$. Let $\mathfrak{J}_1'$ be a left ideal such that $\mathfrak{J}' \supseteq \mathfrak{J}_1' \neq 0$. If every element of $\mathfrak{J}_1'$ is annihilated on the left by e, $(\mathfrak{J}_1')^2 = (\mathfrak{J}_1'e)\mathfrak{J}_1' = 0$ contrary to assumption. Hence there is an element $u \in \mathfrak{J}_1'$ such that $eu \neq 0$. Hence $eue = eu \neq 0$. Since $e\mathfrak{A}e$ is a division ring we can solve $(exe)(eue) = e$. Thus $e = exeue = (exe)u \in \mathfrak{J}_1'$ and $\mathfrak{J}_1' = \mathfrak{J}'$. Hence $\mathfrak{J}'$ is minimal.

THEOREM 6. *If* $\mathfrak{A}$ *is a ring that has no nilpotent ideals, then the left annihilator of any minimal right ideal is a maximal left ideal.*

Let $\mathfrak{J} = e\mathfrak{A}$, $e^2 = e \neq 0$. The left annihilator of $e\mathfrak{A}$ is the same as the left annihilator of the element e. Hence it is identical with the set $\mathfrak{M}'$ of elements $a - ae$. Now we have the Peirce decomposition $\mathfrak{A} = \mathfrak{A}e + \mathfrak{M}'$, $\mathfrak{A}e \cap \mathfrak{M}' = 0$. Since $\mathfrak{A}e$ is minimal and isomorphic to the left $\mathfrak{A}$-group $\mathfrak{A}/\mathfrak{M}'$, $\mathfrak{M}'$ is maximal.

Now let $\mathfrak{A}$ be a primitive ring that contains minimal ideals. It is known that the join $\mathfrak{F}$ of all the minimal right ideals is a two-sided ideal contained in every two-sided ideal $\neq 0$ of $\mathfrak{A}$.[13] Since $\mathfrak{A}$ is semi-simple $\mathfrak{F}^2 = \mathfrak{F}$ so that for any two-sided ideal $\mathfrak{B} \neq 0$ we have $\mathfrak{B}\mathfrak{F} = \mathfrak{F}\mathfrak{B} = \mathfrak{F}$.

THEOREM 7. *Any primitive ring that contains minimal ideals is left primitive.*

Let $\mathfrak{J} = e\mathfrak{A}$, $e^2 = e \neq 0$ be a minimal right ideal. Then the proof of Theorem 6 shows that the left annihilator $\mathfrak{M}'$ of $e\mathfrak{A}$ is a maximal left ideal. If $(\mathfrak{M}':\mathfrak{A})_l \neq 0$, $(\mathfrak{M}':\mathfrak{A})_l \supseteq \mathfrak{F}$. Hence $\mathfrak{F} = \mathfrak{F}\mathfrak{A} \subseteq \mathfrak{M}'$ and $\mathfrak{F}\mathfrak{J} = 0$. Since $\mathfrak{F} \supseteq \mathfrak{J}$ this implies that $\mathfrak{J}^2 = 0$ contrary to the semi-simplicity.

It is clear by symmetry that $\mathfrak{F}$ can also be defined as the join of all the minimal left ideals of $\mathfrak{A}$.

6. The left vector-space determined by a primitive ring possessing minimal ideals

If $\mathfrak{A}$ is a primitive ring that contains minimal ideals, it is known that $\mathfrak{A}$ has essentially only one $(1 - 1)$ irreducible representation. By this is meant that any two irreducible $\mathfrak{A}$-groups that define $(1 - 1)$ representations are $\mathfrak{A}$-isomor-

[12] For simple rings this theorem is due to Artin and Whaples [1] p. 92.

[13] [4] p. 317.

phic.[14] Let $\mathfrak{R}$ be such an $\mathfrak{A}$-group and let $\bar{\mathfrak{A}}$ denote the ring of representing endomorphisms. We know, of course, that $\bar{\mathfrak{A}}$ is a dense ring of l.t. in $\mathfrak{R}$ regarded as a vector space over the division ring $\mathfrak{D}$ of endomorphisms commutative with the $A \in \bar{\mathfrak{A}}$. It is known also that $\bar{\mathfrak{A}}$ (hence $\mathfrak{A}$) has minimal ideals if and only if $\bar{\mathfrak{A}}$ contains finite-valued l.t. $\neq 0$. Moreover, the minimal two-sided ideal $\bar{\mathfrak{F}}$ of $\bar{\mathfrak{A}}$ coincides with the totality of finite valued l.t. contained $\bar{\mathfrak{A}}$.[15]

Now if F is any finite valued l.t. and $\mathfrak{R}F$ has the basis $y_1, y_2, \cdots, y_r$ then

$$xF = y_1x_1'(x) + y_2x_2'(x) + \cdots + y_rx_r'(x) \tag{2}$$

where $x_i'(x) \in \mathfrak{D}$. Since the y's form a basis the expression (2) is unique and the x_i' are therefore single valued functions from $\mathfrak{R}$ to $\mathfrak{D}$. It is clear that these functions are linear:

$$x'(x + y) = x'(x) + x'(y), \quad x'(x\alpha) = x'(x)\alpha. \tag{3}$$

Since $y_i \in \mathfrak{R}F$ there is a vector x_i such that $x_j'(x_i) = \delta_{ij}$. It follows that the x_i' are left linearly independent and in particular that each $x_i' \neq 0$. If $F \in \bar{\mathfrak{A}}$ and u is any vector $\neq 0$ there exists an $A \in \bar{\mathfrak{A}}$ such that $y_iA = u\delta_{ij}$. Then $xFA = ux_j'(x)$ and $FA \in \bar{\mathfrak{F}}$. Thus we see that $\bar{\mathfrak{F}}$ contains mappings $\neq 0$ that send $\mathfrak{R}$ into any one-dimensional space $[u]$. The totality $\bar{\mathfrak{F}}'_{[u]}$ of these mappings is a left ideal.

Consider now a particular $\bar{\mathfrak{F}}' = \bar{\mathfrak{F}}'_{[u]}$ and a particular $u \neq 0$ in $[u]$. If $B \in \bar{\mathfrak{F}}'$, $xB = ux'(x)$, $x'(x)$ a linear function. Denote the totality of x' determined in this way by $\mathfrak{R}'$. Then $\mathfrak{R}'$ is a left vector space over $\mathfrak{D}$. For if $B_1, B_2 \in \bar{\mathfrak{F}}'$, $xB_1 = ux_1'(x)$ and $xB_2 = ux_2'(x)$. Hence $x(B_1 + B_2) = u(x_1'(x) + x_2'(x))$. Thus $x_1' + x_2' \in \mathfrak{R}'$. If α is any element of $\mathfrak{D}$ there exists an $A \in \bar{\mathfrak{A}}$ such that $uA = u\alpha$. Then $xBA = (ux'(x))A = u\alpha x'(x)$. Thus $BA \in \bar{\mathfrak{F}}'$ and $\alpha x'(x) \in \mathfrak{R}'$.

If v is a second non-zero vector in $\mathfrak{R}$ there is an A such that $uA = v$. Then $xBA = vx'(x)$. Thus $\mathfrak{R}'$ is independent of the choice of u in $[u]$ and of the choice of the one-dimensional subspace $[u]$. Our discussion shows also that if $y_1, y_2, \cdots, y_r$ are any vectors in $\mathfrak{R}$ and $x_1', x_2', \cdots, x_r'$ are any forms in $\mathfrak{R}'$ then the l.t. $x \to y_ix_i'(x) \in \bar{\mathfrak{F}}$. Hence also the l.t. $x \to \sum y_ix_i'(x) \in \bar{\mathfrak{F}}$. Conversely by the argument used at the beginning of this section, if $F \in \bar{\mathfrak{F}}$ then F has this form. These results therefore give a complete analysis of the l.t. of $\bar{\mathfrak{F}}$ in terms of the forms in $\mathfrak{R}'$.

Now let A be any l.t. in $\bar{\mathfrak{A}}$ and x' any form in $\mathfrak{R}'$. Then $x'(xA)$ is a linear form in $\mathfrak{R}'$. For if $xB = ux'(x)$ then $xAB = ux'(xA)$. Thus each A determines a mapping $A' : x' \to x'A' \equiv x'(xA)$ of $\mathfrak{R}'$ into itself. It is clear that A' is linear in $\mathfrak{R}'$. Moreover,

$$\begin{aligned} x'(A + B)'(x) &= x'(x(A + B)) = x'(xA) + x'(xB) \\ &= x'A(x) + x'B(x) \end{aligned}$$

[14] [4] p. 318.
[15] [4] p. 317.

and

$$x'(AB)'(x) = x'(xAB) = x'B'(xA) = x'B'A'(x)$$

so that $(A + B)' = A' + B'$ and $(AB)' = B'A'$. If $A \neq 0$ there exists an x such that $y = xA \neq 0$ and if y is any vector $\neq 0$ the density of $\bar{\mathfrak{F}}$ implies that there is an l.t. $C \in \bar{\mathfrak{F}}$ such that $yC \neq 0$. Hence there is an x' such that $x'(y) \neq 0$. Thus $x'(xA) \neq 0$ and $x'A'(x) \neq 0$ so that $A' \neq 0$. These results show that the set $\bar{\mathfrak{A}}'$ of l.t. A' is a ring and the correspondence $A \rightarrow A'$ is an anti-isomorphism between $\bar{\mathfrak{A}}$ and $\bar{\mathfrak{A}}'$.

Suppose now that $\mathfrak{J}'$ is any left ideal in $\bar{\mathfrak{A}}$ and let $\mathfrak{S} = \mathfrak{R}\mathfrak{J}'$. We have seen that any $u \in \mathfrak{S}$ has the form vB, $B \in \mathfrak{J}'$. If x' is arbitrary in $\mathfrak{R}'$, the l.t. A such that $xA = vx'(x)$ is in $\bar{\mathfrak{A}}$. Hence $AB \in \mathfrak{J}'$ and $xAB = vBx'(x) = ux'(x)$. Thus we see that if u is any vector in $\mathfrak{S}$ and x' is arbitrary in $\mathfrak{R}'$ then $x \rightarrow ux'(x)$ is in $\mathfrak{J}'$. More generally it is clear that if $u_1, u_2, \cdots, u_r$ are arbitrary vectors in $\mathfrak{S}$ and $x_1', x_2', \cdots, x_r'$ are arbitrary in $\mathfrak{R}'$ then the l.t. $x \rightarrow \sum u_i x_i'(x) \in \mathfrak{J}'$. Evidently this l.t. belongs also to $\bar{\mathfrak{F}}$. Conversely, it is readily seen that any l.t. in $\mathfrak{J}_0' = \mathfrak{J}' \cap \bar{\mathfrak{F}}$ has this form. Thus the subspace $\mathfrak{S}$ determines the ideal $\mathfrak{J}_0'$ and we have $\mathfrak{S} = \mathfrak{R}\mathfrak{J}_0'$.

If $\mathfrak{S}$ is any subspace of $\mathfrak{R}$ the totality of l.t. in $\bar{\mathfrak{A}}$ that map $\mathfrak{R}$ into $\mathfrak{S}$ is a left ideal $\mathfrak{J}'$. Since $\mathfrak{J}'$ includes all the l.t. $x \rightarrow \sum u_i x_i'(x)$ where $u_i \in \mathfrak{S}$ and x_1' are arbitrary, $\mathfrak{R}'\mathfrak{J}' = \mathfrak{S}$.

These considerations show that there is a $(1 - 1)$ correspondence between the subspaces of $\mathfrak{R}$ and the left ideals $\mathfrak{J}_0'$ contained in $\bar{\mathfrak{F}}$: Any left ideal $\mathfrak{J}_0'$ consists of the l.t. of the form $x \rightarrow \sum u_i x_i'(x)$, u_i in a subspace $\mathfrak{S}$.[16] Clearly this correspondence is a lattice isomorphism. The minimal ideals are in (1—1) correspondence with the one-dimensional subspaces $[u]$. Any minimal ideal is of the form $\mathfrak{J}_{[u]}'$.

7. Dual vector spaces

In order to emphasize that $\mathfrak{R}$ and $\mathfrak{R}'$ enter symmetrically in the above discussion we now denote the element $x'(x)$ in $\mathfrak{D}$ by (x', x). If x' and x are regarded as variables in $\mathfrak{R}'$ and $\mathfrak{R}$, then (x', x) is a *bilinear form* in the sense that

$$\begin{aligned} &(x', x + y) = (x', x) + (x', y), \qquad (x', x\alpha) = (x', x)\alpha \\ &(x' + y', x) = (x', x) + (y', x), \qquad (\alpha x', x) = \alpha(x', x). \end{aligned} \tag{4}$$

The first line is clear from the linearity of $x'(x)$ and the second follows from the definitions of addition and left scalar multiplication of linear forms. Our bilinear form is *non-degenerate* in the sense that $(z', x) = 0$ for all x implies that $z' = 0$ and $(x', z) = 0$ for all x' implies that $z = 0$. Now if $\mathfrak{R}'$ is any left $\mathfrak{D}$-space and $\mathfrak{R}$ is any (right) $\mathfrak{D}$-space for which there exists a non-degenerate bilinear form (x', x) then we shall say that $\mathfrak{R}'$ and $\mathfrak{R}$ are *dual* vector spaces relative to this form.

[16] Cf. [2] p. 20.

Now suppose that $\mathfrak{R}'$ and $\mathfrak{R}$ are any two dual vector spaces. The vectors y' and y of $\mathfrak{R}'$ and $\mathfrak{R}$, respectively, will be called *incident* if $(y', y) = 0$. If S is a subset of $\mathfrak{R}$ ($\mathfrak{R}'$) we denote the subspace of $\mathfrak{R}'(\mathfrak{R})$ of vectors incident to every vector in S by $j(S)$. The following rules hold:

$$j(S_1 \cup S_2) = j(S_1) \cap j(S_2), \qquad j(j(S)) \supseteq S.$$

Here $\cup$ denotes logical sum and $\cap$ intersection.

We now introduce an *$\mathfrak{R}'$-topology* in $\mathfrak{R}$ by prescribing as neighborhoods of x the cosets $x + j(S)$ where S is any finite subset of $\mathfrak{R}'$. Since $j(S_1 \cup S_2) = j(S_1) \cap j(S_2)$ and $\cap j(S) = 0$ we obtain in this way a T_1-space. It is easily seen that $x\alpha$ is a continuous function of x in $\mathfrak{R}$ and α in the discrete space $\mathfrak{D}$ and that $\mathfrak{R}$ is a totally disconnected topological group.

Our study of the $\mathfrak{R}'$ topology will be based on the following lemmas.

LEMMA 1. *If $x_1', x_2', \cdots, x_n'$ are linearly independent linear forms there exist vectors $x_1, x_2, \cdots, x_n$ in $\mathfrak{R}$ such that $x_i'(x_j) = \delta_{ij}$.*

If $x_1' \neq 0$ there is a y such that $x_1'(y) = \beta \neq 0$. Then $x_1 = y\beta^{-1}$ satisfies $x_1'(x_1) = 1$. Suppose now that $x_1, \cdots, x_r$ have already been determined so that $x_i'(x_j) = \delta_{ij}, i, j = 1, \cdots, r$. For any x we define $y = x - \sum_1^r x_i x_i'(x)$. Then $x_i'(y) = 0$ for $i = 1, \cdots, r$. There exists an x such that $x_{0r+1}'(y) \neq 0$. For otherwise we have

$$x_{r+1}'(x) = x_{r+1}'(x_1)x_1'(x) + \cdots + x_{r+1}'(x_r)x_r(x)$$

for all x and this contradicts the independence of the x_i'. Hence we may choose an x so that $x_{r+1}'(y) = \beta \neq 0$. Set $x_{r+1} = y\beta^{-1}$. This gives a set $x_1, \cdots, x_{r+1}$ of the required type.

LEMMA 2. *Let $x_1', x_2', \cdots, x_m', x'$ be linear forms and suppose that $x'(y) = 0$ for all y such that $x_i'(y) = 0, i = 1, 2, \cdots, m$. Then x' is a left linear combination of the x_i'.*

We may suppose that the x_i' are linearly independent. Then if x' is not linearly dependent on the x_i' the set $x_1', x_2', \cdots, x_m', x'$ is an independent set. Hence by Lemma 1 there exists an x such that $x_i'(x) = 0$ for $i = 1, \cdots, m$ but $x'(x) \neq 0$. This contradicts the assumption.

We consider now an l.t. $\mathfrak{A}$ in $\mathfrak{R}$. Such a mapping is continuous in the $\mathfrak{R}'$-topology if and only if it is continuous at 0. Hence the condition for continuity is that for any given $j(x')$ there exist $x_1', x_2', \cdots, x_n'$ such that if $y \in \cap j(x_i')$ then $yA \in j(x')$. Now for any fixed x', the function (x', xA) is linear in x. For any y such that $(x_i', y) = 0, i = 1, \cdots, m$ we have $(x', xA) = 0$. It follows from Lemma 2 that there exists a $y' = \sum \beta_i x_i'$ such that $(x', xA) = (y', x)$ for all x. Since our bilinear form is non-degenerate y' is uniquely determined, so that the transformation $A': x' \rightarrow y'$ is single valued. Clearly we have $(x', xA) = (x'A', x)$ for all x and x'.

DEFINITION. A transformation A' in $\mathfrak{R}'$ is said to be an *adjoint* of a transformation A in $\mathfrak{R}$ if $(x', xA) = (x'A', x)$ for all x and x'.

If such an A' exists it is unique and it is an l.t. in $\mathfrak{R}'$. The above remarks show that if A is a continuous l.t. then A has an adjoint A'. Conversely if A is an l.t. that has an adjoint A', A is continuous. For, if $j(x')$ is given let $y' = x'A'$. Then clearly $(y', y) = 0$ implies that $(x', yA) = 0$. This completes the proof of

LEMMA 3. *A l.t. in $\mathfrak{R}$ over $\mathfrak{D}$ is continuous in the $\mathfrak{R}'$-topology if and only if it has an adjoint.*

If $y_1, y_2, \cdots, y_r \in \mathfrak{R}$ and $x_1', x_2', \cdots, x_r' \in \mathfrak{R}'$ then the mapping $x \rightarrow \sum y_i(x_i', x)$ is an l.t. F in $\mathfrak{R}$. Clearly F is finite-valued. Moreover we can verify that F has the adjoint $x' \rightarrow \sum (x', y_i)x_i'$. Conversely let F be any finite valued continuous l.t. If $\mathfrak{R}F$ has the basis $y_1, y_2, \cdots, y_r$ then $xF = y_1\phi_1(x) + \cdots + y_r\phi_r(x)$. Choose x_i' so that $(x_i', y_j) = \delta_{ij}$. Then $\phi_i(x) = (x_i', xF) = (y_i', x)$ where $y_i' = x_i'F'$. Hence $xF = \sum y_i(y_i', x)$.

We may now state the results that we established in the preceding section in the following way:

THEOREM 8. *If $\mathfrak{A}$ is a primitive ring that contains minimal ideals there exist a pair of dual vector spaces $\mathfrak{R}'$ and $\mathfrak{R}$ such that $\mathfrak{A}$ is isomorphic to a ring of continuous* l.t. *in $\mathfrak{R}$ (relative to the $\mathfrak{R}'$-topology) that contains the ring of finite valued continuous* l.t.

We have shown also that the minimal two-sided ideal $\bar{\mathfrak{F}}$ of $\bar{\mathfrak{A}}$ is the totality of finite valued continuous l.t. in $\mathfrak{R}$. We have also seen that the totality of adjoints A' of the $A \in \bar{\mathfrak{A}}$ is a ring $\bar{\mathfrak{A}}'$ anti-isomorphic to $\bar{\mathfrak{A}}$. Moreover, $\bar{\mathfrak{A}}'$ is a ring of l.t. that are continuous in the $\mathfrak{R}$-topology of $\mathfrak{R}'$ and $\bar{\mathfrak{A}}'$ contains all of the finite valued continuous l.t. The converse of Theorem 8 is the following

THEOREM 9. *Let $\mathfrak{R}'$ and $\mathfrak{R}$ be dual vector spaces and let $\bar{\mathfrak{A}}$ be a ring of l.t. in $\mathfrak{R}$ continuous relative to the $\mathfrak{R}'$-topology and containing all the finite valued continuous* l.t. *Then $\bar{\mathfrak{A}}$ is a primitive ring that contains minimal ideals.*

Let $x_1, x_2, \cdots, x_r$ be linearly independent in $\mathfrak{R}$ and $y_1, y_2, \cdots, y_r$ arbitrary. Choose $x_1', x_2', \cdots, x_r'$ so that $(x_i', x_j) = \delta_{ij}$. Then the l.t. $x \rightarrow \sum y_i(x_i', x) \in \bar{\mathfrak{A}}$ and maps x_i into y_i. Hence $\bar{\mathfrak{A}}$ is dense and consequently primitive. Since $\bar{\mathfrak{A}}$ contains finite valued l.t. $\neq 0$, $\bar{\mathfrak{A}}$ contains minimal ideals.

EXAMPLE. Let $\mathfrak{R}$ be a vector space with a denumerable basis $x_1, x_2, \cdots$ and let $\phi(x)$ be a linear function defined on $\mathfrak{R}$. Then $\phi(x)$ is determined by the values $\phi(x_i)$ and, moreover, we can choose an arbitrary sequence $(\alpha_1, \alpha_2, \cdots)$ and determine a ϕ such that $\phi(x_i) = \alpha_i$. Thus the space $\mathfrak{R}^*$ of all linear functions is equivalent to the space of all denumerable sequences $(\alpha_1, \alpha_2, \cdots)$, α_i in $\mathfrak{D}$.

1) Let $\mathfrak{R}'$ be the space of linear functions whose corresponding sequences all have 0's after a certain point. $\mathfrak{R}'$ has the basis x_i' where $(x_i', x_j) \equiv x_i'(x_j) = \delta_{ij}$. If $x = \sum x_i\alpha_i \neq 0$ some α_i, say α_r, $\neq 0$. Then $(x_r', x) = \alpha_r \neq 0$. It follows that $\mathfrak{R}'$ and $\mathfrak{R}$ are dual. If A is an l.t. in $\mathfrak{R}$ we write $x_iA = \sum x_j\alpha_{ji}$ and associate with A the matrix (α_{ij}) that has only a finite number of non-zero elements in each column. The correspondence $A \rightarrow (\alpha)$ is an anti-isomorphism between the ring $\bar{\mathfrak{L}}$ of all l.t. in $\mathfrak{R}$ and the ring of all matrices of the type described. If A' is an l.t. in $\mathfrak{R}'$ we write $x_i'A' = \sum \alpha_{ij}'x_i'$ and obtain in this way a matrix

that has only a finite number of non-zero elements in each row. Here the correspondence $A' \to (\alpha')$ is an isomorphism. Suppose now that A and A' are adjoint. Then $(x'_i, x_jA) = (x'_iA', x_j)$ for all i, j. This implies that $\alpha_{ij} = \alpha'_{ij}$. Hence the existence of an adjoint of A implies that there are only a finite number of non-zero elements in each row as well as in each column of (α). This condition is also sufficient. Hence the ring $\bar{\mathfrak{A}}$ of all the l.t. in $\mathfrak{R}$ that are continuous in the $\mathfrak{R}'$-topology is anti-isomorphic to the ring of all the matrices $(\alpha_{ij}), i, j = 1, 2, \cdots$ in which there are only a finite number of non-zero elements in each row and column.

An l.t. F in $\mathfrak{R}$ is finite valued if and only if the expressions $x_iF = \sum x_j\beta_{ji}$ involve only a finite number of the x's, i.e. if $\beta_{ji} = 0$ for $j > m$. This means that all the rows after the mth are 0. If, in addition, F is continuous then all the columns of F after a certain one are 0. Thus the matrix of F has a non-zero diagonal block of finite size while outside of this block all of the elements are 0. The ring $\bar{\mathfrak{F}}$ of continuous finite valued l.t. is therefore anti-isomorphic to the ring of matrices of this form.

2) Let $\mathfrak{R}' = \mathfrak{R}^*$ the complete space of linear functions. If $x' \epsilon \mathfrak{R}'$ and A is any l.t., $x'(xA)$ is a linear function. It follows that every l.t. is continuous in the $\mathfrak{R}'$-topology. Hence $\bar{\mathfrak{F}}$ consists of all the finite valued l.t. As before these l.t. correspond to the matrices that have only a finite number of non-zero rows.

8. One-sided ideals in $\mathfrak{A}$

Let $\bar{\mathfrak{J}}'$ be a left ideal in $\bar{\mathfrak{A}}$ and let $\bar{\mathfrak{J}}'_0 = \bar{\mathfrak{J}}' \cap \bar{\mathfrak{F}}$. We have seen that $\mathfrak{S} = \mathfrak{R}\bar{\mathfrak{J}}' = \mathfrak{R}\bar{\mathfrak{J}}'_0$ and that $\bar{\mathfrak{J}}'_0$ consists of the l.t. $x \to \sum y_i(x'_i, x)$ where $y_i \epsilon \mathfrak{S}$ and the x'_i are arbitrary in $\mathfrak{R}'$. It follows that $\bar{\mathfrak{J}}' \supseteq \bar{\mathfrak{F}}$ if and only if $\mathfrak{S} = \mathfrak{R}$. If $\mathfrak{S}$ is any subspace of $\mathfrak{R}$ the totality of l.t. that map $\mathfrak{R}$ into $\mathfrak{S}$ is a left ideal $\bar{\mathfrak{J}}'$ in $\bar{\mathfrak{A}}$ and $\mathfrak{R}\bar{\mathfrak{J}}' = \mathfrak{R}\bar{\mathfrak{J}}'_0 = \mathfrak{S}$ for $\bar{\mathfrak{J}}'_0 = \bar{\mathfrak{J}}' \cap \bar{\mathfrak{F}}$. It follows that if $\bar{\mathfrak{M}}'$ is any maximal left ideal then either $\mathfrak{R}\bar{\mathfrak{M}}' = \mathfrak{R}$ or $\mathfrak{R}\bar{\mathfrak{M}}'$ is a subspace that has a one dimensional complement. Such subspaces will be called *hyperplanes*. We have also seen that the minimal left ideals in $\bar{\mathfrak{A}}$ are precisely the ideals $\bar{\mathfrak{J}}'_{[u]}$ that map $\mathfrak{R}'$ into one dimensional subspaces $[u]$.

The results on left ideals can be transferred to right ideals by making use of the ring $\bar{\mathfrak{A}}'$ of adjoints of the l.t. in $\bar{\mathfrak{A}}$. Thus we see that if $\bar{\mathfrak{J}}$ is a right ideal then $\bar{\mathfrak{J}}_0 = \bar{\mathfrak{J}} \cap \bar{\mathfrak{F}}$ is the totality of l.t. $x \to \sum x_i(y'_i, x)$ where the x_i are arbitrary and the y'_i are in a subspace $\mathfrak{S}'$ of $\mathfrak{R}'$. Hence $\mathfrak{S}' = \mathfrak{R}'\bar{\mathfrak{J}}'$ where $\bar{\mathfrak{J}}'$ is the set of adjoints of the l.t. in $\bar{\mathfrak{J}}$. It is clear from the form of the l.t. in $\bar{\mathfrak{J}}_0$ that the space $\mathfrak{N}$ of vectors annihilated by all the l.t. in this ideal is the space $j(\mathfrak{S}')$ of vectors incident to all the vectors in $\mathfrak{S}'$. We shall show now that $\mathfrak{N}$ is also annihilated by $\bar{\mathfrak{J}}$. For let $z \epsilon \mathfrak{N}$ and suppose that $zB \neq 0$ for some $B \epsilon \bar{\mathfrak{J}}$. Then there exists an $A \epsilon \bar{\mathfrak{F}}$ such that $zBA \neq 0$. Since $BA \epsilon \bar{\mathfrak{J}}_0$ this contradicts $z\bar{\mathfrak{J}}_0 = 0$. We now prove the following

THEOREM 10. *Let* $\bar{\mathfrak{A}}$ *be a dense ring of* l.t. *that contains the identity and that contains finite valued* l.t. $\neq 0$. *Let* $\bar{\mathfrak{J}}$ *be a maximal right ideal in* $\bar{\mathfrak{A}}$. *Then either*

$\bar{\mathfrak{J}} \supseteq \bar{\mathfrak{F}}$ *the totality of finite valued* l.t. *in* $\bar{\mathfrak{A}}$ *or* $\bar{\mathfrak{J}}$ *consists of the totality of* l.t. *in* $\bar{\mathfrak{A}}$ *that annihilate some* $x \neq 0$ *in* $\mathfrak{R}$.

Let $\mathfrak{A}$ be the abstract primitive ring isomorphic to $\bar{\mathfrak{A}}$. Then $\mathfrak{F}$ the correspondent of $\bar{\mathfrak{F}}$ is the minimal two-sided ideal $\neq 0$ of $\mathfrak{A}$. If $\mathfrak{J}$ is any right ideal in $\mathfrak{A}$, $(\mathfrak{J}:\mathfrak{A}) = 0$ if and only if $\mathfrak{J} \not\supseteq \mathfrak{F}$. For if $(\mathfrak{J}:\mathfrak{A}) \neq 0$ this two sided ideal contains $\mathfrak{F}$ and $\mathfrak{J} \supseteq \mathfrak{A}\mathfrak{F} = \mathfrak{F}$. On the other hand, if $\mathfrak{F} \subseteq \mathfrak{J}$ then $\mathfrak{A}\mathfrak{F} \subseteq \mathfrak{J}$ and $\mathfrak{F} \subseteq (\mathfrak{J}:\mathfrak{A})$ so that $(\mathfrak{J}:\mathfrak{A}) \neq 0$. Now let $\mathfrak{J}$ be the maximal right ideal corresponding to $\bar{\mathfrak{J}}$. If $\bar{\mathfrak{J}} \not\supseteq \bar{\mathfrak{F}}$, $\mathfrak{J} \not\supseteq \mathfrak{F}$ and $(\mathfrak{J}:\mathfrak{A}) = 0$. This implies that the $\mathfrak{A}$-group $\mathfrak{A}/\mathfrak{J}$ of cosets $x + \mathfrak{J}$ gives a $(1 - 1)$ representation of $\mathfrak{A}$. Moreover, since $\mathfrak{J}$ is maximal this $\mathfrak{A}$-group is irreducible. Hence $\mathfrak{A}/\mathfrak{J}$ is $\mathfrak{A}$-isomorphic to the given $\mathfrak{A}$-group $\mathfrak{R}$. Now it is clear that $\mathfrak{J}$ is the annihilator of the element $1 + \mathfrak{J}$. If x is the element of $\mathfrak{R}$ that corresponds to $1 + \mathfrak{J}$ in $\mathfrak{A}$-isomorphism between $\mathfrak{A}/\mathfrak{J}$ and $\mathfrak{R}$, then $\mathfrak{J}$ may also be defined as the annihilator of x in $\mathfrak{R}$. Thus $\bar{\mathfrak{J}}$ is the totality of elements B of $\bar{\mathfrak{A}}$ such that $xB = 0$. This proves that if $\bar{\mathfrak{J}} \not\supseteq \bar{\mathfrak{F}}$ then $\bar{\mathfrak{J}}$ is the annihilator of some vector $x \neq 0$ in $\mathfrak{R}$. Clearly if $\bar{\mathfrak{J}} \supseteq \bar{\mathfrak{F}}$ then $\bar{\mathfrak{J}}$ is dense in $\mathfrak{R}$ and the only element annihilated by all the l.t. of $\bar{\mathfrak{J}}$ is 0. This completes the proof.

9. The finite topology in a primitive ring that contains minimal ideals

For a primitive ring containing minimal ideals the most natural right ideal topology appears to be the one that corresponds to the finite topology determined by any $(1 - 1)$ irreducible representation of $\mathfrak{A}$. The set Φ consists of the ideals $\mathfrak{J}_x$ corresponding to the annihilators $\bar{\mathfrak{J}}_x$ of the vectors x in the representation space $\mathfrak{R}$. Clearly $\bar{\mathfrak{J}}_{x_1} \cap \cdots \cap \bar{\mathfrak{J}}_{x_m}$ is the annihilator of the finite dimensional space $[x_1, x_2, \cdots, x_m]$ generated by the x's. Hence the neighborhoods of 0 can be defined to be the ideals corresponding to the annihilators of finite dimensional subspaces of $\mathfrak{R}$. Now if $\mathfrak{S}$ is any finite dimensional subspace there exists an l.t. F in $\bar{\mathfrak{F}}$ such that $\mathfrak{R}F = \mathfrak{S}$. Clearly the annihilator of $\mathfrak{S}$ coincides with the right annihilator of the element F. Thus the neighborhoods of 0 are the right annihilators of the elements $f \in \mathfrak{F}$.

We have seen that any minimal left ideal $\bar{\mathfrak{J}}'$ of $\bar{\mathfrak{A}}$ consists of the totality of l.t. in $\bar{\mathfrak{A}}$ that map $\mathfrak{R}$ into a one dimensional subspace. It follows that Φ may also be characterized as the set consisting of $\mathfrak{A}$ and the right annihilators of the minimal left ideals of $\mathfrak{A}$.

If we make use of the $\mathfrak{R}'$-topology of $\mathfrak{R}$ we are led to consider another topology in $\bar{\mathfrak{A}}$, namely, the uniform topology in this set of transformations. In this topology the neighborhoods of 0 are the sets of l.t. that map $\mathfrak{R}$ into neighborhoods of 0 in the $\mathfrak{R}'$-topology. We recall that a neighborhood of 0 in $\mathfrak{R}$ is a finite intersection of hyperplanes $j(x')$, $x' \in \mathfrak{R}'$. Hence a neighborhood of the 0-transformation in the uniform topology is the set of l.t. A such that $(x_i', xA) = 0$ for all x and a particular finite set $x_1', x_2', \cdots, x_m'$ in $\mathfrak{R}'$. Now $(x_i', xA) = 0$ is equivalent to $x_i'A' = 0$. Hence the neighborhoods here correspond to the annihilators of the finite dimensional subspaces of $\mathfrak{R}'$. Thus the uniform topology in $\bar{\mathfrak{A}}$ corresponds to the finite topology in $\bar{\mathfrak{A}}'$. Interchanging $\mathfrak{R}$ and $\mathfrak{R}'$ we obtain still another definition of the finite topology in $\mathfrak{R}$, namely, the neighborhoods of 0 in this topology correspond to the images of the neighborhoods of 0 in the uniform topology of $\bar{\mathfrak{A}}'$ acting in the $\mathfrak{R}$-space $\mathfrak{R}'$.

If $\mathfrak{A}$ has an identity two other definitions of Φ are available, namely, Φ consists of $\mathfrak{A}$ and the maximal ideals $\mathfrak{J}$ such that $(\mathfrak{J}:\mathfrak{A}) = 0$. Thus Φ coincides with the set Δ of §4. Finally, Φ consists of $\mathfrak{A}$ and the maximal right ideals that do not contain $\mathfrak{F}$. These results follow from Theorem 10.

10. The closure of a one sided ideal

Let $\bar{\mathfrak{A}}$ be any dense ring of l.t. in $\mathfrak{R}$ topologized by means of the finite topology. If x is any vector in $\mathfrak{R}$, $\mathfrak{J}_x$ is open and closed. Since the annihilator of any subspace is an intersection of ideals $\mathfrak{J}_x$ it is a closed right ideal. We shall now show that any closed right ideal has this form. For this purpose we require the following extension of Theorem 3.

LEMMA. *Let $\bar{\mathfrak{J}}$ be a right ideal in a dense ring of* l.t. $\bar{\mathfrak{A}}$ *and let $\mathfrak{N}$ denote the subspace of vectors annihilated by all the elements of $\bar{\mathfrak{J}}$. Then if $x_1, x_2, \cdots, x_m$ are vectors in $\mathfrak{R}$ that are linearly independent* (mod $\mathfrak{N}$), *there exists an* l.t. C *in $\bar{\mathfrak{J}}$ such that $x_iC = y_i$ for arbitrary given y_i.*

By Theorem 3 there is a $B \epsilon \bar{\mathfrak{J}}$ such that $x_iB = x_i + z_i$, z_i in $\mathfrak{N}$. We may suppose that $z_1, z_2, \cdots, z_r$ is a basis for the set of z's. Then $x_1, \cdots, x_m$; $z_1, \cdots, z_r$ are linearly independent. Hence there exists an A in $\bar{\mathfrak{A}}$ such that $x_iA = x_i$ and $z_jA = 0$ for $j = 1, \cdots r$. Then $z_iA = 0$ for all i and $x_iBA = x_iA + z_iA = x_i$. There exists an A_1 in $\bar{\mathfrak{A}}$ such that $x_iA_1 = y_i$. Hence $C = BAA_1$ has the required properties.

THEOREM 11. *Let $\bar{\mathfrak{J}}$ be a right ideal in a dense ring of* l.t. $\bar{\mathfrak{A}}$, *let $\mathfrak{N}$ be the subspace annihilated by $\bar{\mathfrak{J}}$ and $\bar{\mathfrak{K}}$ the right ideal annihilating $\mathfrak{N}$. Then $\bar{\mathfrak{K}}$* is the closure $\bar{\mathfrak{J}}^*$ of $\bar{\mathfrak{J}}$ in the finite topology.

Since $\bar{\mathfrak{K}}$ is closed and contains $\bar{\mathfrak{J}}$, $\bar{\mathfrak{K}} \supseteq \bar{\mathfrak{J}}^*$. Now let $D \epsilon \bar{\mathfrak{K}}$ and let $\mathfrak{S}$ be any finite dimensional subspace. Set $\mathfrak{N}_1 = \mathfrak{S} \cap \mathfrak{N}$ and write $\mathfrak{S} = \mathfrak{N}_1 \oplus \mathfrak{S}_1$. Then if $x_1, \cdots, x_m$ is a basis for $\mathfrak{S}_1$ these vectors are linearly independent (mod $\mathfrak{N}$). Hence there exists a C in $\bar{\mathfrak{J}}$ such that $x_iC = x_iD$. Since C and D both annihilate $\mathfrak{N}_1$ these l.t. have the same effect on $\mathfrak{S}$. Since $\mathfrak{S}$ is an arbitrary finite dimensional subspace $D \epsilon \bar{\mathfrak{J}}^*$. Thus $\bar{\mathfrak{K}} = \bar{\mathfrak{J}}^*$.

COROLLARY. *A right ideal $\mathfrak{J}$ is closed if and only if it is the complete annihilator of some subspace of $\mathfrak{R}$.*

Theorem 11 also gives the following result on the finite topology in a primitive ring containing minimal ideals:

THEOREM 12. *Let $\mathfrak{J}$ be a right ideal in a primitive ring that contains minimal ideals and let $\mathfrak{J}_0'$ be the intersection of $\mathfrak{F}$ with the left annihilator of $\mathfrak{J}$. Then the closure $\mathfrak{J}^*$ of $\mathfrak{J}$ in the finite topology is the right annihilator of $\mathfrak{J}_0'$.*

Let $\bar{\mathfrak{A}}$ be a dense ring of l.t. isomorphic to $\mathfrak{A}$ and let $\bar{\mathfrak{J}}$ be the ideal corresponding to $\mathfrak{J}$. Let $\mathfrak{N}$ be the space annihilated by $\bar{\mathfrak{J}}$. If $F \epsilon \bar{\mathfrak{J}}_0'$ the ideal corresponding to $\mathfrak{J}_0'$, $\mathfrak{R}F \subseteq \mathfrak{N}$ and conversely. Hence $\bar{\mathfrak{J}}_0'$ is the left ideal contained in $\bar{\mathfrak{F}}$

that maps $\mathfrak{N}$ into $\mathfrak{N}$. The annihilator of $\mathfrak{N}$ coincides with the right annihilator of $\bar{\mathfrak{J}}_0'$. Hence the theorem follows from Theorem 11.

THEOREM 13. *Let $\bar{\mathfrak{J}}_0'$ be a left ideal contained in $\bar{\mathfrak{F}}$ the set of finite valued* l.t. *of the dense ring $\mathfrak{A}$. Then $\bar{\mathfrak{J}}_0'$ is closed in $\bar{\mathfrak{F}}$.*[17]

If C is in the closure of $\bar{\mathfrak{J}}_0'$ for any $F \in \bar{\mathfrak{F}}$ there exists a B_F in $\bar{\mathfrak{J}}_0'$ such that $F(C - B_F) = 0$. Now if C is finite valued there exists an F such that $FC = C$. Then $C = FB_F \in \bar{\mathfrak{J}}_0'$.

THE JOHNS HOPKINS UNIVERSITY

BIBLIOGRAPHY

[1] E. ARTIN AND G. WHAPLES, *The theory of simple rings*, Amer. Jour. Math., vol. 65 (1943), pp. 87–107.

[2] J. DIEUDONNE, *Sur le socle d'un anneau et les anneaux simples infinis*, Bull. Soc. Math. de France, vol. 71 (1943), pp. 1–30.

[3] N. JACOBSON, *Structure theory of simple rings without finiteness assumptions*, Trans. Amer. Math. Soc. vol. 57 (1945), pp. 228–245.

[4] ———, *The radical and semi-simplicity for arbitrary rings*, Amer. Jour Math. vol. 67 (1945), pp. 299–320.

[5] ———, *Structure theory for algebraic algebras of bounded degree*, Annals of Math., vol. 46 (1945), pp. 695–707.

[17] Cf. [2] p. 19.

A NOTE ON DIVISION RINGS.*

By N. Jacobson.

In this note we consider division subrings Φ of a division ring P such that the left dimensionality $(\mathrm{P} : \Phi)_l$ is finite. If $\mathfrak{L}$ denotes the ring of linear transformations in P regarded as a vector space over the set Φ_l of left multiplications by elements of Φ then clearly $\mathfrak{L}$ contains the subring P_r of right multiplications by elements of P. The relation $(\mathfrak{L} : \mathrm{P}_r)_r = (\mathrm{P} : \Phi)_l$ can be established (Theorem 1). Thus $\mathfrak{L}$ has finite right dimensionality over P_r. The principal result of this paper is that the converse of these results holds, namely, if $\mathfrak{A}$ is any ring of endomorphisms of the additive group of P such that 1) $\mathfrak{A}$ contains P_r and 2) $(\mathfrak{A} : \mathrm{P}_r)_r$ is finite, then $\mathfrak{A}$ coincides with a ring $\mathfrak{L}$ of linear transformations of P over Φ_l where Φ is a division subring of P such that P over Φ_l is finite. A complete reciprocity is established between the rings $\mathfrak{A}$ (or $\mathfrak{L}$) and the subrings Φ of P.

There are a number of applications of our results. One of these, the theory of automorphisms in a division ring, is developed here. In this connection we generalize the concept of a closed group of automorphisms introduced by Emmy Noether [1] for groups of inner automorphisms and we establish a Galois correspondence between closed groups that satisfy a certain finiteness condition and their division subrings of invariants. The theorem of Noether [2]-Artin and Whaples [3] on commutators of division subalgebras containing the center Γ and our own results on finite groups of outer automorphisms are contained as special cases of the theorems derived here.

1. Let P be a division ring and Φ a division subring of P of *finite left index* m. By this we mean that the left dimensionality $(\mathrm{P} : \Phi)_l = m < \infty$. Here we are regarding P as a left vector space (or module) over Φ or, what amounts to the same thing, as a group relative to the set Φ_l of left multiplications $\xi \rightarrow \xi\alpha_l \equiv \alpha\xi$, ξ in P, α in Φ. Let $\mathfrak{L}$ denote the ring of linear transformations (l. t.) of P over Φ_l. Thus $A \in \mathfrak{L}$ if and only if A is an endomorphism of the additive group of P and

* Received June 28, 1946.

[1] [5], p. 529.

[2] [5], p. 528.

[3] [1], p. 104.

$$(\alpha\xi)A = \alpha(\xi A)$$

for all $\alpha \varepsilon \Phi$ and all $\xi \varepsilon \mathrm{P}$. Evidently $\mathfrak{L}$ contains the set P_r of right multiplications $\xi \rightarrow \xi\rho_r \equiv \xi\rho$ in P. Since $1 \varepsilon \mathrm{P}_r$, $\mathfrak{L}$ may be regarded as a (right) vector space over P_r. When this is done we obtain the following theorem:

THEOREM 1. *The dimensionalities* $(\mathfrak{L} : \mathrm{P}_r)_r$ *and* $(\mathrm{P} : \Phi)_l$ *are equal.*[4]

Let $\xi_1, \xi_2, \cdots, \xi_m$ be a left basis of P over Φ. Then any l. t. in P over Φ_l is completely determined by its effect on the ξ_i. Now define C_j, $j = 1, 2, \cdots, m$, to be the l. t. such that

$$\xi_j C_j = 1 \quad \text{and} \quad \xi_i C_j = 0 \text{ if } i \neq j.$$

Then $\Sigma C_j \eta_{jr}$ maps ξ_i into η_i. Hence $\Sigma C_j \eta_{jr} = 0$ implies that all the $\eta_j = 0$. Thus the C's are right independent relative to P_r. On the other hand if A is any l. t. in P over Φ_l, let $\xi_i A = \eta_i$. Then A coincides with $\Sigma C_j \eta_{jr}$. Hence the C's form a right basis for $\mathfrak{L}$ over P_r.

2. We assume now that $\mathfrak{A}$ is any ring of endomorphisms acting in the additive group of P that 1) contains P_r and 2) has finite dimensionality $(\mathfrak{A} : \mathrm{P}_r)_r$. Let C be any endomorphism in P that commutes with every $A \varepsilon \mathfrak{A}$. Then since C commutes with every right multiplication, $C = \gamma_l$ a left multiplication.[5] The totality of γ obtained in this way is a division subring Φ of P that we shall call the division ring of $\mathfrak{A}$-*invariants.*

THEOREM 2. *Let* $\mathfrak{A}$ *be a ring of endomorphisms of the additive group of* P *such that* 1) $\mathfrak{A} \supseteq \mathrm{P}_r$ *and* 2) $(\mathfrak{A} : \mathrm{P}_r)_r = m < \infty$ *and let* Φ *be the division subring of* $\mathfrak{A}$-*invariants. Then* $(\mathrm{P} : \Phi)_l = m$ *and* $\mathfrak{A} = \mathfrak{L}$ *the complete ring of* l. t. *of* P *over* Φ_l.

Since $\mathfrak{A} \supseteq \mathrm{P}_r$, $\mathfrak{A}$ is an irreducible ring of endomorphisms in P. Since $(\mathfrak{A} : \mathrm{P}_r)_r < \infty$, $\mathfrak{A}$ satisfies the descending chain condition for right ideals. Now Φ_l is the division ring of endomorphisms commutative with the elements of the irreducible ring $\mathfrak{A}$. It follows from known results on irreducible rings of endomorphisms that $(\mathrm{P} : \Phi)_l < \infty$ and that $\mathfrak{A} = \mathfrak{L}$ the complete ring of l. t. of P over Φ_l.[6] By Theorem 1, $(\mathrm{P} : \Phi)_l = (\mathfrak{A} : \mathrm{P}_r)_r$.

To complete the reciprocity between Φ and $\mathfrak{L}$ we require the following theorem:

---

[4] The present proof of this theorem, which is simpler than my original one, was communicated to me by R. Brauer.

[5] [2], p. 54.

[6] [4], in particular Theorems 6 and 3.

THEOREM 3. *Let Φ be a division subring of finite left index* P *and let* $\mathfrak{L}$ *be the ring of* l. t. *in* P *over* Φ_l. *Then reciprocally* Φ_l *is the complete set of endomorphisms in* P *that commute with all the endomorphisms belonging to* $\mathfrak{L}$.

This result is well known for arbitrary vector spaces.[7] It can also be obtained as a consequence of Theorems 1 and 2. For let Ψ be the division ring of $\mathfrak{L}$-invariants. Then by Theorem 2, $(\mathrm{P} : \Psi)_l = (\mathfrak{L} : \mathrm{P}_r)_r$. On the other hand $\Psi \supseteq \Phi$ and $(\mathrm{P} : \Phi)_l = (\mathfrak{L} : \mathrm{P}_r)_r$ by Theorem 1. Hence $\Psi = \Phi$.

If $(\mathrm{P} : \Phi)_l < \infty$ and $\mathfrak{L}$ is the ring of l. t. of P over Φ_l our results establish a $(1-1)$ correspondence between the division rings Ψ between P and Φ and the rings $\mathfrak{B}$ between P_r and $\mathfrak{L}$. If Ψ is given then $\mathfrak{B}$ is the ring of l. t. of P over Ψ_l. Clearly $\mathrm{P}_r \subseteq \mathfrak{B} \subseteq \mathfrak{L}$. On the other hand if $\mathfrak{B}$ is a ring between P_r and $\mathfrak{L}$ then the set Ψ of $\mathfrak{B}$-invariants is a division subring containing Φ. These two correspondences are inverses of each other.

All of the above results can, of course, be duplicated for division subrings of finite right index. In the remainder of this note we consider some applications of these theorems to the theory of automorphisms in a division ring.

3. Let A be an automorphism in P. Then A is an endomorphism of the additive group of P and $(\xi\eta)A = (\xi A)(\eta A)$. Thus we have the following fundamental commutation rules governing automorphisms and multiplications:

$$\eta_r A = A(\eta A)_r \qquad \xi_l A = A(\xi A)_l. \tag{2}$$

We consider now the possible forms of linear relations with coefficients that are multiplications that can hold for automorphisms in P. First let $I_1, I_2, \cdots I_h$ be inner automorphisms, say, $\xi I_j = \tau_j \xi \tau_j^{-1}$. Then $I_j = \tau_{jl}\tau_{jr}^{-1}$ if τ_{jl} is the left multiplication determined by τ_j and τ_{jr} is the right multiplication determined by τ_j. Suppose that the τ's are linearly dependent over Γ where Γ is the center of P. Then there exist γ_i not all 0 such that $\Sigma\tau_i\gamma_i = 0$. Hence $\Sigma I_i\tau_{ir}\gamma_{ir} = \Sigma\tau_{il}\gamma_{ir} = \Sigma\tau_{il}\gamma_{il} = 0$. Since not every $\tau_{jr}\gamma_{jr}$ equals 0 this shows that the I_j are linearly dependent over P_r.

Conversely suppose that we have a relation $\Sigma I_i\eta_{ir} = 0$ where not every $\eta_{jr} = 0$. Then $\Sigma\tau_{il}\mu_{ir} = 0$ where $\mu_j = \tau_j^{-1}\eta_j$ so that not every $\mu_j = 0$. We assert that this implies that the τ_j are dependent over Γ. This is the following well-known lemma:

LEMMA 1. *If* $\tau_1, \tau_2, \cdots \tau_h$ *are elements of* P *that are linearly independent over* Γ *then the left multiplications* $\tau_{1l}, \tau_{2l}, \cdots, \tau_{hl}$ *are linearly independent over* P_r.

[7] [2], p. 22.

For the sake of completeness we give the proof. If the τ_l's are dependent over P_r there exists a shortest relation which we can suppose to be of the form

$$\tau_{1l}\mu_{1r} + \tau_{2l}\mu_{2r} + \cdots + \tau_{sl}\mu_{sr} = 0 \tag{3}$$

where all the μ_i are different from 0. We may assume also that $\mu_{1r} = 1_r$. Since the τ_j are independent over Γ, one of the μ's, say μ_2, is not in Γ. Hence there exists a ν such that $\mu_2\nu - \nu\mu_2 \neq 0$. Multiplication of (3) on the left and on the right by ν_r followed by subtraction of the resulting relations gives a new relation shorter than (3). This contradiction proves the result.

We suppose next that $A_1, A_2, \cdots, A_u$ are automorphisms that are in distinct cosets of the subgroup of inner automorphisms. Then no $A_iA_j^{-1}$, $i \neq j$, is inner. Let $\mathfrak{U}$ denote the ring of endomorphisms generated by the left and the right multiplications. The elements of $\mathfrak{U}$ have the form $U = \tau_{1l}\mu_{1r} + \tau_{2l}\mu_{2r} + \cdots$. We wish to prove the following lemma.

LEMMA 2. *If $A_1, A_2, \cdots, A_u$ are automorphisms such that no $A_iA_j^{-1}$, $i \neq j$, is inner, then a relation $\Sigma A_iU_i = 0$ where the $U_j \in \mathfrak{U}$ can hold only if every $U_j = 0$.*

We may choose elements $\tau_1, \tau_2, \cdots, \tau_h$ such that the left multiplications τ_{il} are independent over P_r and such that every $U_i = \Sigma\tau_{jl}\mu_{jir}$. Then our relation reads $\Sigma A_i\tau_{jl}\mu_{jir} = 0$. We shall show that every $\mu_{ji} = 0$. For otherwise we may suppose that $\Sigma A_i\tau_{jl}\mu_{jir} = 0$ is a shortest linear relation connecting the hu elements $A_i\tau_{jl}$ in which the coefficients $\mu_{jir} \neq 0$. We may suppose also that $\mu_{11} = 1$. We shall show that no A other than A_1 occurs in this relation. For suppose that

$$A_1\tau_{1l}\mu_{11r} + A_2\tau_{kl}\mu_{k2r} + \cdots = 0. \tag{4}$$

Multiply (4) on the left by ξ_r and on the right by $(\xi A_1)_r$ and subtract the resulting relations. This yields

$$A_2\tau_{kl}[(\xi A_2)_r\mu_{k2r} - \mu_{k2r}(\xi A_1)_r] + \cdots = 0. \tag{5}$$

If the bracket is 0 for all ξ

$$\xi A_2 = \mu_{k2}(\xi A_1)\mu_{k2}^{-1}$$

contrary to the assumption that A_1 and A_2 do not differ by an inner automorphism. Hence there is a ξ such that the bracket is not 0 and this gives a shorter relation than (4). Thus (4) must have the form

$$A_1(\tau_{1l}\mu_{11r} + \cdots) = 0.$$

Since A_1 is an automorphism this implies that $\tau_{1l}\mu_{11r} + \cdots = 0$ contrary to the assumption that the τ_{jl} are linearly independent over P_r.

Suppose finally that we have an arbitrary linear relation with coefficients that are in P_r connecting a finite set of automorphisms. We may write this in the form $\Sigma A_i I_j \mu_{jir} = 0$ where the I_j are inner and the A_i are in different cosets relative to the subgroup of inner automorphisms. If $I_j = \tau_{jl}\tau_{jr}^{-1}$, $\Sigma A_i \tau_{jl}\tau_{jr}^{-1}\mu_{jir} = 0$. Hence by Lemma 2, $\Sigma I_j \mu_{jir} = \Sigma \tau_{jl}\tau_{jir}^{-1}\mu_{jir} = 0$. By Lemma 1, either all the μ's are 0 or the τ_j are linearly dependent over Γ.

4. An element ϕ of P is *invariant* under an automorphism A if $\phi A = \phi$. By (2) ϕ has this property if and only if the left (right) multiplication $\phi_l(\phi_r)$ commutes with A. If G is a group of automorphisms we shall call ϕ a G-*invariant* if ϕ is invariant under every $A \varepsilon G$. The totality of these elements is a division subring Φ of P. Let $\mathfrak{A} = G\mathrm{P}_r$ denote the totality of endomorphisms in the additive group of P that have the form $\Sigma A_i \eta_{ir}$ where the $A_i \varepsilon G$. Clearly $\mathfrak{A}$ is closed under addition and subtraction and by the first equation of (2) $\mathfrak{A}$ is closed under multiplication. Hence $\mathfrak{A}$ is a subring of the ring of endomorphisms of P. Since G contains the identity, $\mathfrak{A}$ contains P_r. Thus an endomorphism commutes with every element of $\mathfrak{A}$ if and only if it is a left multiplication that commutes with every $A \varepsilon G$. It follows that the division ring of G-invariants coincides with the set of $\mathfrak{A}$-invariants in the sense defined above.

Let H be the invariant subgroup of inner automorphisms contained in G and let $I_1 = \tau_{1l}\tau_{1r}^{-1}$ and $I_2 = \tau_{2l}\tau_{2r}^{-1}$ belong to H. Then clearly $I_3 = I_1 I_2 = \tau_{3l}\tau_{3r}^{-1}$ where $\tau_3 = \tau_1\tau_2$. An element ϕ is an I_1 and an I_2-invariant if and only if it commutes with τ_1 and with τ_2. If this is so, ϕ also commutes with every $\tau = \tau_1\gamma_1 + \tau_2\gamma_2$, γ_i in Γ. Hence we do not change the set of G-invariants if we adjoin to G the inner automorphisms by the element τ obtained in this way. This remark leads us to restrict our attention to the study of groups G that are *closed* in the sense that if $I_1 = \tau_{1l}\tau_{1r}^{-1}$ and $I_2 = \tau_{2l}\tau_{2r}^{-1}$ are in G then so is $I = \tau_l\tau_r^{-1}$, where $\tau = \tau_1\gamma_1 + \tau_2\gamma_2$ and the γ_i are arbitrary in Γ. It is readily seen that G is closed if and only if the subgroup H is precisely the set of inner automorphisms that are defined by the non-zero elements of the division subalgebra A of P that contains the center Γ. The algebra A, which is uniquely determined, will be referred to as the division subalgebra *associated* with the closed group G (or H).

We shall say that a closed group G has finite reduced order if 1) $(\mathrm{A} : \Gamma)$ is finite and 2) H has finite index in G. The product of these two numbers will be called *the reduced order* of G. Our aim in the remainder of this note is the establishment of a Galois correspondence between closed groups of automorphisms of finite reduced order and their division rings of invariants. We prove first the following theorem:

THEOREM 4. *Let G be a closed group of automorphisms of finite reduced order in the division ring* P *and let $\mathfrak{A} = G\mathrm{P}_r$ be the ring of endomorphisms of the form $\Sigma A_i \eta_{ir}$, A_i in G. Then $(\mathfrak{A} : \mathrm{P}_r)_r =$ reduced order of G.*

Let $A_1, \cdots, A_u$ be representatives of the cosets of H in G and let $I_1, \cdots, I_h$ be inner automorphisms such that if $I_j = \tau_{jl}\tau_{jr}^{-1}$ then the elements τ_j form a basis for the subalgebra A associated with G. Then any automorphism in G has the form $A_k I$ where I is in H. If $I = \tau_l \tau_r^{-1}$, $\tau = \Sigma \tau_i \gamma_i$. As we have seen this implies that $I = \Sigma I_j \eta_{jr}$ where $\eta_j = \tau_j \gamma_j \tau^{-1}$. Hence $A_k I$ is a linear combination of the automorphisms $A_k I_1, A_k I_2, \cdots, A_k I_h$. Consequently the hu automorphisms $A_i I_j$ are generators over P_r of $\mathfrak{A}$. Now if $\Sigma A_i I_j \mu_{jir} = 0$ then $\Sigma I_j \mu_{jir} = 0$ by Lemma 2. Moreover $\Sigma I_j \mu_{jir} = 0$ and the independence of the τ_j over Γ implies that all the $\mu_{ji} = 0$. Thus the $A_i I_j$ are independent over P_r. Hence $(\mathfrak{A} : \mathrm{P}_r)_r = hu$.

5. We may now establish the Galois correspondence between groups of automorphisms and their division rings of invariants. The basic results are contained in Theorems 5 and 8.

THEOREM 5. *Let G be a closed group of automorphisms of finite reduced order in the division ring* P *and let Φ be the division subring of G-invariants. Then $(\mathrm{P} : \Phi)_l =$ reduced order of $G = (\mathrm{P} : \Phi)_r$ and any automorphism of* P *that leaves the elements of Φ invariant belongs to G.*

Let $\mathfrak{A} = G\mathrm{P}_r$. Then by Theorem 4, $(\mathfrak{A} : \mathrm{P}_r)_r =$ reduced order of G. On the other hand we know that Φ is the division ring of $\mathfrak{A}$-invariants. Hence by Theorem 2, $(\mathrm{P} : \Phi)_l = (\mathfrak{A} : \mathrm{P}_r)_r$. Hence $(\mathrm{P} : \Phi)_l =$ reduced order of G. The equality $(\mathrm{P} : \Phi)_r =$ reduced order of G follows by symmetry since the concept of reduced order is a "two-sided" one. Suppose now that A is any automorphism in P with the property that $\phi A = \phi$ for all $\phi \,\varepsilon\, \Phi$. Then $A\phi_l = \phi_l A$ and A is a linear transformation in P over Φ_l. Hence by Theorem 2, $A \,\varepsilon\, \mathfrak{A}$. Thus $A = \Sigma A_i I_j \mu_{jir}$ where the A_i and I_j are determined as in the proof of Theorem 4. By Lemma 2 this relation must have the form $A \equiv A_1 I = A_1 \Sigma I_j \mu_{jr}$. If $I = \tau_l \tau_r^{-1}$, Lemma 1 implies that τ is Γ-dependent on the τ_j, $j = 1, \cdots, h$. Hence $I \,\varepsilon\, H$ and $A = A_1 I \,\varepsilon\, G$.

The special case of this theorem in which $G = H$ consists of inner automorphisms is due to Artin and Whaples.[8] We note also that if G is a finite

[8] [1], p. 104.

group of automorphisms A_i such that each $A_i \neq 1$ is outer, then G is closed and of finite order. Hence Theorem 5 applies in this case too. The result thus obtained has been given in a previous paper of the author's.[9] A more general result can be obtained by combining Theorem 5 with the following theorem:

THEOREM 6. *Let F be any finite group of automorphisms and let J be its subgroup of inner automorphisms. Then the closed group G generated by F has finite reduced order. If $J = \{\tau_{1l}\tau_{1r}^{-1}, \cdots, \tau_{ml}\tau_{mr}^{-1}\}$ then the reduced order of G is the product of the order of the set $(\tau_1, \cdots, \tau_m)$ over Γ by the index of J in F.*

Since J is closed under multiplication $\tau_i\tau_j = \tau_k\gamma_{ij}$, γ_{ij} in Γ. If A is any automorphism and I is the inner automorphism $\tau_l\tau_r^{-1}$ then $A^{-1}IA$ is the inner automorphism $(\tau A)_l(\tau A)_r^{-1}$. Hence if $A \varepsilon F$, $(\tau_i A)_l(\tau_i A)_r^{-1} \varepsilon J$. It follows that $\tau_i A = \tau_h\delta_h$, δ_h in Γ. Now let A denote the totality of linear combinations $\Sigma\tau_i\gamma_i$, γ_i in Γ. Since the product of any two τ's is a multiple of a τ, A is a subalgebra. Clearly A has a finite basis and the dimensionality $(\mathrm{A} : \Gamma)$ is the order of the set $(\tau_1, \cdots, \tau_m)$ over Γ. Since A has a finite basis, A is a division algebra. Let H denote the group of inner automorphisms determined by the elements of A. Evidently H is contained in the closed group G generated by F. If $A \varepsilon F$ and $I = \tau_l\tau_r^{-1} \varepsilon H$ where $\tau = \Sigma\tau_i\gamma_i$ then $A^{-1}IA$ is the inner automorphism determined by the element $\Sigma(\tau_i A)(\gamma_i A)$. Since $\tau_i A = \tau_h\delta_h$, δ_h in Γ and $\gamma_i A \varepsilon \Gamma$, $A^{-1}IA = I'$ is in H. It follows that the totality G' of automorphisms of the form AI, A in F and I in H is a group. For the product $(A_1I_1)(A_2I_2) = A_1A_2(A_2^{-1}I_1A_2)I_2$ is in this set and $(AI)^{-1} = I^{-1}A^{-1} = A^{-1}(AI^{-1}A^{-1})$ is in the set. If AI is inner, so is A. Hence A is in J. Thus $AI \varepsilon H$. This shows that H is the group of inner automorphisms contained in G'. Since H is a group determined by the elements of an algebra, G' is closed. Hence $G' = G$ the smallest closed group containing F. Since the elements of G have the form AI, A in F, it is easy to see that the index of H in G is the same as that of J in F. It follows that G is of finite reduced order and that its reduced order is the product of $(\mathrm{A} : \Gamma)$ by the index of J in F.

6. Let G be an arbitrary closed group of finite reduced order and let $\mathfrak{A} = G\mathrm{P}_r$. We wish to establish a $(1-1)$ correspondence between the closed subgroups K of G and the subrings $\mathfrak{L}$ of $\mathfrak{A}$ that contain P_r.

We note first that if $\mathfrak{L}$ is any subring of $\mathfrak{A}$ containing P_r, then

[9] [3], p. 4.

3

$K = \mathfrak{L} \cap G$ is a closed subgroup of G. Evidently K is closed under multiplication. If $A \varepsilon K$, A has an inverse. Hence A is not a zero-divisor in $\mathfrak{L}$. Since $\mathfrak{L}$ satisfies the chain conditions this implies that $A^{-1} \varepsilon \mathfrak{L}$. Hence K is a group. Now let $I_1 = \tau_{1l}\tau_{1r}^{-1}$ and $I_2 = \tau_{2l}\tau_{2r}^{-1} \varepsilon K$. Then $\tau_{3l} = I_1\tau_{1r} + I_2\tau_{2r} \varepsilon \mathfrak{L}$ and if $\tau_3 \neq 0$, $I_3 = \tau_{3l}\tau_{3r}^{-1} \varepsilon \mathfrak{L}$. Since G is closed $I_3 \varepsilon G$. Thus $I_3 \varepsilon K$. It follows that K is a closed group of automorphisms.

Next let K be any closed subgroup of G and let $\mathfrak{L} = K\mathrm{P}_r$. Clearly $\mathfrak{L}$ is a subring of $\mathfrak{A}$ containing P_r. Let $K' = \mathfrak{L} \cap G$. Then $K' \supseteq K$. On the other hand the elements of K' leave invariant every K-invariant. Hence, by Theorem 5, $K' \subseteq K$. Thus $K' = \mathfrak{L} \cap G = K$.

Finally let $\mathfrak{L}$ be any ring between $\mathfrak{A}$ and P_r. Let $K = \mathfrak{L} \cap G$ and let $\mathfrak{L}' = K\mathrm{P}_r$. We wish to show that $\mathfrak{L}' = \mathfrak{L}$. Let $L = \mathfrak{L} \cap H$. Then L is the subgroup of inner automorphisms in K. We may choose representatives $A_1, A_2, \cdots, A_u$ of the cosets of H in G such that $A_1, \cdots, A_t$ are representatives of the cosets of L in K. Also we may choose the inner automorphisms $I_1 = \tau_{1l}\tau_{1r}^{-1}, \cdots, I_h = \tau_{hl}\tau_{hr}^{-1}$ in such a way that $(\tau_1, \cdots, \tau_h)$ is a basis for the algebra $\mathbf{A}$ associated with G while $(\tau_1, \cdots, \tau_f)$ is one for the algebra $\mathbf{B}$ associated with K. Now suppose that the element $V = \Sigma A_i I_j \mu_{jir} \varepsilon \mathfrak{L}$. We wish to show that $V \varepsilon \mathfrak{L}'$. Since $A_i I_i \varepsilon K$ if $i \leq t$, and $j \leq f$ it suffices to assume that $\mu_{ji} = 0$ for $j \leq f$, $i \leq t$ and to prove under this assumption that if $V \varepsilon \mathfrak{L}$, $V = 0$.

Suppose then that there exists a $V \neq 0$ in $\mathfrak{L}$ such that if $V = \Sigma A_i I_j \mu_{jir}$ then $\mu_{ji} = 0$ for $j \leq f$, $i \leq t$. We may assume also that V is an element satisfying these conditions that has the fewest number of non-zero coefficients μ_{jir}. Then by the argument used to prove Lemma 2 we see that $V = A_i \sum_j I_j \mu_{jir} = A_i \Sigma \tau_{jl}\tau_{jr}^{-1}\mu_{jir}$. By the argument used to prove Lemma 1 we can see that we may assume that the elements $\gamma_j = \tau_j^{-1}\mu_{ji} \varepsilon \Gamma$. Hence $V = A_i \Sigma \tau_{jl}\gamma_{jl}$ and since $V \neq 0$, $\tau = \Sigma \tau_j \gamma_j \neq 0$. Thus $V = A_i I \tau_r$ where $I = \tau_l \tau_r^{-1}$. Evidently $A_i I \varepsilon K = \mathfrak{L} \cap G$. Hence $i \leq t$ and τ is a linear combination of the τ_j with $j \leq f$. This contradicts the assumption on the form of V. We have therefore proved the following theorem:

THEOREM 7. *Let G be a closed group of automorphisms of finite reduced order and let $\mathfrak{A} = G\mathrm{P}_r$. Then any subring $\mathfrak{L}$ of $\mathfrak{A}$ containing P_r has the form $K\mathrm{P}_r$ where K is a closed subgroup of G. Moreover, distinct K's give rise in this way to distinct rings $\mathfrak{L}$.*

Let Φ be the division ring of G-invariants. Then we know that $\mathfrak{A}$ is the complete ring of l. t. of P over Φ_l. We have also seen that there is a $(1-1)$ correspondence between the rings $\mathfrak{L}$ between $\mathfrak{A}$ and P_r and the division rings between Φ and P. If we combine this $(1-1)$ correspondence with the $(1-1)$ correspondence between the rings Ω and the closed subgroups K of G we obtain the following result which includes the basic theorem of the ordinary Galois theory:

THEOREM 8. *Let* P *be a division ring and let* G *be a closed group of automorphisms in* P *of finite reduced order. To each closed subgroup* K *of* G *we associate the division subring* $\Psi = \mathrm{P}(K)$ *of elements that are invariant under every* $B \varepsilon K$ *and to each division subring* Ψ *of* P *containing* $\Phi = \mathrm{P}(G)$ *we associate the closed subgroup* $K = G(\Psi)$ *of automorphisms of* G *that leave the elements of* Ψ *invariant. Then these two correspondences are inverses of each other, and each is* $(1-1)$ *between the closed subgroups of* G *and the division rings between* P *and* Φ.

7. It is an unsolved problem to determine conditions on a division subring Φ of P in order that there exist a closed group G of finite reduced order whose ring of invariants is Φ. If P is commutative we know that necessary and sufficient conditions are that P be finite, separable and normal over Φ.

Suppose now that $(\mathrm{P} : \Gamma) < \infty$ and that Φ is a subfield of Γ such that $(\Gamma : \Phi) < \infty$ and let G be a closed group of automorphisms of P that has Φ as ring of invariants. Let H be the closed subgroup of G of automorphisms that leave the elements of Γ invariant. Now it is well known that an automorphism that produces the identity effect in Γ is inner. Hence H coincides with the subgroup of inner automorphisms of G. If A is the subalgebra associated with H we know that $(\mathrm{A} : \Gamma) = (\mathrm{P} : \Gamma)$. Hence $\mathrm{A} = \mathrm{P}$. Thus G includes all the inner automorphisms. The automorphisms A of G induce automorphisms $\bar{A}$ in Γ and the correspondence A to $\bar{A}$ is a homomorphism of G on a group $\bar{G}$ of automorphisms in Γ. The kernel of this homomorphism is the set of automorphisms that leave the elements of Γ invariant. Hence the kernel is H and $\bar{G} \cong G/H$. Now Φ is the field of invariants of the group G acting in Γ. Hence Γ is finite, separable and normal over Φ.

Conversely suppose that P is finite over Γ and that Γ is finite, separable and normal over Φ. Assume, moreover, that every automorphism $\bar{A}$ of the Galois group of Γ over Φ can be extended to an automorphism A in P. Let G be the totality of automorphisms of the form AI where I is inner. If $\bar{A}\bar{B} = \bar{C}$ and A, B, C, respectively, are the extensions of $\bar{A}, \bar{B}, \bar{C}$ then AB and C differ by an inner automorphism. Hence $AB = CI$. It follows easily that G is a group of automorphisms containing all of the inner automorphisms and that Φ is the ring of G-invariants.

Thus we have proved the following result which is in essence due to

Teichmüller:[10] If Φ is a subfield of the center Γ of P and $(\mathrm{P} : \Phi)$ is finite then necessary and sufficient conditions that there exist a closed group G of automorphisms whose set of invariants is Φ are 1) Γ is separable and normal over Φ and 2) every automorphism of the Galois group of Γ over Φ can be extended to an automorphism in P.

When these conditions hold we have seen that G is a group extension of the group H of inner automorphisms of P by the finite Galois group $\bar{G}$ of Γ over Φ. It is natural to seek conditions that this extension split in the sense that G contains a subgroup G' whose elements are representatives of the cosets of $G/H \cong \bar{G}$. It is easy to see that this is equivalent to the requirement that P be a direct product $\Sigma \times \Gamma$ where Σ and Γ are algebras over Φ. Conditions for this have been given by Teichmüller.[11].

THE JOHNS HOPKINS UNIVERSITY.

BIBLIOGRAPHY.

[1] E. Artin and G. Whaples, "The theory of simple rings," *American Journal of Mathematics*, vol. 65 (1943), pp. 87-107.

[2] N. Jacobson, *The Theory of Rings*, Mathematical Surverys, vol. 2, New York, 1943.

[3] ———, "The fundamental theorem of the Galois Theory for quasi-fields," *Annals of Mathematics*, vol. 41 (1940), pp. 1-7.

[4] ———, "Structure theory of simple rings without finiteness assumptions," *Transactions of the American Mathematical Society*, vol. 57 (1945), pp. 228-245.

[5] E. Noether, "Nichtkommutative Algebra," *Mathematische Zeitschrift*, vol. 37 (1933), pp. 514-541.

[6] O. Teichmüller, "Über die sogenannte nichtkommutative Galoissche Theorie etc.," *Deutsche Mathematik*, vol. 5 (1940), 138-149.
See also the review in the *Zentralblatt*, vol. 23, p. 198.

Reprinted from
American Journal of Mathematics
January 1947.

[10] [6].
[11] [6].

ISOMORPHISMS OF JORDAN RINGS.*

By N. Jacobson.

In a recent paper G. Ancochea [1] has defined a semi-isomorphism S of an associative ring $\mathfrak{A}$ on an associative ring $\mathfrak{B}$ to be a $1-1$ mapping of $\mathfrak{A}$ on $\mathfrak{B}$ such that

$$(a+b)S = aS + bS \tag{1}$$

$$(ab+ba)S = (aS)(bS) + (bS)(aS). \tag{2}$$

Moreover, Ancochea showed that if $\mathfrak{A}$ and $\mathfrak{B}$ are simple algebras with finite bases over their centers of characteristic $\neq 2$ then any semi-isomorphism of $\mathfrak{A}$ on $\mathfrak{B}$ is either an isomorphism or an anti-isomorphism. The starting point of the present paper is the observation that semi-isomorphisms are nothing more nor less than ordinary isomorphisms of the non-associative Jordan ring determined by the given associative ring. For the sake of simplicity we assume that our rings have the property that the mapping $a \to 2a$ is $1-1$. Then for each a, $\frac{1}{2}a$ is uniquely defined. Hence we can introduce the Jordan multiplication

$$a \cdot b = \tfrac{1}{2}(ab + ba)$$

If we use this in place of the ordinary multiplication we obtain the Jordan ring $\mathfrak{A}_j$ determined by the associative ring $\mathfrak{A}$. It is immediate that S is a semi-isomorphism of $\mathfrak{A}$ on $\mathfrak{B}$ if and only if this mapping is an isomorphism of $\mathfrak{A}_j$ on $\mathfrak{B}_j$.

If $\mathfrak{A}$ is a finite simple algebra it is known that $\mathfrak{A}_j$ is a simple Jordan ring. Thus from the non-associative point of view Ancochea's theorem gives a solution for one case of the general problem of determining the isomorphisms between simple Jordan rings that are algebras with a finite basis over a field. Besides the rings $\mathfrak{A}_j$ we note also the following type of simple Jordan ring. As before let $\mathfrak{A}$ be a finite simple associative algebra. Assume, moreover, that $\mathfrak{A}$ possesses an involution J, i. e. an anti-automorphism of period two, and let $\mathfrak{H}(\mathfrak{A}, J)$ be the set of elements that are J-symmetric $(aJ = a)$. Then it can be shown that $\mathfrak{H}(\mathfrak{A}, J)$ is a simple (Jordan) subring of $\mathfrak{A}_j$. In this paper we determine the isomorphisms between any two simple Jordan rings of the types enumerated here ($\mathfrak{A}_j$ or $\mathfrak{H}$). As will be shown

* Received May 2, 1947.

[1] [2] in the Bibliography.

elsewhere [2] these systems together with one other type obtained from Clifford number systems give all of the simple Jordan rings that are algebras with a finite basis over a field of characteristic 0. Since it is trivial to determine the isomorphisms for Clifford systems our results give a complete solution of the isomorphism problem for simple Jordan algebras of characteristic 0.

It should be observed that the results and methods of the present paper are quite similar to our former ones on Lie rings.[3] The Jordan theory is, in fact, simpler in one important respect, namely, simple Jordan algebras have identities. Hence in studying the isomorphisms one can get along without the concept of the multiplication centralizer (extended centrum). It suffices to consider the ordinary centers for these rings. We remark finally that the results in the Lie case could also have been formulated in terms of another kind of "semi-isomorphism" of an associative ring, namely, a $1-1$ mapping S such that

$$(a+b)S = aS + bS \tag{3}$$

$$(ab)S - (ba)S = (aS)(bS) - (bS)(aS). \tag{4}$$

1. Types of Jordan rings. We shall consider the following classes of Jordan rings.

Type A: Let $\mathfrak{A}$ be a simple associative ring that has finite dimensionality over its center Φ. We assume throughout that the characteristic is not two. Let $\mathfrak{A}_j$ be the Jordan ring obtained from $\mathfrak{A}$ by replacing ordinary multiplication by the Jordan multiplication $a \cdot b = \frac{1}{2}(ab + ba)$. We shall say that $\mathfrak{A}_j$ is a Jordan ring of *type* A_I.

Assume again that $\mathfrak{A}$ is simple and finite over its center, which we now denote as P. Assume further that $\mathfrak{A}$ possesses an involution J of second kind. Thus J is an anti-automorphism of period 2 and J induces a non-trivial automorphism in P. Let $\mathfrak{H}(\mathfrak{A}, J)$ be the subring of $\mathfrak{A}_j$ of J-symmetric elements and let $\Phi = \mathfrak{H} \cap P$. Then it is known that P is a quadratic extension of Φ. The rings $\mathfrak{H}(\mathfrak{A}, J)$, J of second kind, will be called Jordan rings of *type* A_{II}.

Types B-C. The distinction between these types will be given later. Both are obtained by starting with a simple associative ring finite over the center Φ that possesses an involution J of first kind. This means that J acts

---

[2] This will be proved in a forthcoming joint paper on representation theory of Jordan algebras by the author and F. D. Jacobson.

[3] [3] and [4].

as the identity in Φ. Let $\mathfrak{H}(\mathfrak{A}, J)$ be the totality of J-symmetric elements. Then $\mathfrak{H}$ is a subring of $\mathfrak{A}_j$ that we shall call a Jordan ring of *type B* or *C*.

2. The extended algebras. Simplicity. If Φ is the field specified in the various cases above then the given Jordan ring $\mathfrak{U}$ contains Φ. If $\alpha \varepsilon \Phi$ the ordinary products αa and $a\alpha$ are equal. Hence

$$\alpha \cdot a = \alpha a = a\alpha. \tag{5}$$

The associative ring $\mathfrak{A}$ from which $\mathfrak{U}$ is obtained can be regarded as an algebra over Φ since

$$\alpha(ab) = (\alpha a)b = a(\alpha b) \tag{6}$$

for all $a, b \varepsilon \mathfrak{A}$. This condition also implies that

$$\alpha(a \cdot b) = (\alpha a) \cdot b = a \cdot (\alpha b) \tag{7}$$

for all $a, b \varepsilon \mathfrak{A}$. Hence $\mathfrak{U}$ can be regarded as a Jordan algebra over Φ. We now let Ω be the algebraic closure of the field Φ and we consider the extended algebras $\mathfrak{U}_\Omega$.

Type A_I. Since Ω is algebraically closed $\mathfrak{A}_\Omega = \Omega_n$ the full matrix algebra over Ω. Since $\mathfrak{U} = \mathfrak{A}_j, \mathfrak{U}_\Omega = \mathfrak{A}_{j\Omega} = (\mathfrak{A}_\Omega)_j = \Omega_{nj}$.

Type A_{II}. Here $\mathfrak{U} = \mathfrak{H}(\mathfrak{A}, J)$, J an involution of second kind. It is known that if $x_1, x_2, \cdots, x_m$ is a basis for $\mathfrak{H}$ over Φ then it is also a basis for $\mathfrak{A}$ over P. Hence m is the dimensionality n^2 of $\mathfrak{A}$ over P. This shows also that $\mathfrak{U}_P = \mathfrak{A}_j$. Hence if Ω is the algebraic closure of Φ chosen to contain P then $\mathfrak{U}_\Omega = \Omega_{nj}$.

Types B and C. In this case $\mathfrak{U} = \mathfrak{H}(\mathfrak{A}, J)$, J of first kind. The involution J can be extended to an involution $\bar{J}$ in $\mathfrak{A}_\Omega = \Omega_n$.[4] As is known we can suppose that $\bar{J}$ is either the usual mapping $a \rightarrow a'$ the transposed or $\bar{J}$ is the mapping $a \rightarrow q^{-1}a'q$ where

$$q = \begin{pmatrix} 0 & 1_\nu \\ -1_\nu & 0 \end{pmatrix}. \tag{8}$$

Of course the latter can hold only if $n = 2\nu$ is even. In the first case we say that J and $\mathfrak{H}(\mathfrak{A}, J)$ are of *type* B and in the second that these are of *type* C. If J is of type B the set $\mathfrak{H}(\Omega_n, \bar{J})$ of $\bar{J}$-symmetric elements is just the set of ordinary symmetric matrices. The dimensionality over Ω is $n(n+1)/2$. If J is of type C the set $\mathfrak{H}(\Omega_n, \bar{J})$ is the set of matrices that satisfy

[4] Cf. [3] pp. 535-537.

$$(9) \qquad q^{-1}a'q = a.$$

These are of the form

$$(10) \qquad a = \begin{pmatrix} a_{11} & a_{12} \\ a_{21} & a_{22} \end{pmatrix}$$

where the $a_{ij} \varepsilon \Omega_\nu$ and satisfy

$$(11) \qquad a_{22} = a'_{11} \qquad a'_{12} = - a_{12}, \qquad a'_{21} = - a_{21}.$$

Hence the dimensionality over Ω is $n(n-1)/2 = \nu(2\nu - 1)$. For both types it is easy to see that $\mathfrak{H}(\Omega_n, \bar{J}) = \mathfrak{H}(\mathfrak{A}, J)_\Omega$.

The algebras Ω_{nj}, $\mathfrak{H}(\Omega_n, \bar{J})$, $\bar{J}$ of either type, are known to be simple.[5] It follows that the Jordan rings $\mathfrak{U}$ are simple when regarded as algebras over Φ. This implies also that *the rings* $\mathfrak{U}$ *are simple.*

3. The centers. The center Γ of any non-associative ring $\mathfrak{U}$ is defined to be the totality of elements γ that commute with every $a \varepsilon \mathfrak{U}$ and that associate with every pair a, b in the sense that

$$(12) \qquad \gamma(a \cdot b) = (\gamma \cdot a) \cdot b, \quad a \cdot (\gamma \cdot b) = (a \cdot \gamma) \cdot b,$$
$$a \cdot (b \cdot \gamma) = (a \cdot b) \cdot \gamma.$$

It is easy to see that the middle condition is a consequence of the other two. Moreover, if the ring is commutative the first condition is sufficient that a be in the center. If we now consider a Jordan ring $\mathfrak{U}$ and we use the definition $a \cdot b = \frac{1}{2}(ab + ba)$ then we can verify that

$$(13) \qquad [[a\gamma]b] = 0$$

is equivalent to the condition $\gamma \cdot (a \cdot b) = (\gamma \cdot a) \cdot b$.

It is clear from this condition that if γ is in the field Φ specified in the various cases above then $\gamma \varepsilon \Gamma$. Hence $\Phi \subseteq \Gamma$. We shall now show that $\Gamma = \Phi$. The proof will not make use of the explicit form of (13) but rather it will depend on some general principles that are applicable to other non-associative simple rings. We observe first that because of the simplicity of $\mathfrak{U}$ the center Γ is an ordinary field.[6] The field Γ as well as $\mathfrak{U}$ is finite dimensional over Φ. Now let Ω be the algebraic closure of Φ and form the extension algebra $\mathfrak{U}_\Omega$ of $\mathfrak{U}$ regarded as an algebra over Φ. Then it is clear that $\mathfrak{U}_\Omega$ contains Γ_Ω and that the latter is in the center of $\mathfrak{U}_\Omega$. Since $\mathfrak{U}_\Omega$ is simple its center is a field. On the other hand if $(\Gamma : \Phi) > 1$ then Γ_Ω has zero-divisors. Hence we must have $(\Gamma : \Phi) = 1$ and $\Gamma = \Phi$.

---

[5] [7] pp. 7-10 or [1] p. 553.
[6] [6] p. 239 or [9] p. 62.

We can now state the results on the dimensionalities obtained in **2** in the following way. The dimensionality $(\mathfrak{U} : \Phi)$ of $\mathfrak{U}$ over its center Φ is given by the table:

$(\mathfrak{U} : \Phi) = n^2$ if $\mathfrak{U} = \mathfrak{A}_j$ and $(\mathfrak{A} : \Phi) = n^2$

$(\mathfrak{U} : \Phi) = n^2$ if $\mathfrak{U} = \mathfrak{H}(\mathfrak{A}, J)$, J of second kind and $(\mathfrak{A} : \Phi) = 2n^2$

$(\mathfrak{U} : \Phi) = n(n+1)/2$ if $\mathfrak{U} = \mathfrak{H}(\mathfrak{A}, J)$, J of type B and $(\mathfrak{A} : \Phi) = n^2$

$(\mathfrak{U} : \Phi) = n(n-1)/2$ if $\mathfrak{U} = \mathfrak{H}(\mathfrak{A}, J)$, J of type C and $(\mathfrak{A} : \Phi) = n^2$.

We shall assume in the sequel that $n > 1$ for the types A_I and B and that $n > 2$ for type C rings.

4. The enveloping rings. By the *enveloping ring* of a subset $\mathfrak{U}$ of a ring $\mathfrak{A}$ we mean the subring of $\mathfrak{A}$ generated by $\mathfrak{U}$. We shall now show that in all of the cases considered above the enveloping ring of the Jordan ring $\mathfrak{U}$ is the ring $\mathfrak{A}$ from which $\mathfrak{U}$ is constructed. This is, of course, clear if $\mathfrak{U} = \mathfrak{A}_j$ is of type A_I. Suppose next that $\mathfrak{U} = \mathfrak{H}(\mathfrak{A}, J)$ is of type A_{II} and let $\mathfrak{B}$ denote the enveloping ring. Since $P\mathfrak{H}$, the totality of linear combinations of elements of $\mathfrak{H}$ with coefficients in P, is $\mathfrak{A}$ we have also that $P\mathfrak{B} = \mathfrak{A}$. Hence if $\mathfrak{B} \supseteq P$ then $\mathfrak{B} = \mathfrak{A}$. Evidently $\mathfrak{B} \supseteq \Phi$. We assert that $\mathfrak{B}$ is a simple algebra over Φ. For $\mathfrak{B}$ is semi-simple since a nilpotent ideal $\mathfrak{N}$ in $\mathfrak{B}$ determines a nilpotent ideal $P\mathfrak{N}$ in $\mathfrak{A}$. Hence if $\mathfrak{B}$ is not simple, $\mathfrak{B} = \mathfrak{B}_1 + \mathfrak{B}_2$ where the $\mathfrak{B}_i$ are two-sided ideals. Then $\mathfrak{B}_1\mathfrak{B}_2 = 0 = \mathfrak{B}_2\mathfrak{B}_1$. If $\mathfrak{A}_i = P\mathfrak{B}_i$ then also $\mathfrak{A} = \mathfrak{A}_1 + \mathfrak{A}_2$ and $\mathfrak{A}_1\mathfrak{A}_2 = 0 = \mathfrak{A}_2\mathfrak{A}_1$. This contradicts the simplicity of $\mathfrak{A}$. Thus $\mathfrak{B}$ is simple. Now suppose that $\mathfrak{B}$ is a proper subring of $\mathfrak{A}$. Then $\mathfrak{B} \cap P = \Phi$ and it follows that Φ is the center of $\mathfrak{B}$. On the other hand J induces an anti-automorphism of period 2 in $\mathfrak{B}$. Since J leaves the elements of Φ invariant J is of first kind in $\mathfrak{B}$. Hence $\mathfrak{H} = \mathfrak{H}(\mathfrak{B}, J)$ is of type B or C. But then the dimensionality of $\mathfrak{H}$ over its center woold be $n(n+1)/2$ or $n(n-1)/2$. Since we know it is n^2 and $n > 1$ this is impossible.

Next let $\mathfrak{B}$ be the enveloping ring of $\mathfrak{U} = \mathfrak{H}(\mathfrak{A}, J)$ of type B or C. Since $\mathfrak{U}$ contains Φ, $\mathfrak{B}$ is a Φ-subalgebra of $\mathfrak{A}$. It is easy to see that the enveloping ring of $\mathfrak{U}_\Omega$ is $\mathfrak{B}_\Omega$. Also $\mathfrak{B}_\Omega = \Omega_n$ implies that $\mathfrak{B} = \mathfrak{A}$. Hence it suffices to show that the enveloping ring of $\mathfrak{U}_\Omega = \mathfrak{H}(\Omega_n, J)$ is Ω_n.

To prove this suppose first that $\tilde{J}$ is of type B. Then $\mathfrak{H}(\Omega_n, \tilde{J})$ has the basis $h_{ij} = (e_{ij} + e_{ji})$, $i < j$, and e_{ii} if the e_{ij} are matrix units. Then $e_{ii} \in \mathfrak{B}_\Omega$ and also e_{ij}; $e_{ii}h_{ij}e_{jj}$ and $e_{ji} = e_{jj}h_{ij}e_{ii} \in \mathfrak{B}_\Omega$. Since $\mathfrak{B}_\Omega$ contains Ω this implies that $\mathfrak{B}_\Omega = \Omega_n$. Next let $\tilde{J}$ be of type C. Here $\mathfrak{H}(\Omega_n, \tilde{J})$ has the basis

$$
\begin{array}{lll}
f_{ij} = e_{ij} + e_{j+\nu,i+\nu}, & i,j = 1, 2, \cdots, \nu & \\
k_{i,j+\nu} = e_{i,j+\nu} - e_{j,i+\nu}, & i,j = 1, 2, \cdots, \nu, & i < j \\
k_{i+\nu,j} = e_{i+\nu,j} - e_{j+\nu,i}, & i,j = 1, 2, \cdots, \nu, & i < j.
\end{array}
$$

It follows that the enveloping ring contains $e_{ij} = f_{ii}f_{ij}$, $i \neq j$, $e_{ii} = e_{ij}e_{ji}$, $e_{j+\nu,i+\nu} = f_{ij}f_{ii}$, $i \neq j$, $e_{j+\nu,j+\nu} = e_{j+\nu,i+\nu}e_{i+\nu,j+\nu}$, $e_{i,j+\nu} = e_{ii}k_{i,j+\nu}$, $i < j$, $e_{j,i+\nu} = - e_{jj}k_{i,j+\nu}$, $i < j$, $e_{i+\nu,j} = k_{i+\nu,j}e_{jj}$, $i < j$, $e_{j+\nu,i} = - k_{i+\nu,j}e_{ii}$, $i < j$, $e_{i,i+\nu} = e_{i,j+\nu}e_{j+\nu,i+\nu}$, $e_{i+\nu,i} = e_{i+\nu,j}e_{ji}$. This shows that $\mathfrak{B}_\Omega = \Omega_n$ and completes the proof.

5. Non-isomorphim of the extended rings. We consider now two Jordan rings $\mathfrak{U}_1$ and $\mathfrak{U}_2$ or types A, B or C with enveloping rings $\mathfrak{A}_1$ and $\mathfrak{A}_2$ and centers Φ_1 and Φ_2 respectively. Let S be an isomorphism of $\mathfrak{U}_1$ on $\mathfrak{U}_2$. Then S induces an isomorphism s of Φ_1 on Φ_2. Moreover, since Jordan multiplication by elements of Φ_i coincides with ordinary multiplication, s is an isomorphism between the (ordinary) fields Φ_i. Also

$$
(14) \qquad (\gamma a_1)S = (\gamma \cdot a_1)S = \gamma^s \cdot a_1 S = \gamma^s (a_1 S).
$$

By constructing a new system $\mathfrak{A}_2$ if necessary we can suppose that $\Phi_1 = \Phi_2 = \Phi$. Then s is an automorphism in Φ and by (14) S is a semi-linear transformation between $\mathfrak{U}_1$ and $\mathfrak{U}_2$ regarded as vector spaces over Φ.

As before let Ω denote the algebraic closure of Φ. Then s can be extended to an automorphism $\tilde{s}$ in Ω and S can be extended in one and only one way to a semi-linear transformation $\tilde{S}$ of $\mathfrak{U}_{1\Omega}$ on $\mathfrak{U}_{2\Omega}$ having $\tilde{s}$ as associated automorphism. It is easy to see that $\tilde{S}$ is an isomorphism between the Jordan rings $\mathfrak{U}_{i\Omega}$.

We have seen that if $\mathfrak{U}$ is one of our Jordan rings then $\mathfrak{U}_\Omega$ is either of the form Ω_{nj} or of the form $\mathfrak{H}(\Omega_n, \tilde{J})$ where J is either of type B or of type C. Now it has been proved by Albert[7] that distinct systems of the forms noted are not isomorphic. We give a new proof of this fact here.

We shall define the *degree* of a simple Jordan ring over its center to be the maximum dimensionality of any subspace A generated by the powers of an element a in the ring. It is clear from the definition of Jordan multiplication that Jordan powers coincide with ordinary powers. Also we know that Ω is the center of Ω_{nj} and of $\mathfrak{H}(\Omega_n, \tilde{J})$ and that Jordan multiplication by elements of Ω is the same as ordinary multiplication. Hence the degree of Ω_{nj} and of $\mathfrak{H}(\Omega_n, \tilde{J})$ is the maximum degree of the minimum polynomials of the elements of the system.

[7] [1] p. 554.

By the Hamilton-Cayley theorem it is clear that the degree $\deg \Omega_{nj} \leq n$ and $\deg \mathfrak{H}(\Omega_n, \tilde{J}) \leq n$. It is known also that if a is a $2\nu \times 2\nu$ matrix such that $a = q^{-1}a'q$ where q is skew-symmetric then a satisfies an equation of degree ν.[8] Hence $\deg \mathfrak{H}(\Omega_n, \tilde{J}) \leq \nu$ if $n = 2\nu$ and J is of type C. Since Ω is infinite, Ω_n contains a matrix

$$\text{diag}\ \{\rho_1, \rho_2, \cdots, \rho_n\}$$

with $\rho_i \neq \rho_j$ for $i \neq j$ and $\mathfrak{H}(\Omega_n, \tilde{J})$ of type C contains

$$\text{diag}\ \{\rho_1, \rho_2, \cdots, \rho_\nu; \rho_1, \rho_2, \cdots, \rho_\nu\}.$$

Hence the bounds indicated for the degrees are attained:

$$\deg \Omega_{nj} = n,\ \deg \mathfrak{H}(\Omega_n, \tilde{J}) = n \text{ or } \nu = n/2$$

according as $\tilde{J}$ is of type B or of type C.

Now suppose that two of these Jordan rings are isomorphic. Then the degrees are the same. Hence the isomorphic pair is in the triple: Ω_{nj}, $\mathfrak{H}(\Omega_n, \tilde{J})$, $\tilde{J}$ of type B; $\mathfrak{H}(\Omega_{2n}, \tilde{J})$, $\tilde{J}$ of type C. The dimensionalities over the center are, respectively, n^2, $n(n+1)/2$, $n(2n-1)$. Since we have assumed $n > 1$ and $2n > 2$ respectively in the last two cases, these numbers are all different. Hence the isomorphic pair coincide.

Our result shows that if $\mathfrak{U}_1$ and $\mathfrak{U}_2$ are isomorphic Jordan rings of types A, B or C then they have the same type and we can suppose that $\mathfrak{U}_{1\Omega} = \mathfrak{U}_{2\Omega}$ for Ω the algebraic closure of the center Φ. In this case, of course, the extended isomorphism $\tilde{S}$ is an automorphism.

6. Isomorphisms between Jordan rings of type A. We suppose now that $\mathfrak{U}_1$ and $\mathfrak{U}_2$ are of type A. We assume first that $\mathfrak{U}_1$ is of type A_I and $\mathfrak{U}_2$ is of type A_{II}. Then $\mathfrak{U}_1 = \mathfrak{A}_j$ for a simple ring $\mathfrak{A}$ with center Φ and $\mathfrak{U}_2 = \mathfrak{H}(\mathfrak{B}, J)$ where $\mathfrak{B}$ is simple with center P a quadratic extension of Φ and J is of the second kind. Now $\mathfrak{A}_\Omega = \Omega_n$. Hence we can suppose that $\mathfrak{A}$ is a subring of the associative ring Ω_n. Similarly since $(\mathfrak{B}$ over $P)_\Omega = \Omega_n$ we can suppose that $\mathfrak{B}$ is a subring of Ω_n.

As above let S be an isomorphism of $\mathfrak{U}_1$ on $\mathfrak{U}_2$ and let $\tilde{S}$ be the extended automorphism in $\mathfrak{U}_{1\Omega} = \mathfrak{U}_{2\Omega} = \Omega_{nj}$. The form of $\tilde{S}$ can be deduced from the following

[8] [5] p. 748 or [4] p. 497.

LEMMA 1 (Ancochea). *Any algebra automorphism of Ω_{nj} over Ω is either an automorphism or an anti-automorphism of the associative algebra Ω_n over Ω.*[9]

By an algebra automorphism T we mean a ring automorphism such that $(\omega a)S = \omega(aS)$ for all $\omega \varepsilon \Omega$. The proof is a simple computation with matrix units and will be omitted.

We consider now the automorphism $\tilde{S}$ in Ω_{nj}. We have $(\omega a)\tilde{S} = \omega^{\bar{s}}(aS)$. The mapping $(\alpha_{ij}) \rightarrow (\alpha_{ij}^{\bar{s}})$ is an automorphism G in the associative system Ω_n. Hence it is an automorphism in Ω_{ni}. Since $(\omega a)G = \omega^{\bar{s}}(aG)$, $T = \tilde{S}G^{-1}$ is an algebra automorphism in Ω_{ni}. By Lemma 1, T is an automorphism or an anti-automorphism of Ω_n. Hence $\tilde{S} = TG$ is either an automorphism or an anti-automorphism of Ω_n. Clearly $\tilde{S}$ induces either an isomorphism S_1 or an anti-isomorphism S'_1 of the enveloping rings of $\mathfrak{U}_1$ on the enveloping ring of $\mathfrak{U}_2$. We know that these rings are respectively $\mathfrak{A}$ and $\mathfrak{B}$. Since S_1 or S'_1 coincides with S on $\mathfrak{U}_1$ and $\mathfrak{U}_1 = \mathfrak{A}$ the image $\mathfrak{U}_1 S$ is closed under multiplication. Hence $\mathfrak{U}_1 S = \mathfrak{B}$. Since $\mathfrak{U}_1 S = \mathfrak{U}_2 = \mathfrak{H}(\mathfrak{B}, J)$ this implies that the elements of $\mathfrak{B}$ are invariant under J contrary to assumption. We therefore have the following

THEOREM 1. *A simple Jordan ring of type A_I can not be isomorphic to one of type A_{II}.*

We suppose next that both $\mathfrak{U}_1$ and $\mathfrak{U}_2$ are of type A_I, say, $\mathfrak{U}_1 = \mathfrak{A}_j$ and $\mathfrak{U}_2 = \mathfrak{B}_j$. Also we can suppose that $\mathfrak{A}$ and $\mathfrak{B}$ are subrings of Ω_n. Then the argument just given shows that S is either an isomorphism or an anti-isomorphism of $\mathfrak{A}$ on $\mathfrak{B}$. The converse is clear: If $\mathfrak{A}$ and $\mathfrak{B}$ are isomorphic (anti-isomorphic) and S is an isomorphism (anti-isomorphism) of $\mathfrak{A}$ on $\mathfrak{B}$ then $\mathfrak{A}_j$ and $\mathfrak{B}_j$ are isomorphic under S. This proves

THEOREM 2 (Ancochea). *If $\mathfrak{A}$ and $\mathfrak{B}$ are simple associative rings that have finite dimensionalities over their centers then the Jordan rings $\mathfrak{A}_j$ and $\mathfrak{B}_j$ are isomorphic if and only if $\mathfrak{A}$ and $\mathfrak{B}$ are either isomorphic or anti-isomorphic. Any automorphism in $\mathfrak{A}_j$ is either an automorphism or an anti-automorphism of $\mathfrak{A}$.*

We assume finally that both $\mathfrak{U}_1$ and $\mathfrak{U}_2$ are of type A_{II}, say. $\mathfrak{U}_1 = \mathfrak{H}(\mathfrak{A}, J)$ and $\mathfrak{U}_2 = \mathfrak{H}(\mathfrak{B}, K)$. Here we can suppose that $\mathfrak{A}$ and $\mathfrak{B}$ are subrings of Ω_n. Our argument shows that if S is an isomorphism of $\mathfrak{U}_1$ on $\mathfrak{U}_2$ there exists either an isomorphism S_1 or an anti-isomorphism S'_1 of $\mathfrak{A}$ on $\mathfrak{B}$ that induces

[9] [2] pp. 151-152.

S in $\mathfrak{U}_1$. We note that in reality both S_1 and S'_1 exist. For if S_1 is an isomorphism of the required type then S_1K and JS_1 are anti-isomorphisms having the required properties. Similarly if S'_1 is given then S'_1K and JS'_1 are isomorphisms. Since $\mathfrak{A}$ and $\mathfrak{B}$ are the enveloping rings of $\mathfrak{U}_1$ and $\mathfrak{U}_2$ respectively there is only one isomorphism and only one anti-isomorphism that coincides with S in $\mathfrak{U}_1$. Hence we see that $S_1K = JS_1$ and $S'_1K = JS'_1$.

We summarize our results as follows: If $\mathfrak{H}(\mathfrak{A}, J)$ and $\mathfrak{H}(\mathfrak{B}, K)$ of type A_{II} are isomorphic then $\mathfrak{A}$ and $\mathfrak{B}$ are isomorphic. In this case we can identify $\mathfrak{A}$ and $\mathfrak{B}$ and consider $\mathfrak{H}(\mathfrak{A}, J)$ and $\mathfrak{H}(\mathfrak{A}, K)$. Then a necessary condition for isomorphism is that the involutions J and K be *cogredient* in the sense that $K = S_1^{-1}JS_1$ where S_1 is an automorphism in $\mathfrak{A}$. It is easy to see that this condition is also sufficient for isomorphism. Finally we see that any automorphism S of $\mathfrak{H}(\mathfrak{A}, J)$ is induced by an automorphism S_1 of $\mathfrak{A}$ that commutes with J. For in this case we have $J = S_1^{-1}JS_1$. The automorphism S_1 is uniquely determined by S. Conversely any S_1 that commutes with J induces an automorphism in $\mathfrak{H}(\mathfrak{A}, J)$.

THEOREM 3. *Let $\mathfrak{A}$ and $\mathfrak{B}$ be simple associative rings that have finite dimensionalities over their centers and that possess involutions J and K, respectively, of second kind. Then if the Jordan rings $\mathfrak{H}(\mathfrak{A}, J)$ and $\mathfrak{H}(\mathfrak{B}, K)$ are isomorphic, $\mathfrak{A}$ and $\mathfrak{B}$ are isomorphic. A necessary and sufficient condition that $\mathfrak{H}(\mathfrak{A}, J)$ and $\mathfrak{H}(\mathfrak{A}, K)$ be isomorphic is that J and K be cogredient. The group of automorphisms of $\mathfrak{H}(\mathfrak{A}, J)$ is the subgroup of the group of automorphisms of $\mathfrak{A}$ that commute with J.*

7. Isomorphisms between Jordan rings of types B and C. The isomorphism theory for these rings is similar to that of the rings of type A_{II}. Let $\mathfrak{U}_1 = \mathfrak{H}(\mathfrak{A}, J)$ and $\mathfrak{U}_2 = \mathfrak{H}(\mathfrak{B}, K)$, J and K of first kind (types B or C) and let S be an isomorphism of $\mathfrak{U}_1$ on $\mathfrak{U}_2$. Then we can suppose that $\mathfrak{A}$ and $\mathfrak{B}$ have the same center Φ. Let the dimensionality of $\mathfrak{A}$ over Φ be n^2. Then $\mathfrak{A}_\Omega = \Omega_n$ and $\mathfrak{U}_{1\Omega} = \mathfrak{H}(\Omega_n \tilde{J})$, $\tilde{J}$ of type B or type C. Similarly if the dimensionality of $\mathfrak{B}$ over Φ is m^2 then $\mathfrak{B}_\Omega = \Omega_m$ and $\mathfrak{U}_{2\Omega} = \mathfrak{H}(\Omega_m, \tilde{K})$. It follows that $m = n$ and that $\tilde{J} = \tilde{K}$. Thus we can suppose that $\mathfrak{U}_{1\Omega} = \mathfrak{U}_{2\Omega} = \mathfrak{H}(\Omega_n, \tilde{J})$ and that the enveloping rings $\mathfrak{A}$ and $\mathfrak{B}$ are subrings of Ω_n.

The isomorphism S can be extended to an automorphism $\tilde{S}$ of the system $\mathfrak{H}(\Omega_n, \tilde{J})$. We can prove that $\tilde{S}$ is induced by an automorphism of Ω_n by using the following

LEMMA 2 (Kalisch). *Any algebra automorphism of the Jordan algebra*

$\mathfrak{H}(\Omega_n, \bar{J})$ *over* Ω *can be extended to an automorphism in the associative algebra* Ω_n.[10]

As before this result can be used to show that $\tilde{S}$ can be extended to an isomorphism in Ω_n. It follows that $\tilde{S}$ induces an isomorphism S_1 between the enveloping rings $\mathfrak{A}$ and $\mathfrak{B}$. The rest of the argument is a duplication of the one given in the A_{II} case. The following result which has been proved for algebras by Kalisch is therefore readily established.

THEOREM 4. *Theorem 3 holds also for involutions J and K of the first kind.*

THE JOHNS HOPKINS UNIVERSITY.

BIBLIOGRAPHY

1. A. A. Albert, "On Jordan algebras of linear transformations," *Transactions of the American Mathematical Society*, vol. 59 (1946), pp. 524-555.
2. G. Ancochea, "On semi-automorphisms of division algebras," *Annals of Mathematics*, vol. 48 (1947), pp. 147-154.
3. N. Jacobson, "Simple Lie algebras over a field of characteristic zero," *Duke Mathematical Journal*, vol. 4 (1938), pp. 534-551.
4. ———, "Classes of restricted Lie algebras of characteristic *p*," I, *American Journal of Mathematics*, vol. 63 (1941), pp. 481-515.
5. ———, "An application of E. H. Moore's determinant of a hermitian matrix," *Bulletin of the American Mathematical Society*, vol. 45 (1939), pp. 745-748.
6. ———, "Structure theory of simple rings without finiteness assumptions," *Transactions of the American Mathematical Society*, vol. 57 (1945), pp. 228-245.
7. G. Kalisch, "On special Jordan algebras," *Transactions of the American Mathematical Society*, vol. 61 (1947), pp. 482-494.
8. I. Kaplansky, "Semi-automorphisms of rings," *Duke Mathematical Journal*, vol. 14 (1947), pp. 521-527.
9. T. Nakayama, "Über einfache distributive Systeme unendlicher Ränge," *Proceedings of the Imperial Academy, Tokyo*, vol. 20 (1944), pp. 61-66.

Reprinted from
American Journal of Mathematics
April 1948.

[10] A more general result will be proved in the paper referred to in [2].

THE CENTER OF A JORDAN RING[1]

N. JACOBSON

If $\mathfrak{A}$ is an arbitrary associative ring we can symmetrize and antisymmetrize the multiplication defined in $\mathfrak{A}$ to obtain two non-associative rings. We set

$$\{ab\} = ab + ba, \qquad [ab] = ab - ba \tag{1}$$

and call the former the *Jordan product* and the latter the *commutator* or *Lie product* of a and b. If we use $\{ab\}$ as product in place of the originally defined ab we obtain the *Jordan ring* $\mathfrak{A}_j$ determined by $\mathfrak{A}$. Similarly the *Lie ring* $\mathfrak{A}_l$ is obtained by using $[ab]$ in place of ab. Naturally if $\mathfrak{A}$ has characteristic 2 then $\mathfrak{A}_j = \mathfrak{A}_l$. It is customary to exclude this case from consideration but in most of our discussion we shall not find it necessary to do so. Clearly $\{ab\} = \{ba\}$, $[ab] = -[ba]$. Also we recall the following well known identity of Jacobi's:

$$[[ab]c] + [[bc]a] + [[ca]b] = 0. \tag{2}$$

If $\mathfrak{R}$ is any non-associative ring one defines the center of $\mathfrak{R}$ to be the totality of elements c that commute,

$$c \cdot a = a \cdot c, \tag{3}$$

and associate,

$$\begin{gathered}(a \cdot b) \cdot c = a \cdot (b \cdot c), \qquad (a \cdot c) \cdot b = a \cdot (c \cdot b), \\ (c \cdot a) \cdot b = c \cdot (a \cdot b),\end{gathered} \tag{4}$$

with all a, b in $\mathfrak{R}$.[2] It is known that the center is a subring of $\mathfrak{R}$. Clearly this subring is associative. It is also known that the center of a simple ring is either 0 or a field. It is easy to see that the middle condition in (4) is a consequence of (3) and the other conditions in (4). Also it is clear that if $\mathfrak{R}$ is commutative then the first condition of (4) characterizes the center.

We consider now the centers $\mathfrak{C}_j$ and $\mathfrak{C}_l$ respectively of $\mathfrak{A}_j$ and $\mathfrak{A}_l$.

Received by the editors May 6, 1947, and, in revised form, July 3, 1947.

[1] I am indebted to A. H. Clifford for a number of valuable conversations on the subject of this note.

[2] See Jacobson, *Structure theory of simple rings without finiteness assumptions*, Trans. Amer. Math. Soc. vol. 57 (1945) p. 239, or T. Nakayama, *Über einfache distributive Systeme unendlicher Ränge*, Proc. Imp. Acad. Tokyo vol. 20 (1944) p. 62 for this definition and for the results quoted in this paragraph.

First let $c \in \mathfrak{C}_l$. Then by (1) and (3), $2[ca]=0$. By (4) and Jacobi's identity, $[[ca]b]=0$. Hence

$$2[ca] = 0, \qquad [[ca]b] = 0 \tag{5}$$

holds. It is also easy to see that these conditions are sufficient that $c \in \mathfrak{C}_l$.

Next let $c \in \mathfrak{C}_j$. We introduce the *Jordan associator*

$$A(a, b, c) = \{\{ab\}c\} - \{a\{bc\}\}$$

and we can verify that

$$A(a, b, c) = [[ca]b]. \tag{6}$$

Hence

$$[[ca]b] = 0 \tag{7}$$

is a necessary and sufficient condition that $c \in \mathfrak{C}_j$. Thus we see that $\mathfrak{C}_l \subseteq \mathfrak{C}_j$. If we denote the center of $\mathfrak{A}$ by $\mathfrak{C}$, $c \in \mathfrak{C}$ if and only if $[ca]=0$. Hence $\mathfrak{C} \subseteq \mathfrak{C}_l \subseteq \mathfrak{C}_j$.

Let $c \in \mathfrak{C}_j$ and a, b, d be arbitrary in $\mathfrak{A}$. Then $[[c, ab]d]=0$. Since

$$[c, ab] = [ca]b + a[cb]$$

we obtain

$$[ca][bd] + [ad][cb] = 0. \tag{8}$$

This simple relation has a number of interesting consequences. In the first place if we set $d=c'$, a second element in $\mathfrak{C}_j$, we obtain $[ca][bc'] + [ac'][cb]=0$. If we use the fact that $[cb] \in \mathfrak{C}$ this reads

$$[ca][c'b] + [cb][c'a] = 0. \tag{9}$$

It is easy to see that this implies

$$[[cc', a]b] = 0. \tag{10}$$

Thus $cc' \in \mathfrak{C}_j$. It is clear also that if $2[ca]=0$ and $2[c'a]=0$ then $2[cc', a]=0$. Hence we have the following theorem.

THEOREM 1. *The Jordan center* $\mathfrak{C}_j$ *and the Lie center* $\mathfrak{C}_l$ *are (ordinary) subrings of* $\mathfrak{A}$.

If we specialize $d=a$ and $b=c$ successively in (8) we obtain

$$[ca][ba] = 0, \qquad [ca][cd] = 0. \tag{11}$$

In particular

$$[ca]^2 = 0. \tag{12}$$

This implies the following theorem.

THEOREM 2. *If $\mathfrak{A}$ is an associative ring whose center $\mathfrak{C}$ contains no nilpotent elements not equal to 0 then the Jordan center $\mathfrak{C}_j$ coincides with $\mathfrak{C}$.*

For if $c \in \mathfrak{C}_j$, $[ca]^2 = 0$ for any $a \in \mathfrak{A}$. Since $[ca] \in \mathfrak{C}$ this implies that $[ca] = 0$. Hence $c \in \mathfrak{C}$.

It is easy to see that if a ring contains a nilpotent element in its center then it contains a nilpotent two-sided ideal. Hence we have the following corollary.

COROLLARY. *If $\mathfrak{A}$ is a ring that has no nilpotent two-sided ideals then its Jordan center $\mathfrak{C}_j$ coincides with the ordinary center.*

It is clear from Jacobi's identity that if $c \in \mathfrak{C}_j$ then $[[ab]c] = 0$ for all a, b. Thus c commutes with every commutator. Clearly the elements that have this property form a subring $\mathfrak{B}$ of $\mathfrak{A}$. It will be shown in a forthcoming paper by Kaplansky that if $\mathfrak{A}$ is a semi-simple ring then $\mathfrak{B} = \mathfrak{C}$.[3] Thus Kaplansky's result has both a stronger hypothesis and a stronger conclusion than our Theorem 2. The following examples will serve to illustrate these results.

EXAMPLE 1. Let $\mathfrak{A}$ be the ring of triangular matrices

$$\begin{pmatrix} \alpha_{11} & & * \\ & \alpha_{22} & \\ & & \ddots \\ 0 & & & \alpha_{nn} \end{pmatrix} \tag{13}$$

with elements in a field Φ. It is easy to see that $\mathfrak{C}$ is the set of scalar matrices. Hence $\mathfrak{C}_j = \mathfrak{C}$. On the other hand it can be seen that $\mathfrak{B}$ is the set of matrices

$$\begin{pmatrix} \alpha & 0 & \cdots & 0 & \beta \\ & \ddots & & & 0 \\ & & \ddots & & 0 \\ & & & \ddots & \\ & & & & \alpha \end{pmatrix}. \tag{14}$$

EXAMPLE 2. Let $\mathfrak{A}$ be the set of triangular matrices for which $\alpha_{11} = \alpha_{22} = \cdots = \alpha_{nn} = \alpha$. Here $\mathfrak{C}$ is the set of matrices of the form

[3] *Added in proof.* This paper has now appeared: *Semi-automorphisms of rings*, Duke Math. J. vol. 14 (1947) pp. 521–527.

(14) and $\mathfrak{B}$ is the set of matrices of the form

$$
(15) \qquad \begin{bmatrix} \alpha & 0 \cdots 0 & \gamma & \beta \\ & \ddots & \epsilon & \delta \\ & & \ddots & 0 \\ & 0 & \ddots & \vdots \\ & & & 0 \\ & & & \alpha \end{bmatrix}.
$$

Finally $\mathfrak{C}_j$ is the subset of $\mathfrak{B}$ of the matrices of the form (15) in which $\epsilon = 0$.

We derive next another property of the element c in $\mathfrak{C}_j$. If a, b, d are arbitrary in $\mathfrak{A}$,

$$
\begin{aligned} [[ca, b]d] &= [(a[cb] + c[ab]), d] \\ &= [ad][cb] + [cd][ab] + c[[ab]d]. \end{aligned}
$$

If we interchange a and d in (8) we see that

$$
[ad][cb] + [cd][ab] = 0.
$$

Hence

$$
[[ca, b]d] = c[[ab]d].
$$

This implies also that $[[a, cb]d] = c[[ab]d]$. By Jacobi's identity we obtain finally $[[ab]cd] = c[[ab]d]$. Thus

$$
(16) \qquad c[[ab]d] = [[ca, b]d] = [[a, cb]d] = [[ab]cd].
$$

This can also be written in the following form in terms of the associator $A(b, d, a) = [[ab]d]$:

$$
(16') \qquad cA(b, d, a) = A(b, d, ca) = A(cb, d, a) = A(b, cd, a).
$$

In the remainder of this note we consider rings $\mathfrak{U}$ that are subrings of a Jordan ring of the form $\mathfrak{A}_j$. By the enveloping (associative) ring of $\mathfrak{U}$ in $\mathfrak{A}$ we mean the subring of $\mathfrak{A}$ generated by $\mathfrak{U}$. If $\mathfrak{E}$ is the enveloping ring of $\mathfrak{U}$ clearly $\mathfrak{U}$ is also a subring of the Jordan ring $\mathfrak{E}_j$. Hence we can suppose that $\mathfrak{E} = \mathfrak{A}$.

If c is in the Jordan center $\mathfrak{C}_j(\mathfrak{U})$ of $\mathfrak{U}$ then $[[ca]b] = 0$ holds for all $a, b \in \mathfrak{U}$. Since $\mathfrak{U}$ is a Jordan ring

$$
[[c, \{ab\}]d] = 0
$$

for all a, b, d in $\mathfrak{U}$. Also we have $[c[ab]] = 0$ so that $[[c, [ab]]d] = 0$. Hence by addition

$$2[[c, ab]d] = 0. \tag{17}$$

We shall assume now that the ring $\mathfrak{A}$ contains no element that has order two in the additive group. Then by (17), $[[c, ab]d]=0$ for all a, b, d in $\mathfrak{U}$. As before this leads to the relation (8) and to its consequence $[ca]^2=0$. On the other hand since $[[ca]b]=0$, $[ca]$ commutes with every b in $\mathfrak{U}$. Hence it commutes with every b in the enveloping ring $\mathfrak{A}$. Thus $[ca]$ is in the center $\mathfrak{C}$ of $\mathfrak{A}$. This shows that if some $[ca]\neq 0$ then the center of $\mathfrak{A}$ possesses nilpotent elements. Hence we have the following theorem.

THEOREM 3. *Let $\mathfrak{A}$ be an associative ring that has no elements of order 2 and that has no nilpotent elements ($\neq 0$) in its center. Let $\mathfrak{U}$ be a subring of the Jordan ring $\mathfrak{A}_j$ such that the enveloping ring of $\mathfrak{U}$ is $\mathfrak{A}$. Then the Jordan center $\mathfrak{C}_j(\mathfrak{U})$ coincides with the totality of elements c of $\mathfrak{U}$ that commute with every a in $\mathfrak{U}$ (or in $\mathfrak{A}$).*

A. A. Albert has recently studied the structure of the subalgebras $\mathfrak{U}$ of a Jordan algebra $\mathfrak{A}_j$, $\mathfrak{A}$ a matrix algebra Φ_n over a field Φ.[4] In the course of this study he has defined the center of $\mathfrak{U}$ to be the totality of elements c such that $[ca]=0$ for all a in $\mathfrak{U}$. We shall show in an example that Albert's definition is unsatisfactory for arbitrary Jordan algebras since it is not invariant under isomorphism. On the other hand it will also be shown that in the case of simple algebras over a field of characteristic 0—and this case is the only one for which Albert uses his definition—Albert's center coincides with the Jordan center as defined in the present paper. We remark that Theorems 2 and 3 give other cases for which this identity holds. In all of these cases Albert's definition can not lead to any difficulties. We consider now the following example.

EXAMPLE 3. Let $\mathfrak{A}$ be the algebra over Φ that has the basis z_1, z_2 such that $z_iz_j=0$. The Jordan ring $\mathfrak{A}_j$ has the basis z_1, z_2 with the multiplication table

$$\{z_1z_1\} = 0, \qquad \{z_1z_2\} = 0, \qquad \{z_2z_2\} = 0.$$

Let $\mathfrak{B}$ be the algebra over Φ that has the basis Z_1, Z_2, Z_3 with the multiplication table

$$Z_1^2 = Z_2^2 = Z_3^2 = 0, \qquad Z_1Z_2 = -Z_2Z_1 = Z_3, \qquad Z_2Z_3 = Z_3Z_2 = 0,$$
$$Z_3Z_1 = Z_1Z_3 = 0.$$

Since any product of three Z's is 0, $\mathfrak{B}$ is an associative algebra. We

[4] *On Jordan algebras of linear transformations*, Trans. Amer. Math. Soc. vol. 59 (1946) p. 540.

have the Jordan multiplication table

$$\{Z_1Z_1\} = 0, \qquad \{Z_1Z_2\} = 0, \qquad \{Z_2Z_2\} = 0.$$

Hence $\mathfrak{U}=(Z_1, Z_2)$ is a subalgebra of $\mathfrak{B}_j$ isomorphic to $\mathfrak{A}_j$. The totality of elements of $\mathfrak{A}_j$ that commute with all the elements of $\mathfrak{A}_j$ is $\mathfrak{A}_j$ itself. However, the elements Z_1 and Z_2 do not commute with all the elements of $\mathfrak{U}$. Evidently the Jordan center of $\mathfrak{A}_j$ is $\mathfrak{A}_j$.

We shall now consider the Jordan center of any simple Jordan algebra with a finite basis over a field of characteristic 0. At first we consider any associative algebra $\mathfrak{A}$ over a field Φ. As before let $\mathfrak{A}_j$ denote the Jordan algebra obtained from $\mathfrak{A}$ by using the operations $a+b$, αa for α in Φ and $\{ab\}=ab+ba$. If $\mathfrak{U}$ is a subalgebra of $\mathfrak{A}_j$ it is clear that the center $\mathfrak{C}_j(\mathfrak{U})$ is a subalgebra. Let $c\in\mathfrak{C}_j(\mathfrak{U})$ and let a be any algebraic element of $\mathfrak{U}$. Then there exists a polynomial $\phi(\lambda)\neq 0$ in $\Phi[\lambda]$ such that $\phi(a)=0$. Since $[ca]$ commutes with a, $[c, a^k]=ka^{k-1}[ca]$. Hence

$$0 = [c, \phi(a)] = \phi'(a)[ca] \tag{18}$$

where $\phi'(\lambda)$ is the derivative of the polynomial $\phi(\lambda)$. We assume next that c is algebraic and that a is arbitrary. Then a similar argument shows that if $\psi(\lambda)$ is a polynomial not equal to 0 such that $\psi(c)=0$ then

$$\psi'(c)[ca] = 0. \tag{19}$$

We suppose now that $\mathfrak{A}=\Phi_n$, Φ of characteristic 0. Let $\mathfrak{U}$ be a simple subalgebra of $\mathfrak{A}_j$. Then it has been shown by Albert that $\mathfrak{U}$ has an identity e.[5] Evidently e is in the Jordan center $\mathfrak{C}_j(\mathfrak{U})$. Hence $\mathfrak{C}_j(\mathfrak{U})\neq 0$ and it follows that $\mathfrak{C}_j(\mathfrak{U})$ is a field. Since $ea+ae=a$, $e^2=e/2$. If we set $e'=2e$ then $(e')^2=e'$ and $(e'a+ae')/2=a$. Hence

$$e'a/4 + e'ae'/2 + ae'/4 = (e'a + ae')/2$$

and

$$a = (e'a + ae')/2 = e'ae'.$$

It follows that $e'a=a=ae'$. Thus e' acts as an identity for all the elements of $\mathfrak{U}$ and therefore for all the elements of the enveloping algebra $\mathfrak{E}$ of $\mathfrak{U}$.

Suppose now that $c\in\mathfrak{C}_j(\mathfrak{U})$ and that $\mu(\lambda)$ is the minimum polynomial of c regarded as an element of $\mathfrak{C}_j(\mathfrak{U})$. Then if $\mu(\lambda)=\lambda^m$

[5] Albert (loc. cit., footnote 3) uses the multiplication $a\cdot b=\{ab\}/2$. Clearly an algebra is simple relative to the dot multiplication if and only if it is simple relative to $\{\ \}$. Moreover, if e' is an identity relative to $\cdot$, then $e=e'/2$ is one relative to $\{\ \}$.

$+\gamma_1\lambda^{m-1}+\cdots+\gamma_m$,

$$\{c\}^m+\gamma_1\{c\}^{m-1}+\cdots+\gamma_{m-1}c+\gamma_m e=0 \tag{20}$$

where $\{c\}^r$ is defined inductively by $\{c\}^r=\{\{c\}^{r-1}, c\}$. Since $\{c\}^r=2^{r-1}c^r$, (20) yields

$$c^m+\frac{\gamma_1}{2}c^{m-1}+\cdots+\frac{\gamma_{m-1}}{2^{m-1}}c+\frac{\gamma_m}{2^m}e'=0. \tag{21}$$

Thus if $\psi(\lambda)=2^{-m}\mu(2\lambda)$ then $\psi(c)=0$. Since $\mathfrak{C}_j(\mathfrak{U})$ is a separable field, $\mu'(\lambda)$ is prime to $\mu(\lambda)$. Hence $\psi'(\lambda)$ is prime to $\psi(\lambda)$. It follows that

$$\psi'(c)=mc^{m-1}+(m-1)\frac{\gamma_1}{2}c^{m-2}+\cdots+\frac{\gamma_{m-1}}{2^{m-1}}e'$$

has an ordinary inverse relative to e'. Hence by (19), $[ca]=0$ for all a. This proves the following theorem.

THEOREM 4. *Let $\mathfrak{A}=\Phi_n$ the ring of $n\times n$ matrices over a field of characteristic 0 and let $\mathfrak{U}$ be a simple Jordan subalgebra of $\mathfrak{A}_j$. Then the center of $\mathfrak{U}$ coincides with the totality of elements c of $\mathfrak{U}$ that commute in the ordinary multiplication with every a in $\mathfrak{U}$.*

THE JOHNS HOPKINS UNIVERSITY

Reprinted from
Bulletin of the American Mathematical Society
April 1948.

Permissions

Birkhäuser Boston thanks the original publishers of the papers of Nathan Jacobson for granting permission to reprint the papers in this collection.

[1], [2] Reprinted from *Annals of Mathematics* **35** © 1935 by Princeton University Press.

[3] Reprinted from *Proceedings of the National Academy of Sciences* **21** 1935.

[4] Reprinted from *Annals of Mathematics* **36** © 1935 by Princeton University Press.

[5] Reprinted from *Proceedings of the National Academy of Sciences* **21** 1935.

[6] Reprinted from *American Journal of Mathematics* **58** © 1936 by Johns Hopkins University Press.

[7] Reprinted from *Proceedings of the National Academy of Sciences* **23** 1937.

[8], [9] Reprinted from *Annals of Mathematics* **38** © 1937 by Princeton University Press.

[10] Reprinted from *Duke Mathematical Journal* **3** © 1937 by Duke University Press.

[11] Reprinted from *Transactions of the American Mathematical Society* **42** © 1937 by the AMS.

[12] Reprinted from *Bulletin of the American Mathematical Society* **43** © 1937 by by the AMS.

[13] Reprinted from *American Journal of Mathematics* **59** © 1937 by Johns Hopkins University Press.

[14] Reprinted from *Annals of Mathematics* **39** © 1938 by Princeton University Press.

[15] Reprinted from *Duke Mathematical Journal* **3** © 1938 by Duke University Press.

[16] Reprinted from *American Journal of Mathematics* **61** © 1939 by Johns Hopkins University Press.

[17] Reprinted from *Bulletin of the American Mathematical Society* **45** © 1939 by the AMS.

[18] Reprinted from *Annals of Mathematics* **40** © 1939 by Princeton University Press.

[19] Reprinted from *Duke Mathematical Journal* **5** © 1939 by Duke University Press.